ISBN 978-1-5281-0092-2
PIBN 10921501

This book is a reproduction of an important historical work. Forgotten Books uses
state-of-the-art technology to digitally reconstruct the work, preserving the original format
whilst repairing imperfections present in the aged copy. In rare cases, an imperfection in
the original, such as a blemish or missing page, may be replicated in our edition. We do,
however, repair the vast majority of imperfections successfully; any imperfections that
remain are intentionally left to preserve the state of such historical works.

1 MONTH OF
FREE
READING

at

www.ForgottenBooks.com

By purchasing this book you are eligible for one month membership to ForgottenBooks.com, giving you unlimited access to our entire collection of over 1,000,000 titles via our web site and mobile apps.

To claim your free month visit:

www.forgottenbooks.com/free921501

English
Français
Deutsche
Italiano
Español
Português

www.forgottenbooks.com

Mythology Photography **Fiction**
Fishing Christianity **Art** Cooking
Essays Buddhism Freemasonry
Medicine **Biology** Music **Ancient
Egypt** Evolution Carpentry Physics
Dance Geology **Mathematics** Fitness
Shakespeare **Folklore** Yoga Marketing
Confidence Immortality Biographies
Poetry **Psychology** Witchcraft
Electronics Chemistry History **Law**
Accounting **Philosophy** Anthropology
Alchemy Drama Quantum Mechanics
Atheism Sexual Health **Ancient History**
Entrepreneurship Languages Sport
Paleontology Needlework Islam
Metaphysics Investment Archaeology
Parenting Statistics Criminology
Motivational

THE
BIOCHEMICAL
JOURNAL

EDITED FOR THE BIOCHEMICAL SOCIETY

BY

SIR W. M. BAYLISS, F.R.S.

AND

ARTHUR HARDEN, F.R.S.

EDITORIAL COMMITTEE

PROF. G. BARGER	PROF. F. G. HOPKINS
PROF. V. H. BLACKMAN	SIR F. KEEBLE
MR J. A. GARDNER	PROF. W. RAMSDEN
SIR E. J. RUSSELL	

VOLUME XVI 1922

CAMBRIDGE
AT THE UNIVERSITY PRESS
1922

PRINTED IN GREAT BRITAIN

CONTENTS

No. 1

No. 2

No. 3

No. 5

CONTENTS

No. 6

I. NOTE ON TANNASE.

By DONALD RHIND and FRANCIS EDWARD SMITH.

*From the Biochemical Laboratory, Chemical Department,
University of Bristol.*

(*Received December 14th, 1921.*)

SINCE the middle of last century, this enzyme has been the subject of many investigations, but it is notable that no method has been described whereby its hydrolysing power may be estimated. The investigation described in this paper was undertaken at the suggestion of Dr Nierenstein, since it was thought that some such method might be of assistance to other work at present in progress in this laboratory.

This enzyme was prepared from *Aspergillus Luchuensis* Inui.[1] by a slight modification of Freudenberg's method [1913]. It was used without further purification, a 1 % solution in distilled water being employed in each experiment.

The gallotannin used was Schering's leviss. puriss. which had been further purified according to Fischer's method [1912]. An approximately 0·3 % solution in distilled water was prepared, and the exact amount of gallotannin present estimated by Nierenstein's caseinogen method [1911] as modified by Spiers [1914].

The hydrolyses were carried out in 250 cc. Erlenmeyer flasks, 100 cc. of gallotannin solution, to which various quantities of tannase solution were added, being employed in each experiment. During the reaction the flasks were plugged with cotton wool, and toluene was added to the reaction mixture to prevent the growth of fungi or bacteria, the whole being kept in a dark incubator at 23°. The amount of gallotannin present in the reaction mixture was estimated immediately after the tannase solution had been added and again after the hydrolysis had proceeded for a certain time.

The details of the estimation are as follows. A solution of approximately 0·1 % $KMnO_4$ was prepared and standardised against pure ammonium oxalate of known strength; 20 cc. of a solution of 0·5 % of indigo carmine in dilute sulphuric acid (A-R reagent, British Drug Houses), being used as indicator. As the indigo carmine reacts with potassium permanganate, it is at first necessary to determine the quantity of the latter used by the indicator solution. To this end 20 cc. of this solution were placed in a large porcelain basin, 750 cc. of distilled water added, and titrated with the $KMnO_4$ solution. The end-point is reached when a faint pink colour can be seen round the edge of the solution. This titration gives the amount of $KMnO_4$ solution taken up by the indicator, which must be allowed for in all calculations. The gallo-

---

[1] We are indebted to Mr F. W. Mason of the Bureau of Biotechnology at Leeds for a pure culture of this fungus.

tannin present is estimated as follows: 25 cc. of the reaction mixture are withdrawn and 4 cc. of this solution placed in a porcelain dish together with 20 cc. of indigo carmine solution and 750 cc. of distilled water. The whole is then titrated with permanganate solution until the end-point as previously described is obtained. This titration represents both the tannin and non-tannins present. The remainder of the solution is shaken for 15 minutes with 1 g. of fat-free caseinogen and then passed through a barium sulphate filter. This process is repeated twice. When all gallotannin is removed [cf. Nierenstein, 1911], 4 cc. of this filtrate are then titrated as before. The difference between the two titration readings represents the amount of gallotannin present. The actual quantity may be calculated by means of Spier's ammonium oxalate gallotannin ratio, viz. 1 g. of ammonium oxalate is equivalent to 0·4648 g. of gallotannin.

In estimating the amount of gallotannin present after hydrolysis, the procedure is the same as that just described except that it is necessary first to filter off the precipitate of gallic acid formed.

The following table gives the results obtained.

In each case three titrations were made, the mean of these being recorded; also as a check each worker carried out an independent investigation.

Sample	Age in days	Cc. of 1 % tannase solution added to 100 cc. of gallotannin solution	Duration of hydrolysis in hours	Percentage of gallotannin hydrolysed
1	112	10·0	23·75	14·70
				14·40
	2	5·0	19·50	10·52
				11·02
	2	5·0	45·00	16·60
				15·70
	~	5·0	19·50	13·48
				13·25
		7·5	19·50	13·00
				12·17
		7·5	45·00	18·25
				17·30
5		10·0	19·50	15·00
				15·35
~	2	10·0	45·00	20·00
				20·51

From the results given in the above table it is evident that (1) this method represents a moderately accurate procedure for estimating the hydrolysing power of any sample of tannase; (2) the bulk of the hydrolysis takes place during the first 24 hours.

It is proposed by Dr Nierenstein to continue this work at a later date, but since the authors are unable to pursue this research at present, it was thought advisable to publish this preliminary note.

REFERENCES.

Fischer (1912). Ber. 45, 923.
Freudenberg (1913). Ber. 52, 177.
Nierenstein (1911). Chem. Zeitg. 36, 31.
Spiers (1914). J. Agric. Sci. 6, 77.

II. THE ACTION OF HYPOPHYSIN, ERGAMINE AND ADRENALINE UPON THE SECRETION OF THE MAMMARY GLAND.

By ERNEST ROTHLIN, ROBERT HENRY ADERS PLIMMER
AND ALFRED DENNIS HUSBAND.

From the Biochemical Department of the Rowett Research Institute of Animal Nutrition, University of Aberdeen and the North of Scotland College of Agriculture, and the Physiological Institute, University of Zurich.

(*Received December 20th, 1921.*)

THE action of extracts of various organs, especially that of the pituitary gland (hypophysin) upon milk secretion has been frequently investigated. The results of these investigations have been summarised by Hammond [1913], who made an extensive study of the action of hypophysin upon the milk flow in the goat. He found that there was an immediate action and that the effect soon passed off, and he concluded that the action was not upon the muscular tissue of the gland but rather upon the glandular epithelium. The action depended upon the state of nutrition of the animal and upon the stage of lactation. The composition of the milk was not appreciably altered, though that produced by the action of the drug had a higher percentage of fat than other samples. The milk analyses were limited to estimations of fat and total solids and did not extend over long periods.

Hammond and Hawk [1917] added some further details respecting the action in different states of nutrition and tested adrenaline, which was found to have no effect.

As two goats under normal conditions, on a constant diet, but in different stages of lactation were available, there arose the opportunity of once more testing the action of hypophysin upon the stage of lactation and of extending Hammond's results. Complete analyses of the milk of these goats had been previously made, and by continuing this procedure and including extra samples some further details could be ascertained. At the same time it was of interest to compare its action with that of adrenaline and of ergamine or histamine. The latter substances, as is well known, have an action corresponding with that of excitation of the sympathetic or para-sympathetic nervous systems. Their action, if any, would give probably some information upon the innervation of the mammary gland with nerve fibres of the involuntary nervous system. Hammond's results with hypophysin have been confirmed and extended; ergamine and adrenaline, as might have been expected, had no action.

EXPERIMENTAL.

The hypophysin, ergamine as acid phosphate, and adrenaline used for injection were the commercial preparations supplied by Burroughs, Wellcome and Co. The injections were always subcutaneous in the lumbar region of the body. They were made at different times of the day, usually from one to two hours before milking and generally in both goats about the same time. Before the injection of the drugs an injection of 0·9 % saline was made and once again later; it had no effect and thus the effect of the actual injection was excluded.

The experiments were carried out in two periods; during the first period there was an interval of only one or two days between the injections; during the second period there was an interval of about seven days between the injections of the three drugs in order to ascertain if the total volume of the milk differed in the period of injection from that before and after. The injections were then made more frequently, sometimes twice in the day.

The two goats used in these experiments were in different periods of lactation. Goat A had two kids on June 14th, goat N one kid on Feb. 2nd, 1920. Goat A was in a good state of lactation, goat N in a poor state.

Goat A during the first ten days was outdoors at grass by day and indoors by night; subsequently she was kept indoors permanently, fastened by a chain in a ground floor room of the building. Goat N was always kept in the building in the same room, but in a cage in which she could freely move and turn about. Both goats were given 112 g. of oatmeal at about 8 a.m., noon and 5 p.m. This was eaten rapidly. Except for the first ten days, when goat A had grass during the day, after the oatmeal both had 454 g. of hay, which they were allowed to eat at their leisure. They were thus under fairly normal conditions and upon a constant diet. Their weights varied very little during the course of the experiment which lasted about three months, and as far as could be noticed they were in excellent health.

The milk was not drawn by suction, but always by hand by the same experienced milker, who had drawn the milk previously over a long period. During this time the goats were milked twice daily, morning and evening; the two samples were mixed and analysed. To observe the effect of the drugs it became necessary to draw the milk more frequently and it was taken at 9.30 a.m., 11.30 a.m. and 5 p.m. and occasionally at 6.30 or 7 p.m., depending upon the time of injection.

Complete daily analyses of the milk of these goats had been previously made. Owing to the number of samples now taken (six to eight) it was not possible to analyse each separately and two or more samples were combined. The combination of two samples was made in such a way that the particular sample after injection was kept separate. The analyses are shown in the table, numbered 1, 2 or 3 according to the milking time of 9.30 a.m., 11.30 a.m. and 5 p.m.

The methods of analysis were those usually employed in milk analysis. Fat was estimated by Soxhlet's method, total protein by Kjeldahl's method: caseinogen by this method after its precipitation from the sample by dilute acetic acid and washing; albumin in the filtrate from the caseinogen by precipitation with tannic acid, washing and Kjeldahl's method. The nitrogen figures by Kjeldahl's method were multiplied by the factor of 6·38 to give the protein. Lactose was estimated in a sample, precipitated by "dialysed iron," by Fehling's method. Ash was determined by incineration in a small crucible. The data are expressed in grams per 100 cc. milk.

The figures in the following table show the volumes at each milking, and the fat analyses, and the calculations to give the figures for the output of the whole day, the times of injection, etc. The data are so numerous that they are greatly abbreviated, and only those showing the essential features are given. The amounts of lactose, total protein, caseinogen and albumin except on the first two days are omitted and none of the figures of the second period is shown.

RESULTS.

(a) Changes of Volume.

1. *Hypophysin.* The injection of hypophysin generally about two hours before the milking time produced a flow of milk only in goat *A*, which was in an early stage of lactation. The effect is distinctly marked in the volume of the second sample collected after an interval of two hours from the first sample. The volumes of 250 and 260 cc. were obtained in comparison with normal volumes varying from 35 to 145 cc. The volume of the next milking was then smaller than the normal, 230 and 150 cc. against 255 and 325 cc. An increase was not definitely noticed in the volume of the third sample collected after an interval of about six hours from the previous one. A volume of 350 cc. was obtained between volumes of 325 and 410 cc., but the average volume for this milking time was between 200 and 300 cc. No increase in volume was observed in the first sample collected at 9.30 a.m., *i.e.* about 16 hours after the sample at 5 p.m. The total daily volume of milk was not appreciably altered. A change over the total period could not be observed as the goat *A* gave a gradually diminishing volume of milk.

No change in volume of the samples was observed in goat *N* after an injection of hypophysin. This goat gave a more regular daily volume and the effect was to make a slight general increase of the total volume; previous to the injection the average volume was 423 cc.; during the period it was 448 cc. and after 468 cc. A similar slight increase was again noted later, the volumes being 395, 424 and 509 cc. The increase may however be due to other circumstances, as the goat *N* had at the time of this injection been in lactation for eight to nine months. The injection in goat *A* at the later period did not produce a rapid flow as at the former time, but there was a very slight increase of total volume; the averages of the periods were 710, 724, 730.

PROTOCOLS.

Date Sept.		Vol.	Ash	Fat	Lact.	T. prot.	Cas.	Alb.	Vol.	Ash	Fat	Lact.	T. prot.	Cas.	Alb.
				Goat *A*							Goat *N*				
7		1100	0·83	6·22	4·06	3·64	3·00	0·47	450	0·91	5·75	4·55	3·57	2·57	0·75
8		1200	0·80	7·80	4·23	3·59	2·95	0·48	390	0·90	5·07	4·64	3·73	2·75	0·75
*9		1200	0·76	7·02	4·26	3·61	2·95	0·51	430	0·97	5·50	4·60	3·52	2·53	0·77
†10	1	900	0·85	4·20	4·36	3·77	2·90	0·51	315	0·91	5·31	4·60	3·63	2·53	0·75
	2	250	0·80	11·21	4·00	3·61	2·75	0·46	10	—	12·07	4·37	3·68	2·50	0·80
	3	230	0·82	9·12	4·11	3·85	2·90	0·47	140	0·90	6·25	4·64	3·69	2·59	0·70
per total		1380	0·83	6·29	4·26	3·75	2·87	0·49	465	0·91	8·47	4·61	3·65	2·55	0·74
‡11	1	900 } 0·87		5·75	4·31	4·00	3·07	0·53	308 } 0·91		5·22	4·70	3·77	2·64	0·75
	3	215							90						
	2	55	0·80	11·35	4·08	3·78	3·02	0·47	12	—	8·67	4·38	3·73	2·26	0·52
		1170	0·87	6·01	4·30	3·99	3·07	0·53	410	0·91	5·28	4·66	3·77	2·63	0·74

* Both goats injected with 1 cc. 0·9 % NaCl after first milking.

† Both goats injected with 0·5 cc. hypophysin after first milking; milk collected one hour after injection (10.25 a.m.).   ‡ No injection.

Date Sept.		Vol.	Ash	Fat	Vol.	Ash	Fat	Remarks
			Goat *A*			Goat *N*		
12	1	920	—	—	310	—	—	No injection.
	2	100	—	—	35	—	—	
	3	255	—	—	85	—	—	
		1275	0·84	7·80	430	0·90	4·99	
13	1	840	0·83	5·00	300	0·91	4·89	Both injected with 1 cc. hypophysin
	2	260	0·74	14·09	37	0·82	9·29	after first milking; milk collected
	3	150	0·90	7·37	83	0·94	4·77	one hour after injection.
		1250	0·82	7·09	420	0·91	5·25	
14	1	940 } 0·85		4·13	335 } 0·96		4·31	No injection.
	3	325			118			
	2	95	0·74	10·35	42	0·90	10·74	
		1360	0·84	4·57	495	0·95	4·86	
15	1	680	0·86	4·65	325	0·92	4·50	Goat *A* injected with 1 cc. 0·9 %
	2	70	0·86	10·62	10	—	9·65	NaCl at 3 p.m. Goat *N* injected
	3	350	0·82	9·00	130	0·90	6·29	with 1 cc. hypophysin at 3 p.m.
		1100	0·85	6·41	465	0·91	5·11	
16	1	520	0·84	4·00	280	0·96	4·10	No injection.
	2	95	0·80	6·30	40	0·90	7·52	
	3	410	0·84	6·39	115	0·96	4·69	
		1025	0·84	5·37	435	0·95	4·57	
17	1	800	—	—	305	0·94	3·89	Both goats injected with 1 cc. hypo-
	2	145	—	—	45		5·98	physin at 8.15 a.m.; milk collected
	3	265	—	—	95	0·92	4·00	one hour later.
		1210	0·83	5·54	445	0·93	4·12	
18	1	750	—	—	325	0·95	4·12	No injection.
	2	35	—	—	40	0·88	7·40	
	3	210	—	—	110	0·88	4·69	
		995	0·81	5·51	475	0·93	4·53	
19	1	680	0·80	5·15	315	—	—	No injection.
	2	50	0·80	9·30	35	—	—	
	3	230	0·80	8·40	110	—	—	
		960	0·80	6·15	460	0·95	4·15	
	4	30	0·83	9·40	35	0·94	5·53	
20	1	690	0·85	5·50	300	—	—	Goat *A* injected with 1 cc. hypo-
	2	65	0·82	8·10	30	—	—	physin at 8.15 a.m.; milk collected
	3	175	0·88	6·52	105	—	—	one hour later.
		930	0·85	6·00	470	0·94	4·39	
21	1	620 } 0·87		5·82	305 } 0·93		3·94	Goat *N* injected with 1 mgm. orga-
	3	145			45	0·75	8·21	mino phosphate at 9.45 a.m.; milk
	2	70	0·82	10·25	100	0·94	4·97	collected one hour later.
		835	0·87	6·49	450	0·93	4·60	

Date Sept.		Goat A Vol.	Ash	Fat	Goat N Vol.	Ash	Fat	Remarks
22	1	570	0·85	5·60	250 ⎫	0·94	4·36	Goat A injected with 1 mgm. ergamine phosphate at 9.45 a.m.; milk collected one hour later.
	3	120	0·88	6·30	100 ⎭			
	2	110	0·74	9·38	35	0·90	7·45	
		800	0·84	6·23	385	0·94	4·64	
23	1	730	0·90	5·72	320	0·94	4·83	Both goats injected with 1 mgm. ergamine phosphate at 3.45 p.m.; milk collected at 5 p.m.
	2	140	0·82	10·02	18	—	7·82	
	3	160	0·91	5·80	142	0·90	5·26	
		1030	0·89	6·41	480	0·93	5·07	
24	1	660	0·85	5·15	315	0·92	4·73	No injection.
	2	80	0·83	9·22	30	0·86	9·85	
	3	200	0·86	6·50	105	0·93	4·95	
		940	0·86	5·78	450	0·92	5·12	
25	1	640	0·90	6·05	300	0·92	4·56	Goat A injected with 0·5 mgm. ergamine phosphate at 8.15 a.m. Goat N injected with 1·0 mgm. ergamine phosphate at 8.15 a.m. Milk of both goats collected one hour later.
	2	50	0·80	9·54	20	—	8·54	
	3	140	0·87	7·40	100	0·91	5·18	
		830	0·89	6·49	420	0·92	4·90	
26	1	570	—	—	295	—	—	
	2	70	—	—	15	—	—	
	3	205	—	—	120	—	—	
		845	0·89	6·73	430	0·98	4·96	
	4	33	lost	7·43	33	0·93	6·57	
27	1	622	0·88	5·03	282	0·94	4·78	Both goats injected with 1 mgm. ergamine phosphate at 3.30 p.m. Milk collected at 5 p.m.
	2	20	—	—	10	—	—	
	3	225	0·87	7·04	130	0·96	6·46	
		900	0·88	5·67	455	0·94	5·40	
28	1	600	0·87	5·68	300	0·92	4·67	Both goats injected with 1 mgm. ergamine phosphate at 6.15 p.m. Milk collected at 9.15 a.m. next day.
	2	50	—	—	15	—	—	
	3	130	0·86	7·32	125	0·93	7·94	
		780	0·87	6·66	440	0·92	5·71	
29	1	630	0·88	7·64	295	0·95	5·15	No injection.
	2	30 ⎫	0·86	8·76	10 ⎫	0·93	5·72	
	3	220 ⎭			130 ⎭			
		880	0·87	7·96	435	0·94	5·33	
30	1	540	0·87	6·80	295	0·93	5·62	Both goats injected with 1 mgm. adrenaline at 1.15 p.m.; milk collected at 5 p.m.
	2	45	—	—	20	—	—	
	3	110	0·87	7·63	100	0·93	8·01	
		695	0·87	7·21	415	0·93	6·31	
Oct.								
1	1	620	0·90	6·92	285	0·97	5·28	Both goats injected with 1 cc. 0·9 % NaCl at 1.30 p.m.; milk collected at 5 p.m.
	2	40	—	—	35	—	—	
	3	210	0·86	8·02	110	0·94	6·51	
		870	0·89	7·24	430	0·96	5·69	
2	1	590	0·81	5·63	300	0·93	5·00	Both goats injected with 1 mgm. adrenaline at 9.45 a.m.; milk collected at 11 a.m.
	2	30	—	—	15	—	—	
	3	200	0·86	8·28	100	0·91	7·01	
		820	0·84	6·37	415	0·92	5·56	
	4	24	—	9·24	20	—	8·34	
3	1	536	0·87	6·60	265	0·94	5·46	No injection.
	2	40	—	—	35	—	—	
	3	140	0·87	8·38	100	0·94	6·66	
		740	0·87	7·12	420	0·94	5·98	
4	1	650	0·86	5·61	315	0·91	4·87	Both goats injected with 1 mgm. adrenaline at 1.15 p.m.; milk collected at 5 p.m.
	2	30	—	—	10	—	—	
	3	170	0·88	8·90	110	0·92	6·21	
		850	0·86	6·27	435	0·91	5·24	
5	1	480	0·91	6·00	260	0·95	4·75	No injection.
	2	75	—	—	10	—	—	
	3	190	0·89	7·49	115	0·94	5·70	
		745	0·90	6·53	385	0·95	5·06	

2. *Ergamine*. The injection of ergamine had no pronounced effect on the separate volumes of the samples of either goat and the average volume of the total milk per day was about the same as when no injection was made. A general effect over the whole period was not noticeable in the case of goat *A* which was giving a diminishing flow of milk, but a slight diminution can be made out in the case of goat *N*. During the second period of injection the flow of milk in goat *A* was more constant, the average daily volumes being 645, 656, 657 cc. for the pre-, actual, and post times; the average figures of goat *N* at this time were 532, 493, 576. In both cases the changes are not essentially different from the ordinary daily volumes.

3. *Adrenaline*. During the first period no effect could be seen in the case of either goat, both the separate volumes and the average daily volumes remaining unchanged. During the second period there was an apparent increase in goat *A*, but the increase occurred at a time when the daily flow was lessening; the reverse effect can be made out in goat *N*. The general effect is not sufficient to indicate any definite action of adrenaline.

(b) Composition of the Milk.

No real change in the chemical composition of the total milk per day occurred in either goat after the injections of hypophysin, ergamine and adrenaline. If any change did occur, it was not greater than the normal variations in composition. These normal daily variations are most marked in the fat content. Possibly the fat content was lowered over the total period during which hypophysin was injected and raised during the ergamine period, but again the alterations were not more than those which occur daily.

The three separate samples of milk taken at different times of the day after different intervals had normally a distinctly different composition. The fat content was always highest in the second sample taken two hours after the first, and the early morning sample had the lowest amount of fat; it was taken after an interval of 16 hours. The higher fat content was observed in both goats and is thus independent of the period of lactation. The second sample had less protein and lactose than the other samples. On the average the first sample at 9.30 a.m. had more protein and lactose than the third sample, but the differences are not distinctly marked off.

If allowance be made for the ordinary daily variations, the separate samples do not show any appreciable difference from the normal.

Discussion.

Since a marked increase in the volume of milk was only observed in goat *A* at the time of the second milking, and since it was followed by a diminished volume at the third milking, it appears that the action of hypophysin is powerful, rapid and of short duration. The phase of hyper-secretion after injection is followed by a compensatory phase of hypo-secretion. This effect explains the absence of an increase in volume at the time of the first milking and third milking. The interval between the milkings was here 16 hours; an increase followed by a decrease would not be noticed. The interval between the first and second milkings was two hours. The gland secretes normally (Sept. 12) about 50 cc. every hour; after hypophysin the secretion was 125 cc. (Sept. 10) and 130 cc. (Sept. 13) per hour. Hypophysin had no effect in goat N and at the later period in goat A. The action of hypophysin is therefore probably not directly upon the gland. If the smooth muscle were stimulated, a flow of milk should follow at either period of lactation. If the secreting cells were affected a flow of milk should also follow in any case. It is most probable that it may act indirectly through the reproductive organs, which have been proved to contain substances acting upon the secretion of the mammary gland. At the earlier period of lactation the reproductive organs will be in a state of activity, whereas in the later stage their state of activity will have disappeared. Hypophysin thus may act upon an active organ which produces the galactogogue; further work will be required to determine whether this active organ be the ovary, corpus luteum, uterus, or placenta.

In practice, the injection of hypophysin will only be of value at early stages of lactation and it may be able to bring a gland into activity when it is not already in that state. It must be remembered that the total volume produced by the gland is not appreciably increased, as shown above and previously described by other investigators.

The composition of the total milk per day is not altered by the injection of hypophysin. It has been previously considered that hypophysin causes a flow of milk with a higher fat content, but, as the data show, it is followed by a milk with a lower fat content. The high fat content of a sample taken at a short interval between milkings is normal and is not really due to the action of hypophysin.

As milk contains more fat if collected at a short interval after a previous longer interval, it appears that the secreting mechanism of the mammary gland is of a two-fold character; a mechanism producing fat and a mechanism producing protein and lactose. The fat-producing mechanism begins to act first and a milk of high fat content results; the protein and lactose mechanism acts later and dilutes the fat content to the normal value. The fat mechanism is probably more easily influenced by other conditions, since the amount of fat in milk is very variable, whilst those of protein and lactose are fairly constant.

Former investigators have observed that adrenaline has no action upon the mammary gland; these results confirm their work.

Ergamine also has no action. If these substances have a positive or negative action, it is rapidly followed by compensation.

The inaction of adrenaline points to the absence of sympathetic nerve fibres from the mammary gland, if the general law of the relationship between the action of adrenaline and the presence of sympathetic fibres is here correct.

It has been suggested from analogy to other glands (stomach) that ergamine perhaps acts on para-sympathetic fibres. The inaction points to the absence of such fibres from the mammary gland.

SUMMARY.

Hypophysin produces a flow of milk only in the early stages of lactation. Its action is rapid and powerful, but of short duration. The flow (hyper-secretion) is followed by a smaller quantity than normal (hypo-secretion). The total volume of milk per day is not altered.

Since hypophysin does not act at later stages of lactation, it is probable that its action is indirect through the organs of reproduction.

The quality of milk secreted after the action of hypophysin is not different from the normal. Normal milk has a high fat content, if it be collected at a short interval after the last milking.

The secretory activity of the mammary gland is not influenced by the subcutaneous injection of adrenaline or ergamine. The gland probably does not contain sympathetic or para-sympathetic nerve fibres for the secretory mechanism. If they be present, the inaction of adrenaline and ergamine is exceptional.

REFERENCES.

Hammond, J. (1913). Quart. J. Exp. Physiol. 6, 311.
Hammond, J. and Hawk, J. C. (1917). J. Agric. Sci. 8, 147.

III. THE REARING OF CHICKENS ON THE INTENSIVE SYSTEM. PART I. THE VITAMIN REQUIREMENTS.

PRELIMINARY EXPERIMENTS.

BY ROBERT HENRY ADERS PLIMMER & JOHN LEWIS ROSEDALE

WITH THE ASSISTANCE OF

ARTHUR CRICHTON & ROBERT BAYNE TOPPING.

From the Biochemical Department, Rowett Research Institute, University of Aberdeen and North of Scotland College of Agriculture.

(*Received December 20th, 1921.*)

THE attempts of previous workers to rear chickens in confinement under laboratory conditions have not been altogether successful. Drummond [1916] considered that there were other factors than an adequate diet which must be supplied to the growing chick. Osborne and Mendel [1918] commented upon various statements by Drummond and remarked "if the conditions under which chickens continue to grow normally in confinement can be learned, it will be possible to obtain much information of practical use in poultry husbandry." Further they say "There is a widespread belief among poultry raisers that young chickens cannot be reared under the artificial conditions of housing and diet which many other experimental animals tolerate without detriment. The current ideas are expressed in the statement that the birds must be kept 'on the ground,' that they must have exercise, outdoor life, and green food."

In their previous experiment [1916] with four chicks from 3 to 4 weeks old, two died and two were observed to the age of 77 days. Two other birds 81 days old at this time were actually kept for a total of 309 days in which time they reached maturity in apparent full vigour. The birds at the ages of 77 and 81 days had very ruffled feathers; this appearance was attributed to frequent handling. Osborne and Mendel beheved the success was largely due to the supply of blotting paper as roughage. Out of another group of ten birds, two only reached maturity; the others failed to grow and showed leg weakness. We may record the better success of Buckner, Nollau and Kastle [1915] and of Palmer and Kempster [1919]. Hart, Halpin and Steenbock [1920] have emphasised the frequent occurrence of leg weakness in chickens reared in confinement and, like Osborne and Mendel, attribute the cause to the absence of roughage from the diet. Drummond's birds were fed upon ordinary commercial chicken food; Osborne and Mendel fed their birds upon the foods which they supplied to rats, containing butter fat and protein-free milk as sources of the A and B accessory food factors, a diet quite satisfactory for their growth. Buckner, Nollau and Kastle used grain mixtures and supplied some "green" food. Hart, Halpin and Steenbock used a diet similar to that of Osborne and Mendel.

It is now known that, though rats and a few other animals can be reared on a diet containing only A and B factors, most animals, including man, require the inclusion of C factor for the maintenance of health. It was thus possible that the poor success was due to the absence of this third factor, although it was added to the diets fed by Hart, Halpin and Steenbock. With reference to this third factor, the work of Chick and her colleagues [1919] insists upon the inclusion of a definite daily quantity which must be above a certain minimal amount. The necessity for a daily amount of the other vitamins has not been so clearly established, but at any rate it is indicated by the experiments of Cooper [1913] and of Chick and Hume [1917].

In these experiments the birds have been supplied with the usual protein, carbohydrate and fat; salts other than those contained in the foodstuffs were not added. Accessory food factors have been given in the form of cod liver oil for A, autolysed yeast or marmite for B, and lemon or orange juice for C. The choice of these three substances was made on account of the general experience that cod liver oil is the most concentrated in A factor, marmite in B, and lemon or orange juice in C, as well as the fact that these substances are very fairly constant in their respective vitamin content. The diet was composed essentially of oatmeal and milk and no account was taken of the presence of vitamins in these foods. Water was always freely provided.

Arbitrary amounts of the vitamin foods had to be chosen as the requirements for chickens are unknown. Later work is designed to determine the minimal daily amount of each vitamin, and subsequently these amounts will have to be translated into the terms of natural foodstuffs as supplied by the poultry keeper. It must be pointed out that the natural foods are very unequal in their vitamin content and that at present, in the absence of chemical knowledge of vitamins, only comparable data are capable of investigation. Also, it must be clearly stated that the diet now supplied is not one that would be given under ordinary conditions to poultry.

With the food supplied 21 chickens out of 24 have been reared to maturity; two were lost from illness apparently due to insufficient vitamin B in the diet and one was lost accidentally; in other terms 91 % were successfully reared. In Part II [1922] it will be seen that nine birds out of nine, or 100 % were brought to maturity.

Experimental.

Housing. The chicks up to the age of about four weeks were housed in an ordinary type of foster mother, heated by a paraffin lamp[1]. After this age they were kept in wooden poultry houses measuring $6 \times 4\frac{3}{4} \times 6\frac{1}{2}$ feet high with an entrance door at one end and an opening into a run at the other end. Ventilation hatches were provided at each end at the top. The run of wire netting on wooden supports on a wood floor had the size of $6 \times 4\frac{3}{4} \times 3$ feet high. Altogether there were four identical houses, two on the roof of the building, two in a room measuring $30 \times 20 \times 15$ feet high on the ground floor of the Institute. The foster mother was kept in the ground floor room.

The wooden floors of the foster mother, poultry houses and runs were covered with broken peat moss litter, except that during wet weather, sand and coal ashes were used in the houses on the roof. The peat moss litter was scraped about but, as far as could be ascertained, very little was eaten. Grit was supplied in the form of coal ashes from the age of about two months. The birds were weighed separately every two days.

Diet. There is very little information about the daily quantity of food consumed by little chicks; it is stated [Board of Agriculture and Fisheries, 1907] that chicks eat about 3 lbs. of dry food during the first eight weeks of their life. Consequently definite amounts were prepared every day; portions were given at intervals of 2–3 hours to the little chicks and 2–4 times daily as they grew older. In this way a record could be kept of the amount eaten at the different ages. It was found that the total consumption corresponded to that stated.

Kilned oatmeal of pin head size was selected as the basal diet, mainly on account of its method of manufacture which would exclude the presence of vitamins, but their entire absence could not be expected. A small quantity of milk was also given to provide extra protein; the amount was at first rather variable, but later was always measured. Towards the end caseinogen was substituted for milk in the case of half the birds and at the same time B factor was excluded, since a comparison between caseinogen and marmite and at the same time an estimate of the vitamin content of the caseinogen, were wanted.

The accessory food factors were added to the daily quantity of oatmeal. According to the health of the birds the original amounts required alteration. The course of the experiment is most conveniently divided into periods from which the variations in food consumption and vitamin needs are more easily ascertained.

Twenty-four chicks were comprised in the group, 12 white and 12 black Leghorn. They were hatched in an incubator and were from Miss Fraser's strains at the North of Scotland College of Agriculture. At the time of their arrival their age was approximately three days; several were suffering from diarrhoea and were very weak and some mortality was expected.

[1] Kindly lent by Miss Fraser of the North of Scotland College of Agriculture.

Period 1 *from July* 13*th to Aug.* 10*th,* 1920. The birds were housed in the foster mother and supplied daily with 120 g. oatmeal, 5 cc. cod liver oil, 0·5 g. marmite, 30 cc. lemon juice.

Extra food, as required, after the above mixture had been eaten, was given in the form of dry oatmeal, or a mixture of dry oatmeal and biscuit meal; at first 30 g. was eaten daily, but it gradually increased until at the end of the period 200 g. was eaten. Including the basal mixture, the total food consumption per bird increased from 6 to 13 g. per day.

In addition the birds were given 200 to 250 cc. of milk every day; it was mostly spilt and thus no record of its consumption could be made. On two occasions the yolk of an egg was given and lemon skins were put into the run for the birds to peck at and for their amusement; as far as could be seen, none was eaten.

On July 21st, *i.e.* on the ninth day, one of the weak birds was found dead in the outer part of the foster mother.

The other chicks recovered from the chicken diarrhoea and no signs of illness were noticed in any of the other birds.

The average weight of the birds on July 27 was 78 g.; on Aug. 20 it was 150 g. The increase averaged 5·1 g. per day.

Period 2 *from Aug.* 11*th to Sept.* 3*rd,* 1920. The 23 birds were now divided into two groups. Group I had six black and six white; group II had five white and six black chicks. The selection was made so that the average weight of the two groups was approximately the same: group I weighed 154, group II 156 g. per bird. The two groups were housed in the poultry houses in the ground floor room.

Each group received as basal diet 120 g. of oatmeal. The vitamins were added as before, but the amounts were divided between the two groups, except in the case of C factor. Each group received 2·5 cc. of cod liver oil and 0·25 g. of marmite per day. The difference in the amount of C factor was made on account of the stress that poultry keepers lay on green food for birds, and because, presumably, C factor is the most abundant in these foods. Group I was given 15 cc., group II 20 cc. of lemon juice per day; in the case of group I it was thus the same as before.

For the first three days extra food was given as oatmeal and biscuit meal; each group consumed about 44 g. per day. On Aug. 14th, the extra food was given as a mixture of oatmeal and milk, 10 cc. of milk being mixed with 30 g. of oatmeal. The proportion of milk was raised to 20 cc. on Aug. 26th, since the protein ratio was very low on the former mixture. The protein ratio of the total thus became 1 : 6 approximately. No account of the presence of vitamins in the milk was taken as it is now proved that the vitamin content of milk is variable and the quantity so supplied was expected to make little difference to that supplied directly.

Five days after the separation into two groups, on Aug. 15th, illness was observed amongst the birds of group I. There was no particular characteristic sign; the birds were less active and drooped. The cause was attributed to insufficiency of C factor. The quantities were therefore increased. Group I was given 35 cc. and group II 30 cc. per day, *i.e.* in the reverse order to that previously. A little improvement was noticed, but the birds still remained unwell. Consideration was then given to the amount of B factor. The quantity of 0·25 g. to each group was very small and the amount was raised to 0·5 g. per day to each group on Aug. 19th.

After this increase some birds of group II showed similar signs of illness to those of group I, and the quantity of lemon juice was again increased to 40 and 50 cc. per day respectively to groups I and II.

Group H. One bird of group I was so ill on Aug. 24th that it was removed and housed in the foster mother (not heated). The next day two other birds from group I and one bird from group II were also put into the foster mother. This so-called hospital group H was given the basal diet of oatmeal and vitamins supplied to group II, but with the cod liver oil increased to 5 cc.; so far no alteration had been made in the quantity of A factor and the illness was possibly due to its insufficiency. The increase had no effect; the birds got no better and on Aug. 27th the first bird of group I died. Its death was diagnosed by Dr W. Taylor as due to chicken cholera.

Some of the birds now showed distinct signs of leg weakness, drooping wings and ruffled feathers. These symptoms have been described in the case of birds suffering from polyneuritis gallinarum in its early stages. The cause of the illness was thus indicated to be due to insufficiency of B factor. Extra marmite was consequently given, but owing to the poor appetite of the birds it could not be given in the food; on Sept. 1st, it was therefore supplied in the form of a 2 % marmite drink. This was greatly relished and on the third day unmistakable signs of improvement in health were observed. Just as this improvement was noticed two more birds from group II had to be admitted to the hospital group (Sept. 3rd). With the marmite drink they also got better in three days. The drink was stopped on the 9th Sept. and as the birds were now quite well they were returned to their groups on Sept. 11th.

Period 3 from Sept. 4th to 10th, 19). The birds remaining in groups I and II were supplied from Sept. 1st to 3rd with the same daily diet consisting of 300 g. oatmeal, 120 cc. milk, 2·5 cc. cod liver oil, 0·5 g. marmite, 50 cc. lemon juice. Since there was no object in continuing the same diet to both groups, a change was made on Sept. 4th in the quantity of A factor. Group II was given 10 cc. Loss of appetite was noticed in both groups and several more birds in group II were ill. A drink of lemon juice had no effect. The remarkable effect of the marmite drink had just been observed in group H and it appeared clear that too little B factor was being supplied. The quantity of marmite in the basal diet of both groups was therefore doubled on Sept. 6th. Group I relished the increase by showing an improved appetite, but group II still appeared ill. On Sept. 8th group II was given a drink of 2 % marmite; its effect was again striking. It thus appeared clear that too little B factor had been given and that chickens are very susceptible to an insufficiency; it was very noticeable that the leg weakness disappeared with the increase. Guinea-pigs show a similar behaviour, but their susceptibility is to C factor. The daily amount of marmite in the diet was doubled for the last two days of this period.

Period 4 from Sept. 11th to 16th, 1920. On Sept. 11th the birds of the hospital group were returned to their respective groups. On account of the death each group had now eleven birds.

The effect of the cod liver oil on group II had had the noticeable effect of producing a loss of appetite and signs of ill-health; another period of a similar diet was therefore tried. Both groups were given more marmite, 2·5 g. per day and the same amount of lemon juice 30 cc. gradually reduced from 50 cc. Oatmeal remained as before. Group I had 5 cc. of cod liver oil; group II 12·5 cc. The birds of group I (except one which was given 0·5 g.. marmite, dissolved in water, by hand on Sept. 14th, after which it recovered) appeared well and had a good appetite. The birds of group II showed marked loss of appetite, and the quantity of marmite was doubled on Sept. 14th; by Sept. 17th their appetite had again become normal. The extra cod liver oil was thus not itself the cause, but rather the extra fat required B factor for its assimilation. Braddon and Cooper [1914] have made similar observations in cases of beri-beri, and have suggested that the amount of B factor ran parallel to the caloric value of the diet.

During this period a bird of group I swallowed a nail and died.

Period 5 from Sept. 19th to Nov. 10th, 1920. The above observations indicating that the amount of B factor needed bore a constant ratio to the amount of carbohydrate and fat in the diet led to a general rearrangement of the diet. For every 30 g. of oatmeal a quantity of 0·5 g. of marmite was given; cod liver oil and lemon juice were given in a constant quantity per day. The only difference in the two groups was the extra cod liver oil for group II and its balancing by 2 g. of marmite. The daily food consumption for the first part of this period was the following:

	Group I				Group II			Extra for groups I and II		
	Oatmeal g.	Cod liver oil cc.	Marmite g.	Lemon juice cc.		Cod liver oil cc.	Marmite g.	Lemon juice cc.	Oatmeal g.	Marmite g.	Milk cc.
Sept. 19–21	180	5	3	30	180	12·5	5	30	120	2	80
„ 22–Oct. 17	180	6	3	30	180	12·5	5	30	240	4	160
Oct. 18–21	180	5	3	30	180	12·5	5	30	360	6	240

The food was supplied in two portions, and in consequence of the greater size of the birds the milk portion was rapidly increasing above the original quantity. It was therefore decided to reduce the milk to 100 cc. per day and for greater convenience all the items were made into a single mixture:

	For group I					For group II			
	Oatmeal g.	Milk cc.	Cod liver oil cc.	Marmite g.	Lemon juice cc.	Oatmeal g.	Milk cc.	Cod liver oil cc.	Marmite g.	Lemon juice cc.
Oct. 22–Nov. 10	600	100	5	11	30	600	100	12·5	13	30

These quantities were eaten daily by the birds of group II, but on two occasions they were not eaten by group I, due to the fact that this group had 10 birds against 11 in group II.

There were no signs of illness throughout this period of 54 days; the birds grew steadily, increasing on the average 25 g. every two days. Group I increased from 406 to 1019 or 613 g., group II from 329 to 1011 or 682 g. It would appear that the extra cod liver oil had a stimulating action, but as two birds of group II were small at the start and two were cockerels of a large strain, it was not a real effect.

An alteration in the housing was made on Oct. 12th, both groups being moved to the roof. The move made no difference to the regular weight increase nor to the daily food consumption, so that housing is of minor importance provided diet and cleanliness are satisfactory. During this period the cocks began to crow, the first one crowing on Oct. 13th.

Period 6 from Nov. 11th to Jan. 13th, 1921. The quantities of the several accessory factors supplied in the previous period were evidently sufficient to keep both groups in perfect health. Group III (see Part II) was receiving a diet containing caseinogen, accwa and 0·25 g. marmite per 30 g. oatmeal per day, *i.e.* half the amount of marmite. The caseinogen and secwa would thus together contain 0·25 g. of B factor reckoned as marmite. It was possible that the caseinogen alone contained B factor equivalent to that in the marmite. Caseinogen was therefore substituted for marmite in group I. It was also of interest to compare the protein of caseinogen with that contained in marmite. The diets thus became:

	For group I					For group II				
	Oatmeal g	Milk cc.	Caseinogen g	Cod liver oil cc.	Lemon juice cc.	Oatmeal g.	Milk cc.	Marmite g.	Cod liver oil cc.	Lemon juice cc.
Nov. 11–Dec. 6	660	100	13	5	30	660	100	13	5	30
Dec. 7–24	990	150	19	7·5	45	990	150	19	7·5	45

The increase at the later date was an extra half ration on account of the greater daily consumption; as before, group I with ten birds did not require food on two occasions.

Both groups increased in weight: group I from 1018 to 1489 or 471 g., group II from 1045 to 1515 or 470 g. per bird. No signs of ill-health were noticed. It was necessary to separate the cocks and hens at the last date. Two main groups were made: group I a white cock and four white hens, group II a black cock and six black hens. The extra cock birds were kept in separate houses. The same diet was continued to the end of the period. The hens then began to lay and another rearrangement of groups was necessary. The white birds (group IV) were compared with group III (see Part II) and the blacks were divided into two groups (I and II) consisting of three hens and a cock, the extra cock being taken from the extra set.

Period 7 from Jan 14th to Feb. 28th, 1921. Numerous experiments on the effect of vitamins on egg laying are conceivable. Since it had been found that birds are most susceptible to insufficiency of B factor, the effect of extra marmite on egg laying was first tried.

The birds (groups I and II) were given an identical diet except that group II was given double the amount of B factor. To compensate for any extra protein in the marmite the two groups were supplied with the same total quantity, but in the case of group I half was heated in the autoclave at 120° C. for one hour to diminish the B factor. The daily diets were:

Group I						Group II				
Oatmeal g.	Milk cc.	Marmite g.	Heated Marmite g.	Cod liver oil cc.	Lemon juice cc.	Oatmeal g.	Milk cc.	Marmite g.	Cod liver oil cc.	Lemon juice cc.
360	40	6	6	3·5	15	360	40	12	3·5	15

No signs of ill-health were noticed in either group and the average weights remained constant at about 1609 g. for group I and 1638 g. for group II. There was a slight difference in the number of eggs:

	Group I	Group II
Jan. 14–31	10 weighing 45·5 g. on average	4 weighing 47 g. on average
Feb. 1–28	27 ,, 51·5 g. ,,	36 ,, 51·5 g. ,,

There were two double eggs weighing 84 and 73 g.

The egg laying period was too short to yield a definite result, but if any recognition can be given to the actual birds in the groups, those in group I were better and started laying earlier than those in group II, and yet more eggs were laid by group II.

The experiment had to be discontinued at this period.

CONCLUSIONS.

Chickens can be reared in confinement on a diet of oatmeal and milk if the three vitamins are all contained in the diet.

Chickens are very susceptible to an insufficiency of B factor in the diet. The amount of B factor required in the diet seems to run parallel to the amount of carbohydrate and fat. It may be represented by 0·5 g. of marmite per 30 g. of oatmeal and per 5 cc. of cod liver oil for 11 birds.

A daily amount of A factor as contained in cod liver oil of 5 cc. and a daily amount of C factor as present in lemon juice of 30 cc. suffice for 11 birds.

The above figures are not to be regarded as minimal figures.

It is probable that the necessary "green" food for fowls supplies extra B factor rather than the expected A and C factors.

It would appear that the leg weakness of chicks is due to insufficiency of B factor in the diet.

REFERENCES.

Board of Agriculture and Fisheries (1907). Leaflet No. 114.
Braddon and Cooper (1914). *J. Hygiene*, 14, 331.
Buckner, Nollau and Kastle (1915). *Amer. J. Physiol.* 39.
Chick and colleagues (1919). Quoted in *Med. Research Committee, Report No.* 38 on Vitamines.
Chick and Hume (1917). *J. Roy. Army Med. Corps*, August.
Cooper (1913). *J. Hygiene*, 13, 436.
Drummond (1916). *Biochem. J.* 10, 77.
Hart, Halpin and Steenbock (1920). *J. Biol. Chem.* 43, 421.
Osborne and Mendel (1916). *J. Biol. Chem.* 26, 293.
Osborne and Mendel (1918). *J. Biol. Chem.* 33, 433.
Palmer and Kempster (1919). *J. Biol. Chem.* 39, 299.
Plimmer and Rosedale (1922). *Biochem. J.* 16, 19.

IV. THE REARING OF CHICKENS ON THE INTENSIVE SYSTEM. PART II. THE EFFECT OF "GOOD" PROTEIN.

By ROBERT HENRY ADERS PLIMMER & JOHN LEWIS ROSEDALE

WITH THE ASSISTANCE OF

ARTHUR CRICHTON & ROBERT BAYNE TOPPING.

From the Biochemical Department, Rowett Research Institute, University of Aberdeen and North of Scotland College of Agriculture.

(Received December 20th, 1921.)

OSBORNE and MENDEL [1912, 1914] have shown that the growth of rats depends upon the presence of the amino acid, lysine, in the protein of the diet and that normal or average growth follows if 2 or 3 % of this compound be present in the protein. They have shown further [1916, 1] that proteins vary greatly in their capacity for promoting growth. Lactalbumin was superior to caseinogen and caseinogen to edestin; cereal proteins were not of "good" quality[1]. Buckner, Nollau and Kastle [1915] found that a mixed grain diet of high lysine content produced more rapid growth than one of low lysine content in the case of chicks and this was confirmed by Osborne and Mendel [1916, 2]. Buckner, Peter, Wilkins and Hooper [1919] repeated the experiments with a larger number of birds kept under better conditions at a poultry farm and again noted the effect of a high lysine content. In all these experiments there was a considerable mortality amongst the birds, probably to be accounted for by an insufficient supply of vitamins. In Part I [Plimmer and Rosedale, 1922] it was shown that chickens could be reared in confinement if an adequate quantity of each of the three vitamins were added to the diet. It seemed that the B factor must be present in preportion to the amount of carbohydrate and fat in the diet.

The present experiment was designed partly to confirm the previous one and partly to ascertain the effect of "good" protein upon the rate of growth. A mixture of lactalbumin and caseinogen was chosen; both these proteins contain a high amount of lysine; the main deficiency is that of cystine in caseinogen. Vitamins, additional to those in the commercial foods, were added and oatmeal formed the basis of the diet. The effect of the "good" protein produced the expected result of rapid growth. The cocks reached the average weight of 1828 g. in 122 days; after this time they increased more slowly reaching the average weight of 2002 g. in 144 days; they were really suitable for marketing at the former age. The hens started laying at the age of 139 days.

---

[1] A summary on "Quality of Protein in Nutrition" is given by Plimmer [1921].

Experimental.

The group (No. III) consisted of nine white Leghorn chicks of the same strains as those in the previous experiment hatched under hens on Aug. 21st, 1920. They were housed for the first three weeks in the foster mother and then in one of the poultry houses in the ground floor room of the Institute, except for a period of 21 days from Sept. 23rd to Oct. 13th, during which time they were kept in a house on the roof of the building. This change made no difference to their rate of growth, which was rapid throughout the experiment.

The diet, except for the first five days, was composed of a mixture of equal parts of caseinogen[1] and secwa[2] or dried whey, mixed with oatmeal. This choice was made so as to give some comparison with the milk used previously. No allowance was made for vitamins in these foods: B factor would be present in the lactalbumin preparation, and both A and B factors in the caseinogen. To ensure an adequate supply they were added in the form of cod liver oil, marmite and lemon or orange juice. The cod liver oil and lemon juice were given, except for a short time at the beginning and again later, in constant daily amounts, but the marmite was increased with the increase of oatmeal required as the birds grew bigger. It was always added in the proportion of 0·25 g. to 30 g. of oatmeal. The food was prepared daily in the same way: the required amounts of oatmeal, caseinogen and secwa were weighed out; marmite was measured from a freshly prepared 25 % solution and mixed with the lemon or orange juice and cod liver oil; some extra water was added and the three ingredients shaken together and well stirred into the food mixture. A wet mash was thus the only food provided; dry food only consisted of the dried surface of the mixture. No extra salts were given and no grit, but on one day the birds were given by mistake some broken pieces of porcelain. The examination of the gizzards of the surplus cocks, after they were killed, revealed the presence of small black stones; they were presumably picked up from the peat moss litter used to cover the floor of the cage.

The quantity of food eaten, protein ratio, average weight and average daily increase are shown in the following table:

	No. of days	Oatmeal g.	Casein-ogen g.	Secwa g.	Cod liver oil cc.	Marmite g.	Lemon juice cc.	Protein ratio	Average weight g.	Average daily increase g.
Aug. 21–25	5	30	0	30	2·5	0·25	15	1 : 6	39–69	6
,, 26–29	4	30	15	15	2·5	0·25	15	1 : 2·5	69–80	3
,, 30–Sept. 3	5	60	30	30	5·0	0·5	30	1 : 2·5	80–111	6
Sept. 4–11	8	90	45	45	7·5	0·75	45	1 : 2·5	111–181	9
,, 12–21	10	120	45	45	7·5	1·0	45	1 : 2·5	181–299	12
,, 22–30	9	180	60	60	7·5	1·5	45	1 : 2·2	299–322	3
Oct. 1–5	6	210	60	60	7·5	1·75	45	1 : 2·5	322–443	20
,, 6–15	10	240	60	60	7·5	2·0	45	1 : 2·5	443–714	27
,, 16–26	11	300	60	60	7·5	2·5	45	1 : 3	714–904	17
,, 27–Nov. 18	23	180	36	36	7·5	1·5	45	1 : 3	904–1233	14
Nov. 19–Dec. 6	18	240	48	48	7·5	2·0	45	1 : 3	1233–1445	12
Dec. 7–26	20	360	72	72	11·0	3·0	67	1 : 3	1445–1668	11
,, 27–Jan. 13	18	360	36	36	7·5	3·0	45	1 : 4	1668–1767	5·5

The diminution in the daily food supply on Oct. 27th was due to the fact that the group consisted of five cocks and four hens and that it was necessary

[1] Caseinogen: moisture 9·9, ash 0·5, fat 0·8, protein 82·7 %.

[2] Secwa: moisture 1·2, ash 0·8, fat 0·3, lactose 74·5, sol. lactalbumin 14·2 %.

to separate off the surplus four cocks (groups III *A* and *B*, below). The figures from this date thus refer to one cock and four hens.

The birds in three weeks reached the weight of the birds in groups I and II [Part I] at the age of six weeks. The cocks in this group began to crow about Oct. 9th at the age of 49 days. The hens began to lay at the age of 139 days on Jan. 7th, 1921. It is recorded that Leghorns have started to lay at 122 days, but these birds were hatched in the spring and had a free existence.

Throughout the period the health of the birds was excellent and they moulted rapidly during October. As might be expected from the work of Palmer and Kempster [1919], the legs and beaks were not pigmented. These investigators showed that the yellow pigmentation depended on the presence of yellow pigments in the food. The food here supplied was almost colourless and the result is a confirmation of their work.

Groups III A and III B. The four cocks after removal from the main group were kept in four separate small cages and were arranged in two groups of two birds as regards food. Both groups were supplied with the same food items as the main group, but with a higher proportion of caseinogen and secwa; this proportion was reduced on Dec. 3rd, so that the protein at first 1 : 1·8 became 1 : 1·3. They were also given more cod liver oil and orange juice per bird than the main group, group III *B* more than III *A*. The marmite was always kept at 0·25 g. per 30 g. of oatmeal. The food consumption, average weight and daily average increase were as follows:

Group III A, each bird.

	No. of days	Oatmeal g.	Casein-ogen g.	Secwa g.	Cod liver oil cc.	Marmite g.	Lemon juice cc.	Protein ratio	Average weight g.	Average daily increase g.
Oct. 27–Dec. 3	39	60	30	30	2·5	0·5	15	1 : 1·8	1053–1730	17
Dec. 4–Jan. 11	39	60	15	30	2·5	0·5	15	1 : 2·9	1730–2202	15

Group III B, each bird.

	No. of days	Oatmeal g.	Casein-ogen g.	Secwa g.	Cod liver oil cc.	Marmite g.	Lemon juice cc.	Protein ratio	Average weight g.	Average daily increase g.
Oct. 27–Dec. 3	39	60	30	30	5·0	0·5	30	1 : 1·8	1015–1587	15
Dec. 4–Jan. 11	39	60	15	30	5·0	0·5	30	1 : 2·9	1587–1804	6

Sometimes the cocks would eat an extra daily allowance during the course of one or more days.

The housing in the smaller cages made no appreciable difference to the daily weight increase, though it was slightly higher than that of the main group. In the first part of the period the greater amount of cod liver oil given to group III *B* had a lessening rather than the augmenting effect which might have been expected. This effect was very marked in the second part. It would support the suggestion made in Part I that the amount of B factor required in the food bears a definite ratio to the consumption of carbohydrate and fat. Cod liver oil was not given for the last few days as the birds were to be killed on Jan. 11th for the purpose of examining the flesh. The flesh of the breast was pale, that of the legs about the colour usual with Leghorns. The flesh was also lacking in flavour, evidently due to the "tasteless" food.

Group III, egg laying period. The hens began to lay on Jan. 7th, 1921; seven eggs weighing 286·5 g. or 40·9 g. per egg were produced up to Jan. 13th.

On Jan. 14th a comparison of egg laying on this diet was made with the same number of hens and a cock (group IV) from the birds of groups I and II. Their diet contained blood meal as protein in exactly the same protein ratio of 1 : 4. Group IV was given less lemon juice, and marmite was omitted from their diet, since it was possible that sufficient of this B vitamin was present in the blood meal. The following tables show the diets, weights etc.:

Group III.

Oatmeal g.	Caseinogen g.	Secwa g.	Cod liver oil cc.	Marmite g.	Lemon juice cc.	Average weight g.	Increase g.
360	36	36	7·5	3·0	45	1767–1815	48

Group IV.

	Oatmeal g.	Blood meal g.	Cod liver oil cc.	Lemon juice cc.	Average weight g.	Decrease g.
	360	87	7·5	45	1537–1456	81

A hen of group III suddenly died on Feb. 4th apparently from choking. No signs of ill-health were noticed beforehand nor could any reason be found at post mortem examination. The other birds were perfectly well throughout.

Group IV did not relish the blood meal; on several occasions it was not necessary to prepare the daily food for them.

Grit was not supplied to either group; the eggs had always a sound shell.

Group III gave 6 eggs during January weighing 253·5 g. or 42·5 g. per egg.
    „   „   „   17   „   February   „   702·5 g. , 41·3 g.   „
    „   IV   „   7   „   January   „   316·5 g. „ 45·2 g.   „
    „   „   „   3   „   February   „   134·4 g. „ 44·8 g.   „

The very great difference in the number of eggs from the two groups may have been due to the blood meal in the food of group IV, but from the results recorded in Part I, it is more probably due to absence of B factor. Unfortunately it was not possible to test this supposition, as the experiment had to be discontinued.

CONCLUSIONS.

On a diet containing sufficient of each of the three vitamins, chickens can be raised in confinement. This confirms the results of Part I.

Rapid growth results on a diet having an albuminoid ratio of 1 : 3 and containing "good" proteins, caseinogen and lactalbumin. Cock birds began to crow in 49 days and hens to lay in 139 days; the cocks had an average weight of 1853 g. and hens of 1815 g.

REFERENCES.

Buckner, Nollau and Kastle (1915). *Amer. J. Physiol.* **39**, 162.
Buckner, Peter, Wilkins and Hooper (1919). *Kentucky Agric. Exp. Stat. Bull.* No. 220.
Osborne and Mendel (1912). *J. Biol. Chem.* **12**, 473.
Osborne and Mendel (1914). *J. Biol. Chem.* **17**, 325.
Osborne and Mendel (1916, 1). *J. Biol. Chem.* **26**, 1.
Osborne and Mendel (1916, 2). *J. Biol. Chem.* **26**, 293.
Palmer and Kempster (1919). *J. Biol. Chem.* **39**, 299, 313, 331.
Plimmer (1921). *Proc. Roy. Inst.*; *Nature*, **107**, 664; *J. Soc. Chem. Ind.* **40**, 227.
Plimmer and Rosedale (1922). *Biochem. J.* **16**, 11.

V. DISTRIBUTION OF ENZYMES IN THE ALIMENTARY CANAL OF THE CHICKEN.

By ROBERT HENRY ADERS PLIMMER
AND JOHN LEWIS ROSEDALE.

From the Biochemical Department, Rowett Research Institute for Animal Nutrition, University of Aberdeen and North of Scotland College of Agriculture.

(*Received December 20th, 1921.*)

THE presence of lactase in the intestines of animals and the non-adaptation of the pancreas and intestine to lactase by feeding with lactose was investigated by Plimmer [1906]. Lactase was always found to be absent from the intestine of chickens. A diet containing lactose had been used by us [1921] in feeding chickens from birth for a period of over three months. Examination of the birds' excreta showed that reducing sugar was absent therefrom, a fact which indicated that the sugar was assimilated. Assimilation of disaccharides is usually preceded by hydrolysis to monosaccharides, which would imply the presence of lactase in the alimentary canal, either in the intestine by adaptation or in some other part. The intestines of the cockerels in this group of birds were therefore examined, after they were killed, for the presence of lactase: it was not found to be present, and the non-adaptation of this organ was verified. If hydrolysis of lactose previous to assimilation occur, it must take place in some other part of the gut. The crop, pancreas and proventriculns were tested and lactase in small amount was detected in the crop. The investigation was then extended to the presence of other enzymes, as no information could be found in the literature about their occurrence in the alimentary canal of birds. The enquiry did not extend to the detection of all known enzymes, but was limited to those concerned in the digestion of the common foodstuffs.

EXPERIMENTAL.

The methods of preparing the enzyme solutions and detecting the presence of enzymes were in general in accordance with those usually adopted; in many cases a longer time of action (up to seven or ten days) was allowed, and in the case of the sucroclastic enzymes, proteins etc. were removed before testing for the reducing sugar formed by their action.

The various parts of the alimentary canal were always taken from chickens killed the same day, or not later than the day previously; on account of the small size of the crop, proventriculus and pancreas, the organs from four to eight birds were collected and examined together. A single small intestine provided sufficient material, but in most experiments several were combined as the whole series of sucro- or proteo-clastic enzymes were tested for simultaneously. Separate tests were made for lactase. At least two experiments were made with each part, except the caeca.

Preparation of enzyme solutions.

The pancreas, on removal, was cut up into small pieces and ground with sand in a mortar; the ground mass was put into glycerol in which it was kept for several days in the presence of a few cc. of toluene. The solution was then prepared by diluting with rather more than an equal volume of water and filtering from sand, etc.

The other parts of the alimentary canal were cut open and washed with running water to remove the contents. The mucous membrane was scraped off, ground up with sand and water and extracted for 24–48 hours with water in the presence of a little toluene to prevent putrefaction. The aqueous portion was strained off through cloth to remove sand and larger pieces and used for testing for enzymes.

It was not possible to scrape off mucous membrane from the inside of the proventriculus. The organ is glandular, covered with numerous small teats, which, on pressing with a scalpel, emit a yellowish, viscous, distinctly acid secretion. This secretion was the material actually used after grinding with sand and mixing with water. Nothing could be scraped off the gizzard, the interior surface of which resembled parchment.

Detection of enzymes.

(a) *Diastase and invertase.* As substrates 100 cc. of 1 % starch solution and 50 cc. of 3 % cane sugar solution were used. Two portions were measured out with a pipette in separate flasks; a known volume of enzyme solution was added to one, and the same volume of boiled enzyme solution, after cooling, to the other; 2 or 3 cc. of toluene were added to each, the flasks corked and put into an incubator at 37° for one or more days. A test for starch by the iodine reaction was made from time to time with a drop removed from the mixture. At the end of the reaction time, the mixtures were washed into a 250 cc. measuring flask, a slight excess of colloidal ferric hydroxide added, any excess of the latter removed by a few crystals of magnesium sulphate, the volumes made up to the mark, the solutions filtered and reducing sugar tested for by the complete reduction of 10 cc. of Fehling's solution. The control solutions containing boiled enzyme did not reduce, or only gave a slight reduction due to sugar present in the extract.

(b) *Lactase.* The detection of lactase was carried out in a similar way to that of diastase and invertase, using 50 cc. of 4 % lactose solution as substrate. The enzyme and control mixtures were put directly into 250 cc. measuring flasks and made up to volume after clearing with colloidal ferric hydroxide and magnesium sulphate. The reducing sugar was estimated by the reduction of 10 cc. of Fehling's solution. The observed difference in reading indicated whether hydrolysis had or had not occurred. No difference in reading was observed in the case of the intestine or proventriculus, but a small though distinct difference was always noticed in the case of the crop extract; it varied from 0·2 to 0·5 cc. in a total of 10 or 10·1 cc. This slight difference indicated an hydrolysis of 10–20 % of the lactose.

(c) *Lipase.* This enzyme was not looked for except in the case of the pancreas. Two exactly equal portions of oil in separate test tubes were made just alkaline to phenolphthalein with 0·1 N caustic soda. Enzyme and boiled enzyme solution were added. On keeping at 37° and occasionally shaking, the pink colour of the tube containing enzyme solution disappeared and it was restored by adding a few drops of the soda. This could be repeated several times and altogether from 1–2 cc. of alkali were added; the control tube did not change colour.

(d) *Proteoclastic enzymes.* Proteoclastic enzymes were detected by their action on Congo-red fibrin in neutral, acid and alkaline media. In the first case, a definite volume of enzyme solution and the same volume of boiled enzyme solution were put into separate flasks; in the other cases the same volumes of enzyme and boiled enzyme solutions were mixed with an equal volume of 0·2 N hydrochloric acid or 0·2 N sodium carbonate solution in separate flasks; 1 g. of Congo-red fibrin and 2 cc. of toluene were added to each and the several flasks were put in an incubator at 37° for one to seven days. Solution of Congo-red fibrin, which, in the case of hydrolysis, generally occurred in one or two days, was taken as indication of the presence of proteoclastic enzyme; solution did not occur in those flasks with boiled enzyme solution. No investigation was made of the products of the hydrolytic action.

RESULTS.

The presence or absence of enzymes in the various parts of the alimentary canal is most easily seen from the following table:

	Crop	Proven-triculus	Pancreas	Intestine whole	Duo-denum	Ileum	Caeca
Invertase	0	0	.	+	.	.	0
Diastase	+	0	+	+	.	.	+
Lactase	+	0	.	0	.	.	.
Lipase	.	.	+	.	.	.	.
Proteoclastic in neutral	0	0	+slight	0	0	0	0
„ acid	+slight	+	+less rapid	+	+	+	0
„ alkaline media	0	0	+rapid	+slight	+	+slight	0

The distribution of the sucroclastic enzymes corresponds in most particulars with that in the animal; most animals have invertase in the intestine, lactase is present in some, absent in others: diastase and lipase are generally

present in the pancreas of animals. The proteoclastic enzymes show a difference: the animal has trypsin acting in alkaline media; the chicken in both
alkaline and acid media. The intestine of the chicken has an enzyme acting most
rapidly in acid medium, less rapidly in alkali. The proteoclastic enzyme of
the proventriculus acts only in acid medium; the organ corresponds to the
stomach of animals. The caeca, as expected, had no enzyme of this group,
but contained diastase.

We wish to/thank Prof. J. A. MacWilliam, F.R.S., for kindly allowing us
to carry out these experiments in his laboratory.

REFERENCES.

Plimmer (1906). *J. Physiol.* **34**, 93; **35**, 20.
Plimmer and Rosedale (1921). *J. Agric. Sci.*

VI. THE AMINO-ACIDS OF FLESH.

THE DI-AMINO-ACID CONTENT OF RABBIT, CHICKEN, OX, HORSE, SHEEP AND PIG MUSCLE.

By JOHN LEWIS ROSEDALE.

From the Biochemical Department, Rowett Research Institute for Animal Nutrition, University of Aberdeen and North of Scotland College of Agriculture.

(*Received December 20th, 1921.*)

A LONG series of food analyses has recently been made by Plimmer [1921, 1], who points out that by the ordinary routine method of analysis, in which the amount of protein is estimated by multiplying the nitrogen content by 6·25, no discrimination is made between the flesh of different animals. The protein of one animal is regarded as being the same as that of another. The work of Emil Fischer and Kossel and their pupils has definitely proved that the various proteins differ very widely in their composition as regards the amino-acids, and this difference is emphasised by the experiments on the food value of the individual amino-acids by Hopkins in conjunction with Willcock and Ackroyd, by Osborne and Mendel and other American investigators[1]. These chemical and biological differences are sufficient evidence that quality of protein in nutrition must be taken into consideration.

Complete analyses of the protein of the muscle of the ox, chicken, halibut and scallop have been made by Osborne and Heyl [1908] and Osborne and Jones [1909], and Drummond [1916] has made some analyses of muscular tissue by Van Slyke's method. Both the more complete analyses by Osborne and co-workers and those by Drummond do not show any marked difference in the amino-acid content of the various muscle proteins. The flesh of various animals shows such distinct appearances, different both to the eye and palate, that it seems probable that greater differences may exist, and that there may be smaller differences in the flesh from various parts of the same animal's carcase, such as back and leg. Some further amino-acid analyses have therefore been made.

The methods of protein analysis are far from perfect: Fischer's ester method for the mono-amino-acids, as he pointed out, is not quantitative: Kossel and Patten's method for the di-amino-acids, in spite of the numerous manipulations, is generally considered to be fairly accurate, but it has been largely superseded by Van Slyke's method which gives higher values for these amino-acids. Van Slyke's method also possesses the advantage of requiring only small amounts of protein and is more rapidly carried out. This method of protein analysis has been used in these experiments, since it was chiefly

---

[1] See summary by Plimmer [1921, 2].

desired to compare the muscle protein of several animals with a view to more complete data at a later time. A comparison of these results with those by Kossel's method has been made in a few cases. The results indicate that differences exist in the amino-acid content of the various muscle proteins. Duplicate analyses were always carried out; frequently these analyses were not so concordant as was expected. This inconsistency of the results was under investigation by Plimmer [1916] who tested the arginine determination; other details of the method are now being studied.

EXPERIMENTAL.

In the case of the smaller animals (rabbit, chicken) opportunity was taken of comparing the flesh of different parts of the body of the same animal. In other cases the flesh was taken from the thigh. The mode of operation was the same throughout. The flesh (about 350 g.) was freed from inside fat, minced and put into about 2 litres of boiling water containing 0·1 % acetic acid and heated for about ten minutes so as to coagulate the protein and remove the extractives. The liquid was poured off and the coagulated protein squeezed dry in a cloth. This procedure was repeated twice. The coagulated protein (about 200 g.) was then digested with 1 g. pepsin in 2 litres of 0·1N HCl, so as to separate nucleins, indigestible matter, etc. After digestion, which usually occupied about ten days at 37° the liquid was filtered off and the total nitrogen estimated. A portion containing about 6 g. of protein was then hydrolysed by boiling with hydrochloric acid added to the liquid so as to make a concentration of 20 %. The hydrolysis was carried on for 36 hours. The hydrolysed solution was evaporated to dryness *in vacuo*, made up to 250 cc. and two samples of 100 cc. were analysed by Van Slyke's method. This was performed as described except for the arginine estimation which was effected by Plimmer's modification [1916]. In the earlier experiments it was impossible to make determinations of amide N owing to the facilities for vacuum distillation not being adequate. The analyses were made in duplicate and the percentage has been calculated from the average.

Table I. Nitrogen percentages.

	Amide	Humin N	Total N	Di-amino-acids Amino N	Non-amino N	Argi-nine N	Histi-dine N	Lysine N	Total N	Mono-amino-acids Amino N	Non-amino N	Total N of hydro-lysed solution
Rabbit, back	—	—	45·7	21·5	24·0	15	19	11·5	49	—	—	94·7
,, fore limb	—	—	44·1	17·9	27·7	8·8	30·9	5·5	50·7	—	—	94·8
,, hind limb	—	—	44·8	18·4	26·6	13	25	5·8	56·3	—	—	101·1
Chicken, breast	6·9	3	27	9	18	10	13	2	61·1	49·7	11·4	98·0
,, legs	5·5	1·3	25·5	15	10·5	8	7	11	68·5	66·6	1·9	100·8
Beef	6·3	0·5	28·5	15	13·5	13·3	5	11·2	55	26·8	28·2	90·3
Horse	2·9	0·9	37·1	18·8	18·3	14·9	10·5	11·6	70	58	11·9	110·9
Mutton	6·5	0·5	38·3	22·3	15·6	15	18	4·3	54	52	2	99·3
Pork	6·4	1·2	28·2	13·3	15	14	7	7	57	53	4	92·8

Relatively little difference can be observed from the figures for the different meats. The amide N is almost similar, in each case averaging about 6·0 % of the total N.

Table II. *Percentages of amino-acids. Giving the amount of amino-acids in 100 g. of protein.*

	Arginine	Histidine	Lysine	Total di-amino N
Rabbit, back	8	10	13	31
„ fore limb	5	19	5	29
„ hind limb	7	15	5	27
Chicken, breast	6	8	1	15
„ legs	4	4	10	18
Beef	7	3	10	20
Horse	7	6	9	22
Mutton	7	11	4	22
Pork	7	4	6	17

Humin N shows a difference. It is, if anything, higher in the white meats, *e.g.* breast of chicken 3 %, legs 1·3 %, pork 1·2 %, than in the red meats, where the average is 0·5 %, except in the horse, where 0·9 % was found. The explanation of this slightly higher value may be that the animal was not properly bled on slaughter.

Lysine figures are, with the exception of mutton, higher for the red meats, averaging about 11 %, while, of the white meats, rabbit limbs show only 5·5 %, chicken breast 2 % and pork 7 %.

Gortner and Holm [1920, 1], working with mixtures of pure amino-acids, have shown that tryptophan, and in the presence of aldehyde also tyrosine, and their analogues are the only known amino-acids which go to form humin. There is therefore no connection between the humin content and the lysine content of the meats; this is exemplified especially in the chicken, where the humin is high, and the lysine is low in the breast; and humin is low and lysine high in the legs. It may perhaps be mentioned that in the preparation of the di-amino-acids by the method of Kossel and Patten a distinct yellowish colouring adheres to the lysine portion.

At the same time too much reliance must not be placed on the humin as an estimation of tryptophan and tyrosine. Gortner and Holm [1920, 2] and Thomas [1921] have shown that tyrosine and tryptophan which go to form humin are not necessarily the only substances giving a reaction with the phenol reagent of Folin and Denis. Estimations of substances giving the blue colour with this reagent were made during the progress of this work, both before the removal of the humin and afterwards. In the case of chicken breast, a white meat, the readings before removal of the humin represented 4 % "tyrosine" whereas after its removal the readings represented 3·5 %. In the case of beef however—a red meat—the difference was greater, the former reading being 3·5 % and the latter 2·1 %, yet the humin N was much lower in the case of beef.

The arginine figures are more constant at about 14 or 15 % except in rabbit fore limb and chicken legs where the average is 8 %.

The histidine figures are less satisfactory, and exhibit perhaps a weak point in the method. In this connection it is of interest to point out that in the cases of abnormally high histidine the figures for the non-amino N are lower than normal and *vice versa, e.g.* beef 5 % histidine, 28 % non-amino N, mutton 18 % histidine, 2 % non-amino N, while in other cases this observation cannot be made. This may be due, either to incomplete precipitation of the histidine by the phosphotungstic acid, or to washing. Work in this connection is in progress.

It is not possible to draw any conclusions from the figures of the mono-amino fraction, which account for about 55 to 60 % of the total N.

The average percentage of the di-amino N is 35.

Comparison with former work on the hydrolysis of meat is difficult, because, with the exception of Drummond [1916] on chicken meat, the other figures relate to the method of Kossel, which generally gives lower results than the Van Slyke method.

The above figures for chicken breast agree in the main with those of Drummond, his total hexone bases N 27·26 being the same as that above. The arginine figures are within 1 % and he records having used the same modification of that process as mentioned above. The figures for histidine and lysine are discordant, Drummond finding 8·45 and 9·81 respectively, while the total N of the mono-amino fraction is 4 % higher than that found by Drummond.

In order to compare the figures of Osborne and co-workers with the above it is necessary to refer to the percentages not of total N but of actual arginine, histidine and lysine. The figures for arginine are generally constant within 1 %, those for histidine are higher than Osborne's, while the lysine figures, owing to the calculation in Van Slyke's method, are dependent on the histidine values. Apart from the arginine values, only the beef of the present sets has given results comparable with those of Osborne, who found 7·5 % arginine, 1·8 % histidine and 7·6 % lysine against 6·8 % arginine, 2·6 % histidine and 9·6 % lysine in this experiment.

SUMMARY.

1. Determinations have been made of the di-amino-acids of the protein of the flesh muscle of rabbit, chicken, ox, horse, sheep, pig by Van Slyke's method.

2. The red meats show a higher lysine content than the white meats.

I wish to take this opportunity of expressing my gratitude to Dr Plimmer, who suggested this work, for his kindness and guidance throughout the time I was under him, and also to Professor J. A. MacWilliam, F.R.S., for so kindly placing his laboratory at my disposal.

REFERENCES.

Drummond (1916). *Biochem. J.* **10**, 473.
Gortner-Holm (1920, 1). *J. Amer. Chem. Soc.* **42**, 821.
Gortner-Holm (1920, 2). *J. Amer. Chem. Soc.* **42**, 1682.
Osborne and Heyl (1908). *J. Biol. Chem.* **22**, 433; **23**, 81.
Osborne and Jones (1909). *J. Biol. Chem.* **24**, 161, 437.
Plimmer (1916). *Biochem. J.* **10**, 115.
Plimmer (1921, 1). *Analyses and Energy Values of Foods* (Stationery Office).
Plimmer (1921, 2). *Proc. Roy. Inst.*; *Nature*, **107**, 664; *J. Soc. Chem. Ind.* **40**, 227.
Thomas (1921). *Bul. Soc. Chim. Biol.* **3**, 197.

VII. XYLENOL BLUE

AND ITS PROPOSED USE AS A NEW AND IMPROVED INDICATOR IN CHEMICAL AND BIOCHEMICAL WORK.

BY ABRAHAM COHEN.

From the Cooper Laboratory, Watford.

(*Received December 24th, 1921.*)

DURING the course of some studies on the most important of the sulphonephthaleins synthesised by W. M. Clark and Lubs [1917], the writer was led to synthesise a new sulphonephthalein which has practically the same properties as thymol blue but with the important advantage of being twice as intense and therefore more economical.

Lubs and Acree [1917] in their earlier quests for evidence on the quinonephenolate theory, now practically firmly established, noticed various relationships between the chemical constitution of the sulphonephthaleins and the corresponding working ranges of utility. Thus the introduction of negative (bromine) groups shifts the working ranges to a more acid region, *e.g.* cresol red with a working range from P_H 7·2 to P_H 8·8 on bromination yields brom cresol purple with a working range from P_H 5·2 to P_H 6·8, a shift of two units for either limit. The same shift occurs on brominating thymol blue to produce brom thymol blue. On brominating phenol red to produce brom phenol blue, a tetra-brominated instead of a dibrominated sulphonephthalein results, and the shift is four units for either limit of the working range.

Whilst then it would appear that negative group-substituents induce fairly regular changes of P_H ranges, another interesting aspect of the matter is presented by the shifting of the ranges towards the more alkaline region, by the introduction into the phenol residue of phenol red of positive groups like methyl and *iso*-propyl as in cresol red and thymol blue, which is a 1-methyl-4-*iso*-propyl-3-phenolsulphonephthalein. Here, however, the well defined acid range of thymol blue complicates matters.

It occurred to me to compare the case of diethyl red and dipropyl red. These are azo indicators and homologues of the well-known methyl red, or more strictly dimethyl red, dimethylaminoazobenzene-*o*-carboxylic acid. The two homologues show the same P_H range, viz. P_H 4·5–6·5, in spite of the increased positive character of the propyl groups. The following show similar anomalous behaviour.

Table I.

1. Methyl Orange: Sodium p-dimethylaminoazobenzenesulphonate } P_H 3·1–4·4
2. Mono-ethyl Orange: Sodium p-monoethylaminoazobenzenesulphonate }
3. Sórensen's: α-naphthylaminoazobenzene } P_H 3·7–5·0
   α-naphthylaminoazo-p-toluene. }
4. Sodium p-benzylanilineazobenzenesulphonate } P_H 1·9–3·3
5. Sodium p-aminoazobenzenesulphonate }

One would suspect therefore the possibility of a similar anomalous un-shifted P_H range when the thymol residue in thymol blue is replaced by p-xylenol; the difference here is in the replacement of the *iso*-propyl group in thymol by another methyl radical.

In view of this, it occurred to me, that owing to the smaller molecular weight, the resulting new sulphonephthalein should be used like phenol and cresol reds in 0·02 % solution to produce the same intensity as 0·04 % thymol blue or in strength between the two percentages. We should then have a more symmetrical list in which only the brominated dyes are used in the higher concentration of 0·04 % (or 1·2 % concentrated), alcoholic or aqueous monosodium salt, whenever phenol and cresol reds are chosen in half these strengths. These relative concentrations among the sulphonephthaleins have been recommended largely on the basis of equality of *depth* of colour, as far as can be judged by the eye with the reds and blues, though the reds and blues do not appear of the same *quality*, owing mainly to difference in degree of dichroism.

By condensing the chloride or anhydride of o-sulphobenzoic acid with p-xylenol, a product was ultimately obtained having the same double P_H ranges as thymol blue and similar colour changes.

EXPERIMENTAL.

The method of preparing 1·4-dimethyl-5-hydroxybenzenesulphonephtha-lein or xylenol blue is as follows [Chemicals & Bye-products, Ltd., and A. Cohen, 1921]:

A mixture of ten parts of o-sulphobenzoic dichloride, or of the acid anhy-dride; ten parts of fused zinc chloride; and fifteen parts p-xylenol, M.P. 74·5°, B.P. 211·5°, is heated in a brine bath for six hours. The melt is heated with forty parts water till disintegrated. It is then filtered hot, washed with hot water, and afterwards with a little alcohol. Next, it is dissolved in excess caustic soda and precipitated with hydrochloric acid whilst being stirred. The mixture is filtered and crystallised from alcohol in the form of a brown solid.

The reactions are:

Samples of the new indicator were titrated against 0·01 N NaOH using a micro-burette up to 5 cc. and reading to 0·01 cc.

Sample 1. 0·0205 g. in 150 cc. water (yellowish orange) required 4·90 cc. instead of the theoretical 5·00 cc. or 98 % molecular equivalent alkali.

Two other samples gave identical results.

Carbon dioxide must of course be rigidly excluded in such titrations involving high dilutions.

As in the cases of phenol and cresol reds, a 0·6 % aqueous solution of the monosodium salt can be made by gently boiling, not merely warming, for 3 minutes, 0·6 g. of the solid powdered dye in about 80 cc. water containing 1·47 cc. of normal alkali, and then diluting when cold to 100 cc.

If preferred, a strong alcoholic solution can be made by boiling 0·2 g. of the solid in 100 cc. absolute alcohol.

A 0·02 % solution was prepared by a thirty-fold dilution of the 0·6 % aqueous monosodium salt solution, and added to cordite tubes (or test tubes of hard white glass, unflanged and of uniform diameter) each containing 10 cc. of a standard buffer solution. Using a concentration of 0·05 cc. or 1 drop indicator per cc. buffer solution, the double series P_H 1·2–2·8 and P_H 8·0–9·6 (each in consecutive steps of 0·2 P_H) with xylenol blue, were similar to those with thymol blue added as a 0·04 % solution in the same concentration of 0·05 cc. per cc. The tints were, in fact, rather more distinct with xylenol blue in the alkaline range, appearing a purer blue in the daylight. Further, unlike dipropyl red in the anomalous case mentioned above, xylenol blue does not tend to precipitate out from buffer solutions even after three months' standing. On the contrary, xylenol blue stands in this respect to thymol blue, as the preferred diethyl red stands to dipropyl red.

Like thymol blue, it can be used for the estimation of the P_H of gastric juice and ammoniacal fluids; and in the differential acidimetric and alkalimetric titrations as described by A. B. Clark and Lubs [1918].

In Table II some results are given, obtained with mixtures of hydrochloric and benzoic acids. A 5 cc. micro-burette graduated in hundredths of a cubic centimetre was used, and I have found it convenient to use a special easily constructed titration flask in which a cordite tube is fused horizontally to the side of the flask to facilitate colorimetric comparison with the standard coloured benzoic acid. The cordite tube attachment also serves to stir the liquid when the flask is tilted. Such a flask can be heated without any cracks developing in the joint, provided the heating is done at the side remote from the joint. I wish here to thank one of my colleagues Mr G. H. Wallis for his kindness in constructing this flask.

0·5 cc. of 0·02 % aqueous xylenol blue was added to 50 cc. of an unknown benzoic acid solution, and 10 cc. of this standard kept in a cordite tube. The 0·5 cc. indicator is arbitrary and others may find it better to use more or less, but the less indicator is used the better; the differentiating power of the eye being an inadequately known function of subjective colour differences.

The mixtures titrated had, of course, the same concentration of indicator as the standard.

Table II. *Titration of mixtures of hydrochloric and benzoic acids.*

	Benzoic acid cc. used	1 N HCl cc. used	0·993 N NaOH		N HCl found cc.	G. of benzoic acid per cc. found
			to standard	to blue		
1	10	nil	nil	0·39	nil	0·0047
2	10	3·03	3·06	3·46	3·04	0·0049
3	25	4·04	4·05	5·00	4·02	0·0046

Experiments are in progress on the conditions for brominating xylenol blue.

SUMMARY.

p-Xylenolsulphonephthalein or xylenol blue, a new indicator having two working ranges of utility from P_H 1·2 (red) to 2·8 (yellow) and from P_H 8·0 (yellow) to 9·6 (blue), can be successfully employed in all work for which thymol blue has heretofore been recommended.

The fact, that *p*-xylenol is easily prepared from diazotised *p*-xylidene and that only half as much xylenol blue as thymol blue is required, should render its use eminently preferable.

REFERENCES.

Clark, W. M. and Lubs (1917). *J. Bact.* 21, 109, 137.
Clark, A. B. and Lubs, H. A. (1918). *J. Amer. Chem. Soc.* 40, 1443.
Lubs and Acree (1917). *J. Amer. Chem. Soc.* 38, 2772.
Chemicals & Bye-products, Ltd. and A. Cohen (1921). British Specification, 30119.

VIII. THE EXAMINATION OF SOME INDIAN FOODSTUFFS FOR THEIR VITAMIN CONTENT.

By SUDHINDRA NATH GHOSE.

From the Institute of Physiology, University College, London.

(*Received January 3rd, 1922.*)

THE subject of this paper was taken up at the suggestion of Dr J. C. Drummond, with a view to ascertaining whether the principal foodstuffs consumed by the mass of people in Bengal contain adequate amounts of different vitamins or not.

Among the previous workers in this field mention should be made of Chick and her collaborators [1917, 1919], who tested certain cereals and flours and samples of dried fruits for the presence of vitamin *B*, etc. and Greig [1917, 1918], who examined the anti-scorbutic value of Indian grains and lentils. Shorten and Roy [1921] also examined some vegetables for vitamins *B* and *C*. Chick had found that pure white flour is deficient in vitamin *B*. Chick, Greig and Shorten found that vegetables and lentils under certain conditions contained good amounts of anti-scorbutic vitamin.

The problem of supplying foodstuffs which should contain liberal amounts of different vitamins is very important in Bengal. Beri-beri occurs to some extent, and scurvy and rickets are encountered but are not widespread. From this but more from the low resistance of the people to various infectious diseases, and the presence of certain forms of eye-diseases, it may be suggested that the mass of the industrial classes may be living at or below the danger level of vitamin deficiency.

The table given below—quoted from MacCay [1912]—shows typical diets ordinarily in use among the different classes in Bengal.

Table I.

	Cultivators ozs.	Middle class (not above indigence) ozs.	Middle class (above indigence) ozs.	Well-to-do ozs.
Rice	20	16	6	4
Dal	1	1	1	1½
Vegetables	4	4	4	4
Oil	1	—	1	1
Fish	1	1	2	4
Wheat flour	—	—	6	6
Milk	—	—	8	16
Butter	—	—	1	4
Calories (about)	2390	2310	2350	2950–3450
			(Meat and eggs also added)	(Meat and eggs are taken, also sweets freely)

From this table it will be seen that so far as calorific values, etc. are concerned, the diets of the people are fairly in compliance with accepted standards. Low resistance to disease, therefore, is probably due to vitamin deficiency—vitamin A specially—although MacCay is of opinion that this low resistance is due to want of sufficient protein in food.

The above table may be regarded as roughly accurate, but MacCay omits to emphasise that all classes of the people eat large quantities of fruit during the spring and summer months; also that the amount of lentils consumed by the poorer classes is much higher than that in the table.

The principal "fatty foods" consumed are "ghee," various preparations of butter, mustard oil, and, to a limited extent, coconut oil. These are mainly used in cooking.

<h2 style="text-align:center">EXPERIMENTAL.</h2>

This enquiry has been confined to an examination of different samples of "ghee," mustard oil and coconut oil for the presence of vitamin A; "dals," *i.e.* lentils, and some samples of Indian flours, for vitamin B.

"Ghee" is made either from cows' milk or buffalo milk. As a rule, the raw milk is first warmed over a low flame for about half-an-hour, and then cooled and inoculated with sour milk. After about 12 hours, when curdling is complete, water is removed and the curd churned until the fats separate out. The fat is removed, clarified, and then cooled into a semi-solid mass of buttery consistency. In some cases this product is allowed to turn rancid and then heated and rendered.

Fresh "ghee" remains good for months if well stored in the dark.

The following table gives the analysis of different "ghees." The results are similar to those quoted by Bolton and Revis [1910].

Table II.

	Reich.-M. value	Sapon. V.	Iodine N.	Free fatty acids
Pure cows' "ghee" (yellow)	28·7	230	31	2·2
Pure cows' "ghee" (remelted) eight months old sample (yellow)	26	229	29	2·6
Pure buffalo "ghee" (white)	38·8	224	27·2	
Adulterated "ghee" (white)	14·4	213	29·3	—

"Ghee" is frequently adulterated with lard, mutton fat, coconut oil, and almond oil.

Rats about one month old, which were over 50 g. in weight, were selected and put on a diet deficient in vitamin A [Drummond and Coward, 1920]. Generally it took about three weeks for the rats to become steady in weight and fit for the feeding experiment. Then each rat was given 0·2 g. of "ghee" every morning in addition to its usual basal diet.

In the majority of cases the rats made steady increase in weight for about ten days (sometimes varying from 10 to 21 days) when they again became stationary. The amount of "ghee" was then raised to 0·4 g. a day, when the rats again resumed growth.

It is to be noted that the remelted sample did not induce growth, but some of the adulterated samples were fairly active when given as 5 % in basal diet.

It has been shown by Drummond, Coward and Watson [1921] that 0·2–0·4 g. per day of average samples of butter is sufficient to maintain growth for rats.

From the examination of the table of weighings it will be seen that both pure cows' "ghee" and buffalo "ghee" may be as good as European butters.

Table III.

(Showing the growth of rats on A-deficient diet and 0·2 g. of pure "ghee." The weighings of the rats during the preparatory period are omitted.) g.

	931, male	835, male
First weighing	91	58
After 10 days	102	74
„ 15 „	104	83
„ 30 „	124	100
„ 40 „	138	110

Table IV.

(Rats on A-deficient diet + 0·2 g. of pure "ghee." Sample 2.) g.

	832, male	834, male	940, male
First weighing	80	89	56
	85	111	77
	90	120	76*
	120	150	105
	129	162	112

* Amount of "ghee" doubled

Table V.

(Rats on remelted cows' "ghee.") g.

	833, male	836, female	828, male
First weighing	90	55	80
After 10 days	94	60	66
„ 15 „	92	52	60
„ 30 „	95	—	—
„ 40 ,			

Table VI.

(Rats on pure buffalo "ghee.") g.

	933, male	1176, male	932, female
	104	76	95
	120	105	109
	—	—	—
	136	150	136
	151	—	148

Table VII.

(Rats on adulterated "ghee." Sample I.) g.

	1510, female	1177, male	1176, male
First weighing	72	65	68
After 10 days	78	68	73
„ 30 „	75	72	72
„ 40 „	80	76	74

Table VIII.

(Rats on adulterated "ghee." Sample II, given as 5 % in basal diet.) g.

	976, female	1300, male
	80	60
	—	—
	110	95
	125	109

Table IX.

(Rats on coconut oil and A-deficient diet. The oil was given in the basal diet 10–20 %.) g.

	1457, male	1456, male	1463, female
First weighing	82	64	76
After 10 days	95	69	94
„ 20 „	118	77	105
„ 30 „	130	91	102

Table X.

(Rats on mustard oil: 10–20 % in the basal diet.) g.

	1508, male	1483, female	1511, male
	77	57	64
	84	62	76
	88	64	85
	98	72	92

Some of the adulterated samples gave good results, but as this depends on the adulterants used, the different samples gave widely varying results.

Daniels and Loughlin [1920] and Drummond, Golding, Zilva and Coward [1920] have shown that lard may contain a certain amount of vitamin A, the latter authors having shown that this occurs when the pigs have been grass-fed, and it should be borne in mind that lard is a most widely used adulterant in "ghees."

Contrary to the results of previous workers [cf. Jansen, 1920], the samples of coconut oil examined by me did produce fair growth in rats. It should be mentioned here that coconut oil was not administered in the same way as "ghees," but was incorporated in the basal diet in 10 % proportion and replaced hardened fat.

The results obtained with pure mustard oil are also interesting. The oil was administered in large quantities (10–20 % in diet) to the rats, and in some cases actual growth was obtained contrary to the results of previous workers. Both mustard oil and coconut oil were those prepared in India and most probably not so refined as in Europe or America.

The buffalo "ghee" is pure white, and the presence of the growth-promoting vitamin A in it confirms Drummond and Coward's work on vitamin A and colour [1920].

That renovation under certain conditions does not affect the vitamin content is also in accord with the results found by Drummond, Coward and Watson [1921].

Examination of different samples of lentils.

The procedure for examining the samples of lentils was similar to that adopted in the case of "ghees," except that in these experiments the rats had been previously fed on diet deficient in water-soluble B. Each rat, after it had begun to lose weight on this diet (B-deficient diet), was given 1 g. of the lentil to be tested, daily.

The B-deficient diet was similar to the Drummond-Coward diet (A-deficient) except that no yeast extract was given and that butter replaced hardened fat.

The weighed amount of lentil (1 g.) was kneaded with a small portion of the diet and given in the morning before the rat had the main bulk of its food.

The different lentils thus examined included:

1. "But" (yellow variety)—*Cicer arientinum* Linn.
2. "Sona Moong" (golden-yellow in colour)—*Phaseolus Mungo* Linn.
3. "Arhar" (small size, yellow-ochre variety)—*Cejanus Indicus* Spreng.
4. "Lal Moong" (red variety)—*Phaseolus radiatus* Linn.
5. "Mash Kalai" (greenish variety)—*Phaseolus Mungo* Linn.
6. "Bara Lal But" (red variety)—*Cicer arientinum* Linn.

It should be mentioned here that as far back as 1900 Grijns [1901] working in Java found "Katjang Hidjoe"—a variety of *Phaseolus radiatus*—a good cure for both human and avian beri-beri. Hulsoff [1917] also worked with the anti-neuritic property of Katjong (*Phaseolus radiatus*). A similar result has also been found by Pol [1917].

Table XI. *Composition of the various "Dals" (lentils).*

Name	H₂O %	Ether extract %	Protein %	Soluble carbo- hydrates %	Woody fibre %
5. *Phaseolus Mungo* Linn.	8·5	0·9	19·5	59·3	4·5
2. „ „ Yellow	10·99	0·83	20·5	59·0	4·8
4. *Phaseolus radiatus.* Red	10·0	0·93	22·9	58·9	3·38
1. *Cicer arientinum.* Yellow	11·4	4·9	18·6	56·3	6·1
6. „ „ Red	8·6	5·3	15·5	60·1	7·2
3. *Cejanus Indicus* Spreng.	14·3	1·9	15·8	57·2	5·8

Examination of the tables showing the growth of the rats on the lentils; shows that all the above varieties of lentils contain appreciable amounts of vitamin *B*.

Table XII.

Rats on *B*-deficient diet (1 g. of lentil a day). (*Phaseolus Mungo*—greenish variety.) g.

	1262, female	1267 a, female	1267, female
First weighing	90	102	75
After 10 days	117	122	95
„ 20 „	133	138	113

Table XIII.

Rats on *Phaseolus Mungo*—golden-yellow variety. g.

	1282, male	1284, male	1285, male
First weighing	92	100	80
After 10 days	109	130	100
„ 20 „	122	151	115

Table XIV.

Rats on *Phaseolus radiatus.* g.

	1258, male	1283, male	1259, female
First weighing	72	100	63
After 10 days	82	140	85
„ 20 „	86	162	104

Table XV.

Rats on *Cejanus Indicus.* Spreng. g.

	1260, female	1261, male
First weighing	86	95
After 10 days	100	131
„ 20 „	115	146

Table XVI.

Rats on *Cicer arientinum.* Yellow variety. g.

	1264 male	1265 female	1266 female
First weighing	76	76	90
After 10 days	68	94	115
„ 20 „	78	110	128

Table XVII.

Rats on *Cicer arientinum.* Red variety. g.

	1274 male	1277 male	1276 male	1275 male
First weighing	82	96	80	95
After 10 days	98	130	101	115
„ 20 „	105	146	112	127

Table XVIII.

Rats on *B*-deficient diet and 1 g. of pure unbleached Indian flour. g.

	1278, male	1279, female
First weighing	106	88
After 10 days	120	115
„ 20 „	135	120

Table XIX.

Rats on crude "Attah." g.

	1266a, female	1268a, female	1269a, male
First weighing	83	75	65
After 10 days	101	105	98
„ 20 „	111	117	117

Table XX.

Rats on bleached Indian flour. g.

	1292a, male	1291a, female	1293a, female
First weighing	81	69	76
After 10 days	80	65	73
„ 20 „	79	Dead	Dead

Examination of different samples of Indian flour.

Here the samples examined were only those that are widely used in Bengal. They were:

1. Crude "attah." ("Attah" is a mixture of various flours, *e.g.* wheat, pea and maize, etc., and contains the husks of the cereals in it.)

2. Pure Indian flour. This flour has a finer texture than the "attah." It is machine-ground, but it contains the ground-up husks of the cereals. This flour is not bleached.

3. White flour. This is machine-ground like the pure Indian flour, but it does not contain the husks, and moreover it is bleached perfectly white. Sometimes chlorine is added in the bleaching process.

1 g. of each of the above samples was given daily to the different sets of rats, in the same way as in the case of lentils.

The composition of the different flours is given below.

Table XXI.

	H_2O	Ether extract	Carbo-hydrate	Protein $N \times 5.7$	Crude fibre
1. Crude "Attah"	14·6	2·9	67·1	11·5	3·9
2. Pure unbleached flour	15	2	71·2	11	0·8
3. Bleached white	12	2	71·3	13·5	0·6

The results found are in accord with those of Chick as well as with the facts announced by Col. Hehir [1919, 1, 2] in the Mesopotamia Commission Report on the presence of deficiency diseases in the besieged garrison in Kut. Willcox [1917] has also discussed "attah" as a protective against beri-beri.

Johns and Finks [1920] found that a mixture of pea-flour and wheat-flour produces better growth in rats than simple wheat-flour, and these facts are also corroborated since "attah" is found to be better in promoting growth than pure (unbleached) flour. From the weighings of the rats, fed on the different samples of the flours, it will be seen that with the exception of bleached flour they contain appreciable amounts of vitamin *B*.

CONCLUSION AND SUMMARY.

Samples of pure Indian "ghee" proved to be as good a source of vitamin *A* as pure butter.

Samples of remelted "ghee" ("makkum galana") failed to restore growth in rats whose weights had become stationary on a shortage of fat-soluble *A*.

Two samples of "ghee" whose history is unknown failed to promote growth.

Some samples of adulterated "ghee" only gave growth in rats when administered in large quantities (*e.g.* 5 % of the basal diet).

Certain edible vegetable oils, *e.g.* ("narikel") coconut oil, and pure mustard-seed oil were examined, and these were found to give growth when administered in 10–20 % in the basal diet.

Various samples of lentils were examined and these showed good content of vitamin B.

Bleached (pure white) Indian flour is found to be deficient in vitamin B, but both crude "attah" and unbleached Indian flour have considerable quantities of water-soluble vitamin B.

In conclusion I take the opportunity of expressing my thanks to Dr J. C. Drummond and Miss Katharine H. Coward, for kind help and advice.

My thanks are also due to Mr Suvaga for kindly giving me some samples of remelted "ghee," and to my brother, Mr B. N. Ghose, for various samples of lentils, flours, oils and "ghees."

REFERENCES.

Bolton and Revis (1910). Quoted in *Analyst*, **35**, 343; from "Fatty Foods" (Churchill).
Chick and others (1917). Proc. *Roy. Soc. B.* **90**, 44.
—— —— (1919). *Biochem. J.* **13**, 199.
Daniels and Loughlin (1920). *J. Biol. Chem.* **42**, 359.
Drummond and Coward (1920). *Biochem. J.* **14**, 661.
Drummond, Coward and Watson (1921). *Biochem. J.* **15**, 540.
Drummond, Golding, Zilva and Coward (1920). *Biochem. J.* **14**, 742.
Greig and others (1917). *Indian Journ. Med. Res.* **4**, 818.
—— —— (1918). *Indian Journ. Med. Res.* **6**, 56.
Grijns, quoted by Hulsoff (1901). *Genees. Tijds. v. Ned. Ind.* **41**, 3.
Hehir (1919, 1). Mesopotamia Commission Report. Appendix. *Proc. Asiatic Soc. Bengal*, **15**, 212.
—— (1919, 2). *Indian Journ. Med. Res.* Indian Science Congress, pp. 44–59 and 79–82.
Hulsoff (1917). *J. Physiol.* **51**, 432.
Johns and Finks (1920). *J. Biol. Chem.* **42**, 491.
Jansen (1920). *Genees. Tijd. v. Ned. Ind.* **58**, 173.
MacCay (1912). "The Protein Element in Nutrition" (Arnold), p. 78.
Pol, D. H. (1917). *Ned. Tijd. v. Genees.* **11**, 806.
Shorten and Ray (1921). *Indian J. Med. Res.* Annual Science Congress. *Biochem. J.* **15**, 274.
Willcox (1917). *Lancet*, ii, 77.

IX. CONDITIONS OF INACTIVATION OF THE ACCESSORY FOOD FACTORS.

By SYLVESTER SOLOMON ZILVA.

From the Biochemical Department, Lister Institute.

(Received January 4th, 1922.)

In some preliminary communications [Zilva, 1920, 1921] attention was drawn to the fact that the fat-soluble and the anti-scorbutic factors were inactivated on exposure to ozone in the case of the former and ozone and air in the case of the latter. It was further pointed out that the anti-scorbutic factor could withstand the temperature of 100° for two hours when air was excluded. These observations have found confirmation in the results obtained independently by other workers. Hopkins [1920] has shown that when air is bubbled through butter for four hours at 120° the fat-soluble factor becomes inactivated. Drummond and Coward [1920] have also demonstrated that the destruction of this factor takes place in butter-fat in presence of air rapidly at high temperatures and more slowly at lower temperatures. Hess [1920] and Hess and Unger [1921] found that milk incubated with hydrogen peroxide lost its anti-scorbutic properties, whilst orange juice, milk, and tomato juice treated with oxygen lost some of their anti-scorbutic potency. Dutcher, Harshaw, and Hall [1921] have since confirmed Hess and Unger's observations concerning the destruction of the anti-scorbutic factor by hydrogen peroxide and the writer's observation on the stability of this principle at high temperatures in the absence of air.

The object of this communication is to describe a series of experiments on the action of ozone and air on the three accessory factors, observations having been made at various temperatures. These experiments owe their origin to the fact that the inactivation of the fat-soluble factor in butter by exposure to a mercury-quartz lamp [Zilva, 1919] was found by the author to be due to the action of the ozone generated by the lamp and not to the action of the ultra-violet rays.

THE INFLUENCE OF OZONE AND AIR ON THE FAT-SOLUBLE FACTOR.

In the preliminary communication [Zilva, 1920] it was pointed out that when cod liver oil was exposed to the mercury-quartz lamp for 16 hours in the absence of air the activity of the oil was not destroyed by the action of the light. On the other hand six hours of exposure of the same oil to ozone in the dark entirely inactivated it.

In testing the above oils for the fat-soluble factor 2 g. of the oil were added to the daily basal diet of 20 g. when the rats had ceased to grow owing to the deficiency in the diet. The following are the results obtained:

*With the addition of oil exposed to the mercury-quartz lamp
in the absence of air.*

Rat I ♀ gained 30 g. in 17 days.
Rat II ♀ gained 24 g. in 17 days.
Rat III ♀ gained 39 g. in 17 days.

With the addition of oil exposed to ozone for six hours in the dark.

Rat I ♂. No gain in weight in four weeks. On addition at the end of that time of a similar dose of the unexposed oil the animal resumed growth and gained 37 g. in three weeks.

Rat II ♂. No gain in weight in four weeks. On addition at the end of that time of a similar dose of unexposed oil the animal resumed growth and gained 23 g. in 16 days.

Rat III ♂. Declined in weight and died 26 days after the commencement of the administration of the oil. Developed keratomalacia during the period of treatment with the oil.

Rat IV ♀. No gain in weight in four weeks. On addition at the end of that time of a similar dose of unexposed oil the animal resumed growth and gained 29 g. in three weeks.

From the above experiments it is quite evident that ultra-violet light as such has no deleterious action on the fat-soluble factor. On the other hand ozone does inactivate it. Moreover the action of ozone must be very drastic, when one bears in mind that cod-liver oil is very potent. The oil employed in the above experiments has been found capable of inducing growth in rats in daily doses of 1·7 mgm. The experimental rats on the exposed oil consumed 60–70 % of their diet, in other words 1·2–1·4 g. of the oil or about 750 times as much as the minimum dose necessary to induce growth, without showing any tendency to grow. Possibly a much shorter exposure would have also produced inactivation of this factor.

In the earlier part of this work whale oil was used instead of cod-liver oil. The former is much less active than the latter and was given up for that reason. At the time the experiments were carried out it was not definitely settled that the fat-soluble factor was associated with the unsaponifiable fraction of the oil and it was therefore considered advisable to study the change in the iodine value of the oil during exposure to the mercury-quartz lamp. The differences observed were of a low magnitude. The following are the figures of a representative experiment:

Iodine value of whale oil before exposure 119
Iodine value of whale oil exposed to the lamp for eight
hours in the absence of air ... ... 117·6
Iodine value of whale oil exposed for eight hours in the
presence of the generated ozone ... ... 114·5

The change is not significant and the results are analogous to those obtained by Hopkins [1920] with butter heated in a stream of air at 120°. No change was recorded in the iodine value of the butter after the treatment. It is very doubtful whether a chemical change such as the iodine value of the oil can have any bearing on the accessory factor.

Some experiments in connection with the inactivation of the fat-soluble were carried out in connection with another inquiry. It was necessary to inactivate a large batch of very potent cod-liver oil and Hopkins' method was adopted for that purpose. This method has an advantage over the employment of ozone because it does not change the consistency of the oil to such a great extent. Air was aspirated through the oil heated at 120° for 18 hours. By this means a very active oil was inactivated to the extent that a daily dose of about 0·75 g. did not induce any growth in rats.

THE INFLUENCE OF OZONE ON THE WATER-SOLUBLE FACTOR.

Autolysed yeast juice was employed as a source for the water-soluble factor in this experiment. The juice was exposed to the ozone with constant shaking in the same apparatus and for the same time as the cod-liver oil. After the exposure the juice was very much altered in taste, smell and colour. It was incorporated in the basal diet free from the water-soluble factor after the rats had been on the restricted diet between three and four weeks. Three animals received 1, 2 and 3 cc. of autolysed yeast juice in their daily basal diet of 20 g. All the three animals responded by resuming growth, gaining 19, 24 and 15 g. respectively in 17 days. As 1 cc. of autolysed yeast juice is not much above the minimum dose capable of inducing growth there could not have been much destruction of the water-soluble factor in the juice by this very drastic method and we may conclude that this factor is decidedly more resistant to oxidation than the other two accessory factors. This resistance explains why on exposing autolysed yeast juice in a Petri dish to a mercury-quartz lamp no inactivation of the water-soluble factor takes place [Zilva, 1919]. Whether drastic oxidation inactivates the water-soluble factor at higher temperatures still remains to be ascertained.

THE INFLUENCE OF OZONE AND AIR ON THE ANTI-SCORBUTIC FACTOR.

The influence of ozone in the dark on the anti-scorbutic factor was next investigated. Decitrated lemon juice, namely lemon juice from which the organic acids are removed by neutralisation with calcium carbonate and which retains almost the entire anti-scorbutic potency of the original juice, was exposed to ozone in precisely the same way as the active oils and the autolysed yeast juice. Fresh batches were prepared every two or three days so that no deterioration could take place on storage. The treated juices, as well as the control untreated juices of the same batch, were administered to guinea-pigs which had been kept on a scorbutic diet of oats and bran and autoclaved milk (40 cc.) for 14 days. Daily doses of 3, 5 and 7 cc. were administered with the following results:

Guinea-pig I. Dose 3 cc. Commenced losing weight after three weeks. Declined gradually and died 44 days after the commencement of the experiment. At the post-mortem examination evidence of scurvy was found.

Guinea-pig II. Dose 5 cc. Declined and died 35 days after the commencement of the experiment. At the post-mortem examination evidence of scurvy was found.

Guinea-pig III. Dose 7 cc. Maintained its weight for 53 days. Chloroformed at the end of this period. At the post-mortem examination some evidence of scurvy was found.

A control dose of 3 cc. of the original decitrated juice was found to be adequate to prevent scurvy for 56 days; the animal was still growing when the experiment was discontinued.

One may conclude from the above experiments that the ozone inactivated the juice to a great extent. The writer has previously found [Zilva, 1919] that when decitrated lemon juice was exposed to the mercury-quartz lamp its anti-scorbutic potency was not destroyed or even impaired in spite of the generated ozone. This can be explained by the fact that the juice was exposed in a quartz tube and, unlike the butter in that experiment, was shielded almost entirely from the generated ozone. The anti-scorbutic, like the fat-soluble factor, is, therefore, inactivated by ozone but is not affected by ultra-violet rays.

In view of the results obtained in the above experiments it became of interest to ascertain whether less drastic oxidative treatment such as exposure to air would also inactivate the anti-scorbutic factor. Air was therefore aspirated through decitrated lemon juice for 12 hours. Batches of juice prepared every two or three days were employed. The following is the protocol of the experiment:

Guinea-pig I. Dose 3 cc. Just maintained its weight for 51 days gradually declining in health and showing symptoms of scurvy during this period. The animal was chloroformed owing to its distressed condition. At the post-mortem examination evidence of advanced scurvy was found.

Guinea-pig II. Dose 5 cc. Commenced declining in weight after 17 days. Chloroformed 48 days after the commencement of the experiment. At the post-mortem examination evidence of scurvy was found.

Guinea-pig III. Dose 7 cc. Gained 57 g. in 51 days. Chloroformed 51 days after the experiment commenced. Evidence of mild scurvy was found at the post-mortem examination in this case.

A control dose of 3 cc. of the original juice was found quite adequate to prevent scurvy for 51 days. The control animals were still growing when the experiment was discontinued.

This experiment affords evidence that the exposure of decitrated lemon juice to air for 12 hours diminishes its anti-scorbutic potency very appreciably.

The above observations suggested the possibility that the inactivation of the anti-scorbutic factor by heat may be due in part if not entirely to the presence of air during the heating. Such a possibility was conjectured by Delf [1920] who found that orange juice and swede juice retained their anti-scorbutic potency to a marked extent after being heated for a considerable time at temperatures higher than 100°. In order to test this theory the following experiments were devised. Two lots of decitrated lemon juice were heated at 100° under a reflux condenser. One was heated for one hour during which time air was bubbled through. The second was heated for two hours, air being all the time rigorously kept out by bubbling carbon dioxide gas through the solution. In this case also fresh batches of decitrated lemon juice were employed every two or three days. The following results were recorded:

Decitrated lemon juice heated in air for one hour.

Guinea-pig I. Dose 1·5 cc. Commenced declining in weight after a month. Chloroformed 43 days after the commencement of the experiment. Evidence of scurvy was found at the post-mortem examination.

Guinea-pig II. Dose 4 cc. Commenced to decline after three weeks and died 41 days after the commencement of the experiment. Evidence of acute scurvy was found at the post-mortem examination.

Guinea-pig III. Dose 6 cc. Maintained weight for 36 days and then suddenly declined. Died 43 days after the commencement of the experiment. Evidence of scurvy was found at the post-mortem examination.

The same juice heated in an atmosphere of carbon dioxide for two hours.

Guinea-pig I. Dose 1·5 cc. Gained 122 g. in 74 days. Chloroformed. No evidence of scurvy was found at the post-mortem examination.

Guinea-pig II. Dose 4 cc. Gained 48 g. in 74 days. Chloroformed. No evidence of scurvy was found at the post-mortem examination.

Guinea-pig III. Dose 6 cc. Gained 138 g. in 74 days. Chloroformed. No evidence of scurvy was found at the post-mortem examination.

This experiment affords proof that the anti-scorbutic factor is quite stable when heated in the absence of air.

Decitrated lemon juice contains a considerable quantity of reducing sugar. On hydrolysis, however, the reducing capacity of the juice increases slightly, a reducing substance, most probably a sugar, being liberated. An experiment was devised to ascertain whether the anti-scorbutic potency is destroyed when the juice is hydrolysed under anaerobic conditions. Decitrated lemon juice was therefore hydrolysed on a water-bath under a reflux condenser for five hours with 2 N hydrochloric acid. The air was displaced during hydrolysis by carbon dioxide. After hydrolysis the solution was neutralised with sodium carbonate and tested out for its anti-scorbutic potency on guinea-pigs. A fresh preparation of lemon juice was hydrolysed every day for feeding purposes. The following results were obtained:

Guinea-pig I. Dose 1·5 cc. Declined after three weeks and died 32 days after the commencement of the experiment. Evidence of scurvy at the post-mortem examination.

Guinea-pig II. Dose 1·5 cc. Declined after 23 days. Died 30 days after the commencement of the experiment. Evidence of scurvy at the post-mortem examination.

Guinea-pig III. Dose 1·5 cc. Declined after three weeks and died 26 days after the commencement of the experiment. Evidence of scurvy at the post-mortem examination.

Guinea-pig IV. Dose 3 cc. Gained 30 g. in 37 days. Chloroformed. At the post-mortem examination the animal was found in excellent condition, but showed evidence of extremely mild scurvy.

Guinea-pig V. Dose 3 cc. Gained 28 g. in 37 days. Chloroformed. At the post-mortem examination the animal was found in excellent condition showing no evidence of scurvy.

Guinea-pig VI. Dose 3 cc. Gained 16 g. in 37 days. Chloroformed. At the post-mortem examination the animal was found in excellent condition showing no evidence of scurvy.

Guinea-pig VII. Dose 5 cc. Gained 18 g. in 37 days. Chloroformed. At the post-mortem examination the animal was found in excellent condition showing no evidence of scurvy.

Guinea-pig VIII. Dose 5 cc. Gained 71 g. in 37 days. Chloroformed. At the post-mortem examination the animal was found in excellent condition showing no evidence of scurvy.

Guinea-pig IX. Dose 5 cc. Gained 30 g. in 37 days. Chloroformed. At the post-mortem examination the animal was found in excellent condition showing no evidence of scurvy.

No control was tested out with this experiment owing to some mishap, but numerous experiments performed by the writer with decitrated lemon juice show that the minimum dose capable of preventing scurvy in guinea-pigs is about 1·5 cc. The conclusion one can draw from this experiment is that the hydrolysis diminishes the activity of the juice, but in spite of the very drastic treatment very significant activity is retained. This further confirms the view that the anti-scorbutic factor is stable towards heat providing anaerobic conditions are observed.

Summary.

1. The fat-soluble factor was destroyed in cod-liver oil by exposing it to ozone in the dark at ordinary temperature and by bubbling air through it at 120°.

2. The exposure of autolysed yeast juice to ozone in the dark did not destroy the activity of its water-soluble factor to any appreciable extent.

3. The exposure of decitrated lemon juice to ozone destroyed its anti-scorbutic activity. This was also destroyed by passing air through the solution at ordinary temperature.

4. Ultra-violet rays had no deleterious action on the accessory food factors in the absence of air.

5. Two hours' boiling did not destroy the anti-scorbutic potency of decitrated lemon juice in an atmosphere of carbon dioxide, whereas one hour's boiling in the presence of air destroyed its potency almost entirely.

6. Hydrolysis of decitrated lemon juice with hydrochloric acid for five hours impaired the anti-scorbutic activity of the juice. The juice however retained its activity to a very considerable extent.

In conclusion I wish to express my best thanks to Professor Sensho Hata and Miss E. M. Low for help in some of the experiments in this inquiry. My thanks are also due to Dr J. S. Edkins for having kindly permitted me to use his ozone generator.

The expenses of this research were defrayed from a grant made by the Medical Research Council, to whom my thanks are due.

REFERENCES.

Delf (1920). *Biochem. J.* **14**, 211.
Drummond and Coward (1920). *Biochem. J.* **14**, 734.
Dutcher, Harshaw and Hall (1921). *J. Biol. Chem.* **47**, 483.
Hess (1920). *Brit. Med. J.* 147.
Hess and Unger (1921). *Proc. Soc. Exp. Biol. Med.* **18**, 143.
Hopkins (1920). *Biochem. J.* **14**, 725.
Zilva (1919). *Biochem. J.* **13**, 164.
Zilva (1920). *Biochem. J.* **14**, 740.
Zilva (1921). *Lancet*, i, 478.

X. AN IMPROVISED ELECTRIC THERMOSTAT CONSTANT TO 0·02°.

By SAMUEL CLEMENT BRADFORD.

From the Science Museum.

(*Received January 17th, 1922.*)

In these days of elaborate physical apparatus it may be of interest to describe a simple thermostat which can be made easily from material at hand and is sufficiently constant for most purposes. The apparatus, shown in the photograph, was required for an investigation expected to extend over twelve months or more and needing only moderate constancy.

The principle of the stirrer, on which the success of the thermostat depends, was suggested by Mr S. Bradshaw, to whom the author is indebted, also, for much kind assistance in making the apparatus (Plate I). The stirrer comprises a solenoid, A, with an iron plunger, B, mounted on a glass tube carrying an ordinary cork bung, shown in dotted outline, at its lower extremity. The upper part of the glass shaft bears two rubber corks, C and D, that strike the left arm of the brass lever, E, whose right arm carries a cylindrical plumbago contact. As the stirrer reaches the lower end of its stroke, the cork, C, pulls over the lever, causing the plumbago contact to engage with a similar contact, F, above. The current from the 110 volt supply now flows through the solenoid, raising the plunger until the cork, D, throws the lever back again, breaking the circuit and allowing the plunger to fall. To render the lever stable in either position, it is made top-heavy by means of the weight, G, carried on a stout wire soldered to the lever near its pivot.

For temperatures up to 25°, a 16 c.p. carbon lamp is sufficient as a heater. For higher temperatures a more powerful lamp may be employed, or an auxiliary one continually alight. The brass parts of the lamp are protected from the water by a short piece of 1 inch rubber tubing.

As pointed out by Lewis [1922], the heater should be placed near the thermo-regulator, M, which is similar to an ordinary toluene thermo-regulator for gas. This was made from a stout glass test-tube of about 50 cc. capacity when sealed to about 3 inches of 4 mm. glass tubing. To this was joined about 6 inches of 1 mm. bore capillary tube, having a short wider tube at its open end. About 2 inches from the upper end of the capillary, a short piece of 4 mm. tube, N, was sealed at right angles. The wider tube, joining the capillary and bulb, was then bent through 180° near its junction with the bulb. The bulb contains toluene and some mercury which also fills the capillary tube. The height of the mercury in the latter is adjusted by sliding, within the side tube, a short piece of capillary tube bearing a platinum contact at its sealed end. The joint between the side tube and its piston is closed by a slieve of rubber tubing. The other platinum contact is carried at the end of a piece of glass tube, O, drawn out so as to pass vertically within the wider end of the capillary tube and rest on the constriction, with the contact projecting into the capillary tube.

These two contacts are connected, by mercury seals, in series with a 4 volt accumulator and the magnetic cut-out for the heater. For the cut-out, a discarded high-resistance telegraph relay, P, was fitted with iron wire dippers and two mercury cups in circuit with the heater. However, almost any high-resistance electro-magnet and armature could be adapted. By using an accumulator to actuate the cut-out, there can be no fouling of the mercury in the thermo-regulator, and it is unnecessary to provide for an up-and-down motion of the upper platinum contact. The capacity of the accumulator is immaterial, as it can be charged while in use through a single lamp, the potential difference across its terminals remaining at 4 volts. Actually a 40 ampere-hour accumulator will last a month without recharging. The capacity of the bath is about 5 litres. A constant level is maintained by the Marriott's bottle shown on the left.

Had the temperature remained constant within $0.1°$, expectations would have been fulfilled. Consequently, it was astonishing to find that the variations were limited to $0.02°$. The observations recorded below extended over a period of nearly 24 hours. The maximum variation occurs within a short period corresponding to, but lagging behind, the heating cycle. By bringing the heater right up to the thermo-regulator, probably even this small variation would be diminished. The disadvantage of doing so, when working for very long periods, lies in the chance of the mercury in the cut-out being fouled on account of the greater number of breaks. During the period under observation only one reading was made differing by more than $0.01°$ from the mean, and this was only $0.05°$ greater.

The solenoid, $4\frac{1}{2}$ inches long, with a pasteboard tube, allowing $\frac{1}{16}$ inch clear space round the iron core, is wound with about eight layers of No. 27 single cotton covered wire insulated with shellac varnish. The core is made from gas pipe, $\frac{3}{4}$ inch diameter, the same length as the solenoid, turned down to about $\frac{7}{12}$ inch thickness. Or it could be made from sheet iron. The glass shaft passes through two rubber corks in the ends of the core and is supported by guides, Q and R, constructed of sheet brass. It is convenient to make the glass shaft in two pieces joined by a cork, shown at the level of the heating lamp.

The contact lever, E, is the only part of the stirrer needing careful construction. It is fashioned from 18 gauge sheet brass. The pivot bearing is a piece of brass tube, $\frac{1}{2}$ inch long, sweated through the lever at its centre of mass, and ground with a little emery and oil to fit smoothly over the shank of a round-headed brass screw forming the pivot. The mass of the weight, G, is about 25 g. One end of the bearing abuts against a brass plate screwed to the base-board, the other against the flange of the screw, which is adjusted so that the lever just swings easily, without looseness, between the two stops, K and L, jacketed with rubber tube. The thin steel springs, H and J, are carefully bent so that their pressures increase or diminish very gradually and evenly throughout the stroke. They engage the under side of the lever and the lower one should project upwards to the level of the end of the upper one. The pressures of the springs are adjusted by means of the screws seen near their centres, and must be only sufficient just to absorb the energy of the lever as it is brought to rest against the stops. To prevent sparking at the pivot, with consequent roughening of the bearing, the current is conveyed to the lever through a short piece of flexible wire soldered to it, as shown in the photograph.

(a) Thermometer remote from thermo-regulator		(b) Thermometer near thermo-regulator	
θ	T (h. m. s.)	θ	T (h. m. s.)
23·52°	3 5 0 p.m.	23·515°	12 30 0 p.m.
·51	7 0	·505	30 30
·50	7 30	·505	31 0
·51	8 0	·515	31 30
·52	9 0	·52	32 0
		·515	32 10
23·50	3 17 0	·51	32 30
		·51	33 0
23·50	3 45 0		
		23·51	1 10 0
23·50	4 0 0	·515	11 0
		·51	11 20
23·51	4 15 0	·51	12 0
		·505	12 20
23·52	10 30 0 a.m.	·51	13 0
23·52	10 50 0	23·515	2 12 30
·52	53 0	·51	12 50
·53	54 0	·50	13 10
		·50	13 30
23·515	11 5 0	·505	14 0
		·51	14 10
23·52	11 39 0	·515	14 30
·51	39 30	·51	14 50
·505	40 0	·505	15 0
·51	40 10		
·52	40 20		
·525	41 0		
·52	41 20		
23·52	12 26 0 p.m.		
·51	26 30		
·50	27 0		
·51	27 20		
·52	27 30		

The induction in the solenoid causes a powerful spark at each break and a great deal of trouble was experienced with the contacts. When first set up, the apparatus worked for several hours, but the metal contacts used soon became corroded. Mercury contacts in oil lasted only about 48 hours. Finally,

4—2

cylindrical plumbago contacts, 8 mm. in diameter, as used for magneto brushes were adopted. The lower one fits firmly in a clip soldered to the lever. The upper contact slides tightly in a thin split metal cylinder, which fits loosely in a short vertical piece of glass tube, clamped to the base-board, and has a projection at the top to prevent the contact holder dropping through the glass tube. The flexible wire bringing the current is soldered to this projection. The position of the upper plumbago rod in its holder is adjusted so that, when the lever is in the contact position, the lower rod has raised the upper one about $\frac{3}{16}$ inch above its resting position. Thus, at each make and break, the lower contact pushes the upper one up and down, and upon this motion the efficiency of the apparatus appears to depend. Probably carbon contacts would serve nearly as well as plumbago ones.

  Even with this arrangement, however, the apparatus could hardly have been regarded as satisfactory, owing to excessive sparking. Ultimately the device was adopted of allowing a small current, just insufficient to raise the plunger, to flow through the coil constantly. Only the extra current, necessary to actuate the stirrer, is passed through the contacts. With this arrangement, the coil is always on a closed circuit, and the smaller induced current, due to breaking the extra current, discharges through the closed circuit, almost entirely avoiding the spark. The constant current flows through an 8 c.p. carbon lamp, and the size of the bung is then adjusted so that its buoyancy is sufficient just not to float the plunger. The extra current passes through an 8 c.p. lamp or preferably, a 16 c.p. lamp with a resistance (which may be made of iron wire), about 150 ohms, adjusted so that the plunger moves regularly with the required power. The position of the corks, C and D, is adapted to the length of stroke, about $2\frac{1}{2}$ inches.

The only attentions the apparatus requires are to give a drop of oil to the pivot, the brass springs and the sliding contact holder, and to sand-paper the surfaces of the plumbago contacts about every 48 hours. Occasionally the pivot screw, or those of the springs, may need a fraction of a turn. The guides of the shaft are best lubricated with vaseline. At the time of writing the thermostat has been running smoothly for over a month.

REFERENCE.

Lewis (1922). *Trans. Farad. Soc.* **17,** 1.

XI. TETHELIN—THE ALLEGED GROWTH-CONTROLLING SUBSTANCE OF THE ANTERIOR LOBE OF THE PITUITARY GLAND.

By JACK CECIL DRUMMOND

AND

ROBERT KEITH CANNAN (*Beit Memorial Research Fellow*).

*From the Biochemcial Laboratories, The Institute of Physiology,
·University College, London.*

(*Received January 17th, 1922.*)

CERTAIN conditions of disordered development, such as acromegaly and gigantism, have long been associated with abnormal conditions of the pituitary body and in consequence the function of this gland in the growth of the animal has been the subject of much investigation. The suggestion that the pituitary body supplies a hormone which regulates the growth of the body, particularly of the skeleton, arose directly from the pathology of the diseases in question and has directed research along two definite lines. On the one hand studies of the effect of extirpation of the gland have been made, on the other, light has been sought in experiments based on the administration of the gland or its extracts to a normal animal.

The extirpation of one or other of the lobes of the pituitary body without injury to the remainder or to surrounding tissues is attended with extreme difficulty owing to the intimate association of the parts and, in consequence, results of such experiments have been conflicting. It is generally agreed that removal of the posterior lobe gives rise to no marked symptoms, whereas the balance of opinion points to the conclusion that with extirpation of the anterior lobe in young animals growth is checked, fatty deposits increase, metabolism is reduced and persistent infantilism results.

The results of experiments on the administration of the gland to normal animals are, again, far from unanimous. Cerletti [1907] injected pituitary emulsion into young animals and reported retardation of bone growth and Cushing [1909] by repeated injection of anterior lobe extracts obtained a fall in weight of young animals. The feeding of pituitary by Sandri [1909] to young guinea-pigs arrested growth whilst Schafer [1912] found that oral administration of anterior lobe was without appreciable effect on the growth and metabolism of young rats. Aldrich [1921] from experiments on puppies came to the same conclusion.

The general opinion at this period was, therefore, that oral administration of anterior lobe did not appreciably accentuate any regulating influence which the gland might normally have upon growth. The value of much of this earlier work is however depreciated by the fact that various forms of administration and different preparations of the gland were adopted by the different workers.

The whole question was reopened by a series of papers by Brailsford Robertson in 1916. He claimed to have demonstrated that the feeding of the fresh anterior lobe of the ox to white mice leads to a "marked retardation of growth in the beginning of the third growth cycle (6–10 weeks) followed by a pronounced acceleration from the 20th to 60th weeks" [1916, 1]. Further, he isolated from the anterior lobe of the ox a substance to which he gave the name *tethelin*, and which he regards as the active growth-controlling principle [1916, 2].

We were, at first, impressed by the care with which Robertson's feeding experiments had been carried out and it did appear that, by the employment of large groups of animals and by a statistical control of his data, he had guarded against all sources of error common to such experiments. A closer inspection of his work, however, revealed many inconsistencies which in our opinion seriously detract from the validity of his conclusions. In particular, it appeared that nothing but a substance of very impure character could possibly be obtained by the method he described for the isolation of tethelin [1916, 3].

The communications referred to were followed by others in which the chemical and physiological properties of tethelin were described and its effects in increasing the rate of tumour growth [1916, 4], in accelerating the healing of wounds and recovery after inanition [1916, 5] were reported. Meanwhile other workers were obtaining varying results. Goetsch [1916] and Marinus [1919] independently observed accelerated growth in rats and Wulzen [1916] accelerated division of planarian worms whilst Pearl [1916] obtained retarded growth of chickens fed upon anterior lobe.

The conflicting nature of all these results is not dispelled by Robertson's later papers in which, from a consideration of his accumulated data, he modifies his former views and comes to the conclusion that the influence of tethelin upon growth consists, as far as the whole animal is concerned, in retardation. His latest explanation of his former conclusions and the development of these views into a theory of growth catalysers is ingenious but quite unconvincing [1919, 1921]. We were led by the conflicting nature of all the evidence to undertake the re-investigation the results of which are reported in this paper. For the opportunity to carry out the work we are indebted to Mr T. O. Kent, who most kindly supplied us throughout with quantities of fresh anterior lobe of the pituitary of the ox.

Isolation of tethelin.

The method for the isolation of tethelin described by Robertson [1916, 3] is briefly as follows. The mixed anterior lobes dried by intimate mixture with an anhydrous inorganic salt are extracted for 48 hours with boiling absolute alcohol. The extract is concentrated *in vacuo* until separation of a precipitate begins. After cooling, one and one half volumes of dry ether are added and a voluminous white precipitate separates, which, after three hours, is filtered, washed with alcohol-ether mixture and dried in an incubator at 30–35°. Robertson appears to claim chiefly from constancy of nitrogen and phosphorus content that the substance is a chemical individual. To anyone familiar with such methods for the extraction of tissues it must be obvious that only a crude product, consisting largely of lipoid substances would be yielded by his technique and, further, that no such product would preserve its characteristics unless carefully protected from oxidation throughout the preparation and drying. We have made several attempts to prepare this substance and to obtain a white precipitate such as Robertson describes by following the detail of his method. Our yield varied from 0·4 to 0·8 % of the moist gland, but the products obtained did not show constancy of nitrogen and phosphorus content as reported by Robertson. The analyses of two samples which agree in their content of nitrogen are given in Table I. In all cases our preparation was carried through in an atmosphere of carbon dioxide.

Table I. *Analyses.*

| | Our preparations | | Robertson | Protagon |
	I	II		
Nitrogen	2·88	2·80	2·58	2·39
Phosphorus	1·76	0·81	1·41	1·07

Our preparations resembled in many ways the mixture of lipoid substances which one can prepare from animal tissues by following a method essentially that of Robertson, and the analytical figures as well as the appearance and properties of the products show a resemblance to admittedly impure mixtures of this type such as protagon. Robertson described a number of properties of tethelin which are typical of such lipoid mixtures but we were unable to confirm the colour reactions upon which he is rash enough to suggest the presence in the molecule of inositol and of an iminazolyl radicle.

On the assumption that tethelin represents an impure mixture of substances of the lipoid class, perhaps contaminated by other cell constituents—no purification of his product appears to have been attempted by the author—we submitted our preparations to a simple fractionation. By extraction with hot alcohol they could be divided into a fraction readily soluble in that solvent and an insoluble fraction.

The soluble fraction deposited, on standing, a white pseudo-crystalline product of which 1·68 g. were obtained (T_1). The mother liquor of T_1 yielded on evaporation a sticky brown residue, T_3, whilst the fraction insoluble in alcohol was designated T_4. On analysis these products gave the figures:

	T_1	T_3	T_4
Nitrogen	1·63	3·43	3·23
Phosphorus	0·88	1·54	3·94

We endeavoured to carry the fractionation further but were led to abandon it owing to the small amount of material at our disposal and to our conviction that the matter was not worthy of further consideration.

We did, however, examine the fraction T_1 which appeared to have the properties of an impure mixture of galactosides.

A determination of the melting point as described by Rosenheim [1914, 1] showed the following:

	Phrenosin (Rosenheim)	T_1
Softens	130–140°	—
Shrinks	170–190°	190°
Darkens	—	209°
Melts	205–215°	214°

In order to determine if the substance were mainly galactosides we hydrolysed 0·5 g. of T_1 by the method described by Rosenheim [1913]. 0·34 g. of impure esters was obtained and this upon hydrolysis yielded 0·23 g. of fatty acids, having a melting point of 58°. Recrystallisation from alcohol gave a crop of white needles melting at 70° which after a second recrystallisation melted at 76°.

A few oily drops were separated from the products of hydrolysis of T_1 and may have been the sulphates of the bases, but there was insufficient for examination.

The hydrolysate, freed from esters, was examined in the polarimeter. The rotation, if any, was slight and laevo in direction. A minute amount of a methylphenylosazone which melted at 182° was obtained (methylphenylgalactosazone M.P. 190°). A further amount of T_1 was extracted with pyridine following the method of Rosenheim [1914, 2] and yielded fractions of varying nitrogen and phosphorus content. There was insufficient of these for further examination.

There appeared, from all the data, to be little doubt that the product obtained by following the method described by Robertson for the isolation of tethelin is a very impure mixture of substances of the lipoid class and that this product is capable of separation by simple means into fractions of variable composition.

Feeding experiments.

Robertson appears to be well aware of the high variability of the growth curves of white mice but it is his claim that, by the employment of large groups of animals, a composite growth curve can be constructed which will to a high degree of probability reproduce the characteristics of growth of the individual. On this statistical basis he feels entitled to draw definite and far reaching conclusions from small differences in such composite curves for normal and tethelin-fed groups. An examination of his methods and data led us to query the justice of this argument.

For example, it is significant that, in his experiments on the effect of feeding egg-lecithin [1916, 6], two batches of normal mice chosen at random at the age of four weeks showed a difference of 1·5 g. in mean weight, *i.e.* 13 % of body weight—a difference of the same order as the defection of weight of tethelin-fed mice at the maximum divergence of this group from the normal. Yet this, in the former case, is explained away as being due merely to the high variability at that age, whilst in the latter case it is reported as being a marked defection due to the effect of tethelin. Many other examples of such biassed interpretation occur throughout this series of papers but one further example will suffice. In support of his contention that mice fed upon pituitary are, size for size, heavier than those normally fed. Robertson reproduces a photograph of two mice of the same age and linear dimensions, one a normal mouse weighing 30 g. the other a pituitary-fed mouse weighing 37 g. [1916, 1]. When it is pointed out that the weight of the normal was about the mean for his batch whereas the other weighed 27 % more than the mean for his batch—*i.e.* was an extravagantly abnormal member—the fallacy of the argument is apparent.

On the basis of such criticisms we submitted the author's papers to Professor Karl Pearson under whose direction, Mr H. Soper, of the Department of Applied Statistics, University College, London, undertook the examination of Robertson's statistical arguments, and who has kindly permitted us to publish extracts from his report. He says "if some mice are apt to be 3 days in advance of or behind the others in stages of growth as seems indicated, then the curve of mean growth will not reproduce the characteristics of individual growths. The addition of curves at different epochs or phases will tend to distort or obliterate the detail." It is significant that just such a possible variation in stage of growth was introduced into the experiments by the empirical grouping together of animals for weighing. Thus animals weighed between the 25th and 31st days inclusive were regarded as having been weighed upon the 28th day. A similar bracketing of ages was employed throughout. Mr Soper sums up in these words: "The conclusion to be drawn is that on the figures presented and statements made there is a significant defect in weight in all the treated groups from the normal group. The normal group was not changed and was not large in numbers and it has been assumed that it was a random sample of the whole community under observation. The figures given are not so complete as they might have been and statements are made and procedures taken which could only be supported by data not detailed."

It is significant that the normal group was not strictly a "random sample of the whole community under observation" as their growth curve antedates that of the pituitary mice by three weeks and that of the tethelin mice by three months. The stages of growth of the contrasted batches were not contemporary and there is no evidence that undetected variations in environment incident to the two groups at different ages were not responsible for the differences in growth. Robertson himself [1919] has acknowledged that the normal growth curve varies year by year.

Any experiment, claiming acceptance upon statistical considerations, which in the methods employed introduces a possibility for variability from that statistical basis must be open to criticism. Finally an inherent source of error vitiates all the conclusions. The animals were in possession of their own functioning pituitary glands, the activity of which could be neither measured nor controlled.

EXPERIMENTAL.

It was not possible with the material at our disposal, nor in view of our previous findings was it considered worth while, to carry out feeding experiments on the scale of those of Robertson, but all the other precautions employed by him in respect to weighing, elimination of sickly mice and general care were taken and the errors which it is suggested existed in his méthods have as far as possible been avoided.

The mice were placed as soon as weaned in groups of six of the same sex in special wooden boxes and fed on a basal diet of mixed grains and dry bread with meat and vegetables once a week and water *ad libitum*. To the treated animals dried anterior lobe was administered in a dose of 50 mg. per mouse per day (equal to about 3 mg. of tethelin). This was given in a paste of dried milk powder and starch in equal parts and was greedily consumed before the day's ration of ordinary food was given. The same quantity of the paste, without pituitary, was given to the normal animals.

In the first experiment mice were taken at an age between 3–5 weeks and divided into batches of 25 normal and 25 pituitary-fed mice of each sex. The individual growth curves were plotted until maturity. In view of the uncertainty of age no composite growth curves were constructed but it was thought that a careful inspection of such groups of individual curves would reveal the general characteristics of the two rates of growth and the distortion of the normal curve found in Robertson's tethelin curve would be detectable. No general tendency towards a differentiation of the shapes of the two batches of curves was to be observed, a close inspection merely emphasising the high variability of growth.

In a second experiment mice were observed from birth so that composite curves might be drawn. The mothers of the treated mice were fed with pituitary until the young were weaned, when the latter were removed and treated as in the first experiment. The experiment was abandoned at 13 weeks owing to pituitary supply having become exhausted but as Robertson reports marked effects from 7 weeks onwards a comparison was possible.

These growth curves show no characteristic differences. Admittedly the curves are built up from a statistically too limited number of mice but, in conjunction with the other work reported their negative result is significant.

	Male mice (ten in each group) Mean weight. g.		Female mice (ten in each group) Mean weight. g.	
Age	Normals	Pituitary-fed	Normals	Pituitary-fed
Birth	1·60	1·44	1·53	1·50
1 week	3·92	4·20	4·00	4·10
2 weeks	5·55	5·90	5·80	5·75
3 ,,	7·44	8·00	8·02	8·30
4 ,	9·70	10·0	10·1	10·1
5 ,	12·0	12·1	12·4	12·0
6 ,	14·3	14·0	14·3	13·4
7 ,	15·1	15·6	15·4	14·8
8 ,	16·0	16·9	16·4	16·0
9 ,	17·2	18·0	17·6	17·6
10 ,,	18·7	19·5	19·0	18·4
11 ,	19·8	20·4	19·6	19·0
12 ,	21·1	21·4	20·3	19·8

CONCLUSIONS.

1. The product obtained by the method described by Robertson for the isolation from the anterior lobe of the pituitary gland of the substance he has called *tethelin* is a very impure mixture of substances of the lipoid class.

2. Criticisms are advanced of the deductions made by Robertson as to the effect of tethelin and of anterior lobe of the pituitary gland upon growth and experiments are reported which fail to point to any influence upon the growth of mice of oral administration of anterior lobe of the pituitary gland.

REFERENCES.

Aldrich, T. B. (1912). *Amer. J. Physiol.* **30**, 320 and 437.
Cerletti, U. (1907). *Arch. Ital. Biol.* **47**, 123.
Cushing, H. (1909). *J. Amer. Med. Assoc.* **53**, 251.
Goetsch (1916). *Bull. John Hopkins.* **27**, 29.
Marinus, C. J. (1919). *Amer. J. Physiol.* **49** 238.
Pearl, R. (1916) *J. Biol. Chem.* **24** 123.
Robertson, T. B. (1916, 1). *J. Biol. Chem.* **24**, 385.
—— (1916, 2). *J. Biol. Chem.* **24**, 397.
—— (1916, 3). *J. Biol. Chem.* **24**, 409.
—— (1916, 4). *J. Exp. Med.* **23**, 631.
—— (1916, 5). *J. Amer. Med. Assoc.* **66**, 1009.
—— (1916, 6). *J. Biol. Chem.* **25**, 647.
—— (1919). *J. Biol. Chem.* **37**, 377.
—— (1921). Biochemistry.
Rosenheim, O. (1913). *Biochem. J.* **8**, 604.
—— (1914, 1). *Biochem. J.* **7**, 121.
—— (1914, 2). *Biochem. J.* **8**, 110.
Sandri, O. (1909). *Arch. Ital. Biol.* **51**, 337.
Schafer, E. A. (1912). *Quart. J. Exp. Phys.* **5**, 203.
Wulzen, R. (1916). *J. Biol. Chem.* **25**, 625.

XII. THE ESTIMATION OF PECTIN AS CALCIUM PECTATE AND THE APPLICATION OF THIS METHOD TO THE DETERMINATION OF THE SOLUBLE PECTIN IN APPLES.

By MARJORY HARRIOTTE CARRÉ and DOROTHY HAYNES.

Department of Plant Physiology and Pathology, Imperial College of Science and Technology.

(*Received January 18th, 1922.*)

IN order to follow the changes which take place in the pectic constituents of stored apples, a number of preliminary investigations have been found to be necessary. In the first place a method was required by which pectin could be estimated accurately in dilute solution. This having been established, it has become possible to proceed to an investigation of the means by which the various pectic constituents of the fruit can be separately extracted. The present communication, in so far as the extractions are concerned, is confined to an examination of methods for extracting the soluble pectin of fruits. The difficulties which are encountered in the course of this extraction are almost entirely mechanical—difficulties inseparable from the process of washing out this soluble colloidal material from the insoluble colloids of the cell wall with which it is intimately associated. The clearing up of these mechanical difficulties is a necessary preliminary to any further investigation of methods of extraction, for unless suitable methods of washing out are adopted much soluble pectin may remain behind to form a mixture with the pectin dissolved out by subsequent treatment.

It has been the general practice hitherto to estimate the pectin content of solutions by precipitating the pectin with alcohol. This method at its best is inconvenient and lacking in accuracy. The pectin obtained is necessarily of variable composition, since pectin probably exists in a number of forms intermediate between neutral pectin and pectic acid. Moreover, as dilution increases precipitation becomes increasingly difficult, until at very low concentrations no precipitate is obtained, even after prolonged standing with a large excess of alcohol. By the method now adopted the difficulty of precipitation is avoided, for calcium pectate can be made to flocculate from solutions of very low concentration, and a product of definite chemical composition is obtained. A comparison of the results of precipitating pectin with alcohol and as calcium pectate is given in the sequel.

THE PRECIPITATION OF CALCIUM PECTATE.

Since calcium pectate is a colloid, its state of aggregation varies greatly with the conditions under which it is precipitated, and in order to obtain a product which can be easily manipulated it is requisite to adjust these conditions carefully. Great difficulties were experienced in the early stages of this investigation by the occurrence of a sticky condition in which the precipitate could only be washed and filtered with extreme difficulty. This has been found to be the result of precipitating in alkaline solution, in which case calcium hydroxide is absorbed by the gel [Haynes, 1914]; if the period of hydrolysis with sodium hydroxide is too prolonged, or if the alkali is too concentrated, a similar result tends to be produced. On the other hand, too small an excess of alkali above the theoretical amount produces unsatisfactory coagulation, and as the concentration of pectin is increased the excess of alkali required increases also. In the opinion of the writers, both effects are probably a consequence of the tendency which alkalies exhibit to replace the loosely combined water molecules of the pectin sol.

To avoid these difficulties the precipitation of calcium pectate is carried out in the following stages:

(1) Hydrolysis with sodium hydroxide.
(2) Acidification with acetic acid.
(3) Addition of calcium chloride.

Since no independent method exists for checking the results of pectin estimations, it has been necessary to carry out series of experiments and to ascertain how the weight of the calcium pectate varies when variations are made from a standard procedure arbitrarily adopted. In most cases determinations of the ash content of the precipitate calculated as calcium were also carried out. Since the number of interdependent factors is considerable, this has involved a large amount of work. No useful purpose would be served by giving a complete account of the details of this preliminary investigation, but some remarks on the successive stages of the process are required, and the results of a few experiments will be quoted in support of the statements made. The results of ash determinations require more detailed discussion, and will be treated separately below. The pectin solutions used for the investigation were in almost every case obtained from apple. Many determinations were made directly on the liquid obtained by expressing and washing out apple pulp, previously killed by freezing; but confirmatory series were also carried out on pectin from a similar source which had been precipitated by alcohol, re-dissolved and allowed to stand under toluene until suspended impurities were deposited. The juice in each case was heated immediately after pressing, to destroy pectase. It was found convenient to work with a quantity of solution giving from 0·02–0·03 g. of calcium pectate, and to dilute to a concentration of approximately 0·01 % before precipitation. Larger quantities give so great a bulk of precipitate that it is difficult either to wash thoroughly or to dry satisfactorily.

Hydrolysis. Fellenberg [1918] has stated that pectin hydrolyses very rapidly in the presence of a small excess of alkali. The following series show the general results obtained in the course of the present investigation.

	Weight of calcium pectate obtained			
	SERIES I (Concentration of NaOH) $N/100$		SERIES II (Concentration of NaOH) $N/50$	
Time				
10 mins.	Very little		0·029	0·026
30 ,,	0·031	0·0285	0·0315	0·032
60 ,,	0·028	0·0285	0·0305	0·0305
4 hrs.	0·028	0·028	—	—
24 ,,	0·028	0·028	0·0295	0·030
48 ,,	0·028	0·0275	—	—

It will be observed that the rate of hydrolysis increases very rapidly with the concentration of alkali. As it is advisable to keep this concentration low, it is necessary to allow the mixture to stand at least an hour. The best results have been obtained by leaving it overnight, in which case the precipitate flocculates readily and is easy to wash and filter. It is noticeable that rather high results were obtained in both series for the half-hour period. This is a very frequent result of incomplete coagulation which renders effective washing extremely difficult.

Acidification and addition of calcium salt. Calcium pectate can be boiled with water without change, but with dilute acids it tends to pass into colloidal solution. The conditions of this change have not been fully investigated, but it appears that prolonged boiling reverses the peptising effect of acid. This is shown by the results of the following series of experiments carried out on equal quantities of pectin precipitated as calcium pectate.

Treatment		Weight of calcium pectate	
Boiled with water		0·0275	0·028
Boiled with $N/2$ acetic acid	5 mins.	0·0205	0·0225
,, ,, ,,	30 ,,	0·0215	0·024
,, ,, ,,	60 ,,	0·026	0·0265
Boiled with $N/10$ acetic acid	5 ,,	0·0245	0·027
,, ,, ,,	30 ,,	0·0285	0·028
,, ,, ,,	60 ,,	0·026	0·023

The precipitates which had been boiled with acid were found to contain a low percentage of ash. The action of acid must therefore be primarily attributed to the decomposition of calcium pectate; this being so, it is not surprising to find that an excess of calcium salts tends to reverse the effect and that calcium pectate may be boiled with dilute solutions of weak acids in the presence of a sufficient excess of calcium chloride without change of weight. The amount of calcium salt required increases very rapidly with the amount of pectin present. This is shown by a comparison of the two following series of precipitations:

Precipitation carried out with: 100 cc. $N/10$ NaOH; 100 cc. $N/1$ acetic acid; x cc. $M/1$ CaCl$_2$:

Cc. $M/1$ CaCl$_2$ added	Series I 200 cc. apple pectin		Series II 400 cc. apple pectin as used in Series I	
	(1)	(2)	(1)	(2)
2	0·0300	0·0300	0·0565	0·0510
5	0·0300	0·0290	0·0560	0·0575
10	0·0295	0·0300	0·0600	0·0555
25	0·0305	0·0310	0·0605	0·0600
50	0·0300	0·0300	0·0605	0·0600
100	—	—	0·0605	0·0615

Effect of acid concentration.

Precipitation carried out with: 100 cc. $N/10$ NaOH; x cc. acetic acid; 50 cc. $M/1$ CaCl$_2$:

Cc. acetic $N/10$	Series I 200 cc. dilute apple pectin		Series II 400 cc. dilute apple pectin as in I	
	(1)	(2)	(1)	(2)
2	0·1042	0·1270	—	—
10	0·0375	0·0365	—	
20	0·0300	0·0305	—	
50	0·0305	0·0290	—	
100	0·0300	0·0295	—	
$N/1$				
15	0·0310	0·0290	0·0645	0·0660
30	0·0300	0·0295	0·0615	0·0620
50	0·0300	0·0300	0·0610	0·0615
100	0·0300	0·0300	0·0605	0·0610

It will be seen that in the more concentrated series II the lowest limit of acid is only reached at 50 cc. of $N/1$ acetic acid, whereas in series I, 20 cc. $N/10$ acetic acid are seen to be sufficient. The higher values in both series are due to the carrying down of calcium carbonate and occluded impurities which are only removed by excess of acid.

It appears from these results that for ordinary concentrations of pectin the optimum concentration of calcium chloride is approximately $M/1$, and of acid $N/5$. Where the quantity of pectin is very small it has been found advisable to reduce the acid concentration to $N/10$, since the larger amount tends to produce stickiness. In this case the quantity of calcium chloride should be correspondingly reduced.

METHOD OF ESTIMATING PECTIN AS CALCIUM PECTATE.

As a result of these investigations the following method of estimating pectin has been adopted:

A quantity of pectin is taken which will yield from 0·02 to 0·03 g. of calcium pectate; this is neutralised and then diluted to a volume such that after addition of all reagents the total volume measures about 500 cc. 100 cc. of $N/10$ NaOH are then added and the mixture is allowed to stand at least an hour, but preferably overnight. 50 cc. of $N/1$ acetic acid are then added, and after five minutes 50 cc. of $M/1$ calcium chloride. The mixture is then allowed to stand for an hour, after which it is boiled for a few minutes and filtered through a large fluted filter. If the precipitation has been properly carried out, filtration should take place very rapidly and subsequent washing should be easy. The washing is continued with boiling water until the filtrate is free from chloride, after which the precipitate is washed back into the beaker, boiled, and filtered again. It is then again tested for chloride, and these processes are repeated until the filtrate from the boiled precipitate gives no indication of chloride with silver nitrate. It is then filtered into a small fluted filter, from which it can be transferred to a dish and finally to a Gooch crucible which has been previously dried at 100°. The precipitate is dried to constant weight at 100° which has been found to require about 12 hours.

If the quantity of pectin is increased, the quantities of soda and calcium chloride should be correspondingly increased. If very small quantities of pectin are dealt with, the acid and calcium chloride should be reduced.

Example of estimations using dilute apple pectin.

Precipitations carried out with: 50 cc. pectin solution; 50 cc. $N/10$ soda; 100 cc. $N/10$ acetic acid; 100 cc. $M/1$ CaCl$_2$:

SERIES I		SERIES II	
Purified pectin solution		Unpurified pectin solution	
0·0236	0·0237	0·0280	0·0285
0·0237	0·0236	0·0287	0·0300
0·0240	0·0231	0·0265	0·0275
0·0240	0·0230	0·0280	0·0280
0·0229	0·0233	0·0280	0·0285
Mean = 0·0235		Mean = 0·0282	

The quantity of acid used for these estimations is less, and the quantity of CaCl$_2$ greater, than prescribed in the text. It has been found that the alteration of these amounts facilitates the washing of the precipitate.

If precipitated according to the method given above, calcium pectate is so insoluble that quantitative precipitation can be carried out in solutions of extreme dilution, the limiting factors being rather those governing the drying and weighing of very small quantities of colloidal material. It is thus possible to carry out estimations by the calcium pectate method over a large range of concentrations at which precipitation by alcohol is impossible. The lower limit of concentration at which pectin can be even partially precipitated by alcohol has been found to be 0·06 %.

Example of method applied to very dilute solutions.

Precipitations carried out with: 50 cc. dilute pectin solution (apple extract); 50 cc. $N/10$ NaOH; 50 cc. $M/1$ CaCl$_2$; 50 cc. $N/1$ acetic acid:

Weight of calcium pectate	
0·0053	0·0055
0·0050	0·0055
0·0050	0·0053
0·0055	0·0050
Mean = 0·00525	

A few examples of estimation by precipitation with alcohol are given below for comparison. They are all carried out on similar quantities (10 cc.) of apple juice to which a small quantity of water extract has been added.

		Weight of precipitate	
		(1)	(2)
25 cc. of alcohol		0·019	0·021
50 „ „	(1)	0·018	0·015
	(2)	0·012	0·015
100 „		0·018	0·020

The weights of calcium pectate obtained from 10 cc. of this extract were: (1) 0·023; (2) 0·023. It is probable that all the alcohol precipitations are too low. The solubility of pectin in alcohol is shown clearly by the following series of results, carried out on the same material as above:

(1) The weight of precipitate obtained by adding 50 cc. of alcohol to 10 cc. of extract diluted with 25 cc. of water was found to be (1) 0·019; (2) 0·020.

(2) With 50 cc. of water no precipitate was obtained in two days in either sample. After this a small precipitate was obtained from one, but not from the other.

(3) With 100 cc. of water no precipitate was obtained after standing for a week or even after doubling the amount of alcohol.

These experiments have not been repeated on purified pectin owing to lack of material. Such experiments would probably give more regular results, since in ordinary juice sugar acids and other substances are present which may themselves tend to favour coagulation.

In the foregoing account it has been assumed that the method of pectin estimation there described is dependent upon the production of a definite chemical compound—calcium pectate. This assumption obviously needs justification, more especially since pectin is a colloidal substance which is likely to form adsorption compounds and is known to form solid solutions with alkalies and akaline earths [Haynes, 1914].

Pectic acid was among the first of the pectic compounds to be described; indeed the formation of pectic acid by hydrolysis still serves to define the class of substances known as pectins. The chemical nature of pectic acid is still obscure, but it is universally recognised as possessing definite acidic properties, and it has been found to possess the same chemical composition expressed by the empirical formula $C_{17}H_{24}O_{16}$ [Schryver and Haynes, 1916][1] when derived from very different sources (strawberry, turnip, and rhubarb stalks).

The pectic acid from apples alone showed a slight variation, and the present work indicates that this may very possibly have been due to the presence of impurities.

Fellenberg [1918] assigns the formula $C_{62}H_{96}O_{52}(CO_2CH_3)_8$ to pectic acid on somewhat arbitrary grounds, and his own analyses of barium salts agree more nearly with the empirical formula given above.

The number of decomposition products obtained by Ehrlich and others from pectin indicates, however, that the molecule of pectin is large, and it is therefore hardly likely that pectic acid can be represented by a simple multiple of $C_{17}H_{24}O_{16}$. It probably differs from a simple multiple by some small

[1] In this paper the name pectin was used as the equivalent of pectic acid. There is however an increasing tendency to reserve the term pectin for soluble products, especially those occurring naturally in fruits, and to denote by pectic acid the last term in a series of pectic compounds obtained by hydrolysis of these soluble substances. This usage is advocated by Fellenberg, and is followed in the present paper.

quantity but the true formula cannot be ascertained with any certainty until the constitution of pectic acid is established; in the meantime it seems advisable to adopt with reservation the formula given above, and it is to this formula, assumed to be that of a dibasic acid that the results of the analyses given below are referred.

The Composition of Calcium Pectate obtained from Apples.

Two methods of analysis have been employed:

    (1) Ignition to CaO.

    (2) Conversion of this to sulphate.

The following results were obtained by the ignition of calcium pectate obtained from a purified solution of apple pectin which contained at most $0.01-0.03 \%$ of ash when precipitated by alcohol. When filtered carefully before precipitation the amount of ash was negligible.

Weight of Ca pectate	Weight of CaO	Percentage Ca
0·0160	0·0017	7·59
0·0226	0·0025	7·90
0·0270	0·0029	7·66
0·0266	0·0028	7·51
0·0746	0·0080	7·65
0·0308	0·0033	7·64

Mean = 7·658

In order to afford a comparison of the results of the two methods of ash determination, the following series are given. These were carried out on material similar to that used for the previous series. The calcium pectate was obtained by exactly similar methods of estimation in each case.

As CaO			As CaSO$_4$		
Weight of Ca pectate	CaO	Percentage Ca	Ca pectate	CaSO$_4$	Percentage Ca
0·0746	0·0080	7·65	0·0240	0·0062	7·60
0·0266	0·0028	7·52	0·0250	0·0065	7·65
0·0270	0·0029	7·60	0·0236	0·0062	7·70
0·0260	0·0017	7·59	0·0234	0·0062	7·65
0·0226	0·0025	7·70	0·0240	0·0064	7·67

Mean = 7·612   Mean = 7·654

Mean of the two series = 7·63.

It is evident that these results agree very closely with the theoretical percentage, 7·66, and the same percentage has been obtained repeatedly from calcium pectate precipitated under widely different conditions, it may therefore be regarded as established that calcium pectate obtained by this method is a definite chemical compound, the empirical formula of which approximates closely to $C_{17}H_{22}O_{16}Ca$.

Since estimations of pectin have usually to be made in a water extract containing sugars, salts and other compounds, it has been necessary to ascertain how nearly the composition of the precipitate from this unpurified material corresponds with theory. It has been found that the ash is usually rather high, but that if sufficient precautions are taken the error can be reduced to an amount which probably seldom exceeds 1 % of the total ash.

The following are representative examples, each set being estimated on different material:

	Ca pectate	CaSO$_4$	Percentage Ca
-	0·0243	0·0062	7·50
	0·0293	0·0076	7·63
II	0·0553	0·0144	7·66
	0·0577	0·0173	8·30
	0·0294	0·0077	7·70
III	0·0514	0·0138	7·90
	0·0151	0·0040	7·79
	0·0503	0·0129	7·54
	0·0425	0·0111	7·68

In the second set, which gives the highest ash, much iron was found to be present. It must be borne in mind that an increase in the weight of ash in the calcium pectate precipitate may be due either to the carrying down of suspended mineral matter or to the occlusion of organic substances such as the salts of organic acids. These latter if present may introduce considerable error, but if the juice is carefully filtered and if the precipitation is carried out in the presence of sufficient acid, it has been found that the difference can be reduced to a very small amount, which as is shown below may almost certainly be regarded as mainly of mineral origin.

It is significant to find that the impurities associated with pectin vary with different stages of pressing. This is illustrated by the following estimations:

Ash content of calcium pectate obtained from successive extractions of apple.

Number of pressings	Proportion of total Ca pectate present	Ash (calculated as percentage Ca)
1–10	79·5	7·62
10–20	14·0	9·20
20–30	6·5	9·70

In other cases final pressings were found to contain as much as 10–12 % of ash calculated as calcium. There seems much reason to think that this excess of ash is due to colloidal mineral matter, possibly associated with the protoplasm, which can only be pressed out when the cell is completely disintegrated. This, together with a variable amount of colloidal iron, appears to constitute the principal source of the impurities carried down with the calcium pectate precipitated from apple juice. The mineral nature of this impurity cannot be completely demonstrated without elementary analysis, but there seems to be very little doubt that it is largely, if not entirely, composed of inorganic material, in which case the error introduced into the weight of the calcium pectate precipitate will be very small.

It may accordingly be stated as a general rule that pectin should be estimated on material obtained by exhaustive extraction, but that if ash determinations are made in order to ascertain whether or not the material affords calcium pectate in a state of comparative purity the later pressings should be rejected.

A few words may also be added here as to the precautions which it is advisable to take if satisfactory ash determinations are to be carried out. It must be remembered that pectin absorbs and carries down other substances very readily, and that these—once occluded—are removed only with very great difficulty.

Thus if calcium carbonate is contained in the precipitated gel, it is practically impossible to remove it by subsequent treatment with acid. Pectin should therefore be precipitated in dilute solution and in the presence of considerable excess of acid. The excess of $CaCl_2$ should not be greater than is sufficient for satisfactory coagulation, and at each stage of the process sufficient time should be allowed for the reagents to penetrate the large particles of the pectin sol.

THE EXTRACTION OF SOLUBLE PECTIN FROM APPLES.

A considerable amount of investigation has been necessary in order to work out a practicable method for the complete extraction of the soluble pectin. After the first few extractions have been carried out, further extraction gives a very dilute solution, and it is easy to imagine that the whole of the pectin has been extracted at this stage. It has been found, however, that a large proportion—sometimes more than 50%—comes out at so great a dilution that no precipitate is obtained from the solution even after prolonged standing with a large excess of alcohol.

The process is of necessity laborious. The following has been found the best method of procedure: 50 g. of finely minced apple are frozen in an efficient freezing mixture, and maintained at a low temperature for a considerable number of hours to ensure that the pulp is completely killed. The pulp is then warmed to the temperature of the air, and pressed through a cloth in a small hand press. After this the residue, which should be as dry as possible, is ground with fine purified sand in a mortar, and repeatedly washed with cold water and pressed. The residue should be well mixed after each addition of water. Sixty to eighty extractions are usually required, and the bulk of extract obtained is about two litres. Warm water should not be used, as this has been found to increase the amount of pectin obtained—presumably by promoting hydrolysis.

After extraction the liquid is boiled to destroy the action of pectase, and before estimations are carried out it is filtered, first through muslin and finally through a fluted filter paper. An aliquot part, containing a suitable quantity of pectin, is then taken for estimation. The amount varies from about one-half in the case of rather immature apples to one-tenth in the case of apples juicy or over-ripe.

The following estimations have been carried out on equal quantities of apple previously finely minced and mixed to as uniform a condition as possible. They show the order of accuracy with which the process of extraction can be carried out.

Weight of calcium pectate	
In 1/10 of extract of 50 g. of apples	From 100 g. of apples
0·0257	0·514
0·0246	0·492
0·0271	0·542
0·0252	0·504
0·0245	0·490
Mean = 0·0256	Mean = 0·508

[*Note* added Feb. 17, 1922.—Other series of results have been obtained which show similar differences. These differences appear to be due rather to the difficulty of mixing the material uniformly than to incomplete extraction. Thus in the following series, in which pressings were collected in successive fractions, there is no indication that later fractions tend to contain more pectin where the first fraction is relatively low.

Weight of Ca pectate from 50 g. of apples			
Main portion	2nd fraction	3rd fraction	Total
0·339	0·0050	—	0·344
·345	·0100	0·0060	·361
·305	·0085	·0030	·3165
·328	·0070	—	·3345
·325	·0040	—	·329]

Summary.

The precipitation of pectin as calcium pectate is described. It is shown that by a careful adjustment of conditions this can be used as an accurate method of analysis, and that the precipitate corresponds in composition closely to the empirical formula $C_{17}H_{22}O_{16}Ca$. A method is given for the extraction of soluble pectin from apples, and the results of a number of similar estimations are compared.

REFERENCES.

Fellenberg (1918). *Biochem. Zeitsch.* 85, 118.
Haynes (1914). *Biochem. J.* 8, 553.
Schryver and Haynes (1916). *Biochem. J.* 10, 539.

XIII. ON THE OCCURRENCE OF MANGANESE IN THE TUBE AND TISSUES OF *MESOCHAETO-PTERUS TAYLORI*, POTTS, AND IN THE TUBE OF *CHAETOPTERUS VARIOPEDATUS*, RENIER.

BY CYRIL BERKELEY.

From the Marine Biological Station, Nanaimo, B.C.

(*Received January 18th, 1922.*)

A NUMBER of marine annelids excrete a mucous substance which hardens to form a tube in which they live. Many of these tubes are partially soluble in cold dilute alkaline solutions.

In the course of an attempt to determine the nature of the material thus dissolved from the tube of an annelid which occurs commonly on the British Columbia coast, *Mesochaetopterus Taylori*, it was found that the substance precipitated on acidifying the alkaline extract yielded a bright blue ash on burning. Qualitative tests showed that this was due to manganese and led to the examination of the tube material and the tissues of the animal for that element. Preliminary quantitative tests indicated that it was present in both cases in unexpectedly large amounts and more careful determinations seemed warranted.

Traces of manganese are known to occur very generally in animal tissues and Bradley [1910] has shown that the element is normally present in considerable amounts in the tissues of fresh-water mussels. It was also observed by that author that the nacre of the shells of both living and fossil Unionidae is rich in manganese and this, so far as the writer is aware, is the only case of its occurrence in quantity in an animal secretion at all comparable with the tube of a worm.

All the more recent methods for the determination of manganese in organic material depend upon the extraction of the manganese salt from the ash and its oxidation to permanganic acid, which is estimated colorimetrically or by titration with arsenious acid or an arsenite. Reiman and Minot [1920] give an excellent summary of the literature of the subject and describe a method for determining the very small amounts of manganese occurring in human blood and tissues. In the present case the amount of manganese present was comparatively so large that the somewhat elaborate method of ashing the material and leaching the ash described by these authors was unnecessary. In working with the tube material it was, indeed, not found necessary to use any oxidising agent to complete the destruction of the organic matter as is recommended by Bradley. Direct burning at a moderate temperature produced an ash, which, excepting some sand, was completely soluble in 35 % nitric acid, to which a few drops of hydrogen peroxide were added. Subsequent fusion of the undissolved residue with potassium nitrate rendered no more manganese soluble. In the case of the tissues the addition of a small quantity of potassium nitrate was found necessary. With this modification and substitution of a solution of sodium arsenite for that of arsenious acid for titration of the permanganic acid, the method used has been substantially that of Bradley. The details of procedure have been as follows:

0·2 to 1·0 g. of material is weighed into a procelain crucible. The bulk of the organic matter is burnt off at a low temperature. Potassium nitrate is then cautiously added if necessary. The crucible is then gradually brought to a bright red heat and maintained there for about half-an-hour. When cool 2 cc. of 35 % nitric acid and 5 drops of hydrogen peroxide are added to the contents of the crucible which is cautiously heated over a small flame. The acid is then diluted and decanted from the small amount of sand remaining undissolved, and the latter is repeatedly leached with small quantities of hot water until the combined leachings amount to about 50 cc. 0·5 cc. of 10 % silver nitrate is added to the solution which is then brought to the boiling point. The vessel is removed from the flame and 10 cc. of 20 % ammonium persulphate cautiously added. The permanganic acid colour develops immediately. The solution is then boiled gently for five minutes and cooled in running water. It is then titrated with a very dilute solution (not stronger than 0·1 g. per litre) of sodium arsenite which has been previously standardised against a solution of pure potassium permanganate reduced by sulphurous acid. An aliquot of this reduced permanganate solution is put through the foregoing oxidation procedure immediately before the sodium arsenite solution is standardised.

In application to aliquots of the standard permanganate solution evaporated to dryness this method gave results of a high degree of accuracy for quantities of manganese as small as 0·00001 g. It is important to boil the solution sufficiently long after addition of the ammonium persulphate to decompose the excess, otherwise too high readings are obtained. The presence of chlorides is stated by some authors to affect the accuracy of determinations of manganese depending upon the principle involved in this method. This has not proved to be the case in the work to be described. Traces of chloride were invariably indicated by the precipitation of silver chloride on adding the silver nitrate to the solution under analysis, but this dissolved on treating with ammonium persulphate and did not affect the determination. No appreciable difference in result was obtained whether the ash were heated with sulphuric acid, to drive off hydrochloric acid, before analysis or not. The figures recorded in Table I illustrate this point and indicate the degree of accuracy attained in duplicate samples.

Table I. *M. Taylori.*

Material	Weight taken g.	Manganese found %	Treatment
Tube	0·1592	0·0276	Ash not treated with sulphuric acid
,,	0·3426	0·0305	,, ,,
,,	0·2000	0·0300	Ash treated with sulphuric acid
,,	0·2176	0·0253	,, ,,
Posterior region	1·0690	0·0015	Ash not treated with sulphuric acid
,, ,,	0·7098	0·0015	Ash treated with sulphuric acid

Closer agreement than is indicated could not be expected owing to the difficulty of obtaining uniform samples. The material of the tube does not lend itself to very fine grinding and, as is subsequently shown (Table III), the different regions of the tube differ considerably in manganese content, so that such fluctuations as appear may easily be due to the regions being represented in varying proportion in the samples.

Determinations in individual tubes of Mesochaetopterus Taylori.

It is extremely difficult to dig up the tube of *M. Taylori* entire. It usually runs down more or less vertically through the fine sand for one to two feet and then takes abrupt turns in a more horizontal direction through the coarser material underneath. Almost invariably, therefore, unless very special care is taken, the tube is cut off short of the closed end. The tubes of which the analyses are given in Table II were the longest which could be selected from a large number dug, but only in the case of No. 1 was the terminal portion present. They were freed from adherent sand as far as possible, but varying quantities remained imbedded in the tube material. This was determined by weighing the residue after the ash had been completely extracted with nitric acid and water.

Table II. Tube of *M. Taylori.*

Serial number	Weight of material g.	Sand %	Manganese found mg.	Manganese in material %	Manganese in material − sand %
1	0·8402	45·9	0·297	0·0353	0·0654
2	0·3209	18·4	0·0715	0·0223	0·0273
3	0·2600	13·9	0·0132	0·0508	0·0589
4	0·2248	15·7	0·0385	0·0171	0·0203
5	0·1427	22·0	0·0495	0·0347	0·0445

The results show a considerable variation. It is noteworthy that although No. 3 gives the highest result on the entire material, this is in part due to its low content of sand and that, if calculation be made on the material minus sand, No. 1, in which the largest proportion of the basal end of the tube was present, gives the highest figure. This suggested that the basal end of the tube might be richer in manganese than the rest and that the variation in results was to some extent due to the varying proportion of this end of the tube represented in the material analysed.

Determinations in various regions of the tube of M. Taylori.

The tube can be easily divided transversely into three distinct regions corresponding roughly to the anterior, median and posterior portions of the animal. The uppermost, most of which projects above the surface of the sand bed, is composed of very thin papery material and is almost white. In the median region the wall is thicker and pale brown, whilst that of the posterior region is comparatively thick and frequently two or three layered. It is much darker in colour than the rest of the tube and often contains thickenings where the tube has been broken and repaired. Material from each of these regions was separated from a number of tubes and duplicate determinations made from the bulked samples. The results are given in Table III.

Table III. Regions of tube of *M. Taylori.*

Region	Weight of material g.	Ash %	Manganese found mg.	Manganese in material %	Manganese in material – ash %
Anterior	0·2520	49·2	0·011	0·0043	0·0086
,,	0·2126	46·4	0·011	0·0051	0·0096
Median	0·2746	39·9	0·0265	0·0096	0·0160
,,	0·3584	36·7	0·0385	0·0107	0·0169
Posterior	0·3426	26·7	0·1199	0·035	0·0475
,,	0·4250	27·2	0·1375	0·0326	0·0444

The duplicate samples agree as well as could be expected having regard to the difficulty of drawing uniform samples of material. The results show a marked difference in the three regions, the posterior being the highest and the anterior the lowest. The ash content varies in the inverse sense. The greater part of the ash was sand, but separate determinations of sand were not made in these cases. Calculated to the material minus ash there is still a very large difference between the manganese content of the three regions. There is no doubt this accounts in some measure for the irregularities shown in the determinations in individual tubes (Table II) since, as has been pointed out, these were not complete and it was in the proportion of material drawn from the posterior portion of the tube that they chiefly differed.

Determinations in tissues of M. Taylori.

The question whether the tissues of the worm itself contain manganese in notable quantity arises naturally out of its occurrence in the tube. The body of the worm is divided transversely into three distinct regions like the tube. For analysis these three regions were treated separately. They were removed from a large number of worms and kept in alcohol until hardened. The material was then dried, ground as finely as possible and duplicate samples from each region taken for analysis. The results are given in Table IV.

Table IV. Tissues of *M. Taylori.*

Region	Weight of material g.	Manganese found mg.	Manganese in material %
Anterior	0·7067	0·044	0·0062
,,	0·7653	0·0495	0·0065
Median	0·4985	0·077	0·015
,,	0·4493	0·0385	0·0086*
Posterior	1·069	0·016	0·0015
,,	0·7098	0·011	0·0016

* This result is certainly too low. No potassium nitrate was used in ashing and combustion was therefore not complete.

Manganese is present throughout the body of the worm, but in smaller proportion than in the tube. It is interesting to note that the median portion of the body contains most manganese since the chief glands which secrete the mucous material from which the tube is constructed are situated in this region.

Determinations in tube of Chaetopterus variopedatus.

C. variopedatus is the best known member of the family Chaetopteridae, but does not occur on our coast. A few tubes of this worm were obtained from Woods Hole, Massachusetts, for comparison with those of *M. Taylori* in respect of manganese content. The tube is a great deal wider in relation to the size of the inhabitant than in the case of the latter species and has both ends open. These both protrude above the surface of the muddy sand in which the animal lives. With the exception of these small and narrowed ends the material of the tube is very similar to the posterior portion of that of *M. Taylori.* It was found possible to divide the tube material longitudinally into two layers and it seemed of interest to determine manganese in the two layers separately. The inner one was smooth and clean and of a somewhat darker colour than the outer which was rough and heavily coated with sandy mud. This was cleaned as carefully as possible and the two lots of material dried separately. The results are given in Table V.

Table V. Tube of *C. variopedatus.*

	Weight of material g.	Ash %	Manganese found mg.	Manganese in material %	Manganese in material – ash %
Inner layer	0·3278	24·95	1·501	0·457	0·610
,, ,,	0·3042	28·27	1·314	0·431	0·602
Outer layer	0·2806	49·14	0·66	0·235	0·462
,, ,,	0·3021	45·81	0·676	0·224	0·410

The tube of *C. variopedatus* is thus more than ten times as rich in manganese as even the posterior portion of that of *M. Taylori.* The inner layer contains about twice as much as the outer one, but this is in part due to there being more sand in the latter. Expressed on ash-free material the ratio is about 3 : 2. This, coupled with the observation that manganese occurs in comparable quantity in the tissues of the allied species *M. Taylori,* suggests very strongly that the manganese in the tube is secreted by the worm and is not due to the direct action of sand or sea water on it. It is probable, on the contrary, that these agencies tend to remove manganese from the outer layers of the tube since the sand lining the cavity from which a tube has been dug is invariably stained dark brown and looks precisely as if a rusty iron bar had been withdrawn. This rustiness is almost certainly due to oxides of iron and manganese resulting from the decomposition and wearing of the outside of the tube by sand and water.

In every case in which the test has been made iron has been found accompanying manganese in the ashes of the materials dealt with in this paper.

Examination of tubes of other worms for manganese.

The tubes of three other worms which occur commonly in this neighbourhood were examined for manganese; those of two *Sabellids* and a *Spiochaetopterus*. It was present in just sufficient quantity to be determined in the material from one of the sabellid tubes. No trace could be detected in that of the other sabellid or of the spiochaetopterus using quantities of material up to 0·75 g. The last case is particularly interesting because, not only is the animal nearly allied genetically to the chaetopterids in the tubes of which manganese occurs, but also it is always found in the sand beds living immediately alongside *M. Taylori*. We seem to have here a very marked instance of a physiological divergence in species taxonomically and ecologically similar.

DISCUSSION.

As an outcome of the observations recorded the question of the source and function of manganese in the tubes of worms naturally arises. It is fairly clear, for reasons which have already been advanced, that the manganese arrives in the tube from the worm outwards and not from the sea water or sand inwards. Whilst it is conceivable that the worms concentrate manganese from the sea water direct it is not probable, because the quantity of the element in sea water is relatively very small indeed and the worms do not pass large quantities of water through their bodies, but strain their food material out of it. Enders [1909], who has carried out a detailed study of the life history and habits of *C. variopedatus*, holds that, in the case of that species, any sand which is suspended in the water taken by the worm into its tube is also discarded and that only diatoms and small animals are taken into the digestive tract. This does not, however, seem to be the case with *M. Taylori*, since a considerable quantity of sand was found in the posterior region of the animal. A sample of the diatoms which are normally to be found in the water covering the sand beds in which this worm lives, and which probably constitute a large part of its food, was obtained by dragging a fine net over beds of eel-grass growing in the immediate neighbourhood. The diatoms were separated by fractional sedimentation from the small animals and debris accompanying them, dried at 100° and ash and manganese determined. 60·8 % of ash containing 0·007 % of manganese was found. The manganese was all readily extractable from the ash by 35 % nitric acid containing hydrogen peroxide. No further manganese could be detected after fusing the residual silica with sodium carbonate, so that it appears to be contained in the organic portion of the diatom. Here, therefore, we have an immediate source of manganese readily available to the worm. It is, however, not impossible that it is also derived from the sand which, in the case of *M. Taylori*, undoubtedly passes through the digestive tract in considerable quantity.

Manganese was determined in the sand in which the worms were living by sodium carbonate fusion followed by application of the method previously used, and 0·158 % found present. About a fifth of this was directly soluble in 35 % nitric acid and was therefore not in silicate combination. The sand passed through the animal therefore provides an ample source of manganese and its utilisation does not necessitate the assumption of any power of silicate digestion.

The presence of manganese in organisms has generally been associated with respiration. No evidence is at hand to show that it is to be connected with this activity in worms. It is difficult to see how its presence in the tube can have any connection with respiration, and, in the absence of any evidence to the contrary, it seems most rational to assume that its presence there is of no physiological significance, but is due to the worm excreting its superfluous manganese with the tube-building material, a theory which is supported by the observation of manganese in larger quantity in the lining than in the outer layers of the tube of *C. variopedatus*, and in the mucous secreting region of the body of *M. Taylori*, more than in the other regions. The presence of more manganese in the posterior than in the other portions of the tube of the latter species is also explained on this assumption since this is the oldest part of the tube and has been most frequently relined and repaired. Bradley [1910] has shown that the excreta of fresh-water mussels living on their normal manganiferous diet contain the element in about the same proportion as it occurs in their food, so that a state of manganese balance is maintained—only if they are deprived of manganiferous food is the metal definitely retained by the animal. It is likely that a similar state of affairs exists in the case of worms having manganiferous tubes, but that some of the superfluous manganese passes into the tube material. The deposition of manganese in the nacre of the shell of the Unionidae observed by Bradley is no doubt a parallel case.

SUMMARY.

1. Manganese occurs throughout the tube of *M. Taylori*, but in much greater quantity in the posterior portion than in the other parts.

2. Manganese is present throughout the body of *M. Taylori*, but in smaller quantity than in the tube. The median region of the body, in which the chief mucous secreting glands are situated is richest in manganese.

3. Manganese occurs in far larger quantity in the tubes of *C. variopedatus* than in those of *M. Taylori*. The lining of the tube contains much more manganese than the outer layer.

4. The tubes of local species of *Sabellid* and *Spiochaetopterus* contain no manganese.

5. Manganese is probably present in worm tubes as a waste material excreted by the worm with its tube building substance. It may be derived either from diatomaceous food or from sand passing through the digestive tract with the food, or from both.

REFERENCES.

Bradley (1910). *J. Biol. Chem.* **13**, 237.
Enders (1909). *J. Morphology*, **20**, 479.
Reiman and Minot (1920). *J. Biol. Chem.* **42**, 329.

XIV. MAMMARY SECRETION. III.

1. THE QUALITY AND QUANTITY OF DIETARY PROTEIN.
2. THE RELATION OF PROTEIN TO OTHER DIETARY CONSTITUENTS.

By GLADYS ANNIE HARTWELL.

*From the Physiological Laboratory, Household and Social Science Department,
King's College for Women, Kensington, London.*

*Thesis approved for the Degree of Doctor of Science in the
University of London.*

(*Received December 21th, 1921.*)

INTRODUCTORY AND HISTORICAL.

IN a previous paper [Hartwell, 1921, 2], it was shown that large quantities of protein in the diet of a lactating rat were detrimental to the young. The diet consisted of bread and protein in the proportion 15·0 g. bread to 5·0 g. protein. An obvious criticism was that such a diet is deficient in fat and vitamins, but it was pointed out that it was unlikely the bad effects were due to deficiency since the rat could bring up a healthy litter on a diet of white bread alone, a diet equally deficient.

It is generally held that the diet of a nursing mother should be rich in protein, because young growing animals need large amounts of protein in their diet. It is probable that if large quantities of protein are ingested, some other factor must be present in the diet to safeguard the young from such harmful effects as those previously noted. The object of the present investigation was, therefore:

1. To find out what proportion of protein can be fed to the mother without danger to the offspring.

2. To determine what other factor or factors should be included in the diet in order that the large quantity of protein may be effectively metabolised.

It is probable that not only should the diet be physiologically complete, but that a greater quantity of one or more constituents might be necessary when the intake of protein is high. For instance, when large quantities of fat are ingested, it is essential that carbohydrate should be included in the diet; otherwise the fat is not properly metabolised and acetone bodies are excreted.

It is also probable that the large amount of protein deemed necessary has been somewhat overestimated, and that less amounts would give equally good results provided that the diet was satisfactory in other respects.

That excess of protein in the mother's diet has detrimental effects on the young is evident from the experiments of other workers, but such results have been explained differently.

The experiments of McCollum and Simmonds [1918] are on almost identical lines with those described in this paper. These observers, however, were working with diets which were inadequate for growth of the young, and the main object of their research appears to be to show how far "the mother can serve as a protective agent in producing milk suitable for the nutrition of the young," when she herself is getting a deficient diet. A great difference between the work of McCollum and Simmonds and that of the research now being described is the duration of the experiments. From the former's curves it appears that the mother was left with the young for 40 days or more, which seems rather long, considering that a normal baby rat is perfectly capable of living alone from the 21st day. In one or two of their experiments the symptoms of the litter closely resemble those described in this series of experiments. In McCollum's experiment 738 the mother rat was fed on rolled oats alone, and certain symptoms were noted in the young. "On about the 20th day a few of the young would throw themselves about the cage, and scream during this performance. Several of the young went into coma and died."

In some experiments where the mother appeared to have a high tolerance for protein, the babies behaved in the way McCollum describes about the 20th–24th days; they had shown no symptoms at the earlier stage. Again in experiment 948 he says that "the young all died at a very early date. The intestines were filled with gas. We are unable to account for the high mortality of these young." The diet used in this case was rolled oats 82·0 and caseinogen 18·0, i.e. a greater proportion of protein than when the rats had rolled oats only. The young died at intervals from the 3rd to the 23rd day. Seven experiments were made.

The condition of the young certainly suggests a similar state to that found in this work [1921, 2], but only a few of the typical symptoms are recorded by McCollum and Simmonds. In their paper they state that the babies were weighed "frequently" and the curves are plotted at intervals of eight days. Unless daily weighings were made, and the babies carefully examined, it is highly probable that the spasms would be overlooked, since this stage is frequently of short duration, sometimes lasting only a few hours. The screaming fits exhibited by the older babies are much more obvious, the noise being distinctly heard in the next room.

It is possible that these results are due to a deficiency of some dietary factor, but this deficiency is only evident when the diet is too rich in protein and therefore it may be that they are directly due to the excess of protein. A similar diet, but with less protein, would not be deficient.

GENERAL CONDITION OF THE ANIMALS.

Only normal healthy animals were used in these experiments. If any rat became ill the results were discarded. There was no trouble with scab, pneumonia, or epidemic diseases. McCollum [1920] says that the "palatability of the ration is the most important factor in animal nutrition," and it is a well-known fact that many animals will starve rather than eat food which is distasteful to them. The rats used in this work readily ate any of the mixtures offered to them, and in some cases consumed enormous quantities. For instance, many rats were taking as much as 40·0 g. bread (+ added protein and other constituents) on the ninth day of lactation. The initial ration of 15·0 g. bread etc. was usually adequate for the first two days, after which the animals' appetites increased rapidly. The cages were examined for any hidden food, the "food-hoarding" instinct being especially marked in rats.

The males and females were kept together until a few days before the birth of the litters, when the mothers were removed and placed in separate cages. After lactation the females were kept from the males for about two weeks to avoid over-strain of the maternal organism, and to give the mother a chance of recovery, should the special diet have upset her in any way. This method of procedure was adopted merely as a safeguard, because even when the litters died, the mothers appeared fit and well. Except during the actual period of lactation (and with one exception pointed out later) the males and females were fed on a diet of kitchen scraps, supplemented with small quantities of whole milk and white bread. This mixture proved a very good one and, as is shown later, the growth curves of the young animals fed on such a diet, are better than those given by Donaldson [1906] as normal for the white rat.

Some white rats, and some white and black were used. It is possible that the latter should be somewhat heavier, since Donaldson states that black rats are heavier than white ones, yet there appeared practically no difference in size and weight.

METHODS.

The technique employed was the same as that previously described [Hartwell, 1921, 1].

The litters, as before, were reduced to six. In a few cases less than six were born, and this is noted in the text and on the curves. The mothers and babies were weighed daily and the weights recorded graphically. In the "foster" experiments to be described later, the number in the litter varies, but is always stated.

The foodstuffs used. (a) Pure caseinogen was not obtainable when these experiments were in progress, and a commercial form, "food casein,"[1] was therefore used. Experiments showed that the results obtained with "food casein" were exactly similar to those previously found with pure caseinogen.

---

[1] I am indebted to Casein, Ltd., for supplying me with the composition of "food casein."

The "food casein" was 83 % pure, and the amounts given were calculated in terms of pure protein.

(b) Pure egg albumin (as supplied by Baird and Tatlock) was used. It was very finely ground, in order to obtain an even mixture of the foodstuffs fed.

(c) A commercial preparation of edestin was used. It was tested at the beginning of the experiments and found to contain only slight amounts of impurities. A new sample was tested lately and found to contain cane sugar. Every sample was not analysed, and as that obtained lately was less pure, it is possible that the results obtained with 1 g. and 2 g. edestin are due to the fact that smaller amounts of protein were actually fed.

To find what amount of protein is really excess, it would be essential to work with absolutely pure foodstuffs.

(d) The lactose was pure, as supplied by Baird and Tatlock.

(e) The salt mixture was that recommended by Mottram, and is a modification of the original one used by Hopkins.

Sodium chloride NaCl ...	...	...	...	46·25 g.
Magnesium sulphate (anhyd.) $MgSO_4$		...		71·20
Acid sod. phosphate NaH_2PO_4, H_2O ...		...		92·68
Di. pot. phosphate K_2HPO_4 ...		...		254·60
Calcium tetrahydrogen phos. $CaH_4(PO_4)_2H_2O$				144·20
Calcium lactate Ca $(C_3H_5O_3)_2 5H_2O$ ...		...		347·00
Ferric citrate Fe $(C_7H_6O_7) 1\frac{1}{2} H_2O$ ...		...		31·52
Sodium fluoride NaF ...	...	...	...	·55
Manganese sulphate $MnSO_4 5H_2O$ ...		...		2·00
Potassium iodide KI ...	...	...	...	10·00
			Total...	1000·00

Water, *ad lib*, was always given unless the rats were having large quantities of milk.

1. THE QUALITY AND QUANTITY OF DIETARY PROTEIN.

In some recent experiments [Hartwell, 1921, 2] the initial diet consisted of 15·0 g. bread to 5·0 g. protein, both being increased proportionately. Only a very few babies survived and those which lived were not normal, therefore in these experiments less protein was given to the mother, in order to see if the litter would then survive. Small quantities of commercial yeast extract, marmite (about 0·2 g. per day) were added to the diet to make it more palatable. It has been found that small amounts make no difference to the final results and the animals ate such a mixture more readily.

Three experiments were made with each protein. This involved 9 mothers and 54 young. The diet was not physiologically complete except in two of the experiments, when 0·5 g. butter, 0·7 cc. lemon juice and 0·7 g. salt were added to the mixture; this made no difference to the final results.

(i) *Proportion of 15 g. bread to 1 g. protein.*

(a) Caseinogen. Fig. 1 a.

Exps. 160 and 173. 15·0 g. bread, 1·0 g. caseinogen, marmite.

Exp. 175. 15·0 g. bread, 1·0 g. caseinogen, marmite, butter, lemon juice and salt mixture.

Litter 160 were perfectly normal in every respect. One was weakly and died three days after weaning, but no typical symptoms (previously described [1921, 2]) were exhibited.

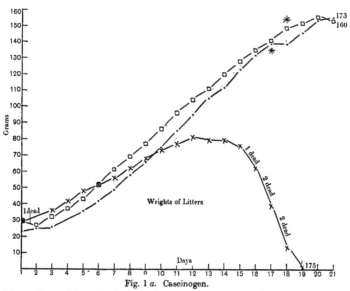

Fig. 1 a. Caseinogen.

Litter 173 exhibited slight symptoms. However they recovered rapidly, the mother was removed on the 21st day, when all the babies appeared normal. They all survived weaning. On the 21st day, the average weight of babies 160 was 30·6 g. and that of litter 173 was 26·0 g. From the normal curve according to Donaldson [1906] the weight at 21 days is about 21·0 g. Hewer [1914] states that she used for subsequent feeding experiments any young rats which weighed at least 23·0 g. at weaning, so it would appear that both litters described above were well above average weight, in spite of the fact that slight symptoms developed in 173. It is possible that "breed" and number in the litter are factors influencing the actual weight of the babies. This is discussed later in the paper.

Litter 175 showed bad symptoms and were all dead by the 19th day.

The mothers all remained well. They were not full grown at the beginning of the experiment and put on a considerable amount of weight.

(b) Edestin.

Exps. 141 and 195. 15·0 g. bread, 5·0 g. edestin, marmite.

Exp. 163. 15·0 g. bread, 5·0 g. edestin, marmite, butter, lemon juice and salt mixture.

All litters were normal in every respect, none of the typical symptoms were noticed in any one baby, not even excitability. The rate of growth in the middle part of lactation was extraordinarily good and apparently maximal (*i.e.* as good as when the mother is fed on bread and milk, a diet previously shown to give the best results).

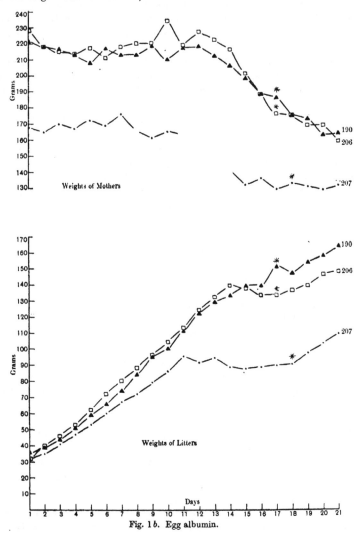

Fig. 1 *b*. Egg albumin.

The mothers, however, showed a slight loss of weight during the first two weeks and this loss became very marked during the last week of lactation.

This seems rather extraordinary because during the last part the young were able to eat for themselves, and were not entirely dependent upon the mother.

(c) Egg albumin. Fig. 1 b.

Exps. 190, 206, 207. 15·0 g. bread, 1 g. egg albumin, marmite.

In these experiments no litter was quite normal, but the mothers were removed on the 21st day and all the babies were successfully weaned.

Litter 190 showed only slight spasms, and recovered rapidly. Two days after the mother was removed they were normal, but small and somewhat thin.

Litter 206 were very similar to 190, but their condition was a little worse and they were not normal till a week after weaning. It will be seen from Fig. 1 b that this litter lost weight for three days before they began to eat for themselves.

Litter 207 showed worse symptoms than 190 and 206 but eventually all the babies recovered.

At first the mothers maintained their weights or showed a slight loss, but from the 11th or 12th day they lost weight considerably. In ten days two of them lost approximately one-quarter of their body-weight and the third lost about one-fifth body-weight. The significance of this is discussed later (p. 86). Thus when 1·0 g. protein to 15·0 g. bread was fed to the mothers, slight symptoms developed in the litters, except with edestin. This is probably due to the fact that commercial edestin was used and, therefore, the mothers were receiving less amounts than with the other proteins.

(ii) *Proportion of* 15·0 g. *bread to* 2·0 g. *protein.*

(a) Caseinogen. Fig. 2.

Exps. 161 and 198. 15·0 g. bread, 2·0 g. caseinogen, marmite.

Exp. 176. 15·0 g. bread, 2·0 g. caseinogen, marmite, butter, lemon juice and salt mixture.

Litter 198 were normal in all respects except that they were somewhat excitable and, as will be seen from the curves, gained only small amounts the two days before they were able to eat for themselves.

Litter 161 showed the typical "excess protein" symptoms, but two lived till the 21st day and survived weaning, although they were not normal till two weeks after.

Litter 176 were in a worse state than 161 and were all dead by the 18th day.

The mothers 161 and 176 put on weight, whilst 198 lost weight. In this connection it is interesting to note that 198 litter were normal, but 161 and 176 suffered badly. This is in agreement with a suggestion previously made that there is some correlation between the weights of the mother and babies.

(b) Edestin.

Exps. 143 and 215. 15·0 g. bread, 2·0 g. edestin, marmite.

Exp. 197. 15·0 g. bread, 2·0 g. edestin, marmite, butter, lemon juice, salt mixture.

Litter 215 were normal except that they developed slight "toe walking" towards the end of lactation. They recovered rapidly and were successfully weaned. Two of the babies died the 2nd day and, therefore, the mother had only four to suckle. Since the average weight of these (see table on p. 87) is so much better than that of the two other litters, it is possible that it can be accounted for in this way. The fact that this rat may have had a greater tolerance for protein is also a factor to be considered.

Litter 197 exhibited all the usual symptoms to some extent, but were never very bad. They soon recovered and were successfully weaned.

Litter 143 developed very bad spasms accompanied by the usual symptoms. One died the 19th day and the others were all dead two days after the mother was removed.

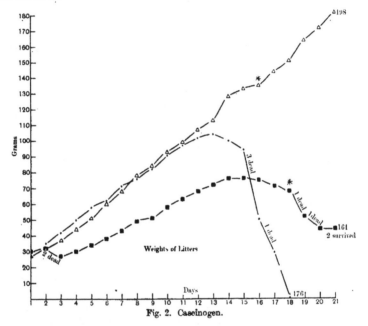

Fig. 2. Caseinogen.

The mothers' curves are somewhat irregular, but on the whole show a maintenance of weight.

(c) Egg albumin.

Exps. 189, 208, 209. 15·0 g. bread, 2·0 g. egg albumin, marmite.

All the litters survived and none showed any spasms.

On the other hand none of the babies was quite normal. Litter 189 did best, but were somewhat excitable. Litter 208 were very excitable, and showed "toe walking."

Litter 209 had similar symptoms to a greater degree and were cold and very weak.

The mothers appeared well, but, as in the case of those which received 1·0 g. egg albumin, they lost weight rapidly.

DISCUSSION OF RESULTS.

The average weights of the young rats at 21 days are given below:

Caseinogen	Edestin	Egg albumin
1·0 g. protein:		
30·6	31·16	27·33
26·0	31·4	24·66
All died	31·16	18·16
2·0 g. protein:		
30·33	33·0 (after 2nd day only four)	29·16
22·0 (only two survived)	24·33	23·5
All died	20·40	19·33

(i) *Irregular growth of the young.*

It has been previously pointed out [Hartwell, 1921, 1] that on any given diet for the mothers, similar growth curves could be obtained for the litters, and that such curves could be regarded as characteristic for that specific diet. This statement now needs some modification.

The experiments described in this paper show that on any given diet the growth curves of the sucklings are by no means identical. It seems probable that when sufficient protein is given to produce bad symptoms in the young, then the above statement no longer holds good. This is well illustrated by the figures above. The average weights of a 21 day rat when the mother received 1·0 g. edestin are practically constant. These litters were normal in every way, so it appears that this amount of protein was not excessive. It is possible that less than 1 g. protein was given in these experiments, since recent samples of the commercial protein have been shown to contain cane sugar as impurity.

In all the other experiments, varying degrees of abnormality were produced and the weights are by no means constant.

This point is again illustrated by experiments to be described later on in this paper.

Thus with sufficient protein in the mother's diet to produce bad symptoms in the young, the individual metabolism of the mother rat seems to play an important part.

The mother's initial weight is not a factor to be considered, since on any given diet the babies of a heavy rat may die while those of a lighter rat may survive, or *vice versa*.

(ii) *Correlation between weights of mothers and offspring.*

It was previously suggested that there was some correlation between the weight of the mother and condition of the offspring, and that if the mother lost weight she might be able to spare the babies some of the symptoms. This suggestion is confirmed by these experiments.

In Fig. 2 litter 198 have a fair growth curve, and showed no symptoms. The mother lost weight. Litters 176 and 161, on the other hand, exhibited typical symptoms and only two of the 12 survived. These mothers put on weight.

In both sets of experiments (*i.e.* 1·0 g. and 2·0 g. protein) it is very obvious that the mothers fed on egg albumin lost weight considerably.

Since albumin is a less good protein than either edestin or caseinogen, it may be that the mother is supplying the necessary amino-acid constituents from her own tissues.

Even 5·0 g. egg albumin to 15·0 g. bread results in some loss in weight of the mother, in a deficient diet [Hartwell, 1921, 2], but practically no loss is found when that proportion is given in a physiologically complete diet (experiments described later).

From the following figures it can be seen that, in some cases, the mother rat lost as much as one-third of her body weight; moreover this loss was very rapid and occupied only ten days.

	Exps.	Weight at beginning of lactation. g.	Weight at end. g.	Loss g.
1·0 g. egg albumin:				
	190	221	164	57
	206	228	159	69
	207	168	142	26
2·0 g. egg albumin:				
	189	247	165	82
	208	214	174	40
	209	149	128	21

Rat 207 lost less than 190 and 206 and her litter were in a worse condition and gained less weight than those of 206 and 190.

Again, in the other series, rat 209 lost less and her litter were the worst.

(iii) *The amount of protein constituting excess.*

It is not possible to give a definite statement as to the proportion of protein which should be fed to a lactating animal, since with different proteins, a different amount will constitute excess. For instance, good growth curves were obtained when the mother was fed with 15·0 g. bread and 1·0 g. edestin, but in the three litters whose mothers had 1 g. egg albumin, there were no normal babies; with caseinogen, one litter was normal, one had slight symptoms and the babies of the third litter all developed bad symptoms and died. As previously explained (p. 81) there is a possibility of impurity in the edestin and hence the animals might be getting less than 1·0 g. of protein, which might account for the slightly better results obtained with the edestin.

It is probable that the amount of protein constituting excess will vary with the presence or absence of other dietary constituents, and that such constituents must be considered *quantitatively* in relation to the protein. This point will be dealt with later in the paper. With caseinogen and edestin, a diet which is complete physiologically from a qualitative point of view gives just as bad results as one not physiologically complete. For example rat 176 received butter, lemon juice and salt mixture in addition to bread, caseinogen and marmite, thus completing the diet qualitatively, yet her litter were worse than those of 198 and 161, whose mother had only bread, caseinogen and marmite.

(iv) *The quality of the protein.*

Certain typical symptoms, already described, were characteristic of the litters, irrespective of the protein fed to the mothers. With egg albumin, however, some extra symptoms were noticed. The weakness appeared to come on at an earlier stage, the babies were thinner and were very cold to the touch. This coldness was observed at an early age and was frequently the first indication that something was wrong. It was evident even when the symptoms were slight and the babies recovered. It is possible that the thinness of the babies is due to less good milk being produced by the mother. It is suggested that egg albumin is deficient or poor in certain amino-acids which are essential for the production of milk of adequate nutritive value. To prove such a point it would be necessary to feed with pure amino-acids.

2. THE RELATION OF PROTEIN TO OTHER DIETARY CONSTITUENTS.

(i) *The effect of large quantities of protein in a physiologically complete diet.*

Hartwell [1921, 2] showed that an initial diet of 15·0 g. bread and 5·0 g. protein fed to nursing rats was detrimental to the young. It was, however, pointed out that such a diet was deficient physiologically, yet it was unlikely the bad symptoms were due to such deficiencies, since the rat can bring up a healthy litter when fed on white bread alone [Hartwell, 1921, 1], a diet also lacking in fat, vitamins and salts.

In these experiments the proportion of protein to bread was kept the same as that used in the previous investigation (*i.e.* 15·0 g. bread to 5·0 g. protein), but 0·5 g. butter (fat and vitamin *A*), 0·2 g. marmite (vitamin *B*), 0·7 cc. lemon juice (vitamin *C*) and 0·7 g. salt mixture (composition given on p. 81) were added to complete the diet.

Three proteins were tried, caseinogen, edestin, and egg albumin and in each case two series of experiments were done.

1. The diet started as soon as possible after the birth of the litter; never more than 24 hours.

2. Diet started before the birth of the litter, and at different stages of pregnancy.

The bread and protein were increased proportionately as required, but the other constituents remained constant.

(a) Caseinogen. Fig. 3.

      Exps. 113, 104, 171. Diet started *at* birth of litter.
      Exp. 120. Diet started 10 days *before* birth of litter.
      ,,   133.   ,,   13   ,,   ,,
      ,,   134.   ,,   16   ,,   ,,
      ,,   166.   ,,   27   ,,   ,,

The gestation period in the rat is given by some observers as 21 and by others as 23 days, so in the second series of experiments the rats were getting the excess protein diet for at least half the gestation period. Rat 166 received this diet during the whole of gestation. In this animal, either the gestation period was longer than normal, or else fertilisation was delayed, because she was separated from the males when the special feeding was started and the litter were not born for 27 days.

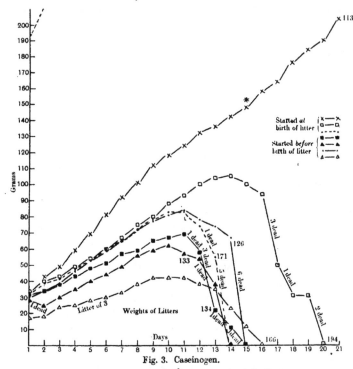

Fig. 3. Caseinogen.

A. *Feeding started after the birth of the litter.*

Litter 113 became excitable on the 9th day and from that time onwards showed signs of being hungry.

No bad symptoms became evident until the 19th day when four of the litter had very bad fits, screamed, rushed round their cage, bit each other and hung on to the cage with firmly fixed jaws. These fits lasted from one to five minutes. Afterwards the babies were very exhausted and would sprawl on the floor of the cage. At this stage the mother frequently removed them to the nest and covered them up. The mother was taken away on the 21st day and all the litter survived, in spite of the fact that bad screaming fits were again noticed on the 22nd day.

Litters 194 and 171 gave the typical "excess protein" curves and all died. The condition of the alimentary canals of the babies was exactly the same as previously described [Hartwell, 1921, 2].

B. *Feeding started before the birth of the litter.*

The actual duration of the feeding before birth seems to be an unimportant factor, since all four curves are similar and the babies were all dead either the 14th, 15th, or 16th day, irrespective of when the feeding was started. In some experiments to be described later (p. 92) the mothers were fed on the caseinogen diet from the time they themselves were weaned and had, therefore, been eating this mixture for 70 days before the birth of the litter.

(b) Edestin.

Exps. 177, 112. Feeding started *at* birth of litter.
Exp. 124. Feeding started   2 days *before* birth of litter.
„   146.   „   11   „   „
„   132.   „   21   ,,   „

A. *Feeding started at birth of litter.*

Litter 177 showed quite good growth up to the 11th day, after which they only gained small amounts. They exhibited all the typical symptoms, though not so markedly as usual. They ate for themselves on the 16th day and were all successfully weaned. They were not normal for about ten days.

Litter 112 grew much more rapidly than did 177, but they developed bad symptoms earlier and were much worse in every way. Only one survived and it was weakly for a month after being taken from its mother.

B. *Feeding started after the birth of the litters.*

In these experiments the time of feeding before birth bears a definite and inverse relation to the time the babies lived, but this does not apply to the litters whose mothers were fed with caseinogen and egg albumin.

There is no reason to describe the experiments separately, or in detail, as the results are similar to those already described for caseinogen.

(c) Egg albumin.

Exps. 111, 193, 221. Feeding started *at* birth of litter.
Exp. 125. Feeding started   4 days *before* birth of litter.
„   129.   „   11   „   „
„   145.   „   9   „   „
„   130.   „   15   „   „

A. *Feeding started at birth.*

Litter 111 had the best growth curve in this series, but bad spasms were seen on the 13th day and after this the usual bad symptoms were marked.

Only one of the litter survived and it was very weak and backward in every way. It did not eat for itself till the 21st day and therefore the mother was not removed until the 24th day. After this the young rat was weakly for some time, but survived.

Litter 193 were all dead the 17th day, having suffered from all typical symptoms, though not to a marked degree. It has been noticed that if extreme weakness comes on at an early stage, the acute spasms are frequently absent.

Litter 221 suffered badly. They showed spasms and all typical symptoms. They were all dead the 16th day.

B. *Feeding started before birth of litter.*

These results again agree with those obtained with edestin and caseinogen.

With all three proteins the mothers were very fit and in most cases put on weight.

From these experiments one would conclude that feeding the mother rat with protein in the proportion 15·0 g. bread to 5·0 g. protein, even in a physiologically complete diet is harmful to the litter. If the feeding is started before the birth of the litter, then no babies survive; also it appears to make very little difference if the feeding is started a few days before the birth or maintained during the greater part of gestation.

Failure in rearing the babies is of much more frequent occurrence when the mother is given an excess of protein in the diet (*at* birth of litter) than when she is fed on other mixtures. It is very easy to see if the young rats have been fed, and with an excess protein diet it was not infrequently found that the whole litter died in the first two or three days, and had hardly had any food.

The symptoms develop at approximately the same time and this seems to suggest a correlation with some stage of development of the young. This possibility is supported by the fact that in a foster litter the symptoms also develop about the same time (experiment to be described later).

If the feeding is started *after* the birth of the litter, some babies may survive, but they are never normal; the majority die.

The results with egg albumin are less good than those with caseinogen and edestin, but this as already suggested is probably due to the different composition of the proteins. Other differences will be commented on later (p. 98).

The fact that some litters do better than others (*e.g.* litter 113 show a better growth curve than any other on the same diet) is possibly due to the greater tolerance of certain rats for high proportions of protein. In this connection it is also probable that the loss of weight of the mother is a factor to be considered; rat 113 lost weight, while other rats on that diet gained and their litters died.

Thus the bad effects on the litters are obtained even when the diet is physiologically complete. It should be pointed out that the amounts of butter (fat and vitamin *A*), marmite (vitamin *B*), lemon juice (vitamin *C*) and salt mixture (inorganic constituents of diet) used were adequate for a normal diet (see experiment described below, when weanlings were fed on a similar mixture).

The mothers remained well during the experiments and when the diet was started before the birth of the litter, the period of feeding lasted as long as five weeks.

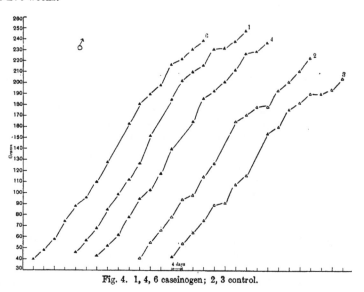

Fig. 4. 1, 4, 6 caseinogen; 2, 3 control.

Growth of young rats on a diet similar to that fed to the mother. Fig. 4.

One criticism which might be offered is that although the mother rats appeared well, they might not have remained so had the feeding been continned longer. To test the adequacy of this diet, young rats were given a similar mixture. Two normal healthy litters were chosen and fed on bread and milk for two days after weaning. Then half of each litter was put on this diet, the protein used was caseinogen, and the other halves of the litters were used as controls, being fed on kitchen scraps supplemented with small quantities of bread and milk.

The young rats were weighed separately every day at first and later every four days. The increase in weight of the males is shown in Fig. 4.

The caseinogen-fed males did better than the controls, but the females grew equally on both diets.

The experiment lasted 90 days, which should prove a suitable length of time for testing the diet.

The protein-fed animals grew rapidly and were very fit, their coats were thick and silky and decidedly better than that of the average animal. The females became pregnant and the litters were born at about 13 weeks, which is recognised as the normal time for the appearance of the first litter. In the controls the first litter appeared about two weeks later. The three caseinogen-fed rats produced healthy litters, which compared very favourably both in size and weight with the average first litter.

The litters were not reared. Their growth curves were typically those described previously when the mothers received excess protein in the diet.

The weights of the animals at 70 and 90 days are compared with those of Donaldson [1906] and Osborne and Mendel [1911].

The average weights given by Osborne and Mendel are slightly less than those of Donaldson. The authors suggest that this is due to differences of "breed."

The rats used in these experiments were black and white and as before suggested they may be of a bigger breed. In other experiments both white, and black and white animals were used, and practically no differences in weight were noticed. The only difference was that the white rats do not make such good mothers and, therefore, for the later experiments, they were rarely used.

	Weight of rat at 70 days Average weight		Weight at 90 days Average weight	
	Male g.	Female g.	Males g.	
Donaldson	107	100	150	
Osborne and Mendel	120	100	135	
Caseinogen-fed	(3){198, 201, 216}	(3){114, 143, 165}	(3){231, 242, 258}	Females not given because they were pregnant
Controls. Fed on kitchen scraps and small amount bread and milk	(2){178, 182}	(3){103, 158, 162}	(2){204, 223}	pregnant

This experiment proves the diet to be adequate in all respects except during lactation. It is extraordinary that a diet, so obviously good for a growing animal, suitable for fertilisation and production of offspring, should cause such disastrous effects on the suckling babes.

Experiments with foster-mothers.

It was suggested on p. 91 that the bad symptoms in the babies were in some way connected with development, since the symptoms usually came on at about the same time. It does not seem at all feasible to suppose that the milk contained toxic substances at a definite time, and then lost these properties again. In support of this view it was found that if a rat could be induced to suckle more babies after her own had all died and the excess protein feeding continued, then the new litter gained for a time though not to any great extent, but finally died. The growth curve for the young is very similar to that obtained with the rat's own litter, except that the increase of weight of the babies is less than with the original litter. (This fact applies also to a good diet, see experiment 107.)

Several "foster" experiments were made. Rat 152 was fed on bread edestin, butter, marmite, lemon juice and salt mixture (the same diet as previously used) while nursing her own babies (experiment 112), and the diet continued with the foster litter, five in number. The foster babies gained weight until the 9th day, then ceased to gain, finally lost weight and died. No spasms were noticed, but weakness was a prominent feature. It was frequently found that slower-growing babies developed weakness earlier and showed only slight spasms, or none at all; however, this was not always the case. The mother, 152, lost weight considerably, but in experiment 112 she had gained a good deal. At the beginning of experiment 112 she weighed 212 g. and at the end of experiment 152 she was 205 g. In this instance loss of weight of the mother did not spare the litter, but the mother's loss is hardly comparable to that of other experiments, because in this case the rat had just put on what she now lost.

Rat 107 brought up foster babies when she was having a diet of bread and milk. Previously she was fed on a diet of egg albumin and bread, and all her own litter had died, with typical symptoms. These two experiments are representative of others, and it is inferred that after her own babies have died as a result of too much protein in her diet, the rat is capable of bringing up another litter. If a good diet is given in the second experiment, the babies are normal but small. They do not develop so rapidly as a rat's own babies, and are not able to eat for themselves until about the 21st day. If the excess protein diet is continued, the babies gain small amounts for a time, but eventually die, having shown the usual symptoms, except that they do not suffer from spasms.

In the experiments in which the mothers received excess protein, the foster babies did not develop symptoms until a few days later than the original babies. It has been pointed out that on a good diet the foster babies do not eat till later. This is in agreement with the suggestion that the development of the young is a factor to be considered.

Quite a number of rats were unable to bring up a foster litter after their own had died as a result of too much protein in the diet. The rats took to the babies and mothered them, but were apparently unable to feed them. The babies lived from two to four days, during which time they lost weight. They did not appear to have been fed, although the rat took care of them in the nest. These babies were rarely eaten by the foster mothers.

It was previously suggested [Hartwell, 1921, 1] that excess protein caused an alteration in composition of the milk, and finally a cessation of the flow. From these experiments it would appear that in some rats the secretion could be re-staited, but not in others.

It is possible in some cases to alter the diet, either by adding milk, omitting the excess protein, or entirely changing the food to bread and milk, and thus to obtain recovery of the babies.

Experiments on changing litters.

That the milk of the rat on excess protein contains some toxic substance is further demonstrated by changing the litters.

Rat 136 received bread and milk. Rat 135 had a diet with excess protein. The litters were born on the same day. As soon as the babies 135 developed spasms (the 12th day) the litters were changed. Litter 135 recovered rapidly when fed by rat 136, which was having bread and milk, and were successfully weaned at the normal time. Two of litter 136 developed screaming fits and bad spasms the 20th day, but recovered. In another exchange the normal babies given to a mother, whose own litter were showing bad symptoms, lost weight for three days, after which they ate for themselves. In this experiment no typical symptoms were noticed.

Thus it is evident that with large amounts of protein in the diet, which is complete physiologically, the milk is affected, and the babies show abnormal and typical symptoms. It is suggested that

(i) The *quality* of the milk is altered, in that it possesses some toxic properties responsible for the spasms. This may be due to excess of amino-acids, or possibly to some deficiency in the milk.

(ii) The *quantity* of the milk is inadequate, since the babies lose weight. .

Since large quantities of protein are usually recommended for a lactating animal, it is probable that definite and large amounts of one or more other constituents should be included in the diet.

With a view to throwing some light on this, the following experiments were made.

(ii) *Effect of adding whole milk to a diet containing excess protein.*

The bread and protein were fed in the same proportion as in the last series of experiments. The milk was heated to 70° C. In these experiments the rats had no water to drink. (In some earlier experiments, when 100 cc. whole milk were given, it was noticed that the animals drank no water, and usually upset the containing vessel as soon as it was given to them.)

The control experiment D was done at the same time of the year as the other experiments, though not actually the same year. So far the growth curves of the young when the mother is fed on bread and milk are practically constant, irrespective of the time of year, so long as the animals are kept in a well-heated room. Other observers find different growth curves according to the time of year. It is possible that this difference is minimised when the diet is exceptionally good, *e.g.* bread and whole milk.

In these experiments no abnormal symptoms of any kind were shown, but in all cases the growth was depressed, as compared with the standard (bread **and milk**).

This is well shown by the following figures:

Average weight of young rat at twenty-one days.

Exp.	Diet			Weight in g.
D	Bread and milk			39·2
41	„	„	edestin	31·3
108				29·6
J	„	„	caseinogen	27·8
C				30·6
201	„	„	gelatin	30·4
202				26·0
203	„	„	egg albumin	29·0
204				34·3 (litter of three only)

In these experiments the rats were eating even larger quantities of protein, since milk contains from 2·5 %–4·0 % protein [Matthews, 1921]. The exact amount of milk required to prevent the babies from developing the typical symptoms has not been determined, but other experiments showed 50 cc. to be inadequate. In all probability the individuality of the rat would be a factor conditioning the amount required. Later in this paper (p. 101) it is demonstrated that the whey from 100 cc. milk is equally efficient in preventing bad results with caseinogen and edestin, and that the growth is equal to that obtained with bread and milk.

Thus milk contains some dietary constituents which render it possible for the rat to ingest even larger quantities of protein and yet the litter is not affected by bad symptoms. The only detrimental effect is that the babies do not grow at a maximal rate.

The next experiments were started with a view to finding the nature of such substance or substances.

Two experiments were made with each protein and the average weight of the two litters does not correspond very closely. In spite of the fact that no symptoms developed in the babies, the mothers were ingesting very large amounts of protein and it is probable (as before suggested) that the individual metabolism of the mother is a partial explanation of these discrepancies.

(iii) *Effect of adding other constituents to a diet containing excess protein.*

In this series the proportion of protein to bread was kept constant as in the previous experiments, *i.e.* 5 : 15.

(a) *Salts. Calcium* given as lactate, 1·0 g. per diem (Fig. 5).

Exps. 114, 172, 156. Bread, caseinogen, marmite, calcium lactate.
Exp. 138. Bread, egg albumin, marmite, calcium lactate.
„ 140. Bread, edestin, marmite, calcium lactate.

Litter 114 grew well the whole time; they were excitable and exhibited toe-walking, but otherwise no typical symptoms. It is probable that the mother had a greater tolerance for protein, as before suggested. Apart from this litter, the results were in agreement and the same for each protein tried, *i.e.* caseinogen, edestin, and egg albumin. All the litters showed typical and very bad symptoms, but the babies lived longer than usual.

Thus the addition of calcium lactate to a diet containing excess protein does not alleviate the symptoms but causes a slight prolongation of the life of the sucklings.

(*b*) *Salt mixture.* Bread, caseinogen, marmite, salt mixture. No improvement in the litter was obtained when 1·0 g. salt mixture was added to the daily ration. The symptoms were bad, and the young were all dead on the 15th day.

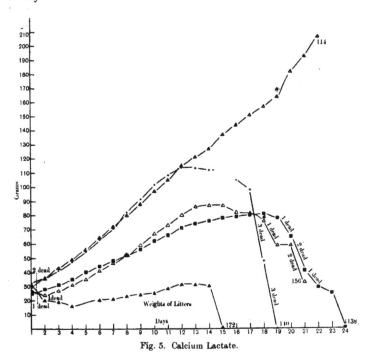

Fig. 5. Calcium Lactate.

(*c*) *Milk ash*[1]. The milk was evaporated to dryness, the residue scraped up and strongly heated in a crucible until a constant weight was obtained; 0·7 g. were given per day (*i.e.* the amount in 100 cc. milk).

These results were very similar to those obtained on adding calcium lactate to the excess protein diet. The symptoms were in no respect alleviated, but the life of the young was slightly prolonged.

(*d*) *Fat and vitamin A.* No beneficial effects were obtained on adding 3·5 g. butter per day to the excess protein diet. Edestin was used.

---

[1] I am indebted to Miss W. M. Clifford for the preparation of the milk ash.

(e) *Lactose*. 4·0 g. lactose were added to a diet containing excess edestin, but the spasms and bad effects were in no way improved.

(f) *Fat, vitamin A, and lactose*. Caseinogen was used and the diet was physiologically complete in all respects, fat and lactose being added in approximately the same amounts as found in 100 cc. milk.

The litter suffered very badly, two eventually survived, but they were weakly for some time and were not in a fit condition to leave the mother until the 26th day.

Vitamin C. The diet consisted of bread and egg albumin in the usual proportions and 0·7 cc. lemon juice per day. Great difficulty was experienced in getting a rat to bring up a litter on this diet. In many experiments the babies died within the first three days. The mothers were well, the babies were average weight, and apparently all right at birth. This may be pure coincidence. No explanation is offered to account for this failure.

Finally two experiments were successful. The litter all exhibited typical symptoms, and were dead by the 13th day.

(iv) *Effect of adding large quantities of yeast extract (commercial preparation, marmite) to a diet containing excess protein* (Fig. 6).

Two experiments were made with each protein (caseinogen, egg albumin and edestin).

1. The diet consisted of 15·0 g. bread, 5·0 g. protein, and 3·0 g. marmite, *all* constituents increased proportionately as necessary.

2. The bread, protein, and marmite were the same as above, but butter, lemon juice, and salt mixture were added to complete the diet.

These curves are very similar to those obtained when 100 cc. whole milk were added to the excess protein diet. To bring out this point curve D (bread and milk standard) is included in the figure. With edestin and caseinogen a depression of the growth curve was obtained, very similar to that obtained when the mother had these proteins + 100 cc. whole milk. The babies were practically normal, but slight toe-walking and excitability were noticed; these features, however, were very quickly recovered from. With egg albumin, on the other hand, the results were not so good. No spasms were noticed, but the babies were far from normal. They were all right at first, but on the 14th day they felt cold, were weak, inclined to roll over, thin and lethargic. They walked very badly and their movements were jerky and incoordinate. Two of the litter 180 died but the rest survived, as did all litter 182. The survivors were not normal for two weeks after weaning.

The average weights of the young rats are given below.

					Average weight at 21 days. g.
Exp. D	Bread and milk (standard)				39·2
„ 164.	„	edestin, marmite			34·7
„ 165.	„	„	„	butter, etc.	25·7
„ 179.	„	caseinogen, marmite			34·0
„ 181.	„	„	„	butter, etc.	30·7
„ 180.	„	egg albumin, marmite			22·0
„ 182.	„	„	„	butter, etc.	26·3

, From these experiments it is clear that "something" in marmite is capable of obviating the bad results in the litters usually obtained with excess of caseinogen or edestin in the mother's diet. Also the actual weights of the babies are better when the mother's diet is lacking in fat, vitamin A, salt mixture and the antiscorbutic vitamin. This may be merely coincidence; a far greater number of experiments would be required before any conclusions could be drawn. It is interesting to note that similar results were obtained

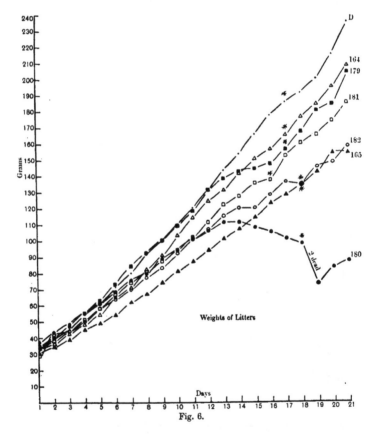

Fig. 6.

when less amounts of protein were given. Since the results are as good when the mother's diet is deficient in fat, vitamins A and C and salt mixture, it appears that the "something" in marmite is the primary factor. It may be of interest that the marmite contained only traces of calcium salts.

With egg albumin the improvement was not nearly so good. The litters certainly did better than when small amounts of marmite were fed to the mother, but they could not compare at all favourably with the caseinogen and edestin babies. These differences are discussed later (p. 102).

The mothers maintained or increased their weights, except 180 (fed on egg albumin in a deficient diet) which lost appreciably.

(v) *The importance of protein.* (Fig. 7).

Exp. D.   Control. Bread and milk (same as in Fig. 6).
Exps. 118, 120. 15·0 g. bread, 5·0 g. edestin ⎫
Exp. 167.   ,,   ,,   3·0   caseinogen ⎪
   ,, 168.   .,   ,,   3·0   edestin   ⎬ + whey from 100 cc. milk.
Exps. 159, 162.   ,,   ,,   0·0 protein   ⎭
Exp. L.   Bread alone. [Taken from Hartwell, 1921 ,1].

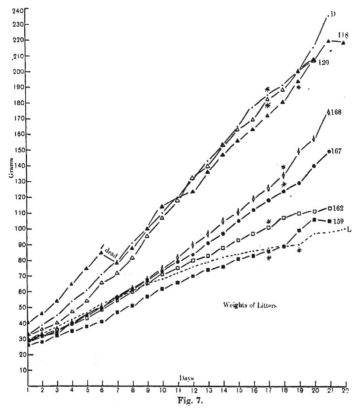

Fig. 7.

Preparation of the whey. 100 cc. milk were warmed to 40° C., and a few drops of rennet added. When the junket was firm, it was hung up in a muslin bag and the whey separated by dripping through. Each rat was given daily as much as could be obtained from 100 cc. milk. The protein and bread were increased proportionately as before.

It has already been mentioned that the whey was as effective as milk in preventing the bad symptoms in the litters. Also when the proportion of bread to protein was 15 : 5 the growth of the litter was practically equivalent to that obtained with a bread and milk diet, which up to the present time is regarded as maximal. It is extraordinary that such good growth should be obtained when the mother's diet was so poor in fat and vitamin A[1].

In this connection it is interesting to note that all the mothers that were killed had large deposits of subcutaneous and peritoneal fat. It may be that with large reserves of body fat, the amount in the diet is immaterial from the point of view of milk production.

In experiments 159 and 162 in which the mothers were fed on bread and whey, the litters did very little better than in experiment L when the mother received bread alone (taken from Hartwell [1921, 1]). Comparison of the protein rations in these experiments shows that the bread and whey contains only slightly more protein than bread alone. When more protein was added to the diet (experiments 167 and 168) a considerable improvement in growth of the young was obtained, though all the other constituents were the same as before. A further addition of protein (experiments 118 and 120) gave a still greater improvement in the growth of the litters. In this way it is seen that adding protein to the diet can raise the growth curve from almost the minimal to the maximal. Thus protein itself does play a very important part in the production of milk of good nutritive value, yet from other experiments described in this paper, it is obvious that the diet must be rich in other constituents as well or else harmful results occur in the litter.

One of litter 118 was killed accidentally on the 7th day and the average weight of these babies is rather greater than that of 120, whose mother received a similar diet. It is quite possible that the actual number in the litter is a factor of importance. Experiments to investigate this point are now in progress.

DISCUSSION OF RESULTS.

(i) Prevention of the bad symptoms in the young.

It has been conclusively proved that large quantities of protein are by no means harmful to the rat itself, since the young and adult animals can live normally on a diet containing excess protein for considerable periods. The proportion of 15·0 g. bread to 5·0 g. protein proved an excellent diet for growing rats, but much less protein, even as little as 15·0 g. bread to 1·0 g. protein produced bad symptoms in the suckling litter. Therefore, it appears that the rat can metabolise the protein effectively from the point of view of its own growth, but is unable to produce normal milk, unless definite quantities of some other constituent (or constituents) are supplied. Our knowledge

---

[1] Since the caseinogen and edestin used were not absolutely pure, it is possible that traces of vitamin A may have been present.

of the metabolism of individual amino-acids is very limited, but the experiments described in this paper show that there is some quantitative relation between the amino-acids and other dietary constituents. What these constituents are is a very difficult problem. The fact that whole milk can prevent the bad symptoms throws very little light on the subject. Owing to the complicated nature of milk, it merely proves that large quantities of protein in the mother's diet produce no harmful effects in the litter, as long as "something" else is included. It is even possible that the necessary factor will vary with the nature of the protein.

It has been pointed out that 3·0 g. marmite to 5·0 g. protein is adequate in preventing spasms, etc., in the case of caseinogen and edestin, but with egg albumin, although the symptoms are somewhat reduced, the litter still suffers badly. Also with egg albumin much better results were found when small quantities of butter, lemon and salt mixture were added to the diet, but with caseinogen and edestin, it was immaterial if the diet was complete or not as long as large quantities of marmite were fed. Since caseinogen and edestin are somewhat similar in constitution, and egg albumin is different there seems some evidence for suggesting that the factor will vary with the constitution of the protein.

Some experiments with egg albumin now in progress show that the litter exhibit still greater improvements (but are *not* normal) when more butter is added to the ration, the diet being physiologically complete.

It is quite obvious that some other constituents are vitally essential for the production of normal milk, if large quantities of protein are ingested.

These are all present in milk, and in the case of caseinogen and edestin, present also in commercial yeast extract ("marmite"). In the light of present-day investigations the most obvious explanation is that vitamin B is the factor, but this conclusion is by no means justified from the experiments described here. Again "marmite" alone was not wholly effective with egg albumin. It might be that insufficient was given, but this is hardly likely, because 3·0 g. to 5·0 g. protein is a large proportion. A more probable explanation is that some other constituent is also needed for the proper metabolism of this protein in a lactating animal.

The fact that large quantities of milk and marmite were necessary suggests an actual chemical relation between the protein and some other factor, and not a catalytic or hormone action. It is possible that less marmite would be required if the right proportion of other constituents could be arrived at. Experiments have shown 1·0 g. marmite per day to be ineffective, yet the amount necessary to supply vitamin B to a rat is estimated as 0·2 g. per day.

It is possible that the animal needs a greater amount of vitamin B for the effective metabolism of large amounts of protein; hence it is used by the mother and she has none to spare for her milk. On the other hand the mother is usually regarded as being a safeguard for the young and it is thought that she can sacrifice her own tissues for the benefit of her offspring. Also it has been proved that in a non-lactating animal quite small amounts of vitamin B are adequate even when the intake of protein is high.

The experiments of Karr suggest that vitamin B is an essential constituent of a dog's diet, otherwise the animal refuses its food. Since the dog is a carnivor, it might be assumed that this vitamin plays some part in connection with protein metabolism. Karr, however, states that the intermediary meta-holism of nitrogen is unaltered by adding vitamin B to the diet, but he appears to deduce such a conclusion from analysis of excreta and food ingested. It is possible that the nitrogen excretion might remain constant and yet the intermediary metabolism be altered.

Thus there seems a certain amount of superficial evidence for associating vitamin B with protein metabolism. In the experiments described in this paper, all other recognised dietary constituents have been eliminated except vitamin B. This is only a negative proof and therefore of little value. The commercial preparation marmite is a somewhat crude product and contains many other substances besides vitamin B. It is even possible that extractives are of use in obviating the bad symptoms resulting from excessive amounts of protein in the mother's diet.

(ii) *The amount of protein.*

At present it is impossible to state what amount of protein is harmful and what amount is beneficial, because it appears to depend largely on the other dietary constituents. The experiments described in this paper show that too much protein depresses the growth curve, even if no bad symptoms are shown. For example when the mother is fed on bread and protein (15 : 5) and whole milk, the litter show a less good growth curve than when the mother is fed on bread and milk only, although the babies are normal in each case. In the former experiment the rat was taking at least 8·0 g. protein per day, but no bad symptoms occurred, yet in a less complete diet 1·0 g. protein can produce a very bad condition in the young.

As the results differ with different proteins it is probably certain amino-acids which are primarily responsible for the bad effects on the babies. If this is so, a diet of mixed proteins (such as taken by man) should prove less harmful, unless the protein is taken in very great excess. The actual composition of foods themselves is a safeguard, because it is rare to find only one protein.

(iii) *Development of the young.*

It has been pointed out that the time of onset of the symptoms is re-markably constant and it is therefore suggested that the stage of development of the young rat is a factor to be considered. In foster litters, the babies develop a little more slowly and their eyes open about two days later than those of normal litters. They do not eat for themselves until a few days later than the original babies and therefore cannot be weaned at the normal time.

(It has been found advisable to leave the mother with the young for at least three days after they can eat for themselves.) In general they seem more backward and less developed than normal babies of the same age. In foster litters, when the mother is fed on an excess protein diet, the symptoms usually develop about 2–3 days later, than in a litter left with its own mother. Also the symptoms come on at the same time, even if the mother has been fed with excess protein during gestation, or if she has first fed her own litter when she has had the excess protein diet. If the composition of the milk changed towards the middle and end of lactation, then the symptoms in the foster-litter should come on at an earlier stage, provided that the excess protein diet were continued throughout. But this is not so.

(iv) *The value of milk in the diet of a lactating animal.*

"Milk in *liberal amounts* should always be included in the diet of the lactating mother." This statement is made by McCollum [1920] and is probably a very true one. It was shown [Hartwell, 1921, 1] that bread and milk fed to the mother produced maximal growth of the young. Various other diets have been tried but as yet no curves have been obtained which are as good (except 15·0 g. bread to 5·0 g. protein + whey from 100 cc. milk— and this is a milk derivative). It has been proved in this paper that large quantities of milk entirely obviate the bad symptoms produced by excess protein in the diet. It is, therefore, possible that by taking large amounts of milk, the mother safeguards herself from any errors in diet.

The value of milk is, no doubt, due to the fact that it contains all the necessary constituents of a good diet, and not excessive amounts of any. It has been suggested by some physiologists that milk contains a specific galactogogue. Some experiments now in progress show that there are other diets which will give as good results as bread and milk (at any rate with rats) provided that the *right proportion* of the various constituents can be ascertained.

Another advantage of milk is that the proteins are adequate and it is quite obvious that the quality of the protein fed to a lactating animal is of primary importance.

SUMMARY.

1. In a lactating rat, the amount of dietary protein constituting excess varies with the type of protein and with the individuality of the rat. The proportion 15·0 g. bread to 1·0 g. protein (egg albumin or caseinogen) fed to the mother results in abnormal symptoms in the litters.

2. On a diet of bread and egg albumin (either 1·0 g. or 2·0 g. egg albumin to 15 g. bread) the mother loses a considerable amount, in one case as much as one-third of her body weight.

3. When excess protein is fed to the mother, the growth curves of the litters are not constant. The variations obtained are probably due to the individual metabolism of the different rats.

4. Large quantities of protein fed to a nursing rat are detrimental to the young, even when the diet contains all the essential constituents.

5. If such a diet is started at birth, some babies may survive, but they are not normal; the majority die.

6. If the diet is begun before or during gestation, none of the young survive.

7. The litters die at approximately the same time, irrespective of the length of time the mother has been fed on the excess protein diet.

8. Foster babies suffer in exactly the same way as the original family if the same diet is continued.

9. The typical symptoms can be induced in a healthy litter by giving them to a mother whose own litter has developed this condition.

10. It is suggested that the toxic substances in the milk affect the baby rat at some special stage of development.

11. A diet containing large amounts of protein (15·0 g. bread to 5·0 g. protein) in a physiologically complete mixture is adequate for growth, fertilisation and reproduction in the rat, and is unsuitable only during lactation.

12. The bad effects are entirely obviated by adding 100 cc. whole milk to the mother's diet, but the growth of the litter is not maximal. This applies to caseinogen, edestin, egg albumin and gelatin. With caseinogen and edestin, 100 cc. whey are equally effective.

13. With caseinogen, edestin and egg albumin the addition of calcium lactate to the diet prolongs the life of the young, but the symptoms are just as severe. Milk ash has a similar effect when added to a diet containing excess egg albumin or edestin.

14. The addition of butter and lactose causes no improvement in the young when the mother is eating large quantities of edestin or caseinogen.

15. Yeast extract (marmite) in large amounts added to the excess protein diet prevents all bad symptoms in the case of caseinogen and edestin; with egg albumin the bad condition is improved, but not entirely cured.

I wish to express my thanks to Prof. V. H. Mottram for his interest in this work, the expenses of which were defrayed by a grant from the Medical Research Council.

REFERENCES.

Donaldson (1906). *Boas Anniv. Vol.*, 5.
Hartwell (1921, 1). *Biochem. J.* 15, 140.
Hartwell (1921, 2). *Biochem. J.* 15, 563.
Hewer (1914). *J. Physiol.* 47, 480.
McCollum and Simmonds (1918). *Amer. J. Physiol.* 46, 275.
McCollum (1920). Newer Knowledge of Nutrition, 129.
Matthews (1921). Physiological Chemistry, 306.
Osborne and Mendel (1911). Feeding experiments.

XV. EFFECT OF SEVERE MUSCULAR WORK ON COMPOSITION OF THE URINE.

By JAMES ARGYLL CAMPBELL
AND THOMAS ARTHUR WEBSTER.

From the Department of Applied Physiology, National Institute for Medical Research, Hampstead.

(*Received January 23rd, 1922.*)

A FAIRLY complete series of observations was undertaken to determine the effect of severe muscular work on the composition of the urine, using recent standard methods. These observations form part of a research connected with industrial fatigue, the subject being employed at the same time for observations on respiratory exchange.

In a previous paper [1921] we recorded the results of observations on day and night urine under various routines—five days' complete rest, five days' ordinary laboratory work, and five days' light muscular work, 67,500 kilogrammeters (kgm.) per day. In the present paper we give results obtained from the same subject, under similar conditions as regards diet, time of meals, etc., during five days' severe muscular work. The diet was controlled as regards quality. About ½ litre more fluid per diem was swallowed during the severe muscular work than during the other routines.

Our subject was accustomed to do 13,500 kgm. per hour, on a bicycle ergometer, working three hours before dinner—10 a.m. to 1 p.m.—and two hours in the afternoon—2 p.m. to 4 p.m.—the total, 67,500 kgm. producing only a slight degree of fatigue. For severe muscular work he attempted to do 100,000 kgm. per day at the rate of 20,000 kgm. per hour with the same interval for dinner. 100,000 kgm. represents a full day's work for a vigorous labourer, and, as might be expected, our subject—a laboratory assistant—found this amount of work a severe task. Table I shows the details of the work actually performed. On the first day the subject had to give up the attempt at the end of the third hour, the first hour's work of 20,000 kgm. affecting him very much. On the second day he was able to do more and on the third, fourth and fifth days he completed the full amount in the five hours. It was obvious from the subjective and objective symptoms that the experiment produced severe strain. The average composition of the 24 hours' urine for the five days is shown in Table II; the significance of any of the five days' results was similar. For comparison we have given the figures for the 24 hours' urine obtained in our previous research [1921]. It will be observed that the undetermined nitrogen, creatinine, neutral sulphur—and, therefore, the total sulphur—and lactic acid were distinctly increased during the severe muscular work. Acetone bodies were present during the third and fourth days and the third night, probably indicating complete consumption of available carbohydrate. Purine nitrogen was also increased, but we had not complete figures for comparison. Other observers have obtained somewhat similar results. Cathcart [1921] reviews the most important of the recent and older researches in his monograph on protein metabolism.

Table I.

Date	Work done in kgm.	Working period in hours	Remarks
10. x. 21	44,482	3	Felt very hot, perspired freely; very much fatigued; severe headache; sore all over
11. x. 21	72,440	5	Felt very hot, perspired freely; very much fatigued; felt stiff all over
12. x. 21	104,760	5	Felt very hot, perspired freely; very much fatigued; felt stiff all over; acetone bodies in urine
13. x. 21	100,000	5	Not so hot; not so fatigued; not so stiff as on previous days; acetone bodies in urine
14. x. 21	100,000	5	Not so hot; not so fatigued; not so stiff as on previous days

Osterberg and Wolf [1907] showed that the undetermined nitrogen was increased during activity. Shaffer [1908] found that muscular activity—walking ten miles—did not increase the excretion of creatinine. Pekelharing and Harkink [1911] obtained a similar result, but observed that after prolonged tonic contraction the output rose. Our subject exhibited muscular stiffness and this may have been a cause for the increased excretion of creatinine. Hoogenhuyze and Verploegh [1908] found that in fevers and other pathological conditions in which there is an increased breaking down of tissue, the creatinine excretion is increased. In our subject there was the possibility that some muscle fibres were injured. The brain contains a fair proportion of creatine [Janney and Blatherwick, 1915], and was a possible seat of production of creatinine during the severe fatigue produced. Weinberg [1921] considers that an important influence on the excretion of creatinine in the urine should be assigned to the mind, high excretion being connected with emotion. Scott and Hastings [1920], Garrat [1898] and others noted that there is a rise in output of inorganic sulphate during activity. Although we observed a similar increase of sulphate during the light muscular work it was the neutral sulphur that was most increased during the severe work. This may have been due to the want of oxygen, so that the sulphur was not so completely oxidised during the severe work. With regard to lactic acid, many observers have found that this acid is increased by severe muscular exercise. Fletcher and Hopkins [1907] clearly demonstrated that lactic acid is produced in excised muscle only when the muscular contraction occurs in a deficiency of O_2. When it occurs in an adequate supply of O_2, CO_2 instead of lactic acid is produced. Hill and Flack [1909] showed that in the fatigue of athletes oxygen inhalation increases the lasting power and decreases fatigue, probably by maintaining or restoring the vigour of the heart. This leads to the question whether creatinine, neutral sulphur and undetermined nitrogen were increased because of insufficient oxidation. This was a possibility, but in one respect these substances differed greatly from lactic acid; lactic acid was excreted in much greater amount during the day (Table II) than at night, whereas creatinine, neutral sulphur and undetermined nitrogen were more evenly distributed between day and night. With regard to the latter substances we did not notice in our previous researches [1921] that there was any parallelism in these excretions; but, nevertheless, there may be the same explanation for

Table II.

	Average 24 hours results				Average hourly day and night results during seve[re] muscular work	
	Complete rest	Laboratory work	Light muscular work (67,500 kgm. in five hours)	Severe muscular work (100,000 kgm. in five hours)	Day	Night
ount cc.	1112	1116	971	686	32·4	25·9
dity %	49·5	53·7	58·0	65·0	60·0	70·0
ratable acidity	309	312	332	342	14·0	14·4
olin) cc. $N/10$						
al acidity cc. $N/10$	703	673	723	722	29·0	30·7
al N g.	8·010 (100)*	9·015 (100)	9·465 (100)	10·250 (100)	·446 (100)*	·414 (100
a N g.	5·990 (74·80)	7·131 (79·10)	7·647 (80·80)	7·680 (74·93)	·336 (75·3)	·308 (74·8
monia N (A) g.	·439 (5·48)	·385 (4·27)	·442 (4·67)	·412 (4·02)	·015 (3·4)	·019 (4·6)
monia N (B) g.	·548	·505	·552	·533	·021	·023
iino-acid N g.	·109 (1·36)	·120 (1·33)	·110 (1·16)	·121 (1·18)	·006 (1·4)	·004 (1·0)
atinine N g.	·741 (9·25)	·736 (8·17)	·695 (7·34)	·925 (9·02)	·036 (8·2)	·040 (9·6)
ic acid N g.	·154 (1·92)	·141 (1·56)	·133 (1·40)	·134 (1·31)	·006 (1·4)	·005 (1·2)
determined N g.	·578 (7·19)	·502 (5·57)	·438 (4·63)	·978 (9·54)	·047 (10·3)	·038 (9·1)
loride (NaCl) g.	9·410	7·207	8·801	6·736	·394	·200
osphate (P_2O_5) g.	1·550	1·703	1·711	1·811	·070	·080
tal S (SO_3) g.	1·594 (100)	1·749 (100)	1·770 (100)	2·069 (100)	·084 (100)	·087 (100
organic S (SO_3) g.	1·148 (72·0)	1·400 (80·0)	1·432 (80·9)	1·422 (68·7)	·057 (67·9)	·061 (70·1
hereal S (SO_3) g.	·212 (13·3)	·132 (7·5)	·140 (7·9)	·209 (10·1)	·007 (8·3)	·009 (10·4
utral S (SO_3) g.	·234 (14·7)	·227 (12·5)	·198 (11·2)	·438 (21·2)	·020 (23·8)	·017 (19·8
ctic acid g.	trace	trace	trace	·061	·0040	·0014
lcium (CaO) g.	—	·281	·252	—	—	—
gnesium (MgO) g.	—	·173	·156	—	—	—
rine N g.	·035	—	—	·058	·0023	·0025

(A) Van Slyke's method.   (B) Malfatti's method.

* Figures in brackets are percentages. Total acidity is Ammonia (B) + titratable acidity.

the increase of each during severe work. It is well known that creatinine and neutral sulphur resemble one another in that they are of endogenous origin, and that they are practically constant in amount for each individual, being evidently connected with the mass of active living protoplasm. In our subject both creatinine and neutral sulphur were higher on the first day than on the succeeding days of severe work. We consider that the probable explanation of the increased excretion of creatinine, neutral sulphur and undetermined nitrogen was damage to protoplasm, both nervous and muscular by the excessive fatigue and strain.

Cathcart [1921] states that no one has been able to demonstrate clearly that muscular work affects greatly the total nitrogen excretion, provided the supply of food, particularly of carbohydrate and of oxygen be sufficient. Under our conditions of experiment we found a slight and steady increase in total nitrogen from complete rest to severe work (Table II). We also found that during the five days' severe work the total nitrogen rose from 8 g. on the first day to 10 g. on the second day, to 12 g. on the third day and fell to 10 g. for the fourth and fifth days, whereas the increments in undetermined nitrogen, neutral sulphur, creatinine were marked from and including the first day. Garratt [1898] found that a rise of sulphate preceded a rise in total nitrogen as a result of exercise. Whilst the figures for total nitrogen are of interest and of some significance and show that the excretion of total nitrogen was increased, we cannot conclude that there was an increased katabolism of nitrogen in our subject as the result of exercise, since we did not estimate the total nitrogen in the food and in the faeces. Under ordinary conditions our subject's total nitrogen averaged only 9 g. per diem during six months and was never as high as 12 g. The large amount of nitrogen, 12 g. on the third day of severe work, was probably due to lack of carbohydrate as indicated by the presence of acetone bodies on that day. Many observers have found an increase of nitrogen when there is an inadequate supply of nitrogen-free food available.

Because it was necessary to increase greatly the amount of work for our subject before any definite change in composition of urine was detected we consider that under equally good dietary and atmospheric conditions a healthy industrial worker would not show any such change in urinary composition unless undertaking work far more severe than his customary daily task. We have done a few experiments with some subjects sitting at rest under hot (36° C.) and close atmospheric conditions and found that the day urine showed higher figures for titratable acidity and ammonia than when under comfortable atmospheric conditions. We hope to extend these observations. One of us [J. A. C.] has noted high excretion of ammonia in a hot climate [1919, 1920].

Table II also gives the average hourly day and night results during the five days' severe work. As in our previous research [1921] we found that the acidity, ammonia and phosphate were higher at night than during the day; also that the sulphur was evenly distributed between day and night, whilst the total nitrogen was higher during the day than at night. We consider that during the night there is an excretion of fixed acids which have been formed in the cells during the day and that these fixed acids are concerned in the production of fatigue and sleep. For all routines examined we observed that the differences between day and night urine were most marked during rest in bed and less marked as the activity increased, activity probably hastening the processes concerned.

We are indebted to Mr C. Pergande, who was the subject in this research.

SUMMARY.

A healthy subject, who was accustomed to do 67,500 kgm. of work, on a bicycle ergometer, in five hours, developed symptoms of muscular strain on attempting to do 100,000 kgm. of work in five hours and showed pathological changes in urinary composition. Creatinine, undetermined nitrogen, neutral sulphur and lactic acid were much increased, whilst acetone bodies were present during part of the experiment.

Our previous results regarding composition of day and night urine were confirmed.

REFERENCES.

Campbell (1919). *Biochem. J.* **13**, 239.
—— (1920). *Biochem. J.* **14**, 603.
Campbell and Webster (1921). *Biochem. J.* **15**, 660.
Cathcart (1921). The Physiology of Protein Metabolism, 129.
Fletcher and Hopkins (1907). *J. Physiol.* **35**, 247.
Garratt (1898). *J. Physiol.* **23**, 150.
Hill and Flack (1909). *J. Physiol.* **38**, *Proc.* xxviii.
Hoogenhuyze and Verploegh (1908). *Zeitsch. physiol. Chem.* **57**, 161.
Janney and Blatherwick (1915). *J. Biol. Chem.* **21**, 567.
Osterberg and Wolf (1907). *J. Biol. Chem.* **3**, 169.
Pekelharing and Harkink (1911). *Zeitsch. physiol. Chem.* **75**, 207.
Scott and Hastings (1920). Quoted in *J. Indust. Hyg.* Feb. 1921, 203.
Shaffer (1908). *Amer. J. Physiol.* **22**, 445.
Weinberg (1921). *Biochem. J.* **15**, 306.

XVI. THE VALUE OF GELATIN IN RELATION TO THE NITROGEN REQUIREMENTS OF MAN.

By ROBERT ROBISON.

From the Lister Institute.

(*Received January 23rd, 1922.*)

THE story of the earliest attempts to discover the value of gelatin as a food-stuff has been told by Carl Voit [1872] in the introduction to his paper on this subject. It commences in 1682 when Dionys Papin prepared gelatin from bones by means of his digestor, and from the gelatin made soup with which he fed the poor. Such attempts were zealously renewed during the French Revolution by Cadet de Vaux, d'Arcet and others, and were supported by the Government, who issued official instructions extolling the nourishing properties of gelatin soup above those of beef tea. The approbation of the Institute of France and of the Academy of Medicine was also forthcoming, but in spite of d'Arcet's attempts to improve the flavour of the soup with spices, it did not meet with very great approval from the poor, who were expected to consume it.

Gannal, a manufacturer of gelatin, fed himself and his family on gelatin, with and without bread, for some weeks until compelled to desist owing to the unsupportable nausea caused by the diet. The effects on the health of these people led him to conclude that gelatin is not only valueless as a food but actually harmful.

Magendie's Report in 1841 to the Paris Academy on the results of the investigations of the second Gelatin Commission was scarcely more favourable. Gelatin was considered to have no food value by itself and to reduce the value of other foodstuffs when fed in combination with them. A similar opinion was expressed by the Academy of Medicine in 1850 but less extreme views were held by some physiologists among whom were Boussingault [1846] and Frerichs [1845]. The latter ascribed to gelatin the same significance as that of the "Luxus" protein, *i.e.* the excess protein in the diet over the requirements of the body as represented by the protein decomposition during starvation. Though unable to replace the body protein gelatin could, he held, be utilised in the same way as the nitrogen-free foodstuffs ("Respirationsmitteln").

Somewhat similar views were held by Bischoff [1853] while Donders [1853] considered that gelatin might reduce the body's needs for protein since these are not restricted merely to the replacement of tissues.

Voit's own experiments and those carried out in conjunction with Bischoff on dogs form the first systematic study of the nitrogen balances on diets containing varying quantities of meat, gelatin and fat. As the result of a large number of experiments he concluded that gelatin always spares protein and in a greater degree than fat or carbohydrate, but that gelatin plus fat reduces the protein decomposition more than does gelatin alone. On the other hand, however much gelatin and fat are given, body protein will be lost.

The energy requirements of the animal were not sufficiently considered in these investigations and a large part of the so-called sparing action of gelatin can be ascribed to its ability to furnish energy and so to reduce the use of body protein for this purpose. I do not suggest that the whole effect is to be explained in this way, but to what extent gelatin can satisfy any portion of the dog's specific nitrogen requirements cannot be ascertained from Voit's results.

During the next thirty years experiments upon dogs by feeding with gelatin were also carried out by Oerum [1879], Pollitzer [1885], Munk [1894], Kirchmann [1900] and Krummacher [1901]. In Oerum's experiments the dog received a basal diet of starch, butter and meat extract equivalent to over 80 calories per kilo body weight. During successive periods of from four to eight days this diet was supplemented by meat or the equivalent amount of gelatin. Unfortunately only the urea nitrogen was determined (by Liebig's titration method) and the faeces were not analysed, so that a nitrogen balance sheet cannot be made out, but the results appear to indicate that gelatin can save about half the amount of nitrogen excreted by a dog when receiving a carbohydrate diet of sufficient calorie value.

Pollitzer also gave his dog abundant carbohydrate to which, during successive periods, were added equivalent quantities of meat, digestion products of meat (peptone, etc.) and gelatin. Positive nitrogen balances were obtained with all except gelatin.

Kirchmann determined the amount of body protein spared by different amounts of gelatin, no other food except water being given. Taking the nitrogen output during starvation as 100 he found a saving of 25 % when the gelatin given was sufficient to satisfy only 7·5 % of the energy requirements of the animal, while eight times this amount was required to save 35 %. He estimated that a maximum saving of 39 % might be expected if the amount of gelatin could be increased to meet these energy requirements in full.

Krummacher continued these experiments with still greater quantities of gelatin, and obtained a result closely agreeing with Kirchmann's calculated figures. It is clear however that unless we also know the minimum nitrogen output when the energy requirements are fully met by nitrogen-free foodstuffs, the above relationships offer no evidence as to the capacity of gelatin to satisfy any of the specific nitrogen requirements of the animal. This value was not determined by either Kirchmann or Krummacher.

Munk's experiments were on a different plan from those mentioned above.

He gave a dog a diet of rice, fat and meat equal to 58 calories per kilo and containing 0·6 g. nitrogen per kilo body weight, which was more than twice the starvation output. He was then able to replace five-sixths of this protein by the equivalent amount of gelatin and still keep the animal in nitrogen equilibrium.

Kauffmann [1905] carried out a series of experiments on a similar plan but with the precaution of reducing the nitrogen intake of the standard diet to a much smaller amount than that given by Munk. This standard diet consisted of milk, rice, caseinogen (plasmon) and fat and was given in amount equal to 0·32–0·39 g. N and 63–72 calories per kilo body weight. Not more than one-fifth of the nitrogen of this diet could be replaced by gelatin nitrogen without an increase in the nitrogen output occurring. Kauffmann also investigated the possibility of improving the value of gelatin by supplementing it with tyrosine, tryptophan and cystine and concluded from his experiments on dogs and on himself that with these additions gelatin becomes of equal value with caseinogen.

At a much earlier date Escher [1876] had fed dogs and pigs with gelatin supplemented by tyrosine and found that their body weight was maintained, but Lehmann [1885] was unable to obtain this result in experiments on rats.

Rona and Müller [1906] carried out a series of very careful experiments with dogs on the same plan as those of Kauffmann but were unable to confirm the latter's conclusions. Their standard diet gave 0·2 g. N and 91 calories per kilo body weight. With this the animal was in nitrogen equilibrium, but when a portion of the milk was replaced by gelatin plus tyrosine and tryptophan a negative balance was found.

The value of gelatin fed in conjunction with other proteins has also been investigated by Murlin [1907, 1] both by experiments on dogs and on himself. With dogs on a diet containing one-fourth more than the fasting requirement of nitrogen, half of this being in the form of cracker meal and half in the form of caseinogen, it was not possible to replace the caseinogen nitrogen by gelatin nitrogen without increased loss of body protein. With other diets however, in which the protein was in the form of meat, up to 58 % could be replaced without loss of body protein. The fuel value of all diets was greater than the energy requirements of the animal but Murlin attributes the high replacement value obtained in some diets largely to the greater proportion of calories supplied by carbohydrate in place of fat.

It is possible that this factor may have influenced the result, though according to Zeller [1914] the nitrogen requirements are not affected by the proportion of fat to carbohydrate in the diet so long as this does not become greater than about 4 : 1.

There seems however to be insufficient reason for assuming that the amount of meat given in some of these diets was the minimum required for nitrogen equilibrium, and unless this were so the fact that a part could be replaced by gelatin without affecting the balance would prove nothing.

On the other hand, on the cracker meal diets a negative balance was always obtained, which was recognised by Murlin as evidence of the lower availability of this form of protein. On these diets it was not possible to replace any part of the protein by gelatin without increasing the relative loss of body nitrogen.

The criticism that the protein in the diet after part of it had been replaced by gelatin may have still been in excess of the minimum required, applies even more forcibly to the experiment on himself, in which the basal diet contained 14·25 g. N, *i.e.* about 10 % more than his nitrogen output during starvation. When two-thirds of this had been replaced by gelatin nitrogen he was still receiving 5·33 g. N (0·076 g. per kilo) derived from eggs, cream, butter and cereals. During the two days on which this diet was taken a positive balance was obtained, but this cannot be accepted as convincing evidence of the value of gelatin nitrogen.

In a later paper Murlin [1907, 2] brought forward satisfactory proof that in a dog the reduction (about 30 %) of the fasting nitrogen output produced by small amounts of gelatin, was much greater than could possibly be accounted for by the dextrose which might be synthesised in the body from this gelatin[1].

From the investigations so far considered it may be taken as definitely established that:

1. Gelatin when given as the sole source of nitrogen is unable to maintain the animal body in nitrogen equilibrium.

2. With dogs gelatin is able to reduce the loss of body nitrogen considerably below that occurring during starvation, and this effect is not proportional to the amount of potential energy thus supplied and cannot therefore be simply explained on these grounds.

3. Some of the experiments indicate that when gelatin is mixed with other proteins, they may complement one another so that a proportion of the nitrogen of gelatin is utilisable.

A critical examination of the results of these experiments does not enable us to form any definite conclusions as to the capacity of gelatin alone to satisfy any part of the specific nitrogen needs of the body in man, although some of the results obtained with dogs indicate a limited capacity in this direction if the nitrogen output on an abundant nitrogen-free diet is taken as representing these specific requirements. There is however a difficulty in accepting this since the results obtained with dogs do not fall into line with those obtained with man and some other animals, and suggest that the nitrogen metabolism of the carnivora varies from that of the omnivora and herbivora in some details.

The fasting output of a man is equal to about 0·2 g. N per kilo body

---

[1] A brief account of other researches by Ganz, Gerlach (1891), who investigated the value of gelatin peptones, Gregor (1901), who used gelatin for feeding infants, and by Brat (1902) and Mancini (1905), who fed it to convalescents, will be found in Murlin's paper [1907, 1].

weight, that of a large dog is of the same order. On an abundant carbohydrate diet the nitrogen output of man can be reduced to one-quarter of this amount, *i.e.* 0·05 g. N per kilo whereas according to most observations under the same circumstances the nitrogen output of a dog is only reduced by 10 % to 20 %.

The difference in detail between the nitrogen metabolism of man and dog also emerges on comparison of the ratios of the total nitrogen to that excreted in the form of creatinine during starvation and on abundant nitrogen-free diets (see Table I). The constancy of the creatinine output and its probable relationship to the endogenous metabolism has been noted by Folin [1905], McCollum [1911], Zeller [1914] and others.

Table I

Observer	Animal	Weight Kg.	Diet	Total nitrogen in urine per kilo body weight	Creatinine nitrogen per kilo body weight	Total urine N / Creatinine N
Cathcart [1907]	Man V.B.	62·0	Fasting, 4th day	0·221	0·0056	39
,, ,,	,, ,,	60·0	,, 8th ,,	0·159	0·0053	30
Benedict and Osterberg [1914]	Dog 39	7·6	,, 3rd ,,	0·360	0·0099	36
,, ,,	,, 33	12·7	,, 3rd ,,	0·280	0·0112	25
Towles and Voegtlin [1912]	,, 3	9·0	,, 2nd ,,	0·294	0·0129	23
Murlin [1907, 2]	,, C	13·0	,, 4th ,,	0·257	0·0080	32
Folin [1905]	Man H.B.H.	85·7	Starch, cream, 1 g. N	0·0420	0·0070	6·0
Graham and Poulton [1912]	,, G.G.	62·4	Starch, cream, ·912 g. N	0·0445	0·0093	4·8
,, ,,	,, E.P.P.	72·4	Starch, cream, 1·23 g. N	0·0468	0·0107	4·4
af Klercker [1907]	,, a.K.	88·0	Low N	0·0319	0·0079	4·0
Robison [1922]	,, C.J.M.	60·5	Carbohydrate, fat, ·3 g. N	0·0352	0·0072	4·9
,, ,,	,, R.R.	58·0	,,	0·0355	0·0084	4·2
McCollum [1911]	Pig	10·9	Carbohydrate	0·0495	0·0095	5·2
,, ,,	,,	68·4	,,	0·0387	0·0069	5·6
Mendel and Rose [1911]	Rabbit	1·74	..	0·126	0·0172	7·3
Murlin [1907, 2]	Dog C	11·3	,,	0·158	0·0104	15

There is a close parallelism between the figures for men and dogs during starvation and between those for men, pigs and rabbits on abundant nitrogen-free diets. The creatinine excretion for Murlin's dog C on such a diet is also in good agreement with the corresponding figures for men and pigs but the ratio $\frac{\text{Total urine N}}{\text{Creatinine N}}$ is about three times as high as the same ratio for other animals. It is of course not possible to state on such evidence alone that the real endogenous metabolism of this dog should be represented by a nitrogen output of one-third the observed amount, but it is clear that the nitrogen metabolism of dogs differs in some way from that of man, and that caution must be used in applying conclusions from experiments with these carnivora to other animals and man.

These criticisms however do not apply to the experiments of McCollum [1911] on pigs, for in these the constancy of the proportion of the endogenous metabolism represented by creatinine nitrogen was recognised and was used as a criterion for judging when the minimum nitrogen excretion of the animals had been reached.

The pigs were fed on a basal nitrogen-free diet of ample fuel value consisting of starch, a salt mixture and water, until the nitrogen output had reached the minimum, whereupon an amount of the protein under examination equivalent to this minimum (urine nitrogen only) was added to the diet during a further period, after which the basal diet alone was fed until the output had again fallen to the minimum, the nitrogen excreted during this last period being also included in the calculation. In the experiment recorded by McCollum 2·62 g. of gelatin nitrogen was given daily during eight days, i.e. 20·96 g. in all. The total output during these eight and the following four days on which no nitrogen was given, amounted to 41·71 g. in the urine and 12·48 g. in the faeces, i.e. 54·19 g. in all, making a negative balance of 33·23 g.

Had the pig received no nitrogen at all its total output during these twelve days would have amounted to 31·44 g. in the urine and 12·48 g. in the faeces, making 43·92 g. in all, so that a saving of 10·69 g. nitrogen has been effected by 20·96 g. of gelatin nitrogen. This implies a utilisation of 50 % of the nitrogen given in this form, which was confirmed by five other similar experiments the details of which are not given. If the result is stated in terms of body protein saved, this amounts to 1·34 g. per day (if reckoned on eight days), i.e. 37 % of the minimum output in urine and faeces or 51 % of that in the urine only, which is taken by McCollum as representing the essential tissue metabolism of the animal.

Boruttau [1919] has recently attempted to determine the biological value of gelatin by two experiments on dogs, using the method and formulae adopted by Karl Thomas [1909] and has obtained the figures 49·1 % calculated by formula I and 67·3 % calculated by formula II. These values would agree much more closely had Boruttau not made an error in his use of formula I by taking the total food nitrogen as denominator in place of this amount less the nitrogen of the faeces, as intended by Thomas. In any case however such figures have no real significance in the case of gelatin since they will necessarily vary with the amount of the intake, and moreover the experiments were of too short duration to possess much value.

Apart then from the experiments of McCollum no very satisfactory evidence has been produced regarding the value of gelatin alone to satisfy any of the nitrogen requirements of the animal body. Most of the investigations have in fact been concerned with its value when fed in conjunction with other proteins and this introduces the possibility of complementary effect, about which very little is definitely known. That such effect is possible is shown by the experiments of Osborne and Mendel [1912] on rats. With gelatin as the sole protein the animals rapidly declined in weight but recovered when half of the gelatin was replaced by gliadin, a protein incapable of inducing more than a very slight growth when fed as the sole protein constituent of the diet. Further, almost all the previous work, including that of McCollum, has been carried out on animals, and the results might not necessarily apply to man. The whole question is of very great theoretical and practical interest because of its bearing on protein metabolism in general and the nature of the body's requirements for particular compounds of nitrogen.

EXPERIMENTAL.

The investigation about to be described was an attempt to obtain more light on the problem by direct experiments on man.

The subject of the experiment was myself, age 37 years, medium build, weight 59 kilo, height 173·5 cm. My minimum nitrogen output had been determined by previous experiments which will be discussed in another paper.

In the second of these experiments, in which a diet containing about 0·3 g. N and equivalent to 2600–3000 calories (45–52 cals. per kilo) was taken for a period of seven days, the nitrogen output in the urine fell to a fairly constant level of 2·06 g., while the average amount of nitrogen excreted in the faeces was 1·13 g. per day.

In the present investigation the basal diet supplemented by different quantities of gelatin was taken for periods of ten days, the nitrogen intake being kept absolutely constant during each period.

Profiting by the experience of the previous experiments the basal diet was somewhat altered, the original attempt to introduce some variety and palatability being given up in favour of greater simplicity and uniformity of the food intake. The proportion of calories supplied by fat and the total nitrogen in the diet were both reduced. In the later experiments the process of simplification was carried to its furthest extent, the diet consisting of corn starch, lactose, sucrose and a salt mixture. Minimal quantities of lemon juice and cod liver oil were added to supply the antiscorbutic and fat soluble A accessory factors and agar-agar was taken to increase the bulk of the faeces and prevent constipation. The corn starch, lactose, salt mixture and agar for each day's ration were weighed out and mixed together before the experiment began. The mixture was taken in the form of a cream made with cold or warm (but not boiling) water and washed down with more water. The uncooked starch grains were very well absorbed, extremely few being found in the faeces. Usually a third of the day's ration was taken at 8 a.m., 1 p.m. and 7 p.m., but sometimes it was found necessary to increase the number of meals in order to consume the prescribed amount. The gelatin was dissolved in warm water and taken either by itself or mixed with some of the starch and lactose. The lemon juice, sweetened with cane sugar, was taken as a drink and a little weak tea with lemon was also permitted. The very small amount of nitrogen in the tea was assumed to be due to caffeine and to the excreted unchanged in the urine. It was therefore always subtracted from the total nitrogen intake and from the output.

The salt mixture had the following composition:

Calcium diacid phosphate $CaH_4(PO_4)_2, H_2O$	20	Na	2·5 %
Calcium lactate $(C_3H_5O_3)_2Ca, 5H_2O$	30	K	13·5
Potassium hydrogen phosphate K_2HPO_4	30	Ca	7·1
Sodium dihydrogen phosphate NaH_2PO_4, H_2O	15	P	13·6
Magnesium carbonate $MgCO_3$	3	Mg	·85
"Iron carbonate"	2	Fe	·7

It was intended that 10 g. of this mixture with the addition of 5 g. sodium chloride should be taken daily. The amounts of calcium and phosphorus would then correspond with those recommended by Sherman as 50 % above the minimum requirements of the body for these elements [Sherman, Wheeler, and Yates, 1918; Sherman, 1920]. It was found necessary however to reduce these quantities to 6 g. and 4 g. respectively on account of the diarrhoea caused by the diet. The ash of the above salt mixture is markedly alkaline, a point of importance in view of the observations by McCollum and Hoagland [1913] on the increased nitrogen output caused by diets having an acid ash.

The urine was collected from 8 a.m. to 8 a.m. and stored under toluene. The faeces were collected over the whole period and mixed with dilute sulphuric acid, those passed during the morning being considered as belonging to the previous day. Owing to the fluid consistency of the faeces the use of markers was found to be impracticable, but in view of the regular evacuation of the intestines and the length of the experiment, no serious errors can have been introduced in this way. Estimations of nitrogen in urine, faeces and in all components of the diet were carried out by the Kjeldahl method in duplicate. Creatinine was estimated by the method of Folin.

The percentages of nitrogen found in the constituents of the diet and their fuel values are given in Table II.

Table II.

	Nitrogen per 100 g.	Calories per 100 g.
Gelatin (Coignet's "Extra." Gold Label)	14·16	324
Corn starch (a)	0·039	360
„ „ (b)	0·027	360
Dextrin	0·065	360
Agar-agar	0·242	—
Lactose	0·013	370
Sucrose	—	395
Butter	0·080	775
Cod liver oil	Not determined	930
Lemon juice (per 100 cc.)	0·067	40
Vermouth „	0·005	140
Tea infusion* „	0·008	—

* The strength of the tea infusion was kept as nearly constant as possible but the total nitrogen intake from this source was checked by removing an aliquot portion of all tea drunk during an experiment and estimating the nitrogen in the whole quantity.

Concerning the purity of the gelatin used in the experiments.

The source from which commercial gelatin is obtained and the methods employed in its manufacture are not likely to produce a pure product. One would expect to find it contaminated with traces of other animal proteins or their decomposition products. Such included impurities being colloids could not be removed by washing, and might conceivably possess a high value for the replacement of body nitrogen.

Kirchmann drew attention to the fact that the best French gelatin gave a slight precipitate with Millon's reagent and with potassium ferrocyanide and acetic acid, and claimed to have succeeded in removing the impurities to which these reactions were due. The unoxidised sulphur was also reduced from 0·387 % to 0·263 %. He considered that the difference between his own results and those of previous workers was to be attributed largely to the presence of this protein in their gelatin. One of his methods consisted in soaking the gelatin first in water, then in 10 % sodium chloride solution, again in water and finally in alcohol. Murlin [1907, 1], using the same methods, was unable to detect any improvement in the purity of the product.

The gelatin used in the present investigation also gave a slight positive reaction with Millon's reagent and with potassium ferrocyanide and acetic acid, and an attempt was therefore made to purify it by soaking it for 24 hours in $N/20$ HCl followed by $N/20$ NaOH, then for some days in running water. No appreciable reduction in the intensity of the colour produced with Millon's reagent was observed after such treatment.

Folin and Denis [1912] have recorded finding a trace of tyrosine in gelatin, using Folin's colorimetric method, and Dakin [1920] has recently obtained a similar result using a gravimetric method. He estimates the amount of tyrosine at about 0·01 % and considers that it cannot be an integral part of the gelatin molecule.

I attempted to estimate the amount of tyrosine present by means of Folin's method, using relatively large quantities of gelatin. The tyrosine in a sample of dried ox muscle was also estimated by the same method. The results are shown in Table III. Millon's reaction is not well adapted for colorimetric measurement but under suitable conditions it was found possible to make approximate determinations by comparing the colour with that developed by different amounts of pure tyrosine, and the results agree reasonably well with those obtained by Folin's method.

Table III.

	Tyrosine estimated by Folin's method	Tyrosine estimated with Millon's reagent
Gelatin (Coignet's extra)	0·57 %	0·6 % to 0·7 %
„ after purification	0·45	—
„ Swiss	0·55	—
Glue	1·47	
Ox muscle	5·8	

The accuracy of Folin's method has been called in question by Abderhalden [1913, 1, 2] who has suggested that other amino-acids, tryptophan, hydroxy-tryptophan and hydroxyproline give the same colour reaction. Of these the first two are not present in gelatin but Dakin estimated the amount of hydroxy-proline as 14·1 %. Through the kindness of Prof. Leathes, F.R.S., who supplied me with a specimen of this amino-acid, I was able to test its behaviour with Folin's reagent and found that a slight colour developed under the conditions laid down by Folin and Denis for the estimation of tyrosine, but that the intensity was about $\frac{1}{315}$ of that produced by the latter compound.

A slight colour, similar to that given by hydroxyproline, was also obtained from a specimen of phenylalanine.

In the face of the results given by gravimetric methods it would be rash to assert that the gelatin actually contained 0·57 % of tyrosine, but the colour produced with Folin's reagent does not appear to be due to any of the other amino-acids known to be present. It is also probable that the same compound, tyrosine or other amino-acid, is the cause of the colour produced with Millon's reagent.

If the percentage of tyrosine is correct and if it is present as a constituent of another protein similar to ox muscle, the proportion of the latter in the gelatin would be about 10 %. This calculation however is based on too many assumptions to be of more than speculative interest.

Up to the present neither cystine nor any other compound containing sulphur has been isolated from gelatin though the presence of such unidentified compounds has been noted by Dakin [1920].

The gelatin used in these experiments after purification in the manner described above, contained 0·24 % of total sulphur, calculated on the dry substance. Krummacher [1903] after purifying gelatin by Kirchmann's method found 0·28 % S (of which 0·02 % was in the form of SO_3 and SO_4). The original commercial product used by Krummacher contained 0·62 % total S, of which 0·4 % was present as SO_3 and SO_4. Such an amount (0·24 %) of unoxidised sulphur would correspond with 0·9 % of cystine (or other compound containing a like proportion of S) and this can hardly be ascribed to impurities in the gelatin.

RESULTS OF THE EXPERIMENTS ON GELATIN DIETS.

It was proposed to carry out three diet experiments in which low, medium and high amounts of gelatin nitrogen should be given in addition to the basal diet, in order to determine

(1) whether the minimum nitrogen loss on abundant nitrogen-free diet can be still further reduced by gelatin, and if so,

(2) what relation the amount of this reduction bears to the amount of gelatin ingested.

Two experiments were completed during the early part of 1921, but the third had to be broken off through illness shortly after it was begun. It was repeated in September 1921 the conditions being somewhat modified on account of certain results that had in the meantime been obtained from other experiments carried out with Prof. C. J. Martin. These appeared to indicate that the amount of certain proteins required for nitrogen equilibrium could be greatly reduced if a carbohydrate diet very much in excess of the energy requirements was taken. In this last experiment, therefore, I increased the fuel value of the diet to the maximum that could be tolerated, so that the body weight was maintained and even increased during the first half of the period (see Fig. 1, Curve C). Synthesis of fat from the carbohydrate of the food was also indicated by the high respiratory quotient.

The diets for the three experiments are given together in Table IV. The diet taken during the experiment in which my minimum requirements were determined is also included for purposes of comparison.

Table IV.

Food	Nitrogen minimum 29/11/20—5/12/20 Total calories=2605 Calories per kilo=45[1] Calories supplied as Fat=31%.		Gelatin I 28/1/21—6/2/21 Total calories=2525 Calories per kilo=44 Calories supplied as Fat=6·4%.		Gelatin II 18/4/21—27/4/21 Total calories=2757 Calories per kilo=47 Calories supplied as Fat=6·5%.		Gelatin III 8/9/21—17/9/21 Total calories=3256 Calories per kilo=54 Calories supplied as Fat=0·86%.	
	Wt g.	N g.	Wt g.	N g.	Wt g.	N g.	Wt g.	N g.
Gelatin	—	—	66·67[2]	12·000	27·14[2]	4·885	41·89[2]	7·540
Corn starch	280	0·118	350	0·137	340	0·143	500	0·135
Dextrin	50	0·033	16	0·010	—	—	—	—
Butter and margarine	105	0·074	20	0·016	20	0·002	Cod oil 3	—
Honey	55	0·013	—	—	—	—	—	—
Sucrose	25	—	—	—	70	—	30	—
Lactose	65	0·008	180	0·023	250	0·032	300	0·039
Lemon juice	30 cc.	0·020	25	0·017	25	0·017	20	0·013
Vermouth	25 cc.	0·001	50	0·002	—	—	—	—
Tea	1200 cc.	0·072	350	0·028	250	0·020	600	0·043
Agar-agar	15	0·036	13	0·031	10	0·024	10	0·024
Salt mixture	5	—	10[3]	—	10[3]	—	10[3]	—
	—	0·375	—	12·264	—	5·123	—	7·794
Total fluid	—	—	2200–2500	—	2000	—	2200	—

[1] During the first five days of the experiment the fuel value of the diet was equal to about 52 cals. per kilo.
[2] Weight calculated as dry gelatin.
[3] Includes 4 g. sodium chloride.
No purification was attempted for experiment I. For II and III the gelatin, after purification by the method described in the text, was dissolved in hot water. Weighed amounts of this solution were transferred to bottles and sterilised in the autoclave. The figures for the nitrogen intake are based on a number of analyses of samples from different bottles.

Gelatin I (28th Jan. to 6th Feb. 1921).

During the three days previous to the experiment a mixed diet containing about 12 g. N was taken.

The experimental diet is given in Table IV. It included:

N in the form of gelatin   ...   12·00 g.
N in accessories[1] (excluding tea)   0·23
Total N ...   .   12·23

[1] The nitrogen in the constituents of the basal diet.

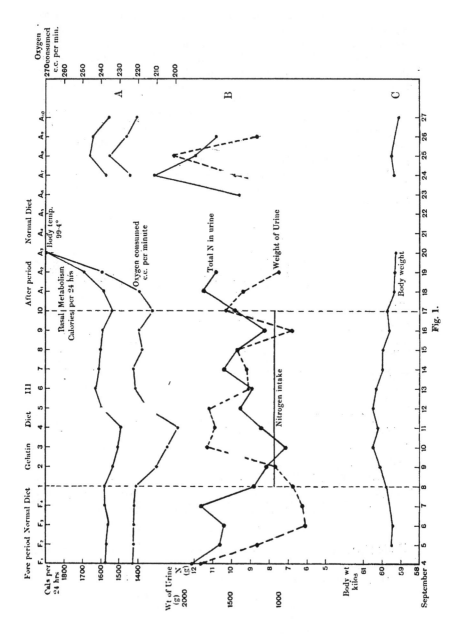

Fig. 1.

A few nitrogen-free biscuits made from starch, dextrin and agar were chewed at each meal to increase the flow of saliva. The rest of the starch etc. was taken in the raw state as already described.

The results of the experiments are shown in Table V. It will be seen that the nitrogen output in the urine remained almost stationary for three days, then rose and a series of oscillations set in.

Table V.

Date	Day of exp.	Body weight k.	Weight of urine g.	Sp. gr. of urine g.	Creatin- ine N g.	Total N in urine[1] g.	N in faeces g.	Total N (U + F) g.	Balance
Jan. 28	1	58·75	1447	1·0175	0·44	12·37	(1·38)[2]	(13·75)	(−1·50)
29	2	—	1790	1·014	0·53	12·42	1·38	13·80	−1·57
30	3	—	714	1·023	0·46	12·38	1·38	13·76	−1·53
31	4	—	1236	1·015	0·48	14·05	1·38	15·43	−3·20
Feb. 1	5	—	1550	1·013	0·52	12·84	1·38	14·22	−1·99
2	6	—	1496	1·015	0·53	14·42	1·38	15·80	−3·57
3	7	57·80	1200	1·016	0·51	13·13	1·38	14·51	−2·28
4	8	—	1300	1·015	0·53	14·20	1·38	15·58	−3·35
5	9	—	1601	1·014	0·50	12·27	1·38	13·65	−1·42
6	10	57·55	1914	1·014	0·54	14·44	1·38	15·82	−3·59
Average for whole period					0·50	13·25	1·38	14·63	−2·40
Average for the last six days					0·52	13·55	1·38	14·93	−2·70

[1] The "caffeine N" = 0·03 g. (from tea) has been subtracted from the total urinary nitrogen.
[2] The faeces were only collected for the last nine days.

Gelatin II (15th–27th April 1921).

During the two days previous to this experiment the diet consisted of eggs, milk, bread, potatoes, butter and apples, and contained 11·2 g. nitrogen. The experimental period was divided into two parts.

During the first three days the basal diet was supplemented by 250 g. of egg-white while during the last ten days this was replaced by gelatin. The two diets contained:

	Three days 15th–17th April	Ten days 18th–27th April
N in form of egg-white ...	5·0	—
„ gelatin ...	—	4·88
N in accessories (excluding tea)	·21	·22
Total N	5·21	5·10
Fuel value	2580 cals = 44 cals per kilo	2757 cals = 47 cals per kilo (6·5 % supplied by fat)

The results are shown in Table VI.

A pronounced negative balance occurred on the egg-white diet and the nitrogen output during the first few days of the gelatin period is slightly above the average of the last six days.

Gelatin III (8th–17th Sept. 1921).

During the preceding four days a mixed diet containing about 12 g. N and equal to about 2400 calories was taken. On the 7th Sept. 100 g. lactose and 100 g. starch were consumed in addition to the above, making the total calories 3130.

Table VI.

Date	Day of exp.	Body weight k.	Weight of urine g.	Sp. gr. of urine g.	Creatin- ine N g.	Total N in urine[1] g.	N in faeces g.	Total N (U + F) g.	Balance
First period:									
April 15	E 1	59·3	1351	1·017	—	7·30	—	—	—
16	E 2	—	1155	1·019	—	5·91	—	—	—
17	E 3	—	962	1·019	—	6·09	—	—	—
Second period:									
April 18	1	58·85	1517	1·013	0·52	7·42	1·28	8·70	−3·60
19	2	—	1779	1·012	0·56	6·32	1·28	7·60	−2·50
20	3	—	904	1·018	0·55	6·85	1·28	8·13	−3·03
21	4	—	1125	1·015	0·53	6·73	1·28	8·01	−2·91
22	5	—	833	1·019	0·55	5·68	1·28	6·96	−1·86
23	6	—	1148	1·017	0·54	7·12	1·28	8·40	−3·30
24	7	—	1523	1·013	0·54	6·18	1·28	7·46	−2·36
25	8	58·35	817	1·021	0·53	6·12	1·28	7·40	−2·30
26	9	—	1209	1·015	0·52	6·77	1·28	8·05	−2·95
27	10	—	1020	1·017	0·53	6·56	1·28	7·84	−2·74
28	—	58·15	—			—	—	—	—
Average of whole period	... ...				0·54	6·57	1·28	7·85	−2·75
Average of last six days	... ...				0·53	6·41	1·28	7·69	−2·59

[1] The "caffeine N" = ·02 g. (from tea) has been subtracted from the total urinary nitrogen.

The experimental diet is given in Table IV. It included:

N in the form of gelatin	...	...	...	7·54 g.
N in accessories (excluding tea)	...	...	·21
Total N	...	...	...	...	7·75

From Sept. 14th the fuel value was increased by 100 calories taken in the form of starch biscuits and honey. The increased nitrogen is negligible.

The daily exercise consisted of a walk of from five to seven miles.

The results are shown in Table VII.

Here as in Gelatin I the nitrogen output during the first four days is below the average for the whole period.

Table VII.

Date	Day of exp.	Body weight k.	Weight of urine g.	Sp. gr. of urine g.	Creatin- ine N g.	Total N in urine[1] g.	N in faeces g.	Total N (U + F) g.	Balance
Sept. 8	1	59·8	850	1·0225	0·59	8·77	1·54	10·31	−2·56
9	2	60·1	1038	1·0175	0·59	8·12	1·54	9·66	−1·91
10	3	60·5	1764	1·011	0·60	7·11	1·54	8·65	−0·90
11	4	60·25	1680	1·013	0·58	8·43	1·54	9·97	−2·22
12	5	60·5	1735	1·013	0·60	9·49	1·54	11·03	−3·28
13	6	60·35	1320	1·0165	0·59	8·87	1·54	10·41	−2·66
14	7	60·0	1347	1·0155	0·57	10·37	1·54	11·91	−4·16
15	8	59·93	1439	1·015	0·59	9·68	1·54	11·22	−3·47
16	9	59·65	859	1·0215	0·59	8·20	1·54	9·74	−1·99
17	10	59·78	1555	1·0145	0·60	9·84	1·54	11·38	−3·63
Average for the whole period	... ...				0·59	8·89	1·54	10·43	−2·68
Average for the last six days	... ...				0·59	9·41	1·54	10·95	−3·20

[1] The "caffeine nitrogen" = ·04 g. (from tea) has been subtracted from the total urinary nitrogen.

BASAL METABOLISM.

My basal metabolism was determined each day while on the experimental diet and during the periods immediately before and after.

The method adopted was that of the Douglas bag, the expired air being analysed in Haldane's gas analysis apparatus. From the 4th to the 7th of September the estimation was made at 9 a.m., fasting, after a walk of one mile followed by a resting period of at least 30 minutes. All the remaining determinations were made between 7.30 a.m. and 8 a.m. on waking. The results are shown in Table VIII.

A slight decrease in the basal metabolism occurred at the commencement of the experimental diet but the normal level was regained by about the fourth day. This decrease coincides with the lower nitrogen output, as may be seen in Fig. 1 A and B, but it is not possible to say whether the two are in any way connected. No corresponding decrease occurred in the creatinine output which was constant throughout the period.

In calculating the basal metabolism in calories per 24 hours the protein oxidation has been ignored but the error thus introduced is less than $+ 1 \%$. For those days in which the R.Q. is greater than 1 a correction has been made for the nett heat production due to synthesis of fat from carbohydrate by adding 0·3 of the calories equivalent to the excess of the CO_2 output over the oxygen consumed. I am aware that this is an arbitrary estimate, but the possible error involved is only slight.

Table VIII.

Date	Day of exp.	Diet during previous 24 hours (calories per kilo)	Oxygen consumed per minute (cc)	R.Q.	Calories per 24 hours
Sept. 4	F 1	Normal (40)	223·0	0·831	1575
5	F 2	,, ,,	222·5	0·830	1571
6	F 3	,, ,,	222·4	0·803	1561
7	F 4	,,	222·3	0·866	1580
8	G 1	Normal + 200 g. starch (54) lactose mixture	221·2	0·904	1582
9	G 2	Carbohydrate + gelatin (54)	210·1	1·012	1537
10	G 3	,, ,,	204·8	1·043	1510
11	G 4	,, ,,	199·0	1·093	1492
12	G 5	,, ,,	215·3	1·065	1600
13	G 6	,, ,,	221·6	1·051	1633
14	G 7		222·8	0·967	1614
15	G 8	,, ,,	218·0	1·040	1604
16	G 9	,, ,,	219·8	0·967	1592
17	G 10	,, ,,	212·9	0·968	1542
18	A 1	,, ,,	218·7	0·979	1587
19	A 2	Normal diet	239·4	0·823	1691
20	A 3	,,	269·3[1]	0·805	1893
24	A 7		224·3	0·801	1574
25	A 8		235·5	0·809	1660
26	A 9		226·3	0·823	1596
27	A 10	,,	220·9	0·830	1559

[1] The body temperature was 99·4° when this determination was made. Several attacks of vomiting had occurred during the previous night. The condition became worse and necessitated some days' rest.

The normal *basal metabolism* calculated from Harris and Benedict's formula for a man of age 37, weight 59·8 kilo, height 173·5 cm. would be
$$66·4730 + 13·7516 \times 59·8 + 5·0033 \times 173·5 - 6·755 \times 37 = 1506 \text{ calories.}$$

GENERAL CONSIDERATIONS.

My usual mode of life was followed throughout these experiments, eight or nine hours of each day being occupied with laboratory work. The only form of exercise was a walk of from two to seven miles. No great difficulty was found in consuming the diet although a feeling of nausea was frequently experienced. This was greatly intensified by the excessive quantity of food in the third experiment; the tongue became furred and more or less headache was common. Slight diarrhoea occurred in all three periods, faeces of fluid consistency being passed two or three times a day. The diet was however very well assimilated scarcely any starch being found in the faeces. Traces of reducing sugar were regularly present in the urine during the third experiment but only occasionally during the first two.

It was intended that a period on nitrogen-free diet should follow immediately on the third gelatin diet in order to determine my minimum requirements once more. This was not possible, and indeed much difficulty was experienced in completing the ten days proposed for the above experiment.

In the two experiments (G I, G III) which followed a normal mixed diet the average nitrogen output during the first four days was lower than that for the remainder of the period while the reverse of this was observed when the previous diet had been insufficient to satisfy the protein requirements (G II) and a considerable negative balance had occurred. These differences are probably due to the influence of the protein of the preceding diet, a diminishing store of which, perhaps in the form of amino-acids, remains in the body for some days. The first four days have therefore been excluded in considering the results. Another disturbing factor is the variation in the urine nitrogen from day to day. This frequently amounted to more than 20 % of the average output and showed no relationship whatever to the volume of the urine. Consequently it cannot be explained by diuresis.

SUMMARY AND DISCUSSION OF RESULTS.

The results obtained in the three experiments are summarised in Table IX, the average figures being given for the last six days of each ten day period. The average amounts of the nitrogen intake and output on the last three days of the earlier experiment on "nitrogen-free" diet are also included.

Table IX

Date	Body weight (kilos)	Fuel value of diet, cals per kilo	Nitrogen intake		Nitrogen output				
			Gelatin g.	Accessories g.	Urine g.	Faeces g.	Total g.	Creatinine g.	Balance
3. xii. 20–5. xii. 20	58·0	52–44	—	·30	2·06	1·13	3·19	0·49	−2·89
1. ii. 21–6. ii. 21	57·8	44	12·00	·23	13·55	1·38	14·93	0·52	−2·70
21. iv. 21–27. iv. 21	58·4	47	4·88	·22	6·41	1·28	7·69	0·53	−2·59
11. ix. 21–17. ix. 21	60·2	54	7·54	·21	9·41	1·54	10·95	0·59	−3·20

In attempting to calculate the amount of body protein spared by the gelatin from these results we are met by two difficulties, namely what is to be done with that part of the intake due to the nitrogen of the accessories, and with that part of the nitrogen output due to the faeces.

Nitrogen of the accessories. The nitrogen from the tea does not appear in the table, having been subtracted both from intake and output. The assumption that this is all "caffeine N" is not strictly correct but as the total amount is very small, usually under 0·05 g. any error involved in this mode of treatment must be negligible. The greater part of the accessory nitrogen comes from the corn starch. In McCollum's account of his experiments on pigs, no mention is made of any nitrogen arising from this source although large quantities (1700 g.?) of corn starch were given. If this starch contained as much nitrogen as the samples used by me the nitrogen intake from this source may easily have been ·0·5 g. or more. We do not know in what form this nitrogen is present nor its value in the human body and cannot therefore estimate its effect on the nitrogen output. If the amount and nature of such accessories are the same during the determination of the nitrogen minimum and experiments with gelatin, the following argument might be applied. If the value of this nitrogen in the accessories is zero then the real nitrogen requirements will be less than the observed output by the full amount of such nitrogen intake since the latter must be excreted in addition to the nitrogen resulting from the protein metabolism of the body. If the value of this nitrogen for the replacement of body nitrogen is 100 % then it will spare an equal amount of the latter and the observed output on the so-called "nitrogen-free" diet will represent the actual minimum requirements. But in this case an equal amount of gelatin will also be spared when gelatin is taken, and the apparent sparing effect of the gelatin will thus be increased by the same amount.

In either case the real saving of body nitrogen due to the gelatin will be less than the apparent saving, *i.e.* the difference between the negative balance on the gelatin and the minimum nitrogen output on the "nitrogen-free" diet, by an amount equal to the nitrogen of the accessories. This will also hold for all values of the latter between 0 and 100 °/$_o$. Unfortunately the proviso that the accessories should be the same on both diets does not strictly hold in the above experiments but if the butter nitrogen be subtracted from the total intake on the "nitrogen-free" diet the remainder is practically the same as the accessory nitrogen on the gelatin diets. The butter nitrogen would probably have a high value and I have therefore assumed that it does not appreciably increase the nitrogen output. The rest of the accessory nitrogen (·22 g. average) has been deducted in calculating the amount of body nitrogen saved by the gelatin. This is not strictly accurate but is probably the best that can be done with the figures. The above argument however ignores the possibility of complementary action of the accessories and the gelatin. It has been already pointed out that such action occurs when gelatin is fed with certain cereal proteins. McCollum, Simmonds and Pitz [1917] have shown that a mixture of oat protein and gelatin has a higher value than either alone or than oat protein plus caseinogen. It is impossible to say whether the results in my experiments were affected by such complementary action but if this did occur the real value of the gelatin alone is still less than the calculations appear to show.

Faecal nitrogen. The problem of how to treat the nitrogen excreted in the faeces is even more difficult. In McCollum's experiments the urinary nitrogen is alone considered, that in the faeces being estimated merely as a check on the complete absorption of the food protein. He considers that the nitrogen of the urine represents the essential tissue metabolism, while that of the faeces represents losses which may be termed accidental in character.

It has been shown by Rubner [1919] that the increased amount of nitrogen in the faeces of men, when fed on various diets, above that found on a carbo-hydrate diet cannot be taken as entirely due to undigested food protein, but that a considerable proportion of the increase comes from the body. In my experiments somewhat large variations were observed in the nitrogen of the faeces and these may have been due to the slight diarrhoea which occurred. Probably most of the nitrogen came from the body but the possibility that a small amount of gelatin escaped absorption must not be overlooked. We know very little about the relationship between loss of body nitrogen through the intestines and that excreted in the urine, but there seems to be no evidence that an increase in the former is accompanied by a decrease in the latter. The reverse of this is perhaps more probable.

This question remains at present the limiting factor for the accuracy of such experiments. I have attempted to define the limits between which the true conclusion from my results is to be found, by calculating the amount of body nitrogen saved by the gelatin in two ways. In the first (A) I have assumed that my minimum nitrogen requirements are represented by the sum of the nitrogen in the urine and that in the faeces on the nitrogen-free diet, and that the difference between the latter amount and the corresponding excretion on the gelatin diets represents unabsorbed gelatin nitrogen.

Table X.

	N Intake				A		B		
					Body Nitrogen saved			Body Nitrogen saved	
	Gelatin	Accessories	N Balance	N minimum		% of	N minimum		% of
Experiment	g.	g.	g.	g.	g.	minimum	g.	g.	minimum
R.R. II	4·88	0·22	−2·59	3·19	0·38	11·9	3·34	0·53	15·9
R.R. III	7·54	0·21	−3·20	3·19	0	0	3·60	0·19	5·3
R.R. I	12·00	0·23	−2·70	3·19	0·26	8·1	3·44	0·51	14·7
McCollum's pig average of last six day	2·62	?	−2·35	3·68	1·33	36·1	3·66	1·31	36·3

In the second (B) I have assumed that the gelatin is completely absorbed and that the minimum requirements for such periods are represented by the output in the urine on "nitrogen-free" diet plus the nitrogen in the faeces on the gelatin diet under consideration. The truth probably lies somewhere between these two extremes. The results of these calculations are given in Table X. McCollum's figures are also included for comparison. The accessory nitrogen has been in each case deducted from the apparent amount of body nitrogen saved.

The maximum saving in terms of the nitrogen minimum is thus 11·9 % if the first method of calculation is employed and 15·9 % if the second is used.

The fact that this was obtained with the lowest amount of gelatin would appear to prove that the effect is not due to any impurity, in which case there should be a proportionality between the amount saved and the gelatin intake. In this connection however the possibility that the increased protein in the diet entails an increased loss of body nitrogen must not be overlooked, although the creatinine excretion does not lend any support to such an hypothesis.

The results of the first and second experiments agree well between themselves, but in the third a much lower value was apparently indicated. It will be noticed however that the creatinine excretion in the experiment was higher than the normal and the body weight was also higher. This nitrogen minimum may therefore have been higher than the amount shown (perhaps owing to the excessive amount of food taken) in which case the calculated result would be too low.

All these values are much lower than those found by McCollum and this discrepancy cannot be explained except by assuming a difference in the metabolism of man and pig. If the values are calculated in terms of the urinary output alone they are all proportionately increased—the value for my experiment G II then becoming 25·7 % of the minimum, but the difference between the man and the pig still persists.

The creatinine output has been shown to bear some close relationship with the minimum nitrogen output. I therefore attempted to compare my results with McCollum's in terms of the amounts of creatinine excreted in the several experiments. I ignored the faeces and calculated the negative balance by subtracting the nitrogen intake from the output in the urine. The ratio of this balance to the average amount of creatinine nitrogen excreted during the same period is shown in the last column of Table XI. The last six days of each period have been alone considered.

Table XI.

Experiment	Nitrogen intake g.	Nitrogen in urine g.	Nitrogen balance g.	Creatinine N g.	Nitrogen balance / Creatinine nitrogen
R.R. II	5·10	6·41	− 1·31	0·53	2·47
R.R. III	7·75	9·41	− 1·66	0·59	2·81
R.R. I	12·23	13·55	− 1·32	0·52	2·54
McCollum's pig	2·62	3·89	− 1·27	0·48	2·65

There is obviously no discrepancy between our results when they are considered in this way, though what this agreement implies is not easy to state. It has been shown in many papers by Grafe (1912–1914), Abderhalden [1915], Underhill and Goldschmidt [1913] and others, that many nitrogen compounds other than amino-acids, namely organic ammonium salts, urea etc., have the capacity to spare a certain proportion of the loss of body nitrogen occurring on a carbohydrate diet. The results obtained by these workers do not agree in all points but the amount of nitrogen thus spared appears to be of the same order as that spared with gelatin in my experiments. It may well be that the action of the gelatin is of the same nature as that of these simpler

9

compounds and consists essentially in the reduction of the waste of amino-acids derived from body protein through deaminisation and subsequent oxidation in the body. The amino-acids of the gelatin and the ammonia and urea produced from them can play a part in the reversible reactions that are constantly proceeding in the body and thus influence the resulting equilibrium.

$$\text{Amino-acid} \rightleftarrows \text{Non-nitrogenous compound} + NH_3 \rightleftarrows \text{urea.}$$
(Ketonic or hydroxy acid)

If this is true the loss of body nitrogen when both carbohydrate and gelatin are fed may represent a "N-minimum" that corresponds more closely with the specific nitrogen requirements of the body than does the output on carbohydrate diet alone. This may perhaps be the explanation of the close agreement between the ratios of such loss to the creatinine nitrogen shown by my experiments and those of McCollum.

In conclusion I would express my very sincere thanks to Prof. C. J. Martin for his constant encouragement and advice throughout this investigation.

REFERENCES.

Abderhalden (1913, 1). *Zeitsch. physiol. Chem.* **85**, 91.
—— (1913, 2). *J. Biol. Chem.* **15**, 357.
—— (1915). *Zeitsch. physiol. Chem.* **96**, 1.
Benedict and Osterberg (1914). *J. Biol. Chem.* **18**, 195.
Bischoff (1853). Der Harnstoff als Maass des Stoffwechsels, 70.
Boruttau (1919). *Biochem. Zeitsch.* **94**, 194.
Boussingault (1846). *Ann. chim. phys.* **18**, 444.
Cathcart (1907). *Biochem. Zeitsch.* **6**, 109.
Dakin (1920). *J. Biol. Chem.* **44**, 499.
Donders (1853). Die Nahrungstoffe, 72.
Escher (1876). *Vierteljahrsschr. nat. Ges. Zurich*, 36.
Folin (1905). *Amer. J. Physiol.* **13**, 117.
Folin and Denis (1912). *J. Biol. Chem.* **12**, 245.
Frerichs (1845). *Handwörterbuch Physiol.* **3** (i), 683.
Graham and Poulton (1912). *Quart. J. Med.* **6.** 82.
Kauffmann (1905). *Pflüger's Archiv.* **109**, 440.
Kirchmann (1900). *Zeitsch. Biol.* **40**, 54.
af Klercker (1907). *Biochem. Zeitsch.* **3**, 45.
Krummacher (1901). *Zeitsch. Biol.* **42.** 242.
—— (1903). *Zeitsch. Biol.* **45**, 310.
Lehmann (1885). *Sitzungsber Münchener morph-phys-gesell.*
McCollum (1911). *Amer. J. Physiol.* **29**, 210.
McCollum and Hoagland (1913). *J. Biol. Chem.* **16**, 299.
McCollum, Simmonds and Pitz (1917). *J. Biol. Chem.* **29**, 341.
Mendel and Rose (1911). *J. Biol. Chem.* **10**, 475.
Munk (1894). *Pflüger's Archiv.* **58**, 309.
Murlin (1907, 1). *Amer. J. Physiol.* **19**, 285.
—— (1907, 2). *Amer. J. Physiol.* **20**, 234.
Oerum (1879). *Nordiskt medicinskt Arkiv.* 11.
Osborne and Mendel (1912). *J. Biol. Chem.* **13**, 233.
Pollitzer (1885). *Pflüger's Archiv.* **37**, 301.
Robison (1922). *Biochem. J.* **16**, 131.
Rona and Müller (1906). *Zeitsch. physiol. Chem.* **50**, 263.
Rubner (1919). *Arch. Physiol.* 73.
Sherman (1920) *J. Biol. Chem.* **41**, 173.
Sherman, Wheeler and Yates (1918). *J. Biol. Chem.* **34**, 383.
Thomas (1909). *Arch. Physiol.* 219.
Towles and Voegtlin (1912). *J. Biol. Chem.* **10**, 479.
Underhill and Goldschmidt (1913). *J. Biol. Chem.* **15**, 341.
Voit, C. (1872). *Zeitsch. Biol.* **8**, 297.
Zeller (1914). *Arch. Physiol.* 213.

XVII. DISTRIBUTION OF THE NITROGENOUS CONSTITUENTS OF THE URINE ON LOW NITROGEN DIETS.

BY ROBERT ROBISON.

From the Lister Institute.

(*Received January 17th, 1922.*)

DURING the course of experiments carried out in conjunction with Prof. C. J. Martin on ourselves, with the object of determining our minimum nitrogen requirements and the biological value of certain proteins, the opportunity was taken of investigating the distribution of the nitrogenous constituents in the urine when the total nitrogen output had reached a very low level.

In the particular experiments referred to in this paper, a diet consisting of carbohydrate and fat (corn starch, sucrose, lactose, honey, butter) together with a little lemon juice, inorganic salts and agar-agar was taken during seven days. The fuel value of this diet was equal to 44–52 calories per kilo body weight and the nitrogen content was about 0·3 g. A little tea and coffee was also taken but the nitrogen in these beverages was assumed to be "caffeine nitrogen" and to be excreted as such in the urine during the same 24 hours. It was therefore subtracted from the total nitrogen before calculating the percentage amounts of the other constituents. The above assumption is of course not strictly accurate, but in the one experiment (R.R.) the total amount of such nitrogen is very small, less than 0·1 g., so that any error thus introduced may be considered negligible.

The period on this low nitrogen diet was immediately followed by another of five days during which the same basal diet with the addition of a little milk was taken, the nitrogen intake being thus raised to about 3 g.

The total nitrogen in the urine and in all constituents of the diet was estimated by Kjeldahl's method, urea by van Slyke's urease method, ammonia by Folin's aeration method, amino-acids by formal titration after removal of the ammonia, creatinine by the method of Folin, and uric acid by that of Hopkins as modified by Folin and Schaffer.

The results are set out in Table I. In Table II the distribution of nitrogen on the days for which the total nitrogen output reached its lowest values has been compared with corresponding figures obtained by other investigators.

It will be seen that these observations are in complete agreement with Folin's [1905, 2] generalisations respecting the variation in the distribution of the urinary nitrogen. The creatinine nitrogen is practically constant and is

Table 1.

Subject: C.J.M. Weight Nov. 28th—61.9 kilos; Dec. 5th—60.5 kilos

Date 1920 Dec.	Day of exp.	Intake N g.	Volume of urine cc. Sp. gr. of urine	Total N g.	Total N less caffeine N g.	Urea N g.	Ammonia N g.	Amino-acid N g.	Creatinine N g.	Uric acid N g.	Undetermined N g.	Urea	Ammonia	Urea+ ammonia	Amino-acid	Creatinine	Uric acid	Undetermined
												In percentage of total N (less caffeine N)						
4	6	·35 / ·29¹	1421 / 1·014	2·69	2·40	1·27	·29	·05	·45	·07	·27	52·9	12·1	65·0	2·1	18·7	2·9	11·3
5	7	·29¹ / ·33¹	908 / 1·020	2·39	2·13	1·08	·21	·05	·43	·07	·29	50·7	9·9	60·6	2·3	20·2	3·3	13·6
6²	8	·26¹ / 3·87	1708 / 1·011	2·67	2·42	1·24	·30	·11	·47	·02	·28	51·2	12·4	63·6	4·5	19·4	·8	11·5
7	9	3·36 / ·25¹	1207 / 1·0165	2·74	2·57	1·39	·23	·08	·42	·06	·39	54·1	9·0	63·1	3·1	16·3	2·3	15·2
8	10	3·42 / ·17¹	1060 / 1·020	3·30	3·01	1·91	·24	·06	·45	·06	·29	63·4	8·0	71·4	2·0	15·0	2·0	9·6
9	11	3·40 / ·34¹	1128 / 1·0185	3·46	3·12	1·96	·28	·06	·44	·08	·30	62·8	9·0	71·8	1·9	14·1	2·6	9·6

Subject: R.R. Weight Nov. 28th—58.6 kilos; Dec. 5th—57.7 kilos.

Date 1920 Dec.	Day of exp.	Intake N g.	Volume of urine cc. Sp. gr. of urine	Total N g.	Total N less caffeine N g.	Urea N g.	Ammonia N g.	Amino-acid N g.	Creatinine N g.	Uric acid N g.	Undetermined N g.	Urea	Ammonia	Urea+ ammonia	Amino-acid	Creatinine	Uric acid	Undetermined
3	5	·32¹ / ·09¹	1886 / 1·007	2·08	1·99	·74	·35	·04	·50	·07	·29	37·2	17·6	54·8	2·1	25·1	3·5	14·5
4	6	·30¹ / ·07¹	1703 / 1·010	2·24	2·17	·98	·27	·05	·50	·10	·27	45·2	12·4	57·6	2·3	23·0	4·6	12·4
5	7	·27¹ / ·06¹	1547 / 1·009	2·07	2·01	·91	·25	·04	·47	·07	·27	45·3	12·4	57·7	2·0	23·3	3·5	13·4
6²	8	2·98 / ·10¹	1783 / 1·010	2·59	2·49	1·26	·31	·04	·50	·07	·31	50·6	12·5	63·1	1·6	20·1	2·8	12·4
7	9	2·94 / ·09¹	1991 / 1·011	3·04	2·95	1·71	·29	·04	·50	·03	·38	57·9	9·8	67·7	1·4	17·0	1·0	12·9
8	10	3·09 / ·16¹	1403 / 1·012	3·38	3·22	2·06	·26	·04	·45	·05	·36	64·0	8·1	72·1	1·2	14·0	1·5	11·2
9	11	2·91 / ·15¹	1134 / 1·014	3·01	2·86	1·60	·29	·04	·45	·09	·30	59·1	10·1	69·2	1·4	15·8	3·1	10·5
10	12	2·97 / ·18¹	1260 / 1·015	3·01	2·83	1·89	·26	·03	·46	·02	·17	66·8	9·2	76·0	1·1	16·2	·7	6·0

¹ N from tea and coffee—taken as "caffeine N."
² Milk diet began on Dec. 6.

Table II.

Observer	Subject	Body weight	Day of exp.	Intake N g.	Total N g.	Urea N g.	Ammonia N g.	Amino-acid N g.	Creatinine N g.	Uric acid N g.	Undetermined N g.	Urea	Ammonia	Urea+ Ammonia	Amino-acid	Creatinine	Uric acid	Undetermined
												Percentages of total N						
Folin [1905, 1]	H.B.H.	85·7	7	1	3·6	2·2	·42	—	·60	·09	·27	61·7	11·3	73·0	—	17·2	2·5	7·3
,,	E.S.A.	55·7	7	1	2·8	1·7	·28	—	·41	·08	·32	60·7	10·0	70·7	—	14·0	3·0	11·6
,,	A.H.	70·7	7	1	3·5	2·1	·27	—	·50	·14	·52	59·3	7·8	67·1	—	14·1	4·0	14·8
Zeller [1914]	—	67·9	17	1·04	2·88	1·42	·34	·22	·63	·15	·05	49·3	11·7	61·0	7·5	21·9	5·1	1·8
Graham and Poulton [1912]	G.G.	61·4	9	·91	2·25	1·047	·343	—	·598	·072	·19	46·54	15·25	61·79	—	26·58	3·2	8·44
Robison [1912]	C.J.M.	60·5	7	·33¹ / ·26¹	2·13²	1·08	·21	·05	·43	·07	·29	50·7	9·8	60·6	2·3	20·2	3·3	13·6
,,	R.R.	57·8	5	·32									9·1		2·1	25·1	3·5	14·5

equal to 7·3 mg. (C.J.M.) and 8·2 mg. (R.R.) per kilo body weight. These amounts like those of the other constituents are very similar to those found by Folin and others. The values for the total nitrogen are amongst the lowest on record and correspond with still further reductions in the percentage of urea nitrogen, this falling in one instance to 37 % of the total. The last figure was however accompanied by a somewhat high percentage of ammonia, but the sum of the urea and ammonia nitrogen was only 54·6 % of the whole amount. It would have been interesting to discover whether this sum (urea + ammonia) could have been further reduced by the ingestion of alkalies or whether any decrease in the ammonia nitrogen would have been accompanied by an increase in the urea.

The question as to whether any part of the urea nitrogen represents what Folin has termed the endogenous metabolism remains open.

I wish to express my indebtedness to Prof. C. J. Martin for his interest and help in this work.

REFERENCES.

Folin (1905, 1). *Amer. J. Physiol.* **13**, 66.
—— (1905, 2). *Amer. J. Physiol.* **13**, 117.
Graham and Poulton (1912). *Quart. J. Med.* **6**, 82.
Zeller (1914). *Arch. Physiol.* 213.

XVIII. THE ESTIMATION OF TOTAL SULPHUR IN URINE.

By ROBERT ROBISON.

From the Lister Institute.

(*Received January 27th, 1922.*)

THE methods for the estimation of total sulphur in urine are less satisfactory than those available for determining many of the other urinary constituents. A high degree of accuracy is specially important when we are concerned with the amount of unoxidised or "neutral" sulphur in the urine. This can only be determined by subtracting the amount of sulphates from that of the total sulphur and any error in the latter value falls on the relatively small difference.

Benedict's [1909] method in which the oxidation is effected with a mixture of copper nitrate and potassium chlorate has been criticised on the ground that considerable losses frequently occur through spattering during the initial stages of the reaction. Denis [1910] has stated that out of forty analyses mechanical loss resulted in every case. She has proposed to substitute a solution of copper nitrate, ammonium nitrate and sodium chloride in place of Benedict's oxidising reagent, thereby reducing the vigour of the reaction.

In my hands Denis' modification has not proved so reliable as the original method. Decrepitation is certainly less troublesome but the results are frequently too high, the errors amounting at times to more than 10 mg. The following series taken from nearly a hundred analyses are fairly representative:

25 cc. urine (1) gave 0·0711, 0·0719, 0·0783, 0·0826 g. $BaSO_4$,

    ,,   ,, (2) ,,   0·0864, 0·0901, 0·0865, 0·0899, 0·0877 g. $BaSO_4$.

The residue, even after prolonged ignition always contained appreciable amounts of nitrate, this being probably due to sodium nitrate formed from the sodium chloride and copper nitrate since the latter is readily decomposed at temperatures lower than those used. Whether the presence of this nitrate would alone account for the high results I am not able to say. Considerable errors in this direction were observed by Kolthoff and Vogelensang [1919] using higher concentrations of both sulphate and nitrate.

Consistent results were usually obtained by Benedict's original method, using an electric hot plate as suggested by Givens [1917] although in spite of this precaution vigorous spattering sometimes occurred. Only slight traces of nitrate were found in the residues after ignition and it is probable that these arose from the small amounts of alkali salts in the reagent and in the urine itself. The necessity of igniting the residue at a red heat was however a frequent cause of spoiled analyses through the cracking of the porcelain basins. Probably those used by Benedict were of better quality than are at present obtainable.

Errors due to the use of coal gas for igniting the residue.

Neither Benedict nor Denis appears to have recognised the possibility of errors from the presence of sulphur in the coal gas used for the ignition of the residue. Folin [1903] advised spirit burners for his original method and according to Liebesny [1920] errors of 8–10 % in the amount of neutral sulphur may arise from the use of gas. My own experience confirms this. In determining the blank of the oxidising solution referred to below, 2·5 cc. was found to yield 0·7 mg. BaSO$_4$, using a spirit burner, and 2·4 mg. using an argand burner with coal gas. If this error were constant in amount it would be cancelled in subtracting the blank but such was not found to be the case. Probably the absorption of oxides of sulphur depends on the surface area of the residue as well as on the total quantity present in the gas.

The use of a spirit burner is therefore essential if accurate results are to be obtained.

Since copper nitrate, which forms the basis of Benedict's reagent, is decomposed at comparatively low temperatures it seemed possible that the necessity of igniting the residue at a red heat might be avoided if neither chlorate nor alkali salts were introduced. Other substances were therefore tried as a diluent for the copper nitrate and of these copper chloride was found to be the most satisfactory. In the presence of a suitable proportion of this salt spattering never occurs, while the nitrate can be completely decomposed by heating the residue over a small spirit stove of the simplest kind. The temperature of a good argand burner is sufficiently high but gas cannot be used.

The oxidising reagent finally adopted has the following composition:

Copper nitrate (cryst.)	40 g.
Copper chloride (cryst.)	15 g.
Water	to 100 cc.

2·5 cc. of this solution are added to 10 cc. of the urine in a 4-inch porcelain basin and evaporated to dryness on a water-bath or electric hot plate. The oxidation can be started on the hot plate, or over a very small spirit flame. It proceeds rapidly but smoothly, leaving a coherent residue which frequently swells up. The dish is then heated over a broad spirit flame for 20 minutes. A spirit stove of the common kind is suitable but a sound tin, half filled with methylated spirit answers very well. A better flame is obtained if a number of holes are punched about half-way up the tin. The residue is dissolved in 10 cc. of 2N HCl and diluted with 300 cc. distilled water. The sulphate is precipitated in the boiling solution with 10 cc. of a 5 % solution of barium chloride, dropped in very slowly by means of a dropping tube. The precipitate is allowed to stand overnight before being filtered.

In the following analyses the precipitates were weighed in platinum crucibles.

For 25 cc. urine, 6·25 cc. of the oxidising solution, 20 cc. $2N$ HCl, 400 cc. H_2O and 15 cc.–20 cc. of 5 % barium chloride solution were used. A blank determination must be carried out with the reagents.

Two series, each of eight estimations, of the total sulphur in a normal urine yielded the following results.

$BaSO_4$ from 10 cc. urine		$BaSO_4$ from 25 cc. urine	
1.	0·0451	1.	0·1140
2.	0·0454	2.	0·1143
3.	0·0460	3.	0·1135
4.	0·0454	4.	0·1141
5.	0·0456	5.	0·1138
6.	0·0457	6.	0·1133
7.	0·0461	7.	0·1141
8.	0·0454	8.	0·1148
Average 0·0456 g.		Average 0·1140 g. equivalent to 0·0456 g. for 10 cc.	

It is doubtful whether any gain in accuracy is achieved by the use of 25 cc. instead of 10 cc. of urine.

In all the above estimations the filtrates were examined for nitrates by the very sensitive diphenylamine test, slight traces being found in one or two cases only.

The accuracy of the method was checked by estimating the total sulphur in two solutions of cystine containing also 2 % of urea and a definite volume of $N/10$ H_2SO_4 neutralised by an equal amount of $N/10$ NaOH.

The sample of cystine, for which I am indebted to the kindness of Dr J. C. Drummond, was analysed by the Carius method:

0·1878 g. gave 0·3679 $BaSO_4$; calculated 0·3681 g. $BaSO_4$.

$BaSO_4$ from 10 cc. solution A	$BaSO_4$ from 10 cc. solution B
0·0671	0·0662
0·0670	0·0659
0·0675	0·0661
0·0675	0·0659
—	0·0659
—	0·0663
0·0673 g.	0·0661 g.
Calculated 0·0678 of which 0·0212 is from cystine.	Calculated 0·0668 of which 0·0202 is from cystine.

This method has been employed for some time by myself and others in connection with metabolism experiments in this laboratory and has proved satisfactory.

I wish to thank Miss M. Tazelaar and Miss M. H. Carr for much valuable help in carrying out the numerous analyses, only a very small proportion of which have been recorded above.

REFERENCES.

Benedict (1909). J. Biol. Chem. 6, 363.
Denis (1910). J. Biol. Chem. 8, 401.
Folin (1903). Amer. J. Physiol. 9, 265.
Givens (1917). J. Biol. Chem. 29, 15.
Kolthoff and Vogelensang (1919). Pharm. Weekblad, 56, 122.
Liebesny (1920). Biochem. Zeitsch. 105, 43.

XIX. THE ACTION OF YEAST-GROWTH STIMULANT.

By OSWALD KENTISH WRIGHT.

From the Biochemical Department, Lister Institute, and the Bacteriological Laboratory, Household and Social Science Department, King's College for Women.

(*Received January 30th, 1922.*)

WILLIAMS [1919] and Bachman [1919] have confirmed the observation of Wildiers [1901] that certain yeasts are only able to grow at the expense of ammonium salts provided that a heavy inoculation is employed or a small amount of organic material (termed "bios" by Wildiers) is added to the medium. After a study of the ability of various substances to produce this effect, they suggest that the essential substance, which we will continue to call "bios" for the sake of convenience, is identical with the water-soluble *B* or anti-beri-beri vitamin. Williams [1920] has gone further, and has elaborated a method for measuring quantitatively the amount of anti-beri-beri vitamin present in any substance by observing its effect on the growth of yeast.

This suggestion has given rise to much discussion [see Eddy, 1921] and at present it is generally considered that the case for the identity of the two principles is not proven.

Lemon juice freed from citric acid by the method of Harden and Zilva [1918] when added to a mineral nutrient solution in small quantities enables a yeast to grow which could not grow in its absence. As animal feeding experiments show that the amount of water soluble *B* vitamin in lemon juice is very small relative to that in yeast extract [see Osborne and Mendel, 1920], it was thought that an investigation of its effect on the growth of yeast in mineral nutrient solutions might throw some light on the way in which the stimulation is brought about.

Preliminary experiments were performed to ascertain the smallest amount of lemon juice capable of producing growth in a nutrient solution made up as follows (Williams):

Saccharose	...	20 g.
$(NH_4)_2SO_4$	...	3 g.
KH_2PO_4 ...	...	2 g.
$CaCl_2$ ...	...	0·25 g.
$MgSO_4$ ...	...	0·25 g.
Distilled water ...		1000 cc.

A series of tubes was prepared containing this solution with the addition of increasing percentages of lemon juice. The yeast employed was a pure culture of a baker's yeast isolated by Dr Harden. This was grown for 24 hours in yeast water glucose and the growth was washed three times with sterile distilled water by means of the centrifuge. A loopful of a suspension of this washed yeast was put into each tube of the series, the amount being judged so that a fair proportion of drops made, according to Lindner's method, with a mapping pen on a cover slip contained one, two, or three cells each. The coverslip was sealed over a moist chamber with vaseline and, after noting the number of cells in each drop, was incubated for 24 hours at 25° and again examined for growth.

By this method it was found that no growth took place in the mineral nutrient solution unless 5 % or more of lemon juice was added. It was also noticed however that some of the tubes containing smaller amounts of lemon juice than 5 %, which had been standing in the laboratory, contained an amount of yeast appreciably larger than the amount originally inoculated. This suggested that growth did not take place as readily in the case of one or two cells in a minute drop as in a larger amount of medium. Subsequent investigation confirmed this, and the drop method was abandoned in favour of estimating the number of cells per cc. by means of a counting chamber.

In order to avoid adding any appreciable amount of "bios" with the seed yeast, it is necessary to inoculate each tube with as small a quantity of yeast as possible. Attempts to inoculate tubes by picking up a drop of distilled water containing one, two, or three cells from a coverslip by means of sterile filter paper proved unsuccessful. The method adopted was to make a light emulsion in sterile distilled water of yeast cells prepared and washed as described above. A slide was marked with several circles with a grease pencil and a drop of the emulsion transferred to the centre of each by means of a platinum loop. The drops were dried and stained and the number of cells in each counted, and by this means an emulsion was prepared which contained somewhere between 50 and 500 cells in a loopful. One loopful was then used to inoculate each tube.

A series of ten tubes was prepared with a mineral nutrient solution containing $(NH_4)_2SO_4$ with increasing percentages of lemon juice. A similar series was prepared without the $(NH_4)_2SO_4$.

Each tube contained 7 cc. of the following solution (Williams' solution without the $(NH_4)_2SO_4$, see p. 137):

Saccharose	...	20 g.
KH_2PO_4 ...	...	2 g.
$CaCl_2$ ...	...	0·25 g.
$MgSO_4$ ...	...	0·25 g.
Distilled water	...	1000 cc.

The tubes containing $(NH_4)_2SO_4$ received 1·5 cc. of a solution containing 2 g. of the salt in 100 cc. The saccharose and the $(NH_4)_2SO_4$ were purified by several recrystallisations. Repeated attempts to grow the yeast in tubes of this solution with $(NH_4)_2SO_4$ inoculated with from 50 to 500 cells in the manner described above and without the addition of "bios" failed, no growth being observed even after a lapse of several weeks.

. The lemon juice was prepared by the method of Harden and Zilva [1918], namely as follows: calcium carbonate was stirred into fresh lemon juice in excess of the amount required to neutralise all free acid. Absolute alcohol to twice the volume of the original lemon juice was added, and the whole filtered. The filtrate was evaporated down almost to dryness *in vacuo* at 35°. and the residue made up to the original volume of the lemon juice with distilled water. The sample used contained 0·0868 mg. of nitrogen per cc. of which 0·0112 mg. was ammoniacal nitrogen. The desired amount of this lemon juice was added to each tube of the series which was then made up to 10 cc. with distilled water. The tubes were sterilised by steaming for 30 minutes on each of three successive days. Each was then inoculated with from 50 to 500 cells as described above, the tubes were aerated by blowing air into them through a sterile glass tube plugged with wool and having its end drawn out to a fine point, and were incubated at 25°.

The tubes were examined each day and the number of cells counted by means of a Thoma haemacytometer. At least four drops were counted from each tube and if any considerable discrepancy was found between them the counting was continued until the margin of error was reduced below 5 %. Owing to the number of tubes it was not possible to count all of them every day, but the different rates at which the yeast grew in the various tubes rendered this unnecessary. The results are set out in Table I.

Table I.

Number of tube	Percentage of lemon juice	(NH₄)₂SO₄	Days: growth in millions of cells per cc.						
			1	2	3	4	5	6	9
1	0·1	0	s.g.	s.g.	s.g.	s.g.	—	s.g.	1·6
11	0·1	+	,,	,,	,,	,,	—	1·5*	29·0†
2	0·25	0	,,	,,	,,	,,	—	1·8	1·0
12	0·25	+	,,	,,	,,	,,	—	4·3*	32·0†
3	0·5	0	,,	,,	,,	1·3	—	3·5	1·0
13	0·5	+	,,	,,	,,	1·7	—	2·1*	34·2†
4	0·75	0	,,	,,	,,	0·9	—	2·5	1·8
14	0·75	+	,,	,,	,,	1·0	—	4·9*	26·7†
5	1·0	0	,,	,,	2·8	3·1	—	2·7	6·1
15	1·0	+	,,	,,	1·1	1·4	—	4·6*	22·6†
6	2·5	0	,,	2·8	4·3	—	6·3	—	—
16	2·5	+	,,	3·1	5·5	—	21·2	—	—
7	5·0	0	,,	4·9	7·1	—	—	—	—
17	5·0	+	,,	4·8	27·4	—	—	—	—
8	7·5	0	,,	6·7	9·2	—	—	—	—
18	7·5	+	,,	21·1	19·4	—	—	—	—
9	10·0	0	,,	12·0	12·1	—	—	—	—
19	10·0	+	,,	26·5	29·4	—	—	—	—
10	15·0	0	,,	15·9	19·6	—	—	—	—
20	15·0	+	,,	24·3	40·6	—	—	—	—

S.g. means some growth but too slight to count.
* Involution forms present.
† Films formed. Accurate counting impossible owing to clumping of cells. Many involution forms.

It is well known that the growth of yeast is susceptible to slight variations of conditions which cannot be controlled absolutely, such as the degree of aeration and the formation of products of metabolism and fermentation. Further, the accuracy of the counting is affected by the smallness of the drop examined in the low counts and by the tendency of the cells to form clumps in the higher counts or in the later stages of growth, so that below 3,000,000 and above 15,000,000 cells per cc. the counts of different drops of the same sample are found to be noticeably less uniform than between those figures, and the counts made after more than six days of growth can only be regarded as approximate. In considering the results obtained regard must be paid to these facts, and account taken only of large differences in the numbers. The table shows, however, that, until the yeast reaches a concentration of somewhere in the neighbourhood of five or six million cells per cc., the rate of growth is independent of the presence or absence of $(NH_4)_2SO_4$ and depends on the concentration of the lemon juice.

It does not appear unreasonable that it should take longer for an individual cell to collect sufficient nutriment to enable it to multiply when the nutriment is dilute than when it is concentrated, and possibly the fact that cells were unable to multiply in a minute drop in dilutions that enabled growth to take place in larger quantities of medium is explained by the fact that the whole drop did not contain sufficient nutriment.

After the yeast reaches a concentration of five or six million cells per cc. it is able to continue growing freely in the $(NH_4)_2SO_4$ tubes. Further it was observed that after six days involution forms, as described by Will [1895] in his investigation of film formation, begin to appear in all tubes containing $(NH_4)_2SO_4$ where the growth had not reached the critical point of five or six million cells per cc., and by the ninth day there was a heavy growth of film yeast on all these tubes. All the remaining tubes after a lapse of three weeks showed only very rare involution forms and no film formation, but most of the cells were highly refractile with many large vacuoles and granules and a thick membrane, similar in appearance to the permanent cells described by Will. It would appear, then, that after an interval of six or seven days the cells in smaller concentrations than five or six million per cc. are able to adapt themselves to the use of $(NH_4)_2SO_4$ and that film formation results.

A similar series of tubes was prepared with increasing percentages of aqueous yeast extract instead of lemon juice. This was prepared by Osborne and Wakeman's method [1919]; about 250 g. of wet brewer's yeast was washed and boiled for two minutes with 1 litre of distilled water containing 0·01 % of acetic acid and the liquid was separated off in the centrifuge. The solid residue was again heated with 500 cc. of 0·01 % acetic acid and the extracts united and concentrated to 500 cc. This extract contained 2·856 mg. of nitrogen per cc., of which 0·2273 was ammoniacal nitrogen. As this is rather more than 30 times the amount contained in the lemon juice, the amounts added to the various tubes of the yeast extract series were much smaller than in the lemon juice series. The results are set out in Table II.

Table II.

Number of tube	Percentage of yeast extract	$(NH_4)_2SO_4$	Days: growth in millions of cells per cc.						
			1	2	3	4	5	6	11
21	0·01	0	s.g.	s.g.	s.g.	s.g.	—	—	3·1
31	0·01	+	,,	,,	,,	,,	—	—	10·6*
22	0·02	0	,,	,,	,,	,,	—	—	3·0
32	0·02	+	,,	,,	,,	,,	—	—	10·5*
23	0·05	0	,,	,,	,,	,,	—	—	4·0
33	0·05	+	,,	,,	,,	,,	—	—	15·0*
24	0·1	0	,,	,,	,,	,,	—	1·6	10·0*
34	0·1	+	,,	,,	,,	,,	—	15·1*	20·0*
25	0·15	0	,,	,,	,,	6·1*	—	10·0*	—
35	0·15	+	,,	,,	,,	3·6	—	11·5*	—
26	0·2	0	,,	,,	0·5	3·6	5·1	—	—
36	0·2	+	,,	,,	0·5	3·7	15·5	—	—
27	0·5	0	,,	4·9	7·6	—	—	—	—
37	0·5	+	,,	4·4	23·7	—	—	—	—
28	1·0	0	,,	5·3	14·5	—	—	—	—
38	1·0	+	,,	11·8	39·5	—	—	—	—
29	2·0	0	,,	13·6	27·2	—	—	—	—
39	2·0	+	,,	19·9	43·0	—	—	—	—
30	3·0	0	,,						
40	3·0	+	,,						

(rows 30 and 40) } Very beavy growth. Not counted.

S.g. means some growth but too slight to count.
* Involution forms present. Very slight film formation.

On comparing the results obtained with the yeast extract with those obtained with the lemon juice it will be seen that their general trend is similar, and that the yeast extract is approximately ten times as effective as the lemon juice although its nitrogen content is more than 30 times as great. The results obtained are somewhat less uniform than those obtained with the lemon juice, and tubes Nos. 24 and 25 fall out of line on the 11th and 4th days respectively. Film formation does not occur so readily, only very slight traces being present in tubes Nos. 31, 32, 33, 34 and 35 by the 11th day.

These differences may be due to differences in the nature of the nitrogen in the two substances under examination or to the presence of toxic substances in the yeast extract which are absent from the lemon juice. The yeast experiment, however, affords confirmation of the fact that the rate of growth is at first independent of the presence or absence of $(NH_4)_2SO_4$ and depends on the concentration of the "bios" until the yeast has reached a concentration of somewhere in the neighbourhood of five or six million cells per cc., after which it proceeds further in the presence of the $(NH_4)_2SO_4$.

General Conclusions.

The suggestion that "bios" may be of the same nature as the vitamins, which are widely held to be necessary for the nutrition of higher animals, and the further suggestion that it may be identical with one of the recognised vitamins, open up possibilities for investigating the subject of vitamins which are particularly attractive, owing to the ease with which a biological process can be studied in a yeast as compared with a similar study in the case of a more complex organism. Before we can proceed to investigate the general question of vitamins by such methods it is necessary, firstly, to be certain that "bios" is actually a vitamin, i.e. a substance whose presence in the food in relatively small quantities is necessary to enable the remainder of the foodstuffs to be properly assimilated and utilised by the organism, and secondly that the substance in question is identical with one of the recognised vitamins. The present investigation points to the fact that "bios" does not enable the yeast to assimilate $(NH_4)_2SO_4$ simply by its presence or by being consumed at the same time, but that the yeast grows solely at the expense of the "bios" until it reaches a certain degree of concentration, and after that it is able to use the $(NH_4)_2SO_4$. No explanation is offered of this phenomenon which requires further investigation.

I wish to thank Prof. A. Harden, who suggested the investigation, for his advice and assistance in carrying it out.

REFERENCES.

Bachman (1919). *J. Biol. Chem.* **39**, 235.
Eddy (1921). The Vitamine Manual, Chap. iv (Baltimore).
Harden and Zilva (1918). *Biochem. J.* **12**, 259.
Osborne and Wakeman (1919). *J. Biol. Chem.* **40**, 383.
Osborne and Mendel (1920). *Ibid.* **42**, 465.
Wildiers (1901). *La Cellule*, **18**, 313.
Will (1895). *Zeitsch. ges. Brauwesen*, **18**, 1.
Williams (1919). *J. Biol. Chem.* **38**, 465.
—— (1920). *J. Biol. Chem.* **42**, 259.

XX. THE FUNCTION OF PHOSPHATES IN THE OXIDATION OF GLUCOSE BY HYDROGEN PEROXIDE.

BY ARTHUR HARDEN AND FRANCIS ROBERT HENLEY.

From the Biochemical Department, Lister Institute.

(*Received January 25th, 1922.*)

THE conditions under which the oxidation of glucose by H_2O_2 can take place in aqueous solution have been studied by W. Löb and his co-workers [1910, 1911, 1912, 1915]. He believes that the oxidation depends on the hydroxyl ion concentration of the solution, and further that phosphate ions exercise a specific accelerating effect on the reaction. He was unable to produce any oxidation of glucose by H_2O_2 in presence of water alone or when other substances were substituted for phosphates.

For instance neither Sörensen's mixtures of glycocol and NaOH nor of borate and HCl could be used to replace the phosphate mixture. Witzemann [1920] confirmed Löb's observations and agreed that phosphate mixtures exercise a specific effect. He was unable to produce the same effect with mixtures of $NaHCO_3$ and Na_2CO_3 as with phosphate mixtures.

Witzeman [1920] suggested that the phosphates might produce an intermediate compound with glucose of the same nature as the hexosephosphate which has been shown by Harden and Young [1910] to be produced during alcoholic fermentation, and which appears from the work of Embden and others to be connected with the production of lactic acid in muscle.

Witzemann was unable to detect the formation of any compound of glucose and phosphate in absence of H_2O_2.

As more exact knowledge of the mechanism of this reaction is desirable in view of the part played by phosphate in alcoholic fermentation, further experiments were made to try and ascertain whether an intermediate compound of the nature of a phosphoric ester of glucose was formed during this oxidation. For this purpose free phosphate was estimated at intervals in a mixture similar to one of those used by Witzemann for oxidation of glucose by H_2O_2 at 37°, both by the method of Schmitz [1906] and by the ordinary magnesium citrate method. No change in the amount of free phosphate was observed.

Besides the possibility of the formation of an intermediate compound between phosphate and glucose there are at least two other possible explanations of the effect produced by phosphate in this reaction:

(1) Phosphate may act as a peroxidase, thus rendering possible the oxidation of glucose by H_2O_2. If this were so phosphate might act in the same way towards mixtures of H_2O_2 and benzidine or guaiacum. No evidence could be obtained of such an action, but it does not follow from this negative result that no such action occurs with H_2O_2 and glucose.

(2) The phosphate mixtures may simply act as buffer substances, the maintenance of the P_H within certain points being essential. If this hypothesis is correct it should be possible to replace the phosphate mixtures by other substances, provided the P_H be maintained within the same limits.

We have carried out experiments on the oxidation of glucose by H_2O_2 with solutions the P_H of which was never over 7·3 and find that glucose is oxidised by H_2O_2 in presence of 0·125 M $NaHCO_3$ saturated with CO_2; or 0·25 M sodium arsenate saturated with CO_2; or 0·25 M sodium acetate.

The phosphate ion cannot therefore be regarded as playing a specific part in the reaction. It appears more likely that the buffer action of the salts employed is the important factor, whether as providing and maintaining the most suitable conditions for oxidation or as protecting the H_2O_2 from too rapid decomposition. In this connection the experiment of Witzemann [1920, p. 12, sec. 7] with a mixture of sodium carbonate and bicarbonate should be referred to. This shows that in presence of these salts, at a P_H slightly over 9·3, H_2O_2 is rapidly decomposed and little oxidation takes place, whereas in our experiment in presence of 0·125 M $NaHCO_3$ saturated with CO_2 (P_H 6·8) a considerable amount of oxidation is produced.

Experiments were accordingly made roughly to test the stability of H_2O_2 in solutions the P_H of which was maintained at various levels by the use of buffer solutions. The rate of decomposition was found to increase with rise of alkalinity. In those cases in which the buffer solutions consisted of phosphate mixtures the rate of decomposition of the H_2O_2 was not so rapid as in the corresponding experiments with other buffer solutions; so that in this sense the phosphate ion may be said to have a specific action.

In the experiments of both Löb and Witzemann the P_H of the solutions was only carefully measured in those experiments in which Sörensen mixtures were used. In most if not all of the experiments in which the effects of other substances were compared to that of phosphate mixtures the reaction of the solution was tested with litmus paper only. The glycocol + NaOH and borate + HCl mixtures used are only efficient buffers between P_H 8·0 and 10·2, and at these higher degrees of alkalinity the H_2O_2 is rapidly decomposed. The same applies to the $NaHCO_3$ + Na_2CO_3 mixture used by Witzemann which must have been much more alkaline than the phosphate mixture.

The experiments in which he compared the effect of Na_2CO_3 and NaOH were not carried out at the same P_H. The reaction of the solution in the latter case was suitable for the oxidation, whereas the alkalinity of the solution used in the former case was too high.

EXPERIMENTAL.

I. *Experiments to determine if mixtures of* Na_2HPO_4 *and* NaH_2PO_4 *will act as a peroxidase towards* H_2O_2 *and benzidine or guaiacum.*

Phosphate mixtures of P_H 9·0, 7·4, 6·9, 6·6, 6·2, 4·8 were tested with H_2O_2 and benzidine or guaiacum. No action was observed in any of the experiments.

To test whether the optimum P_H for the action of a peroxidase lies within the above range, similar experiments were made with the same phosphate mixtures, H_2O_2 and guaiacum with the addition of potato juice. The maximum effect was produced at room temperature from P_H 7·4 to 6·9. A similar result was obtained when benzidine was substituted for guaiacum.

II. *Oxidation of glucose by* H_2O_2 *in presence of various buffer mixtures.*

Seven flasks were made up as follows:

	Glucose g.	NaHCO₃ g.	Na₂HAsO₄. 12 H₂O g.	NaC₂H₃O₂. 3 H₂O g.	K₂HPO₄ g.	KH₂PO₄ g.	5·7 % sol. H₂O₂ cc.	
1	8·75	2·6	—	—	—	—	62	
2	8·75	—	25·1	—	—	—	62	
3	8·75	—	—	8·5	—	—	62	to 250 cc. with
4	8·75	—	—	—	8·8	2·2	62	distilled water
5	11·25	2·6	—	—	—	—	—	
6	11·25	—	25·1	—	—	—	—	
7	11·25	—	—	8·5	—	—	—	

In Nos. 1, 2, 5, 6 the solution was saturated with CO_2.

The P_H of each solution was estimated colorimetrically, and the glucose estimated by Bertrand's method and by the polarimeter.

All the flasks were then kept in the incubator at 37° for 24 hours and the P_H and glucose again estimated in each. The results are shown below:

	At start of experiment glucose g. per 100 cc.			After 24 hours			Apparent loss of glucose in g. per 100 cc.	
	P_H	Bertrand A	Polarimeter B	P	A	B	by A	by B
1	6·8	4·16	3·50	6·7–6·8	3·39	2·99	0·77	0·51
2	6·8	3·96	3·55	6·7–6·8	2·96	2·38	1·00	1·17
3	6·8–7·0	3·64	3·54	5·2	3·01	2·50	0·63	0·53
4	7·0	3·65	3·20	5·6–5·8	2·24	1·85	1·41	1·35
5	6·8	—·	4·55	7·6	—	4·55	—	0
6	6·8	—·	4·50	7·0	—	4·50	—	0
7	7·3	—·	4·25	7·3	—	4·25	—	0

Note. The sample of glucose employed for these experiments showed a 5 % higher content of glucose as estimated by Bertrand's method than when the polarimeter and $[a]_D$ 52·5° were used. This, and the fact that the Bertrand estimations of glucose at the start of the experiment were carried out before the polarimeter readings were made, may account for the discrepancies between the two sets of figures shown above.

Only traces of H_2O_2 remained in 1, 2 and 4 after 24 hours in the incubator.

No. 3 still contained a considerable amount of H_2O_2 and the action was allowed to proceed for a second period of 24 hours. After this the P_H was found to be 4·6–4·8 and glucose estimated polarimetrically amounted to 1·76 g. per 100 cc.: a total apparent loss of glucose of 1·78 g. per 100 cc.

These experiments show that in absence of H_2O_2 no loss of glucose has apparently taken place (Nos. 5, 6, 7). Whereas in presence of H_2O_2 (Nos. 1, 2, 3, 4) an apparent loss of glucose has occurred in every case, greatest in the two solutions containing phosphates and arsenates, but considerable in the other two solutions.

III. *Effect of change of* P_H *on the stability of* H_2O_2.

All solutions contained 5·0 cc. of 5·7 % H_2O_2 per 100 cc. solution. The H_2O_2 was estimated at the start and again after 24 hours and 48 hours at 37° by titration with standard permanganate.

No.	Mixture	P_H	% H_2O_2 lost after 24 hours	after 48 hours
1	HCl	1·0	—	1
2	Sörensen's glycocol + HCl ...	1·9	1	2·1
3	Na acetate + acetic acid ...	4·6	2	2·1
4	Phosphates	6·8	2	4·2
5	,, ...	7·1*	4·9	7·5 (43 hours)
6	,, ...	7·8*	6·1	7·4 (43 hours)
7	,, ...	8·0*	7·5	13·1
8	Palitzsch's borax + boric acid	6·9	20·2	36·1
9	,, ,, ,,	7·1	20·2	37·2
10	,, ,, ,,	7·7	22·3	39·3
11	,, ,, ,,	7·9	37·2	9·5
12	,, ,, ,,	8·2	55·3	74·4
13	,, ,, ,,	8·4	65·9	84·0
14	,, ,, ,,	8·6	65·9	84·0
15	,, ,, ,,	8·9	67·0	85·1
16	Sörensen's borate + HCl ...	8·8	—	78·5
17	,, + NaOH ...	9·2	—	99·0
18	,, ,, ...	12·8	100	—

* In these cases the concentration of the hydrogen peroxide used was 5 % instead of 5·7 %.

IV. *The* P_H *of some of the solutions tested by Witzemann.*

A solution was made up as described by Witzemann [1920, p. 12, sec. 7]: 2·43 g. $Na_2CO_3 . 10H_2O$; 0·72 g. $NaHCO_3$; 35 cc. H_2O; 20 cc. (0·2 g.) glucose solution and 20 cc. of 3 % H_2O_2.

The P_H, estimated colorimetrically, was over 9·3. At this degree of alkalinity H_2O_2 is rapidly decomposed and it is therefore not surprising that little oxidation took place.

To compare the effect of NaOH and Na_2CO_3 Witzemann [1920, p. 14, sec. 9 (2) and sec. 10 (2)] prepared solutions as follows:

P. 14, sec. 9: 32·0 cc. 0·33 M Na_2HPO_4; 3·0 cc. H_2O; 20·0 cc. 1 % glucose; 20·0 cc. 3 % H_2O_2; 0·61 g. $Na_2CO_3 . 10H_2O$;

and sec. 10 (3) 25·6 cc. 0·33 M Na_2HPO_4; 6·4 cc. 0·33 M NaH_2PO_4; 20·0 cc. 1 % glucose; 20·0 cc. 3 % H_2O_2; 3 cc. H_2O; 0·1714 g. NaOH.

The P_H of the former was 9·3–9·6 and of the latter 7·1. The amount of oxidation was greater in the second than in the first case and Witzemann argues that NaOH exercises a less harmful effect on the reaction than does Na_2CO_3.

It does not seem that this inference can fairly be drawn unless the experiments are carried out under the same conditions of P_H. The acidity of the H_2O_2 solution used by Witzemann is not stated in his paper. It is therefore not possible to say what was the exact P_H in his experiments. But in any case the alkalinity of the solutions used in experiments 7 and 9 (2) must have been much higher than in experiment 10 (3).

SUMMARY.

1. The oxidation of glucose by H_2O_2 takes place in presence of the following buffer substances: $NaHCO_3 + CO_2$; $Na_2HAsO_4 + NaH_2AsO_4$; $NaC_2H_3O_2$; $K_2HPO_4 + KH_2PO_4$, provided the P_H does not rise much above 7·3.

2. The stability of H_2O_2 in aqueous solution is increased by the presence of phosphates which do not however show a specific action in other respects.

REFERENCES.

Harden and Young (1910). *Proc. Roy. Soc. B.* **82**, 321.
Löb, W. (1911). *Biochem. Zeitsch.* **32**, 43.
Löb, W. and Beysel (1915). *Biochem. Zeitsch.* **68**, 368.
Löb, W. and Gutman (1912). *Biochem. Zeitsch.* **46**, 288.
Löb, W. and Pulvermacher (1910). *Biochem. Zeitsch.* **29**, 316.
Schmitz (1900). *Zentr. anal. Chem.* **45**, 513.
Witzemann (1920). *J. Biol. Chem.* **45**, 1.

XXI. VITAMIN REQUIREMENTS OF *DROSO-PHILA*. I. VITAMINS *B* AND *C*.

BY ARTHUR WILLIAM BACOT AND ARTHUR HARDEN.

From the Departments of Entomology and Biochemistry, Lister Institute.

(*Received January 30th, 1922.*)

LOEB [1915] ascertained the fact that flies of the genus *Drosophila* could be reared on a culture medium containing as mineral constituents only K, Mg, PO_4 and SO_4 ions. In these experiments micro-organisms were not excluded, and the growing larvae undoubtedly lived at the expense of these. Loeb also found that when Na was substituted for K no growth took place and concluded that K was essential for the development of the insect. It is however possible that the absence of K stopped the growth of the micro-organisms and thus deprived the larvae of nitrogen in an available form. Loeb's experiments therefore only prove that the mineral constituents named above are adequate, but not that they are essential.

Subsequently Loeb and Northrop [1916] succeeded in sterilising the eggs of *Drosophila* by exposure for 6–7 minutes to 0·1 % aqueous $HgCl_2$ or a saturated alcoholic solution of the same salt and found that the larvae hatched out from the few eggs which survived the disinfection were sterile. They ascertained that these sterile larvae could not be reared on sterile media consisting only of purified proteins, carbohydrates, fats and salts, but that when yeast killed by heat was added, normal development occurred. Alcoholic extracts of yeast were however inactive. Northrop [1917] pursued the subject further and found that the addition of caseinogen and similar substances along with cane sugar to yeast greatly increased the number of flies which could be reared at the expense of a definite amount of yeast and concluded that they acted as foodstuffs when properly supplemented by yeast. Similar results were obtained by the addition of kidney, liver and pancreas from the dog, but many other organs of the dog were inactive. The author was inclined to interpret these results in the sense that yeast supplies one or more accessory substances, necessary for the growth of the organism.

The question was again investigated by Baumberger [1919] who found that the pupae and eggs of flies reared in presence of a pure culture of yeast could readily be sterilised by immersion for 20 minutes in 85 % alcohol. The sterile larvae obtained from such eggs could easily be reared in sterile culture media provided the latter contained killed yeast or nucleo-protein prepared from yeast. Baumberger came to the conclusion that the essential requisite was a sufficient concentration of yeast nucleo-protein and that accessory food factors were not concerned.

The following experiments were instituted with the object of ascertaining more definitely whether any of the recognised vitamins are essential for the nutrition of these flies. The subject is of general interest as regards the extent to which the need for accessory food factors is distributed in the animal kingdom. Special interest attaches to the possibility of introducing a simplified technique by which results could be obtained more rapidly and economically than by the tedious and costly feeding experiments which are at present necessary.

TECHNIQUE OF THE EXPERIMENTS.

It was soon found that the eggs could be sterilised much more readily and certainly than the pupae and this method was always employed.

Following Baumberger's instructions the flies were reared in banana-agar (approximately equal parts of mashed banana and 1·5 % water-agar) which had been autoclaved and subsequently inoculated with yeast (*S. cerevisiae*). After a few sub-cultures on this medium the flies were found to be free from bacteria, both anaerobic and aerobic, and the pure stock was readily maintained by occasionally passing a few flies into a fresh banana-agar-yeast tube.

When eggs were desired for experimental purposes, freshly emerged flies were transferred to a special apparatus. This consisted of an inverted wide necked bottle, standing in a Petri dish and having a strip of moistened filter paper on the wall. A small circular glass dish with vertical walls was also placed on the Petri dish, so as to be covered by the inverted bottle. The walls of this dish were lined with hardened filter paper moistened with water. The bottle was in turn covered by an inverted cylindrical tinned iron can and the whole apparatus was autoclaved at 120° for half-an-hour. A few cc. of a 24–48 hour culture of yeast in yeast extract containing glucose were then placed in the small dish, the flies introduced into the inverted bottle and the whole apparatus placed in an incubator at 30°. Under these conditions the eggs were laid within 24–48 hours almost exclusively on the hardened filter paper lining the small circular glass dish containing the yeast culture. The strips of paper were removed, placed on a sterile slide under a dissecting microscope and the eggs picked off singly by means of sterilised dissecting needles and placed in 85 % alcohol as directed by Baumberger. They were then divided up into collections of the number required for each tube by means of a sterile pipette provided with a rubber teat, and were finally, after half-an-hour's exposure to the alcohol, sucked up into the pipette and delivered into the appropriate tube of medium. It was found most convenient to place 20 eggs in each test tube (6″ × ¾″) containing 5–5·5 cc. of medium in the form of an agar slope.

The agar jelly in the medium tube must be so soft that the larvae can readily penetrate it in their search for food. As a rule 3 cc. of 1½ % agar were added to 2 cc. of medium and the whole then steamed for 20 minutes on three successive days, and cooled in the form of slopes.

The inoculated tubes were incubated at 30° and examined daily, note being made of the number hatched, the general progress and the time of appearance of pupae and flies.

At the close of each experiment the tubes were tested for sterility by making cultures on beef broth-peptone-agar and incubating at 37°. Contamination by yeast usually showed itself in the experimental tube, but was when necessary tested for by culture on beer wort agar at 25°.

DIET.

A basal diet consisting of

Caseinogen (purified)	...	...	...	...	0·15 g. per tube
Starch and salts[1] (94 % starch and 6 % salts)					0·05 g. ,,
Cane sugar	...	...	...	...	0·1 g. ,,

was employed, to which various additions of butter-fat (vitamin A) yeast-extract (vitamin B) and lemon juice freed from citric acid (vitamin C) were made as required, 2 cc. of liquid in all being added, and then 3 cc. of $1\frac{1}{2}$ % agar. The caseinogen had been extracted with alcohol and light petroleum or in some cases heated at 120° for many hours and repeatedly stirred.

RESULTS.

The experiments were directed in the first instance to ascertaining which, if any, of the three recognised accessory food factors are necessary for the growth and metamorphoses of these insects.

The results, which are summarised below, showed definitely that:

(1) in the presence of butter-fat and yeast extract (vitamins A and B) growth occurred irrespective of the presence or absence of 1 cc. of lemon juice (vitamin C);

(2) in the presence of butter-fat (vitamin A) growth occurred in presence of 1 cc. of yeast extract (vitamin B), but not in its absence, this result being independent of the presence or absence of lemon juice;

(3) in the presence of 1 cc. of yeast extract (vitamin B), with or without 1 cc. of lemon juice (vitamin C), growth occurred in the presence of two drops of butter-fat, but not in its absence.

In the following summary only results of experiments in which the tubes remained sterile are quoted.

Additions to basal diet

	Butter-fat	Yeast extract	Lemon juice	Tubes inoculated	Eggs hatched	Pupae formed	Flies emerged
1	−	+	+	10	100	0	0
2	+	+	−	11	115	92	63
3	+	−	+	10	57	0	0
4	+	+	+	10	68	48	42
5		+	−	2	19	0	0
6	−	−	−	2	16	0	0

[1] As used in this laboratory for rats.

Control experiments were carried out to ascertain whether the yeast extract alone was capable of supplying sufficient nitrogen for the growth of the larvae, but with entirely negative results. This was done by omitting the caseinogen from mixtures 2 and 4, which gave positive results in its presence. Even the addition of 2 cc. of the yeast extract gave entirely negative results in the absence of the caseinogen.

Other sources of vitamin B.

As flies of this group are specially associated with yeast, which probably forms the chief food of the larvae in their natural condition, it was thought that in order to prove the necessity for vitamin *B*, it would be essential to show that it could be supplied from other sources.

The first experiments were made with milk, but without success. Recourse was then had to wheat germ, the alcoholic (80 %) extract of which is known to contain a considerable amount of vitamin *B*.

50 g. of wheat germ were heated on the water-bath for one hour with 200 cc. of 80 % alcohol, pressed out and the clear liquid evaporated to dryness at 50°. The residue was shaken up with 50 cc. of water and filtered, the clear filtrate being used. This contained 10·47 g. total solids, 0·21 g. mineral matter, and 0·196 g. N per 100 cc.

Tubes were made up with 2, 1 and 0·5 cc. of this extract in addition to the usual basal diet of caseinogen, starch, salts, cane sugar and butter-fat. Some difficulty was experienced in adjusting the stiffness of the agar jelly, but finally 0·5 cc. H_2O and 3 cc. of 1·25 % agar were added to each tube, so that the total volume was 5·5 cc. as shown below.

	Caseinogen, starch and salts g.	Cane sugar g.	Butter-fat	Wheat germ extract cc.	H_2O cc.	1·25 % agar cc.
7	0·2	0·1	2 drops	2	0·5	3
8	0·2	0·1	,,	1	1·5	3
9	0·2	0·1	..	0·5	2·0	3
	Starch and salts					
10	0·05	0·1	,,	2	0·5	3

The following are the summarised results of three quite independent and consistent experiments:

	Wheat germ extract	Eggs known to have hatched	Pupae	Flies
11	2	34	18	9
12	1	56	24	13
13	0·5	39	0	0
14	2 cc. as sole source of N	9	0	0

The experiments were somewhat hindered by the hot weather prevailing at the time, which caused unduly rapid drying of the tubes, but they show quite conclusively that, like yeast extract, wheat germ extract is capable of supplying some factor necessary for the growth and metamorphosis of the fly.

It is at the same time obvious that the wheat germ extract (1 cc. of which corresponds with 1 g. of wheat germ) is much less efficacious than the yeast extract (1 cc. of which corresponds with 0·5 g. yeast) and this is in accordance with what is known of the vitamin B content of these two materials, 1 g. of yeast being approximately equivalent to 3·6 of wheat germ.

Taken together the experiments with yeast extract and wheat germ extract leave little doubt that these insects require vitamin B for their development, thus confirming Northrop's view. Experiments are in progress on the amount of yeast extract required and the effect of varying concentrations on the duration of the period of development.

THE FAT-SOLUBLE FACTOR (VITAMIN A).

The striking fact that in presence of 1 cc. yeast extract growth occurred in the presence of two drops of clarified butter-fat, but not in its absence, may be taken to show either that a fat of some kind is necessary or that the fat-soluble factor is required or that both are required. To decide between these possibilities it is necessary to ascertain whether or not growth occurs in presence of a fat devoid of vitamin A. Attempts of this kind are being made, but no definite result has yet been obtained. The preliminary results however indicate that the fly is certainly able to develop in the presence of very small amounts of the vitamin if not in its entire absence.

SUMMARY.

The complete development of *Drosophila* requires the presence of vitamin B but not of vitamin C.

REFERENCES.

Baumberger (1919). *J. Exp. Zoology*, 28, No. 1.
Loeb (1915). *J. Biol. Chem.* 23, 431.
Loeb and Northrop (1916). *J. Biol. Chem.* 27, 309.
Northrop (1917). *J. Biol. Chem.* 30, 181.

XXII. STUDIES ON CARBONIC ACID COMPOUNDS AND HYDROGEN ION ACTIVITIES IN BLOOD AND SALT SOLUTIONS.

A CONTRIBUTION TO THE THEORY OF THE EQUATION OF LAWRENCE J. HENDERSON AND K. A. HASSELBALCH.

By ERIK JOHAN WARBURG.

From the Laboratory of the Finsen Institute (Copenhagen).

(*Received March 3rd, 1922.*)

INTRODUCTION

K. A. HASSELBALCH in 1916 introduced a formula with the help of which it is possible to determine the hydrogen ion concentration of the blood from the amount of its combined and dissolved carbonic acid, basing his equation upon the work of L. J. Henderson carried out from 1908–1910. Henderson employed the then generally known equation for the first dissociation of carbonic acid, namely,

$$\gamma \frac{C_H \cdot C_{HCO'_3}}{C_{H_2CO_3}} = K_1.$$

Applied to the determination of the $C_{\text{H}^\cdot}$ of the blood he got

$$C_{\text{H}^\cdot} = \frac{C_{\text{H}_2\text{CO}_3}}{C_{\text{HCO}'_3}} \times \frac{K_1}{\gamma}.$$

By substituting the best known values at the time for combined and dissolved CO_2, for the first dissociation constant of carbonic acid and for the dissociation of bicarbonate, he found a value for $C_{\text{H}^\cdot}$ which corresponded fairly well to the reaction of the blood as it was thought to be at that time. Henderson [1909–1910] concluded from his calculations that the major portion of combined carbonic acid in the blood was in the form of bicarbonate.

Hasselbalch [1916] determined in the same sample of blood (and serum) $C_{\text{H}^\cdot}$, $C_{\text{HCO}'_3}$ and $C_{\text{H}_2\text{CO}_3}$, and found that $C_{\text{H}^\cdot}$ was the same as calculated by substituting in the equation the value of $\frac{K_1}{\gamma}$ for a solution of pure sodium bicarbonate. He thought he could infer from this that all the combined carbonic acid of the blood was in the form of bicarbonate and that the $C_{\text{H}^\cdot}$ of the blood could be estimated by means of the equation when $C_{\text{H}_2\text{CO}_3}$ and $C_{\text{HCO}'_3}$ were known.

As investigations on kindred subjects progressed it became clear the theory could only be recognised as a first approximation—a view which a knowledge of S. P. L. Sörensen's *Studies on Proteins* [1915–1917] strongly supported—and the question immediately arose, how is it possible the calculated and experimental values of $C_{\text{H}^\cdot}$ can coincide when the suppositions of the calculation are inadequate?

It further became evident from experience gradually accumulated regarding the $C_{\text{H}^\cdot}$ measurements by the Hasselbalch-Michaelis principle (shaking method combined with minimal immersion), that experimental difficulties were present which had not been entirely overcome, since measurements with various electrodes gave constant differences. And as the Henderson-Hasselbalch equation seemed to be destined to play an important rôle in physiology and pathology it was necessary to subject both the measurements and the theories which formed the basis of the equation to a further test.

The present work therefore has a double object, viz.:

(1) to investigate whether the Henderson-Hasselbalch equation gives correct results when the measurements are carried out by another method than that of Hasselbalch;

(2) to develop a theory connecting the amount of carbonic acid in the blood with its reaction.

In connection with this a number of points have naturally arisen concerning the distribution of ions between blood cells and serum, and a series of experiments on the reactions in simple watery bicarbonate solutions in equilibrium with different tensions of CO_2 have been made.

CHAPTER I

SOME GENERAL PHYSICO-CHEMICAL CONCEPTIONS RELATING TO THE
NEW ELECTROLYTIC DISSOCIATION THEORY OF BJERRUM.

Osmolar Concentration.

By the molecular or molar concentration of a solute in a solution we understand the number of gram-molecules of solute in a litre of the solution. We
also speak of the molar concentration of the ions and molecular aggregates,
looking upon the particles in question as if they were molecules when determining the gram-molar weight. In physiology a distinction has often been
made between molecular and molar concentrations (cf. Hamburger [1902],
Hedin [1915], Ege [1920]), by imagining all solutes to exist in molecular form
(thus Na·Cl′ as NaCl) when reckoning the total molecular concentration—a
conception which is also made use of in this work in stating the molecular
concentration—while in estimating the molar concentration the solute is considered to be in its actual state of aggregation. This distinction has been of
especial significance in physiology for questions of osmotic pressure as it is
the total number of dissolved particles that is the determining factor (corrected
for secondary effects), but as this terminology is not recognised in physical
chemistry it must be abandoned. In place of the molar concentration of
some physiologists I have therefore employed the expression *osmolar concentration.*

Activity of a Solute.

In 1908 G. N. Lewis formed the conception activity. As regards the
thermodynamical definition of this conception I shall confine myself to referring
to the article of Lewis [1908], however I shall here quote slightly modified
the very perspicuous explanation which Lewis gives in the paper mentioned.

The activity of an ideal gas is equal to the concentration of the gas.

The activity of a solute which forms an ideal solution is equal to its
concentration[1].

If a substance has the same activity in two phases, the substance will
not by itself go from the one phase to the other.

If the activity of a substance is greater in one phase than in the other, the
substance will go from the first phase to the second as soon as it is possible.

By the *activity coefficient* of a solute we mean the factor the concentration of the solute must be multiplied by to give its activity. For
uncharged mols in aqueous solution the coefficient will nearly always be
greater than unity as it is then merely an expression for what we call depression
of solubility, salting out effects, action of salt, etc. For charged mols, "ions,"
in weak aqueous solutions the coefficient will be less than unity, in strong
solutions often greater than unity, as will shortly be shown in discussing
electrolytic dissociation. Following Bjerrum [1916, 1918, 1919], we distinguish

[1] Lewis writes ...is by constant temperature and pressure proportional to the concentration.

between the apparent and true activity coefficients, the apparent being the relation between the activity of the desolvated solute and the concentration of the solvated solute, while the true coefficient eliminates solvatation effects. In so far as we do not know whether a solute is solvated we will use the expression *apparent activity coefficient*.

Solubility and Activity of Gases.

In the case of a solution of *the ordinary gases* at room temperature (and low pressure), the activity of the gas can be estimated by determining its tension in a gaseous phase in equilibrium with the solution. This will be understood from what is said later. The gases mentioned in this work, oxygen, hydrogen, carbon dioxide and nitrogen, for the ranges of pressure and temperature used, can in practically every respect be regarded as ideal[1]. The tension of such a gas will thus be a direct measure of its activity. For instance if Henry's law applies, the amount of gas which is *dissolved* in a solution in a state of equilibrium will be proportional to its pressure. The factor which expresses the amount of gas dissolved at a given pressure is called the *absorption coefficient*. Bunsen's absorption coefficient is used here which is called a, and which gives the number of cc. of gas reduced to $0°$ C. and 760 mm. Hg, that are dissolved in 1 cc. of solution at a pressure of 760 mm. of the gas itself. At a pressure of P mm. Hg, a cc. of gas will then be dissolved in 1 cc. of the solution

$$\frac{Pa}{760} = a \dots\dots\dots(1)$$

That we use Bunsen's absorption coefficient and not Ostwald's, is because we want to estimate the concentration of the gas in the solution and the former way of expressing it is most suitable for this purpose. A gram-molecule of an ideal gas at 0° C. and 760 mm. Hg occupies a volume of *22393* cc. A solution containing a cc. of gas (0° C. and 760 mm. Hg) in 1 cc. will contain 1000 a cc. per litre and the gas will therefore have a concentration of

$$\frac{1000a}{22393} = 0.044656 \times a = C_{\text{ideal gas}}. \dots\dots\dots(2)$$

Expressing the dissolved gas in volumes per cent. (Vols. %) this will be equal to 100 a and the concentration will be

$$0.00044656 \times (100\,a) = C_{\text{ideal gas}}. \dots\dots\dots(3)$$

The constant 22393 is valid for an ideal gas; for CO_2 the corresponding constant is *22263*[2]. The difference between these two values is a little over $\frac{1}{2}$ % (the deviation of CO_2 from Avogadro's law at 0° C. and 760 mm. Hg). The last constant is of course used for the calculation of the concentration of CO_2.

For *carbon dioxide* therefore we have the corresponding equations

$$0.044917 \times a = C_{CO_2} \dots\dots\dots(4)$$

and

$$0.00044917 \times (100\,a) = C_{CO_2}. \dots\dots\dots(5)$$

[1] An ideal gas obeys the "gas laws."

[2] These constants are calculated according to F. W. Küster, *Logarithmische Rechentafeln für Chemiker u.s.w.* 19th edition, Leipzig, p. 40, 1918.

If we desire to express the concentration in terms of the absorption coefficient and the tension we have, for an ideal gas,

$$Pa \times 0\cdot000058756 = C_{\text{ideal gas}} \quad \dots\dots\dots\dots\dots\dots(6)$$

and for CO_2

$$Pa \times 0\cdot000059101 = C_{CO_2}. \quad \dots\dots\dots\dots\dots(7)$$

Activity Coefficients and Depression of Solubility of Gases.

We will now consider the value a a little more closely. Equation (1) $\frac{Pa}{760} = a$ is valid without exception for pure solvents. In this work we shall assume the same is the case for all solutions if the correct absorption coefficient is used but it must be mentioned there may be small deviations in heterogeneous solutions. If a gas dissolves in a pure solvent,

$$C_{\text{gas}} = A_{\text{gas}} \text{ (Activity of the gas). } \dots\dots\dots\dots(8)$$

If other solutes are already present in the solvent a correction must be introduced. In such solutions the absorption coefficient is

$$a_{\text{corrected}} = \frac{100 - Y}{100} \times a \quad \dots\dots\dots\dots\dots(9)$$

and the concentration of the gas will then be

$$P \frac{100 - Y}{100} a \times k = C_{\text{gas}}, \quad \dots\dots\dots\dots\dots(10)$$

while the activity of the gas is

$$P \times a \times k = A_{\text{gas}} \dots\dots\dots\dots\dots\dots(11)$$

and

$$C_{\text{gas}} \times \frac{100}{100 - Y} = A_{\text{gas}}. \quad \dots\dots\dots\dots\dots(12)$$

Y is called the *relative depression of solubility*[1],

$\frac{100 - Y}{100}$ is called the *relative absorption coefficient*, and

$\frac{100}{100 - Y}$ is the *apparent activity coefficient* of the gas.

It will be noticed that the apparent activity coefficient is the reciprocal of the relative absorption coefficient. The apparent activity coefficient for a gas is called F_a (gas), the value a is reserved for the absorption coefficient in the pure solvent.

For watery solutions the relative absorption coefficient has been determined in a large number of experiments, in different ways, for numerous gases, and it has been shown that its value in a given solution is about the same for all gases.

The relative absorption coefficient has also been determined for various solid substances which we have reason to think do not unite with other solutes, by estimating the partition of the substance between the solution and a liquid which does not mix with the latter and in which the solute under investigation may be assumed to form a "true" solution. It has been found the same laws apply as for gases.

[1] This applies to homogeneous solutions, whilst the symbol Ψ is used for this quantity in a heterogeneous solution.

Lastly it has been discovered that the relative absorption coefficient, within the experimental error, is independent of temperature. The solutes that most markedly depress solubility in water are the electrolytes, but non-electrolytes can also have a large effect, e.g. cane sugar. There is an extensive literature on the solubility of gases in water which is exhaustively dealt with in Rothmund's monograph [1907] and cited in Landolt and Börnstein's tables, to which the reader is referred.

Bohr and Bock's [1891] values for the absorption coefficients have been used.

Table I. *The Absorption Coefficients of CO_2 (Chr. Bohr and J. C. Bock).*

Temp.	α	Temp.	α
15	1·019	20	0·878
16	0·985	21	0·854
17	0·956	22	0·829
18	0·928	37	0·567
19	0·902	38	0·555

Reduction to 0° C. and 760 mm. Hg has been done in the usual way according to Boyle's and Gay Lussac's laws.

Mass Action Law.

We have now arrived at the point where it will be advisable to discuss the question of the chemical equilibrium which takes place between substances which react with one another in solution.

When the mass action law is expressed as a formula and the apparent activity of the mol B is designated by a_B the following holds good:

The reaction[1] $bB + cC + dD \rightleftarrows mM + nN + oO$ proceeds from left to right with the rate

$$a_B^b \cdot a_C^c \cdot a_D^d \cdot k_\mathrm{I}. \quad \ldots\ldots\ldots\ldots\ldots\ldots(13)$$

and from right to left with the rate

$$a_M^m \cdot a_N^n \cdot a_O^o \cdot k_\mathrm{II}, \quad \ldots\ldots\ldots\ldots\ldots(14)$$

therefore $\quad a_B^b \cdot a_C^c \cdot a_D^d \cdot k_\mathrm{I} = a_M^m \cdot a_N^n \cdot a_O^o \cdot k_\mathrm{II} \quad \ldots\ldots\ldots\ldots(15)$

and $\qquad \dfrac{a_M^m \cdot a_N^n \cdot a_O^o}{a_B^b \cdot a_C^c \cdot a_D^d} = K. \quad \ldots\ldots\ldots\ldots\ldots(16)$

The expression for a reaction like the following (a binary dissociation) is especially simple, $B \rightleftarrows M + N$, namely

$$\dfrac{a_M \cdot a_N}{a_B} = K. \quad \ldots\ldots\ldots\ldots\ldots(17)$$

---

[1] That is to say b mols of B react with c mols of C and d mols of D, giving m mols of $M + n$ mols of $N + o$ mols of O. But the reaction can also be expressed thus—m mols of M react with n mols of N and o mols of O giving b mols of $B + c$ mols of $C + d$ mols of D.

If we wish to express the law for the reacting (solvated) mols in terms of their concentrations we get

$$\frac{C_M \cdot F_{a(M)} \cdot C_N \cdot F_{a(N)}}{C_B \cdot F_{a(B)}} = K, \quad\quad\quad\quad\quad (18)$$

in which the activity of a mol is defined as the concentration multiplied by the apparent activity coefficient. If the apparent activity coefficient is 1, or if

$$\frac{F_{a(M)} \cdot F_{a(N)}}{F_{a(B)}} = 1,$$

(18) reduces to

$$\frac{C_M \cdot C_N}{C_B} = K. \quad\quad\quad\quad\quad (19)$$

When the reaction is in equilibrium and when it is dissociated so much that $xC_{B(T)}$ mols of M and N are formed, (19) then becomes

$$\frac{x^2}{1-x} C_{B(T)} = K, \quad\quad\quad\quad\quad (20)$$

where $C_{B(T)}$ is the concentration B would have if the reaction proceeded completely from right to left.

Osmotic Pressure.

In watery solutions which only contain non-electrolytes the "ideal" conditions for the osmotic pressure are very nearly fulfilled, particularly at higher temperatures, but here also certain corrections must be introduced for hydration (see Findlay [1913]) which will be more pronounced if several solutes are present simultaneously in the solution some of which appropriate considerable quantities of the solvent.

In solutions of electrolytes the conditions will be still more complicated when we consider the osmotic activity of the charged particles (ions). It has until recently been generally accepted, though not without hesitation, that the osmotic activity of the ions was the same function of their concentration as in the case of non-electrolytes, but latterly it has been appreciated that the electrical forces between the ions must diminish their osmotic activity.

Milner (quoted from Bjerrum [1916, 1918, 1919]) has attempted to calculate the magnitude of this effect, which Bjerrum calls the *Milner effect*, without, according to Bjerrum, having actually succeeded in doing so, but as regards the questions dealt with here this is of no importance because we accept the empirical formula given by Bjerrum which will be discussed later on. But first we must recall some of the principal points in Arrhenius' theory of electrolytic dissociation.

Electrolytic Dissociation.

In the following sketch of the fundamental conceptions of the theory of electrolytic dissociation aqueous solutions are exclusively referred to.

Water dissociates to a slight extent into hydrogen and hydroxyl ions,

$$HOH = H^{\cdot} + (OH)'. \quad\quad\quad\quad\quad (21)$$

According to the law of mass action,

$$\frac{a_{\mathrm{H}^{\cdot}} \times a_{(\mathrm{OH})'}}{a_{\mathrm{H_2O}}} = k. \quad\dots\dots\dots\dots\dots\dots\dots(22)$$

In weak solutions the following equation is very nearly true,

$$a_{\mathrm{H_2O}} = K \quad\dots\dots\dots\dots\dots\dots\dots\dots(23)$$

From (22) and (23) we get

$$a_{\mathrm{H}^{\cdot}} \times a_{(\mathrm{OH})'} = K_w, \quad\dots\dots\dots\dots\dots\dots(24)$$

where K_w is of the order of 10^{-14}. [See also Lewis, 1916, **1**, p. 325, and S. P. L. Sörensen, 1912, **12**, p. 400.]

When an electrolyte is dissolved in water it will be split up—dissociated— and positively and negatively charged ions will be formed.

The ions carry one, two or more electrostatic units and the dissociation can therefore take place according to the following schemes:

$$M_{\mathrm{I}} A_{\mathrm{I}} \rightleftharpoons M^{\cdot}{}_{\mathrm{I}} + A'{}_{\mathrm{I}} \quad\dots\dots\dots\dots\dots\dots\dots\dots(25)$$
(electrolyte)   (cation) (anion)

$$M_{\mathrm{I}} M_{\mathrm{I}} A_{\mathrm{II}} \rightleftharpoons M^{\cdot}{}_{\mathrm{I}} + M^{\cdot}{}_{\mathrm{I}} + A''{}_{\mathrm{II}} \quad\dots\dots\dots\dots\dots(26)$$

$$M_{\mathrm{I}} M_{\mathrm{I}} M_{\mathrm{I}} A_{\mathrm{III}} \rightleftharpoons M^{\cdot}{}_{\mathrm{I}} + M^{\cdot}{}_{\mathrm{I}} + M^{\cdot}{}_{\mathrm{I}} + A'''{}_{\mathrm{III}} \quad\dots\dots(27)$$

or $\quad\quad M_{\mathrm{I}} M_{\mathrm{I}} M_{\mathrm{I}} A_{\mathrm{III}} \rightleftharpoons M^{\cdot}{}_{\mathrm{I}} + M_{\mathrm{I}} M_{\mathrm{I}} A'{}_{\mathrm{III}} \quad\dots\dots\dots\dots(28)$

$$M_{\mathrm{I}} M_{\mathrm{I}} A'{}_{\mathrm{III}} \rightleftharpoons M^{\cdot}{}_{\mathrm{I}} + M_{\mathrm{I}} A''{}_{\mathrm{III}} \quad\dots\dots\dots\dots\dots(29)$$

or $\quad\quad M_{\mathrm{II}} A_{\mathrm{I}} A_{\mathrm{I}} \rightleftharpoons \dot{M}_{\mathrm{II}} A^{\cdot}{}_{\mathrm{I}} + A'{}_{\mathrm{I}} \quad\dots\dots\dots\dots\dots(30)$

$$M_{\mathrm{II}} A^{\cdot}{}_{\mathrm{I}} \rightleftharpoons M^{\cdot\cdot}{}_{\mathrm{II}} + A'{}_{\mathrm{I}} \quad\dots\dots\dots\dots\dots\dots(31)$$

and so on.

$M^{\cdot}{}_{\mathrm{I}}$ is a monovalent cation, $M^{\cdot\cdot}{}_{\mathrm{II}}$ a divalent cation, etc.; $A'{}_{\mathrm{I}}$ and $A''{}_{\mathrm{II}}$ respectively monovalent and divalent anions. There is reason to distinguish between three kinds of electrolytes.

Electrolytes can be divided into:

Totally dissociated electrolytes $\begin{cases} \text{Alkali salts and similar salts} \\ \text{Strong acids} \\ \quad,, \quad \text{bases} \end{cases}$

Electrolytes the dissociation of which varies with the concentration $\begin{cases} \text{Dissociation varying} \\ \quad \text{with the reaction} \begin{cases} \text{Weak acids} \\ \quad,, \quad \text{bases} \\ \text{Ampholytes} \end{cases} \\ \text{Complex salts} \end{cases}$

The acids dissociate according to the following schemes:

For a monovalent acid

$$HA_{\mathrm{I}} \rightleftharpoons H^{\cdot} + A'{}_{\mathrm{I}}, \quad\dots\dots\dots\dots\dots(32)$$

and for a divalent acid

$$H H A_{\mathrm{II}} \rightleftharpoons H^{\cdot} + H A'{}_{\mathrm{II}} \text{ (partial dissociation)}, \quad\dots\dots(33)$$

$$H A'{}_{\mathrm{II}} \rightleftharpoons H^{\cdot} + A''{}_{\mathrm{II}}. \quad\dots\dots\dots\dots\dots\dots(34)$$

The bases dissociate as follows:

$$M_1 (OH) \rightleftarrows M^{\cdot}{}_I + (OH)', \quad\dots\dots\dots\dots\dots\dots\dots(35)$$

$$M_{II} (OH) (OH) \rightleftarrows M^{\cdot\cdot}{}_{II} + (OH)' + (OH)'. \quad\dots\dots(36)$$

For an ampholyte which can split off one hydrogen ion and one hydroxyl ion per molecule the following hold:

for the hydrogen ion dissociation

$$H R (OH) \rightleftarrows H^{\cdot} + R (OH)', \quad\dots\dots\dots\dots\dots(37)$$

for the hydroxyl ion dissociation

$$H R (OH) \rightleftarrows H R^{\cdot} + (OH)', \dots\dots\dots\dots\dots(38)$$

and for the simultaneous dissociation of $H^{\cdot}$ and $(OH)'$

$$H R (OH) \rightleftarrows H^{\cdot} + (OH)' + {}'R^{\cdot}. \quad\dots\dots\dots\dots(39)$$

Although the dissociation of weak acids and bases, and ampholytes follows the mass action law, difficulties are encountered with the strong acids and bases which compel us to believe that they are completely dissociated (*i.e.* the reaction progresses completely from left to right), as will now he more fully explained in speaking of salts.

"Strong" Electrolytes.

The question which has caused most trouble in connection with the electrolytic dissociation theory is whether the "strong" electrolytes follow the "mass action law."

We will not discuss further the extremely wide development the subject has undergone; in all the larger textbooks of physical chemistry it is dealt with, and in Arrhenius' [1912] *Theories of Solution* a whole chapter is reserved for it to which the reader is referred. Suffice it to say that in whatever way the dissociation is estimated—whether by depression of the freezing point or by conductivity methods, the mass action law is not fulfilled. In using the method involving the depression of the freezing point which is proportional to the osmotic pressure it was assumed the depression caused by ions and non-electrolytes was the same function of their concentration. As an example of such an estimation the dissociation (25)

$$M_I A_I \rightleftarrows M^{\cdot}{}_I + A'{}_I$$

may be considered. The solution has a freezing point depression which is i[1] times greater than a non-electrolyte of the concentration of $M_I A_I$ (if the reaction proceeded completely from right to left). If we regard the freezing point depression as proportional to the osmolar concentration without corrections for the Milner effect, the following equations will apply:

$(C_{M_I A_I (T)}$ is the concentration of the salt when the reaction proceeds completely from right to left.)

$$C_{M^{\cdot}{}_I} = (i - 1) C_{M_I A_I (T)} = C_{A'{}_I} \quad\dots\dots\dots\dots(40)$$

and
$$C_{M_I A_I} = (2 - i) C_{M_I A_I (T)}; \quad\dots\dots\dots\dots\dots(41)$$

[1] i is van 't Hoff's coefficient.

substituting in (19) we get

$$\frac{(i-1)^2}{2-i} \cdot C_{M_I A_I (T)} = K. \qquad \qquad \text{.........................(42)}$$

In calculations made from conductivity experiments it was assumed the velocity of the ions was independent of the other ions present in the solution. It was thought therefore that

$$\frac{\text{the equivalent conductivity at the concentration } C_{M_1 A_1 (T)}}{\text{the equivalent conductivity at infinite dilution}} = \frac{\mu_v}{\mu_\infty} = f_\mu$$

was equal to the value x in (20) and it was subsequently substituted in the equation, but corrections for altered viscosity were often introduced.

As it appeared to be impossible to obtain any concordance between the experiments and the requirements of the mass action law—whether the dissociation was estimated by the methods described or in any other way—either the assumptions which formed the basis of the calculation were erroneous, or it must be concluded the strong electrolytes do not follow the mass action law, a conclusion which under these conditions would be destructive to the law. .

It has for many years been known that the dissociation determined by conductivity methods (see for example Jahn [1900]) was not quite accurate while that determined by depression of the freezing point has as a rule been regarded with greater confidence, at any rate by physiologists, although there has been some suspicion about it also.

The problem of elucidating the nature of the complications mentioned seems first to have been mastered by the Dane, Niels Bjerrum [1909, 1916, 1918, 1919].

Dissociation Theory of Bjerrum.

Bjerrum's theory can aptly be called *the theory of the complete dissociation of the strong electrolytes.*

From a number of experiments he carried out in 1906 on the chromium salts Bjerrum came to the conclusion the strong electrolytes were completely dissociated. It was found that they possessed the colour of the ions both in weak and in strong solutions if complex combinations were not formed (Bjerrum)[1]. The fact that the colour of a strong electrolyte was independent of its concentration led him in 1909 to formulate the hypothesis that the strong electrolytes must be taken to be completely dissociated into their ions in solution provided complex compounds were not produced.

In 1919 Bjerrum developed this theory further and showed that a large number of the difficulties of the electrolytic dissociation theory disappeared on his supposition.

It can thus be said that Bjerrum showed in some very characteristic cases

---

[1] I shall follow extremely closely Bjerrum's exposition which to a large extent is quoted word for word.

that it was not necessary to attribute any catalytic effect to the undissociated acid (in systems where hydrogen ions were the catalyst), which was the case when accounting for the dissociation determined by conductivity.

Bjerrum and Gjaldbaek [1919] succeeded in getting agreement between the calculated and experimental reaction constants for calcium carbonate and their investigations will be briefly discussed later in this work.

They also succeeded in calculating the potential for the concentration cell Hg/HgCl, $0 \cdot 1n$ KCl/$0 \cdot 01n$ KCl, HgCl/Hg; after the elimination of the diffusion potential the potential was

$$\epsilon = 0 \cdot 0548 \text{ volt.}$$

Reckoned with the Nernst formula

$$\epsilon = 0 \cdot 0591 \log \frac{c_2}{c_1} \text{ volt,} \qquad \dots\dots\dots\dots\dots\dots(43)$$

by actual concentration $\epsilon = 0 \cdot 0591$ volt,

(corrected) by the degree of conductivity

$$\epsilon = 0 \cdot 0569 \text{ volt,}$$

(corrected) by the activity coefficient

$$\epsilon = 0 \cdot 0553 \text{ volt.}$$

The following is a translation of a part of Bjerrum's paper in the reports of the Nobel Institute, those sections only being included which are necessary for the understanding of the theory, as it will be indispensable for those wishing to work at the subject to consult the original.

"In the year 1887 Arrhenius enounced his famous hypothesis according to which the ions in a solution of electrolytes are present in the solution in the free state. During the 26 years passed since then this hypothesis has become an indispensable part of physics and chemistry, and even to-day it acts as a fruitful working hypothesis thus showing the powerful richness and force of the original conception. In the course of time, however, several difficulties have attended this hypothesis especially in explaining the properties of strong electrolytes. But after the work of later years these difficulties appear to vanish when considering the effects that must be produced on the properties of the solutions of electrolytes by the electric forces between the ions. The greatness of these effects may be indicated by means of coefficients expressing the relation between the real value of the property in question and the value the property should have held supposing the electric forces did not act between the ions. Thus the activity-coefficient f_a indicates the effect of the interionic forces on the activity of the ions; the conductivity-coefficient f_μ indicates the influence of the interionic forces on the conductivity; the osmotic coefficient f_0 indicates the influence of the interionic forces on the osmotic pressure etc.

If we suppose that the strong electrolytes are completely dissociated into ions we may determine the value of the osmotic coefficient by means of freezing-point determinations. From Noyes's and Falk's excellent exposition it thus results that the osmotic coefficient of all electrolytes with monovalent ions may, with fair approximation, be rendered by the formula

$$1 - f_0 = k \sqrt[3]{c}, \quad \dots\dots\dots\dots\dots\dots\dots(44)$$

in which k varies between 0·146 and 0·225 the mean being 0·17.

This formula, no doubt, does not agree with Milner's values, but within the limits of concentration where both experiments and Milner's calculation are fairly reliable, viz. $0\cdot01m - 0\cdot1m$, the agreement is a rather fair one. This fact must be looked on as a support of the assumption that the strong electrolytes are completely dissociated or, at any rate, are much more dissociated than originally assumed.

When the formula of the osmotic coefficient is known, the formula of the activity-coefficient may be deduced by means of the following thermodynamic relation:

$$f_0 + c\,\frac{\delta f_0}{\delta c} = 1 + c\,\frac{\delta ln f_a}{\delta c}. \quad \dots\dots\dots\dots\dots\dots(45)$$

If this be combined with (44) the following expression is obtained:

$$ln f_a = -\,4k\,\sqrt[3]{c^1}. \quad \dots\dots\dots\dots\dots\dots\dots(46)$$

When trying to determine experimentally the value of the activity-coefficient f_a for an ion, various difficulties are thrown in one's way.

First, it is difficult to make sure that the electrolyte is really completely ionised or, if that is not the case, to determine its degree of dissociation. If the molar conductivity of the electrolyte increases with decreasing concentration according to Ostwald-Walden's valence rule it will however be natural, as I have developed formerly, to presume complete dissociation. In many cases this point may be investigated by means of spectro-photometric measurements the colour of the ions being, to a rather high degree, independent of the interionic forces and, on the other hand, greatly changing when the ions combine chemically together.

Secondly, it is not possible, even if the ion concentration is known, to state with certainty what the activity of the ion would have been if the interionic forces did not exist. This difficulty particularly manifests itself in concentrated solutions. As the measure of the activity one should, most likely, take neither the mol number per litre nor the mol number per 1000 g. of water but what is known as the mol fraction,

$$x = \frac{n}{n+n'}, \quad \dots\dots\dots\dots\dots\dots\dots(47)$$

where n is the mol number of the solute and n' the mol number of the solvent. We shall designate this mol fraction as x-concentration. As to the justification for putting the activity equal to the x-concentration we may for instance refer to the excellent exposition of ideal solutions in Alex. Findlay's book: The osmotic pressure.

Thirdly, the water content of the ions in the solution must be considered. When the ions form hydrates in solution the consequence is that the x-con-

---

[1] Table 11 gives the corresponding values of the constant factors in equations (44) and (46).

Table 11.

$\dfrac{1-f_0}{\sqrt[3]{c}}$	$\dfrac{ln f_a}{2\cdot3026\sqrt[3]{c}}$
0·146	0·25
0·17	0·30
0·225	0·39

centrations of the ions will be smaller than they would have been without the hydration. The effect of hydration is well known and frequently discussed. There is, however, another effect of hydration which has not as yet been taken into account; although it is rather of greater importance than the former effect. If we want to determine the activity of an ion by means of potential measurements, say with metal or hydrogen electrodes or with mercurous chloride electrodes, *we obtain for a hydrated ion by means of the ordinary Nernst formula not a measure of the activity of the ion itself but only of the activity of the ion without water. If the ion contains m H_2O we should to obtain the measure of the activity of the ion-hydrate multiply the activity of the ion without water, which we may call A, by the activity of the water to the mth power.* We have a good measure of the activity of the water in the vapour pressure of the solution p divided by the vapour pressure of pure water p_0. So the activity of the actual ion may be put at

$$a = A \left(\frac{p}{p_0}\right)^m. \quad \ldots\ldots\ldots\ldots\ldots\ldots\ldots\ldots\ldots\ldots(48)$$

Strange to say no account has till now been taken of this correction in spite of the fact that in the electrometric determination of the hydrogen-ion-concentration, so frequently used of late, it is of considerable importance, the hydrogen-ion being very strongly hydrated. The measurement of the potential of a hydrogen-electrode in a solution does not directly give us either the hydrogen-ion-concentration or the hydrogen-ion-activity, but only a calculative quantity which may be designated as the activity of the dehydrated hydrogen-ion. It is only for want of due consideration of this fact that the conclusion that the hydrogen-ion-activity increases by plentiful addition of neutral salts to hydrochloric and other strong acids has been arrived at (comp. for instance Harned, *Journ. Amer. Chem. Soc.* **37**, 2460 (1915), **38**, 1986 (1916)). This correction should also be made in the case of activity determinations founded on the solubility of salts.

Fourthly, the exact determination of ion-activities in concentrated solutions is rendered difficult by the association of the water, which is of importance in calculating the x-concentration, and which is not properly known."

By making calculations based on the above Bjerrum has succeeded in estimating the amount of hydrate water of the ions.

F_a is the apparent activity coefficient and f_a the true activity coefficient. $- \log F_a$ is the apparent activity exponent.

In Fig. 1 which is reproduced here Bjerrum shows the relation existing between F_a and c for different degrees of hydration of the ions.

While it is the hydrogen ion concentration, as Bjerrum [1916, 1918] has shown for the catalytic action of hydrogen ions (esterification of organic acids), which is the determining factor, it is the apparent hydrogen ion activity which is the important factor in the mass action law relating to those chemical actions in which the hydrogen ions take part. The apparent hydrogen ion activity can be determined by the electrical method in the same way as the "hydrogen ion concentration" has hitherto been estimated.

In measuring the potential difference between a hydrogen-platinum elec-

[1] It may be noted that Bjerrum is using the symbols a and A with a significance differing from that which we have accepted in these articles.

trode and a solution, and a constant electrode we have the following form of
Nernst's equation,

$$p_{\mathrm{H}^{\cdot}} = - \log a_{\mathrm{H}^{\cdot}} = \frac{E - E_0}{0.0577 + 0.0002\,(t - 18°)}, \qquad\qquad(49)$$

E being the measured potential and E_0 a constant dependent upon the
electrode used for comparison. If a $0.1n$ KCl−HgCl electrode ($0.1n$ calomel
electrode) is used, S. P. L. Sörensen on the basis of conductivity experiments
at $18°$ obtained $E_0 = 0.3377$ volt (Sörensen). ....(50)

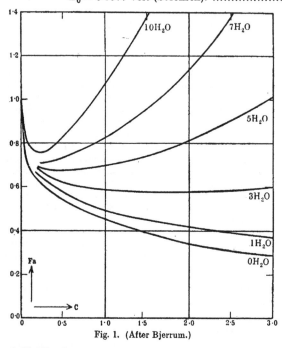

Fig. 1. (After Bjerrum.)

Bjerrum and Gjaldbaek [1919] from calculations based on the activity coefficient obtained $E_0 = 0.3348$ (Bjerrum)   ....(51)

from which E_0 (Sörensen) $- 0.0029 = E_0$ (Bjerrum)[1]. ....(52)

Both results are valid for a hydrogen pressure of $760 - f$ mm. Hg where f
is the pressure of saturated aqueous vapour at the temperature in question.
If the hydrogen pressure is different, e.g. P_{H_2}, the following value (c) must be
added to E,

$$c = \frac{0.0001983}{2}\,(273 + t°)\,\log\frac{P_{\mathrm{H}_2}}{760 - f}, \qquad\qquad(53)$$

where $t°$ is the temperature on the centigrade scale, P_{H_2} is the H_2 pressure
in the electrode, and f is the pressure of saturated aqueous vapour at $t°$.

[1] At $38°$ E_0 (Bjerrum) $= 0.3313$.

It will be observed that Bjerrum's E_0 is 2·9 millivolts less than Sörensen's. The corresponding difference in p_{H}· from 5·00 to 8·00 at room temperature or body temperature can with sufficient accuracy be expressed by

$$p_{\text{H}}· \text{ (Bjerrum)} = p_{\text{H}}· \text{ (Sörensen)} + 0·048. \quad\text{...............(54)}$$

It is thus easy to transfer the values from one scale to another.

As will be seen from Bjerrum's reasoning,

$$F_a \text{ (Ion) } C_{\text{Ion}} = a_{\text{Ion}}·\quad\text{...........................(55)}$$

In weak solutions, $\quad \log f_a \text{ (Ion)} = \log F_a \text{ (Ion)} \quad\text{.......................(56)}$

and the concentration of an ion can be calculated when its activity is known.

For hydrogen ions according to Bjerrum [1916, 1918] the following equation holds,

$$- \log f_a \text{ (H)}· = 0·2 \sqrt[3]{c}. \quad\text{...........................(57)}$$

For the bicarbonate ion k is not known but it will be shown later that k for sodium bicarbonate is about 0·46.

For an n-valent ion, according to Bjerrum and Gjaldbaek [1919, p. 79],

$$- \log f_a = nk \sqrt[3]{c}, \quad\text{...........................(58)}$$

where k has the same value as for a monovalent ion in the same solution, and where it varies within the same limits as in the case of monovalent ions. Table III, according to Bjerrum, gives the values of the deviation coefficients for KCl.

Table III.

Molec. conc.	f_0	f_a	$f_\mu = \dfrac{\mu_v}{\mu_\infty}$
0·001	0·985	0·943	0·979
0·01	0·963	0·882	0·941
0·1	0·932	0·762	0·861
1·0	0·854	0·552	0·755

Since $f_a \text{(KCl)} = \sqrt{f_a \text{(Cl')} f_a \text{(K·)}}$.

It is a matter of fact that all salts do not follow the same law of dissociation. Thus while the salts of the alkali metals and presumably also the majority of those of the alkaline earths are completely dissociated (Bjerrum and Gjaldbaek [1919], Gjaldbaek [1921]), this is not the case for many of the salts of the heavy metals. The latter form complex compounds, that is to say they are only partially dissociated—often only very slightly dissociated. The equilibrium between the ions and the undissociated salts is determined in a similar manner to that between hydrogen or hydroxyl ions and the electrolytes the dissociation of which varies with the reaction, and the remarks anent these can almost be directly applied to the complex salts (electrolytes the dissociation of which varies with the concentration) by substituting metal ions for hydrogen ions.

As it is by no means a matter of course that a salt is completely dissociated it will be one of the objects of the present work to investigate whether protein salts are complex or not. The working hypothesis that alkali albuminates and protein salts with Cl' and HCO'$_3$ are completely dissociated has been adopted throughout, and it has been investigated whether any facts following from this contradicted the hypothesis.

Complications arising from Heterogeneity.

The following equation holds for a heterogeneous solution,

$$\frac{C_{S_y}(100-Q)}{100} + \frac{C_{S_y}DQ}{100} = C_{S_t}, \qquad \qquad (59)$$

where Q is the volume of the internal phase expressed as a percentage of the total volume of the solution; C_{S_y} is the true concentration of the solute in the external phase, the possibility of an aggregation being neglected, and C_{S_i} is the concentration in the inner phase; D is a proportionality factor determined by

$$D = \frac{C_{S_i}}{C_{S_y}}, \qquad \qquad \qquad (60)$$

and C_{S_t} is the concentration of the solute in the whole system provided it is evenly distributed over the whole volume. We shall call C_{S_t} the *mean concentration.*

In biological chemistry only scant attention has been paid to the conditions expressed by (59) but there are a number of investigators who have been cognizant of them. Thus Hamburger [1902] was the first to recognise the importance of the volume of the disperse phase in experiments on the osmotic conditions of the blood corpuscles.

Rona [1910, 1913, 1, 2] and his collaborators also corrected for the volume of the disperse phase in dialysis of serum but they did not discuss any more than Hamburger the equation of the solubility of the solute in the continuous phase.

S. P. L. Sörensen in his monumental work *Studies on Proteins* was the first thoroughly to appreciate these conditions and the author of this essay willingly admits he has been greatly influenced by the above work.

After the appearance of *Studies on Proteins* Rich. Ege [1920, 1, 2, 3, 4; 1921, 1, 2, 3, 4] published an excellent piece of work on the partition of glucose between the blood cells and serum in which he pays the proper attention to the characteristic qualities of colloid solutions just mentioned.

In Henderson's and Hasselbalch's equations the heterogeneity of the solution is not allowed for and it was this that originally prompted me to investigate the theory of these equations.

T. R. Parsons in 1917 called attention to the difficulties caused by the different distribution of bicarbonate between the blood corpuscles and serum. L. S. Fridericia [1920] as well as Joffe and Poulton [1920] have recently determined the conditions of distribution but none of these authors regarded the serum and the fluids of the blood corpuscles as heterogeneous systems separate from one another. The work of all these authors is extremely valuable in relation to the questions before us and it will be fully dealt with later in this work.

CHAPTER II

THE HYDRATION OF CARBON DIOXIDE AND THE DISSOCIATION OF CARBONIC ACID.

The Derivation of Henderson's and Hasselbalch's Equations and the Modification of these in accordance with Bjerrum's Theory. The Limits within which the Derived Equations are Valid.

In 1 cc. of water in equilibrium with a gas mixture in which the tension of CO_2 is P_{CO_2} mm. Hg, there will be dissolved according to Henry's law (equation (1)),

$$a = \frac{P_{CO_2}}{760} a \text{ cc. } CO_2 \text{ (at } 0° \text{ and } 760 \text{ mm. Hg).}$$

Some of the dissolved carbon dioxide is hydrated, forming carbonic acid,

$$CO_2 + H_2O \rightleftharpoons H_2CO_3, \quad \dots\dots\dots\dots\dots(61)$$

and the mass action law gives for this reaction

$$\frac{A_{CO_2} a_{H_2O}}{a_{H_2CO_3}} = k_0. \quad \dots\dots\dots\dots\dots(62)$$

Since a_{H_2O} can be regarded as constant under the experimental conditions,

$$\frac{a_{H_2CO_3}}{A_{CO_2}} = K_0, \quad \dots\dots\dots\dots\dots(63)$$

and therefore from (7) and (11),

$$a_{H_2CO_3} = P_{CO_2} a K_0 \, 0.00005910. \quad \dots\dots\dots\dots(64)$$

This equation in accordance with what was said on p. 157 applies to watery solutions also, with the proviso that K_0 is the same for pure water and for watery solutions. How far this is the case cannot be estimated at present, but as from (62) and (63)

$$K_0 = \frac{a_{H_2O}}{k_0} \quad \dots\dots\dots\dots\dots(65)$$

and we regard a_{H_2O} as constant, K_0 will only vary from one solution to another if k_0 varies.

Such a variation under the conditions investigated is improbable, but if it should happen it would be without importance for the following argument and would be included in the total constants. Only in the investigation of the extent to which the activity of the bicarbonate ion conforms to Bjerrum's [1916, 1918, 1919] equation for the calculation of the activity coefficient, would such a condition be of real importance and invalidate the proof. Assuming however that Bjerrum's theory is correct the above-mentioned investigation can be used as a proof that K_0 does not vary appreciably with the ionic concentration within the limits of concentration investigated.

In equation (63)

$$\frac{a_{H_2CO_3}}{A_{CO_2}} = K_0,$$

$a_{H_2CO_3}$ is the apparent activity of carbonic acid, and in so far as it is not hydrated (*i.e.* it is not H_2CO_3, H_2O for example), we have $a_{H_2CO_3} = A_{H_2CO_3}$. The error involved by putting

$$\frac{a_{H_2CO_3}}{A_{CO_2}} = \frac{G_{H_2CO_3}}{C_{CO_2}} \quad \dots\dots\dots\dots\dots\dots(66)$$

will in all cases be small, as the activity coefficient of non-electrolytes is an expression of the salting out effect which ought to be about the same for all non-electrolytes. K_0 indicates therefore how much of the dissolved CO_2 is hydrated. Experiment has shown that K_0 is small. Thus Thiel and Strohecker [1914] have calculated from the rapidity with which CO_2 neutralises bases that only $\frac{1}{2}$ % of the dissolved CO_2 is hydrated and therefore our K_0 is roughly 0·005.

Walker and Cormack [1900] make some remarks in their experiments on conductivity in CO_2 solutions about the value of K_0. They show that such solutions of carbonic acid fulfil the requirements of the mass action law when the total amount of dissolved CO_2 is substituted for the carbonic acid in the equation. If only a small part of the dissolved CO_2 was hydrated the mass action law would not be satisfied in the form employed and they think therefore they are right in concluding about half the dissolved CO_2 is hydrated (which means that our K_0 should be 0·5). Walker and Cormack's determinations however certainly do not justify them in drawing these conclusions.

Carbonic acid dissociates in the following manner:

$$H_2CO_3 \quad \rightleftharpoons \quad H^{\cdot} \quad + \quad HCO'_3 \quad \text{(1st dissociation)}, \quad \dots\dots(67)$$
$$\text{(Carbonic acid)} \quad \text{(Hydrogen ion)} \quad \text{(Bicarbonate ion)}$$

$$HCO'_3 \quad \rightleftharpoons \quad H^{\cdot} \quad + \quad CO''_3 \quad \text{(2nd dissociation)}. \quad \dots\dots(68)$$
$$\text{(Bicarbonate ion)} \quad \text{(Hydrogen ion)} \quad \text{(Monocarbonate ion)}$$

From the mass action law

$$\frac{a_{H^{\cdot}} \cdot a_{HCO'_3}}{a_{H_2CO_3}} = k_1 \text{ (1st dissociation of carbonic acid)}, \quad \dots\dots(69)$$

$$\frac{a_{H^{\cdot}} \cdot a_{CO''_3}}{a_{HCO'_3}} = K_2 \text{ (2nd dissociation of carbonic acid)}. \quad \dots\dots(70)$$

Multiplying equations (63) and (69) we get

$$\frac{a_{H^{\cdot}} \cdot a_{HCO'_3}}{A_{CO_2}} = k_1 K_0, \quad \dots\dots\dots\dots\dots\dots(71)$$

and therefore

$$\frac{a_{H^{\cdot}} \cdot a_{HCO'_3}}{A_{CO_2}} = K_1. \quad \dots\dots\dots\dots\dots\dots(72)$$

Since, almost without exception before the appearance of Bjerrum's theory, we put

$$a_{H^{\cdot}} = C_{H^{\cdot}}, \quad \dots\dots\dots\dots\dots\dots(73)$$

$$a_{CO_2} = C_{CO_2}, \quad \dots\dots\dots\dots\dots\dots(74)$$

$$a_{HCO'_3} = C_{HCO'_3} \gamma = (C_{HCO'_3} f_\mu), \quad \dots\dots\dots\dots(75)$$

where γ was the "conductivity dissociation" of the bicarbonate in the concentration under investigation, (72) became

$$\frac{C_{H^{\cdot}} \cdot C_{HCO'_3} \gamma}{C_{CO_2}} = K_1, \quad \dots\dots\dots\dots\dots\dots(76)$$

which was transformed into

$$C_{H^{\cdot}} = \frac{K_1}{\gamma} \frac{C_{CO_2}}{C_{HCO'_3}}. \quad \dots\dots\dots\dots\dots\dots(77)$$

This is Henderson's equation.

Equations like (76) and (77) have repeatedly been used in physical chemistry, for instance in W. Nernst's experiments on the dissociation constants of weak monovalent acids (cited from Datta and Dhar [1915]), in Datta and Dhar's [1915] determinations of the second dissociation constant of weak divalent acids, and in investigations on the second dissociation constant of carbonic acid by McCoy and Smith [1911] and McCoy and Test [1911], Auerbach and Pick [1912], and Seyler and Lloyd [1917, 1, 2], and lastly in work upon the solubility of the carbonates of the alkaline earths, for example, by Johnston [1916].

An equation analogous to (72) is one derived by Bjerrum and Gjaldbaek [1919, p. 82, equation (24)].

In logarithmic form (77) becomes

$$p_{\mathrm{H}^{\cdot}} = pK_1 + \log C_{\mathrm{HCO}'_3} - \log C_{\mathrm{CO}_2} + \log \gamma. \quad \ldots\ldots\ldots\ldots(78)$$

As will be shown later in considering the limits between which an equation similar to (77) is valid, $C_{\mathrm{HCO}'_3}$ in the cases treated by Hasselbalch and Henderson (as always in these investigations) can be assumed to have originated from the combined carbonic acid, as the amount of HCO'_3 derived from the dissociation of the dissolved carbonic acid is negligible in comparison with it. If we know the total quantity of CO_2 in a solution and its pressure in a gaseous phase in equilibrium with the solution, the temperature and the relative absorption coefficient of the solution, it is easy to calculate the quantity of combined carbonic acid by subtracting the amount dissolved from the total amount. *If now it is assumed all the combined carbonic acid is in the form of bicarbonate,* and the CO_2 is expressed in volumes per cent., (77) compared with (5) becomes

$$C_{\mathrm{H}^{\cdot}} = \frac{K_1}{\gamma} \frac{\text{vol. \% dissolved } CO_2}{\text{vol. \% combined } CO_2} \frac{0.0004492}{0.0004492}, \quad \ldots\ldots\ldots\ldots(79)$$

$$C_{\mathrm{H}^{\cdot}} = \frac{K_1}{\gamma} \frac{\text{vol. \% dissolved } CO_2}{\text{vol. \% combined } CO_2}. \quad \ldots\ldots\ldots\ldots\ldots(80)$$

This is Henderson's equation in a simplified form, expressed in the same way as that of Hasselbalch.

Hasselbalch [1916, 2] regarded carbonic acid as a divalent acid and put the concentrations in (77) in terms of normality, assuming a solution which contained a gram-molecule of CO_2 in a litre to be twice normal as regards carbonic acid, while he regarded a solution which contained a gram-molecule of CO_2 bound as bicarbonate in a litre, as normal with respect to bicarbonate[1].

Equation (77) under these circumstances becomes

$$C_{\mathrm{H}^{\cdot}} = \frac{K_1}{\gamma} \frac{\text{vol. \% dissolved } CO_2}{\text{vol. \% combined } CO_2} 2. \quad \ldots\ldots\ldots\ldots(81)$$

And since Hasselbalch substitutes K_I for $\frac{K_1}{\gamma}$ his equation becomes

$$C_{\mathrm{H}^{\cdot}} = K_I \frac{\text{vol. \% dissolved } CO_2}{\text{vol. \% combined } CO_2} 2. \quad \ldots\ldots\ldots\ldots(82)$$

[1] Parsons [1919] and L. Michaelis [1920] have recently pointed out that Hasselbalch uses this unusual convention and Michaelis has protested against it.

If we now compare (80) with (82) it will be seen that

$$\frac{K_1 \text{(Henderson)}}{\gamma} = 2K_I \text{ (Hasselbalch)}. \quad \dots \dots \dots (83)$$

Hasselbalch's equation in logarithmic form becomes

$p_{H^{\cdot}} = p_{K_I} + \log$ vol. % combined $CO_2 - \log$ vol. % dissolved $CO_2 - 0.3010$,

and when (83) is put in logarithmic form we get

$$pK_1 \text{ (Henderson)} + \log \gamma = pK_I \text{ (Hasselbalch)} - 0.3010 \dots \dots (84)$$

Although the above appears to be correct it should be noted that Hasselbalch *determined his total constant* by the potentiometrical method. If this is recognised (83) becomes

$$\frac{K_1 \text{(Henderson)}}{F_a \text{(HCO}_3')} = 2K_I \text{ (Hasselbalch)} \quad \dots \dots \dots (85)$$

and (84) is transformed accordingly, but at the same time a correction ought to be introduced into all Hasselbalch's calculations so that p_{K_I} always becomes 0.048 times larger (given by the difference between Sörensen's and Bjerrum's E_0).

If we abide by what is the usual custom in the literature of physical chemistry there is no doubt we ought to use molar concentration and not normality in such equations, and as a similar convention also has priority in physiological literature, Hasselbalch's mode of expression should be given up.

As the assumptions underlying (73), (74) and (75) are no longer tenable the equation should be modified in accordance with Bjerrum's requirements. When the mols in the equation with the exception of $a_{H^{\cdot}}$ are expressed by their concentrations, (72) becomes

$$\frac{a_{H^{\cdot}} \cdot C_{HCO_3'} F_a \text{(HCO}_3')}{C_{CO_2} F_a \text{(CO}_2)} = K_1. \quad \dots \dots \dots \dots (86)$$

It will be seen from (54) that

$$1.117 \, a_{H^{\cdot}} = C_{H^{\cdot}} \text{ (Sörensen)}, \quad \dots \dots \dots (87)$$

and we thus obtain

$$\frac{C_{H^{\cdot}} \text{ (Sörensen)} C_{HCO_3'}}{C_{CO_2}} = K_1 \frac{F_a \text{(CO}_2) \, 1.117}{F_a \text{(HCO}_3')} . \quad \dots \dots \dots (88)$$

Substituting

$$K'_1 = \frac{K_1}{F_a \text{(HCO}_3')} \quad \dots \dots \dots \dots (89)$$

we get from (88) and (85)

$$K'_1 = \frac{1.793 \, K_I \text{ (Hasselbalch)}}{F_a \text{(CO}_2)}, \quad \dots \dots \dots \dots (90)$$

and from (83) and (90)

$$K'_1 = \frac{0.896 \, K_1 \text{ (Henderson)}}{\gamma \, F_a \text{(CO}_2)} . \quad \dots \dots \dots \dots (91)$$

If the relative absorption coefficient for blood is taken as 0.92 (Bohr [1905]), $F_a \text{(CO}_2)$ becomes 1.087 and we get for blood

$$\text{Blood} \begin{cases} K'_1 = 1.65 \, K_I \text{ (Hasselbalch)}, & \dots \dots \dots \dots (92) \\ pK'_1 = pK_I \text{ (Hasselbalch)} - 0.218. & \dots \dots \dots (93) \end{cases}$$

Now that we have developed the relations between the various constants we will revert to (72) and express it in the same way as Hasselbalch's equation by using (89)

$$a_{H^{\cdot}} = K'_1 \, F_a \, (CO_2) \, \frac{\text{vol. \% dissolved } CO_2}{\text{vol. \% combined } CO_2} \quad \dots\dots\dots\dots(94)$$

or

$$a_{H^{\cdot}} = K'_1 \, \frac{P_{CO_2}a}{7{\cdot}60 \, \text{vol. \% combined } CO_2}. \quad \dots\dots\dots\dots(95)$$

Putting (94) in logarithmic form we get

$$p_{H^{\cdot}} \text{ (Bjerrum)} = pK'_1 + \log \text{vol. \% combined } CO_2 - \log \text{vol. \% dissolved } CO_2 \\ - \log F_a \, (CO_2). \quad \dots\dots\dots(96)$$

We have now worked out the relations between Henderson's, Hasselbalch's and our own equations as expressed in (90) and (91), but a considerable qualification of their significance must be made.

Henderson's and Hasselbalch's equations were evolved, as will be remembered, on the assumption that the degree of dissociation measured by conductivity gave the degree of activity of the bicarbonate, but since this is incorrect as Bjerrum later has shown, (73), (74), (75), (76), (77), (78), (79), (80), (81), (82), (83), (84), and (91) are only approximations to the true equations and have *only mathematical significance*, while (88) and (90) are correct actually as well as mathematically.

It is perhaps also worth pointing out that (90) only holds good if K_I (Hasselbalch) and K'_1 are calculated from the same experiments or from experiments which give the same results with the same method of calculation, and that Sörensen's E_0 (50) should be used for the calculation of K_I (Hasselbalch), while E_0 (Bjerrum) (51) should be used for the calculation of K'_1. If (90) is not satisfied by a correct calculation from two series of experiments it is proof that the experimental results do not agree.

In the equations evolved up to the present we have assumed without hesitation that all the combined carbonic acid is present as bicarbonate. *We will now inquire how far the equations are valid if the second dissociation of carbonic acid* (68) *takes place to any considerable extent.* We will first find out, however, the reaction at which this happens.

Equation (70)

$$\frac{a_{H^{\cdot}} \cdot a_{CO''_3}}{a_{HCO'_3}} = K_2$$

can be transformed into

$$\frac{C_{CO''_3}}{C_{HCO'_3}} = \frac{K_2}{a_{H^{\cdot}}} \cdot \frac{F_a(HCO'_3)}{F_a(CO''_3)}. \quad \dots\dots\dots\dots(97)$$

The value of the right side of the equation is still not accurately known in spite of a great deal of work on the subject, since McCoy's [1911], Auerbach and Pick's [1912], Shield's [1893], and Seyler and Lloyd's [1917, 1] results must be recalculated according to the new ideas. Bjerrum and Gjaldbaek [1919] give $K_2 = 10^{-10.22}$ from which we may expect to get a good approximation by putting

$$K_2 \frac{F_a(HCO'_3)}{F_a(CO''_3)} = 10^{-10} = K'_2. \quad \dots\dots\dots\dots(98)$$

It should however be noted that the value may vary somewhat with the ionic concentration.

If with this value $\dfrac{C_{CO''_3}}{C_{HCO'_3}}$ is calculated from (97) the table below is obtained.

Table IV.

$a_{H\cdot}$	$p_{H\cdot}$	$\dfrac{C_{CO''_3}}{C_{HCO'_3}}$ room temp.	$a_{H\cdot}$	$p_{H\cdot}$	$\dfrac{C_{CO''_3}}{C_{HCO'_3}}$ room temp.
1×10^{-7}	7·00	0·001	0.25×10^{-8}	8·60	0·040
0·5	7·30	0·002	0·2	8·70	0·050
0·3	7·52	0·003	0·17	8·77	0·059
0·2	7·70	0·005	0·15	8·82	0·067
0·15	7·82	0·007	0·13	8·89	0·077
1×10^{-8}	8·00	0·010	0·11	8·96	0·091
0·5	8·30	0·020	1×10^{-9}	9·00	0·100
0·3	8·52	0·033			

It will be seen from the table that at $p_{H\cdot}$ 7·00 the ratio is only $\frac{1}{1000}$, at $p_{H\cdot}$ 8·00, $\frac{1}{100}$ and at 9·00 it is $\frac{1}{10}$. It may be concluded from this that in reactions of physiological importance all the carbonate may be considered to be in the form of bicarbonate—a result which numerous workers have previously arrived at in a similar way (e.g. L. Meyer [1857], Heidenhain and L. Meyer [1863], Zuntz [1868], Chr. Bohr [1909], van Slyke and Cullen [1917], Parsons [1919], Bayliss [1915] and others). We shall revert briefly to this subject in discussing the variations of volume of the blood corpuscles.

It will now be shown that with a little alteration the equations evolved are valid even if the reaction is so alkaline that a large amount of monocarbonate ions are present in the solution. Instead of the concentration, the corresponding amounts of CO_2 are substituted expressed in volumes per cent., and the amount

corresponding to C_{CO_2} is called S_0,

,,　　　　　　$C_{HCO'_3}$,,　 S_1,

,,　　　　　　$C_{CO''_3}$,,　 S_2.

(94) then becomes

$$a_{H\cdot} = K'_1 F_a (CO_2) \frac{S_0}{S_1} \qquad \dots\dots\dots\dots\dots(99)$$

And (97) combined with (98)

$$a_{H\cdot} = K'_2 \frac{S_1}{S_2} \qquad \dots\dots\dots\dots\dots(100)$$

Calling the total amount of combined CO_2, B, we get

$$S_1 + S_2 = B, \qquad \dots\dots\dots\dots\dots(101)$$

and from (100) and (101)

$$S_1 = \frac{a_{H\cdot} B}{K'_2 + a_{H\cdot}}. \qquad \dots\dots\dots\dots\dots(102)$$

When this is substituted in (99) we get a quadratic equation

$$a_{H\cdot} = \frac{S_0 K'_1 F_a (CO_2)}{2B} \pm \sqrt{\frac{(S_0 K'_1 F_a (CO_2))^2 + 4 S_0 K'_1 F_a (CO_2)^2 K'_2 B}{4B^2}}.$$

The numerator under the square root sign is equal to

$$(S_0 K'_1 F_a (CO_2) + 2 K'_2 B)^2 - 4 K'^2_2 B^2.$$

$4 K'^2_2 B^2$ can be neglected as K'^2_2 is very small, and as the negative sign is without physical significance, we get

$$a_{H\cdot} = \frac{K'_1 F_a (CO_2) S_0}{B} + K'_2. \qquad \dots\dots\dots\dots\dots(103)$$

The equation (103) *thus developed is valid generally* (independent of hydrolysis of the ions), *so long as the concentration of the combined carbonic acid is large compared with the concentration of the dissolved CO_2, and so long as all the combined carbonic acid is present as monocarbonate and bicarbonate.*

It will be noticed that it is identical with the earlier equation (94) (the modified Henderson-Hasselbalch equation) when K'_2 is small compared with

$$K'_1 F_a (CO_2) \frac{\text{vol. \% dissolved } CO_2}{\text{vol. \% combined } CO_2}.$$

In the following short table the first column contains values of

$$K'_1 F_a (CO_2) \frac{\text{vol. \% dissolved } CO_2}{\text{vol. \% combined } CO_2};$$

the second column contains the corresponding values of

$$K'_1 F_a (CO_2) \frac{\text{vol. \% dissolved } CO_2}{\text{vol. \% combined } CO_2} + K'_2,$$

K'_2 being again 10^{-10}; in the last column the corresponding p_H values are given.

Table V.

$K'_1 F_a (CO_2) \frac{S_0}{S_1}$	$K'_1 F_a (CO_2) \frac{S_0}{S_1} + K'_2$	p_H
10^{-8}	$1 \cdot 010 \times 10^{-8}$	7·996
10^{-9}	$1 \cdot 100 \times 10^{-9}$	8·959
10^{-10}	$2 \cdot 000 \times 10^{-10}$	9·699

It will be seen from the table *that at reactions more acid than 10^{-8} the correction is negligible and the equation becomes identical with the modified Henderson-Hasselbalch equation.*

We have still to investigate how small the quantity of combined carbonic acid can be without being small in comparison with the amount of bicarbonate arising from the dissolved CO_2.

The hydrogen ions in a solution containing dissolved CO_2 may either

(1) exclusively be derived from the dissociated carbonic acid (except for the minimal amount due to the dissociation of water), or

(2) partly come from the carbonic acid and partly from another acid, or

(3) the reaction may be so acid that the carbonic acid is undissociated, in which case there can be no carbonate present in the solution and the equation does not apply.

We will only consider the case when carbonic acid is the sole source of hydrogen ions, because the field is more restricted than in the other cases.

Let us call the concentration of bicarbonate ions which is equivalent to the amount of available base, C_m, the bicarbonate ions which are derived from the dissociation of carbonic acid being, as already mentioned, equal to C_H. Equation (86) taken in conjunction with (89) and (85) then gives

$$\frac{C_H \cdot F_a(H') (C_m + C_H')}{C_{CO_2}} = K'_1 F_a (CO_2), \quad \dots\dots\dots\dots (104)$$

from which $\qquad C_H' = -\frac{C_m}{2} \pm \sqrt{\frac{C_m^2}{4} + \frac{K'_1 F_a (CO_2) C_{CO_2}}{F_a(H')}} \quad \dots\dots\dots\dots (105)$

in which the negative sign is without actual meaning. If C_m is equal to 0, (105) reduces to

$$a_{H^{\cdot}} = \sqrt{K'_1 \, F_a \, (CO_2) \, F_a \, (H^{\cdot}) \, \overline{C_{CO_2}}}. \qquad \dots\dots\dots\dots(106)$$

(106) applies to reactions in solutions of CO_2 where there is no combined carbonic acid.

A hard and fast rule for using (105) instead of the modified Henderson-Hasselbalch equation is difficult to give, but a good approximate rule may be arrived at in the following way.

(104) is compared with (86) and (89) in the form

$$\frac{C_{H^{\cdot}} \, F_a \, (H^{\cdot}) \, C_m}{C_{CO_2}} = K'_1 \, F_a \, (CO_2). \qquad \dots\dots\dots\dots(107)$$

The difference between (104) and (107) is due to the difference between the factors C_m and $(C_{H^{\cdot}} + C_m)$. If $C_{H^{\cdot}}$ is small in comparison with C_m, the equations become the same. Making the approximate assumption that $C_{H^{\cdot}}$ is equal to $a_{H^{\cdot}}$ we will evaluate $\frac{C_{H^{\cdot}}}{C_m}$. $a_{H^{\cdot}}$ is calculated from (94) which only differs from (105) in form, and an estimation of the value of $\frac{a_{H^{\cdot}}}{m}$ is made. If this does not exceed 0·01 the use of equation (94) is admissible.

Table VI.

$a_{H^{\cdot}}$	C_m	Vol. % combined CO_2	mm. Hg, CO_2
10^{-6}	10^{-4}	0·22	0·563
10^{-5}	10^{-3}	2·24	53·2
10^{-4}	10^{-2}	22·39	4710

In the table the values which C_m must have at different reactions so that the above requirements for the applicability of (94) may be fulfilled, are given. In the third column C_m is expressed in volumes per cent. of CO_2 and in the fourth column the corresponding tension of CO_2 calculated with the help of the constants determined for sodium bicarbonate in Chapter VIII.

RÉSUMÉ.

From the above results it may be said that in homogeneous solutions

I. Henderson's equation is in agreement with the equation evolved on the basis of Bjerrum's new dissociation theory when $K_1 \, \frac{F_a \, (CO_2)}{F_a \, (HCO'_3)}$ is substituted for $\frac{K_1}{\gamma}$ in the former.

II. Hasselbalch's $K_I = K_1 \dfrac{F_a \, (CO_2) \, . \, 0 \cdot 557}{F_a \, (HCO'_3)}$.

III. The value found then becomes $a_{H^{\cdot}}$ (Bjerrum), not $C_{H^{\cdot}}$ (Sörensen), and instead of $p_{H^{\cdot}}$ (Sörensen) we have $p_{H^{\cdot}}$ (Bjerrum).

IV. The limits for the modified Henderson-Hasselbalch equation are given.

V. The complete equations for the relation between the hydrogen ion activity, the amount of dissolved CO_2 and the combined carbonic acid are evolved.

CHAPTER III

THE DEVELOPMENT OF A MODIFIED HENDERSON-HASSELBALCH EQUATION FOR USE IN THE CASE OF HETEROGENEOUS SOLUTIONS.

As already hinted at several times the conditions in heterogeneous solutions are complicated by the fact that the concentration of the solute in the individual phases is not directly given by its mean concentration, as the solute cannot be assumed to be evenly distributed over the whole system. Further we cannot regard the activity coefficients in the single phases as already known, and lastly there is the possibility that the solute may be concentrated by adsorption in the interfaces between the different phases.

The influence of the first condition on the determination of reaction by Henderson's and Hasselbalch's equations has latterly been recognised by Fridericia [1920], Parsons [1917] and by Joffe and Poulton [1920], but they have not realised the full significance of it, only regarding serum-blood cells as a heterogeneous system, and failing to appreciate that serum (plasma) and the fluid of the blood cells are likewise heterogeneous systems.

Fridericia's and Parsons' contributions appeared while the experimental part of this research was in progress, but Joffe and Poulton's work only came to my notice after I had embarked upon the literary section. These investigations are of the greatest importance for the questions here considered and they will be fully dealt with in a later chapter of this work.

We will proceed with our calculations *under the assumption* that all the combined carbonic acid is present as bicarbonate (that no CO_2 is bound by complex combinations or adsorbed), and we will investigate how the experimental results accord with this assumption.

Let   ·   Vol. % combined carbonic acid $= \beta$,

β being the mean concentration in the heterogeneous solution. Let us first consider a system of two watery phases in mutual equilibrium, and in equilibrium with a gaseous phase containing CO_2.

100 parts of the system consist of q parts of phase (2) and $100 - q$ parts of phase (1).

From this we get

$$100 \times \frac{\text{vol. of phase (2)}}{\text{vol. of phase (1)} + \text{vol. of phase (2)}} = q. \quad \ldots\ldots\ldots\ldots(108)$$

Also

$$\frac{\text{Vol. \% combined } CO_2 \text{ in phase (2)}}{\text{Vol. \% combined } CO_2 \text{ in phase (1)}} = d. \quad \ldots\ldots\ldots\ldots(109)$$

Seeking now the amount of combined CO_2 in the single phases expressed by β, d and q,

$$\frac{V_{(1)}qd}{100} + \frac{V_{(1)}(100-q)}{100} = \beta, \quad \ldots\ldots\ldots\ldots\ldots(110)$$

$$V_{(1)} = \frac{100\beta}{100 - q(1-d)} \text{ (vol. \% combined } CO_2 \text{ in phase (1)), } \ldots(111)$$

and

$$V_{(2)} = \frac{100d\beta}{100 - q(1-d)} \text{ (vol. \% combined } CO_2 \text{ in phase 2)). } \ldots(112)$$

From (111) in conjunction with (95) we get

$$a_{H'(1)} = K'_{1(1)} \frac{100 - q(1-d)}{100} \times \frac{P_{CO_2} a}{7 \cdot 60 \beta}, \dots\dots\dots\dots(113)$$

where $a_{H'(1)}$ is the apparent hydrogen ion activity in phase (1) and $K'_{1(1)}$ is K_1 divided by the apparent activity coefficient of HCO'_3 in phase (1).

Analogous to this we have

$$a_{H'(2)} = K'_{1(2)} \frac{100 - q(1-d)}{100d} \times \frac{P_{CO_2} a}{7 \cdot 60 \beta}. \dots\dots\dots\dots(114)$$

$K'_{1(1)}$ and $K'_{1(2)}$ are not different from K'_1 in a homogeneous solution and the indices only signify that $F_a(HCO'_3)$ refers to a particular solution.

Comparing (95) with (113) and (114) it will be seen that the last two can be expressed in the same form as (95), viz. (113) in the form

$$a_{H'(1)} = \lambda_{(1)} \frac{P_{CO_2} a}{7 \cdot 60 \beta} \dots\dots\dots\dots(115)$$

and (114) in the form

$$a_{H'(2)} = \lambda_{(2)} \frac{P_{CO_2} a}{7 \cdot 60 \beta}. \dots\dots\dots\dots(116)$$

Since

$$\lambda_{(1)} = K'_{1(1)} \frac{100 - q(1-d)}{100}, \dots\dots\dots\dots(117)$$

$$\lambda_{(2)} = K'_{1(2)} \frac{100 - q(1-d)}{100d}, \dots\dots\dots\dots(118)$$

$\lambda_{(1)}$ and $\lambda_{(2)}$ occur in (115) and (116). These are only constants as long as the factors in them do not vary or the variations neutralise one another.

$K'_{1(1)}$ and $K'_{1(2)}$ will only vary with the reaction if the activity coefficient of the bicarbonate ion varies with it.

L. Michaelis [1920] has lately drawn attention to this condition, regarding it however from the classical standpoint of Arrhenius, and therefore assuming that K'_1 varies with the degree of dissociation of the bicarbonate. His views appear in the form of a criticism on the theory which H. Straub and Klothilde Meier [1918–1920] put forward for the carbonic acid combining power of the blood on the strength of a series of experiments carried out by them. Straub and Meier's experiments and theories are however subject to an elementary error, and as will later be shown they are quite fallacious, but Michaelis' comments naturally do not lose their interest on this account.

According to Bjerrum's[1] theory there is reason to expect that the activity coefficient of an ion will decrease in a solution containing ions with an opposite charge and this will be especially the case if these ions are polyvalent or have a large molecular volume.

If there are already rather many ions present in the solution the effect of the new-comers will be small, and possibly in special cases it will not be noticed at all, but it is not right, without due consideration, to regard $K'_{1(1)}$ and $K'_{1(2)}$ as constants, as it is a fact that the ionic concentrations in phases (1) and (2) vary with the reaction.

The factor $\frac{100 - q(1-d)}{100}$ in (117) and the corresponding factor in (118)

[1] I have repeatedly had the opportunity of discussing this condition with Prof. Bjerrum about which he has up to the present only published a rough sketch.

will vary provided that q and d vary. From S. P. L. Sörensen's [1915–1917] experiments on egg albumin there is every reason to expect they will vary with the reaction, but what influence the variations will have on the "total constants" experimental investigations alone can inform us at present.

As will be shown in one of the subsequent chapters $\lambda_{(1)}$ in serum and the liquid phase of the blood corpuscles is constant within the limits of the reactions investigated (experimental errors are not very small in experiments with the fluids of blood cells), and we will therefore continue our calculations under this assumption.

Let $\quad a_{H^{\cdot}}$ in the liquid phase of serum $\quad = a_{H^{\cdot}(s)}$,

and $\quad a_{H^{\cdot}}$ in the liquid phase of blood cells $= a_{H^{\cdot}(c)}$,

$$\lambda_{(1)} \text{ in serum} \quad = \lambda_{(s)},$$

$$\lambda_{(1)} \text{ in blood cells} = \lambda_{(c)}.$$

The volume of the blood cells expressed as a percentage of the total volume of the blood is Q. The partition of HCO'_3 between blood cells and serum is D, so that

$$\frac{c_{HCO'_3} \text{ in blood cells}}{c_{HCO'_3} \text{ in serum}} = D,$$

and lastly the mean concentration of combined carbonic acid in the blood expressed in volumes % $= B$, and

$$Q = \frac{\text{vol. of blood cells}}{\text{vol. of blood}} \times 100.$$

From (115) and (116) we then get

$$a_{H^{\cdot}(s)} = \lambda_{(s)} \frac{100 - Q(1-D)}{100} \times \frac{P_{CO_2} a}{7\cdot 60\,B}, \quad \dots\dots\dots\dots(119)$$

$$a_{H^{\cdot}(c)} = \lambda_{(c)} \frac{100 - Q(1-D)}{100\,D} \times \frac{P_{CO_2} a}{7\cdot 60\,B}. \quad \dots\dots\dots\dots(120)$$

(119) and (120) can be put in a form analogous with (95), viz.

$$a_{H^{\cdot}(s)} = \Lambda_{(s)} \frac{P_{CO_2} a}{7\cdot 60\,B}, \quad \dots\dots\dots\dots\dots\dots(121)$$

$$a_{H^{\cdot}(c)} = \Lambda_{(c)} \frac{P_{CO_2} a}{7\cdot 60\,B}, \quad \dots\dots\dots\dots\dots\dots(122)$$

$$\Lambda_{(s)} = \lambda_{(s)} \frac{100 - Q(1-D)}{100}, \quad \dots\dots\dots\dots\dots(123)$$

$$\Lambda_{(c)} = \lambda_{(c)} \frac{100 - Q(1-D)}{100\,D}. \quad \dots\dots\dots\dots(124)$$

It follows from the above that in $\Lambda_{(s)}$ and $\Lambda_{(c)}$ four variables take part, of quite a different kind from those which determine K'_1 in (95), and it will thus be understood that we cannot immediately draw conclusions regarding the combination of carbonic acid in the blood from a numerical agreement between $\Lambda_{(s)}$ and K'_1 in any bicarbonate solution.

Fridericia [1920], Henderson [1908–1910], Hasselbalch [1916, 2], Parsons [1917–1920], Joffe and Poulton [1920], Michaelis [1920] and many others have not realised the full consequences of this condition (which in an almost analogous way can be developed by the classical dissociation theory), and we are therefore compelled to investigate anew the questions mentioned in the introduction:

I. Can the hydrogen ion concentration of the blood be calculated by an equation similar to Henderson's and Hasselbalch's?

II. Is all the combined carbonic acid of the blood ionised?

I have attacked the first problem from a purely empirical standpoint. For practical work it is convenient to employ (121) in logarithmic form

$$p_{H^{\cdot}(s)} = p\Lambda_{(s)} + \log B - \log \frac{P_{CO_2}a}{7 \cdot 60}, \dots\dots\dots\dots(125)$$

and transform it thus

$$p_{H^{\cdot}(s)} = p\Lambda'_{(s)} + \log B - \log \int_0, \quad \dots\dots\dots\dots(126)$$

where $\int_0$ is the mean concentration of dissolved CO_2 in the blood expressed in volumes %, and

$$p\Lambda'_{(s)} = p\Lambda_{(s)} - \log \Phi_a (CO_2), \quad \dots\dots\dots\dots(127)$$

while $\Phi_a (CO_2)$ is the reciprocal of the relative absorption coefficient in blood.

The relative absorption coefficient in serum is 0·975 according to Bohr [1905] and 0·81 in blood corpuscles, from which $\Phi_a (CO_2)$ can be calculated by the following equation:

$$\Phi_a (CO_2) = \frac{100}{100 - \frac{19Q}{100} - \frac{2 \cdot 5 (100 - Q)}{100}}. \quad \dots\dots\dots\dots(128)$$

Résumé.

I. The modified Henderson-Hasselbalch equation is adapted for use with heterogeneous solutions.

II. It appears from the resulting equation that nothing can be concluded from the size of the total constant about the way the carbonic acid is combined, without taking the heterogeneity into account.

CHAPTER IV

METHODS FOR THE ELECTROMETRIC DETERMINATION OF THE APPARENT HYDROGEN ION ACTIVITY IN SOLUTIONS CONTAINING CO_2 WITH THE HELP OF THE PLATINUM HYDROGEN ELECTRODE, AND THE TECHNIQUE EMPLOYED.

The general principles and methods of estimation by the potentiometrical method have been often described and are to be found in every handbook of physical chemistry. The method used here is that described in S. P. L. Sörensen's *Études enzymatiques*, II [1909-1910] and those interested are referred to this work for further information. Within the limits of reaction obtaining in the following experiments the reaction of a solution will be a function of the CO_2 tension with which it is in equilibrium. What effect a change of the CO_2 tension will have on the apparent hydrogen ion activity cannot be expressed in a general way, because the buffer action of the solution will determine it.

If it is desired to determine the reaction in a solution in equilibrium with a certain CO_2 tension, care must be taken that the CO_2 content does not vary during the measurement. There are two ways of avoiding this, either by saturating the solution with a hydrogen-CO_2 mixture (Höber's principle, 1903), or by bringing a very limited amount of hydrogen in equilibrium with the solution and taking care by suitable means that the CO_2 tension in the hydrogen only very slightly departs from that existing originally in the solution (Hasselbalch's principle, 1910), the potential of a platinum plate in contact with the solution being then measured.

For didactic reasons it is convenient to first describe Hasselbalch's principle. It is a development of methods with "stationary hydrogen" first used by Farkas in 1903 and later modified in different ways, see for example Botazzi in Neuberger's *Die Harn*.

Farkas allowed the liquid, the reaction of which he wished to estimate, to come into equilibrium with a quantity of confined hydrogen by diffusion through its surface, and he then measured the potential of a platinum plate dipping into the liquid. K. A. Hasselbalch showed however in 1910 that equilibrium is not reached in the time occupied by the experiment by this method, and as it was impossible to increase the duration because changes would occur in the blood he modified the method in such a way that the hydrogen was quickly brought into equilibrium with the liquid by rocking the electrode vessel containing the hydrogen and liquid a certain number of times. The liquid was then replaced by a fresh supply and the process repeated. Hasselbalch combined his method with the device of Michaelis and Rona [1909] which consisted in minimal immersion of the platinum in the liquid. Michaelis and Rona had demonstrated that by this refinement the depolarising action of the oxygen could be avoided or at any rate reduced to a minimum.

Konikoff [1913], who also worked at the "oxygen error," realised that the rocking must be continued until all the oxygen is reduced by the hydrogen (with the help of the platinised platinum). This is however hardly practicable in blood on account of the slow speed of the reduction (both electrode and blood become changed during the experiment). But there is no great difficulty in removing the free oxygen from inorganic salt solutions, (human) urine and similar liquids if a sufficiently large platinum plate is used, but a platinum wire of the size usually employed (Michaelis [1914], Hasselbalch [1916, 2] electrode at the bottom of Fig. 3) cannot remove the oxygen from solutions. Höber [1914], Peters [1914] and T. R. Parsons [1917] have also pointed out the possibility of depolarisation by O_2 in Hasselbalch's measurements and they recommended estimations on Höber's principle.

It is easy to convince oneself of the truth of the above by measuring the reaction of a phosphate mixture by Hasselbalch's method. In a vessel con-

taining a platinum plate electrode the liquid can be renewed several times
without the potential practically varying at all, as Hasselbalch [1910 and
1911] showed a long time ago. But if the estimation is carried out in a vessel
containing a platinum wire as electrode the potential[1] will fall on renewal
(in the electrode vessel illustrated in Fig. 3 the potential fell about 2 millivolts
for each renewal of liquid). This is due to the fact that besides H_2 and CO_2
the gas becomes contaminated with O_2 which depolarises the H_2-Pt electrode.
No information is known to me in the literature about the activity of small
quantities of O_2 in this connection. In measurements with an electrode vessel
on Höber's principle, to be described later, I have obtained a rough idea of
the magnitude of this factor. O_2 below 2 mm. Hg does not give an error of
1 millivolt, 3 mm. gives about $1\frac{1}{2}$ millivolt, and 8 mm. gives a large error if
the potential otherwise remains constant. I can also confirm the frequent
experience that no conclusion from the way in which the potential becomes
established can be drawn about the presence of traces of O_2 in the electrode
hydrogen.

Until 1916 Hasselbalch employed a small platinum plate as an electrode,
while Michaelis used a platinum wire. In 1916 Hasselbalch adopted a hook-
shaped wire for blood estimations, the point of which was allowed to touch
the side wall of the vessel just over the surface of the liquid, and in this way
he obtained minimal immersion.

According to experience at the Finsen Institute therefore wire electrodes
should be used for liquids containing haemoglobin, and platinum plates for
other liquids. In estimations of blood there are four points in particular to
be noted:

(1) Contact between platinum and liquid should be minimal.

(2) The electrode must not be moved while waiting for the potential to
become constant.

(3) The three-way tap must not become fixed.

(4) There must be no air bubbles in the connecting tube to the KCl
solution.

We will now quite shortly consider a few points of importance in making
measurements by Hasselbalch's method, neglecting the depolarising action of
the oxygen, but remembering that if we did not do this the potential would
be lower (and p_H smaller). We will assume also that the volume and pressure
of the gas do not vary during the determination.

If a limited quantity of liquid, saturated with a gas at a given tension,
which forms no combinations that split off the gas under the conditions of
experiment, e.g. bicarbonate, is allowed to come into equilibrium with a
limited space occupied by a gas, it can easily be shown[2] that at a given tem-

---

[1] By the potential is understood, here and henceforth when not otherwise stated, the total
potential of the chain of concentration cells.

[2] See for example D. D. van Slyke and Cullen's [1917] equation for the calculation of the
amounts of CO_2 in the apparatus of van Slyke.

perature the same proportion of the dissolved gas always goes over into the space, so that its partial pressure in the space will always be the same fraction of the pressure with which the liquid was originally in equilibrium. If the space is now allowed to come into equilibrium with a fresh amount of liquid, the partial pressure of the gas will rise further, but it will again be a definite fraction of the original pressure irrespective of its absolute magnitude.

It will be seen from equation (95) that $a_{H^.}$ is proportional to the CO_2 pressure, from which it follows that the error, occasioned by the CO_2 tension in Hasselbalch's electrode being too low, will be independent of the absolute CO_2 tension and will only be dependent on the relation between the volume of liquid and gas and on the absorption coefficient of the investigated liquid for CO_2.

The statement of Hasselbalch [1910, 1911] that more renewals of the liquid should be made in the case of higher CO_2 tensions than with lower to obtain a constant potential, is due to the fact that a slight alteration of the H_2 tension has hardly any effect on the potential, while a large change greatly affects it (see equation (53)).

Neither Hasselbalch, Michaelis nor any other person who has worked with the method, corrected for the decreased H_2 tension (the correction is usually small), but Hasselbalch came to the conclusion that the best results are obtained by omitting this correction. It follows from what has gone before that we ought to be very cautious about admitting measurements by Hasselbalch's method as standard ones, but they can legitimately be used for comparison with one another if they are done with the same electrode and the CO_2 tensions do not differ very much from one another.

Similar considerations apply to blood, as the CO_2 split off does not greatly disturb the conditions but rather tends in favour of equilibrium being attained more easily than in bicarbonate solution. It appears that Hasselbalch and his collaborators found different values for $p_{H^.}$ at different times. This is perhaps most evident from the fact that Hasselbalch and also Michaelis until 1916 held that blood with 40 mm. CO_2 had about 0.20 $p_{H^.}$ greater at 18° than at 38°, but in 1916 when he changed the electrode he found the same $p_{H^.}$ value at 18° as at 38°.

It is also worth while recounting a few experiences we have had at the Finsen Institute with regard to measurements of blood in the Hasselbalch electrode, which have also been noticed in numerous other laboratories. When the change of potential is followed after rocking it will be noted that it rises to a maximum and then slowly falls a little (the destruction of the electrode by the proteins). We have always taken the maximum reading as the correct potential.

When the electrode vessel is ready for an estimation the blood corpuscles sink, leaving the platinum in contact with serum alone. This is unquestionably of no small importance in the estimations, a view which is shared by many others. McClendon [1917] has even devised an electrode in which the blood corpuscles and serum are separated by centrifuging before the estimation is made; and T. R. Parsons [1917] has saturated the blood with a mixture of H_2 and CO_2 and then centrifuged at the temperature of saturation (38°), keeping the saturation constant, and he showed that the potential of the separated serum was the same as that of the blood.

The other technical points are the same as those met with in measurements with Höber's electrode, to which the reader is referred.

Höber's principle, as already mentioned, depends on first bringing the liquid into equilibrium with a mixture of hydrogen and CO_2 of known constitution, and it is tacitly assumed that the oxygen is expelled by the saturation.

Numerous electrode cells have been constructed on this principle (e.g. by Bjerrum and Gjaldbaek [1919], Höber [1903], Peters [1914], McClendon [1917], Milroy [1917], Walpole [1913] and others). There can be no doubt that the measurements made on this principle should be corrected for diminished hydrogen tension, but there are no difficulties arising from the depolarising action of the oxygen (see however later measurements of the blood cell fluids).

GENERAL TECHNIQUE.

The gases used were stored in the ordinary commercial metal cylinders. The hydrogen was prepared electrolytically and the cylinders contained as a rule about 99 % H_2 and about 0·5 % O_2. Once it was strongly contaminated with atmospheric air.

The hydrogen was led through palladium-asbestos kept red-hot in an electric oven. On account of the war I had only about 2 g. of 25 % palladium-asbestos at my disposal. It was packed in a combustion tube between asbestos corks. With this device I could get rid of the oxygen in 1 litre of hydrogen in a minute. The hydrogen was first washed with 5 % sulphuric acid and then with concentrated aqueous sublimate solution. Hydrogen prepared in this way contained about 99·5 % H_2.

The gases were mixed in a spirometer. For the hydrogen mixtures a 20-litre spirometer with water sealing was used, and care was taken that oxygen never gained access to it, a precaution it is absolutely necessary to observe. If the spirometer was not used for several days it had to be washed with O_2-free air several times. In addition to this spirometer two others of 125 litres were employed. In using a spirometer with water sealing it is important to recognise that gas mixtures containing CO_2 slowly change because CO_2 diffuses out through the water. The best way to avoid error is to take care that the CO_2 content of the gas mixture only differs slightly from the mixture that was previously in the spirometer, so that in investigating the effect of different tensions of CO_2 on a solution we start with the weakest or the strongest mixture and let the next nearest mixture follow on.

I have convinced myself a large number of times that the same gas mixture is obtained from the spirometer (20 litres) if 9 litres are removed at the end of half-an-hour or 18 litres at the end of an hour. Only very occasionally have I found slight differences. In working with the large spirometers I have always taken the precaution to prepare 100 l. of the mixture and only withdraw 60 l. for the experiment. In this way the same mixture will be obtained after the lapse of one hour as after the lapse of four hours. Occasionally however I have noticed a slight fall in the CO_2 tension. Mixtures which greatly differed from the one which was last in the spirometer have usually been prepared on the evening before they were to be used. The mixing in the spirometer took place simply by moving the bell up and down. With the 20-litre spirometer, 25 times was sufficient; with the large one, 50 times was necessary. The gas mixture was conducted from the spirometer through a wash bottle which often contained 0·9 % NaCl when I was working with blood and strong salt solutions, otherwise pure water. I have however not strictly adhered to this but no appreciable error can have arisen from this cause. Before reaching the wash bottle in the thermostat the gas passed through a lead tube 1·5 metres long and about 0·3 cm. diameter so that it should attain the correct temperature. A Roux thermostat was employed (see Fig. 2). As the temperature in the upper part was often several degrees higher than in the lower, the perforated copper plates were replaced by widemeshed galvanised iron netting and the netting covering the floor was raised somewhat. Under this a stirrer was installed which was driven by a small electric motor.

The stirrer[1] is illustrated in detail in Fig. 5 (p. 193). Its springs were similar to those of a dentist's drill (Claudius Ash and Co.). When it revolved at the proper speed and had the right inclination the temperature could be kept constant in the thermostat to within 0·5°. The conducting wires to the electrodes were led through the wall of the thermostat in paraffined glass tubes which were fixed in paraffined corks.

For the sake of economy in space it was necessary to place the thermoregulator somewhat lower. In addition to the stirrer, the wires and the thermo-regulator an axle pierced the wall of the upper part to rotate the saturator and another the wall of the lower part to rock the Hasselbalch electrode. Both axles were worked with the help of a small electric motor.

The *saturators* used for small amounts of liquid were composed of 250 cc. bottles like those Hasselbalch [1916, 2] has used, but the inlet tube was not fitted with the piece of rubber tube which acts as a valve. In such an instrument 4–12 cc. of liquid were rotated about 60 times a minute for half-an-hour and about 1 litre of gas blown through every three minutes, at the same time. Under these conditions equilibrium was always established except in a very few cases with high CO_2 tension and very viscous liquids and when preliminary

---

[1] This and various other items of the apparatus used were constructed by the mechanic of the Finsen Institute, the late Mr Eriksen, after consultation with me and my best thanks are due to him.

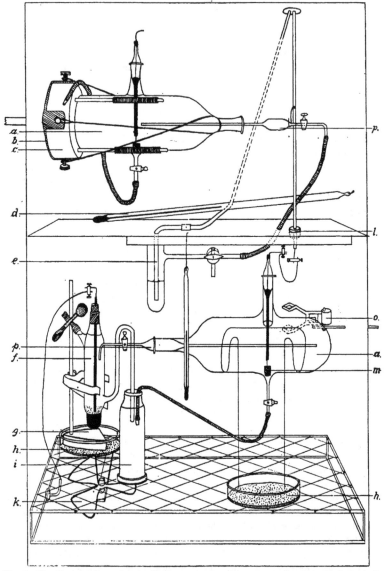

Fig. 2. The thermostat employed. In the upper department a small saturator electrode is rotating; in the lower, a similar one is ready for the estimation. *a*, small saturator electrode vessel; *b*, metal holder for the same during rotation; *c*, hollow wooden block; *d*, thermometer; *e*, wash bottle; *f*, calomel electrode; *g*, stand for calomel electrode; *h*, Petri dish filled with sealing-wax; *i*, vessel containing 3·5*n* KCl solution; *k*, fan for mixing the air in the thermostat; *l*, paraffined cork; *m*, platinum electrode; *o*, stand for Hasselbalch electrode; *p*, ground joint for tap.

treatment with pure CO_2 had been undertaken, but the difference was only slight. The experiment below will show the maximum error in the first case. Preliminary treatment with CO_2 in the small saturator was only employed in experiments where absolute accuracy was unimportant for the conclusions drawn, while it should be remembered that some of the CO_2 combinations may be a few vols. per cent. too low with tensions above 350 mm. CO_2.

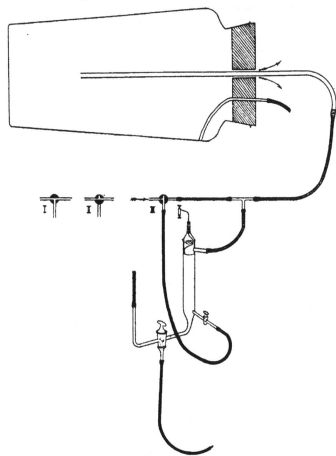

Fig. 3. Above, a large saturator; below, the Hasselbalch electrode E VI used as a Höber electrode. The tap is in position III during saturation and in position I when the liquid is transferred from the saturator to the electrode.

Ox blood containing 22·7 vols. % combined O_2 (concentrated by centrifuging) was treated with pure CO_2; half-an-hour after there were 178·8 vols. % total CO_2 in the blood; after a further treatment with CO_2 for 2½ hours 183·4 vols. % CO_2 were present.

For larger amounts of liquid a saturator of the form illustrated in Fig. 3 was used. The capacity was a litre. A curved tube for extracting samples and a conducting tube for the gas were led through the cork in a similar manner to the small saturator. In a number of the experiments a stirrer was installed in the vessel, rotating with it. The frame of the stirrer was of heavily gilt brass and its blades of wood. It became evident however that the stirrer was superfluous. As equilibrium was always established in half-an-hour with 70 cc. in the saturator without the stirrer I dispensed with it afterwards, but rotated for an hour at the same time blowing about a litre of gas through every three minutes, total amount about 20 litres.

The saturator was rotated by being mounted in a holder of practically the same type as that illustrated in Fig. 2 (without wooden blocks), and fixed in it by means of straps.

The *electrodes* used were on Höber's principle and five kinds were employed.

I. Small saturator electrode[1]. This electrode which rather resembles a form described by McClendon [1917] was constructed as a development of the above saturators. The details of its structure will be understood from Fig. 2 in which an electrode is seen which is rotated in the upper part of the thermostat while gas is blown through it, and in the lower part an electrode ready for an estimation in a little wire basket fastened in a Petri dish with sealing-wax.

The electrode consists of a vessel with a rounded end. Its capacity is about 250 cc. In the middle a small funnel is blown which is continued into a thin glass tube with a small glass tap having a single hole. On the end of the tube a piece of rubber tube is placed which partly serves for taking samples of the liquid and partly for making a connection with the KCl solution. In the plane of the glass tube at right angles to the long axis of the vessel the outer part of a ground joint is blown in the side of the vessel and it is prolonged about a cm. through the wall. The axis of the ground joint and that of the glass tube form an angle of about 150° with one another and the axis of the ground joint cuts off a sector of the cross section (not looking towards the axis of the vessel). In the ground joint there is an inner part which is continued as a glass tube through the end of which a 0·5 mm. platinum wire passes. The wire carries a 0·5 × 1 cm. platinum plate of about 0·5 mm. thickness. Connection with the sealed platinum wire is made by means of mercury or by welding a copper wire on to it. The neck of the vessel also consists of a ground joint the internal half of which inside the vessel is prolonged into a tube and outside bears a single bored tap.

The electrode vessel in the upper portion is fixed in a hollow block of wood in which three 5 mm. strong rubber coated iron wires are fastened. The block of wood is held against an iron plate by means of straps which go round the neck of the vessel and fasten on to brass knobs attached to the iron plate.

---

[1] The electrode vessels were made at Jacob's technical glass works, Copenhagen.

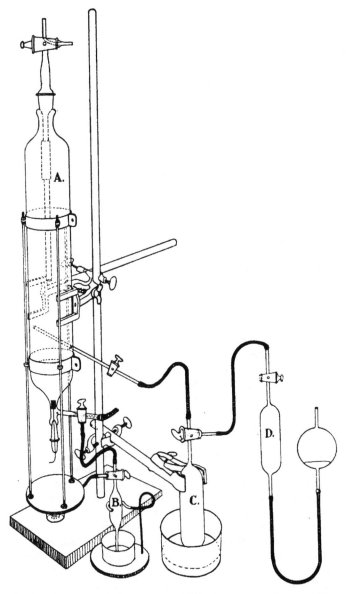

Fig. 4. On the extreme left is the large saturator L II in its stand after the potential has been measured. Then comes the small receiver filled with mercury for taking samples of the liquid. Following this is seen an "Ente" and in connection with its tap there is an ordinary gas receiver.

On the opposite side to the wooden block the iron plate carries a box which fits into a screw thread in the axis of the rotator.

The ground joints are lubricated with vaseline. The exit taps are not greased. In the electrode vessel 10 cc. of liquid are placed and rotated for half-an-hour while blowing through 8–10 litres of the H_2–CO_2 mixture. Samples of the liquid are taken while the introduction of the gas mixture continues, in a little mercury filled receiver which is seen in Fig. 4 (p. 189). The exit tap is then closed and afterwards the inlet tap, and the outer part of the ground joint is put in its place in the neck. The connecting tube is removed and the electrode placed ready for the estimation in the lower portion of the thermostat, the tube dipping into a vessel of $3.5n$ KCl and a connection made with the copper wire by a binding screw.

In constructing such an electrode again it would be an advantage to replace the small straight-bored exit tap by a tap with tail boring, as the conductivity through it is better.

In cases where the carbonic acid binding capacity of the liquid was known no samples were taken, and only 3–5 cc. were therefore introduced into the electrode vessel. In working with this saturator electrode and the saturator described above, samples of the gas blown through were taken immediately after the saturation was completed.

Now that the small saturator electrode has been described in detail it will only be necessary to give a short account of the other electrodes employed.

II. Large Saturator electrodes. (L I and L II.)

The electrode vessel was, on the whole, of the same type as the previously described smaller electrode but it was much longer and had a capacity of 1050 cc. The exit tap was an ungreased 3-way tap with tail boring (the straight boring being filled with paraffin). The inlet tap was a 3-way tap similar to the one fitted to the evacuation receiver (see Fig. 5, p. 193, upper row of taps) and where the neck of the vessel (the ground joint) joined the body a thin tube was sealed in at an angle of 45° with the axis of the vessel. This tube which was used for taking samples of the liquid had a small greased single-bore tap. The electrode vessel was mounted in a similar stand to that shown in Fig. 4, which carries the electrode vessel shown there. This stand had a box with a screw clamp and could be fixed to a rotating axle.

In measuring haemoglobin solutions (haemolysed blood) the following procedure was adopted. 20 cc. of the liquid was put in the electrode vessel and rotation started. Then either a strong stream of CO_2 was sent through it for a quarter to half-an-hour, or this preliminary treatment was dispensed with and 17 to 18 litres of H_2–CO_2 mixture passed through the electrode for half-an-hour. Then the ground joint was put in position and the inlet tap closed, and the electrode was rotated for half-an-hour with closed tap. A similar gas mixture was again passed through the electrode while rotating and finally the electrode was again rotated for half-an-hour with closed tap.

The stand was then removed from the rotating axle and hung up by silk threads, so much liquid being sucked out that the exit tube was filled, and then the rubber tube to the tap was filled with KCl solution making a connection for the measurement in the usual way (the potential was not affected beyond the experimental error (1 millivolt) by further rotation). After the estimation was completed the electrode was put in such a position that its long axis formed an angle of about 45° with the horizon and the inlet tap was lowest. The latter was now put in connection with a receiver for taking samples of gas and its two inlet tubes filled from the receiver with mercury. Through the small glass tube a sample of the liquid was taken; and then a sample of gas for analysis, through the inlet tap. In this way there was a control on the equilibrium by comparing the analysis of the gas mixture used for the saturation with that in the electrode.

The same electrode vessel (L I) was used for an investigation of the liquid in equilibrium with low tensions of CO_2. The electrode was filled with pure H_2 and then the $H_2 + CO_2$ saturated liquid was introduced while H_2 was blown through the vessel, and the ground joint was then placed in position and it was rotated for one hour. As it seemed desirable not to allow contact between the haemoglobin solution and the platinum plate to take place before the measurement was actually begun, the electrode vessel shown in Fig. 4 (L II) was constructed. Its capacity was about 1050 cc. and it consisted of a glass tube with a ground joint in each end. The inner portion of the small ground joint carried the platinum electrodes; on the prolongation of the outer portion a side tube was joined with an ungreased 3-way tap with tail boring, with the help of which communication with the KCl solution and the removal of samples of the liquid could take place. A tube for taking samples of gas was sealed in the wall of the vessel and prolonged almost to the opposite side. This tube had a straight-bored tap. By means of the contrivance which is to be seen on the drawing the tube and the 3-way tap were rinsed with about 10 cc. gas from the electrode before the sample was admitted to the receiver.

It is useful to let the platinum wire in the electrode project 2 cm. on either side of the point where it is sealed into the glass. The other details will be appreciated with sufficient clearness from the diagrams.

The technique of saturation was the same as with the electrode L I.

III.   Electrode vessels of the Hasselbalch type were likewise used but with H_2-CO_2 mixtures. One was the electrode shown by Hasselbalch and Gammeltoft [1915] and was called E III. It was only used for salt solutions. A little liquid was introduced and an H_2-CO_2 mixture bubbled through. Connection with the KCl solution was made by filling the 3-way tap and exit tube with KCl solution. The gas mixture was passed through until the potential became constant, the ground joint was turned and the 3-way tap adjusted in such a way that the bore of the tap was in line with that of the electrode vessel after which the potential was measured again.

The other electrode vessel of the Hasselbalch type called E VI was that illustrated in Fig. 3. It will be seen that there is a $CaCl_2$ tube with a ground joint from which the handle is removed, a platinum wire through the top of the joint being substituted. The wire is coiled with a single turn in the interior of the ground joint and is bent in the form of a hook against its wall. The electrode vessel is washed out before the estimation with the H_2-CO_2 gas mixture and the gas is then led through the saturator as shown in Fig. 3. As soon as saturation is complete the stream of gas is turned off and the gas is conducted, by turning a T-tap to the position marked in the illustration, directly out into the saturator. The electrode vessel is then filled with liquid to the side opening in the ground joint avoiding the introduction of air into it and the measurement made in the usual way.

In using all these electrodes it must be remembered to wash out the tube and taps with the gas mixture in order to displace any air present.

As *connecting liquid* 3·5n KCl was used.

As *standard electrode* a $n/10$ KCl calomel electrode was prepared according to Ostwald and Luther's directions. For a large number of the estimations a calomel electrode and KCl solution was used which with long standing had become too concentrated giving an error of 2 millivolts, but this has been allowed for. Later four new electrodes were made which agreed very well with one another and numerous measurements of S. P. L. Sörensen's [1909–1910] phosphate solutions gave results which did not deviate 1 millivolt from Sörensen's own value. It is however probable that the stated value of p_H· is 0·008 too low which corresponds to an error of $\frac{1}{2}$ millivolt. There is possibly thus a systematic error running through all the results but I have not corrected for it as it falls within the errors due to the method of regulating the temperature which I employed. A *Weston cell* was used as a *standard cell* and was tested twice during the investigations.

The *gas analysis* was carried out partly with the analysis apparatus of Haldane modified by A. Krogh and partly with a Petterson, Bohr, Tobiesen apparatus. CO_2 was absorbed in 50 % KOH, and O_2 in strongly alkaline pyrogallic acid solution. The maximum accuracy with Haldane's apparatus was 0·01 %; with Petterson's 0·02 %.

The gas content of the liquid was determined by exhausting with a Töpler-Hagen-Bohr mercury gas pump and analysing in Petterson's apparatus.

The liquids were collected in the little receiver shown in Fig. 4, the 3-way tap being rinsed with the liquid from which the sample was taken by suction, and the volume of the liquid being determined by weighing the mercury that had run out.

The evacuation receiver had a 3-way tap, like those shown in the uppermost row in Fig. 5. The sample was transferred to it by filling the dead space with distilled water and turning the taps as shown in the diagram (from left to right). It was necessary, as indicated, to have the tap closed with water and mercury during the pumping.

The extraction of gas from blood and amino-acids took place with a saturated solution of boric acid in the receiver while in the case of inorganic salts 5 % sulphuric acid was used as a rule. The receiver was usually heated during the process in a water bath at 38°–45°.

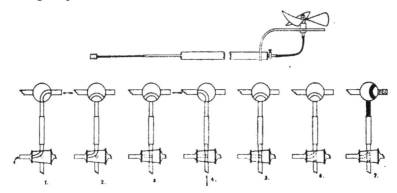

Fig. 5. Below is the small (liquid) receiver in connection with the inlet tap of the evacuation receiver. In the first position (from the left) the dead space is filled with distilled water, in the fourth position the liquid is led into the pump and in the last position the tap of the evacuation receiver is closed and the small receiver can be removed and pumping begun. Above is the stirrrer for the thermostat.

CHAPTER V

THE DIRECT EMPİRICAL DETERMINATION OF $A_{(s)}$ IN THE MODIFIED HENDERSON-HASSELBALCH EQUATION IN ITS APPLICATION TO BLOOD.

The values obtained in the present chapter are calculated from experiments in which the apparent hydrogen ion activity (H_2–Pt electrode) and the combined carbonic acid were estimated on the same sample of blood, and in which the CO_2 tension with which the blood was in equilibrium was known. In the literature similar estimations by K. A. Hasselbalch [1916, 2], T. R. Parsons [1917] and J. F. Donegan and T. R. Parsons [1919] will be found.

Hasselbalch's measurements were made with a small hook-shaped wire electrode. The volume of the hydrogen was about half that of the liquid in the electrode vessel used. Hasselbalch reckoned his constants from these experiments but did not correct for depression of hydrogen pressure. The constants have been recalculated by means of the equations given in chapter III, and the values obtained by correcting for decreased hydrogen pressure are given on the assumption that the CO_2 tension in the electrode vessel was the same as that in the saturation mixture and that CO_2 and H_2 alone were present in the electrode vessel. This proviso as we have seen in a previous chapter is only approximately true as the liquid is usually not renewed in Hasselbalch's measurements in this electrode but it will be observed from the table that the correction is only small. Parsons' and Donegan and Parsons' measurements were done on Höber's principle with Walpole's electrode vessel and are already corrected for diminished H_2 tension.

There is reason to believe that the electrical measurements of Parsons and his collaborators are the best in the literature because they replatinised their electrodes each day and only let them come in contact with the blood when the measurement was about to commence, saturation being accomplished in a special saturator. Parsons made use of a water thermostat and possibly obtained a more constant temperature than we did at the Finsen Institute. But it seems likely that his estimations of combined CO_2 are a little uncertain. This assertion is supported by an examination of his experiments by plotting them as curves, and the view is also shared by E. Jarlöv [1919]. The electrical measurements recorded in Table VIII were carried out on serum which was centrifuged from blood saturated with CO_2, at the saturation temperature, saturation being maintained throughout. In experiments with reduced blood the saturation was carried out with H_2-CO_2 mixture and the serum was resaturated with the same mixture. In experiments with oxygenated blood saturation with CO_2-air mixture was performed and the serum separated by the centrifuge was resaturated with H_2-CO_2 mixture with the same content of CO_2 as in the mixture used for the blood. The measurements given in Table IX were made directly on H_2-CO_2 saturated blood. All Parsons' figures for combined CO_2 refer to blood.

I have myself performed a number of similar experiments most of which were carried out with the small saturation electrode, but some with low CO_2 tension with the large saturation electrode L I. The platinum electrodes were platinised before each measurement, which was found to be necessary (cf. Hasselbalch and Gammeltoft [1915]) otherwise the potential was lower than with freshly platinised ones. It was found unnecessary after a few experiments to heat the electrode to redness before platinising, but later on measuring strong haemoglobin solutions this was found to be essential, for which reason it would be advisable in future to avoid all trouble by always heating the electrode to redness first.

The potential always became constant a few minutes after the electrode was ready for an estimation (except at 38° when the air in the thermostat got cool on account of the unavoidable manipulations connected with the taking of samples etc.), and this was taken as one of the indications that everything was in order.

The relative absorption coefficient in Hasselbalch's, Parsons', Donegan and Parsons' and the author's blood experiments was always taken as 0·92, while in measurements of serum-rich and serum-poor mixtures of blood I have used 0·95 and 0·855 respectively.

When not otherwise stated the blood used in the present experiments came from the cattle market and was not used before the day after the animal was slaughtered (so that "1st day" signifies the first day after slaughter). All the experiments with blood at room temperature are marked series A (and also a series with serum). They were all carried out in the autumn of 1918 and I have reason to believe all the potentials were too low without however being able definitely to state the cause. The mutual agreement within a series of experiments was however sufficiently good and therefore I have considered the experiments should be published; the course of the curve for the combination of carbonic acid is also of some interest.

The experiments marked B are subject to no such uncertainty.

Certain points about the value of $A_{(s)}$ can be surmised in advance. From the very considerable experience accumulated in different quarters on the reaction of the blood (serum) with various CO_2 tensions it can be concluded that the activity of the bicarbonate ion in serum cannot be greatly different from its mean concentration in the corresponding blood, from which it follows that $A_{(s)}$ must be of the same order of magnitude as K_1. Hasselbalch considered that K_I (Hasselbalch) varied with the mean concentration of bicarbonate in the same way as it varied in sodium bicarbonate solutions while Milroy [1917], Parsons [1919] and Michaelis [1920] on theoretical grounds thought the value ought not to vary in blood. I can support the opinion of these authors in so far as I have grounds for believing $\lambda_{(s)}$ does not vary but one cannot predict to what extent the variations of the volume of the blood corpuscles (Q in the equation (123)) react upon $A_{(s)}$ or whether the partition of the bicarbonate ion (D) between serum and blood corpuscles is variable. These conditions will be thoroughly investigated in a subsequent chapter. Haggard and Y. Henderson in 1919 calculated this constant from a number of Hasselbalch's and Parsons' experiments. The results differ somewhat from those about to be given chiefly because Haggard and Henderson have dealt with "hydrogen ion concentration" while I have dealt with "apparent activity" (Bjerrum), but furthermore these two authors have not realised that a number of Hasselbalch's experiments which they used in their calculations are really not valid for this purpose because the "hydrogen ion concentration" was not determined electrically, but reckoned from Henderson and Hasselbalch's equation.

In calculating the mean figures of the experiments I have felt justified in neglecting the values in brackets. This applies to the experiments of Parsons [1917] and Donegan and Parsons [1919] which refer to very low tensions of CO_2 because a small error in the determination of the combined CO_2 will produce a large error in $p\Lambda'_{(s)}$ in these cases. Further I have excluded all the exepriments of Donegan and Parsons where $p\Lambda'_{(s)}$ is under 6·10, since as a matter of experience too low values are much commoner than too high ones with the electrode cells employed, and the low values will therefore not be balanced in the mean value without the adjustment alluded to.

The correction for the temperature in my experiments at room temperature is effected in such a way that 0·005 was subtracted from $p\Lambda'_{(s)}$ at the temperature of the experiment for each degree below 18° and added for each degree above 18°.

The mean of Hasselbalch's 24 determinations at 38° was

$$p\Lambda'_{(s)} = 6\cdot139 \pm 0\cdot0039[1].$$

From Parsons' and Donegan and Parsons' 48 determinations at 37°

$$p\Lambda'_{(s)} = 6\cdot178 \pm 0\cdot0039.$$

From my own 15 determinations at 38°

$$p\Lambda'_{(s)} = 6\cdot147 \pm 0\cdot0079.$$

Parsons [1917] and Donegan and Parsons [1919] have remarked that there is a difference between their p_H· values and those calculated according to Hasselbalch's method. They write:

"It will be seen that the general direction of the curve (p_H-P_{CO_2} curve) is the same but that Hasselbalch's blood appears to be more acid than mine at each CO_2 pressure. The reason for this may be to a certain extent in an individual variation, and without more data of this kind for the blood of a number of individuals the extent to which this factor operates must remain undecided. But the divergence may possibly be partly explained by differences in our experimental procedure. It is a significant fact that practically all errors which are likely to occur in electrometric determinations on blood (with the exception of loss of CO_2, which is out of the question in the experiments here described) tend to produce a reduction in the value of the E.M.F. with a consequent shift of the results to the acid side. The particular points in which Hasselbalch's procedure differs from ours is that he runs oxygenated blood into the electrode vessel, and so his results are liable to be affected by an error due to depolarisation of the electrode."

Parsons has repeatedly emphasised that blood is a heterogeneous system and he also realises that Hasselbalch's constant is a total constant, but in the article written in conjunction with Donegan they draw some conclusions which will not bear criticism. They write:

"There is no doubt of the applicability of Hasselbalch's calculations in homogeneous systems containing CO_2 and sodium bicarbonate in equilibrium. But while aqueous solutions of sodium bicarbonate and also blood plasma[2] represent such systems, the same is not true of whole blood. Here we have a two-phase system, composed of corpuscles and plasma in which the differences of composition of the two phases lead to marked differences of their combining powers for CO_2. Therefore the relation between CO_2-tension and CO_2-content in the whole blood will be different from that in plasma. Now it has been shown previously that it is the relation between free and combined CO_2 in the plasma, which determines the value obtained in an electrometric determination of p_H· of the whole blood, while it is' from the different relation

---

[1] The "mean error" of the mean determination.

[2] It follows from what was said in chapter III, that I am unable to support Donegan and Parsons on this point.

between free and combined CO_2 in the whole blood we shall obtain the calculated value of its p_H. In other words the difference between the observed and the calculated p_H·'s of blood is simply an expression of the difference in behaviour of corpuscles and plasma towards CO_2."

Although I am in agreement with Parsons [1917] that the oxygen error may have caused Hasselbalch's measurements to be too low, he overlooks the fact that the CO_2 tension has not been properly appreciated in Hasselbalch's electrode (no renewal) which will militate against the O_2 error so that it is impossible to predict which factor will be paramount.

Donegan and Parsons also overlook the fact that Hasselbalch's constant was determined by measurements on whole blood so that the constant contains a correction for the different bicarbonate contents of serum and blood corpuscles, and that the calculated and measured p_H· must be the same if Hasselbalch's and Parsons' techniques give concordant results.

That the $p\Lambda'_{(s)}$ value of Parsons and Donegan and Parsons is 0·023[1] greater than mine can only be due to lack of uniformity in the technique of the electrical measurements and as they are all carried out on the Höber principle it seems to me that provided the difference is a real one there can only be two possibilities:

(1) either our calomel electrodes are different,

(2) or my lower value depends upon my not having glowed my platinum electrode before each platinising.

The difference between our constants is really very small, although it seems to be real judging from the mean error, and it would mean a considerable number of new experiments to determine what the true value ought to be.

If the tables are examined more carefully it will appear that $p\Lambda'_{(s)}$ from Donegan and Parsons' measurements seems to increase greatly with reactions more alkaline than p_H· 8·0. This is however doubtful and my own measurements with a marked alkaline reaction though few in number do not show this.

On the face of it there appears to be no certain variation of $p\Lambda'_{(s)}$ caused by change of reaction but a more careful investigation of the question might be useful.

If we divide Parsons' and Donegan and Parsons' measurements in such a way that all the experiments in which p_H· was equal to or over 7·46 are placed in one group, and all those in which p_H· was equal to or below 7·45 are placed in a second group, we get 21 experiments

$$p_H· \gtrless 7·46 \quad p\Lambda'_{(s)} = 6·188 \pm 0·0063$$

with a standard deviation of 0·029; and 27 experiments

$$7·45 \gtrless p_H· \quad p\Lambda'_{(s)} = 6·170 \pm 0·0049$$

with a standard deviation of 0·023. The difference between the constants is 0·018 ± 0·008.

[1] Corrected for the influence of the temperature.

It seems from the above as if $p\Lambda'_{(s)}$ increases a little with the $p_{H^.}$ value. The standard deviation in the two series is of the same order so that we are justified in assuming there is a real deviation of the "constant," and the results group themselves about the mean in the same way in the two groups so that we can compare the mean values.

Dealing with my own experiments at 38° in a similar way, we have 8 experiments

$$p_{H^.} \gtreqless 7\cdot39 \quad p\Lambda'_{(s)} = 6\cdot161 \pm 0\cdot0099$$

with a standard deviation of 0·028; and 9 experiments

$$7\cdot38 \gtreqless p_{H^.} \quad p\Lambda'_{(s)} = 6\cdot133 \pm 0\cdot012$$

with a standard deviation of 0·036. The difference between the constants is 0·028 ± 0·0155; $p\Lambda'_{(s)}$ seems therefore to increase a little with the $p_{H^.}$. It is clear however from the calculations that the change in $p\Lambda'_{(s)}$ is small in all cases and a large number of experiments would still be necessary to accurately determine its magnitude.

At 38° the following is the mean of Parsons' [1917], Donegan and Parsons' [1919] and my own measurements:

$$p\Lambda'_{(s)} = 6\cdot159, \text{ from which } p\Lambda_{(s)} = 6\cdot193$$

and $\quad\quad\quad\quad \Lambda_{(s)} = 6\cdot41 \times 10^{-7}.$

At 18° I found $\quad p\Lambda'_{(s)} = 6\cdot24, \text{ from which } p\Lambda_{(s)} = 6\cdot27$

and $\quad\quad\quad\quad \Lambda_{(s)} = 5\cdot4 \times 10^{-7}.$

Résumé.

I. The constant $\Lambda_{(s)}$ has been calculated at body temperature and room temperature from all the suitable experiments to be found in the literature.

II. A considerable number of new experiments for the determination of the constant have been performed.

III. A small discrepancy has been shown to exist between the constant reckoned from Parsons' and Donegan and Parsons' experiments and my own. and the possible causes of the disagreement are discussed.

IV. It has been shown to be probable that the constant increases a little with increasing hydrogen ion activity.

Table VII. Calculated after K. A. Hasselbalch, *Biochem. Zeitsch.* **78**, p. 123.

	mm. Hg CO_2	Vols. % CO_2 (combined)	$p_{H'(s)}$	$p\Lambda'_{(s)}$	Corrected $p_{H'}$	Corrected $p\Lambda'_{(s)}$
38°. Ox blood I	39·3	42·6	7·31	6·11	7·32	6·12
	28·1	37·2	7·43	6·14	7·44	6·15
	63·1	47·9	7·19	6·14	7·21	6·15
	95·6	62·7	7·09	6·10	7·12	6·13
38°. Ox blood II	10·7	31·6	7·77	6·13	7·77	6·13
	33·6	47·8	7·47	6·15	7·48	6·16
	96·7	65·0	7·11	6·11	7·14	6·14
	61·2	58·7	7·27	6·10	7·29	6·12
	20·1	39·8	7·62	6·15	7·63	6·16
	43·7	51·8	7·36	6·11	7·37	6·12
	74·0	64·7	7·22	6·11	7·24	6·13
38°. Ox blood·III	41·1	48·8	7·37	6·12	7·38	6·13
	41·0	48·5	7·37	6·13	7·38	6·14
38°. Ox blood V	42·7	44·3	7·31	6·13	7·32	6·14
	39·3	44·5	7·33	6·10	7·34	6·11
	40·9	42·2	7·29	6·12	7·30	6·13
38°. Ox blood VI	36·3	45·2	7·39	6·12	7·40	6·13
	54·4	52·0	7·32	6·17	7·34	6·19
						Mean 6·14
38°. Human blood (K.A.H.)	50·8	57·1	7·31	6·09	7·33	6·11
	45·7	52·9	7·37	6·13	7·39	6·15
	32·7	45·1	7·43	6·12	7·44	6·14
	18·5	35·1	7·60	6·15	7·61	6·16
	22·4	36·7	7·54	6·15	7·55	6·16
	80·7	61·3	7·17	6·12	7·20	6·15
						Mean 6·145
18°. Human blood (K.A.H.)	44·8	54·3	7·23	6·20	7·24	6·21
	22·3	44·3	7·47	6·22	7·48	6·23
						Mean 6·22

Table VIII. T. R. Parsons, *J. Physiol.* **51**, p. 448.

No.	mm. Hg CO_2	Vols. % CO_2 (combined) Oxygenated blood	Reduced blood	$p_{H'}$ Oxygenated blood	Reduced blood	$p\Lambda'_{(s)}$ 37° Oxygenated blood	Reduced blood
1	37·4	43·4	47·3	7·37	7·45	6·13	6·18
2	0·79	—	12·4	—	8·55	—	(6·19)
3	19·6	34·4	39·9	7·62	7·69	6·20	6·22
4	72·1	59·9	64·4	7·25	7·28	6·17	6·17
5	10·1	30·0	31·5	7·79	7·86	6·16	6·21
6	8·1	—	27·4	—	7·91	—	6·22
7	55·3	—	55·1	—	7·35	—	6·19
8	33·4	41·5	47·6	7·48	7·53	6·13	6·21
9	5·7	21·3	23·8	7·96	8·01	6·22	6·22
					Mean	6·17	6·20
							6·19

Table IX. Calculated after Donegan and Parsons, *J. Physiol.* **52**, pp. 317–318.

mm. Hg CO_2	Vols. % CO_2 (combined)	p_H·	$p\Lambda'_{(s)}$ 37°	mm. Hg CO_2	Vols. % CO_2 (combined)	p_H·	$p\Lambda'_{(s)}$ 37°
0·7	4·5	8·29	(6·32)	1·0	12·4	8·28	(6·01)
9·7	33·1	7·78	(6·08)	21·3	38·5	7·58	6·16
29·6	44·3	7·51	6·17	40·5	46·8	7·37	6·14
44·0	53·1	7·39	6·14	60·0	50·8	7·26	6·17
65·7	60·7	7·30	6·17	2·0	8·9	8·15	(6·34)
1·3	6·4	8·26	(6·30)	14·2	29·3	7·67	6·19
12·7	30·5	7·75	6·20	30·8	38·9	7·43	6·17
37·0	45·0	7·45	6·20	45·9	45·0	7·34	6·18
67·6	52·6	7·24	6·18	2·0	8·5	8·23	(6·32)
2·1	8·4	8·26	(6·29)	19·1	33·5	7·61	6·20
18·8	36·3	7·67	6·22	40·8	44·6	7·39	6·19
31·5	48·2	7·49	6·14	60·7	54·9	7·27	6·15
2·2	10·4	8·30	(6·36)	10·9	26·0	7·74	6·20
18·1	39·4	7·68	6·16	29·3	37·8	7·47	6·20
38·9	43·8	7·44	6·22	45·3	46·2	7·32	6·20
59·3	53·9	7·30	6·18	0·7	8·8	8·02	(6·75)
1·3	7·7	8·19	(6·25)	11·6	27·4	7·58	(6·03)
20·0	35·5	7·59	6·18	31·3	37·0	7·37	6·13
42·7	45·1	7·36	6·17	46·1	41·3	7·27	6·15
65·9	53·0	7·24	6·17	12·5	23·0	7·61	6·18
1·3	10·0	8·27	(6·22)	25·6	30·3	7·42	6·18
15·9	35·6	7·66	6·15	40·9	36·9	7·28	6·16
38·2	46·9	7·41	6·16			Mean	6·18
51·5	50·7	7·31	6·14				

Table X. Determination of $p\Lambda'_{(s)}$ in blood at 38° with the small saturator electrode, Series B.

mm. Hg		Vols. % combined				
CO_2	O_2	CO_2	O_2	p_H·	$p\Lambda'_{(s)}$	
13·6	0·4	27·7	0·2	7·65	6·18	
74·6	0·5	26·5	0·6	—	—	Human blood (E.J.W. 31. i. 19). Placed
39·5	0·4	42·8	0·6	7·39	6·20	on ice immediately
109·2	0·5	64·7	0·2	7·10	6·15	
544·5	0·5	99·5	0·1	6·60	6·17	In the ice safe overnight
		Ordinary air	24·8		Mean 6·18	
501·9	0·4	84·2	0·0	6·53	6·13	
83·5	0·2	55·7	0·1	7·07	6·08	
36·0	0·5	44·1	0·3	7·34	6·10	Ox blood
8·4	0·3	24·2	—	7·75	6·11	
108·7	0·2	60·5	0·4	7·07	6·15	
39·6 +ord. air		39·6	20·1		Mean 6·10	
33·4	0·3	44·3	0·2	7·43	6·14	Rabbit blood from four animals (quite
120·9	0·4	65·4	0·2	7·04	6·14	fresh, placed on ice immediately)
39·4	0·3	45·5	0·7	7·38	6·15	
37·8	140·0	41·2	13·2		Mean 6·14	
518·1	0·7	99·4	0·0	6·55	6·10	
39·6	0·3	49·8	0·7	7·45	6·18	Horse blood
97·3	0·5	65·7	0·5	7·16	6·19	
31·9	0·5	47·0	0·6	7·51	6·17	In the ice safe overnight
34·1	137·5	40·0	20·0		Mean 6·16	
26·0	1·0	36·3	0·1	7·48	6·16	Human blood, fresh (chronic nephritis)
19·9	0·2		0·1	7·59	—	
96·6	0·4			7·10	—	
36·4	136·8	31·7	20·8			

Mean of all the determinations 6·15.

Table XI. Determination of $p\Lambda'_{(s)}$ in blood at 18° with the small and large (L I) saturator electrodes, Series A.

Temp.	mm. Hg		Vols. % combined		$p_{\mathrm{H}}\cdot$	$p\Lambda'_{(s)}$	
	CO₂	O₂	CO₂	O₂			
17·5	250·4	0·8	112·1	0·1	6·85	6·23	Human blood (E.J.W. 3. x. 18),
18·5	91·8	0·6	89·2	0·4	7·17	6·23	fresh, placed on ice
18	24·1	0·8	59·4	—	7·62	6·28	
18	26·3	0·8	56·5	—	7·60	6·32	2nd day
18	37·7	0·3	—	—	7·50	—	
18	16·6	0·8	48·9	0·2	—	—	
18	73·1	0·7	78·6	0·5	7·23	6·25	
18	58·0	143·8	65·6	22·7	—	—	
18	58·0	143·8	65·9	23·1	—	—	
18	4·2	0·8	26·3	0·8	8·02	6·29	3rd day. Large saturator elect.
						Mean 6·27	(L I)
16·5	147·1	0·8	74·0	0·2	6·84	6·21	Ox blood
17	80·2	0·6	65·3	0·4	6·07	6·22	
17	58·8	0·8	59·8	0·5	6·19	6·24	
17	58·8	0·8	59·6	0·0	6·19	6·24	CO₂ preliminary treatment
16·5	419·6	1·1	85·8	0·4	6·43	6·19 ·	Next day. CO₂ preliminary
17	51·7	1·2	58·4	0·0	7·26	6·27	treatment
17	16·2	0·6	42·3	0·0	7·57?	(6·22)?	„ „
17	22·3	1·0	48·4	0·2	7·50	6·23	„ „
16·5	5·3	0·6	29·1	0·0	7·92	6·24	3rd day. Large saturator elect.
	Air from water blower		14·9			Mean 6·24	(L I)
	„ „		15·1				
19	77·1	0·8	68·5	—	7·15	6·25	Dog blood (0·1 % oxalate)
19	29·3	0·3	53·5	—	7·47	6·26	
19	56·4	0·3	62·4	0·4	7·29	6·29	
19	127·5	0·2	76·0	0·2	6·97	6·24	
19	337·6	0·5	88·8	0·2	6·65	6·26	
19	2·2	0·7	21·8	0·4	8·15	(6·38)	Large saturator elect. (L I)
	Air from water blower		19·2			Mean 6·26	
	„ „		19·2				
18·5	41·5	0·3	63·0	0·4	7·38	6·25	60′ after blood Rabbit blood
18·5	24·2	0·3	53·1	0·0	—	—	90′ was taken. (on ice)
18·5	83·1	0·5	76·5	0·2	7·14	6·23	150′ „
18·5	368·0	0·2	107·0	0·4	6·61	6·20	195′ „
18·5	44·2	0·5	64·1	0·4	7·35	6·24	240′ „ Vol. % O₂
18·5	16·0	0·3	48·6	0·5	7·64	6·21	300′ „ (combined)
18·5	81·7	0·4	73·8	0·1	7·16	6·25	390′ „ 16·6
						Mean 6·23	
19	91·5	0·6	83·5	0·7	—	—	1st day. Horse blood
19	22·0	0·3	56·2	0·7	7·62	6·26	
19	48·7	0·3	69·9	0·7	7·39	6·28	
18·5	3·0	0·4	25·2	0·4	8·14	6·26	L I
19	39·6	0·2	66·3	0·5	7·43	6·25	
18·5	484·7	0·2	112·2	0·9	6·54	6·25	2nd day
19	80·2	0·2	78·2	—	7·20	6·24	3rd day
	Air from water blower		22·3			Mean 6·26	
	„ „		22·9				
18·5	128·6	0·6	82·0	0·6	6·99	6·23	2nd day. Equal parts of horse
19	58·5	0·6	71·9	0·5	—	—	blood and serum. $\Psi = 5$
19	38·8	0·3	66·6	0·4	7·42	6·24	
21	269·8	0·4	90·3	—	6·72	6·24	3rd day
						Mean 6·24	
20	126·9	0·4	90·3	—			3rd day. Horse blood cell sus-
20	42·6	0·4	59·6	—	7·42	6·27	pension
20	72·1	0·2	74·3	—	7·26	6·25	
						Mean 6·26	

Table XI (continued)

Temp.	mm. Hg CO₂	mm. Hg O₂	Vols. % combined CO₂	Vols. % combined O₂	p_{H}	$p\Lambda'_{(s)}$		
19·5	536·4	0·4	102·2	—	6·45	6·22	2nd day	Ox blood
21	91·9	0·3	75·5	—	7·13	6·25	3rd day	
· 19·5	24·3	0·2	53·3	—	7·52	6·23	4th day	
—	—	156·0	—	18·96	Mean 6·23			
—	—	156·0	—	18·97				
19	510·5	0·5	89·6	—	6·44	6·25	2nd day	Equal parts of ox
21	91·9	0·3	72·5	—	7·08	6·23	3rd day	blood and serum.
20	34·3	0·4	60·7	—	7·45	6·25		$\Psi = 5$
20	24·3	0·2	57·8	—	7·53	6·23	4th day	
—	—	156·6	—	8·5	Mean 6·24			
19	508·0	—	115·8	—	6·59	6·24	1st day	Ox blood cell sus-
20	65·7	0·3	70·3	—	7·24	6·23	3rd day	pension. $\Psi = 14·5$
20	34·3	0·4	57·1	—	7·42	6·26		
20	2·8	1·1	34·4	—	7·80	6·25	5th day	L I
—	—	156·6	—	33·8	Mean 6·25			
18	65·3	0·5	73·0	0·1	—	—	1st day	Ox blood
19	51·0	1·2	66·6	0·2	7·29	6·21	2nd day	
19	6·0	—	36·5	0·3	7·96	6·22	3rd day	L I
18	58·0	140·7	—	16·5	Mean 6·215			
18	58·0	140·7	—	16·6				
19	38·6	0·8	64·0	0·5	7·42	6·28	1st day	Blood and serum,
19	51·0	1·2	66·7	0·4	7·29	6·23	2nd day	equal parts. $\Psi = 5$
18	144·5	0·6	81·6	—	6·91	6·22	3rd day	
—	—	146·3	—	6·9	Mean 6·24			
18	53·7	1·0	71·0	0·9	7·33	6·23	2nd day	Ox blood cell sus-
18	64·9	1·3	76·7	0·1	7·24	6·19	3rd day	pension
18	64·9	1·3	75·4	0·6	7·24	6·20		
—	—	146·3	—	32·3	Mean 6·20			

CHAPTER VI

THE DETERMINATION OF THE $p\Lambda'_{(s)}$ VARIATION.

From equation (127) and (123) in logarithmic form we get

$$p\Lambda'_{(s)} = p\lambda_{(s)} - \log \frac{100 - Q(1-D)}{100} - \log \Phi_a (CO_2). \quad \ldots\ldots(129)$$

If we now calculate $p_{\mathrm{H}'(s)}$ for blood by using the following equation

$$p_{\mathrm{H}'(s)} \text{ (uncorrected)} = p\lambda_{(s)} + \log B - \log \int_0 = p\lambda_{(s)} + \log \Phi_a (CO_2)$$

$$+ \log B - \log \frac{P_{CO_2}a}{7·60}\ldots\ldots(130)$$

instead of (126) $\qquad p_{\mathrm{H}'(s)} = p\Lambda'_{(s)} + \log B - \log \int_0,$

where B is as usual the volume % of combined CO_2 in the blood and $\int_0$ is the dissolved CO_2 in blood, the p_H· value is not correct because

$$+ \left(- \log \Phi_a (CO_2) - \log \frac{100 - Q(1-D)}{100} \right)$$

must be added to it.

As will be seen from (129) this gives in addition the $p\Lambda'_{(s)}$ variation provided $p\lambda_{(s)}$ is constant, which *a priori* is very probable as the volume of the protein phase in serum is relatively small and the amount of salt large in comparison with the quantity of electrolytes which vary with the reaction.

$\Phi_a (CO_2)$ has been determined from (128) on the assumption that 100 cc. blood corpuscles combine with 49 cc. O_2. This is undoubtedly incorrect but the error involved in the above approximation is but trifling compared with the uncertainty of the relative absorption coefficient in blood corpuscles (0·81) [Bohr, 1905].

When these investigations were begun there were only a few rather uncertain experiments which allowed of the determination of D.

$$\left(D = \frac{\text{c.c. combined } CO_2 \text{ in } 100 \text{ c.c. blood corpuscles}}{\text{c.c. combined } CO_2 \text{ in } 100 \text{ c.c. serum}} \right).$$

Since then some experiments by L. S. Fridericia [1920] and by J. Joffe and E. P. Poulton [1920] have appeared from which this factor may be determined. The authors have themselves made the calculation which in many ways is similar to the one I am about to put forward, and they have also drawn attention to the difficulties with the Henderson-Hasselbalch equation which arise in connection with it. Joffe and Poulton have moreover shown how the difficulties can be evaded.

I have however made a recalculation of these authors' experiments because I believe the method of calculation I have employed has considerable advantages over that of Joffe and Poulton and it also better admits of a comparison of the various series of experiments than would otherwise be possible. A. Schmidt [1867] and N. Zuntz [1867] showed almost simultaneously in 1867 that the blood corpuscles contain considerable amounts of combined CO_2. They further proved that this amount rose rapidly with the CO_2 tension (although Schmidt on ·the basis of some poorer experiments from a technical standpoint, with dog's blood, considered that the CO_2 combination passed through a minimum with increasing CO_2 tension), and lastly they showed that the combination of CO_2 with serum which was obtained from blood saturated at high CO_2 tensions was considerably greater than with serum of the same CO_2 tension provided the serum was got from (the same) blood at lower tensions. They explained this phenomenon—which I shall repeatedly revert to and which I have called the Schmidt-Zuntz phenomenon—by assuming a transference of sodium from the blood corpuscles to the serum (plasma) under the conditions cited. The experiments do not permit D to be calculated with reasonable certainty. L. Fredericq [1878] has shown that in horse blood corpuscles a little over half as much CO_2 is combined as in the serum at physiological CO_2 tensions, while at 745 mm. about nine-tenths are combined in the blood corpuscles as compared with an equal volume of serum.

From experiments carried out by Fr. Kraus [1898]—experiments which seem to me to be much too little known—it is clear that D in ox blood at physiological CO_2 tensions is about $\frac{3}{4}$, while at a couple of hundred mm. CO_2 it is about 1.

Quite recently W. Falta and Richter-Quittner [1920] have asserted that the partition of Cl, glucose and non-protein nitrogen between blood corpuscles and serum is absolutely different from that between blood corpuscles and plasma because they thought they were able to show that the blood corpuscles, before coagulation sets in, do not contain appreciable amounts of these substances, while after coagulation has taken place they contain considerable quantities—as is well known. They further claim to have proved that different treatment of the blood such as the prolonged action of cold, addition of oxalate, etc. has the same effect as defibrination. If Falta and Richter-Quittner's experiments are really sound they will be of epoch-making importance in the subject of the osmotic conditions of the blood corpuscles and the permeability of their surfaces, and pursued to their ultimate conclusion they will occasion a complete change of the current view concerning the conditions of equilibrium between blood corpuscles and plasma, and thereby influence our ideas on the production of lymph and kindred subjects.

L. Fredericq's [1878] experiments show that the partition of combined CO_2 between blood cells and serum is the same as between blood cells and plasma. Fredericq hindered the coagulation of the blood (horse blood) by preserving it in an excised jugular vein.

As early as 1893 H. J. Hamburger showed that the partition of chloride between blood corpuscles and plasma (obtained under oil) was the same as that between blood corpuscles and serum when the blood was defibrinated without admitting the air.

K. L. Gad Andresen [1920] has raised objection to the experiments of Falta and Richter-Quittner, as he has demonstrated that the partition of urea is independent of coagulation (hirudin-plasma). Rich. Ege [1920] has shown the same for chlorine and glucose. A. Norgaard and H. C. Gram [1921] have shown that the Cl content of the blood is a simple function of that of the blood corpuscles (or amount of haemoglobin) when it is assumed the corpuscles contain about half the amount of Cl present in the serum (per unit volume). They also found this ratio in blood rendered incoagulable by the addition of isotonic sodium citrate solution.

H. C. Hagedorn [1920] has shown in numerous experiments that human blood corpuscles obtained from blood with the addition of hirudin contain considerable amounts of glucose. In a short paper I have myself [Warburg, 1920] shown that chlorine and combined CO_2 are distributed in the same manner between blood corpuscles and hirudin plasma as between blood corpuscles and serum even though the CO_2 tension is varied greatly.

Falta and Richter-Quittner [1921] have maintained their position in opposition to these experiments of Gad Andresen, Ege, Hagedorn and Warburg, or at any rate have only modified it to a very slight extent as regards the main problem. It is however impossible to bring the two views into agreement with one another. There are two possibilities—either Falta and Richter-Quittner are unable to analyse the contents of the blood corpuscles with regard to the substances in question, or Fredericq, Hamburger, Gad Andresen, Ege, Norgaard and Gram, Hagedorn[1] and the author are unable to do it.

It may further be of interest to discuss an erroneous assumption which seems to be widespread.

N. Zuntz [1867, 1868] showed in the paper previously referred to, that the alkali in blood which was estimated by titration decreased on coagulation and concluded from this that acid was formed in the process. This fact has subsequently been confirmed several times, amongst others by Loewy and Zuntz [1894] and by A. Jaquet [1892], but it has been curious that the amount of acid formed varied comparatively greatly from one experiment to another. Quite recently Howard Haggard and Yandell Henderson [1920, 1] have again demonstrated the presence of such an acid, having found, at high CO_2 tensions, several volumes % less CO_2 combined in defibrinated blood than in oxalated blood (a difference which is however small compared with what Zuntz [1868] originally found). J. Joffe and E. P. Poulton [1920] have carried out numerous CO_2 determinations of Joffe's blood, defibrinated and oxalated, and they found no difference in the combined CO_2. A. Krogh and G. Liljestrand [1921] have also found the same amount of CO_2 combined in the defibrinated and oxalated blood of Liljestrand. Dr Chr. Lundsgaard and Dr E. Möller have kindly reported to me that in a number of experiments not yet published they have found no difference in the CO_2 combined in defibrinated and oxalated blood (man).

In the previously mentioned experiments of the author no difference in the combined CO_2 in defibrinated and "hirudin" blood (horse) was found. That some investigators discovered a decreased titrimetric alkalinity or CO_2 combining power after coagulation can undoubtedly be ascribed to the fact that a quantity of blood corpuscles may be removed by the defibrination, the bearing of which on the present problem has not hitherto been appreciated, as these at higher CO_2 tensions or more acid reactions function as a base (see chap. XI) and combine with acid.

L. S. Fridericia [1920] has reported three experiments with 0·1 % oxalated ox blood which permit of the calculation of D at three different reactions. The reactions were calculated from the plasma figures, the value for $p\lambda_{(s)}$ which will be found below being used.

Hasselbalch and Warburg [1918] have reported experiments (see Table XIV) from which a similar calculation can be made, as the carbonic acid combination

[1] E. Wichmann [*Pflüger's Arch.* 1921, **189**, p. 108] has likewise shown that Falta and Richter-Quittner's view regarding the difference of hirudin blood is incorrect.

curve of the blood was determined (which had not been previously reported) simultaneously with the experiment with the ox blood which combined with 13·4 vols. % O_2. The values obtained in this experiment are very uncertain because the blood contained few blood corpuscles and there were also local difficulties with the centrifuging. The other experiment of Hasselbalch and Warburg which allows of the calculation being made should give much better results. I have also carried out a few experiments with defibrinated ox blood which will be found in Table XVII.

In calculating D from ox blood experiments it is always assumed that 100 vols. blood corpuscles combine with 45 vols. O_2.

I have performed a number of experiments with defibrinated horse blood which appear in Tables XVIII and XIX. The details of the *technique* were as follows: 50–70 cc. blood were saturated for one hour in the saturator shown in Fig. 3. Instead of the electrode shown in the diagram, a glass cylinder with a rubber cork bored with two holes was substituted through which a long and a short glass tube were introduced. The long tube could be pushed up and down through the cork, the movement being airtight. The gas was conducted by the rubber tube through the glass cylinder before reaching the saturator.

After the completion of saturation a sample of blood was first taken for testing (B_I), whereupon the T tap was turned to position I and the exit rubber tube was removed from its glass tube. The cylinder was then filled with the blood through the glass tube by siphon action avoiding contact with the air. The position of the tap was finally as shown in position II.

When the blood corpuscles had settled a sample was taken through the long glass tube with the help of a rubber tube for pumping out of the serum (S_I) and then another of the blood corpuscle emulsion (C_I). A portion of the serum (S_{II}) was then saturated again and also a portion of blood corpuscles (C_{II}) in a small saturator (Fig. 5) and their gas content estimated in the usual way. Lastly a sample of blood was saturated afresh which before the beginning of the main experiment was saturated with the gas mixture and had been standing in the thermostat (at room temperature), closed in a small saturator, and the sample was pumped out and analysed (B_{II}).

At higher CO_2 tensions the other gas in the spirometer was O_2.

Haematocrite estimations were performed on B_I, C_I and B_{II} as a rule, usually two at a time, and centrifuging was continued until the column of corpuscles was transparent. In those cases where haematocrite estimations were not done the volume of the blood corpuscles was calculated from the amount of oxygen, which is duly noted in the tables.

In the experiments in which solutions rich in blood corpuscles were employed these were the only ones used for the calculation of D, other experiments being used as controls. The values for the oxygen combining power of the corpuscles given in Fig. 9 are all derived from experiments with blood cell emulsions.

Most of the horse blood came from the cattle market. The blood for experiments 1–3 was fresh and obtained from the serum laboratory of "Landbohöjskolen."

A complete experiment lasted 4–5 hours. Saturation was usually finished after $3\frac{1}{2}$–4 hours.

In experiments with ox blood it was passed from the saturator into a centrifuge tube, the rubber tube being put right down to the bottom. The blood then rose slowly in the tube which was closed immediately with a cork, and melted paraffin poured over it. After centrifuging the serum and blood cell emulsion were resaturated separately.

Hasselbalch [1916, 2] has carried out the determinations[1] of $p\lambda'_{(s)}$ given in Table XII. They were done with the technique of Hasselbalch and the remarks previously made about this technique refer also to these experiments. It should be stated however that the oxygen error is much less in these experiments than with blood.

In Table XIII a number of determinations of $p\lambda'_{(s)}$ are given which were carried out with the small saturator electrode in the usual way. The oxygen in the spirometer gas was always estimated but on no occasion did it exceed 1 mm.; as a rule it was about 0·5 mm.

From the mean of Hasselbalch's four experiments at 38°

$$p\lambda'_{(s)} = 6\cdot13 \text{ and consequently } p\lambda_{(s)} = 6\cdot14.$$

From the 23 estimations given in the table I get at 38°

$$p\lambda'_{(s)} = 6\cdot144 \pm 0\cdot0042.$$

The standard error was 0·018. From the mean of 19 estimations at 18°

$$p\lambda'_{(s)} = 6\cdot283 \pm 0\cdot0042.$$

The standard error was 0·019. From these we get at 38°

$$p\lambda_{(s)} = 6\cdot155 \pm 0\cdot0042,$$

and at 18° $p\lambda_{(s)} = 6\cdot294 \pm 0\cdot0042.$

The tables show that there is no variation of the constants determined by the reaction, which is in harmony with the small standard deviation.

If the estimations at room temperature and at 38° are combined we get

$$p\lambda_{(s)}t° = 6\cdot294 + (18° - t°) \, 0\cdot007. \qquad \ldots\ldots\ldots\ldots(131)$$

In Fig. 6 the values of D are plotted as ordinates and the apparent hydrogen ion activity exponents as abscissae. The individual determinations, as will be seen, are well distributed about the curve and it will be noted D varies considerably in the range of reaction dealt with. The cause of this variation will be the subject of a thorough inquiry in the last chapter.

Some points on the figure are marked with a circle. They refer to the determinations in Table XIX. It will be observed that the addition of a solution isotonic with the blood, 5 cc.–100 cc. of 0·9 % NaCl solution (0·154n), does not change D, while a similar quantity of a hypertonic salt solution, 0·5n NaHCO₃ and 1·0n NaCl, makes D smaller. This last fact is of par-

---

[1] $p\lambda'_{(s)}$ has been formed analogically with $p\Lambda'_{(s)}$ in equation (127).

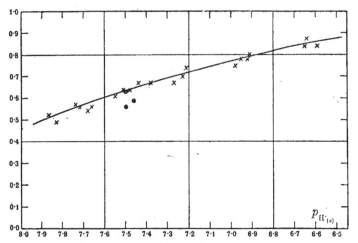

Fig. 6. Horse blood (defibrinated), room temperature.

$$\frac{\text{cc. combined } CO_2 \text{ in 100 cc. blood corpuscles}}{\text{cc. combined } CO_2 \text{ in 100 cc. serum}} = D.$$

ticular interest and is in harmony with what would be expected from theoretical considerations.

In the experiment with the addition of physiological salt solution the blood combines with 62·5 vols. % CO_2 at a CO_2 tension of 37·3 mm., with a serum of p_H· 7·50, while in the corresponding serum 73·5 vols. % CO_2 (combined) are found. In the experiment with hypertonic salt solution the blood combines with the same amount of CO_2 (62·7 vols. %) at practically the same tension (38·5 mm.) and the same p_H· (7·50), but the serum combines with *rather more* CO_2 (76·0 vols. %).

The CO_2 combined by the serum is 4 % greater in the last experiment than in the first. At the same time it can be demonstrated that the volume of the blood corpuscles has diminished by 12 %.

Table XII. $p\lambda'_{(s)}$ calculated from K. A. Hasselbalch's experiments, *Biochem. Zeitsch.* **78**, p. 123.

	mm. Hg CO_2	Vols. % CO_2 combined	$p_{H'(s)}$	$p\lambda'_{(s)}$	Corrected $p_{H'(s)}$	Corrected $p\lambda'_{(s)}$
38°. Ox serum III	41·8	62·0	7·44	6·12	7·45	6·13
	14·5	54·3	7·83	6·11	7·83	6·11
38°. Ox serum IV	40·6	60·9	7·45	6·13	7·46	6·14
	59·8	59·8	7·45	6·12	7·46	6·13
					Mean	6·13
18°.	41·0	64·6	7·36	6·24	7·37	6·25

Table XIII. Determinations of $p\lambda'_{(s)}$ in serum carried out with the small saturator electrode.

Temp.	mm. Hg CO_2	Vols. % CO_2 combined	$p_{H'(s)}$	$p\lambda'_{(s)}$ 18°	$p\lambda'_{(s)}$ 38°	
18·5	512·5	71·1	6·31	6·24	—	Series A
18·5	119·6	64·4	—	—	—	1st day
18·5	18·7	58·2	—	—	—	
19·5	41·2	62·6	7·36	6·24	—	Horse blood
19·5	12·7	55·8	7·82	6·24	—	2nd day
18	545·9	75·9	6·32	6·21	—	Series A
18·5	118·6	69·4	6·95	6·26	—	Ox serum
19	43·1	64·7	7·36	6·24	—	
19	16·0	58·5	7·71	6·22	—	
			Mean	6·23		
38	550·1	69·1	6·43	—	6·18	Series B
20	573·7	71·4	6·28	6·24	—	1st day
38	79·6	—	7·19	—	—	
19·5	83·1	65·3	7·12	6·28	—	Horse serum
38	12·9	50·7	7·89	—	6·14	2nd day
19·5	13·1	55·6	7·85	6·28	—	
38	32·6	58·1	7·50	—	6·10	
19·5	34·1	61·1	7·49	6·29	—	
38	521·7	75·5	—	—	—	Series B
18	546·8	65·1	6·37	6·30	—	Ox serum
38	79·8	68·1	7·24	—	6·16	
18	83·6	76·1	7·16	6·28	—	
21	595·0	75·4	—	—	—	Series B
21	122·9	71·9	—	—	—	1st day
21	40·6	65·3	7·46	6·31	—	Ox serum
21·5	16·1	61·8	7·80	6·28	—	3rd day
21·5	40·5	66·0	7·40	6·25	—	
21·5	70·2	68·5	7·22	6·29	—	
21	558·5	79·5	6·38	6·30	—	Series B
21	96·7	74·4	7·12	6·30	—	Horse serum
21	14·2	63·0	7·86	6·28	—	
22	18·9	65·6	6·75	6·25	—	Series B
21·5	54·3	67·6	7·32	6·29	—	Horse serum
21·5	38·4	67·5	7·48	6·29	—	
19·5	96·3	74·9	7·12	6·30	—	Series B
19	15·4	63·5	7·82	6·28	—	Horse serum
19	9·3	61·8	8·03	6·28	—	
			Mean	6·283	·	
38	24·6	63·4	7·70	—	6·13	Series B
38	23·5	63·8	7·72	—	6·14	Horse serum
38	45·6	67·5	7·45	—	6·15	
38	104·6	71·8	7·14	—	6·16	
38	430·3	77·3	6·56	—	6·15	
38	327·4	71·2	6·64	—	6·15	Series B
38	69·6	61·9	7·24	—	6·15	Ox serum
38	535·6	74·9	6·43	—	6·14	Series B
38	130·7	69·7	7·02	—	6·14	1st day
38	35·2	63·3	7·54	—	6·13	Horse serum
38	10·4	58·1	8·01	—	6·12	3rd day
38	106·7	69·2	7·11	—	6·15	
38	510·7	80·1	—	—	—	Series B
38	103·7	74·7	7·15	—	6·13	Ox serum
38	22·7	62·6	7·74	—	6·16	
38	59·5	69·0	7·34	—	6·14	
38	546·7	78·5	—	—	—	Series B
38	53·8	67·8	7·40	—	6·16	Ox serum
38	562·1	78·7	6·44	—	6·15	Series B
38	84·5	68·7	7·23	—	6·18	Human serum. (Patient
38	15·2	57·2	—	—	—	with chronic nephritis)
38	18·3	60·7	7·81	—	6·14	
				Mean	6·144	

Table XIV. The value D is calculated from Hasselbalch and Warburg's experiments 4 and 6. *Biochem. Zeitsch.* **86**, p. 417. Defibrinated ox blood.

In experiment 4 the volume of the blood cells is estimated at 27·5 % as the blood combined with 12·4 vols. % O_2.

In experiment 6 the volume of the blood cells is estimated at 68 % in the blood cell suspension B; the suspension combined with 31·3 vols. % O_2.

The values for the corresponding carbonic acid combinations in serum and blood cells are determined by interpolation.

Experiment 4.

Blood		Serum		
mm. Hg CO_2	Vols. % CO_2 combined	mm. Hg CO_2	Vols. % CO_2 combined	mm. Hg CO_2 during centrifuging
12·5	43·3	110·4	62·3	12·7
		193·9	64·4	
12·7	44·5			
46·5	61·1			
58·8	65·7	25·0	61·8	58·8
94·1	69·7	96·8	69·9	
181·9	80·0	173·9	74·0	
185·4	79·7			
		29·0	67·3	89·7
377·5	90·2	101·1	75·4	
		182·9	76·3	
		23·4	68·5	181·9
		134·9	79·3	
		401·4	87·7	
430·5	88·0	17·9	75·4	430·5
		160·0	88·2	
		737·0	91·6	
737·5	99·3	26·4	80·6	737·0
		161·1	92·6	
		428·3	95·0	

mm. Hg CO_2	Vols. % combined CO_2		D	p_{H} in serum
	in serum	in blood cells		
12·7	52	23	0·35	7·78
58·8	66	66	1·00	7·22
89·7	74	52	0·70	7·09
181·9	82	65	0·80	6·83
430·5	90	75	0·87	6·49
737·0	96	107	1·11	6·29

Experiment 6.

Serum B. CO_2 tension during centrifuging, 65 mm.

Blood cell suspension B

mm. Hg CO_2	Vols. % CO_2 combined	mm. Hg CO_2	Vols. % CO_2 combined
19·5	61·6	29·8	46·1
79·8	69·6	86·8	67·7
182·0	73·0	159·8	81·2
378·3	73·7	387·7	100·8

from which at 65·0 mm. Hg CO_2 in serum 68 vols. %, in blood cells 55·9, $D = 0·82$, p_{H} in serum $= 7·23$.

Table XV. The value D calculated from J. Joffe and E. P. Poulton's experiments with human blood at 38°. *J. Physiol.* **54**, pp. 148–149, Tables II, III and V.

	Oxygenated blood Vols. % combined CO_2				Reduced blood Vols. % combined CO_2				
mm. Hg CO_2	in serum	in blood cells	D	p_H calc. in serum	in serum	in blood cells	D	p_H calc. in serum	
10	29·2	14·5	0·50	7·77	31·3	21·5	0·69	7·81	
20	38·0	23·3	0·61	7·59	42·1	30·2	0·72	7·64	J.J.'s defibrinated
30	44·8	29·6	0·66	7·49	48·7	36·7	0·75	7·52	blood
40	50·1	35·1	0·70	7·41	54·4	41·6	0·77	7·45	
55	57·3	40·7	0·71	7·31	61·4	48·1	0·78	7·36	Blood cell volume
70	63·1	45·4	0·72	7·27	67·3	54·2	0·81	7·30	50·93
90	68·0	49·5	0·73	7·19	73·4	61·4	0·84	7·23	
156·0	77·2	58·3	0·76	7·01	—	—	—	—	Blood cell volume
180·0	81·5	57·2	0·70	6·97	—	—	—	—	53·5
110·3	73·8	54·2	0·74	7·14	—	—	—	—	
157·0	76·4	65·1	0·85	7·00	—	—	—	—	
376·3	97·5	83·6	0·85	6·72	—	—	—	—	
610·0	108·9	92·9	0·85	6·56	—	—	—	—	
278·0	97·2	73·9	0·76	6·86	—	—	—	—	
477·0	100·5	91·8	0·91	6·63	—	—	—	—	
40	55·3	25·5	0·46	7·45	60·8	36·0	0·59	7·49	J.J.'s oxalated blood
55	65·3	25·7	0·39	7·38	67·8	42·5	0·63	7·40	Blood cell vol. 44·5
15	36·6	22·7	0·62	7·70	40·8	20·6	0·51	7·75	W.R.'s oxalated
25	45·8	24·3	0·53	7·58	49·9	31·8	0·64	7·61	blood
35	52·0	29·6	0·57	7·48	55·9	40·7	0·73	7·51	Blood cell vol. 39·77
45	58·2	33·0	0·57	7·42	59·8	44·5	0·74	7·44	
60	62·2	40·6	0·65	7·32	63·8	45·4	0·71	7·34	
75	63·8	42·4	0·66	7·23	65·0	47·9	0·74	7·25	

Table XVI. The value of D calculated from L. S. Fridericia's experiment 7. *J. Biol. Chem.* **42**, p. 254. Ox blood with 0·1 % sodium oxalate. 38 vols. % blood cells at 17°.

mm. Hg	Vols. % CO_2 combined in plasma	Vols. % CO_2 combined in blood cells	$D (HCO_3')$	p_H' (calc.) in plasma	Gram NaCl in 100 cc. plasma	Gram NaCl in 100 cc. blood cells	$D (Cl')$
0·08	23·4	16·6	0·71	9·46	0·578	0·260	0·45
6·1	42·6	27·6	0·65	8·04	0·533	0·334	0·63
39·1	67·6	62·0	0·92	7·44	0·511	0·369	0·72

Table XVII. The value of D in ox blood. 100 cc. ox blood cells combined with 45 cc. O_2.

Temp.	mm. Hg CO_2	Vols. % combined CO_2	(p_H') (calc.)	Vols. % combined O_2	
18	35·5	64·0	7·46	—	Serum
		58·6	—	16·0	Blood haematocrite determination 35
		50·7	—	35·8	Blood cell suspension
					from serum and blood ... ... $D=0.79$
					from serum and blood cell suspension $D=0.74$
18	43·3	74·4	7·44	—	Serum
		69·1	—	10·9	Blood
		61·6	—	29·6	Blood cell suspension
					from serum and blood ... ... $D=0.79$
					from serum and blood cell suspension $D=0.74$
18	21·9	63·5	7·67	—	Serum
		60·2	—	10·9	Blood
		48·9	—	33·5	Blood cell suspension
					from serum and blood ... ... $D=0.78$
					from serum and blood cell suspension $D=0.69$
18	9·8	56·8	7·97	—	Serum
		50·2	—	10·9	Blood
					from serum and blood ... ... $D=0.52$
18	22·5	62·5	7·65	—	Serum
		56·9	—	15·7	Blood
		49·2	—	24·5	Blood cell suspension

Table XVIII. The value of D in defibrinated horse blood at room temperature.

Temp.	mm. Hg CO_2	Vols. % combined CO_2	Mean of pH calc. in serum	Vols. % combined O_2	Haemato-crite number	Vols. % combined O_2 in blood cells				Vols. % combined CO_2 in blood cells	D
19	34·7	59·3		20·3	41·5	48·9	B_I	from B_I, S_I and haematocrite	B_I	44·4	
		69·9		0·2	—	—	S_I				
		51·1		36·0	—	—	C_I	,, C_I, S_I ,,	B_I	44·4	
		68·7		0·9	—	—	S_{II}				
			7·51								0·64
17	15·4	46·2		20·2	40·0	50·5	B_I	,, B_I, S_I ,,	B_I	30·0	
		57·0		0·6	—	—	S_I				
		40·4		35·1	—	—	C_I	,, C_I, S_I ,,	B_I	33·1	
		56·1		0·6	—	—	S_{II}				
			7·76								0·57
18	82·2	76·1		20·0	41·0	48·9	B_I	,, B_I, S_I ,,	B_I	61·7	
		87·1		1·3	—	—	S_I				
		69·2		34·7	—	—	C_I	,, C_I, S_I ,,	B_I	61·1	
		69·4		33·2	66·5	49·9	C_{II}				
		87·2		1·6	—	—	S_{II}	,, C_{II}, S_{II} ,,	C_{II}	60·8	
		76·9		19·9	—	—	B_{II}	,, B_{II}, S_{II} ,,	B_I	62·2	
			7·23								0·70
18	38·4	59·3		19·8	40·4	49·0	B_I	,, B_I, S_I ,,	B_I	37·2	
		74·1		0·5	—	—	S_I				
		55·1		35·8	—	—	C_I	,, C_I, S_I ,,	B_I	48·1	
		53·5		34·1	66·0	51·8	C_{II}				
		71·8		0·7	—	—	S_{II}	,, C_{II}, S_{II} ,,	C_{II}	45·2	
		60·1		19·2	37·5	51·1	B_{II}	,, B_{II}, S_{II} ,,	B_{II}	40·7	
			7·48								0·64
17	17·3	47·4		19·7	40·3	49·2	B_I	,, B_I, S_I ,,	B_I	32·3	
		57·7		0·4	—	—	S_I				
		39·7		37·9	—	—	C_I	,, C_I, S_I ,,	B_I	34·3	
		39·1		36·2	65·8	(55·0)	C_{II}				
		56·4		0·8	—	—	S_{II}	,, C_{II}, S_{II} ,,	B_I	29·9	
		47·0		18·8	39·5	47·6	B_{II}	,, B_{II}, S_{II} ,,	B_{II}	32·7	
			7·72								0·56
17	86·2	78·4		19·8	41·4	48·5	B_I	,, B_I, S_I ,,	B_I	63·8	
		88·9		1·1	—	—	S_I				
		71·7		35·5	—	—	C_I	,, C_I, S_I ,,	B_I	65·4	
		89·0		1·1	—	—	S_{II}				
		74·8		31·3	—	—	C_{II}	,, C_{II}, S_{II} ,,	B_I	65·8	
		78·1		18·6	36·5	51·0	B_{II}	,, B_{II}, S_{II} ,,	B_{II}	59·2	
			7·21								0·74
19	178·3	90·4		18·4	41·3	44·6	B_I	,, B_I, S_I ,,	B_I	78·0	
		99·1		0·8	—	—	S_I				
		81·9		35·7	—	—	C_I	,, C_I, S_I ,,	B_I	77·6	
		82·2		35·0	75·5	46·6	C_{II}				
		98·7		1·2	—	—	S_{II}	,, C_{II}, S_{II} ,,	C_{II}	76·9	
		89·1		17·9	39·6	45·3	B_{II}	,, B_{II}, S_{II} ,,	B_{II}	74·5	
			6·95								0·78
18	182·7	95·1		23·8	48·9	48·7	B_I	,, B_I, S_I ,,	B_I	80·4	
		109·3		1·4	—	—	S_I				
		88·5		36·6	—	—	C_I	,, C_I, S_I ,,	C_{II}	81·8	
		88·1		36·6	77·4	47·3	C_{II}				
		109·6		0·8	—	—	S_{II}	,, C_{II}, S_{II} ,,	C_{II}	82·4	
		96·4		22·7	48·8	46·5	B_{II}	,, B_{II}, S_{II} ,,	B_{II}	82·6	
			6·98								0·75
18	76·0	75·1		23·0	47·5	48·4	B_I	,, B_I, S_I ,,	B_I	57·9	
		90·7		1·0	—	—	S_I				
		68·6		36·0	—	—	C_I	,, C_I, S_I ,,	C_{II}	61·2	
		67·1		36·6	76·3	48·0	C_{II}				
		89·6		0·5	—	—	S_{II}	,, C_{II}, S_{II} ,,	C_{II}	60·2	
		75·1		22·7	47·0	48·3	B_{II}	,, B_{II}, S_{II} ,,	B_{II}	58·7	
			7·28								0·67

Table XVIII (*continued*)

Temp.	mm. Hg CO_2	Vols. % combined CO_2	Mean of pH calc. in serum	Vols. % combined O_2	Haematocrite number	Vols. % combined O_2 in blood cells		Vols. % combined CO_2 in blood cells	Γ
18	35·5	66·7	—	39·0	—	B_I	from B_I, S_I and haematocrite B_I	46·6	
		79·5	0·5	—	—	S_I			
		55·5	37·6	—	—	C_I	,, C_I, S_I ,, C_{II}	48·4	
		53·8	38·8	79·7	48·7	C_{II}			
		78·7	0·4	—	—	S_{II}	,, C_{II}, S_{II} ,, C_{II}	47·7	
		66·7	—	38·3	—	B_{II}	,, B_{II}, S_{II} ,, B_{II}	47·3	
			7·55						0·61
18	461·2	115·7	19·0	—	—	B_I	,, B_I, S_I ,, C_{II}	98·2	
		128·2	0·4	—	—	S_I			
		113·8	34·7	—	—	C_I	,, C_I, S_I ,, C_{II}	109·3	
		110·1	35·3	77·5	45·6	C_{II}			
		126·5	0·7	—	—	S_{II}	,, C_{II}, S_{II} ,, C_{II}	105·3	
		117·6	18·7	41·8	44·8	B_{II}	,, B_{II}, S_{II} ,, B_{II}	105·2	
			6·65						0·84
17	12·6	60·5	0·4	—	—	S_I	,, C_I, S_I ,, C_{II}	32·9	
		40·1	36·9	—	—	C_I			
		35·8	39·4	79·1	49·8	C_{II}	,, C_{II}, S_{II} ,, C_{II}	29·4	
		60·0	0·4	—	—	S_{II}			
		49·0	18·5	37·4	49·5	B_{II}	,, B_{II}, S_{II} ,, B_{II}	30·5	
			7·87						0·52
18	443·1	99·5	13·4	30·4	44·1	B_I	,, B_I, S_I ,, B_I	78·6	
		97·6	33·2	—	—	C_I			
		108·7	0·2	—	—	S_I	,, C_I, S_I ,, C_{II}	93·8	
		90·9	32·6	73·0	44·7	C_{II}			
		104·7	0·4	—	—	S_{II}	,, C_{II}, S_{II} ,, C_{II}	85·8	
		98·5	13·6	31·0	43·9	B_{II}	,, B_{II}, S_{II} ,, B_{II}	84·4	
			6·59						0·84
19	21·5	52·7	13·7	27·9	49·1	B_I	,, B_I, S_I ,, B_I	32·4	
		60·7	0·2	—	—	S_I			
		40·6	38·5	—	—	C_I	,, C_I, S_I ,, C_{II}	34·9	
		38·4	38·8	78·3	49·6	C_{II}			
		59·8	0·7	—	—	S_{II}	,, C_{II}, S_{II} ,, C_{II}	32·5	
		52·6	13·1	26·5	49·4	B_{II}	,, B_{II}, S_{II} ,, B_{II}	32·8	
			7·66						0·56
18	413·9	108·1	18·2	41·3	44·2	B_I	,, B_I, S_I ,, B_I	98·4	
		114·9	0·4	—	—	S_I			
		103·3	33·9	—	—	C_I	,, C_I, S_I ,, B_I	99·8	
		101·2	33·2	75·0	44·3	C_{II}			
		111·8	0·7	—	—	S_{II}	,, C_{II}, S_{II} ,, C_{II}	97·6	
			6·64						0·87
18	14·3	49·2	18·9	37·8	50·1	B_I	,, B_I, S_I ,, B_I	26·8	
		63·2	0·6	—	—	S_I			
		38·7	38·7	—	—	C_I	,, C_I, S_I ,, B_I	30·0	
		37·7	38·2	76·0	50·2	C_{II}			
		58·9	0·5	—	—	S_{II}	,, C_{II}, S_{II} ,, C_{II}	29·7	
		48·7	17·5	36·8	47·5	B_{II}	,, B_{II}, S_{II} ,, B_{II}	31·3	
			7·83						0·49
18	217·3	99·4	18·8	40·8	46·1	B_I	,, B_I, S_I ,, B_I	81·1	
		111·9	0·5	—	—	S_I			
		94·2	35·8	—	—	C_I	,, C_I, S_I ,, C_{II}	89·1	
		92·5	35·3	76·5	46·1	C_{II}			
		110·1	0·5	—	—	S_{II}	,, C_{II}, S_{II} ,, C_{II}	88·5	
		99·3	18·5	39·9	47·9	B_{II}	,, B_{II}, S_{II} ,, B_{II}	83·3	
			6·91						0·80
17	19·4	51·4	20·8	41·5	50·3	B_I	,, B_I, S_I ,, B_I	39·6	
		59·8	0·9	—	—	S_I			
		40·0	38·9	—	—	C_I	,, C_I, S_I ,, C_{II}	32·7	
		41·8	38·1	75·5	50·4	C_{II}			
		59·3	0·5	—	—	S_{II}	,, C_{II}, S_{II} ,, C_{II}	31·3	
		50·2	19·9	40·0	49·7	B_{II}	,, B_{II}, S_{II} ,, B_{II}	36·0	
			7·68						0·54

Table XVIII (continued)

Temp.	mm. Hg CO₂	Vols. % combined CO₂	Mean of pu' calc. in serum	Vols. % combined O₂	Haematocrite number	Vols % combined O₂ in blood cells		Vols. % combined CO₂ in blood cells	Γ
17	45·1	67·2		20·6	41·8	49·4	B_I from B_I, S_I and haematocrite B_I	50·4	
		79·2		0·6	—	—	S_I		
		58·3		37·7	—	—	C_I ,, C_I, S_I ,, C_{II}	52·1	
		59·9		35·9	72·9	49·3	C_{II}		
		78·1		0·9	—	—	S_{II} ,, C_{II}, S_{II} ,, C_{II}	53·1	
		67·1		19·9	—	—	B_{II} ,, B_{II}, S_{II} ,, C_{II}	52·0	
			7·44						0·67
19	42·3	53·6		21·6	45·0	47·9	B_I ,, B_I, S_I ,, B_I	42·7	
		62·6		0·3	—	—	S_I		
		46·4		37·7	—	·—	C_I ,, C_I, S_I ,, C_{II}	42·0	
		62·3		0·7	—	—	S_{II}		
		46·6		36·0	75·0	48·0	C_{II} ,, C_{II}, S_{II} ,, C_{II}	41·3	
		52·9		20·7	43·4	47·7	B_{II} ,, B_{II}, S_{II} ,, B_{II}	40·3	
			7·38						0·67
19	172·4	89·0		0·7	—	—	S_I ,, C_I, S_I ,, C_{II}	72·0	
		75·7		35·8	—	—	C_I		
		73·2		35·3	77·0	47·1	C_{II} ,, C_{II}, S_{II} ,, C_{II}	68·2	
		90·8		1·0	—	—	S_{II}		
			6·92						0·78

Table XIX. The value of D in defibrinated horse blood with various additions.

Temp.	mm. Hg CO₂	Vols. % combined CO₂	Mean of pu' calc. in serum	Vols. % combined O₂	Haematocrite number	Vols. % combined O₂ in blood cells		Vols. % combined CO₂ in blood cells	D
				100 cc. blood + 5 cc. 0·9 % NaCl sol. (isotonic with blood)					
18	37·3	62·5		19·6	40·3	48·6	B_I from B_I, S_I and haematocrite B_I	46·1	
		73·5		0·4	—	—	S_I		
		53·3		36·6	—	—	C_I ,, C_I, S_I ,, C_{II}	47·8	
		51·3		37·3	75·8	49·2	C_{II}		
		72·7		0·2	—	—	S_{II} ,, C_{II}, S_{II} ,, C_{II}	44·5	
		61·4		18·4	38·8	47·4	B_{II} ,, B_{II}, S_{II} ,, B_{II}	43·4	
			7·50						0·6
				100 cc. blood + 5 cc. n/2 NaHCO₃ (hypertonic to blood)					
19	77·6	118·5		18·6	36·0	51·7	B_I from B_I, SI and haematocrite B_I	83·0	
		138·5		0·3	—	—	S_I		
		97·4		37·1	—	—	C_I , C_I, S_I ,,, C_{II}	81·0	
		96·6		37·5	72·4	51·8	C_{II}		
		136·7		0·7	—	—	S_{II} ., C_{II}, S_{II} ,,, C_{II}	81·2	
		119·3		18·9	36·0	52·5	B_{II} , B_{II}, S_{II} ,,, B_{II}	88·6	
			7·46						0·5
				100 cc. blood + 5 cc. n/1 NaCl (more hypertonic)					
19	38·5	62·7		19·1	34·0	56·2	B_I from B_I, S_I and haematocrite B_I	36·8	
		76·0		0·2	—	—	S_I		
		56·2		35·4	—	—	C_I ,, C_I, S_I ,, C_{II}	44·7	
		56·5		30·6	54·5	56·1	C_{II}		
		75·3		0·4	—	—	S_{II} ,, C_{II}, S_{II} ,, C_{II}	40·7	
			7·50						0·5

The curve with a continuous line in Fig. 7 is the same as that in Fig. 6 (horse blood at room temperature); the "recumbent" crosses represent the values of D in oxygenated blood calculated from the experiments of Joffe and Poulton [1920] with defibrinated human blood at 38°; the "erect" crosses represent the values of D in almost completely reduced blood; the marks enclosed in squares refer to oxalated blood.

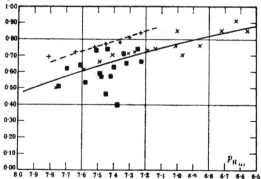

Fig. 7. Joffe and Poulton's results compared with experiments with horse blood.

——— Oxygenated horse blood at room temperature, defibrinated.
— — Reduced human blood at 38°, defibrinated.
× Defibrinated human blood, oxygenated ⎫
+ ,, ,, reduced ⎬ 38°. The value of D.
⊠ Oxalated ,, oxygenated ⎪
⊞ ,, ,, reduced ⎭

Joffe and Poulton's experiments with defibrinated human blood at 38° give values for D which, allowing for experimental error, are identical with those for horse blood cited above, while reduced blood gives values which are fairly accurately represented by a curve situated 0·125 higher in the diagram and which runs parallel with the curve for horse blood. The experiments with oxalated blood all give smaller values for D than corresponding ones with defibrinated blood but the values are more irregular. Apart from this the experiments with oxalated blood agree with those with defibrinated blood as regards all the other factors.

The addition of oxalate therefore causes a change in the distribution of carbonic acid in the blood and it is quite analogous to the alteration which takes place in horse blood on the addition of hypertonic salt solution; sodium oxalate in point of fact introduces fresh cations into the plasma which makes the serum hypertonic and causes the blood corpuscles to shrink.

Referring to J. Joffe's experiments on human blood[1] we find at a pressure of 40 mm. CO_2, 45·0 vols. % CO_2 (total) in defibrinated oxygenated blood and 44·5 vols. % in oxalated blood, while in the corresponding serum there are 52·8 vols. % and in plasma 58·0. As regards reduced blood defibrinated contains 50·4 vols. % total CO_2 and oxalated 52·3, while there are 57·1 vols. % in serum against 63·5 vols. % in plasma. At 55 mm. CO_2 exactly similar conditions prevail. It is seen therefore that the CO_2 content of plasma is

[1] J. Joffe and E. P. Poulton [1920, p. 148, Table II].

· 10 % higher than that of serum and the blood corpuscles have shrunk about 13 %.

Similar conditions would presumably be found in osmotically hypertonic blood (nephritis and sweating baths, diarrhoea, etc.).

D. D. van Slyke and G. E. Cullen [1917] some years ago introduced a method for determining the amount of combined bicarbonate in the blood which is in many ways extremely useful and quite accurate. This method appears to give more variable values for the combined CO_2 than might have been expected from the experiments of J. Christiansen, Douglas and Haldane [1914], Hasselbalch [1916, 2], Donegan and Parsons [1919] and Jarlöv [1919] on the CO_2 combining capacity of whole blood when it is assumed the partition between plasma and blood corpuscles is constant at the same reaction.

In the previously mentioned paper Hasselbalch and Warburg [1918] have cast suspicion on methods based upon the determination of the combined CO_2 in the serum (or plasma) when it is desired to draw conclusions about the "alkali reserve" (van Slyke) or the "reduced hydrogen number" [Hasselbalch, 1916, 1], having pointed out that the Schmidt-Zuntz phenomenon may play some part, provided the blood is not centrifuged at the alveolar CO_2 tension. Recently Joffe and Poulton [1920] in their much cited work have dealt with the same question and put forward a similar explanation of the degree of the dispersion in the experiments of van Slyke and his collaborators.

It has for long seemed strange to the author that the "fortuitous" variations which are associated with van Slyke's method appear to be considerably greater than those we should expect to arise from the Schmidt-Zuntz phenomenon. I think I have already made it clear that the most important cause of the variations is to be sought in the addition of varying amounts of oxalate to the blood before centrifuging it. Van Slyke and Cullen [1917] have shown that oxalate mixed with serum does not alter its combining power with CO_2 but they have overlooked the fact, as far as I am aware, that the oxalate could change the distribution of CO_2 in the blood. Van Slyke and Cullen recommend the addition of 0·5 % oxalate, which is a rather large amount. I have little doubt it would be an advantage to the method if the amount of oxalate employed was always the same and was as small as possible because there will always be a slight want of uniformity on account of the variation in the quantity of blood corpuscles in different experiments.

In the final chapter it will be shown that the Cl' partition between blood corpuscles and serum follows exactly similar laws to that of HCO'_3. It is thus important in determining the amount of Cl' in plasma to pay attention to the amount of oxalate added (the same applies also to fluoride and citrate), just as we should naturally work with a known CO_2 tension as has been particularly pointed out by H. J. Hamburger [1902] and L. S. Fridericia [1920].

After this digression let us revert to the curves. In Fig. 8 the value of D will be found in oxygenated ox blood at room temperature calculated from Fridericia's [1920], Hasselbalch and Warburg's [1918] and my later experiments. The values emanating from the previously mentioned unreliable experiments of Hasselbalch and Warburg are put in parentheses. The other experiments fall evenly on a curve similar to that for reduced human blood according to Joffe and Poulton's [1920] experiments. The experiments are too few to bear further investigation.

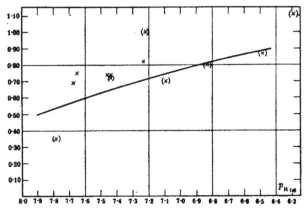

Fig. 8. D in ox blood at room temperature.

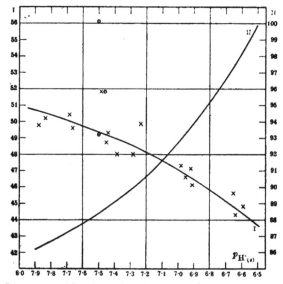

Fig. 9. Curve I. cc. of O_2 in 100 cc. horse blood corpuscles.
    „   II. The volume of blood corpuscles as a % of their volume at p_H· 6·50.

Fig. 9, curve I, and Table XX show how the oxygen in 100 cc. of horse blood corpuscles varies. In the earlier literature we find the statements of H. Nasse [1878] and Hamburger [1902] on the volume of the blood corpuscles at low CO_2 tensions and at tensions about 1 atmosphere. These statements, however important they may have been for the development of our knowledge of the physical chemistry of the blood, are obviously not accurate. Joffe and Poulton have not been able to demonstrate with certainty the change of volume determined by the reaction with greatly varying CO_2 tensions. The variations shown in the diagram are well grouped about the curve with the exception of two points, so that it will have considerable value on account of its novelty. The variations in volume will in chapter XII be the subject of a theoretical investigation. Curve II gives the volume of the blood corpuscles as a percentage of the volume at p_H· 6·50.

At p_H· 7·40 as will be seen horse blood cells combine with 49 cc. O_2 per 100 cc. blood cells. A. Norgaard and H. C. Gram [1921] found values diverging from mine in an examination of human blood at physiological reaction; as a mean of three determinations I found 48 vols. %. The difference is presumably in the haemoglobin determinations (see chapter XI, p. 285 footnote). With regard to the volume of ox blood corpuscles I have only a very few determinations available, the mean of which gives 45 cc. O_2 in 100 cc. blood corpuscles at "physiological" p_H·, but the determination can very well be a couple of vols. % out.

Table XX.

p_H·	Vols. % combined O_2 in blood cells	Vol. of blood cells as a % of vol. at p_H· 6·50	p_H·	Vols. % combined O_2 in blood cells	Vol. of blood cells as a % of vol. at p_H· 6·50
6·50	43·7	100·0	7·30	48·6	89·9
6·60	44·5	98·2	7·40	49·0	89·1
6·70	45·2	96·7	7·50	49·4	88·4
6·80	45·9	95·2	7·60	49·8	87·8
6·90	46·6	93·9	7·70	50·1	87·2
7·00	47·2	92·7	7·80	50·4	86·7
7·10	47·7	91·7	7·90	50·7	86·2
7·20	48·2	90·7			

From the curves in Figs. 6 and 9 we can now calculate the $p\Lambda'_{(s)}$ variations for horse blood at room temperature. This is done in Table XXI and the results are graphically represented in Fig. 10. It will be noted the curves also give a measure of the error committed if p_H· of the blood is estimated by the Henderson-Hasselbalch equation and $p\lambda'_{(s)}$ used in the calculation. The error would then always be 0·011 greater than the correction in the table. It will also be observed that it is very easy by choosing a suitable constant to get a good result in calculating p_H· from the Henderson-Hasselbalch equation, and that a larger (logarithmic) constant should be used in an alkaline reaction than in a more acid reaction. This is in agreement with what we found in the preceding chapters. The rather steep rise of the curves indicates that

Table XXI.

p_{H}	D	Q	$\dfrac{100-Q(1-D)}{100}$	$-\log\dfrac{100-Q(1-D)}{100}$	$-\log\dfrac{100-Q(1-D)}{100}$ $-\log\Phi_a(CO_2)$
		10 vols. % O$_2$	$\Psi = 5.9$	$\Phi_a(CO_2)=1.06$	$\log\Phi_a(CO_2)=0.027$
7·90	0·50	19·7	0·901	0·045	0·018
7·60	0·60	20·1	0·920	0·036	0·009
7·40	0·66	20·4	0·931	0·031	0·004
7·27	0·70	20·7	0·938	0·028	0·001
6·88	0·80	21·5	0·957	0·019	−0·008
6·50	0·88	22·9	0·973	0·011	−0·016
		15 vols. % O$_2$	$\Psi = 8.0$	$\Phi_a(CO_2)=1.09$	$\log\Phi_a(CO_2)=0.036$
7·90	0·50	29·6	0·852	0·069	0·033
7·60	0·60	30·2	0·879	0·056	0·020
7·40	0·66	30·6	0·896	0·048	0·012
7·27	0·70	31·1	0·907	0·042	0·006
6·88	0·80	32·3	0·935	0·029	−0·007
6·50	0·88	34·4	0·959	0·018	−0·018
		20 vols. % O$_2$	$\Psi = 9.9$	$\Phi_a(CO_2)=1.10$	$\log\Phi_a(CO_2)=0.042$
7·90	0·50	39·4	0·803	0·095	0·053
7·60	0·60	40·2	0·839	0·076	0·034
7·40	0·66	40·8	0·862	0·064	0·022
7·27	0·70	41·4	0·876	0·057	0·015
6·88	0·80	43·0	0·914	0·039	−0·003
6·50	0·88	45·8	0·945	0·025	−0·017
		25 vols. % O$_2$	$\Psi = 11$	$\Phi_a(CO_2)=1.12$	$\log\Phi_a(CO_2)=0.049$
7·90	0·50	49·3	0·753	0·123	0·074
7·60	0·60	50·9	0·796	0·099	0·050
7·40	0·66	51·0	0·827	0·082	0·033
7·27	0·70	51·8	0·845	0·073	0·024
6·88	0·80	53·8	0·893	0·049	0·000
6·50	0·88	57·3	0·931	0·031	−0·018
		30 vols. % O$_2$	$\Psi = 12$	$\Phi_a(CO_2)=1.14$	$\log\Phi_a(CO_2)=0.057$
7·90	0·50	59·1	0·705	0·152	0·095
7·60	0·60	60·3	0·759	0·120	0·063
7·40	0·66	61·2	0·792	0·101	0·044
7·27	0·70	62·1	0·814	0·089	0·032
6·88	0·80	64·5	0·871	0·060	0·003
6·50	0·88	68·7	0·916	0·038	−0·019

Donegan and Parsons' estimations in marked alkaline reactions are mostly correct.

As already mentioned the curves are so constructed that the correction for defibrinated horse blood can be directly read when the p_{H} is calculated with (130).

In the calculation of the $p_{\mathrm{H'}(s)}$ of human defibrinated oxygenated blood at 38° the curves are similarly directly applicable. (The condition for this should be that D as well as Q is the same in oxygenated human blood at 38° as in oxygenated horse blood at 18°, but the error due to Q being slightly different will be negligible.) In using the correction curves for calculating the $p_{\mathrm{H'}(s)}$ value of oxygenated defibrinated ox blood at room temperature and of reduced defibrinated human blood at 38°, one proceeds in such a way that having determined the p_{H} value by means of (130), the $p_{\mathrm{H'}(s)}$ value is looked for on the continuous line curve of Fig. 7, which has the same D value as the calculated $p_{\mathrm{H'}(s)}$ on the interrupted line curve, and then the correction corresponding to the new p_H· is read off Fig. 10. This manoeuvre is most easily performed by subtracting 0·40 from the value of p_H· calculated with (130) and then using the resulting number for seeking the correction in Fig. 10. The error committed by this method will hardly be appreciable in the second decimal place.

It will be noticed from the course of the correction curves that $p\Lambda'_{(s)}$ increases with the p_H· and with increasing amount of haemoglobin in the blood. That this can only just be demonstrated with the potentiometer in

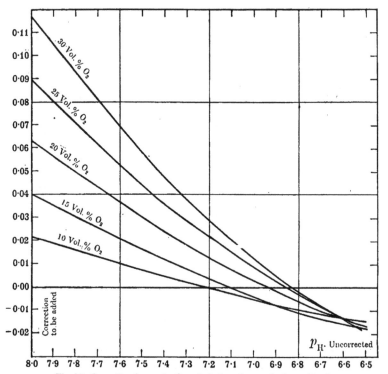

Fig. 10. Correction for use in the calculation of $p_{H'(s)}$ for horse blood
at room temperature and human blood at 38°.

the case of the variation determined by the reaction and not at all in the case of that determined by the concentration is due to the fact that potential measurements, even in the excellent arrangement employed by Parsons whose method can hardly be improved upon at present, are not sufficiently reliable in their application to blood. It follows therefore that the p_H· value ought to be determined according to the principles given here if the best possible results are desired.

It will also be observed that $p\Lambda'_{(s)}$, when calculated as described in this chapter, is larger than the constant found in an earlier one. The reason is presumably that the blood corpuscles in spite of all precautions are to a certain extent "deleterious" to the platinum electrode, even in Parsons' and Donegan and Parsons' experiments where saturation and measurement of potential were undertaken in different vessels.

Lastly it is important to note that the calculations of Hasselbalch using his p_{K_1} curve gave much more certain results than by a measurement of potential, and that the results obtained can easily be rectified with the help of the constants given here.

It has often been attempted to estimate the reaction of blood (and serum) colorimetrically and in later years such methods—partly combined with diffusion processes—have been described by Levy, Rowntree and Marriot [1915], W. M. Bayliss [1919] and quite recently in a very refined form by H. H. Dale and C. Lovatt Evans [1920].

J. Lindhard [1921] has devised a micro-method on Dale and Evans' principle which allows of a determination of the p_{H}· value with a relative accuracy of 0·02.

With the method of calculating p_{H}· given here, which is entirely based on a refinement of the Henderson-Hasselbalch principle, the apparent hydrogen ion activity exponent by using the proper method of CO_2 determination can be reckoned with a relative accuracy of 0·005 in any two determinations in human blood at 38° and horse blood at room temperature, and with an error which does not exceed the absolute value by more than 0·015. In human blood and ox blood at room temperature two determinations can be carried out which do not differ by more than about 0·01 relative to each other and with an error which does not exceed the absolute value by more than 0·03.

In conjunction with the above a few examples of the calculation of $p_{H'(s)}$ with Henderson's and Hasselbalch's equation may be given. Let us examine a sample of human blood with 20 vols. % combined O_2 at 38°. The relative absorption coefficient is 0·91. At 10 mm. CO_2 we find 25 vols. % total CO_2.

The amount absorbed is

$$\frac{10 \times 0.5554 \times 91}{7.60 \times 100} = 0.7 \text{ vol. \%.}$$

The amount combined is thus 24·3 vols. % :

$$\log 24.3 - \log 0.7 = 1.563.$$

Converting the values from Hasselbalch's 1916 curve to p_{H}· (Bjerrum) we get

$$p_{H'(s)} = 7.727.$$

Using Parsons' and Donegan and Parsons' $p\Lambda'_{(s)}$ we get

$$p_{H'(s)} = 7.734.$$

With the author's constant $p_{\dot{H}'(s)} = 7.710,$

and by calculating with (130) and correcting with Fig. 10

$$p_{H'(s)} = 7.751.$$

At 40 mm. we find in a similar blood 50 vols. % total CO_2 and consequently 47·3 vols. % combined CO_2, from which, in a similar manner to the previous example, we get

			Warburg	
			---	---
Hasselbalch	Parsons and Donegan	(126)	(130) and correction
$p_{H'(s)} = 7·374$	7·421	7·397	7·418

and for blood at 300 mm. CO_2 with 90 vols. % total CO_2 and 70·1 vols. % combined CO_2

			Warburg	
			---	---
Hasselbalch	Parsons and Donegan	(126)	(130) and correction
$p_{H'(s)} = 6·719$	6·787	6·763	6·750

CHAPTER VII

THE REACTION OF THE BLOOD CORPUSCLES.

Although the majority of workers who have investigated the reaction of the blood have realised that what is generally called "the hydrogen ion concentration of the blood" is really only "the hydrogen ion concentration of the serum," few have attempted to get some knowledge of the reaction in the interior of the blood corpuscles, a question however which is of the greatest interest because analogies with other cells may be drawn from it if similar factors determine the difference in reaction between serum and blood cells, and serum and tissue cells. Hasselbalch and Lundsgaard [1912] and later J. M. de Corral y Garcia [1914] claim to have shown that blood cells at physiological CO_2 tensions are more acid than the corresponding serum. The evidence produced is however faulty, as the phenomenon discovered by Hasselbalch and Lundsgaard is only an expression of the Schmidt-Zuntz effect as it appears in A. Schmidt's [1867] and N. Zuntz's [1867, 1868] and many other old experiments, and as is very clearly seen in Hasselbalch and Warburg's [1918] experiments and expositions. Hasselbalch and Lundsgaard showed that the reaction at constant CO_2 tension was more acid in blood than in the serum centrifuged from it. The experiments of the above mentioned authors show that the content of the serum in bicarbonate and therefore the reaction of the serum at a given CO_2 tension is a function of the CO_2 tension in the blood at the time of centrifuging, so that we cannot conclude anything about the reaction in the interior of the blood cells from a difference of reaction between serum in blood and separated serum.

The Schmidt-Zuntz phenomenon, which has been the cause of Hasselbalch and Lundsgaard's error, indicates that some kind of equilibrium prevails between the activity of the ions in blood cells and serum but it is not possible at the moment to say how this equilibrium is maintained. In the final chapter an attempt will be made to give a theoretical and experimentally workable solution of the problem on a relatively broad basis.

If the reaction of blood is determined electrically and then haemolysis produced (*e.g.* by freezing) without letting the blood come in contact with any new gas mixture, the reaction after haemolysis will show whether there is any difference of reaction between blood corpuscles and serum. The assumption involved in this reasoning—as for numerous other instances later in the chapter—is that the dissociation of the electrolytes determined by the reaction is the same at the same reaction before and after the haemolysis.

Konikoff [1913] has reported experiments in which he estimated the reaction of blood electrometrically using Hasselbalch's method. He employed a special electrode vessel with a relatively large platinum plate and assumed he avoided the oxygen error in this way (demonstration of the oxygen error was the real object of his work). Having determined the reaction of the blood he haemolysed it by freezing and found that the reaction had become much more acid.

Milroy [1917] determined the reaction in haemolysed blood at a CO_2 tension of about 40 mm. and found a p_H· of roughly 6·60, that is about ten times as large a hydrogen ion activity as in blood serum at the same CO_2 tension.

Although Konikoff devised a special technique to avoid the O_2 error (he did not however make use of minimal immersion) and Milroy employed Höber's principle and was aware of the existence of the O_2 error, I do not hesitate to say that the results of both these investigators are misleading as they undoubtedly had considerable quantities of O_2 in their haemoglobin solutions. Neither of them attempted to prove that the oxygen was actually dissipated during the measurements and it is beyond all question that too low potentials will be obtained if the estimations are carried out with deeply immersed platinum electrodes in strongly oxygenated haemoglobin solutions.

Parsons [1917] has published some electrical determinations in haemolysed blood in which the reaction was almost the same as that usually met with in serum with a similar CO_2 tension.

L. E. Walbum [1914, p. 231] has shown that the reaction in a solution of blood corpuscles (10 blood + 90 physiological NaCl) is the same before and after haemolysis. Although the quantity of blood corpuscles in the experiments was rather small this is counterbalanced by the liquid containing them (serum + NaCl solution) being relatively poor in buffer substances. L. S. Fridericia [1920], and J. Joffe and E. P. Poulton [1920], in the papers extensively referred to in a previous chaper, calculated the reaction in blood corpuscles and serum at the same CO_2 tension in a manner which in essence is identical with that employed in the experiments about to be described, but they made the assumption that $p\lambda_{(c)}$ was the same as p_{K_1} in bicarbonate solutions of the same carbonic acid binding power, an assumption which is to a certain extent supported by Hasselbalch's [1916, 2] rather scanty estimations in dialysed haemoglobin solutions.

It is therefore hardly possible from experiments in the literature to con-elude anything with certainty about the reaction in the blood corpuscles but I believe that Joffe and Poulton's contribution must be looked upon as the most important on this subject even though it is open to objection as $p\lambda_{(c)}$ was not experimentally determined.

If (120)
$$a_{H'(c)} = \lambda_{(c)} \frac{100 - Q(1-D)}{100D} \times \frac{P_{CO_2}a}{7\cdot60B}$$

is divided by (119)
$$a_{H'(s)} = \lambda_{(s)} \frac{100 - Q(1-D)}{100} \times \frac{P_{CO_2}a}{7\cdot60B}$$

we get
$$\frac{a_{H'(c)}}{a_{H'(s)}} = \frac{\lambda_{(c)}}{D\lambda_{(s)}}, \quad\text{...(132)}$$

which in logarithmic form becomes

$$p_{H'(s)} - p_{H'(c)} = p\lambda_{(s)} - p\lambda_{(c)} - \log D. \quad\text{...............(133)}$$

Since we have previously determined $p\lambda_{(s)}$ and D, we only require the value $p\lambda_{(c)}$ for estimating the difference in reaction between blood corpuscles and serum, which we will now attempt to determine. The determination of $p\lambda_{(c)}$ was associated with much greater difficulties than I originally expected. One of the most important was to get rid of the oxygen at the reactions and temperatures dealt with, but this was overcome to a large extent by the technique described in chapter IV. It was easy to haemolyse ox blood by repeatedly freezing so that it became completely transparent and only a trifling amount of blood corpuscles was left, but it was practically impossible by freezing alone to haemolyse horse blood so thoroughly. Even after freezing and thawing three times numerous intact blood corpuscles are present and many amorphous fragments are seen with the microscope. If the volume of the disperse phase is determined by the haematocrite—which can easily be done—it will never be found to be over 5 % of the whole system even when very concentrated blood cell suspensions are used. If a liberal amount of saponin is added to horse blood it will become completely transparent and only a few formed constituents (about 1 % in the haematocrite) can be seen with the microscope. This difficulty of haemolysing horse blood by freezing led me to work with blood haemolysed by saponin as it was found that the combined CO_2 was the same whichever of the two methods was employed as the following experiment shows.

Defibrinated horse blood was concentrated by centrifuging. Haematocrite reading 59·5.

mm. Hg CO_2	Vols. % combined CO_2	
24·3	45·2	Saponin
24·0	45·0	Freezing
79·2	72·7	,,
79·9	72·9	Saponin

In using a concentrated solution of horse blood haemolysed by saponin a new difficulty arose. When it is treated for a long time with high tensions of CO_2 it becomes very viscous and shortly afterwards a large quantity of haemoglobin crystals separate out so that the experiment has to be abandoned. It has been found that this precipitation never takes place in the first quarter of an hour so that a preliminary treatment of the haemoglobin solution with CO_2 may be undertaken for this space of time and the experiment continued with lower CO_2 tensions (lower $a_{H^.}$). This crystallising out of horse haemoglobin will be reverted to in chapter XI.

In horse blood haemolysed by freezing (concentrated in the centrifuge) I have only once seen a similar crystallisation, and the haemoglobin solution was in this instance cooled to $0°$.

That haemoglobin very readily crystallises out at high CO_2 tensions has been known a long time and is mentioned for example by Preyer [1871] without any particular comment. I. Setschenow [1879, p. 48] reported similar observations with strong concentrated frozen horse blood at room temperature (CO_2 and H_2SO_4 addition).

As already repeatedly mentioned the potential in an electrical determination of reaction falls when the platinum electrode has been in contact with protein solutions for some time. The drop is not large and in the course of 2–3 hours an almost constant potential seems to be reached (within $\frac{1}{4}$ millivolt), but I tried nevertheless to avoid any possible error from this cause ("deterioration" of the electrode) by developing the technique employed with L II described earlier in this work. In using this electrode vessel, in which it will be remembered the platinum electrode does not come in contact with the haemoglobin before the potentiometry is started, the potential was found to rise quickly about 10–20 millivolts in the first quarter of an hour after contact (total immersion) was established but quite irregularly. Then it became constant for a time and afterwards slowly declined. When the platinum was heated to redness before platinising the rise was much less, but as a rule a few millivolts. This must be what Parsons [1917] referred to when he wrote that it is essential to heat the electrode red hot before every determination in haemolysed blood.

In Table XXII $p\lambda_{(m)}$ is calculated from Hasselbalch's experiments with dialysed haemoglobin in weak sodium bicarbonate solution the conversion being carried out in the same way as in the preceding chapters.

$p\lambda_{(m)}$ is in agreement with $p\lambda_{(s)}$ and $p\lambda_{(c)}$ and therefore in haemolysed blood we have

$$p\lambda_{(m)} = p_{H^.} + \log \frac{P_{CO_2} a}{7.60} - \log \beta, \quad \ldots\ldots\ldots\ldots(134)$$

where β is the mean concentration of combined CO_2 (expressed in vols. % CO_2) in the haemolysed blood.

In Table XXIII[1] a number of determinations of $p\lambda_{(m)}$ in haemolysed ox blood are given. They were done with the small saturation electrode and within the same period as the experiments with blood designated series A in chapter V.

---

[1] The temperature corrections here and in what follows are made by adding 0·0075 to the value found for each degree over 18°, and subtracting the same amount for each degree under 18°. In the calculations from experiments in chapters V and VI 0·005 was used as the correction, but the difference is so small that I have not found it necessary to recalculate these earlier experiments with the correction employed in this chapter.

It is extremely probable that the apparent hydrogen ion exponent at this time was 0·06 too low, according to which $p\lambda_{(m)}$ in haemolysed ox blood with about 33 vols. % combined O_2 should be about 6·27.

The determinations in Table XXIV of haemolysed ox blood with 29 vols. % O_2 at 20° and 38° give respectively values of 6·26 and 6·15. They belong to series B and are carried out with the small saturation electrode.

The experiments in Table XXV were made in the large saturation electrode L I. In almost all cases a preliminary treatment with CO_2 for $\frac{1}{4}$–$\frac{1}{2}$ an hour was undertaken. Analyses of two gas mixtures are given in the table corresponding to one measurement of potential, the first relating to the last gas in the spirometer and the second to the gas in the electrode vessel.

$p\lambda_{(m)}$ at 18° referring here to horse blood cells haemolysed by saponin with about 38 vols. % combined O_2 is found to be 6·32; for ox blood haemolysed by freezing with 35 vols. % O_2 it is 6·26.

In Table XXVI measurements are given carried out with L II without heating the platinum electrode red hot before platinising. The results are quite in agreement with those reported in Table XXV, as $p\lambda_{(m)}$ in horse serum haemolysed blood cells with about 40 vols. % combined O_2 is also 6·32, while in haemolysed ox blood with about 32 vols. % it is 6·27. It is worth noticing that the final potential is not appreciably different although the platinum plate with one technique was 2–2$\frac{1}{2}$ hours in contact with the protein and with the other technique only $\frac{1}{4}$–$\frac{1}{2}$ hour. I should however expect that with a sufficiently extensive series of measurements a deviation of 1–2 millivolts might be demonstrated.

In Table XXVII are given measurements carried out in L II with platinum plates freshly heated to redness. The results are as will be seen a little different from those just obtained, $p\lambda_{(m)}$ in haemolysed horse blood and ox blood of similar constitution being respectively 6·35 and 6·28 at 18°. These values differ but slightly from those of the first series and the difference is hardly greater than the experimental error, a calculation of the mean error as in the case of the experiments with serum being hardly feasible.

It will be remembered that $p\lambda_{(s)}$ is 6·29 (see chapter VI) and since $p\lambda_{(m)}$ in a mixture of one part serum and three parts haemolysed blood corpuscles is 6·35 we shall not make a large error by assuming $p\lambda_{(c)} - p\lambda_{(s)}$ is 0·07, but as this difference is rather uncertain calculations have been made using (133) with values ranging from 0·05 to 0·09. The results are to be found in Table XXVIII. In the last column the values for (132) are given, the difference here being 0·07.

Table XXII. $p\lambda_{(c)}$ in dialysed haemoglobin with $NaHCO_3$ 0·025n calculated from K. A. Hasselbalch's experiment at 38°.

mm. Hg CO_2	Vols. % combined CO_2	p_H corrected	$p\lambda_{(m)}$
20·2	42·2	7·59	7·13
94·2	55·4	7·06	7·15
7·0	32·9	7·95	7·14

Table XXIII. $p\lambda_{(m)}$ in mixtures of serum and blood cell fluids (freezing) determined by the small saturator electrode. Series A.

Temp.	mm. Hg		Vols. % combined		p_H	$p\lambda_{(m)}$ 18°	
	CO_2	O_2	CO_2	O_2			
19·5	459·9	0·5	117·3	0·6	6·51	6·19	1st day
19·0	77·7	0·5	70·5	0·7	7·09	6·21	
19·0	73·7	0·4	69·8	0·7	7·12	6·23	Ox blood
18·0	50·2	0·9	60·6	—	7·21	6·21	2nd day
18·0	50·2	0·9	63·0	0·8	7·19	6·18	Preliminary treatment with CO_2
18·0	28·6	0·5	50·2	1·8	7·36	6·18	
19·5	28·6	0·5	49·5	0·6	7·35	6·20	,,
			Colorimetric 32				
19·0	37·4	0·5	61·1	0·7	7·33	6·20	1st day
19·5	57·1	0·5	68·6	1·4	7·23	6·23	
19·5	57·1	0·5	70·2	0·5	7·19	6·19	Preliminary treatment with CO_2
19·0	131·9	0·5	90·3	1·6	6·97	6·22	Ox blood
18·0	575·7	0·5	136·2	0·3	6·49	6·20	2nd day
19·0	82·6	0·5	79·1	0·8	7·12	6·22	
18·5	82·6	0·5	82·7	0·7	7·13	6·22	Preliminary treatment with CO_2
18·0	14·3	0·5	40·7	1·3	7·58	6·21	
						Mean 6·21	
20·0	38·6	147·0	51·9	32·2	—	—	3rd day
20·0	314·4	320·8	103·2	34·0	—	—	
19·0	103·5	0·5	77·4	1·1	7·08	6·29	1st day
19·0	40·3	0·2	—	—	7·33	—	
19·0	46·7	·0·3	64·5	—	7·28	6·22	Horse blood
19·0	481·5	0·0	128·2	0·9	6·58	6·24	2nd day
			Colorimetric circ. 32				

Table XXIV. Washed ox blood cells haemolysed by freezing circ. 225 cc. + 6 cc. $n/3$ Na_2CO_3. Determinations of $p\lambda_{(m)}$ in the small saturator electrode. Series B.

Temp.	mm. Hg		Vols. % combined		p_H (measured)	$p\lambda_{(m)}$ 20°	$p\lambda_{(m)}$ 38°	p_H (calculated)
	CO_2	O_2	CO_2	O_2				
27. ii. 19								
38	589·2	0·9	80·8	0·4	—	—	—	6·42
20	615·8	1·0	101·5	0·7	—	—	—	6·41
38	114·3	0·3	47·2	0·6	6·90	—	6·15	6·90
20	119·5	0·3	67·2	0·5	6·95	6·26	—	6·95
38	64·7	0·5	37·3	0·2	7·05	—	6·15	7·05
20	64·6	0·5	55·4	0·7	7·15	6·27	—	· 7·14
28. ii. 19								
38	121·7	0·6	49·1	0·5	6·88	—	6·13	6·90
21	127·2	0·6	69·2	0·4	6·93	6·25	—	6·93
38	12·1	0·5	13·7	0·7	7·32	—	6·14	7·33
20	12·1	0·5	27·4	0·4	7·53	6·24	—	7·55
38	41·0	0·4	29·8	—	7·09	—	6·09	7·15
20	47·4	0·4	49·3	0·8	7·21	6·26	—	7·21
						Mean 6·26	6·15	
1. iii. 19								
38	36·2	140·0	22·8	28·4	—	—	—	7·10
20	37·8	146·5	38·6	29·3	—	—	—	7·20
38	112·6	565·3	42·2	29·0	—	—	—	6·87
20	117·6	590·6	62·8	—	—	—	—	6·92
38	501·3	199·1	78·4	27·2	—	—	—	6·39

Table XXV. $p\lambda_{(m)}$ in mixtures of serum and blood cell fluids estimated with L I. Series B.

Temp.	mm. Hg		Vols. % combined		p_H	$p\lambda_{(m)}$ 18°		
	CO_2	O_2	CO_2	O_2				
—	46·9	0·5	—	—	—	—		
19·5	46·5	0·6	64·4	0·3	7·39	6·33	1st day	
—	106·2	0·7	—	—	—	—		
20·5	105·4	2·3	86·5 ·	1·9	7·13	6·30	2nd day	Horse blood haemo-
—	29·4	0·8	—	—	—	—		lysed with saponin
19·0	29·4	0·9	50·4	0·9	7·48	6·33	3rd day	
			Colorimetric 32					
—	40·2	0·5	—	—	—	—		
20·0	40·4	.1·2	64·7	1·2	7·48	6·35	1st day	Horse blood haemo-
—	16·6	0·2	—	—	—	—		lysed with saponin
19·0	15·3	1·2	40·7	2·1	7·65	6·31	2nd day	
—	12·3	0·5	—	—	—	—		
18·0	12·9	1·1	38·4	1·3	7·67	6·28	3rd day	
			Air from blower 37·6			Mean 6·32		
—	48·4	0·2	—	—	—	—		
19·0	·47·7	0·7	66·7	?	7·40	6·34		Human blood, frozen
—	586·5	0·2	—	—	—	—		
18·0	586·5	0·4	144·9	0·2	6·57	6·27	1st day	Concentrated by
—	60·8	0·5	—	—	—	—		centrifuging
18·0	66·4	0·5	85·3	0·2	7·28	6·26		Ox blood, frozen
—	27·2	0·2	—	—	—	—		
18·0	26·7	0·7	64·1	0·2	7·55	6·26	2nd day	
			Air from blower 35·1			Mean 6·26		

Table XXVI. Determinations of $p\lambda_{(m)}$ in mixtures of serum and blood cell fluids made with L II. The electrode was not heated to redness before each measurement. Series B.

Temp.	mm. Hg		Vols. % combined		p_H	$p\lambda_{(m)}$ 18°		
	CO_2	O_2	CO_2	O_2				
—	47·0	0·1	—	—	—	—		
21·0	47·0	2·9	60·7	1·4	7·35	6·31	1st day	
—	258·3	0·1	—	—	—	—		
20·0	258·8	0·2	113·0	—	6·89	6·33		Horse blood, frozen
—	. 74·2	0·1	—	—	—	—		
19·5	73·7	1·6	75·9	0·5	7·23	6·30	2nd day	
—	45·6	0·1	—	—	—	—		
19·0	45·2	1·2	62·1	0·5	7·40	6·33		
18·0	334·7	0·3	—	—	—	—		
—	335·2	0·3	122·1	0·1	6·78	6·31		
			Air from blower 39·7			Mean 6·32		
20·5	24·6	5·8	43·4	3·3	7·47	6·30	1st day	
—	32·2	0·3	—	—	—	—		
20·0	31·6	3·8	48·8	2·1	7·38	6·27	2nd day	Ox blood, frozen
—	101·3	0·6	—	—	—	—		
20·0	95·0	0·8	74·6	0·4	7·06	6·23	3rd day	
—	305·2	0·1	—	—	—	—		
19·0	305·2	0·7	104·4	0·3	6·73	6·28		
			Air from blower 31·8			Mean 6·27		
			Haematocrite number 72·8					

Table XXVII. Determinations of $p\lambda_{(m)}$ in mixtures of serum and blood cell fluids made with L II. Freshly burnt out platinum electrode.

Temp.	mm. Hg		Vols. % combined		pH·	$p\lambda_{(m)}$ 18°		
	CO_2	O_2	CO_2	O_2				
—	33·7	0·0	—	—	—	—		
19·0	33·3	1·9	53·0	1·5	7·50	6·38	1st day	
—	375·3	0·1	—	—	—	—		
19·0	375·0	0·1	117·6	0·1	6·73	6·32	3rd day	Horse blood, frozen
—	20·1	0·1	—	—	—	—		
19·0	23·6	0·6	54·5	0·0	7·63	6·33	4th day	
—	60·3	0·0	—	—	—	—		
19·0	61·6	0·0	75·1	0·0	7·36	6·37		
			Air from blower 36·3					
—	29·6	0·1	—	—	—	—		
16·0	29·7	0·1	57·5	0·0	7·50	6·31		Horse blood, frozen
			Air from blower 41·6					
—	18·9	0·3	—	—	—	—		
16·0	18·9	0·7	44·3	1·4	7·57	6·30		Horse blood, frozen
—	343·1	1·0	—	—	—	—		
18·5	342·1	2·4	120·7	1·2	6·85	6·39		Horse blood, frozen
—	89·9	0·2	—	—	—	—		
18·5	90·5	1·7	80·8	0·0	7·25	6·39		
			Air from blower 38·2		Mean	6·35		
—	388·9	0·1	—	—	—	—		
17·0	388·4	1·3	110·8	0·6	6·70	6·34		
—	22·8	0·2	—	—	—	—		
16·0	22·6	2·4	51·3	1·2	7·51	6·25		Ox blood, frozen
—	78·4	0·4	—	—	—	—		
16·0	78·9	0·2	76·3	0·1	7·17	6·28		
—	215·1	0·2	—	—	—	—		
16·0	214·5	0·6	99·8	0·1	6·82	6·25		
			Air from blower 27·8		Mean	6·28		

Table XXVIII. The difference of reaction between blood cells and serum (horse blood at room temperature).

$p\lambda_{(s)} - p\lambda_{(c)}$		−0·05	−0·06	−0·07	−0·08	−0·09	−0·07
				pH·$_{(s)} - p$H·$_{(c)}$			$\dfrac{a_{H'(c)}}{a_{H'(s)}}$
pH·$_{(s)}$	$-\log D$						
6·50	+0·055	+0·005	−0·005	−0·015	−0·025	−0·035	0·966
6·88	+0·097	+0·047	+0·037	+0·027	+0·017	+0·007	1·06
7·27	+0·155	+0·105	+0·095	+0·085	+0·075	+0·065	1·22
7·40	+0·180	+0·130	+0·120	+0·110	+0·100	+0·090	1·29
7·60	+0·222	+0·172	+0·162	+0·152	+0·142	+0·130	1·42
7·90	+0·301	+0·251	+0·241	+0·231	+0·221	+0·211	1·70

A dotted line is drawn through the table which indicates the reactions at which blood cells and serum have the same reaction. It will be seen this is the case between 6·50 and 6·88. Fig. 11 is a graphic representation of the table, the apparent hydrogen ion exponents of serum being the abscissae and the differences between the exponents of serum and blood cells the ordinates. For clearness only the curves relating to $p\lambda_{(c)} - p\lambda_{(s)}$ 0·05, 0·07 and 0·09 are given.

`E. J. WARBURG

It will be further seen from the tables and curves that the apparent hydrogen ion activity is larger in blood corpuscles than in serum at serum reactions more alkaline than $p_{H^.} = 6.9$ ($a_{H^.} = 1.26 \times 10^{-7}$) and that the difference increases with the hydrogen ion exponent.

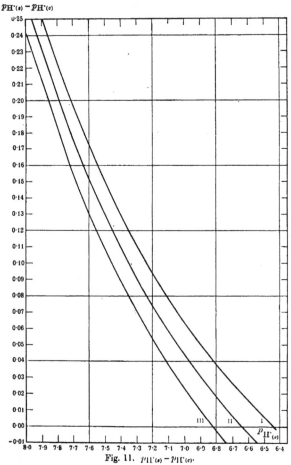

Fig. 11. $p_{H^.(s)} - p_{H^.(c)}$.

As D in human blood at 38° is the same as D in horse blood at 18° the above considerations will also apply to this species of blood if $p\lambda_{(c)} - p\lambda_{(s)}$ is the same. In the tables there is only one measurement of human blood at room temperature to be found but this fits in quite well in the horse blood series. The question can however only finally be settled by many estimations.

According to Fr. Kraus' [1898] experiments D for ox blood appears to be 1·00 at $p_{H^.}$ about 6·8. When this is compared with the determinations of $p\lambda_{(m)}$ in ox blood the same reaction should exist in serum and blood cells at this reaction because $p\lambda_{(c)}$ and $p\lambda_{(s)}$ are very nearly identical in such blood.

It appears also from the few determinations of D for ox blood which are reported in a preceding chapter that the difference of reaction between blood cells and serum increases with the $p_{H'}$.

If we assume that the dissociation of electrolytes which varies with the reaction is the same at a similar reaction before and after haemolysis we can draw conclusions from the combination of CO_2 at the same CO_2 tension, before and after haemolysis, about a possible difference of reaction between blood cells and serum. The combination of CO_2 increases with the apparent hydrogen ion activity in serum and in the fluids of the blood cells but it increases most in the latter case. If the same $p_{H'}$ persists after haemolysis as there was in serum and blood cells before haemolysis (e.g. in horse blood at $p_{H'}$ 7·60) then the CO_2 combination is not altered. If however there was a higher $a_{H'}$ in blood cells than in serum before haemolysis, $a_{H'}$ will be midway between the original reaction of the serum and blood cells under the given conditions after haemolysis and less CO_2 will be combined with the electrolytes varying with the reaction in blood cells but more with those in serum. The result will be that, altogether, less CO_2 will be combined after haemolysis than before if the volumes of blood cells and serum are equal.

I have only made a few experiments (about ten) with ox blood with this object in view but they all go to show that less CO_2 is combined after haemolysis than before, at alkaline reactions. At reactions about 6·30 the combined CO_2 was almost or actually the same in blood before and after haemolysis. In some experiments in which the osmolar concentration of the blood was a little diminished the difference was smaller than in ordinary blood. In an isolated experiment where the osmolar concentration of the blood was rather increased the opposite was the case. These effects of the changes in the osmolar concentration are in accordance with the theory, as D varies with the variations in volume of the blood corpuscles.

As will be noticed there is a disagreement, although a small one, between the results attained by the first and last mentioned principles for the determination of the difference of reaction in ox blood. From the first principle we concluded that the reaction was identical in serum and corpuscles at $p_{H'}$ 6·8 while the last pointed to the fact that there was no difference as far as $p_{H'}$ 6·3. Further experiments are needed to clear up the matter.

In Table XXIX some of the experiments[1] mentioned are given. Haemo-

---

[1] The experiment at 38° was particularly interesting as the relation between oxy-haemoglobin and reduced haemoglobin is not altered by haemolysis. This result was supported by several experiments with saponin haemolysed blood which I hope later to have the opportunity of publishing. The phenomenon itself is not without interest because it indicates that the quantity and kind of salt does not play so great a part in determining the form of the O_2 combination curve of haemoglobin as Barcroft [1914] and his collaborators imagined. That the dilution of the haemoglobin due to the haemolysis plays no great part in the relation between oxy-haemoglobin and reduced haemoglobin was only to be expected because they are both diluted to the same degree and there should thus be no change in the extent of oxygenation of the haemoglobin either according to the interpretation of the process expressed by G. Hüffner [1901] in his later papers or by A. V. Hill [1910, 1913, 1921].

If we could estimate the degree of oxygenation with sufficient accuracy we might however expect to find a little greater oxygenation after haemolysis than before at reactions more alkaline than $p_{H'}$ 6·8 because the extent of oxygenation is a function of the reaction; cf. Chr. Bohr, K. A. Hasselbalch and A. Krogh [1904], R. A. Peters [1914, 2], K. A. Hasselbalch [1916, 2] and L. J. Henderson [1920].

Table XXIX.

mm. Hg		Vols. % combined CO_2		Vols. % combined O_2		PH[(a)]
CO_2	O_2	in blood	in haemoglobin	in blood	in haemoglobin	calculated
	Blood and water-haemolysed blood at 19°
14·9	151·6	24·1	—	15·0	—	7·25
14·9	151·6	—	21·6	—	14·6	—
53·0	143·3	—	35·8	—	14·4	—
53·0	143·3	37·5	—	—	—	6·98
144·3	163·8	—	47·3	—	14·4	—
144·3	163·8	48·6	—	14·5	—	6·73
164·5	313·2	50·5	—	14·6	—	6·29
164·5	313·2	—	48·5	—	14·7	—
467·6	256·2	—	60·5	—	14·5	—
467·6	256·2	61·6	—	14·3	—	6·32
731·5	trace	—	64·9	—	—	—
731·5	,,	67·0	—	1·3	1·4	6·15
	Blood and water-haemolysed blood at 19°
10·7	153·0	25·9	—	8·0	—	7·60
10·7	153·0	—	24·1	—	8·1	—
34·1	148·1	35·4	—	8·0	—	7·22
34·1	148·1	—	32·9	—	7·9	—
67·3	141·3	40·0	—	—	—	6·99
67·3	141·3	—	38·7	—	8·2	—
135·3	127·0	46·1	—	7·7	—	6·75
135·3	127·0	—	44·8	—	7·4	—
133·0	126·4	45·8	—	—	—	6·76
133·0	126·4	—	44·6	—	—	—
405·9	69·9	54·3	—	—	—	6·33
405·9	69·9	—	54·3	—	—	—
736·0	trace	59·4	—	3·3	—	6·13
736·0	,,	—	60·0	—	1·6	—
	Blood and blood haemolysed by freezing at 18°
17·4	152·1	47·1	—	16·0	—	7·65
17·4	152·1	—	45·1	—	15·6	—
98·4	135·1	—	72·8	—	—	—
96·1	135·6	72·1	—	15·8	—	7·07
430·4	64·8	—	96·5	—	12·9	—
429·2	65·3	96·3	—	14·0	—	6·50
	Blood and water-haemolysed blood at 18°
37·2	150·4	—	28·0	—	12·9	—
37·2	150·4	30·1	—	12·9	—	7·11
37·2	150·4	—	28·7	—	13·0	—
37·2	150·4	30·2	—	12·8	—	7·11[1]

[1] CO_2 preliminary treatment for half an hour.

mm. Hg		Vols. % combined CO_2		Vols. % combined O_2		PH
	Blood and water-haemolysed blood at 38°
15·3	144·5	—	12·9	—	8·1	—
15·3	144·5	13·4	—	7·8	—	7·25
42·3	138·6	—	19·5	—	7·7	—
42·3	138·6	20·2	—	7·4	—	6·98
114·6	124·2	—	27·0	—	7·1	—
114·6	124·2	27·5	—	7·4	—	6·62
213·7	100·1	32·6	—	6·2	—	6·46
213·7	100·1	—	32·0	—	6·2	—
455·6	48·0	40·1	—	2·3	—	6·23
455·6	48·0	—	41·3	—	2·2	—

lysis was brought about either by freezing or by the addition of water. When haemolysis by water was completed sufficient NaCl was added to make the salt content up to 0·9 % again.

I. Setschenow [1879, p. 44] reported the following experiment at $37°-37°·5$. Emulsion of dog blood corpuscles:

514·6 mm. CO_2, total CO_2 in 50·18 cc. = 50·31 cc. $(0°, 1$ mtr.$)$.

After freezing:

513·3 mm. CO_2, total CO_2 in 50·18 cc. = 49·98 cc. $(0°, 1$ mtr.$)$.

From this p_{H}· is 6·22 if $\Psi = 15$ and $p\lambda_{(m)} = 6·20$, at which reaction there seems to be the same reaction in dog blood cells and serum.

RÉSUMÉ.

The apparent hydrogen ion activity in horse blood corpuscles has been determined.

CHAPTER VIII

THE DETERMINATION OF THE FIRST DISSOCIATION CONSTANT OF CARBONIC ACID AND THE DEVIATION COEFFICIENTS OF THE BICARBONATE ION.

In the foregoing chapters we have determined the value of $p\Lambda_{(s)}$ and $p\Lambda_{(c)}$, and it will now be interesting to inquire into the factors which control these constants rather more closely.

It will be remembered that in chapter III it was shown that the apparent activity coefficient of the continuous phase of serum and of the blood cell fluid participates in the constants and a rather large number of determinations have therefore been performed of the apparent activity coefficients in salt solutions. I have at several points pursued the investigations further than was absolutely necessary for the problem being dealt with, because it may be of particular interest from a purely physico-chemical standpoint, especially since the appearance of Bjerrum's theory.

The first dissociation constant of carbonic acid has been determined by the conductivity method by J. Walker and W. Cormack [1900] and by J. Kendall [1916] on the basis of experiments carried out by himself, by Pfeiffer[1], by Knox[1] and by Walker and Cormack. The determinations of Walker and Cormack and Kendall are better than the earlier ones (Pfeiffer's and Knox's) and therefore the values calculated from them are the most valuable. The molecular conductivity of the bicarbonate ion at "infinite dilution" comes into the calculation. This value is obtained by extrapolation from conductivity determinations of sodium bicarbonate solutions (and also calcium bicarbonate solutions), but it seems to the author that Walker and Cormack and Kendall have not executed this extrapolation in a satisfactory

---

[1] Cited from Kendall [1916].

manner. A slight error in the extrapolation is of hardly any consequence in the determination of the first dissociation constant of carbonic acid because the molecular conductivity of the bicarbonate ion has to be added to the much greater molecular conductivity of the hydrogen ion and it is the sum of these quantities which is used as a factor in the calculation (see chapter I). But if it is required to determine the conductivity coefficient an error in μ_∞ will have a great effect particularly in weak solutions. The question as to how μ_∞ should be obtained by extrapolation from conductivity determinations of salt solutions has up to the present been the subject of much controversy but I am unacquainted with any well-grounded theoretical method of performing the extrapolation. The best way must therefore be to obtain a relation between a function of the salt concentration and a function of the molecular conductivity which is the equation for a straight line. We can then determine μ_∞ either graphically or by the method of least squares.

It will as a rule be much the most convenient method to perform the extrapolation graphically but the method of least squares has the advantage of giving an estimation of the accuracy of the determinations.

That I only occasionally estimate the error in this chapter is because the value obtained is only of real significance if a systematic error can be excluded and a determination of the error can only be used in comparing results obtained by the same experimentalist with the same technique in a similar process. In the majority of cases the graphic method carried out in a suitable way, coupled with a good idea of the accuracy of the technique, will give more valuable information than the more arduous determination of the error.

Kohlrausch (cited from Lehfeldt [1908, p. 61]) recommends that the cube roots of the concentrations be plotted as abscissae and the molecular conductivities as ordinates. The curve will then be a straight line for many salt solutions. This kind of extrapolation has proved very suitable in the case of Walker and Cormack's determinations but particularly so for Kendall's measurements of conductivity in sodium bicarbonate solutions.

In Table XXX Walker and Cormack's and also Kendall's results are given. In the first column the number of litres in which a gram equivalent is present are recorded; in the second column, the concentration in terms of normality; in the third the cube root of this amount; in the fourth, the equivalent conductivity. In Fig. 12 these values are plotted, $\sqrt[3]{c}$ as abscissae and μ_0 expressed in reciprocal ohms as ordinates. The ordinates which refer to sodium bicarbonate are on the left, the numbers referring to the calcium bicarbonate series are on the right. It will be seen from the figure that the curve which represents Kendall's measurements of sodium bicarbonate solutions corresponds extremely well to a straight line. Extrapolation can therefore be undertaken with very considerable certainty and it gives $\mu_\infty = 99\cdot2$ reciprocal ohms (25°), while Kendall himself made the extrapolation to be 97·5. From this—using the same corrections as Kendall did—μ_∞ at 18° is 86·3 reciprocal ohms, while Walker and Cormack's determinations on extrapolation give 88·1 reciprocal ohms. The figure shows that the extrapolation of Walker and Cormack's measurements is rather uncertain, the two estimations in the most dilute solutions being a little high. The want of agreement between Walker and Cormack's and Kendall's results is presumably due to this cause and I shall regard Kendall's as correct.

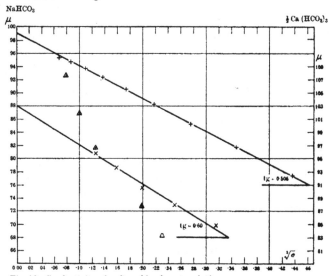

Fig. 12. Conductivity of sodium bicarbonate solutions.
Upper curve from Kendall's measurements; lower curve from Walker and Cormack's measurements.
△ ½ calcium bicarbonate.

The curve of the calcium bicarbonate experiments (triangles) demonstrates that only the three experiments with the strongest solutions lie on a straight line—the experiments with the two weakest solutions lie considerably above. Extrapolation, having regard to these two points, gives quite an unreasonable value for μ_∞, and I therefore think it cannot be performed with these experiments.

If we calculate the molecular conductivity of carbonic acid from Kendall's experiments with sodium bicarbonate (carbonic acid being regarded as monobasic) we get

at 25°   395·5,
„ 18°   355·8,
„ 0°   265·6.

Using these figures, values for the first dissociation constant of carbonic acid are obtained which diverge in the third decimal place from those calculated by Kendall [1916]. In Table XXXI Walker and Cormack's [1900] and Kendall's results are recalculated. Pfeiffer's and Knox's experiments are not given and the reader is referred to Kendall's paper for these. The small difference between Kendall's value for μ_∞ for carbonic acid and mine is without real significance in the calculations from these experiments.

The mean of Walker and Cormack's and Kendall's determinations at 18° of the first dissociation constant of carbonic acid is $3\cdot1 \times 10^{-7}$, from which $p_{K_1} = 6\cdot509$. At 25° we obtain from Kendall's experiments $K_1 = 3\cdot47 \times 10^{-7}$, $p_{K_1} = 6\cdot460$, and at 0°, $K_1 = 2\cdot21 \times 10^{-7}$, $p_{K_1} = 6\cdot656$; p_{K_1} therefore decreases $0\cdot20$ between 0° and 25° which is equivalent to $0\cdot008$ per degree.

By thermodynamic methods it is possible to determine the change in p_{K_1} with temperature when the heat of ionisation of carbonic acid is known. Julius Thomsen has estimated it at 2800 calories, from which the change in p_{K_1} can be calculated by van 't Hoff's equation (cf. Henderson [1909], Hasselbalch [1916, 2] and Kendall [1916]), and turns out to be $0\cdot0065$ per degree, which is in good agreement with Kendall's direct determination. Kendall has pointed out that the heat of hydration of CO_2 participates in Thomsen's determinations and that the calculation of the change of the dissociation exponent with temperature according to van 't Hoff's equation is only correct if the degree of hydration is independent of the temperature. He thinks however he is justified in concluding, from the good agreement between the calculated and experimental values, that the calculation is permissible—that is to say that the change in the degree of hydration is negligible in this connection.

From Fig. 12 it appears that the relation between the molecular conductivity of sodium bicarbonate and the concentration of the salt is expressed by the following equation:

$$- 50\cdot5 \sqrt[3]{c} + 99\cdot2 = \mu_v, \quad\quad\quad\quad\quad (135)$$

from which we get $\quad\quad f_\mu = \frac{\mu_v}{99\cdot8} = 1 - 0\cdot509 \sqrt[3]{c}. \quad\quad\quad\quad (136)$

No general relation is known between the conductivity coefficient of a salt and its activity coefficient but Hasselbalch, in his frequently cited paper [1916, 2], has carried out determinations which permit of the estimation of the apparent activity coefficient of the bicarbonate ion.

The experiments referred to were electrical estimations of hydrogen ion activity in sodium bicarbonate solutions of known concentration with simultaneous estimation of the CO_2 tension with which the solutions were in equilibrium. From these experiments Hasselbalch determined p_{K_1} (Hasselbalch) with the help of the equations developed in chapter III. In Table XXXII these experiments will be found recalculated to give our p_{K_1}' and, as before, the values corrected for depressed hydrogen tension as well as the uncorrected are given. Hasselbalch's experiments were performed by his own method with a plate electrode. For the sake of clearness we will postpone the further examination of these experiments.

Hasselbalch's experiments were carried out, as repeatedly stated, on the principle he evolved himself. I began therefore by repeating these experiments with the earlier described technique on Höber's principle, widening the scope of the investigation by including sodium bicarbonate solutions which in addition contained NaCl or KCl in the concentrations $0.0718n$, $0.145n$, and $0.291n$. These experiments are given in Tables XXXIII–XXXVIII and belong to series A but were done early in this series. I do not think they are subject to any systematic error but I cannot express myself with absolute certainty on this point.

In the first column of the tables the constitution of the solutions is given. The bicarbonate content of the solutions was estimated with the exhaust pump, at least two samples being used and usually several more. The error of this method is negligible compared with that of the electrical method. NaCl and KCl were added to the solutions in measured quantities, the salts being washed from a watch-glass into a measuring flask containing the bicarbonate solution. In a number of cases NaOH was used instead of $NaHCO_3$ and the solutions were therefore treated with CO_2 for a quarter of an hour before the actual saturation. It was impossible to demonstrate any difference between solutions prepared with the bicarbonate and the hydroxide.

In calculating the combined CO_2 (B), dissociation of the bicarbonate into the monocarbonate is taken into account.

In accordance with what was said in chapter II we have

$$S_1 = \frac{a_{H^\cdot}}{K'_2 + a_{H^\cdot}} B,$$

where S_1 is the amount of combined CO_2 corresponding to the bicarbonate expressed in vols. %, and B is the total amount of combined CO_2 expressed in the same way.

Since S_2 is the amount of combined CO_2 corresponding to the monocarbonate, and T is the amount of CO_2 in vols. % which would be combined if the combined CO_2 was in the form of bicarbonate, we have

$$S_1 + 2 S_2 = T, \quad \dots\dots\dots\dots\dots\dots(137)$$

from which with the aid of (102) we get

$$B = \frac{\frac{a_{H^\cdot}}{K'_2} + 1}{\frac{a_{H^\cdot}}{K'_2} + 2} T. \quad \dots\dots\dots\dots\dots\dots(138)$$

When B and $a_{H^\cdot}$ are known, therefore, T can be calculated and inversely B can be calculated from T and $a_{H^\cdot}$. In the pumping out experiments for the determination of $C_{HCO'_3}$, B was put equal to T when $p_{H^\cdot}$ was below 7.70; if it was above, (138) was used in the calculation, K'_2 being taken as 10^{-10}. In calculating $p_{K'_1}$, T was put equal to B when $p_{H^\cdot}$ was less than 7.90; when greater, (138) was used. In these last experiments equation (103) was also used instead of (95) (in logarithmic form) for calculating $p_{K'_1}$.

In Table XXXIX the results of this series of measurements are compared. They show that $p_{K'_1}$ undoubtedly decreases with increasing salt concentration and that the effect of the sodium ion on the constant is more pronounced than that of the potassium ion. The measurements in pure sodium bicarbonate solutions in the case of the two weakest concentrations ($T = 8.88$ and $T = 17.96$) are uncertain because the conductivity of these solutions was so slight that the resistance in the exit tap was very great and the oscillations of the capillary electrometer were too small.

In Table XXXVIII there are a number of determinations of $p_{K'_1}$ at room temperature in sodium bicarbonate solutions with the addition of equivalent amounts of NaCl or KCl. The experiments with the two series of salt solutions were performed on the same day and with the same electrode and they are therefore as comparable as possible, with the technique employed.

In Tables XL and XLIV experiments belonging to series B are given. $p_{K'_1}$ is here determined for sodium bicarbonate solutions partly with the addition of salt, at room temperature and at 38°.

In Table XLI measurements are given, carried out with the Hasselbalch electrodes E III and E VI, employed as Höber electrodes in the manner described in chapter IV, at room temperature. I consider these last measurements to be the best of all from a technical point of view, and I shall examine them a little more in detail. In the first place it will be observed that $p_{K'_1}$ for pure sodium bicarbonate solutions decreases with the concentration.

$$n/100 \text{ NaHCO}_3 \text{ gives } p_{K'_1} \ 6.400$$
$$n/40 \quad ,, \quad ,, \quad ,, \quad 6.393$$
$$n/20 \quad ,, \quad ,, \quad ,, \quad 6.335$$
$$n/10 \quad ,, \quad ,, \quad ,, \quad 6.279$$

In potassium bicarbonate solutions for

$$n/100 \text{ KHCO}_3, \ p_{K'_1} = 6.413$$
$$n/10 \quad ,, \quad ,, \quad = 6.317$$

If we compare the solutions with salt concentrations of $n/10$ we get

$$n/10 \text{ NaHCO}_3, \ p_{K'_1} = 6.279$$
$$n/10 \text{ KHCO}_3, \ p_{K'_1} = 6.317$$
$$n/20 \text{ NaCl} + n/20 \text{ NaHCO}_3, \ p_{K'_1} = 6.297$$
$$n/20 \text{ KCl} + n/20 \text{ KHCO}_3, \ p_{K'_1} = 6.312$$
$$n/10 \text{ NaCl} + n/1000 \text{ NaHCO}_3, \ p_{K'_1} = 6.313$$

It is seen therefore that $p_{K'_1}$ for $n/10$ sodium salt solutions is rather less than the constant for $n/10$ potassium salt solutions. This is still more noticeable in the determinations in $n/5$ salt solutions, and from the latter it also appears that $p_{K'_1}$ for mixtures of Na and K salts lies between those for the pure salts.

$$n/10 \text{ NaCl} + n/10 \text{ NaHCO}_3 \text{ gives } p_{K'_1} = 6.246$$
$$n/10 \text{ KCl} + n/10 \text{ KHCO}_3 \quad ,, \quad ,, = 6.260$$
$$n/10 \text{ KCl} + n/10 \text{ NaHCO}_3 \quad ,, \quad ,, = 6.282$$

Turning again to the $n/10$ solutions it will be noticed that $p_{K'_1}$ seems to be a little greater for solutions containing NaCl than for those containing equivalent amounts of NaHCO$_3$. The difference in $p_{K'_1}$ for solutions containing pure $NaHCO_3$ and for those containing equal parts of $NaCl$ and $NaHCO_3$ is however so small that it certainly falls within the experimental error, but the difference between $n/10$ $NaHCO_3$ and $n/10$ $NaCl + n/1000$ $NaHCO_3$ appears to be real. I think that in this case the method of least squares ought to be used.

From equation (89) in logarithmic form we get

$$p_{K'_1} = p_{K_1} + \log F_a (HCO'_3).$$

According to Bjerrum (cf. equation (58))

$$\log f_a (HCO'_3) = - k \sqrt[3]{c}.$$

If we put $\qquad F_a (HCO'_3) = f_a (HCO'_3),$

which is permissible at $n/10$ concentration, we get

$$p_{K'_1} = p_{K_1} - k \sqrt[3]{c}, \qquad \ldots\ldots\ldots\ldots\ldots(139)$$

where c is the molar concentration of the cations and k is the constant in Bjerrum's above cited equation, while p_{K_1} is the negative logarithm of the first dissociation constant of carbonic acid.

If we have a series of equations of the form

$$ax + y = l, \qquad \ldots\ldots\ldots\ldots\ldots\ldots(140)$$

we can calculate the constants which fulfil the conditions best with the following equations:

$$x = \frac{n \Sigma al - \Sigma a \Sigma l}{n \Sigma a^2 - (\Sigma a)^2}, \qquad \ldots\ldots\ldots\ldots\ldots(141)$$

$$y = \frac{\Sigma l \Sigma a^2 - \Sigma al \Sigma a}{n \Sigma a^2 - (\Sigma a)^2}. \qquad \ldots\ldots\ldots\ldots\ldots(142)$$

If we employ the above equations in connection with all the 26 experiments which only contain $NaHCO_3$ in the series recorded we obtain

$$p_{K_1} = 6.512,$$
$$k = 0.475.$$

An estimate of the mean error of $p_{K'_1}$ can be obtained by putting the values found for it and k in (139), and determining the deviations from the above value, thus

$$M = \sqrt{\frac{7519}{25}} \times \frac{1}{100} = 0.0173.$$

For a series of nine experiments we may therefore expect a mean error of 0.0058. Thus for such a series of measurements with $n/10$ $NaHCO_3$

$$p_{K'_1} = 6.291 \pm 0.0058.$$

In the experiments with the solution of $n/10$ $NaCl + n/1000$ $NaHCO_3$

$$p_{K'_1} = 6.313 \pm 0.0045.$$

The difference is therefore 0.022 ± 0.0073 and in all probability it is a real one.

It is an obvious consequence of Bjerrum's activity theory, as it has been expounded in chapter I and the present chapter, that the activity coefficient in a solution of a binary salt consisting of two monovalent radicles shall not vary[1] if the solution is diluted with another which contains an equivalent amount of a similar salt having no ions in common with the first.

[1] Apart from secondary effects represented by the difference in k in equation (46).

According to a view widely held by physiologists, which for example is put forward by Hamburger [1902], Hedin [1915] and Ege [1920], the dissociation of a monovalent salt will increase if the salt solution is diluted with another monovalent salt having no ions in common with the first, even if the two solutions are equivalent. This view depends however on an incorrect application of Arrhenius' [1888] theory of "isohydric" solutions, but I was myself involved in it until Prof. Bjerrum kindly pointed out the fallacy to me. In view of the general acceptance it has received I will enter a little more deeply into the question regarding for the time being Arrhenius' classical theory of the incomplete dissociation of salts as correct.

Let us consider a $n/1$ solution of NaCl and a $n/1$ solution of KI.

Then according to Arrhenius' theory extended to isohydric dissociation

$$\frac{C_{Na^{\cdot}} + C_{Cl'}}{C_{NaCl}} = k,$$

and

$$\frac{C_{K^{\cdot}} + C_{I'}}{C_{KI}} = k.$$

The equilibria in a $n/1$ KCl solution and a $n/1$ NaI solution will be given by

$$\frac{n/1\,\gamma K^{\cdot}\; n/1\,\gamma Cl'}{n/1\,KCl\,(1-\gamma)} = k,$$

and

$$\frac{n/1\,\gamma Na^{\cdot}\; n/1\,\gamma I'}{n/1\,NaI\,(1-\gamma)} = k,$$

γ being the (common) degree of dissociation of the salts.

If we mix some NaCl solution with some KI solution, e.g. equal parts, an equilibrium will be established which will satisfy all the four equations. This is only possible at a quite definite degree of dissociation of the different salts as the determining equations can only have one solution.

Assuming that the dissociation does not change on mixing the solutions, we have

$$C_{Na^{\cdot}} = C_{Cl'} = C_{K^{\cdot}} = C_{I'} = n/2\gamma,$$

and

$$C_{NaCl} = C_{KI} = n/2\,(1-\gamma).$$

Now NaCl and KI will react with each other and equal quantities of NaI and KCl will be formed if the same laws hold for all the undissociated salts in question. Thus we have the following equilibria:

$$C_{NaCl} = C_{KI} = C_{NaI} = C_{KCl} = n/4\,(1-\gamma),$$

$$\frac{n/2\,\gamma\,Na^{\cdot}\; n/2\,\gamma\,Cl'}{n/4\,NaI\,(1-\gamma)} = k,$$

$$\frac{n/2\,\gamma\,K^{\cdot}\; n/2\,\gamma\,I'}{n/4\,KI\,(1-\gamma)} = k,$$

$$\frac{n/2\,\gamma\,Na^{\cdot}\; n/2\,\gamma\,I'}{n/4\,NaI\,(1-\gamma)} = k,$$

$$\frac{n/2\,\gamma\,K^{\cdot}\; n/2\,\gamma\,Cl'}{n/4\,KCl\,(1-\gamma)} = k,$$

which are correct in accordance with the four original equations. As only one state of equilibrium is possible the above mathematical reasoning is proof that no change in the dissociation of the salts takes place on mixing.

Finally it may be stated that as far as the proof is concerned it is immaterial whether the dissociation equations are really equal to a constant or only to a function of the total salt concentration, provided that all the equations are equal to the same quantity.

Equation (139) is a straight line and corresponds to one in a rectangular coordinate system in which the abscissae are $\sqrt[3]{c}$ and the ordinates $p_{K'_1}$, the straight line cutting the ordinate axis at p_{K_1}, while the tangent of the angle made by the straight line with the abscissa axis measured in the second or fourth quadrant is equal to x. In Fig. 13 $\sqrt[3]{c}$ is plotted as abscissae and $p_{K'_1}$ as ordinates, the total cation concentration in the solutions which contain either sodium salts or potassium salts alone being calculated. Only the experiments for which $\sqrt[3]{c}$ is given in the tables are plotted in the figure.

Hasselbalch's experiments (the two lowest dotted curves) are seen to be about 0·09 lower in the coordinate system than mine, so that Hasselbalch's p_{K_1} (18°) is 6·412, p_{K_1} (38°) is 6·302, p_{K_1} (18°) $- p_{K_1}$ (38°) $= 0·11$, while k is 0·52.

The continuous line refers to the series of pure sodium bicarbonate solutions fully discussed above, while the uppermost dotted line is the one that best represents, as far as one can judge, all the experiments with sodium salts (the experiments with $n/1000$ $NaHCO_3$ are however omitted). This line gives

$$p_{K_1} = 6·514,$$
$$k = 0·46.$$

There is only slight uncertainty in drawing the line and without doubt only a small real error. It is worth noting that the equation

$$- \log F_a (HCO'_3) = 0·46 \sqrt[3]{c} \qquad \qquad (143)$$

appears to hold up to 0·4n with good approximation.

The lowest dotted curve is parallel with the uppermost but 0·11 lower. It represents the determinations at 38° tolerably well. It will be observed that the change in the constant per degree is 0·0055 in Hasselbalch's experiments and also in mine, therefore considerably less than the change found by Kendall [1916], namely 0·008, and rather less than that obtained by thermodynamic methods with Thomsen's value for the heat of reaction, namely 0·0065. The uppermost dotted curve represents the determinations of $p_{K'_1}$ for pure potassium salt solutions. It gives $p_{K_1} = 6·497$ and $k = 0·38$. The value found for p_{K_1} is presumably a little too low because it is hardly possible there can be any real difference between p_{K_1} in sodium and potassium solutions, but there are too few determinations in low concentrations in the series to enable us to attribute the same importance to the constants for potassium as to those for sodium salts.

From their experiments on the reaction of $CaCl_2$ solutions saturated with $Ca (HCO_3)_2$ Bjerrum and Gjaldbaek [1919] calculated the "reaction constant" of calcium carbonate as

$$\log K = - 5·02 \text{ at } 18°.$$

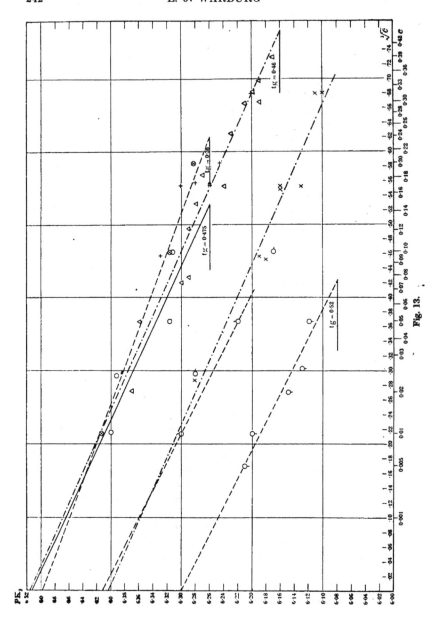

Fig. 13.

This experimental determination they compared with a calculation of the reaction constant by means of the following equation, Bjerrum and Gjaldbaek's equation (28), $\log K = \log K_1 + \log Kp - \frac{1}{2} \log K'$, ...(144)

where K_1 is the first dissociation constant of carbonic acid ($\log K_1$ is therefore equal to $- p_{K_1}$), Kp is the "solubility" constant of CO_2 which is estimated as

$$\log Kp = - 1\cdot38,$$

and K' is a constant, which we shall shortly consider a little more closely, the value used by Bjerrum and Gjaldbaek being

$$\log K' = - 5\cdot68.$$

Bjerrum and Gjaldbaek employed the mean of Walker and Cormack's and Kendall's determinations, namely $- 6\cdot51$, as the value for $\log K_1$, from which they obtained $\log K = - 5\cdot05,$

a figure which therefore was only $0\cdot03$ less than that determined directly.

Now it will be seen that equation (144) reversed can be used for the calculation of $\log K_1$ if $\log K'$ is known, and that by substituting the value found, namely $\log K' = - 5\cdot02$, we obtain

$$\log K_1 = - 6\cdot48.$$

This value agrees extremely well with that previously obtained from Walker and Cormack's and Kendall's experiments, and with those I myself determined, but it will be shown the agreement can be made still better.

The constant K' determines the following equilibrium, Bjerrum and Gjaldbaek's equation (15),

$$\frac{a_{Ca^{..}} \cdot a^2_{HCO_3}}{P_{CO_2}} = K''.$$ ....(145)

Apropos of this they write:

"The ionic activity (a) can be calculated by multiplying the ionic concentration (c) by the ionic activity coefficient (F_a). For the monovalent bicarbonate ion the activity coefficient is approximately given by

$$\log F_a = - 0\cdot3 \sqrt[3]{c_{Ion}},$$

and for the divalent calcium cation it is given by

$$\log F_a = - 2 \times 0\cdot3 \sqrt[3]{c_{Ion}},$$

c_{Ion} standing for the ionic normality of the solution. (145) now assumes the following form:

$$\frac{c_{Ca^{..}} \cdot c^2_{HCO'_3}}{P_{CO_2}} = K' \cdot 10^{1\cdot2 \sqrt[3]{c_{Ion}}}.$$ ....(146)

If we call the equivalent concentration of the calcium carbonate which Schlösing[2] found dissolved C, we have

$$C_{Ca^{..}} = \frac{1}{2}C, \quad C_{HCO'_3} = C, \quad C_{Ion} = C.$$

---

[1] Bjerrum and Gjaldbaek substitute p_{CO_2} for my P_{CO_2}, and naturally their numbering of equations is different from that given here.

[2] Cited from Bjerrum and Gjaldbaek [1919].

By substitution and taking logarithms (146) becomes

$$3 \log C - \log P_{CO_2} - 1 \cdot 2 \sqrt[3]{c} - \log 2 = \log K'. \quad \ldots\ldots\ldots(147)"$$

Bjerrum and Gjaldbaek calculate the values of $\log K'$ with the help of (147) from Schlösing's[1] experiments on the solubility of calcium carbonate at different CO_2 tensions, and their results are recorded in Table XLV in the first three columns. It will be noted they have calculated the activity of the bicarbonate ion from the equation

$$\log F_a (HCO'_3) = -0 \cdot 30 \sqrt[3]{C}, \quad \ldots\ldots\ldots\ldots(148)$$

but as we have previously found the value $0 \cdot 3$ is too low their tables have been recalculated with the equation

$$3 \log C - \log P_{CO_2} - 1 \cdot 52 \sqrt[3]{c} - \log 2 = \log K', \quad \ldots\ldots\ldots(149)$$

having assumed that the constant in the equation for calculating the activity coefficient of the bicarbonate ion in calcium carbonate solutions is the same as in sodium salt solutions (see above). The values obtained have been put in the fourth column of Table XLV. $\log K'$ becomes $-5 \cdot 709$ and it will be observed that the agreement between the individual experiments is better calculated in this way than by Bjerrum and Gjaldbaek's method. Schlösing's[1] experiments were carried out at $16°$ and Bjerrum and Gjaldbaek therefore convert the values to $18°$ using a temperature coefficient derived from experiments of R. C. Wells[1]. Correcting the constant in the same way we obtain $\qquad \log K' = -5 \cdot 749$ at $18°$.

Now it appears that Bjerrum and Gjaldbaek's "reaction constant" for calcium carbonate can be used without conversion if we confine ourselves to the experiments they signify as best suited for the calculation ($0 \cdot 1$ and $0 \cdot 02n$ $CaCl_2$), because a recalculation using $0 \cdot 46$ instead of $0 \cdot 30$ gives no difference in the second decimal place. Calculated by (144) p_{K_1} becomes

$$p_{K_1} = 6 \cdot 515 \text{ at } 18°,$$

which is in surprisingly good agreement with Walker and Cormack's, Kendall's and my own determinations.

At $18°$ the first dissociation exponent is therefore according to

Knox	6·426	Kendall	6·507
Walker and Cormack	6·512	Bjerrum and Gjaldbaek	6·515
Hasselbalch	6·420	Warburg[2]	6·514

Of these values Knox's was obtained with a relatively indifferent technique. In view of what was said in chapter IV Hasselbalch's experiments may very well have a systematic error; all the other determinations give values which approximate closely to $6 \cdot 51$. Therefore

$$K_1 = 3 \cdot 1 \times 10^{-7} \ (18°).$$

T. H. Milroy [1917] has carried out electrical determinations at $37 \cdot 5°$ with pure $0 \cdot 2n$ sodium bicarbonate solution and the same diluted with $0 \cdot 2n$ sodium chloride solution from which $p_{K'_1}$ at the concentration $0 \cdot 2n$ can be calculated.

---

[1] Cited from Bjerrum and Gjaldbaek [1919].   [2] Sodium salts.

I have gone over Milroy's experiments and the technique is hardly as good as could be wished. The dispersion is rather large and $p_{K'_1}$ has a lower value than would be expected, being not as much as 6·00 on an average. J. F. McClendon, A. Shedlov and W. Thomson [1917] have also made measurements in salt solutions containing bicarbonate but they have only published their results graphically so that it is impossible to adjudicate upon the accuracy of the determinations.

Lastly L. Michaelis and P. Rona [1914] have made some measurements in sodium bicarbonate solutions but as Hasselbalch has shown they are subject to a technical error which makes them very doubtful.

The osmotic coefficient of the bicarbonate can be calculated with equation (44),
$$1 - f_0 = k \sqrt[3]{c},$$
where k is determined by (46),
$$\log_e f_a = - 4k \sqrt[3]{c}.$$
By substituting the value $2·303 \times 0·46$ for $4k$ we get
$$f_0 (HCO'_3) = 1 - 0·26 \sqrt[3]{c}. \qquad \ldots\ldots\ldots\ldots\ldots\ldots(150)$$
According to (136) $\quad f_\mu (NaHCO_3) = 1 - 0·51 \sqrt[3]{c},$
and from (143) we have
$$\log f_a (HCO'_3) = (\log F_a (HCO'_3)) = - 0·46 \sqrt[3]{c}.$$

With the help of these equations the deviation coefficients for the bicarbonate ion have been calculated and the results recorded in Table XLVI and graphically displayed in Fig. 14.

When one compares the deviation coefficients given here with those valid for KCl which are recorded in chapter I, Table III one cannot help thinking the marked depression of the activity of the bicarbonate ion and the relatively low conductivity coefficient are an indication that the salt is not completely dissociated as we previously supposed.

It would be of great interest to compare the deviation coefficients of a considerable number of salts with one another and with the dissociation constant of the corresponding acid, but such an investigation lies outside the scope of this work and it is to be hoped it will soon be undertaken by someone more expert in that branch of the subject. It may however be put forward that the fact that the activity of the bicarbonate ion can be determined with (143) over a large range of concentration does not at all well fit in with the idea of an incomplete dissociation of sodium bicarbonate. If we assume that equation (143) only represents the relation between the bicarbonate concentration and the activity of its ions, while the relation between the ionic concentration and activity of bicarbonate and sodium is given by
$$- \log F_a = 0·25 \sqrt[3]{c}, \qquad \ldots\ldots\ldots\ldots\ldots\ldots\ldots(I)$$
we can construct the following equation:
$$c\gamma \cdot 10^{-0·25 \sqrt[3]{c\gamma}} = c \cdot 10^{-0·46 \sqrt[3]{c}}, \qquad \ldots\ldots\ldots\ldots\ldots(II)$$
and therefore $\quad \log \gamma = \sqrt[3]{c} (0·46 - 0·25 \sqrt[3]{\gamma}). \qquad \ldots\ldots\ldots\ldots\ldots(III)$

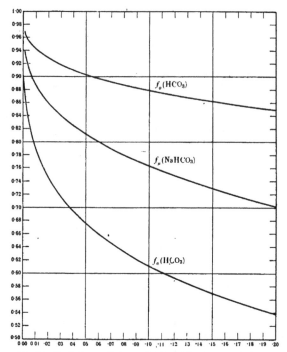

Fig. 14. Deviation coefficients of sodium bicarbonate solutions.

According to the mass action law the following holds:

$$\frac{C_{Na} \cdot F_a(Na^{\cdot}) \cdot C_{HCO'_3} F_a(HCO'_3)}{C_{NaHCO_3} F_a(NaHCO_3)} = K. \quad\ldots\ldots\ldots\ldots(IV)$$

If we put $F_a(NaHCO_3) = 1$ we get

$$\frac{c\gamma \cdot 10^{-0.5\sqrt[3]{c\gamma}}}{1-\gamma} = K. \quad\ldots\ldots\ldots\ldots\ldots(V)$$

If now equation (III) is solved by successive approximations and the values found are substituted in (V) we obtain

c	γ	K
0·001	0·952	$1·7 \times 10^{-2}$
0·01	0·897	$6·1 \times 10^{-2}$
0·1	0·790	$1·8 \times 10^{-1}$

The example given shows therefore that "the constant" in (V) must increase with the salt concentration and that the mass action law cannot therefore be satisfied if the salt is incompletely dissociated, provided that an equation for the relation between the concentration and ionic activity of the form of (143) holds good.

In Table XLVII are given determinations of $p_{K'_1}$ in solutions which contain small amounts of phosphate in addition to NaCl and NaHCO$_3$. It is probable that $p_{K'_1}$ allowing for experimental error is identical with the constant one might expect if phosphate was substituted by chloride.

In Table XLVIII are recorded two series of estimations carried out in about $0{\cdot}01n$ sodium bicarbonate solutions which were at the same time $0{\cdot}501m$ as regards cane sugar.

The solutions were prepared by making stock solutions double the strength of those used in the experiments. These were kept on ice. Equal parts of sodium bicarbonate and cane sugar solutions were measured with Geissler pipettes and mixed half an hour before saturation was commenced during which time they were in the laboratory at room temperature.

The potential established itself—in contrast to the pure salt solutions— not immediately, but only when the electrode had been rocked 10–25 times. The first series of measurements belongs to series A and was performed late in this series so that $p_{K'_1}$ is too low. It shows that the relative absorption coefficient is about $0{\cdot}835$ and y is therefore $16{\cdot}5$. The second series of measurements was carried out with E III and E VI simultaneously with the experiments previously noted as technically the best. It gives $p_{K'_1} = 6{\cdot}461$ ($18°$). As $p_{K'_1}$ in the corresponding pure $NaHCO_3$ solution is $6{\cdot}413$, it appears that the addition of cane sugar increases the constant. If now we regard the depression of solubility of CO_2 as an expression of the amount of water the cane sugar has appropriated (we assume that CO_2 and $NaHCO_3$ are insoluble in hydrated cane sugar), and correct the calculations for this we obtain

$$p_{K'_1} - \log F_a\,(CO_2) = 6{\cdot}383,$$

and at the same time the solution may be regarded as $0{\cdot}012n$ with respect to bicarbonate, which corresponds to

$$p_{K'_1} = 6{\cdot}402.$$

The difference between $6{\cdot}383$ and $6{\cdot}402$ does not lie outside the experimental error for certain, so much the more so because of the peculiarity in the establishment of the potential in cane sugar solutions, just referred to.

The result of this small series of experiments is that the activity of the bicarbonate ion is not affected with any certainty by the cane sugar molecules when a correction for the hydration is made.

RÉSUMÉ.

I. The first dissociation constant of carbonic acid has been recalculated from the best experiments in the literature, and it has been determined afresh by the electrical potential method at $18°$ and $38°$.

II. The apparent activity coefficient of the bicarbonate ion has been determined in sodium and potassium solutions up to $0{\cdot}4n$.

III. The apparent osmotic coefficient of the bicarbonate ion has been calculated from the apparent activity coefficient.

IV. The apparent conductivity coefficient of the bicarbonate ion has been recalculated.

V. It has been rendered probable that the alkaline bicarbonates are completely dissociated.

ADDENDUM

After this chapter was completed a paper by C. Lovatt Evans appeared (*J. Physiol.* 1921, **54**, pp. 353–366) which necessitates a certain amount of criticism. The author himself is rather cautious, as he writes: "Although perhaps some doubt still remains as to the finality of the conclusions which will be presented here etc...."

The object of the paper was to prove that electrical determinations of the reaction of bicarbonate solutions give too acid values (too low p_H·) and that correct results can be obtained by the colorimetric method. The explanation of the reaction being found too acid by potential measurements is given by the author as a consequence of the formation of formic acid in sufficient quantity to set up a reduction potential by the catalytic action of the platinum black on the electrode, according to the equations

$$CO_2 + H_2 \xrightleftharpoons{} HCOOH$$

and
$$HCO'_3 + H_2 = HCOO' + H_2O.$$

(1) It can very easily be shown that the combined CO_2 in an alkali bicarbonate solution does not decrease when the solution is treated for half an hour (or one hour) with hydrogen in the presence of platinum black. I have myself carried out over 50 such experiments. However it would be very probable that the combined amount of CO_2 would decrease if a reduction potential was set up as Evans suggests.

(2) If such a process took place p_{K_1} would be dependent upon the CO_2 tension and HCO'_3 concentration and not, as has been shown, practically only upon the cation concentration.

(3) The calculation of p_{K_1} from Bjerrum and Gjaldbaek's and from my own experiments would give lower values than from conductivity determinations if the formation of formic acid took place.

(4) The agreement between the calculated amount of combined CO_2 and that found experimentally in the phosphate experiments in chapter XI would be bad if Evans was right in his conclusions.

The most important of Evans' experiments are the following (p_{K_1} calculated by the author):

mm. Hg CO_2	Neutral red		Phenol red		Electrically	
	p_H·	p_{K_1}	p_H·	p_{K_1}	p_H·	p_{K_1}
			$0.02n$ NaHCO$_3$ 20°			
8·93	8·23	(6·59)	8·04	6·40	7·96	(6·33)
21·8	7·70	6·45	7·42	6·17	7·55	6·30
33·3	7·57	6·51	7·28	6·22	7·34	6·28
46·0	7·37	6·45	7·17	6·25	7·16	6·24
68·9	7·19	6·34	6·97	6·12	7·07	6·22
		Mean 6·44		Mean 6·19		Mean 6·26
			$0.02n$ NaHCO$_3$ + $0.18n$ NaCl 20°			
6·5	8·13	(6·36)	—	—	—	—
11·5	7·77	6·25	—	—	—	—
13·9	7·66	6·22	7·67	6·23	7·60	6·16
18·8	—	—	—	—	7·34	6·06
26·0	7·48	6·31	—	—	—	—
30·5	7·33	6·23	7·34	6·24	7·15	6·05
42·8	7·23	6·28	—	—	—	—
44·0	7·21	6·27	—	—	—	—
46·1	—	—			7·08	6·16
57·2	7·11	6·28			—	—
58·5	—	—			6·88	6·10
65·4	7·04	6·27			—	—
		Mean 6·26		Mean 6·235		Mean 6·11

From the determinations made in the preceding chapter $p_{K'_1}$ in the first series is 6·38 and in the second series 6·23 at 20°.

The reason Evans did not obtain agreement between the electrical and colorimetric results is probably the fact that his electrode vessel is of an unsuitable type. As far as I can judge it must be almost impossible with his electrode to avoid the O_2 error. Evans himself writes: "Definitive potentials were obtained almost at once provided the gas was free from oxygen; when this was not the case, there was first a somewhat lower potential than that finally obtained after one or two hundred inversions."

The colorimetric determinations of the reaction in salt solutions seem from the above to be rather good (agreement between $p_{K'_1}$ in each series), but the curves of the reaction of the blood indicate too alkaline a reaction possibly because the blood was saturated at body temperature, the dialysis and colorimetry however being performed at room temperature.

The reactions of salt solutions which are "saturated" with smaller CO_2 tensions than 10 mm. are more alkaline than would be expected as Evans himself has noted. He ascribes this to hydrolysis which is incorrect as hydrolysis at the reactions of the experiments is negligible in this connection (it is of the same order as $a_{OH'}$). The strongly alkaline reactions probably arise from equilibrium not being reached during the "saturation" at the low CO_2 tensions (cf. chapter XI).

I shall not enter further into a discussion of the theory of the "degree of dissociation" of the bicarbonate here but refer the reader to the preceding chapters.

Table XXX. Conductivity determinations in bicarbonate solutions.

V	c	$\sqrt[3]{c}$	μ_v	
12·1	0·0826	0·436	77·3	Kendall
24·2	0·0413	0·346	81·7	$NaHCO_3$ 25°
48·4	0·0207	0·275	85·3	
96·8	0·0103	0·217	88·2	
193·6	0·00516	0·173	90·6	
387·2	0·00258	0·137	92·4	
774·4	0·00129	0·109	93·8	
1548·8	0·000646	0·0864	94·8	
3097·6	0·000323	0·0686	95·5	
—	0	0	99·2	
32·0	0·0313	0·315	69·8	Walker and Cormack
64·0	0·0156	0·250	72·9	$NaHCO_3$ 18°
128·0	0·00782	0·198	75·6	
256·0	0·00391	0·158	78·6	
518·0	0·00195	0·125	80·8	
—	0	0	88·1	
64·0	0·0156	0·250	83·3	Kendall
128·0	0·00782	0·198	88·8	1/2 $CaCO_3$ 25°
256·0	0·00391	0·158	93·4	
512·0	0·00195	0·125	96·7	
1024·0	0·000977	0·0992	102·0	
2048·0	0·000488	0·0787	107·7	
—	0	0	?	

Table XXXI. Determinations of the first dissociation constant of carbonic acid calculated from Walker and Cormack's and from Kendall's experiments.

V	μ_v	γ	$K_1 \times 10^{-7}$	
31·25	1·104	0·00310	3·09	Walker and Cormack
62·5	1·570	0·00441	3·13	18° CO_2 from marble
93·7	1·916	0·00539	3·11	
125·0	2·218	0·00623	3·12	
∞	355·8		Mean 3·11	
			p_{K_1} 6·507	
27·5	1·033	0·00291	3·07	Walker and Cormack
55·0	1·454	0·00409	_ 3·05	18° CO_2 from carbonic acid snow
82·5	1·754	0·00502	3·05	
110·0	2·052	0·00577	3·04	
∞	355·8		Mean 3·05	
			p_{K_1} 6·516	
25·4	0·631	0·00237	2·23	Kendall
38·3	0·770	0·00290	2·20	0°
50·0	0·880	0·00331	2·20	
76·3	1·081	0·00407	2·18	
99·8	1·242	0·00468	2·21	
152·6	1·548	0·00584	2·24	
∞	265·6		Mean 2·21	
			p_{K_1} 6·656	
30·9	1·100	0·00309	3·11	Kendall
42·0	1·281	0·00360	3·11	18°
61·2	1·550	0·00437	3·14	
83·4	1·792	0·00504	3·06	
∞	355·8		Mean 3·11	
			p_{K_1} 6·507	
36·4	1·403	0·00355	3·47	Kendall
51·3	1·659	0·00320	3·44	25°
72·8	1·977	0·00500	3·45	
102·4	2·341	0·00592	3·44	
145·5	2·820	0·00713	3·54	
∞	395·5		Mean 3·47	
			p_{K_1} 6·460	

Table XXXII. Calculation of $p_{K'_1}$ in pure sodium bicarbonate solutions from K. A. Hasselbalch's experiments, *Biochem. Zeitsch.* **78**, p. 119.

		$p_{H'}$	$p_{K'_1}$	$p_{H'}$	$p_{K'_1}$
				Corrected for decreased	
Concentration	mm. CO_2	Uncorrected		H_2 pressure	
0·05n	47·2	7·47	6·19	7·48	6·20
$\mathrm{^s/c}=0·368$	403·2	6·53	6·18	6·68	(6·33)[1]
18°	74·0	7·28	6·20	7·30	6·22
Mark ♂	147·7	6·97	6·19	7·01	6·23
		Mean 6·19		Mean 6·22	
0·01n	54·3	6·82	6·29	6·83	6·30
$\mathrm{^s/c}=0·215$	134·3	6·42	6·28	6·46	6·32
18°	101·8	6·51	6·25	6·54	6·29
Mark ♂	84·5	6·59	6·25	6·61	6·27
	68·8	6·73	6·30	6·75	6·32
	92·4	6·59	6·29	6·62	6·32
		Mean 6·28		Mean 6·30	

[1] Not included in the mean value.

Table XXXII (*continued*)

Concentration	mm. CO_2	$p_{H^.}$	$p_{K'_1}$	$p_{H^.}$	$p_{K'_1}$
		Uncorrected		Corrected for decreased H_2 pressure	
$0.05n$	168·9	7·02	6·06	7·08	6·12
$\nu/c = 0.368$	247·8	6·87	6·08	6·97	(6·18)[1]
38°	52·6	7·55	6·09	7·57	6·11
Mark ♀	717·0	6·40	6·07	?	?
		Mean	6·08	Mean	6·12
$0.03n$	40·6	7·48	6·12	7·49	6·13
$\nu/c = 0.311$	80·7	7·15	6·09	7·18	6·12
38°	135·0	6·92	6·09	6·96	6·13
Mark ♀	96·9	7·08	6·10	7·11	6·13
		Mean	6·10	Mean	6·13
$0.02n$	123·3	6·83	6·13	6·87	6·17
$\nu/c = 0.271$	36·7	7·33	6·11	7·34	6·12
38°	62·8	7·13	6·14	7·15	6·16
Mark ♀		Mean	6·13	Mean	6·15
$0.01n$	85·0	6·70	6·14	6·73	6·17
$\nu/c = 0.215$	218·6	6·33	6·18	6·40	6·25
38°	137·4	6·51	6·16	6·55	6·20
Mark ♀	189·2	6·40	6·19	6·46	6·23
	119·8	6·57	6·16	6·61	6·20
	149·0	6·44	6·13	6·48	6·17
		Mean	6·16	Mean	6·20
$0.005n$	23·5	7·02	6·21	7·03	6·22
$\nu/c = 0.171$	41·35	6·75	6·18	6·76	6·19
Mark ♀	71·8	6·53	6·20	6·55	6·22
		Mean	6·20	Mean	6·21

[1] Not included in the mean value.

Table XXXIII. Determinations of $p_{K'_1}$ for 18° in sodium bicarbonate solutions carried out with the small saturator electrode. Series A. Mark △.

Concentration	Temp.	mm. Hg		$p_{H^.}$	$p_{K'_1}$ 18°	
		CO_2	O_2			
$T = 8.88$	16·0	240·2	0·6	5·84	6·37	
$0.00399n$ HCO'_3	16·5	118·6	0·2	6·19	6·42	
$0.00399n$ Na·	16·5	41·6	0·2	6·56	6·33	
	16·5	15·2	0·2	7·04	6·37	
	16·5	5·8	0·2	7·45	6·37	
				Mean	6·37	
$T = 17.52$	19·5	543·4	0·3	5·80	(6·37)	
$0.00786n$ HCO'_3	19·5	141·0	0·5	6·38	6·37	
$0.00786n$ Na·	19·5	55·2	0·5	6·82	6·40	
	19·5	18·0	0·5	7·29	6·38	
	20·0	7·1	0·3	7·70	6·39	
	20·0	4·2	0·3	8·05	(6·51)	
				Mean	6·38	
$T = 55.5$	19·5	486·5	0·7	6·35	6·37	
$0.0249n$ HCO'_3	19·0	124·4	0·5	6·95	6·38	
$0.0249n$ Na·	19·0	51·0	0·5	7·33	6·37	$\nu/c = 0.273$
	19·0	30·9	0·3	7·51	6·35	
	19·0	11·3	0·3	8·01	6·40	
				Mean	6·37	

Table XXXIII (*continued*)

mm. Hg.

Concentration	Temp.	CO_2	O_2	p_H	$p_{K'_1}$ 18°	
$T = 110.0$	20.5	538.0	0.6	6.60	(6.37)	
$0.0494n$ HCO'_3	20.0	267.0	0.5	6.91	6.37	
$0.0494n$ $Na^.$	20.0	67.2	0.5	7.50	6.36	$\sqrt[3]{c} = 0.367$
	20.0	25.2	0.2	7.91	6.35	
	20.0	9.2	0.2	8.36	6.37	
					Mean 6.36	
$T = 221.9$	19.5	126.3	0.5	7.44	6.28	
$0.0997n$ HCO'_3	19.5	37.5	0.5	8.00	6.32	
$0.0997n$ $Na^.$	19.5	14.7	0.3	8.39	6.31	
	19.0	58.8	0.3	7.80	6.31	
	17.0	193.4	0.2	7.28	6.31	$\sqrt[3]{c} = 0.464$
	17.0	85.3	0.9	7.63	6.31	
	17.0	25.1	1.2	8.08	6.33	
	17.0	50.1	0.7	7.88	6.32	
	17.0	130.1	0.5	7.41	6.27	
	17.0	56.6	0.3	7.81	6.32	
					Mean 6.31	

Table XXXIV. Determinations of $p_{K'_1}$ for 18° in sodium bicarbonate solutions with the addition of 0.42 % NaCl ($0.0718n$). Carried out with the small saturator electrode. Series A. Mark △.

mm. Hg

Concentration	Temp.	CO_2	O_2	p_H	$p_{K'_1}$ 18°	
$T = 8.88$	18.5	245.8	0.6	5.78	6.31	
$0.00399n$ HCO'_3	19.0	29.5	0.6	6.70	6.31	
$0.0718n$ Cl'	19.0	40.7	0.6	6.56	6.30	$\sqrt[3]{c} = 0.421$
$0.07479n$ $Na^.$	19.0	63.8	0.4	6.23	6.28	
					Mean 6.30	
$T = 17.52$	19.0	395.5	1.0	5.86	6.31	
$0.00786n$ HCO'_3	19.0	108.9	0.4	6.42	6.28	
$0.0718n$ Cl'	19.5	34.5	0.2	6.91	6.29	$\sqrt[3]{c} = 0.429$
$0.07966n$ $Na^.$	19.5	6.0	0.2	7.68	6.30	
	19.5	14.4	0.2	7.31	(6.31)	
					Mean 6.29	
$T = 55.5$	20.0	36.7	6.6	7.34	(6.24)	
$0.0249n$ HCO'_3	20.0	119.2	1.2	6.86	6.27	
$0.0718n$ Cl'	20.0	106.2	0.7	6.93	6.29	
$0.0967n$ $Na^.$	20.0	7.3	0.5	8.10	6.30	$\sqrt[3]{c} = 0.459$
	20.0	512.0	0.6	6.22	(6.25)	
	20.0	126.7	0.3	6.85	6.29	
	20.0	36.1	0.2	7.40	6.29	
	20.0	3.1	0.2	8.46	(6.31)	
					Mean 6.29	
$T = 110.0$	21.0	144.2	0.2	7.10	6.29	
$0.0494n$ HCO'_3	21.0	48.3	0.2	7.57	6.29	
$0.0718n$ Cl'	21.0	34.7	0.2	7.72	6.29	$\sqrt[3]{c} = 0.495$
$0.1212n$ $Na^.$	20.5	7.7	0.2	8.36	6.29	
	20.5	3.8	0.2	8.77	(6.43)	
					Mean 6.29	
$T = 221.9$	17.0	476.4	1.5	6.76	(6.18)	
$0.0997n$ HCO'_3	17.5	122.5	0.8	7.41	6.23	
$0.0718n$ Cl'	17.0	58.2	0.3	7.77	6.28	$\sqrt[3]{c} = 0.556$
$0.1715n$ $Na^.$	17.0	23.3	0.3	8.15	6.27	
	17.5	22.2	0.3	8.17	6.27	
					Mean 6.26	

Table XXXV. Determinations of $p_{K'_1}$ for 18° in sodium bicarbonate solutions with the addition of 0·85 % sodium chloride (0·145n). Carried out with the small saturator electrode. Series A. Mark △.

Concentration	Temp.	mm. Hg CO₂	mm. Hg O₂	p_H	$p_{K'_1}$ 18°	
$T=8·88$	19·0	238·4	0·9	5·78	6·30	
0·00399n HCO'₃	19·0	111·6	0·0	6·10	6·29	
0·145n Cl'	19·0	30·1	0·3	6·64	6·26	$\sqrt[3]{c}=0·530$
0·1490n Na·	19·0	12·2	0·3	7·03	6·26	
	19·0	63·7	0·3	6·33	6·27	
	19·0	12·6	0·3	7·02	6·27	
					Mean 6·28	
$T=17·52$	20·0	587·6	0·7	5·60	(6·20)	
0·00786n HCO'₃	19·5	149·5	0·3	6·24	6·26	
0·145n Cl'	19·0	43·9	0·5	6·76	6·24	$\sqrt[3]{c}=0·535$
0·1529n Na·	19·0	14·3	0·4	7·23	6·22	
	19·0	6·3	0·4	7·61	6·24	
	19·0	3·6	0·4	7·87	(6·26)	
					Mean 6·24	
$T=55·5$	21·0	152·0	0·4	6·72	6·22	
0·0249n HCO'₃	20·5	47·6	0·4	7·24	6·24	
0·145n Cl'	21·0	14·6	1·6	7·77	6·26	$\sqrt[3]{c}=0·554$
0·1699n Na·	20·5	2·7	0·4	8·55	(6·25)	
	19·5	598·0	1·1	6·13	(6·24)	
	19·5	411·7	0·5	6·27	6·22	
					Mean 6·24	
$T=110·0$	19·5	60·9	0·2	7·44	6·27	
0·0494n HCO'₃	19·5	26·5	0·2	7·81	6·28	
0·145n Cl'	19·5	7·7	0·2	8·34	6·29	
0·1844n Na·	20·5	565·0	1·4	6·42	(6·25)	$\sqrt[3]{c}=0·569$
	20·5	142·0	0·6	7·06	6·23	
	20·5	27·7	0·6	7·79	6·28	
					Mean 6·27	
$T=221·9$	17·0	460·5	0·6	6·73	(6·14)	
0·0997n HCO'₃	17·0	161·2	0·6	7·24	6·19	
0·0997n Cl'	17·0	49·1	0·2	7·79	6·23	
0·2447n Na·	17·0	15·5	0·2	8·28	6·23	$\sqrt[3]{c}=0·625$
	17·0	67·9	0·2	8·64	6·25	
	19·0	45·0	0·6	7·83	6·22	
	19·0	90·9	0·7	7·56	6·26	
	19·0	32·8	0·6	8·00	6·25	
					Mean 6·23	

Table XXXVI. Determinations of $p_{K'_1}$ for 18° in sodium bicarbonate solutions with the addition of 1·7 % NaCl (0·291n). Carried out with the small saturator electrode. Series A. Mark △.

Concentration	Temp.	mm. Hg CO₂	mm. Hg O₂	p_H	$p_{K'_1}$ 18°	
$T=8·88$	18·5	243·4	0·4	5·69	6·22	
0·00399n HCO'₃	19·0	115·5	0·4	6·00	6·20	
0·291n Cl'	19·0	36·1	0·4	6·52	6·22	$\sqrt[3]{c}=0·666$
0·2950n Na·	19·5	11·7	0·4	7·00	6·21	
	19·5	16·1	0·4	6·85	6·19	
					Mean 6·21	
$T=17·52$	18·5	529·3	1·1	5·57	(6·13)	
0·00786n HCO'₃	18·5	149·5	0·8	6·17	6·19	
0·291n Cl'	18·5	46·2	0·2	6·67	6·18	$\sqrt[3]{c}=0·669$
0·2989n Na·	19·0	15·3	0·2	7·17	6·19	
					Mean 6·19	

Table XXXVI (continued)

Concentration	Temp.	mm. Hg CO$_2$	mm. Hg O$_2$	p_H	$p_{K'_1}$ 18°	
$T = 55.5$	19.5	602.7	1.1	6.08	(6.20)	
0.0249n HCO$'_3$	19.5	416.2	0.5	6.27	6.23	
0.291n Cl$'$	19.5	131.1	0.6	6.73	6.19	$\sqrt[3]{c} = 0.681$
0.3159n Na$^{\cdot}$	19.0	62.0	0.5	7.09	6.23	
	19.0	17.2	0.5	7.61	6.18	
	19.0	6.0	0.5	8.08	6.19	
	19.0	2.8	0.5	8.48	(6.19)	
					Mean 6.20	
$T = 110.0$	20.0	518.9	0.4	6.50	(6.25)	
0.0499n HCO$'_3$	20.0	139.1	0.4	7.01	6.19	
0.291n Cl$'$	20.5	48.4	0.4	7.50	6.21	$\sqrt[3]{c} = 0.699$
0.3409n Na$^{\cdot}$	20.0	16.2	0.4	7.94	6.19	
					Mean 6.19	
$T = 221.9$	17.0	442.1	0.6	6.76	6.15	
0.0997n HCO$'_3$	17.0	40.0	0.4	7.84	6.19	
0.291n Cl$'$	17.0	20.7	0.4	8.09	6.15	
0.3907n Na$^{\cdot}$	17.0	40.1	0.4	7.83	6.18	
	17.0	14.5	0.4	8.23	6.14	$\sqrt[3]{c} = 0.731$
	18.0	95.2	0.6	7.50	6.18	
	18.0	106.4	0.6	7.39	6.15	
	18.0	365.2	0.7	6.86	6.17	
	18.0	26.1	0.4	8.04	6.19	
					Mean 6.17	

Table XXXVII. Determinations of $p_{K'_1}$ for 18° in sodium bicarbonate solutions with the addition of varying quantities of KCl. Carried out with the small saturator electrode. Series A.

Concentration	Temp.	mm. Hg CO$_2$	mm. Hg O$_2$	p_H	$p_{K'_1}$ 18°
$T = 44.3$	20.0	80.4	0.3	6.98	6.32
0.0199n HCO$'_3$	20.0	132.0	0.3	6.79	6.34
0.54 % KCl (0.0718n)	20.0	15.4	0.6	7.71	6.33
					Mean 6.33
$T = 110.0$	18.0	58.6	0.5	7.48	6.28
0.0494n HCO$'_3$	18.0	98.7	1.7	7.26	6.29
0.54 % KCl (0.0718n)	18.0	21.2	0.5	7.92	6.29
					Mean 6.29
$T = 44.3$	19.0	130.3	0.6	6.72	6.27
0.0199n HCO$'_3$	19.0	37.1	0.8	7.27	6.28
1.08 % KCl (0.145n)	19.5	13.8	0.6	7.69	6.26
	19.5	56.2	0.6	7.09	6.27
					Mean 6.27
$T = 44.3$	19.0	64.3	0.5	6.98	6.22
0.0199n HCO$'_3$	20.0	95.3	0.7	6.84	6.24
2.16 % KCl (0.291n)	20.0	30.1	0.7	7.34	6.25
	19.5	39.3	0.7	7.20	6.23
					Mean 6.24
$T = 55.5$	17.0	132.2	0.5	6.76	6.20
0.0249n HCO$'_3$	17.0	55.0	0.4	7.14	6.23
2.16 % KCl (0.291n)	17.0	18.1	0.6	7.61	6.22
					Mean 6.22
$T = 111.0$	18.0	58.6	0.6	7.40	6.21
0.0494n HCO$'_3$	18.0	98.7	1.7	7.17	6.21
2.16 % KCl (0.291n)	18.0	21.2	0.5	7.85	6.22
					Mean 6.21

Table XXXVIII. Determinations of $p_{K'_1}$ for 18° in sodium bicarbonate solutions with the addition of 0·145n NaCl or KCl. Carried out with the small saturator electrode. Series A.

| | | mm. Hg | | NaCl | | KCl | |
| | | CO_2 | O_2 | $p_{H^.}$ | $p_{K'_1}$ | $p_{H^.}$ | $p_{K'_1}$ 18° |
Concentration	Temp.						
$T=56·25$	17·0	162·2	0·2	6·66	6·21	6·69	6·24
0·0253n HCO$'_3$	17·5	59·1	0·7	7·11	6·23	7·16	6·28
					Mean 6·22		Mean 6·26
$T=111·0$	17·0	125·3	0·4	7·07	6·21	7·10	6·25
0·0499n HCO$'_3$	17·5	34·2	1·7	7·68	6·23	7·70	6·28
	17·5	45·3	0·4	7·54	6·24	7·57	6·28
					Mean 6·23		Mean 6·27

Table XXXIX. Determinations of $p_{K'_1}$ for 18° in sodium bicarbonate solutions with varying additions of salt. Series A.

| T | No salt added | 0·0718n | | 0·0145n | | 0·291n | |
		NaCl	KCl	NaCl	KCl	NaCl	KCl
8·88	(6·37)	6·30	—	6·28	—	6·21	—
17·5	(6·38)	6·29	—	6·24	—	6·19	—
44·3	—	—	6·33	—	6·27	—	6·24
55·5	6·37	6·29	—	6·24	—	6·20	6·22
110·0	6·36	6·29	6·29	6·27	6·27	6·19	—
111·0	—	—	—	—	6·23	—	6·21
221·9	6·31	6·26	—	6·23	—	6·17	—

Table XL. Determinations of $p_{K'_1}$ for 18° in sodium bicarbonate solutions with the addition of salt. Carried out with the small saturator electrode. Series B. Mark +.

Concentration	Temp.	mm. Hg CO_2	$p_{H^.}$	$p_{K'_1}$ 18°	
$T=45·2$	21·0	50·1	7·18	6·30	
0·0203n HCO$'_3$	21·0	8·7	7·90	6·26	$\sqrt{}/c=0·452$
0·42 % NaCl (0·0718n)	21·0	27·7	7·46	6·32	
0·0718n Cl$'$					
0·0921n Na$^.$				Mean 6·29	
$T=56·5$	15·0	579·4	6·18	6·32	
0·0254n HCO$'_3$	15·5	139·0	6·79	6·31	
0·42 % NaCl (0·0718n)	15·5	30·1	7·50	6·35	$\sqrt{}/c=0·458$
0·0718n Cl$'$	19·5	61·8	7·20	6·34	
0·0962n Na$^.$	19·5	52·0	7·27	6·32	
	19·5	41·6	7·39	6·35	
	20·0	41·6	7·38	6·33	
				Mean 6·33	
$T=45·2$	21·0	51·2	7·16	6·30	
0·0203n HCO$'_3$	21·0	36·0	7·30	6·28	$\sqrt{}/c=0·452$
0·42 % NaCl (0·0718n)					
0·0718n Cl$'$				Mean 6·29	
0·0921n Na$^.$					
$T=55·7$	20·0	36·5	7·42	6·30	
0·0250n HCO$'_3$	20·0	23·2	7·55	6·24	
0·85 % NaCl (0·145n)	20·0	22·2	7·60	6·27	$\sqrt{}/c=0·554$
0·145n Cl$'$	21·0	38·4	7·36	6·26	
0·170n Na$^.$	19·0	142·6	7·76	6·24	
	19·0	53·2	7·22	6·27	
				Mean 6·26	

Table XL (continued)

Concentration	Temp.	mm. Hg CO_2	$p_{H^.}$	$p_{K'_1}$ 18°	
$T = 56.5$	21.5	579.4	6.13	(6.22)	
$0.0254n$ HCO'_3	21.0	111.3	6.92	. 6.29	
0.85 % NaCl (0.145n)	21.0	48.5	7.26	6.27	$\sqrt[3]{c} = 0.558$
$0.145n$ Cl'	21.0	31.8	7.46	6.29	
$0.174n$ Na'	21.0	49.0	7.26	6.26	
	21.0	19.7	7.68	6.30	
				Mean 6.28	
$T = 56.5$	19.5	199.8	6.55	6.19	
$0.0254n$ HCO'_3	19.0	81.3	6.97	6.21	
1.7 % NaCl (0.291n)	19.0	36.8	7.29	6.18	$\sqrt[3]{c} = 0.681$
$0.291n$ Cl'	19.0	20.0	7.58	6.21	
$0.3164n$ Na'	19.0	8.4	7.96	6.22	
	19.0	17.1	7.65	6.21	
				Mean 6.20	
$T = 56.5$	19.5	61.8	7.18	6.30	
$0.025n$ HCO'_3	19.5	52.5	7.25	6.31	$\sqrt[3]{c} = 0.554$
1.08 % KCl (0.145n)	19.5	41.6	7.34	6.30	
$0.145n$ Cl'	20.0	41.6	7.36	6.30	
$0.145n$ K'					
$0.025n$ Na'				Mean 6.30	

Table XLI. Determinations of $p_{K'_1}$ in salt solutions for 18°. Carried out by Höber technique in E VI and E III. Series B. Mark ○.

Concentration	Temp.	mm. Hg CO_2	$p_{H^.}$	$p_{K'_1}$ 18° E VI	E III	
$T = 224.4$	19.0	96.2	7.533	6.247	—	$n/10$ NaHCO₃
$0.1008n$ HCO'_3	19.0	65.9	7.737	6.287	—	
$0.1008n$ Na'	19.0	40.7	7.969	6.299	—	
	19.0	31.6	8.051	6.281	—	$\sqrt[3]{c} = 0.464$
	19.0	31.6	8.061	6.292	—	
	19.0	30.7	8.070	6.288	—	
	19.0	32.7	8.026	6.261	—	
				Mean 6.279		
$T = 112.2$	18.0	86.9	7.364	6.340	—	$n/20$ NaHCO₃
$0.05041n$ HCO'_3	18.0	86.9	7.364	—	6.340	
$0.05041n$ Na'	18.0	61.5	7.506	6.321	—	
	18.0	61.5	7.497	—	6.323	$\sqrt[3]{c} = 0.369$
	18.0	50.7	7.604	6.346	—	
	18.0	50.7	7.597	—	6.349	
				Mean 6.335	6.337	
			,,	6.335		
$T = 56.49$	18.5	52.7	7.324	6.384	—	$n/40$ NaHCO₃
$0.02538n$ HCO'_3	20.0	24.8	7.680	6.400	—	
$0.02538n$ Na'	20.0	14.7	7.890	6.383	—	$\sqrt[3]{c} = 0.294$
	20.0	32.0	7.571	6.402	—	
	19.0	47.8	7.382	6.391	—	
				Mean 6.393		
$T = 22.96$	17.0	104.1	6.644	6.392	—	$n/100$ NaHCO₃
$0.01031n$ HCO'_3	17.0	104.1	6.636	—	6.385	
$0.01031n$ Na'	16.5	43.4	7.038	6.416	—	
	16.5	43.4	7.035	—	6.413	$\sqrt[3]{c} = 0.217$
	19.5	82.2	6.742	6.382	—	
	19.5	35.9	7.133	6.413	—	
	19.5	35.9	7.133	—	6.413	
	19.5	37.3	7.090	6.380	—	
				Mean 6.398	6.403	
			,,	6.400		

Table XLI (*continued*)

Concentration	Temp.	mm. Hg CO_2	$p_{H'}$	pK'_1 18°		
				E VI	E III	
$T = 111.4$	21·0	28·2	7·803	6·291	—	$n/20$ KHCO$_3$
0·05004n HCO'$_3$	20·5	33·3	7·751	6·308	—	
0·0497n Cl'	20·5	33·3	7·742	—	6·299	$n/20$ KCl
0·1000n K·	20·5	49·8	7·600	6·332	—	
	20·5	49·8	7·593	—	6·323	$\sqrt[3]{c} = 0.464$
	22·0	26·2	7·876	6·319	—	
			Mean	6·313	6·312	Mark ⊗
			,,	6·312		
$T = 224.4$	20·0	83·3	7·621	6·268	—	$n/10$ NaHCO$_3$
0·1008n HCO'$_3$	20·0	83·3	7·618	—	6·265	
0·1012n Cl'	20·0	41·1	7·911	6·250	—	$n/10$ KCl
0·1008n Na·	20·0	41·1	7·911	—	6·260	
0·1012n K·						$\sqrt[3]{c} = 0.585$
0·09965n Cl'	22·0	91·7	7·586	6·269	—	
0·09965n K·	22·0	38·3	7·938	6·242	—	
	22·0	38·3	7·931	—	6·235	
$T = 222.7$	21·5	85·1	7·611	6·271	—	$n/10$ KHCO$_3$
0·1000n HCO'$_3$	21·5	85·1	7·604	—	6·264	
0·1012n Cl'	21·5	44·5	7·896	6·275	—	$n/10$ NaCl
0·1000n K·			Mean	6·262	6·256	
0·1012n Na·						
			,,	6·260		
$T = 2.244$	19·5	35·2	6·029	6·310	—	$n/1000$ NaHCO$_3$
0·001008n HCO'$_3$	19·5	35·2	6·026	—	6·310	
0·0994n Cl'	20·0	47·0	5·904	6·303	—	$n/10$ NaCl
0·1004n Na·	20·0	38·6	5·991	6·298	—	
	20·0	38·6	5·991	—	6·298	
	19·0	37·4	6·034	6·334	—	
	19·0	37·4	6·031	—	6·333	
	19·5	39·5	5·987	6·318	—	
	19·5	39·5	5·984	—	6·315	
			Mean	6·313	6·314	
				6·313		
$T = 221.45$	19·5	44·1	7·929	6·322	—	$n/10$ KHCO$_3$
0·09948n HCO'$_3$	19·5	44·1	7·918	—	6·315	
0·09948n K·	19·5	29·2	8·101	6·317	—	
	19·5	29·2	8·094	—	6·310	$\sqrt[3]{c} = 0.464$
	19·5	60·9	7·799	6·324	—	
	19·5	60·9	7·789	—	6·314	
			Mean	6·321	6·313	Mark ⊗
				6·317		
$T = 22.14$	19·5	38·9	7·092	6·422	—	
0·009948n HCO'$_3$	19·5	38·9	7·095	—	6·425	$n/100$ KHCO$_3$
0·009948n K·	19·5	48·4	6·979	6·404	—	
	19·5	48·4	6·979	—	6·404	$\sqrt[3]{c} = 0.215$
	19·5	50·2	6·969	6·410	—	
			Mean	6·412	6·415	Mark ⊗
				6·413		
$T = 224.4$	20·0	57·8	7·737	6·226	—	$n/10$ NaHCO$_3$
0·1008n HCO'$_3$	20·0	57·8	7·737	—	6·226	
0·1002n Cl'	19·5	41·5	7·906	6·267	—	$n/10$ NaCl
0·2010n Na·	19·5	41·5	7·904	—	6·265	
	22·0	81·3	7·623	6·254	—	
	22·0	81·3	7·621	—	6·252	$\sqrt[3]{c} = 0.585$
	22·0	29·5	8·048	6·228	—	
	22·0	29·5	8·058	—	6·238	Mark +
	22·5	78·2	7·642	6·249	—	
	22·5	78·2	7·639	—	6·256	
			Mean	6·245	6·247	
				6·246		

Table XLI (continued)

Concentration	Temp.	mm. Hg. CO_2	$p_H\cdot$	pK'_1 18° E VI	E III	
$T=112.2$	21·0	75·6	7·411	6·314	—	$n/20$ NaHCO$_3$
0·05041n HCO$'_3$	21·0	75·6	7·408	—	6·311	
0·0501n Cl'	22·0	24·3	7·889	6·295	—	$n/20$ NaCl
0·1005n Na·	22·0	24·3	7·871	—	6·277	
	20·0	31·6	7·773	6·300	—	$\sqrt[3]{c}=0.464$
	20·0	31·6	7·761	—	6·288	
			Mean	6·303	6·292	Mark +
			,,		6·297	
$T=222.7$	21·5	80·9	7·647	6·285	—	$n/10$ KHCO$_3$
0·1000n HCO$'_3$	21·5	80·9	7·637	—	6·275	$n/10$ KCl
0·0995n Cl'	21·5	29·2	8·070	6·276	—	
0·1995n K·	21·5	29·2	8·068	—	6·274	$\sqrt[3]{c}=0.585$
	21·5	30·4	8·063	6·286	—	
	21·5	30·4	8·063	—	6·296	
			Mean	6·282	6·282	Mark ⊗
			,,		6·282	

Table XLII. Determinations of pK'_1 at 38°; salt solutions. Carried out by Höber technique in E VI. Series B. Mark o–.

Concentration	mm. Hg CO_2	O_2	$p_H\cdot$	pK'_1 38°	
$T=56.3$	465·7	0·2	6·25	(6·02)	$\sqrt[3]{c}=0.681$
0·0253n HCO$'_3$	108·7	0·2	7·20	6·15	
1·7 % NaCl (0·291n)	44·3	0·2	7·35	6·11	
0·291n Cl'	17·9	0·3	7·22	6·09	
0·3163n Na·	40·1	0·3	7·38	6·09	
			Mean	6·11	
$T=56.5$	587·6	0·0	6·21	(6·09)	$\sqrt[3]{c}=0.681$
0·0254n HCO$'_3$	338·4	0·1	6·45	6·10	
1·7 % NaCl (0·291n)	127·7	0·0	6·89	6·11	
0·291n Cl'	45·8	0·0	7·44	6·11	
0·3164n Na·	8·7	0·6	7·04	6·08	
			Mean	6·10	

Table XLIII. Determinations of pK'_1 at 38° in sodium bicarbonate solutions without the addition of salt. Carried out by Höber technique in E VI. Series B. Mark o–.

Concentration	mm. Hg CO_2	O_2	$p_H\cdot$	pK'_1 38°	
$T=56.5$	54·1	—	7·42	6·27	$\sqrt[3]{c}=0.287$
0·0254n HCO$'_3$	41·4	—	7·56	6·29	
0·0254n Na·	43·4	—	7·55	6·30	
	95·7	—	7·19	6·28	
	32·1	—	7·66	6·27	
			Mean	6·28	
$T=224.4$	74·0	—	7·82	6·20	$\sqrt[3]{c}=0.466$
0·101n HCO$'_3$	59·4	—	7·90	6·18	
0·101n Na·	90·7	—	7·75	6·22	
	59·5	—	7·87	6·15	
	112·0	—	7·58	6·14	
	50·2	—	7·92	6·14	
			Mean	6·17	

Table XLIV. Determinations of $p_{K'_1}$ at 38° in sodium bicarbonate solutions with the addition of salt. Carried out with the small saturator electrode. Series B. Mark ×.

Concentration	mm. Hg		p_H·	$p_{K'_1}$ 38°·1	
	CO_2	O_2			
$T = 45 \cdot 2$	48·8	—	7·31	6·21	$\sqrt{c} = 0 \cdot 452$
0·0203n HCO$'_3$	8·7	—	8·01	6·17	
0·42 % NaCl (0·0718n)	26·5	—	7·56	6·18	
0·0718n Cl′					
0·0921n Na·				Mean 6·18	
$T = 56 \cdot 5$	34·6	0·1	7·55	6·19	$\sqrt{c} = 0 \cdot 458$
0·0254n HCO$'_3$	114·0	0·6	7·02	6·19	
0·42 % NaCl (0·0718n)	52·3	0·3	7·34	6·17	
0·0718n Cl′	39·3	0·0	7·51	6·21	
0·0962n Na·				Mean 6·19	
$T = 45 \cdot 2$	543·5	—	6·15	(6·09)	$\sqrt{c} = 0 \cdot 549$
0·0203n HCO$'_3$	62·7	—	7·10	6·11	
0·85 % NaCl (0·145n)	49·2	0·8	7·26	6·17	
0·145n Cl′	36·2	0·5	7·41	6·17	
0·1653n Na·	34·4	0·1	7·43	6·18	
	19·1	0·5	7·66	6·15	
				Mean 6·16	
$T = 56 \cdot 3$	529·7	0·2	6·28	6·12	$\sqrt{c} = 0 \cdot 554$
0·0253n HCO$'_3$	107·6	0·2	7·00	6·15	
0·85 % NaCl (0·145n)	37·4	0·2	7·44	6·12	
0·145n Cl′	35·3	0·2	7·47	6·13	
0·1703n Na·				Mean 6·13	
$T = 56 \cdot 5$	567·3	0·0	6·27	6·14	$\sqrt{c} = 0 \cdot 554$
0·0254n HCO$'_3$	142·0	0·1	6·89	6·16	
0·85 % NaCl (0·145n)	25·6	0·2	7·61	6·13	
0·145n Cl′	34·3	0·0	7·53	6·18	
0·1704n Na·	34·0	0·6	7·54	6·18	
	56·6	0·6	7·30	6·17	
	39·9	0·6	7·45	6·16	
				Mean 6·16	

Table XLV. Conversion of Bjerrum and Gjaldbaek's Table VI.

P_{CO_2}	$c \, 10^3$	$\log K'$	$\log K'$ converted
0·000504	1·492	−5·619	−5·666
0·000808	1·700	−5·660	−5·698
0·00333	2·744	−5·677	−5·754
0·01387	4·462	−5·688	−5·741
0·0282	5·930	−5·649	−5·707
0·0501	7·200	−5·661	−5·723
0·1422	10·66	−5·635	−5·705
0·2538	13·27	−5·621	−5·697
0·4167	15·75	−5·630	−5·710
0·5533	17·71	−5·613	−5·697
0·7297	19·44	−5·620	−5·705
0·9841	21·72	−5·616	−5·704
		Mean −5·641	Mean −5·709

Table XLVI. Deviation coefficients of the bicarbonate ion
in sodium salt solutions.

C	$\sqrt[3]{}/c$	f_a	f_0	$f\mu$ (NaHCO$_3$)
0·001	0·100	0·900	0·974	0·949
0·005	0·171	0·834	0·955	0·913
0·01	0·215	0·796	0·944	0·890
0·02	0·271	0·751	0·929	0·862
0·03	0·311	0·719	0·919	0·841
0·04	0·342	0·696	0·911	0·827
0·05	0·368	0·677	0·904	0·812
0·06	0·391	0·661	0·898	0·801
0·07	0·412	0·646	0·893	0·790
0·08	0·431	0·634	0·888	0·780
0·09	0·448	0·622	0·884	0·778
0·10	0·464	0·612	0·879	0·764
0·12	0·493	0·593	0·872	0·749
0·14	0·519	0·577	0·865	0·735
0·16	0·543	0·563	0·859	0·723
0·18	0·565	0·550	0·853	0·712
0·20	0·585	0·538	0·848	0·702
0·22	0·604	0·528	0·843	0·692
0·24	0·621	0·518	0·839	0·683
0·25	0·630	0·513	0·836	0·679

Table XLVII. Determinations of $p_{K'_1}$ in salt solutions containing phosphates,
for 18°. Carried out with the small saturator electrode. Series A.

Concentration	Temp.	mm. Hg CO$_2$	Vols. % CO$_2$ combined	p_{H}	$p_{K'_1}$ 18°	
0·85 % NaCl	19·0	49·3	35·8	7·03	6·26	$Y=6$
70 cc. NaHCO$_3$ ($T=43·0$)	19·0	156·9	45·9	6·63	6·24	Phosphate
6 cc. KH$_2$PO$_4$ (0·066m)	19·0	48·2	34·8	6·45	6·27	
24 cc. Na$_2$HPO$_4$ (0·066m)	19·0	16·3	28·9	6·41	6·24	
0·02m phosphate	19·0	38·3	33·6	6·44	6·25	
0·145n Na·					Mean 6·25	
0·004n K·						
0·42 % NaCl	18·0	51·5	38·5	7·06	6·28	$Y=5$
70 cc. NaHCO$_3$ ($T=43·0$)	18·5	28·9	33·4	7·25	6·28	Phosphate
6 cc. KH$_2$PO$_4$ (0·066m)	18·5	53·6	37·5	7·05	6·29	
24 cc. Na$_2$HPO$_4$ (0·066m	19·0	156·9	48·2	6·66	6·25	
0·02m phosphate					Mean 6·28	
0·0718n Cl′						
0·101n Na·						
0·004n K·						

Table XLVIII. Determinations of $p_{K'_1}$ for 18° in cane sugar solutions
(0·501m) with sodium bicarbonate (circ. m/100). Carried out with E VI
and E III.

Concentration	Temp.	mm. Hg CO$_2$	p_{H}	$p_{K'_1}$	Vols. % CO$_2$ combined	
0·501m saccharose	18·0	519·9	6·03	6·36	27·6	$Y=16·5$
$T=27·7$	18·0	117·3	6·66	6·38	28·1	E VI
0·0125n HCO′$_3$	18·5	28·6	7·29	6·39	27·6	Series A
0·0125n Na·	19·0	16·9	7·47	6·34	27·6	
	19·5	523·4	6·03	6·40	27·7	
	18·5	34·7	—	(6·31)	27·9	
			Mean 6·38	Mean 27·7		
0·501m saccharose	20·5	43·9	7·098	6·471	—	$Y=16·5$
$T=22·44$	20·5	43·9	7·092	6·465	—	E VI and E III
0·01008n HCO′$_3$	20·5	41·4	7·102	6·450	—	Series B
0·01008n Na·	20·5	41·4	7·102	6·450	—	
	20·5	43·6	7·095	6·465	—	
	20·5	43·6	7·095	6·465	—	
			Mean 6·461			

CHAPTER IX

PRELIMINARY CONSIDERATIONS CONCERNING THE MODE OF COMBINATION OF CARBONIC ACID IN THE BLOOD

In the first seven chapters of this work the relations existing between the apparent hydrogen ion activity, the CO_2 tension and the combined carbonic acid of the blood have been developed. These relations depended on the assumption that all the combined carbonic acid in the various phases of the blood is present as bicarbonate[1], but on further scrutiny it will be obvious this proviso is by no means proved by the fact that the apparent hydrogen ion activity can be calculated from the modified Henderson-Hasselbalch equations. On the contrary it will be perceived that in the empirically determined "constants," corrections for combined carbonic acid not in the form of bicarbonate can very easily be hidden.

When I started these investigations I thought that the corroboration of Henderson's and Hasselbalch's "proof" that all the combined carbonic acid in the blood was present as bicarbonate, would not involve great theoretical difficulties so that by taking the heterogeneity of the solution into consideration it would be feasible to determine whether the evidence was really sound. But the more I have pondered over the theoretical conditions, the greater the difficulties have seemed to be. I have nevertheless, as far as I have been able, fulfilled my original plan, inasmuch as, it seems to me, questions of considerable interest have arisen but I have furthermore endeavoured to establish the proof in another way in order to avoid the most important difficulties. The evidence will be found in chapter XI but for the present it will be shown how far we can go with the original idea. The value that gives the greatest trouble is the apparent activity coefficient of the bicarbonate ion in the liquid phase of the blood corpuscles, and uncertainty in evaluating this will produce a significant error in the calculations in this chapter and chapter XII. I do not think however the conclusions arrived at can be disputed from a qualitative point of view but I quite realise that quantitatively, especially in this chapter, they may rather diverge from the results that may eventually be obtained.

In chapter III an equation was evolved for the calculation of the apparent hydrogen ion activity in the water phase of a two-phase system. With this equation (115) we can estimate the amount of bicarbonate which will be found in the serum proteins and haemoglobin if the combined carbonic acid is present as bicarbonate:

$$(115) \quad a_{H^{\cdot}} = \lambda_{(1)} \frac{P_{CO_2} a}{7 \cdot 60 \beta},$$

---

[1] No account is taken here or henceforth of the small quantities of monocarbonate found in the blood (cf. chapter II).

the condition for which is

$$(117) \quad \lambda_{(1)} = K'_1 \frac{100 - q(1-d)}{100},$$

from which we obtain $\qquad d = \frac{100\,(\lambda_{(1)} - K')}{K'_1\,q} + 1,$ ....(151)

in which the significance of the constants is discussed in chapters II and III. Let us first consider serum. We found in chapter V

$$\lambda_{(1)} = \lambda_{(s)} = 10^{-6 \cdot 294} = 5 \cdot 20 \times 10^{-7}.$$

K'_1 can be determined from the equation

$$(143) \quad -\log F_a\,(HCO'_3) = 0 \cdot 46 \sqrt[3]{c}.$$

For the determination of the apparent activity coefficient we only lack the value of c.

Numerous and thorough investigations by Hamburger [1902], Hedin [1915], Ege [1920] and many others have shown that serum is almost isotonic (isosmotic) with a 0·9 % sodium chloride solution. For further information respecting this I would refer the reader to Rich. Ege's thesis. It would be erroneous however immediately to conclude that c was the same as in 0·9 % NaCl solution, that is 0·154n, because NaCl mols are not the only osmotically active ones. Now the mols which are not electrolytes are only present in such a small concentration in serum that for this purpose we can neglect them entirely. But at the reactions here in question there are also some protein anions present in serum which neutralise corresponding amounts of cations. The quantity of these never exceeds 0·02n and presumably is usually about 0·01n, so we are justified in estimating c at between 0·16 and 0·18n.

The volume of the disperse phase in serum is certainly very nearly 8 % and putting this in equation (151) we obtain

$$c = 0 \cdot 16 \quad K'_1 = 10^{-6 \cdot 260} = 5 \cdot 50 \times 10^{-7} \quad d = 0 \cdot 32,$$
$$c = 0 \cdot 18 \quad K'_1 = 10^{-6 \cdot 251} = 5 \cdot 61 \times 10^{-7} \quad d = 0 \cdot 07.$$

The calculation is rather uncertain because a slight error in K'_1 or in $\lambda_{(s)}$ has a great effect, but it seems as though some bicarbonate is dissolved in the serum proteins.

Table XLIX. Calculation of the value of d in the blood corpuscles with equation (151).

c	$K'_1 \times 10^{-7}$	$q = 40$	$q = 35$
0·20	5·30	0·51	0·44
0·225	5·41	0·47	0·40
0·250	5·52	0·42	0·35

The concentration of cations in the water phase of the blood corpuscles is not certainly known. It has been estimated here as between 0·2 and 0·25. The reason for this will be discussed later in chapter XII.

$\lambda_{(1)} = \lambda_{(c)}$ in horse blood cell fluid, according to the measurements reported in chapter VII, is about 6·37. The volume of the disperse phase in blood cells according to Ege's determinations is between 35 % and 40 %. If d is calculated with these values we obtain the results given in Table XLIX, using equation (58) with the constant 0·40, as a large part of the cations of blood cells consists of potassium.

From the table it will be seen that the value of d, under the given assumptions, lies between 0·35 and 0·51.

As emphasised above it is quite possible that the calculation may be rather uncertain quantitatively but there is hardly any doubt that d is really a positive quantity. If $d = 0$ and q and $A_{(c)}$ at the same time have the values employed in the calculation above, we can estimate K'_1 and c since in that case the following holds:

$$K'_1 = \frac{100}{100 - q} \lambda_{(c)},$$

and therefore for $q = 40$ we get $K'_1 = 7·12 \times 10^{-7}$ and $c = 0·74n$, and for $q = 35$, $K'_1 = 6·57 \times 10^{-7}$ and $c = 0·55n$. These values for c are undoubtedly too high.

It may therefore be taken as settled that a part of the bicarbonate is dissolved in the haemoglobin provided that all the combined carbonic acid of the blood is really present as bicarbonate, and the calculations also show that the bicarbonate is about half as soluble in haemoglobin as in a roughly $n/4$ KCl solution (the dispersion medium).

This result is by no means antagonistic to what we already know about the solubility of salts in proteins. Thus S. P. L. Sörensen [1915–1917] found that ammonium sulphate was freely soluble in egg albumin but, on the other hand, it compels us to inquire whether a part of the combined carbonic acid is not present in some other form than the bicarbonate ion (or undissociated bicarbonate in the haemoglobin phase). It would be interesting to estimate first what amount of the combined carbonic acid of the blood may be expected to be dissolved in the protein phases according to the above calculations.

As an example let us take horse blood at 18° with a $p_{H'(s)}$ value of 7·40 and at a CO_2 tension of 40 mm., and further we will assume the volume of the blood cells is 40 % of the total volume of blood, which according to Fig. 10 corresponds to 19·6 vols. % combined O_2 (Haldane 106). We now calculate with (130) the combined carbonic acid (B) making the necessary correction with the help of Fig. 10, and we obtain 58·4 vols. %. From Fig. 7 we read that the partition ratio of combined CO_2 at this reaction (D) is 0·66.

In 100 cc. blood therefore there are 58·4 cc. combined CO_2 (0° and 760 mm.), of which 40·6 cc. are present in the 60 cc. serum, equivalent to a content of 67·6 cc. in 100 cc. serum; and in 40 cc. blood cells there are consequently 17·8 cc. combined CO_2 corresponding to 44·6 cc. in 100 cc. blood cells. Of the 60 cc. serum the water phase is 55·2 cc. and the protein phase (8 %) 4·8 cc., and if we assume d is 0·35, we obtain 39·4 in the water phase corresponding to 71·3 cc. in 100 cc. and 1·2 cc. in the protein phase which is equivalent to 25·0 cc. in 100 cc. protein. If we carry out a similar calculation for the blood corpuscles and put $q = 40$ and $d = 0·45$, the water phase becomes 24 cc. and the protein phase 16 cc. There are then 13·7 cc. in the water phase, equivalent to 57·2 cc. in 100 cc. and 4·1 cc. in the protein phase, equivalent to 25·7 cc. in 100 cc. In the two disperse phases there are altogether $4·1 + 1·2 = 5·3$ cc. combined CO_2 which is equivalent to about 9 % of the total quantity of combined CO_2. We have found that 71·3 vols. % combined CO_2 are present in the water phase of serum, while there are only 57·2 vols. % in the water phase of the blood cells. Now the apparent activity coefficient of the bicarbonate ion is very nearly identical in the water phase of serum and blood cells because the greater cation concentration in the blood cells is almost balanced by the smaller Milner effect of the potassium ions when it is compared with the sodium ions, and the relation between the amounts of combined CO_2 in the two water phases will thus directly afford a measure of the relation between the activity of the bicarbonate ion in them, so that

$$\frac{a_{HCO_3} \text{ in the water phase of the blood cells}}{a_{HCO_3} \text{ in the water phase of the serum}} = 0·8.$$

This result is by no means at variance with what we should expect, as it will be shown in chapter XII that on the basis of Ostwald's [1890] and Donnan's [1911] partition law such a distribution is to be looked for.

The above results are condensed in Table L.

Table L.

	Vols. % combined CO_2	Cc. combined CO_2
100 cc. blood at p_H 7·40 and 40 mm.	58·4	58·4
In the serum of the above (60 cc.)	67·6	40·6
„ water phase of the serum (55·2 cc.) ...	71·3	39·4
„ protein phase of the serum (4·8 cc.) ...	25·0	1·2
„ blood cells (40 cc.)	44·6	17·8
„ water phase of the blood cells (24 cc.) ...	57·2	13·7
„ protein phase of the blood cells (16 cc.)	25·7	4·1

The amount of CO_2 combined in a form other than bicarbonate would thus appear to be small and, according to the above calculations, at most 9 % of the total. We will now try to get an idea of what form of combination this might theoretically be.

The combined carbonic acid is determined as the difference between the total dissociable combined CO_2 and the dissolved amount calculated by Henry's law. The relative absorption coefficient of blood and serum has been determined indirectly by Bohr [1905], the assumption being that it is the same for all gases. This is of course only approximately true but the divergence is not so large that a grave error can arise if the CO_2 tension is not too high. It is however possible that CO_2 could be adsorbed at the interface between protein and the dispersion medium. That such an adsorption actually takes place in blood is claimed by Findlay for various gases as will be referred to in the next chapter but he undoubtedly overestimates the significance of his experiments. It must also be noted that $\lambda_{(c)}$ could not be independent of the reaction if such an adsorption took place for reasons identical with those that will be advanced at the end of this chapter against the existence of a protein-carbamino-acid.

The possibility may also be raised that some of the bicarbonate may be present as undissociated protein bicarbonate (complex salt). This assumption is however not quite correct because in alkaline reactions only very small or negligible amounts of protein cations are present, since proteins function almost entirely as acids and so there can be no question here of undissociated protein bicarbonate. At reactions in the region of, or more acid than, $10^{-7 \cdot 0}$ some haemoglobin cations are in all probability formed so that there is the possibility of the existence of undissociated haemoglobin bicarbonate and in such a case $p\lambda_{(c)}$ might decrease with increasing hydrogen ion activity, but in chapter VII we saw this was not so. We can therefore conclude that undissociated bicarbonate does not exist in sufficient quantity to compromise our theories and calculations. Further it is possible that the bicarbonate ion might be adsorbed at the interface between protein and salt solution. As a matter of experience however such an adsorption only takes place when the disperse phase has a charge of opposite sign to that of the adsorbed ions, and so the theory of the adsorption of the bicarbonate ion by proteins is untenable for the same reason as applies to undissociated protein bicarbonate. Furthermore the problem of the adsorption of the ions is presumably identical with the problem of the large Milner effect of the colloid ions, according to the views of which Prof. Bjerrum has kindly given me the benefit. There remains the possibility that some of the CO_2 combines with the protein in some other form than protein bicarbonate. Such a form of combination has been stated by Siegfried to be protein carbamino-acid.

It will be clear to the reader that the combined carbonic acid referred to in this work is only the dissociable combined carbonic acid, that is to say the carbonic acid which can be liberated by the addition of acid or by diminishing the CO_2 tension; that which can only be split off from the organic molecules by such vigorous processes that the latter are destroyed is naturally not included, it can moreover be shown that this dissociation of the carbonic acid combination of the blood is completely reversible. If protein carbamino-acid took part in the reversible carbonic acid combination $A_{(c)}$ (or $\lambda_{(m)}$) could not be constant in the fluid of the blood cells because the non-ionised combined H_2CO_3 would then vary with the CO_2 tension. This evidence that all the combined carbonic acid is really ionised in the blood is well supported by the determinations of $p\lambda_{(mi)}$ in chapter VII, but is weakened to some extent by the determinations being so extremely difficult to carry out. But it also involves the assumption that q and d in equation (117) do not vary appreciably with the hydrogen ion activity. From a close examination of (117) however it appears the variations must be fairly large to make themselves felt in the determinations in question.

As it would certainly involve very great experimental difficulties to proceed further with the questions raised in this chapter there is every inducement to examine the literature for experiments which can throw light on these problems.

In conclusion it may be remarked that the experiments and views put forward give no information as to whether the bicarbonate dissolved in the protein phase is present exclusively as ions, or as undissociated salt as well.

RÉSUMÉ.

It has been shown that a small part of the combined carbonic acid is not present in the dispersion medium of the serum and blood cells, and the manner of combination of this amount of combined acid is discussed. It is probable that it is dissolved in the protein phases.

CHAPTER X

A BRIEF HISTORICAL REVIEW OF THE OLDER THEORIES OF
THE COMBINATION OF CO_2 IN THE BLOOD.

The first to extract gas out of blood was John Mayow of Oxford (quoted from P. Bert), who reported his results in a paper in 1674. He thought the gas obtained was what we now call oxygen.

In 1783 Pietro Moscati of Milan stated he had extracted a gas out of blood and he believed he had proved it to be carbonic acid (fixed air), but his evidence was not good as by allowing blood to stand for 24 hours with lime water he obtained a precipitate [1784].

In 1799 Humphrey Davy proved that CO_2 could be driven out of blood by heating it, and in 1814 Vogel showed that CO_2 could be pumped out of blood. Up to 1837 when Magnus finally established that CO_2 could be pumped out of blood, the question was one of considerable interest as it was closely bound up with the current discussion on where oxidation took place in the organism.

Lavoisier had originally discussed whether oxidation occurred in the peripheral circulation or in the lungs. In his later works he claimed to have proved it took place in the lungs (according to W. Edwards and Ch. J. B. Williams). He believed that the carbon in a state of fine division was carried by the blood to the lungs and burned up there in association with oxygen. It became therefore of prime importance at that time to ascertain whether carbonic acid (CO_2) was free in the blood or only in a state of combination. In those days it was evidently not considered sufficient that carbonic acid could be driven out of the blood by heat, or by treatment with another gas such as hydrogen or atmospheric air, because it was supposed these gases replaced carbonic acid in loose combinations although it was partly realised similar processes took place in the lungs.

The first to have pumped CO_2 out of blood seems to have been, as already mentioned, Vogel [1814] and, apparently almost simultaneously, Vaquelin. This last investigator's experiments were never published by himself but were recorded by W. Edwards [1824].

Between 1814 and 1837, Scudamore, Brande and others succeeded in pumping CO_2 out of blood, while Darwin[1], Davy [1803], Stromeyer [1832], Misterlich, Gmelin and Tiedemann [1834] failed in an attempt to do so.

Stevens' [1832, 1835] experiments and theories are especially interesting. He was a Scotsman who lived the greater part of his life in the then Danish West Indies, and he seems to be almost forgotten in scientific literature, which is partly due to the immense range of his works, but he appears to have introduced a treatment for cholera which although based on faulty experimental material, constituted a real step forward at the time (cf. *Observations on the Nature and the Treatment of the Asiatic Cholera*, London, 1853).

In his book of 1832 Stevens put forward the theory, on pp. 22–26, that the carbonic acid of the blood is present as alkaline carbonates and as free CO_2, comparing the change of reaction of serum on exposure to the air with the change which takes place in the alkaline mineral water highly charged with CO_2 from a spring at Saratoga in New York State, when it is allowed to stand in the air.

In 1835 he communicated a series of experiments to the *Philosophical Transactions* which showed that he was unable to pump CO_2 out of venous blood immediately after it was taken but he succeeded after the blood had stood for some time in contact with hydrogen or air. It had been repeatedly shown that blood could absorb considerable amounts of CO_2—a fact which Stevens himself demonstrated experimentally—but he could not get the CO_2 out again without letting the blood stand some time in the air or hydrogen. Stevens thought therefore that these gases liberated CO_2 allowing it to be pumped out. It is perfectly clear to us now that the disagreement between the different observers was due to their gas pumps being of different degrees of efficiency and those who had the worst pumps could not pump CO_2 out of freshly drawn blood. The one who finally proved with certainty that the blood contained CO_2 which could be expelled by blowing hydrogen through it or by the exhaust pump was Magnus. His article in Poggendorff's *Annalen* of 1837 is still worth reading not only on account of its historical interest but also in view of the great insight of the author, and on perusing it, it is striking, as is the case with so many other works of the first half of last century, how little physiological problems have changed their character and how often the same phenomena are rediscovered.

A sentence in an article in 1834 by R. Hermann of Moscow is of interest. On the 5th September, 1833, Drs Stevens, Markus, Wyllie, Jänchen and R. Hermann met in Moscow for an investigation.—"Nachdem man sich einstimmig darüber ausgesprochen hatte, dass das Serum als eine Flüssigkeit

[1] Quoted from Williams [1835].

zu betrachten sei, die neben doppelt-kohlensaurem Natron noch freie Kohlensäure erhalte (on account of the reaction of the serum), ging man zu Untersuchung des Blutes über."

It was not all however who realised that blood contained free carbonic
acid. Misterlich, Gmelin and Tiedemann [1834], and also Davy [1803] and
others, thought that blood only contained combined CO_2 while one gets the
impression that Magnus [1837] regarded all the CO_2 as free. Even in the middle
of the sixties W. Preyer [1867] could proclaim with success that no free CO_2
was to be found in the blood because it could not be present in an "alkaline"
liquid. When Zuntz [1868] pointed out that free CO_2 might be found in blood
when it was in equilibrium with an atmosphere containing CO_2 the question
was finally decided. P. Bert however in a paper in 1878 opposed this view.

Magnus, whose technique was a considerable advance on that of earlier
workers, contributed analyses which even in those early days gave fairly
correctly the quantity of CO_2 in the blood. These experiments were the best
until 1857 when Fernet in France and Lothar Meyer in Germany published
extensive investigations on the subject. Fernet's technique—which Meyer
also used—consisted in driving the gas out of blood by means of a vacuum
produced with the help of water vapour and afterwards analysing the gas
driven out. Fernet and Meyer thought the combined CO_2 in blood could be
divided into two kinds, namely that which could be liberated with a vacuum
and heat alone, and that which required the addition of acid. About this time
Bunsen's first absorption coefficients appeared and it was then possible to
calculate the amount of combined CO_2 in the blood. The values found by
Fernet and Meyer were however too low—in fact in the case of CO_2 they were
no better than those of Magnus—on account of the relatively bad vacuum.
Fernet proposed the theory that the CO_2 of the blood was combined with
phosphates as a double salt but Heidenhain and Meyer [1863] showed a few
years later that Fernet's analyses with respect to the double salt were not
correct, and Sertoli proved in 1868 that the amount of free phosphates in the
blood was so small that it could be entirely neglected as regards its capacity
for combining with CO_2 in view of the relatively large amount of combined
CO_2 which had been found. Later investigations have fully supported Sertoli's
findings and according to the latest experiments which have been collected
amongst others by Poul Iversen [1920], blood and serum contain about
$m/1000$ dissociated phosphate. The combination of CO_2 in phosphate solutions
has nevertheless had no small influence on our knowledge of CO_2 compounds
in blood, as it is a rather simple example of a CO_2 combination in an electrolyte
the dissociation of which varies with the reaction. Heidenhain and Meyer
[1863], Setschenow [1879], Henderson and Black [1908] and K. A. Hasselbalch
[1910] have all used phosphates as a solution comparable with blood, and the
author will also report some similar experiments later on in this work.

Ludwig [1865] and his collaborators Schöffer [1861, 1868] and Setschenow
[1859], in the latter half of the "fifties" and in the "sixties," carried out a

number of experiments upon the CO_2 compounds in blood, and these investigators also considered that all the CO_2 could not be extracted out of the blood
(with Ludwig's mercury air-pump) without the addition of acid. It was not
until Pflüger [1864] in conjunction with Geissler constructed a new pump
that the extraction was so complete that the yield was not increased by adding
acid. Pflüger also showed that all the CO_2 could not be pumped out of serum
(when it was separated from the blood corpuscles) without adding acid. Pflüger
finally showed that in addition to the CO_2 which blood yielded to a vacuum
a further supply could be liberated from sodium carbonate. Pflüger fully
realised that pumping out in the presence of sodium carbonate was accompanied by a transformation of monocarbonate into bicarbonate and it was
owing to this that the CO_2 was driven out (cf. H. Rose [1835] and Marchand
[1846]). He considered the acids which drove out the CO_2 were haemoglobin
and the serum proteins. Hoppe-Seyler thought that haemoglobin only acted
as an acid under the influence of the exhaust pump, being destroyed during
the pumping, in spite of the fact that such an assumption is unnecessary for
explaining the phenomenon, and, so far as I am aware, a proof has never
been furnished; the hypothetical acid could only drive out a very small
quantity of CO_2, because Gaule [1878] and Zuntz [1882] have shown that
after pumping out, blood combines with almost as much acid as before.

Pflüger [1864] thought the difference between blood and serum consisted
in the fact that less acid was formed in serum on pumping out than in the
blood corpuscles.

Sertoli [1868] and Setschenow [1879] both showed that globulin is an acid
which can expel CO_2 from sodium carbonate, and taking everything into
consideration there is now hardly any doubt that the proteins really possess
an acid character. As it is also an accepted fact that they can combine with
acid their ampholytic nature is proved. It is an observation over 100 years
old that the proteins function both as bases and acids. About the beginning
of the present century they were classified as ampholytes, at which time the
peculiar properties of this group of electrolytes became known, and their
physico-chemical characteristics were thoroughly worked out by Bredig, Walker,
Hardy and Pauli, and later particularly by Sörensen and Michaelis and their
numerous collaborators, while Henderson and Spiro and Hasselbalch have
rather elaborately dealt with the consequences of the ampholytic character
in relation to CO_2 combinations in the blood.

In 1882 Zuntz, in Hermann's handbook, collected the investigations
relating to the combination of carbonic acid in the blood in a deservedly
celebrated monograph. He was able, largely on the basis of his own and
Setschenow's experiments, to enunciate the theory for the combination of
carbonic acid, that CO_2 was combined with the alkali of the blood in such a
way that carbonic acid and the blood proteins divided the alkali between
them, and also the haemoglobin at high CO_2 tensions acted as a base and thus
combined with CO_2 itself.

Some years before, Setschenow [1879] had come to a very similar conclusion about the combination of CO_2, the only difference being that he interpreted CO_2 combinations at high CO_2 tensions not as a production of haemoglobin bicarbonate but as a more specific reaction for haemoglobin.

The view which is held to-day regarding the CO_2 compounds in the blood only slightly differs from Zuntz's, the most important advance being the knowledge that the proteins belong to the group of ampholytes. The premises upon which Zuntz built his theory were pre-eminently the following:

(1) The well established fact that proteins combine with both acids and bases.

(2) Pflüger's observation that haemoglobin can drive CO_2 out of sodium carbonate, in conjunction with Sertoli's and Setschenow's demonstration of the same thing for *globulin* and *paraglobulin*.

(3) Setschenow's demonstration that the blood can combine with more CO_2 than corresponds to the difference between the cations and anions the dissociation of which varies with the reaction (expressed in modern terminology).

Of these premises only Pflüger's has later become open to doubt as Buckmaster tried to show in 1917, but the latter author is certainly wrong. His results can only be explained by his having added too much alkali so that the amount of bicarbonate formed is so small in relation to the amount of monocarbonate that there is not a measurable CO_2 tension in the system and thus CO_2 could not be pumped out. Quite recently Adolph [1920] has also shown Buckmaster was in error, but the latter's experiments have raised so much discussion that I think it worth while reporting one of two experiments I have undertaken which were in good agreement and which well support Pflüger's experiments.

Experiment.

Defibrinated ox blood kept in the ice cupboard from 19. viii. till the beginning of the experiment.

The blood corpuscles were separated by centrifuging.

On 20. viii. 6·98 cc. of the blood corpuscle emulsion in equilibrium with 40·5 mm. CO_2 were pumped out into an acid-free evacuation receiver. It contained 39·1 vols. % CO_2 and 38·2 vols. % O_2 (0° and 760 mm.).

On 21. viii. 7·12 cc. blood cell emulsion in equilibrium with 40·9 mm. CO_2 and 146·1 mm. O_2 at 18° contained 38·5 vols. % CO_2 and 32·3 vols. % O_2 on pumping out in the presence of boric acid.

In an acid-free evacuation receiver were put 6·64 cc. NaOH in equilibrium with 40·9 mm. CO_2. This NaOH, according to previous estimations, combined with 219·7 vols. %, and there was physically absorbed 5·01 vols. %, that is altogether 224·7 vols. % (0° and 760 mm.).

Further there was put in the receiver 10·55 cc. of the blood cell emulsion.

0° 760 mm.

In 6·64 cc. $NaHCO_3$ ... ... 14·92 cc. CO_2
Therefore 10·55 cc. of the above blood cell emulsion in equilibrium
with 40·9 mm. CO_2 and 146·1 mm. O_2 contained ... ... 4·06 „ „
18·98 „

0° 760 mm.

In one hour there was pumped out ... ... 18·35 cc. CO_2

After the addition of boric acid for 7 mins. a further ... ... 0·65 ,, ,,

Each pumping out took place at 38°. 19·00 ,,

Pflüger's [1864] phenomenon that all the combined CO_2 of the blood can be pumped out and further CO_2 can be driven out by addition of sodium carbonate shows beyond doubt that substances are present in the blood which act as acids at reactions more alkaline than the neutral point, but does not prove that all the combined CO_2 in the blood is in the form of bicarbonate because every compound of CO_2 with another substance will be dissociated as soon as the CO_2 tension is diminished.

It is justifiable to assume that the electrolytes of the blood the dissociation of which varies with the reaction are mainly proteins, particularly paraglobulin (Setschenow [1879]) and haemoglobin (Zuntz [1868], Mathieux and Urbain, Setschenow [1879], Bohr [1891, 1898, 1904, 1909], Hasselbalch [1916, 2], Campbell and Poulton [1920] and many others). But Setschenow [1879] has shown that *lecithin* also acts as a weak acid, and it is known that phosphates act in the same way although as already mentioned they are only found in quite small concentrations.

There are some theories in the literature which differ from the Zuntz-Setschenow theory. The first is Bohr's [1904]. He believed the combination of CO_2 took place in the following way:

$$CO_2 + A_B HbA_s \rightleftarrows CO_2 (A_B Hb) + A_s$$

(the symbols being Bohr's own). In other words he imagined haemoglobin contained an acidic and a basic group and that at the same moment in which it combined with CO_2 it split off a new acid. Bohr has worked out this relation on the assumption that there is no alkali in the system and that the process is independent of the reaction of the solution.

So far as I can see the only proof which Bohr has tried to give that the process really proceeds is that the CO_2 compounds in a haemoglobin solution and in the blood can be calculated by an equation derived under the above assumptions, but it is generally accepted a proof of this kind is always very weak. Although I cannot assert that a reaction of the kind in question does not take place particularly when developed according to Siegfried's [1905, 1, 2; 1908, 1, 2; 1910] theory, as will be explained later on, it is probable at any rate that such a process only proceeds to a very slight extent. Bohr has reported numerous experiments on CO_2 compounds in pure haemoglobin and claims to have shown that various modifications of haemoglobin exist which can pass into one another but he is unable to show how. This complicates Bohr's view still more and it seems to me it would be useless to express an opinion on the CO_2 compounds in pure haemoglobin solutions without carrying out fresh experiments.

Siegfried and his collaborators have shown that by treating amino-acids with CO_2 in the strongly alkaline barium hydroxide or calcium hydroxide solution at a temperature about freezing point a complex compound of carbonic acid and amino-acids is formed and he has further shown that the same compound is also formed even if no Ba or Ca is present in the solution. When a solution in which a complex CO_2-amino-acid compound is formed is heated and Ba or Ca is present, a precipitate of the carbonates of these elements will separate out.

Siegfried believes that complex compounds of the amino group and CO_2 are formed according to the following scheme:

$$\underset{\overset{|}{COOH}}{-R-\overset{\overset{\displaystyle H}{\diagup}}{N}-H} + CO_2 \rightleftharpoons \underset{\overset{|}{COOH}}{-R-\overset{\overset{\displaystyle H}{\diagup}}{N}-COOH}$$

and that they are strong acids which form salts as follows:

$$\underset{\overset{|}{COO——Ba}}{-R-\overset{\overset{\displaystyle H}{\diagup}}{N}-COO}$$

(Siegfried seems to imply by this formula that the barium salt is not dissociated; this is however unimportant in the present connection.)

In my opinion Siegfried has only shown that CO_2 is combined in a complex manner with the amino group in the cold and in a very alkaline reaction, and he appears to have failed to supply any proof that the compounds formed are acids, let alone strong acids.

Siegfried [1908, 2] claims to have shown that proteins form with CO_2 complex compounds similar to the amino-acids. The experiments he considers most telling in this respect are certainly not convincing and they are probably quite wrongly interpreted. Siegfried passes CO_2 through a protein solution and shows that the conductivity increases more than can be accounted for by the CO_2 alone. He concludes from this that dissociated carbamino-acids are formed but it seems to me more natural to ascribe the effect to the formation of dissociated protein bicarbonates. Siegfried has *assumed* that protein carbamino-acids are formed at the temperature and reaction of the body and that it is a reversible reaction and they would be formed in greater or less amounts according to the prevailing CO_2 tension, and further that these compounds participate in the transport of CO_2 from the tissues to the lungs. This view has gained some acceptance (cf. for example Hammarsten [1914] and Wilstätter and Stoll [1918]) but in my opinion there is no evidence that the compounds exist or are dissociable at physiological temperatures, reactions and CO_2 tensions. *Per contra* it has naturally not been proved such bodies do not exist.

The third theory relating to CO_2 compounds in the blood is mainly supported by experiments of Findlay and his collaborators [1908, 1910, 1911, 1912, 1913, 1915]. The theory itself has not been actually propounded but Findlay's experiments and his whole method of reasoning in his papers certainly suggest the enunciation of such a theory. Findlay's idea is that gases can be adsorbed by colloids and that it is possible such an adsorption takes place in the blood. The experiments of Findlay and his collaborators have from a physiological standpoint had an unfortunate fate. In 1908 Findlay formulated the working hypothesis referred to and investigated in the following years the adsorption of various gases in several colloid solutions. By this great pioneer work Findlay has brought to light a considerable number of previously unknown conditions which still await further investigation. From his experiments he could conclude that inactive gases are frequently adsorbed in colloid solutions, and that in some cases CO_2 is adsorbed and in others combined (as bicarbonate).

In 1911 Findlay and Creighton reported some experiments on the absorption of oxygen, nitrogen, nitrous oxide, carbon monoxide, and carbon dioxide in blood and serum but these experiments, from which he concluded that oxygen, nitrogen, nitrous oxide and carbon monoxide were adsorbed by the proteins, are from a technical point of view so indifferent that the results are far from convincing. Findlay and Creighton have not taken sufficient notice of the fact that there are large quantities of combined O_2 and CO_2 in the blood which are dissociable and that it is necessary to make sure the compounds of these gases have not changed if one wishes to determine the absorption conditions of another gas by Findlay's technique; but this has certainly not been the case in the experiments referred to. In physiological circles this has been generally known since the days of Fernet and Meyer, and it is only the fact that Findlay has been trained as a chemist that makes it intelligible he should overlook the point.

I will quite briefly show from a single series that Findlay and Creighton's experiments are much too uncertain to allow of any conclusion being drawn. Findlay and Creighton have determined the absorption coefficient (Ostwald's) at a number of different tensions. It is easy with the help of the equation first employed by Fernet [1857] and Meyer [1857] to estimate the absorption coefficient between two tensions. I have performed this and then calculated this value as the relative absorption coefficient at the temperature of the experiment, in pure water.

A few of the experiments will be found recalculated in the following table:

CO in blood which had been exhausted before the saturation[1]		CO in blood not exhausted	
mm. Hg CO_2	Relative absorption coefficient	mm. Hg CO_2	Relative absorption coefficient
751–871	0·32	751–944	1·12
871–1056	1·61	944–1144	1·52
1056–1140	1·08	1144–1371	0·95
1140–1274	1·07	1371–1434	2·75
1274–1543	0·90	1434–1528	1·80

It appears further from other experiments that the absolute values found for the absorption coefficients are on an average much lower than they have

[1] With a water exhaust pump.

been found before, in fact inexplicably low. There is no alternative but to conclude that Findlay's experiments with blood are subject to such a large technical error that they cannot be utilised. The issue itself which Findlay has raised is very interesting. Jolyet and Sigalas [1892] claim to have shown that blood absorbs more nitrogen than it should if the absorption coefficient were the same as for pure water. Bohr [1909] and later Buckmaster have been able to confirm this although Buckmaster [1911–1912] only finds about 10 % too much. There is yet another fact which might be brought forward in support of the contention. Bohr [1905] has found the relative absorption coefficient for hydrogen in blood is 0·92 and Fahr [1912] found approximately the same values.

In serum Bohr estimated 0·975 as the coefficient for hydrogen.

In serum Fahr estimated 0·94 to 0·95 as the coefficient for hydrogen.

In blood cells Bohr estimated 0·81 as the coefficient for hydrogen.

In blood cells Fahr estimated 0·78 to 0·88 as the coefficient for hydrogen.

R. Siebeck [1909] has found values for nitrous oxide in blood, blood cell solutions and haemoglobin solutions which are a little over 1 (for blood 1·03), while Lindhard [1914] found about 0·97 for blood; and for serum Siebeck found the same value with nitrous oxide as Bohr found with the hydrogen[1].

The disagreement between the depression of solubility for hydrogen and nitrous oxide is pronounced. Should this be supported by experiments it points strongly to complicating factors in the solubility of the gases in protein solutions.

Thorup [1887] states he has demonstrated the existence of a CO_2-haemoglobin compound spectroscopically. He has shown that reduced haemoglobin when treated with high tensions of CO_2 develops a different absorption spectrum. From a modern physico-chemical point of view we cannot judge by this method whether complex combinations are formed or whether the haemoglobin becomes ionised, because electrolytes with characteristic spectra usually change their absorption bands on ionisation.

In conclusion I will discuss a view put forward by Jaquet [1892]. He has asserted that the CO_2 combination curve of the blood, when plotted with the CO_2 tensions along one axis and the combined CO_2 along the other, is shifted parallel with itself on addition of acid or base to the blood. If Jaquet's statement were right it would prove that the CO_2 combination with haemoglobin was independent of the reaction and thus was not a bicarbonate compound as it appears to be from the reasoning in the next chapter. But Jaquet's experiments are a long way from proving his contention because in the first place the curves are not truly parallel and secondly they are produced by points corresponding to experiments with different samples of blood.

[1] D. D. van Slyke and W. C. Stadie (*J. Biol. Chem.* 1921, **54**, 1) in accordance with earlier investigators have found greater quantities of N_2 in the blood than is to be expected from Henry's law. A similar observation was made by Smith, Dawson and Cohen (*Proc. Soc. Exp. Biol. and Med.* 1919–20, **17**, 211), quoted from van Slyke and Stadie. At the Finsen Institute I have myself frequently had a similar experience.

RÉSUMÉ.

A search of the old literature does not disclose any facts which would lead us to abandon Zuntz's theory modified in accordance with modern physico-chemical requirements.

CHAPTER XI

FURTHER CONSIDERATIONS AND EXPERIMENTS ON THE NATURE OF THE COMBINATION OF CO_2 IN THE BLOOD, ILLUSTRATED BY THE "CARBONIC ACID COMBINATION" CURVE[1].

In a solution in equilibrium with air containing CO_2, a quantity of CO_2 is found to be dissolved according to Henry's law, as repeatedly stated. This CO_2 can be extracted from the solution by diminishing the CO_2 tension in the gas phase over the solution, and further by this process more CO_2 which must have been reversibly "combined" in the solution can often be liberated.

It has frequently been stated in earlier chapters that the combined CO_2 may be present

(1) as adsorbed CO_2 if the solution is heterogeneous,

(2) as carbonate ions (monocarbonate and bicarbonate),

(3) as undissociated carbonate (complex salt),

(4) as adsorbed carbonate which will presumably be identical with (3), since the molecular groups lying on the surface in a high molecular solute may be expected under certain circumstances to form compounds with the carbonate,

(5) as a dissociable combination in which CO_2 takes part after assumption into the molecule, e.g. carbamino-acids.

At reactions more acid than $a_{H^{\cdot}} = 10^{-8}$ we can regard the carbonate ions present solely as bicarbonate (cf. chapter II).

We will first examine systems in which adsorption combinations and complex salts are not found. In general, the following equation for the equilibrium of the positive and negative charges is valid for any solution whatever:

$$C_{M^{\cdot}{}_{\mathrm{I}}} + 2C_{M^{\cdot\cdot}{}_{\mathrm{II}}} + 3C_{M^{\cdot\cdot\cdot}{}_{\mathrm{III}}} + \dots nC_{M^{\cdot}{}_n} \dots n \text{ charges}$$
$$= C_{A^{\prime}{}_{\mathrm{I}}} + 2C_{A^{\prime\prime}{}_{\mathrm{II}}} + 3C_{A^{\prime\prime\prime}{}_{\mathrm{III}}} + \dots nC_{A^{\prime}{}_n} \dots n \text{ charges}, \quad \dots\dots(152)$$

where $C_{M^{\cdot}{}_{\mathrm{I}}}$ stands for the concentration of monovalent cations and $C_{A^{\prime}{}_{\mathrm{I}}}$ for the concentration of monovalent anions and so on. As hydrogen and hydroxyl ions at reactions not far removed from the neutral point only occur in very small concentrations we can neglect them as terms in the above equation.

---

[1] D. D. van Slyke [1921, 1] has recently reviewed the question of the combination of CO_2 in blood from the theoretical as well as from the clinical standpoint. Those interested are referred to the original articles which only became known to me after the following had been written.

In a solution in which only strong acids and their salts are present combined CO_2 cannot exist as there are no cations to counterbalance the bicarbonate ions. But in a solution containing a free base the hydroxyl ions will react thus:

$$CO_2 + OH' = HCO'_3.$$

Under the given conditions this reaction practically proceeds quantitatively. In such a solution the combination of CO_2 will, as is well known, be independent of the pressure (if monocarbonate is not formed).

If a weak base or the salt of a weak acid is present in a solution, CO_2 will be combined in amounts varying with the CO_2 pressure if the reaction is suitable.

For a weak monovalent acid, as shown in chapter I, the following equation is valid:

$$\frac{a_{H'} \times a_{A'_I}}{a_{HA'_I}} = K_{HA_I}. \quad \ldots\ldots\ldots\ldots\ldots\ldots\ldots(153)$$

On saturation the following reaction

$$CO_2 + H_2O + A'_I \rightleftharpoons HCO'_3 + HA_I \quad \ldots\ldots\ldots\ldots(154)$$

proceeds in such a way that a part of A'_I is substituted by HCO'_3 until the equilibrium

$$K_1 \frac{a_{A'_I}}{a_{HA_I}} = \frac{a_{HCO'_3}}{a_{H_2CO_3}} K_{HA_I} \quad \ldots\ldots\ldots\ldots\ldots(155)$$

is reached.

If a weak monovalent base is present in a solution the reaction

$$CO_2 + M_I OH \rightleftharpoons M'_I + HCO'_3 \quad \ldots\ldots\ldots\ldots(156)$$

proceeds until the following equilibrium is reached:

$$\frac{a_{M'_I}}{a_{M_IOH}} \times \frac{K_W}{K_{M_IOH}} = K_1 \frac{a_{HCO'_3}}{a_{H_2CO_3}}. \quad \ldots\ldots\ldots\ldots(157)$$

The base is thus dissociated with the production of cations and bicarbonate ions. When equilibrium is attained an amount of bicarbonate ions will be present in the solution equivalent either to the cations the dissociation of which does not vary with the reaction (which before equilibrium were balanced by other anions the dissociation of which varies with the reaction) or to newly formed cations the dissociation of which varies with the reaction. The amount of combined CO_2 therefore becomes a measure of the *available cation concentration*. D. D. van Slyke and his collaborators have used the term "alkali reserve" for this amount (in serum) and E. Jarlöv has employed the almost analogous expression "available alkali." The reason I consider these expressions should be given up is because of the possibility of misunderstanding them. Thus Parsons, Davies, Haldane and Kennaway have erroneously assumed that in the "maximum" amount of CO_2 combined we have a measure of the difference between the cations and anions the dissociation of which does not vary with the reaction.

In order to illustrate the above I have performed experiments with sodium hydroxide solutions to which NaCl and phosphate solutions were added. The experiments will be found in Table LI and graphically in Fig. 15.

Table LI.

mm. Hg CO_2 1	Vols. % combined 2	pH calc. 3	$aH· ×10^{-7}$ calc. 4	With equation (171)		pH calc. 7	With equation (172)		With equation (180)	
				Vols. % CO_2 combined 5	Diff. Vols. % 6 (2—5)		Vols. % CO_2 combined 8	Diff. Vols. % 9 (2—8)	Vols. % CO_2 combined 10	Diff. Vols. % 11 (2—10)
I. 30 cc. $m/15$ KH_2PO_4 + 50 cc. 0·04941n NaOH + H_2O to 100 cc. + 0·42 g. NaCl. 20°.										
7·0	16·1	7·585	0·260	17·0	-0·9	7·608	17·6	-1·5	15·9	+0·2
12·7	19·4	7·407	0·385	19·9	-0·5	7·418	20·2	-0·8	19·7	-0·3
23·4	23·5	7·225	0·596	23·7	-0·2	7·229	23·8	-0·3	23·6	-0·1
46·4	28·0	7·004	0·990	28·7	-0·7	7·015	29·1	-0·9	28·3	-0·3
76·5	31·9	6·844	1·43	32·7	-0·8	6·854	33·2	-1·3	32·4	-0·5
172·3	38·4	6·571	2·68	39·3	-0·9	6·582	39·8	-1·4	37·6	+0·8
745·0	48·7	6·038	9·15	48·8	-0·1	6·039	49·2	-0·5	49·1	-0·4
II. 6 cc. $m/15$ Na_2HPO_4 + 24 cc. $m/15$ KH_2PO_4 + 50 cc. 0·04941n NaOH + H_2O to 100 cc. + 0·42 g. NaCl. 19°.										
8·6	49·2	7·954	0·111	49·4	-0·2	7·956	49·4	-0·2	43·1	+6·1
14·4	51·0	7·746	0·180	51·2	-0·2	7·747	51·3	-0·3	47·8	+3·2
20·2	52·1	7·608	0·247	52·9	-0·8	7·614	52·8	-0·7	50·7	+1·4
26·7	53·5	7·498	0·318	54·3	-0·8	7·504	54·6	-1·1	53·2	+0·3
47·3	57·7	7·283	0·521	58·4	-0·7	7·288	58·6	-0·9	57·9	-0·2
82·5	62·0	7·073	0·846	63·0	-1·0	7·080	63·0	-1·0	62·7	-0·7
436·4	77·5	6·446	3·58	78·8	-1·3	6·453	78·1	-0·6	76·4	+1·1
750·2	80·6	6·228	5·92	82·9	-2·3	6·240	82·1	-1·5	81·4	-0·8
III. 30 cc. $m/15$ Na_2HPO_4 + 50 cc. 0·04941n NaOH + H_2O to 100 cc. + 0·42 g. NaCl. 20°.										
5·2	56·3	8·240	0·0576	56·7	-0·4	8·242	56·8	-0·5	48·3	+8·0
12·5	58·8	7·878	0·132	58·8	0·0	7·878	58·8	0·0	54·1	+4·7
19·6	60·4	7·694	0·202	60·7	-0·3	7·696	60·8	-0·4	58·1	+2·3
25·2	62·4	7·599	0·252	61·8	+0·6	7·595	61·9	+0·5	60·2	+2·2
43·5	65·4	7·383	0·415	65·4	0·0	7·383	65·4	0·0	64·9	+0·5
81·0	70·1	7·142	0·721	70·3	-0·2	7·144	70·4	-0·3	70·1	0·0
186·2	77·4	6·824	1·50	78·1	-0·7	6·828	78·1	-0·7	77·0	+0·4
414·4	84·2	6·513	3·07	85·5	-1·3	6·520	85·7	-1·5	83·8	+0·4
731·5	87·5	6·283	5·21	90·1	-2·6	6·296	90·2	-2·7	88·8	-1·3
IV. 24 cc. $m/15$ Na_2HPO_4 + 6 cc. $m/15$ KH_2PO_4 + 0·0243n Na_2CO_3 to 100 cc. + 0·85 g. NaCl. The two first experiments at 18°, the others at 19°.										
6·9	32·8	7·835	0·146	33·3	-0·5	7·841	33·4	-0·6	27·6	+5·2
10·4	34·1	7·673	0·212	34·9	-0·8	7·683	35·0	-0·9	31·4	+2·7
21·1	37·6	7·416	0·384	38·6	-1·0	7·427	38·7	-1·1	37·3	+0·3
42·7	42·4	7·162	0·689	43·6	-1·2	7·174	43·9	-1·5	43·1	-0·7
85·3	48·6	6·921	1·20	49·3	-0·7	6·927	49·5	-0·9	48·7	-0·1
221·8	57·1	6·575	2·66	58·0	-0·9	6·582	58·3	-1·2	56·7	+0·4
414·0	63·0	6·347	4·50	63·0	0·0	6·347	63·0	0·0	62·0	+1·0
747·3	67·0	6·117	7·63	66·8	+0·2	6·117	66·6	+0·4	67·3	-0·3
747·3	66·6	6·115	7·68	66·8	-0·2	6·114	66·8	-0·2	67·3	-0·7
V. 24 cc. $m/15$ Na_2HPO_4 + 6 cc. $m/15$ KH_2PO_4 + 50 cc. 0·04941n NaOH + H_2O to 100 cc. + 0·85 g. NaCl. 19°.										
10·0	49·9	7·859	0·138	50·1	-0·2	7·861	50·1	-0·2	47·4	+2·5
22·7	53·0	7·529	0·296	53·7	-0·7	7·534	53·9	-0·9	53·8	-0·8
47·7	57·8	7·244	0·570	58·9	-1·1	7·252	59·1	-1·3	59·3	-1·5
85·2	62·3	7·023	0·945	63·9	-1·6	7·036	64·1	-1·8	63·6	-1·3
178·6	68·5	6·744	1·80	70·9	-2·4	6·759	71·2	-2·7	69·0	-0·5
375·2	76·9	6·473	3·37	77·6	-0·7	6·476	77·7	-0·8	74·2	+2·7
743·2	81·0	6·198	3·34	82·6	-1·6	6·207	82·6	-1·6	79·5	+1·5

In the first three solutions 0·42 % NaCl was present and 50 cc. 0·04941n NaOH in 100 cc. solution. In the last two solutions 0·85 % NaCl was present and in the first of this series 70 cc. 0·0243n Na$_2$CO$_3$ in 100 cc. In the last, NaOH was of the same concentration as in the first three. In all the solutions there was also 0·02m phosphate (30 cc. of S. P. L. Sörensen's phosphate mixtures to 100 cc. solution), the alkali content of the solutions being varied by adding different amounts of primary and secondary phosphate. The experiments were performed as usual (with the small saturator). In the first column of the table the CO$_2$ tension is given; in the second column the combined CO$_2$ in vols. %; in the third column p_H calculated in the usual way, the values for K'_1 given later being used; the fourth column contains a_H.

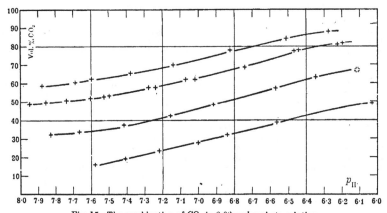

Fig. 15. The combination of CO$_2$ in 0·02m phosphate solution.

We will let C_I stand for the difference between the total cations and total anions the dissociation of which does not depend upon the reaction. The value for K'_1 employed is calculated by equations (139) and (143). The first dissociation of phosphoric acid

$$H_3PO_4 \rightleftharpoons H^{\cdot} + H_2PO'_4 \quad \ldots\ldots\ldots\ldots\ldots(158)$$

progresses, under the conditions of the experiment, from left to right completely, the dissociation constant according to Abbott and Bray [1909] being $10^{-2·0}$. The third dissociation of phosphoric acid under the same conditions does not take place at all, that is to say the reaction

$$HPO''_4 \rightleftharpoons H^{\cdot} + PO'''_4 \quad \ldots\ldots\ldots\ldots\ldots(159)$$

proceeds completely from right to left, the dissociation constant according to Abbott and Bray being $10^{-12·5}$. The only significant reaction therefore is

$$H_2PO'_4 \rightleftharpoons H^{\cdot} + HPO''_4. \quad \ldots\ldots\ldots\ldots\ldots(160)$$

At a given CO_2 tension the following equilibrium holds for (160):

$$a_{H^.} = K'_1 \frac{P_{CO_2} a}{7 \cdot 60 B} = k_2 \frac{C_{H_2PO'_4} F_a (H_2PO'_4)}{C_{HPO''_4} F_a (HPO''_4)}. \qquad \ldots\ldots\ldots\ldots(161)$$

(An almost analogous equation was evolved by L. J. Henderson as early as 1908.)

If we now put

$$k_2 \frac{F_a (H_2PO'_4)}{F_a (HPO''_4)} = k'_2 \qquad \ldots\ldots\ldots\ldots\ldots\ldots\ldots(162)$$

we get

$$a_{H^.} = K'_1 \frac{P_{CO_2} a}{7 \cdot 60 B} = k'_2 \frac{C_{H_2PO'_4}}{C_{HPO''_4}}. \qquad \ldots\ldots\ldots\ldots(163)$$

The total phosphate concentration is expressed by

$$C_{Ph} = C_{H_2PO'_4} + C_{HPO''_4}. \qquad \ldots\ldots\ldots\ldots\ldots(164)$$

The charge expressed in normalities corresponding to the phosphates then becomes

$$u C_{Ph} = C_{H_2PO'_4} + 2 C_{HPO''_4}. \qquad \ldots\ldots\ldots\ldots\ldots(165)$$

From (163), (164) and (165) we get

$$u = \frac{15 \cdot 20 B k'_2 + K'_1 P_{CO_2} a}{7 \cdot 60 B k'_2 + K'_1 P_{CO_2} a} \qquad \ldots\ldots\ldots\ldots\ldots(166)$$

or

$$u = \frac{2 k'_2 + a_{H^.}}{k'_2 + a_{H^.}}. \qquad \ldots\ldots\ldots\ldots\ldots\ldots\ldots(167)$$

Further, C_I is defined by

$$C_I = C_{HCO'_3} + C_{H_2PO'_4} + 2 C_{HPO''_4} \qquad \ldots\ldots\ldots\ldots(168)$$

and therefore from (168) and (165)

$$B = (C_I - u C_{Ph}) \, 2226. \qquad \ldots\ldots\ldots\ldots\ldots(169)$$

From (169) and (166) we obtain

$$B^2 + B \frac{K'_1 P_{CO_2} a + 2226 \times 7 \cdot 60 k'_2 (2C_{Ph} - C_I)}{7 \cdot 60 k_2} + \frac{K'_1 \, 2226 \, P_{CO_2} a \, (C_{Ph} - C_I)}{7 \cdot 60 k'_2} = 0. \ldots(170)$$

Substituting in this the constants given in Table LI a we get, for the first series

$$B = 5 \cdot 2 - 0 \cdot 217 \, P_{CO_2} + \sqrt{0 \cdot 0471 \, P^2_{CO_2} + 21 \cdot 5 \, P_{CO_2} + 27 \cdot 0};$$

for the second series

$$B = 23 \cdot 0 - 0 \cdot 230 \, P_{CO_2} + \sqrt{0 \cdot 0529 \, P^2_{CO_2} + 31 \cdot 6 \, P_{CO_2} + 529};$$

for the third series

$$B = 27 \cdot 5 - 0 \cdot 226 \, P_{CO_2} + \sqrt{0 \cdot 0511 \, P^2_{CO_2} + 32 \cdot 6 \, P_{CO_2} + 756};$$

for the fourth series at 18°

$$B = 14 \cdot 5 + 0 \cdot 252 \, P_{CO_2} + \sqrt{0 \cdot 0635 \, P^2_{CO_2} + 29 \cdot 8 \, P_{CO_2} + 210};$$

at 19°

$$B = 14 \cdot 5 + 0 \cdot 248 \, P_{CO_2} + \sqrt{0 \cdot 0615 \, P^2_{CO_2} + 29 \cdot 3 \, P_{CO_2} + 210};$$

for the fifth series

$$B = 23 \cdot 0 + 0 \cdot 250 \, P_{CO_2} + \sqrt{0 \cdot 0626 \, P_{CO_2} + 33 \cdot 9 \, P_{CO_2} + 529}.$$

$$\ldots\ldots\ldots(171)$$

Table LI a. Constants used in the calculation of (171).

Series	Total cation concentration	C_I	K'_1	C_{Ph}	k'_2	Temp.
I	0·1165	0·0447	$10^{-6.286} = 5 \cdot 18 \times 10^{-7}$	0·02	$10^{-6.86} = 1 \cdot 38 \times 10^{-7}$	19
II	0·1325	0·0607	$10^{-6.278} = 5 \cdot 36 \times 10^{-7}$	0·02	$10^{-6.86} = 1 \cdot 38 \times 10^{-7}$	20
III	0·1365	0·0647	$10^{-6.268} = 5 \cdot 40 \times 10^{-7}$	0·02	$10^{-6.86} = 1 \cdot 38 \times 10^{-7}$	19
IV (18°)	0·1940	0·0530	$10^{-6.244} = 5 \cdot 70 \times 10^{-7}$	0·02	$10^{-6.86} = 1 \cdot 38 \times 10^{-7}$	18
IV (19°)	0·1940	0·0530	$10^{-6.239} = 5 \cdot 77 \times 10^{-7}$	0·02	$10^{-6.86} = 1 \cdot 38 \times 10^{-7}$	19
V	0·2027	0·0607	$10^{-6.235} = 5 \cdot 82 \times 10^{-7}$	0·02	$10^{-6.86} = 1 \cdot 38 \times 10^{-7}$	19

k'_2 is taken as $10^{-6.86}$, which has been very accurately determined by S. P. L. Sörensen [1909–1910] in a mixture of equal parts of $m/15\,KH_2PO_4$ and $m/15\,Na_2HPO_4$. It is probable that the constant in our case only differs slightly from this value. In solutions of 0.42 % NaCl, Y is taken as 6, in other solutions, as 8. The results of the calculations will be found in Table LI in the fifth column; in the sixth column the differences between the values found and calculated are given. If we put aside the experiments with the highest CO_2 tensions (which are subject to the largest experimental error), it will be seen the agreement is as good as could be expected on the whole, but it should be noted the calculated value is always a little greater than that found by experiment. The cause of this is not quite clear but it may be pointed out that the agreement would have been very nearly ideal if k'_2 was taken as $10^{-6.80}$.

Now that we have effected these calculations let us turn again to the equations (171). It will immediately be obvious that it is difficult to form a clear conception of the course of the five carbonic acid combination curves if we only have these equations to go by, but it will be simple if from (167) and (169) we evolve the equation

$$B = \left(C_I - C_{Ph}\,\frac{2k'_2 + a_{H^{\cdot}}}{k_2 + a_{H^{\cdot}}}\right) 2226. \qquad \ldots\ldots\ldots\ldots\ldots(172)$$

This equation indicates that all the CO_2 combination curves in solutions which contain the same phosphate concentration run parallel with one another if they are drawn in a right angled coordinate system where the apparent hydrogen ion activity, or a value which contains no other variable than this, is placed along one axis and the combined CO_2 along the other. If the combined CO_2 is expressed in vols. % the distance between the two curves projected on the vols. % axis will be the difference between the $C_I \times 2226$ (vols. %) of the two solutions.

The above reactions are of general interest in that, in any similar solution whatever we have

$$B = (C_I - C_V f\,(a_{H^{\cdot}}))\,2226, \qquad \ldots\ldots\ldots\ldots\ldots(173)$$

where C_V is the concentration of the electrolytes of which the dissociation varies with the reaction, and where C_I and $C_V f\,(a_{H^{\cdot}})$ can have positive or negative values. Even though we do not know C_I, C_V, or $f\,(a_{H^{\cdot}})$ we can, under the given assumptions, be certain of the shift in the CO_2 combination curve if C_I varies on the addition of alkali (or acid), the volume of the system being constant. If for instance C_I is increased to $C_1 + b$ we get

$$B_I = (C_I + b - C_V f\,(a_{H^{\cdot}}))\,2226. \qquad \ldots\ldots\ldots\ldots(174)$$

Subtracting (173) from this we obtain

$$B_1 - B = 2226b, \qquad \ldots\ldots\ldots\ldots\ldots(175)$$

which is the algebraic expression of the fact that any CO_2 combination curve will be shifted parallel with itself in an $a_{H^{\cdot}} -$ vols. % CO_2 graph provided adsorption combinations of CO_2, or CO_2 compounds in the form of complex bodies (*e.g.* carbamino-acids), are not formed. The distance between the curves is then $2226b$ vols. %.

If the CO_2 in a solution can be adsorbed or combined in the complex manner indicated above and if these combinations are reversible, the amount combined will be a function of the CO_2 pressure. The general expression for the combination will be

$$B = (C_I - C_V f(a_{H^.}) - f_1(P_{CO_2})) \, 2226, \quad \ldots\ldots\ldots\ldots(176)$$

and the difference in CO_2 combined in a solution before and after the addition of alkali at the same reaction will therefore, from analogy with (175), be

$$B_I - B = 2226b - 2226 \, (f_1(P_{CO_2}) - f_1(P_{CO_{2_1}})). \quad \ldots\ldots\ldots(177)$$

In this equation the amount

$$- 2226 \, (f_1(P_{CO_2}) - f_1(P_{CO_{2_1}}))$$

will always be positive because the combination of CO_2 must increase with the CO_2 tension in the given conditions and in order to produce the same reaction a higher CO_2 tension will be necessary after the addition of alkali than before (cf. equation (95)).

It is thus possible to state in general that a CO_2 combination curve, in which the combined CO_2 is set off along the one axis and an expression for the active reaction along the other axis, will be shifted on the addition of alkali in such a way:

(1) that the distance between the original curve and the new curve is always equal to the combination of CO_2 corresponding to the added amount of alkali provided that the variable combination in each curve is only a function of the reaction;

(2) that the distance between the original and new curves will increase with the apparent hydrogen ion activity provided that the variable combination is also a function of the CO_2 tension;

(3) that the distance between the two curves can never be less than the combined CO_2 corresponding to the added alkali.

We will now revert to the phosphate solutions again. In the eighth column of the table the combined CO_2 expressed in vols. % will be found, calculated by (172) with the help of the apparent hydrogen ion activity in column 4 reckoned directly from the experiments. The agreement is, as expected, good. In Fig. 15 the combined CO_2 is plotted as ordinates and

$$- \log a_{H^.} = p_{H^.}$$

as abscissae, however on a hundred times greater scale. In such a coordinate system, as explained, the course of the curves will be parallel. Now it appears from the figure that the CO_2 combination curves are straight lines for a long portion of their courses, and we will therefore, as an example of what is to follow, calculate these sections of the curves by the method of least squares, using the figures in columns 2 and 3, and include all experiments in which

$p_{H^.}$ is less than 7·59. Equations (141) and (142) are employed in the calculation and the mean error of the constants is estimated by the equation

$$M_x = M_y = \frac{M}{\sqrt{n\,(\Sigma a^2 - (\Sigma a)^2)}}, \quad \dots\dots\dots\dots(178)$$

where
$$M = \sqrt{\frac{\Sigma a^2}{n-2}}, \quad \dots\dots\dots\dots\dots\dots\dots(179)$$

Σa being the sum of the squares of the deviations of the found and calculated values and n the number of experiments. The equation of the straight line referred to above is
$$x \log a_{H^.} + y = B, \quad \dots\dots\dots\dots\dots(180)$$

where x is the tangent to the angle made by the line with the axis of the abscissa, measured in the first or third quadrant, and y is the part of the ordinate axis cut off by the line. The calculation gives

Series	x	y	M	No. of experiments
I	21·460±0·057	178·65±0·057	0·51	7
II	22·139±0·17	219·15±0·17	0·91	5
III	21·720±0·16	225·24±0·16	0·87	5
IV	23·076±0·077	208·41±0·077	0·68	7
V	19·301±0·27	199·13±0·27	1·9	6

It will be seen that a determination of the error indicates that x is different n the various series, but as it has been proved already by calculations that the different curves are really parallel, allowing for experimental error, we cannot attribute any real significance to the mean error calculated by this method in the present and similar cases. There are several reasons for this rather unsatisfactory result. In the first place the number of experiments in each series, within the range of reaction employed, is far too few to permit of an accurate determination of the error; in the second place the points determined are not regularly distributed over the various sections of the curve, in fact they do not all cover exactly the same range of reaction, which is of importance as it will be shown the portion of the curve in question is not in reality a straight line. For the appreciation of what is to follow it will be necessary to examine this a little more closely.

By equation (172) it is easy to estimate the true slope of the curve. In the second column of the following table the combinations of CO_2 in a $0·02n$ phosphate solution of $p_{H^.}$ varying from 7·50 to 6·20 are given, 2226 $C_I = 52·3$ vols. %, the combined CO_2 at $p_{H^.}$ 7·50 being here put equal to 0. In the third column will be found the tangent to the straight lines (angle of inclination which is x) which join the nearest points together two by two. If the point referring to $p_{H^.}$ 7·50 is joined with that referring to p_H 6·20 the tangent to the angle of inclination of the line is 21·6. In the fourth column will be found the amount of combined CO_2 expressed in vols. %, which, according to a calculation based on the straight line smoothing formulae (141) and (142) from the figures in columns 1 and 3, should be found combined at the given $p_{H^.}$ when the best constants are used. These constants are

$$x = 22·719 \pm 0·021, \quad y = 169·77 \pm 0·021.$$

In the fifth column the differences between the CO_2 combinations calculated by (172) and by the smoothing are given. The standard deviation is calculated to 0·45 vols. %.

Table LI b.

p_H 1	Vols. % CO_2 combined. Calculated by (172) 2	x 3	Vols. % CO_2 combined. Algebraic smoothing 4	Difference vols. % 5 (2-4)
7·50	0·0		−0·6	+0·6
		17		
7·40	1·7		1·7	0·0
		19		
7·30	3·6		3·9	−0·3
		21		
7·20	5·7		6·2	−0·5
		23		
7·10	8·0		8·5	−0·5
		24		
7·00	10·4		10·7	−0·3
		25		
6·90	12·9		13·0	−0·1
		26		
6·80	15·5		15·3	+0·2
		25		
6·70	18·0		17·6	+0·4
		24		
6·60	20·4		19·8	+0·6
		22		
6·50	22·6		22·1	+0·5
		21		
6·40	24·7		24·4	+0·3
		18		
6·30	26·5		26·6	−0·1
		16		
6·20	28·1		28·9	−0·8

If column 5 is examined it will be observed that the differences change their sign three times, that is to say the straight line cuts the true curve (calculated by (172)) three times. If it should happen, in the determination of a similar CO_2 combination curve, that points were obtained at these places of intersection, it would naturally follow that even an ideal method of analysis would be unable to show that the curve departed from a straight line. It will be realised that a very considerable number of experimental points are required to determine the true course of the curve and that in using the smoothing method of calculation one must be exceedingly cautious about one's conclusions.

As mentioned the constant x was determined by the smoothing method to be 22·719 ± 0·021. If this is compared with the constants found by calculation from the experiments it will be seen that only that from series II appears to be of the same magnitude, which is another plea for the contention that the mean error of the constants is really too sharply defined.

There is still another point in relation to the smoothing method to be touched upon. It is not possible by this method in the form which is used here with the aid of the mean error of the constants to determine the mean error from a single point on the curve. This is due to the choice of the zero point of the coordinate system, because this lies, as will be seen, far outside the experimentally determined points. The reason I have given the numbers for the constants correct to the first decimal place in spite of the objections put forward relating to the mean error, is in order to show them with all the figures used in the calculations and the mean error is therefore only employed roughly to demonstrate the uncertainty of the constants.

It is accepted that a large part of the reversibly combined CO_2 of the blood is present as the bicarbonate ion, the haemoglobin and other protein substances acting as ampholytes. The objections against this view which have been put forward in the last few years have been to some extent negatived in previous chapters and will also be questioned at the end of the present one. It can therefore be concluded that no general condition for the combination of CO_2 in the blood can be formulated which cannot be traced back to an equation of the form of (173) or (176). Provided no CO_2 is adsorbed or combined in a complex manner (173) will be the correct equation.

In 1920 T. R. Parsons put forward a mathematical treatment of the CO_2 combination in the blood but it seems to me he has made a mistake on this point. Firstly Parsons assumed that haemoglobin is a monobasic acid, each haemoglobin molecule however encompassing several such groups. Secondly he assumed that in blood especially there were so many more cations than anions the dissociation of which did not vary with the reaction, that they were able just to neutralise the haemoglobin acid. Thirdly he believed that in the "maximum" of combined CO_2 we have a measure of the concentration of the haemoglobin acid.

The first of these assumptions is quite uncertain; the second is undoubtedly erroneous, because the blood can expel CO_2 from alkaline carbonate as Pflüger [1864] was the first to show. The third assumption is incorrect as there is no maximal CO_2 combination in blood at atmospheric CO_2 tensions. Parsons' final relation evolved as it is from faulty assumptions has no real significance and it is only one of many instances that nearly every equation with a sufficient number of constants can be brought into agreement with an observed fact when the constants chosen are the best possible.

At present there seems to be no possibility—contrary to what was the case with the phosphate experiments—of establishing an *a priori* relation between a_{II}· and the combination of CO_2 with the blood proteins. There is no other course open therefore than to find the simplest empirical relation between these values.

In the previously mentioned paper of Hasselbalch and Warburg some CO_2 combination curves for blood and serum are given in a p_H·—vols. % curve. They are straight lines. This fact appears to have been quite overlooked, but I believe the appreciation of this phenomenon will be of far-reaching importance in the physiology of respiration. In this work I shall content myself with more thoroughly demonstrating the relation and employ it for elucidating the question of the existence of complex CO_2 combinations as well as some other problems.

In tables LII—LVII (pp. 287–290) the combination of CO_2 at various CO_2 tensions in a number of blood samples from different people and animals is given. In the last column but one the results of the algebraic smoothing are recorded, the CO_2 combination curve in the p_H·—vols. % combined CO_2 chart being taken as a straight line.

It is therefore assumed that

$$x \log a_{H^{.}} + y = B \qquad \qquad (181)$$

or

$$- x p_{H^{.}} + y = B. \qquad \qquad (182)$$

That these relations, within the experimental error, are really true seems to follow from the curves and tables, but I have further investigated some rather extensive unpublished material dealing with the CO_2 combination curves in ox blood which has been elaborated of late years for another purpose in the laboratory of the Finsen Institute, without being able to find any sign of a departure from the relations. After what was said about the phosphate curves it will readily be understood I do not wish to urge that the sections of the curves being dealt with are actually straight hues, but it is asserted the deviations from a straight line are small compared with the variations in the combination of CO_2 at different apparent hydrogen ion activities.

In the calculation of $p_{H^{.}(a)}$ (130) was used. In the case of oxygenated human blood at 38° and oxygenated horse blood at room temperature the corrections given in Fig. 10 were added to the values found. In the ease of reduced human blood at 38° and reduced horse blood at room temperature the correction corresponding to $p_{H^{.}}$ (uncorrected − 0·40) was added. In the case of Hasselbalch's [1916] "half reduced" blood 0·20 was subtracted from $p_{H^{.}}$ (uncorrected), while 0·40 was subtracted from oxygenated ox blood at room temperature. Equation (130) was used without correction in the calculation of the other blood samples.

The correction curve corresponding to 18·5 vols. % combined CO_2 was used in the calculations relating to Haldane's blood, from Christiansen, Douglas and Haldane [1914], the 15 vols. % curve for Parsons' [1917] blood, the 20 vols. % curve for Davies' [1920] and Hasselbalch's [1916, 2] blood and the 25 vols. % curve for Joffe's [1920] and Warburg's blood[1]. In the calculations

---

[1] I have not attempted to discover how much combined O_2 was actually present in Davies', Haldane's, Joffe's and Parsons' blood but I have contented myself with estimating their p_H–CO_2 combination curves by means of the corrections mentioned because an error of 5 vols. % combined O_2 only causes a difference in the constant x of about 1. The reason of this is that I have not been able to obtain agreement between the O_2 determinations got by pumping out and the values obtained by Haldane's ferricyanide method. I originally became aware of this by comparing the relation between the O_2 combined by the blood and the volume of the blood corpuscles of the same blood. A. Norgaard and H. C. Gram and later H. C. Gram [1921] found about 38·5 vols. % combined O_2 in 100 cc. human blood corpuscles but V. Bie and P. Möller [1913] found about 45 vols. %, a similar quantity to that I have myself found. Later on, it appears, W. B. Cannon, J. Eraser and A. N. Hooper [1919], in oxygenated blood from patients suffering from shock, found about 43 vols. %. While Norgaard and Gram's haemoglobinometer was corrected by the ferricyanide method, Bie and Möller's apparatus was corrected by Fridericia by pumping out, just as my own results were obtained. Dr Marie Krogh was kind enough to determine the O_2 in some blood samples from man and horse with a Haldane haemoglobinometer which was very accurately calibrated by the ferricyanide method. She constantly found lower O_2 values than I did with the exhaust pump. I have no doubt whatever that too low readings are obtained under special circumstances with Haldane's ferricyanide method as shown lately by van Slyke and Stadie [1921] and I hope soon to take up this question again. F. Müller [1904] previously found the same for dog blood but considered it only applied to quite fresh blood and under unfavourable circumstances.

relating to ox and horse blood the curve which most nearly corresponded to the O_2 determinations was employed.

The value used in the calculations for $p\lambda_{(s)} + \log \Phi_a (CO_2)$ with human, ox, and horse blood at 38° was 6·190; at 37°, 6·185; at 17°, 6·334; at 18°, 6·327; at 19°, 6·320. In the case of pig and pigeon blood at 38°, 6·20 was used; with dog and rabbit blood at 19° and 18·5°, 6·35. With haemolysed blood equation (134) was employed with the following constants: at 17°, 6·325 was employed; at 18°, 6·320; at 19°, 6·315. With serum equation (121) was used and as constants at 38°, 6·165 was used; at 19°, 6·286; at 20°, 6·281; at 21°, 6·274. The results of these calculations and of the smoothing on the curves for some of the experiments carried out in the laboratory of the Finsen Institute will be found in Tables LII, LIII and LVIII and in Figs. 16, 17, 19, 20 and 21. The combination of CO_2 in *human, pigeon, horse, dog, rabbit, pig and ox blood* and in some sera can therefore be expressed by an equation of the form of (181) or (182).

It will be observed, as already mentioned, that all the curves determined in this laboratory are straight lines, allowing for experimental error. In the case of a number of the serum curves however the point corresponding to the highest CO_2 tension is too low. This seems to me only to be an indication that equilibrium had not been quite reached in these experiments as at the time I undertook them I was not aware of the deficiency of the small saturator. I have omitted these experiments in the calculus of smoothing, and in the tables this is denoted by a line drawn through them. It will be easily seen without a special note which experiments are included and which rejected. Those portions of the curves which correspond to the rejected experiments are drawn with a dotted line. In one of the tables an experiment has been excluded because it seems to be subject to an unusually large error. This experiment is put in parentheses.

The experiments with very low tensions were carried out with the large saturator electrodes using the technique described in chapter IV.

In Table LIV I have performed similar calculations for Haldane's blood (taken from Christiansen, Douglas and Haldane's [1914] paper) and in Table LVI for Davies' blood (taken from Davies, Haldane and Kennaway's [1920] paper). The curves are given in Fig. 18.

It will be noticed the CO_2 combinations corresponding to the most alkaline values are too high both in oxygenated and reduced blood—in Davies' blood the most alkaline values are too high and the most acid too low. I believe I am right in saying these deviations from a straight line are due to the fact that the technique of saturation employed is not applicable at high and low CO_2 tensions. The experiments just recorded show this and I myself had a similar failure before I became aware of the relatively great difficulties of saturation. Davies, Haldane and Kennaway's mistake in thinking the combination of CO_2 in the blood is maximal at the highest tensions is only due to equilibrium not being established. Calculations relating to J. Joffe's blood, Table LVII and Fig. 19 (from Joffe and Poulton's [1920] paper) also demonstrate

Table LII.

Pig blood (O_2-combination 21·1 vols. %). Half saturated with O_2, 38°. Calculated after K. A. Hasselbalch. $p\lambda_{(s)} + \log \Phi_s = 6·20$ in every case.

mm. Hg CO_2	Vols. % CO_2 combined	$pH'_{(s)}$ calculated	Vols. % CO_2 combined calculated	Difference
8·4	21·9	7·75	20·8	+1·1
(11·4	22·7	7·64	26·5	−3·8)
15·7	28·6	7·60	28·6	0·0
30·4	36·8	7·42	37·9	−1·1
30·8	36·4	7·41	38·4	−2·0
53·1	46·3	7·28	45·2	+1·1
76·1	51·8	7·17	51·0	+0·8

$x = 51·865 \pm 0·5$. $y = 422·76 \pm 0·5$. $M = 1·4$.

Pigeon blood (O_2-combination 23·7 vols. %). Half saturated with O_2, 38°. Calculated after K. A. Hasselbalch. $p\lambda_{(s)} + \log \Phi_s = 6·20$ in every case.

5·8	21·0	7·90	20·9	+0·1
18·3	35·0	7·62	35·1	−0·1
38·5	45·3	7·41	45·7	−0·4
75·0	56·2	7·21	55·9	+0·3

$x = 50·774 \pm 0·08$. $y = 421·96 \pm 0·08$. $M = 0·4$.

Dog blood (O_2-combination 19·2 vols. %) totally reduced, 19°. $p\lambda_{(s)} + \log \Phi_s = 6·35$ in every case.

2·2	21·8	8·27	22·1	−0·3
29·3	53·5	7·54	53·6	−0·1
56·4	62·4	7·32	63·1	−0·7
77·1	68·5	7·22	67·5	+1·0
127·5	76·0	7·05	74·8	+1·2
337·6	88·8	6·70	89·9	−1·1

$x = 43·230 \pm 0·1$. $y = 379·58 \pm 0·1$. $M = 1·0$.

Rabbit blood totally reduced, 18·5°. $p\lambda_{(s)} + \log \Phi_s = 6·35$ in every case.

16·0	48·6	7·76	44·4	+4·2
24·2	53·1	7·62	52·9	+0·2
41·5	63·0	7·46	62·6	+0·4
44·2	64·1	7·44	63·9	+0·2
81·7	73·8	7·23	76·6	−2·8
83·1	76·5	7·24	76·0	+0·5
368·0	107·0	6·74	106·4	+0·6

$x = 60·838 \pm 0·4$. $y = 516·49 \pm 0·4$. $M = 1·5$.

Reduced horse blood (vols. % combined O_2, 22·6), 19°.

3·0	25·2	8·24	26·1	−0·9
22·0	56·2	7·68	55·7	+0·5
39·6	66·3	7·48	66·3	0·0
48·7	69·9	7·41	70·0	−0·1
80·2	78·2	7·24	79·0	−0·8
91·5	83·5	7·20	81·1	+2·4
484·7	112·2	6·59	113·4	−1·2

$x = 52·908 \pm 0·2$. $y = 462·04 \pm 0·2$. $M = 1·3$.

Oxygenated horse blood (vols. % O_2, 25·4; 25, 1; 25, 2), 20°.

32·7	51·5	7·50	51·4	+0·1
75·2	69·4	7·24	69·4	0·0
194·5	91·1	6·92	91·4	−0·3
364·0	106·1	6·71	105·9	+0·2
558·4	116·2	6·56	116·2	0·0

$x = 68·897 \pm 0·06$. $y = 568·17 \pm 0·06$. $M = 0·2$.

Oxygenated ox blood, 38°. Hasselbalch and Warburg. $p\lambda_{(s)} + \log \Phi_s = 6·20$ in every case.

43·5	41·2	7·31	40·7	+0·5
93·6	54·3	7·10	54·9	−0·6
150·9	63·9	6·96	64·5	−0·6
203·9	70·1	6·87	70·4	−0·3
269·4	76·5	6·79	75·8	+0·7
361·8	83·4[1]	6·70	81·9	+2·5
692·9	101·9[1]	6·50	95·3	+6·6

[1] O_2 deficit.

$x = 67·458 \pm 0·03$. $y = 533·81 \pm 0·03$. $M = 0·7$.

Table LIII.

K. A. Hasselbalch's half reduced blood, 38°.   K. A. Hasselbalch [1916, 2][1].

mm. Hg CO_2	Vols. % CO_2 combined	$p_{H'(a)}$ calculated	Vols. % CO_2 combined calculated	Difference
18·5	35·1	7·63	36·6	−1·5
22·4	36·7	7·56	40·6	−3·9
32·7	45·1	7·48	45·2	−0·1
40·9	48·2	7·41	49·3	−1·1
45·7	52·9	7·40	49·9	+3·0
50·8	57·1	7·39	50·4	+6·7
67·0	57·9	7·27	57·4	+0·5
80·7	61·3	7·21	60·9	+0·4
102·2	62·5	7·11	66·6	−4·1

$$x = 57·858 \pm 0·9. \quad y = 477·98 \pm 0·9. \quad M = 3·6.$$

E. J. Warburg's reduced blood, 38°.

13·6	27·7	7·66	25·9	+1·8
39·5	42·8	7·36	45·9	−3·1
74·6	56·5	7·20	56·5	0·0
109·2	64·7	7·09	63·8	+0·9
544·5	99·5	6·56	99·1	+0·4

$$x = 66·567 \pm 0·5. \quad y = 535·79 \pm 0·5. \quad M = 2·1.$$

[1] The experiments were performed with blood taken on three different days and were not specially designed for the determination of the CO_2 combination curve.

Table LIV.

J. S. Haldane's oxygenated blood, 37°.

mm. Hg CO_2	Vols. % CO_2 combined	$p_{H'(a)}$ calculated	Vols. % CO_2 combined calculated	Difference
2·6	14·9	8·15	3·8	+11·1
8·7	25·7	7·84	23·0	+ 2·7
18·9	35·8	7·64	35·5	− 0·3
37·7	48·6	7·46	46·6	+ 2·0
44·1	49·4	7·40	50·4	− 1·0
41·7	48·2	7·42	49·1	− 0·9
56·1	54·5	7·33	54·7	− 0·2
73·3	58·1	7·25	59·7	− 1·6
78·5	60·3	7·23	60·9	− 0·6
105·2	68·1	7·15	65·8	+ 2·3

$$x = 62·016 \pm 0·5. \quad y = 509·32 \pm 0·5. \quad M = 1·6.$$

J. S. Haldane's reduced blood, 37°.

2·15	15·3	8·24?	4·2	+11·1
7·75	30·3	7·94	23·6	+ 6·7
8·1	30·0	7·92	24·9	+ 5·1
17·9	40·4	7·60	39·8	+ 0·6
37·9	53·5	7·48	53·4	+ 0·1
58·2	60·5	7·34	62·5	− 2·0
74·2	65·3	7·27	67·0	− 1·7
78·7	69·9	7·27	67·0	+ 2·9
111·1	75·4	7·15	74·8	+ 0·6

$$x = 64·731 \pm 0·7. \quad y = 537·60 \pm 0·7. \quad M = 1·8.$$

Table LV.

T. R. Parsons' oxygenated blood, 37°.

mm. Hg CO_2	Vols. % CO_2 combined	$p_{H'(u)}$ calculated	Vols. % CO_2 combined calculated	Difference
5·7	21·3	7·93	21·4	− 0·1
10·1	30·0	7·83	26·2	+ 3·8
19·6	34·4	7·59	37·6	− 3·2
33·4	41·5	7·43	45·2	− 3·7
37·4	43·4	7·40	46·6	− 3·2
72·1	59·9	7·25	53·7	+ 6·2

$x = 47\cdot449 \pm 2\cdot4.\ \ y = 397\cdot70 \pm 2\cdot4.\ \ M = 5\cdot4.$

T. R. Parsons' reduced blood, 37°.

5·7	23·8	7·96	23·0	+ 0·8
8·1	27·4	7·87	27·7	− 0·3
10·1	31·5	7·83	29·6	+ 1·9
19·6	39·9	7·64	39·8	+ 0·1
33·4	47·8	7·48	48·1	− 0·3
37·3	47·3	7·42	51·3	− 4·0
55·3	55·1	7·31	57·0	− 1·9
72·1	64·4	7·26	59·6	+ 4·8

$x = 52\cdot319 \pm 0\cdot7.\ \ y = 439\cdot48 \pm 0\cdot7.\ \ M = 3\cdot0.$

Table LVI.

H. W. Davies' blood, 38°.

mm. Hg CO_2	Vols. % CO_2 combined	$p_{H'(u)}$ calculated	Vols. % CO_2 combined calculated	Difference
8·3	26·3	7·88	6·0	+ 20·3
9·02	29·6	7·90	4·3	+ 25·3
11·2	28·7	7·78	14·5	+ 14·2
19·3	35·5	7·63	27·2	+ 8·3
29·8	41·4	7·50	38·2	+ 3·2
38·9	46·4	7·44	43·3	+ 3·1
39·0	45·5	7·42	44·9	+ 0·6
39·1	43·1	7·39	47·5	− 4·4
46·6	47·3	7·35	50·9	− 3·6
51·3	53·2	7·36	50·0	+ 3·2
64·9	57·4	7·29	55·9	+ 1·5
68·5	60·0	7·28	56·8	+ 3·2
76·7	62·0	7·25	59·3	+ 2·7
89·9	65·7	7·20	63·5	+ 2·2
102·9	66·5	7·15	67·8	− 1·3
202·0	81·8	6·93	86·3	− 4·5
215·0	82·6	6·91	88·1	− 5·5
411·0	97·3	6·69	106·7	− 9·4
455·0	97·6	6·65	110·1	− 12·5
497·0	95·7	6·60	114·4	− 18·7

$x = 84\cdot657 \pm 0\cdot4.\ \ y = 673\cdot09 \pm 0\cdot4.\ \ M = 4\cdot8.$

Table LVII.

J. Joffe's defibrinated oxygenated blood, 4. viii.–28. xi. 1919.

Joffe and Poulton, Tables V and VII.

mm. Hg CO_2	Vols. % CO_2 combined	$p_{H'(a)}$ calculated	Vols. % CO_2 combined calculated	Difference
4·0	13·7	7·93	5·4	+ 8·3
5·0	16·8	7·92	6·1	+10·7
12·9	23·8	7·65	24·0	− 0·2
19·7	29·4	7·54	31·3	− 1·9
24·9	36·0	7·53	31·9	+ 4·1
34·6	40·0	7·42	39·3	+ 0·7
44·4	44·8	7·36	43·2	+ 1·6
50·2	44·1	7·30	47·2	− 3·1
58·0	47·6	7·26	49·9	− 2·3
65·3	52·6	7·25	50·5	+ 2·1
67·8	53·9	7·25	50·5	+ 3·4
82·2	56·0	7·17	55·8	+ 0·2
110·3	63·3	7·10	60·5	+ 2·8
149·0	62·6	6·96	69·8	− 7·2
156·0	67·1	6·97	69·1	− 2·0
157·0	70·3	6·98	68·4	+ 1·9
180·0	68·5	6·91	73·1	− 4·6
376·3	90·0	6·69	87·7	+ 2·3
610·0	100·3	6·53	98·3	+ 2·0

$x = 66·345 \pm 0·16.$ $y = 531·52 \pm 0·16.$ M = 3·2.

J. Joffe's defibrinated reduced blood, 4. viii.–9. x. 1920.

Joffe and Poulton, Table VII.

9·9	27·9	7·81	20·1	+7·8
15·0	34·4	7·72	26·3	+8·1
19·4	35·3	7·61	34·0	+1·3
27·6	40·1	7·50	41·6	−1·5
30·9	44·1	7·50	41·6	+2·5
36·8	47·9	7·45	45·1	+2·8
41·7	46·0	7·37	50·6	−4·6
47·0	50·4	7·36	51·3	−0·9
50·5	49·8	7·33	53·4	−3·6
55·7	52·9	7·31	54·8	−1·9
70·1	55·6	7·23	60·4	−4·8
76·8	62·1	7·23	60·4	+1·7
77·7	64·0	7·24	59·7	+4·3
92·7	67·6	7·19	63·1	+4·5

$x = 69·475 \pm 0·7.$ $y = 562·65 \pm 0·7.$ M = 3·6.

that the most alkaline values are too high but the acid values are correct. Parsons' [1917] blood gives values scattered irregularly about a straight line (Table LV and Fig. 19).

The constant x, which is the tangent to the angle the straight line makes with the axis of the abscissa measured in the first or third quadrant, is for

Davies'	oxygenated blood	...	84·7
Haldane's	,, ,,	... ...	62·0
,,	reduced ,,	... ...	64·7
Hasselbalch's half oxygenated blood			57·9
Joffe's	oxygenated blood	...	66·3
,,	reduced ,,		69·5
Parsons'	oxygenated ,,		47·4
,,	reduced ,,		52·3
Warburg's	,,		66·6

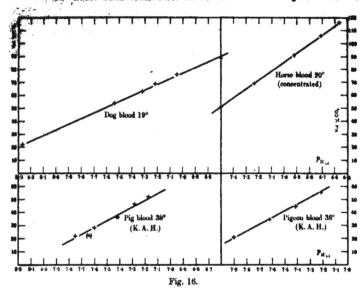

Fig. 16.

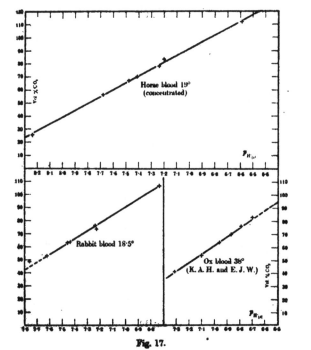

Fig. 17.

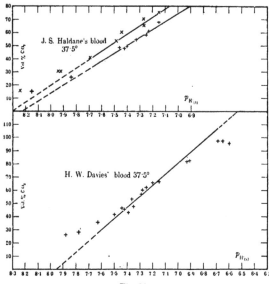

Fig. 18.

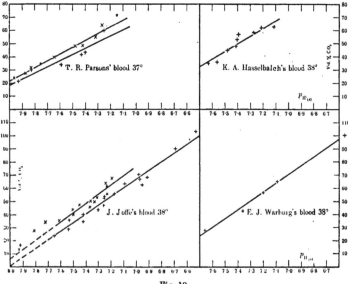

Fig. 19.

It will be observed in the first place that the constant varies from blood to blood. The greater part of the difference is undoubtedly due to variations in the amount of haemoglobin as Hasselbalch [1916, 2] has pointed out and as is evident from the experiments of Schmidt [1867], Zuntz [1867, 1868], Setschenow [1879], Jaquet [1892], Bohr [1905] and many others, but the examination of numbers of curves has shown that this cannot be the sole determining factor, as Peters, Jr. and Barr [1921] have also indicated from a rather different point of view (Peters and Barr's curves are constructed as P_{CO_2},—vols. % total CO_2 curves). It will also be seen that x is a little greater in reduced blood than in the corresponding oxygenated blood. The difference is certainly a real one and is of almost the same magnitude in Haldane's, Joffe's and Parsons' blood. L. J. Henderson [1920, 1] has recently demonstrated the same thing graphically and he has discussed the cause in a paper dealing with the influence of oxygen on the combined CO_2 of the blood. This paper seems to be of very considerable interest although I do not agree with the author in all his views but as I have not yet succeeded in quite clearing up several of the problems I shall only refer those interested to the original article.

By plotting a p_H·—vols. % combined CO_2 curve it is possible to form a good idea of the reliability of the curve. I have done this for a number of Haggard and Henderson's experiments and for some of Peters, Barr and Rule's also. It appears especially in the case of Haggard and Henderson's, that the technique is not so good as the authors believe and it seems to me problematical whether the accuracy is sufficient to draw any conclusion about the tension with which the blood is in equilibrium from the total amount of CO_2 in the blood, as these authors do. Haggard and Henderson's [1920, 1] discovery that there is no difference between the combined CO_2 in oxygenated and reduced oxalate blood, and that the combined CO_2 of the blood decreases irreversibly by blowing air through it vigorously, is presumably due to experimental error.

I must now discuss some experiments published in the last few years by H. Straub and Klothilde Meier [1918–1920]. They have unfortunately gained some recognition, e.g. by L. Michaelis[1] and by Parsons. But their results are so surprising, that Joffe and Poulton [1920], Peters, Barr and Rule [1921] for example have announced that they are sceptical. Straub and Meier's chief claim is that they have shown the combined CO_2 in the blood does not increase before a reaction of about p_H· 6·70 is reached when suddenly (in an interval of p_H· 0·01) an amount of CO_2 as great as the oxygen capacity is combined by the blood and then all further combination ceases. In haemo-

---

[1] Michaelis and Airila have recently [1921] shown by cataphoresis experiments that the change of haemoglobin varies uniformly with the reaction over a very large range of reaction and is thus greatly at variance with Straub and Meier's assertions about the ionisation of haemoglobin. ˙ The conclusion may be drawn from these interesting experiments that the combination of CO_2 in haemoglobin solutions can be represented by a curve which, if not a straight line, is a very close approximation thereto, in a p_H–vols. % diagram over a very considerable range of reaction.

lysed blood the sudden change is at p_H· 7·00 and is of the same extent. They believe serum is quite unable to combine with CO_2 in quantities varying with the pressure. Straub and Meier claim to have shown that various substances shift this sudden change backwards and forwards, and they have formed a theory which for extravagance is only surpassed by the obvious imperfection of their technique. With regard to these authors I cannot suppress my surprise that they really seem to be quite ignorant of the older, and as far as the present question is concerned, of the more recent literature on the combined CO_2 of the blood and it is to me completely incomprehensible that they have not paid more attention to their technique.

As an illustration of Straub and Meier's technique I will give one of their experiments [1920, 2, p. 250].

Experiment 9. 38°.

mm. Hg CO_2	Vols. % CO_2 total
16·8	15·1
50·2	19·6
72·6	30·6
153·3	38·7
41·4	35·9
64·0	34·9

The only possible explanation is, as the authors themselves have realised, that equilibrium has not been reached again at the low CO_2 tension. Instead of examining their technique they employ the experiment in propounding one of their extraordinary theories, and they fail to see that p_H· cannot be calculated by Henderson and Hasselbalch's equation if equilibrium in the system has not been attained. The following similar experiment shows that equilibrium can easily be attained.

Ox blood slightly concentrated by centrifuging. Combined O_2, 22·8 vols. %. 70 cc. blood in the large saturator.

21°.		mm. Hg CO_2	Vols. % CO_2 total		
	First	29·0	47·9 after 30 mins.		
	Then	740	185·5	,, 30	,,
	,,	,,	184·3	,, 65	,,
	Afterwards	29·0	49·8	,, 30	,,
	,,	,,	47·8	,, 60	,,
	,,	,,	47·7	,, 120	,,

In Table LVIII and in Figs. 20 and 21 experiments and algebraic smoothing will be found referring to different sera. These curves also appear to be straight lines although a number of points at high CO_2 tensions fall a little below the curve. It should however be noted that deviations from a straight line course are more difficult to detect than in the case of the blood curves on account of the relatively small value of x.

That serum and plasma really form a dissociable combination with CO_2 is raised beyond all doubt and has already been demonstrated by Setschenow [1879] and by Jaquet [1892], and later by Hasselbalch [1916, 2] and Hasselbalch and Warburg [1918] among others. It is also quite certain that varying carbonic acid is combined as bicarbonate, as is shown by Loewy and Zuntz [1894, 1] and by Gürber's [1895, 2] old diffusion experiments. Gürber even claimed to have proved that CO_2 quantitatively was present as bicarbonate, a conclusion which is strongly supported by the determinations of $p\lambda_{(s)}$ in chapter VI of this work.

Table LVIII.

Ox serum 38°. Hasselbalch and Warburg. I.

mm. Hg CO_2	Vols. % CO_2 combined	$p_{H'(s)}$ calculated	Vols. % CO_2 combined calculated	Difference
41·3	54·4	7·42	54·1	+0·3
74·5	58·5	7·20	58·6	-0·1
140·5	62·2	6·95	63·7	-1·5
203·0	67·5	·6·82	66·4	+1·1
271·5	68·9	6·71	68·6	+0·3
710·0	76·3	6·33	76·4	-0·1

$x = 20.507 \pm 0.2.$ $y = 206.23 \pm 0.2.$ $M = 1.0.$

Ox serum 19° (serum B). Hasselbalch and Warburg. II.

19·5	61·6	7·73	Unsuitable for calculation	
79·8	69·6	7·17	(see Fig. 20)	
182·0	73·0	6·82		
378·3	73·7	6·52		

Horse serum 38°. III.

23·5	63·8	7·73	63·9	-0·1
24·6	63·4	7·71	64·1	-0·7
45·6	67·5	7·47	67·0	+0·5
104·6	71·8	7·14	71·0	+0·8
430·3	77·3	6·57	77·8	-0·5

$x = 12.016 \pm 0.2.$ $y = 156.76 \pm 0.2.$ $M = 0.7.$

Horse serum 38°. IV.

10·4	58·1	8·05	57·9	+0·2
35·2	63·3	7·56	63·6	-0·3
106·7	69·2	7·11	68·9	+0·3
130·7	69·7	7·03	69·8	-0·1
535·6	74·9	6·42	76·9	-2·0

$x = 11.644 \pm 0.1.$ $y = 151.67 \pm 0.1.$ $M = 0.3.$

Ox serum 38°. V.

22·7	62·6	7·74	64·2	-1·6
59·5	69·0	7·37	69·4	-0·4
103·7	74·7	7·16	72·4	+2·3
510·7	80·1	6·50	81·6	-1·5

$x = 14.012 \pm 0.6.$ $y = 172.69 \pm 0.6.$ $M = 2.3.$

Human serum 38°. VI.

15·2	57·2	7·88	58·7	-1·5
18·3	60·7	7·82	59·5	+1·2
84·5	68·7	7·21	68·2	+0·5
562·1	78·7	6·45	79·0	-0·3

$x = 14.237 \pm 0.3.$ $y = 170.83 \pm 0.3.$ $M = 1.4.$

Horse serum 19°. VII.

12·7	55·8	7·86	56·5	-0·7
18·7	58·2	7·71	57·9	+0·3
41·2	62·6	7·40	60·9	+1·7
119·6	64·4	6·95	65·2	-0·8
512·5	71·1	6·36	70·9	+0·2

$x = 9.5901 \pm 0.2.$ $y = 131.87 \pm 0.2.$ $M = 1.2.$

Table LVIII (*continued*)

Horse serum (19·5°) 20°. VIII.

mm. Hg CO_2	Vols. % CO_2 combined	$p_{H'(t)}$ calculated	Vols. % CO_2 combined calculated	Difference
13·1	55·6	7·86	56·7 ·	−1·1
34·1	61·1	7·48	60·5	+0·6
83·1	65·3	7·13	64·0	+1·3
573·7	71·4	6·32	72·2 ·	−0·8

$x = 10·025 \pm 0·3.$ $y = 135·51 \pm 0·3.$ M = 1·4.

Ox serum 19°. IX.

16·0	58·5	7·78	59·3	−0·8
43·1	64·7	7·40	63·9	+0·8
118·6	69·4	6·99	68·9	+0·5
545·9	75·9	6·37.	76·4	−0·5

$x = 12·127 \pm 0·2.$ $y = 153·65 \pm 0·2.$ M = 0·9.

Ox serum 21°. X.

16·1	61·8	7·81	61·4	+0·4
40·5	66·0	7·44	66·0	0·0
40·6	65·3	7·44	66·0	−0·7
70·2	68·5	7·22	68·8	−0·3
122·9	71·9	7·00	71·5	+0·4
595·0	75·4	6·33	79·8	−4·4

$x = 12·412 \pm 0·2.$ $y = 158·38 \pm 0·2.$ M = 0·7.

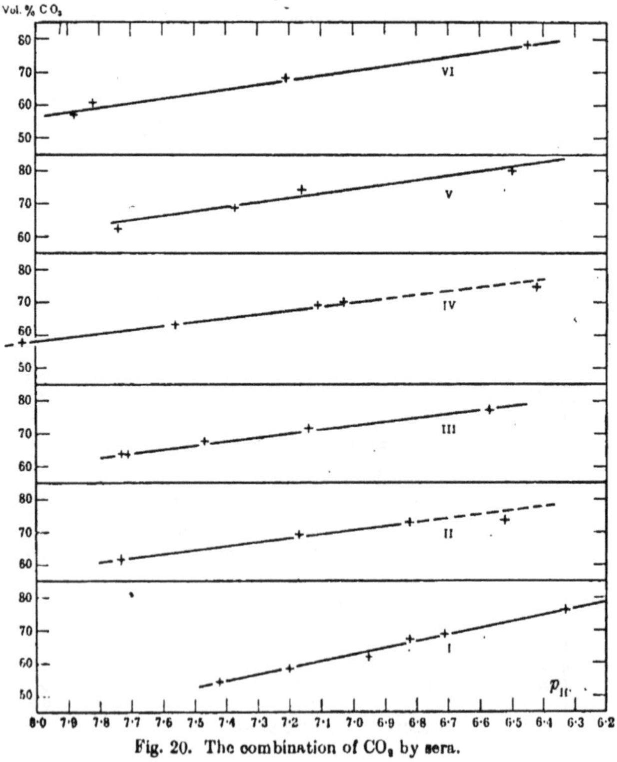

Fig. 20. The combination of CO_2 by sera.

In a paper on the neutrality of the blood W. M. Bayliss [1919] has recently advanced the opinion that the proteins of serum and plasma could not function as ampholytes at the reactions obtaining in my experiments and in the above mentioned papers. I do not agree with Bayliss on this point and as the subject is of prime importance for the problems we are discussing I shall attempt briefly to dispute his statements.

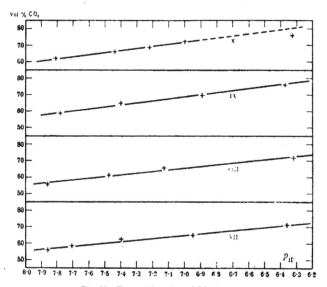

Fig. 21. The combination of CO_2 by sera.

Bayliss[1] asserts that serum does not combine with CO_2 in a reversible manner at the usual CO_2 tensions, supporting this by an experiment without doubt wrongly interpreted. It is obvious from the experiment that the variable combination—which Bayliss considers unimportant—is of the same order as that found by other authors. Bayliss further maintains that proteins do not act as buffers at reactions in the neighbourhood of the neutral point. This result is absolutely opposed to the finding of L. J. Henderson [1909–1910] and T. B. Robertson [1908–1910, 1912] with serum globulin and to the above mentioned experiments. The reason he has come to this erroneous conclusion must be that he has overrated the accuracy with which the reaction of plasma can be determined colorimetrically. In the same paper W. M. Bayliss has concluded by analogy that *haemoglobin* cannot act as a buffer. This analogy is however unjustifiable, quite apart from whether the experiments with plasma are correct or not, because the ampholytic character of the various

---

[1] Mukai (*J. Physiol.* 1921, 55, 356) has quite recently, in Bayliss' laboratory, found that serum combines with variable amounts of CO_2 of a similar order to those found by earlier authors,

proteins is very different as is undoubtedly shown by the work of Robertson [1912] and his collaborators, and by the studies of Pauli [1920] and his co-workers.

In Table LIX and Fig. 22 the calculations from Campbell and Poulton's [1920] experiments with dialysed haemoglobin solution at 38° will be found. The corresponding curves are here also straight lines allowing for experimental error. As $2226C_I$ was 91·5 vols. % in the experiments the isoelectric point is 7·14. This is, as Campbell and Poulton [1920] themselves have remarked, considerably more alkaline than Michaelis and Takahashi [1910] found at room temperature the isoelectric point according to these observers being p_H· (Bjerrum) 6·79. Now it can be shown from a glance at the CO_2 combination curves—as Hasselbalch [1916, 2] has pointed out—that the isoelectric point for haemoglobin must be more acid at body temperature than at room temperature (the difference is about 0·30), so that the difference between Campbell and Poulton's and Michaelis' results is real[1].

Table LIX. Dialysed solution of haemoglobin (Haldane 52) with $0.041n$ $NaHCO_3$, 38°; $p\lambda_{(m)} = 6.23$. Calculated after J. M. H. Campbell and E. P. Poulton.

mm. Hg CO_2	Vols. % CO_2 combined	p_{H}· calculated	Vols. % CO_2 combined calculated	Difference
4·8	52·3	8·40	36·2	+ 16·1
4·9	52·1	8·39	36·6	+ 15·5
11·0	57·5	8·08	50·3	+ 7·2
21·9	63·3	7·83	61·2	+ 2·1
40·7	72·9	7·62	70·4	+ 2·5
43·6	76·5	7·61	70·9	+ 5·6
50·7	75·5	7·54	70·0	+ 5·5
52·5	70·2	7·47	77·0	− 6·8
78·1	74·1	7·34	82·8	− 8·7
85·5	79·8	7·34	82·8	− 3·0
90·0	77·1	7·30	84·5	− 7·4
110·5	92·3	7·29	85·0	+ 7·3
119·8	86·3	7·22	88·0	− 1·7
151·4	93·5	7·16	90·7	+ 2·8
221·9	99·1	7·02	96·8	+ 2·3
226·7	97·8	7·00	97·7	+ 0·1
310·4	101·9	6·88	103·0	− 1·1
338·0	107·0	6·87	103·4	+ 3·6
435·3	110·0	6·77	107·8	+ 2·2
596·4	112·2	6·64	113·5	− 1·3

$x = 43.959 \pm 0.2$. $y = 405.41 \pm 0.2$. M = 4·5.

Some experiments I have performed give an indication of the position of the isoelectric point. According to Hardy the agglutination optimum of

[1] T. R. Parsons and Winifred Parsons (Biochem. Zeitschr. 1921, 126, 108) have recently pointed out that Campbell and Poulton's haemoglobin solutions must have contained inorganic salts because when they incinerated the haemoglobin they found more ash than could have come from the iron alone. Calculating the amount of HCO'_3 which would be balanced by metallic ions on the assumption the excess of ash was K_2CO_3 we get 30 vols. % and the isoelectric point is therefore about 6·5, which agrees with Michaelis' determination when we take the difference of temperature into account.

a colloid is at the isoelectric point. The crystallising out of haemoglobin at high CO_2 tensions (after saponin haemolysis) mentioned in chapter IV never began at reactions more alkaline than $p_H \cdot$ 7·10 at room temperature under the conditions of the experiment, but the ease with which crystallisation was set going increased until $p_H \cdot$ 6·80 was reached which supports Michaelis' result. Should further experiments confirm the difference between the iso-electric point estimated from the combined CO_2 and from the other method, it will indicate that some CO_2 is adsorbed or combined in a complex manner with the haemoglobin.

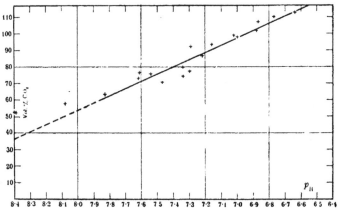

Fig. 22. CO_2 combined by dialysed haemoglobin (with the addition of 0·041n NaHCO$_3$). (J. M. H. C. & E. P. P.)

With the help of the relations developed in this chapter, which found their simplest expression in (173) and (176), it will be possible to investigate whether all the reversibly combined CO_2 is present as bicarbonate. Before reporting the experiments I have made in this connection I would draw attention to the fact that an error in $p\lambda_{(s)}$ or $p\lambda_{(m)}$ or in the correction taken from Fig. 10 is without significance here, because it will influence the estimation of the hydrogen ion exponent in the original and in the curve displaced by the addition of alkali to the same extent and thus the relative positions of the curves will not be altered.

The experiments were carried out as follows: blood or saponin haemolysate ($\frac{1}{2}$ % saponin) which was slightly concentrated by pipetting off serum was sloped in a 100 cc. measuring flask so that it filled about half of it. With a pipette either 5 cc. $n/2$ NaCl solution or 5 cc. $n/2$ NaHCO$_3$ solution was then added during shaking, the titre of which was determined beforehand by the exhaust pump. Two drops of octyl alcohol were next added to prevent frothing after which the flask was filled up with blood (or haemolysate).

With the mixtures thus prepared CO_2 combination curves were constructed
with the help of the large saturator. The calculations were made as before,
and the results are given in Table LX and in Figs. 23, 24 and 25.

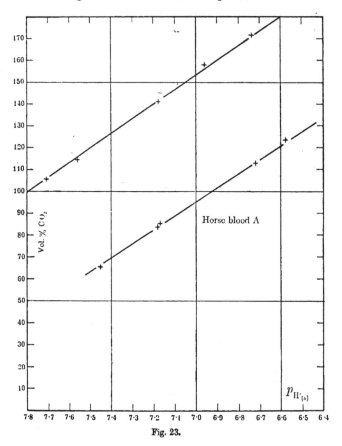

Fig. 23.

Let us first compare the constant x.

		Lower curve	Upper curve	Difference
Horse blood A		$66{\cdot}9 \pm 0{\cdot}3$	$69{\cdot}2 \pm 0{\cdot}3$	$2{\cdot}3 \pm 0{\cdot}4$
Horse blood haemolysate B		$87{\cdot}0 \pm 0{\cdot}6$	$89{\cdot}2 \pm 0{\cdot}5$	$2{\cdot}2 \pm 0{\cdot}7$
Ox ,, ,, C		$57{\cdot}9 \pm 1{\cdot}1$	$57{\cdot}8 \pm 0{\cdot}9$	$-0{\cdot}1 \pm 1{\cdot}4$
Horse ,, ,, D		$81{\cdot}3 \pm 0{\cdot}1$	$83{\cdot}3 \pm 0{\cdot}3$	$2{\cdot}0 \pm 0{\cdot}3$
,, ,, ,, E		$64{\cdot}0 \pm 0{\cdot}1$	$70{\cdot}3 \pm 0{\cdot}3$	$6{\cdot}3 \pm 0{\cdot}3$

Table LX.

Defibrinated horse blood A. Curve I. $\Psi = 11$.

Temp.	mm. Hg CO₂	mm. Hg O₂	Vols. % combined CO₂	Vols. % combined O₂	pH'(s) calculated	Vols. % CO₂ combined calculated	Difference vols. %
17	43.5	147.0	65.6	26.5	7.45	65.6	0.0
"	99.8	135.5	83.8	26.4	7.18	83.7	+ 0.1
"	104.6	—	85.4	26.4	7.17	84.3	+ 1.1
"	358.8	—	112.7	—	6.72	114.4	- 1.7
"	537.8	—	123.5	—	6.59	123.1	+ 0.4

Mean 26.4

$x = 66.883 \pm 0.3. \quad y = 563.87 \pm 0.3. \quad M = 1.2.$ Theoretical distance 57.0.

Curve H.

17	40.4	147.6	105.6	26.9	7.71	104.9	+ 0.7
"	59.9	143.7	114.5	26.5	7.56	115.3	- 0.8
"	168.0	120.5	140.9	26.4	7.18	141.6	- 0.7
"	303.8	—	158.1	—	6.96	156.8	+ 1.3
"	522.4	—	171.5	—	6.74	172.0	- 0.5

Mean 26.6

$x = 69.172 \pm 0.3. \quad y = 638.24 \pm 0.3. \quad M = 1.1.$

Saponin haemolysed horse blood B. Curve I. $\Psi = 11$.

18	14.4	—	33.6	26.2	7.60	31.3	+ 2.3
"	45.2	—	54.2	25.9	7.31	56.5	- 2.3
"	84.8	—	69.3	—	7.15	70.4	- 1.1
"	359.3	—	107.4	—	6.71	108.7	- 1.3
"	536.8	—	121.4	—	6.59	119.2	+ 2.2
"	137.4	—	75.0	—	6.97	86.1	- 11.1[1]
"	359.3	—	92.6	—	6.64	114.8	- 22.2[1]

[1] Crystallisation.

$x = 87.033 \pm 0.6. \quad y = 692.70 \pm 0.6. \quad M = 2.5.$ Theoretical distance 53.0.

Curve II.

18	20.3	—	69.4	—	7.77	68.3	+ 1.1
"	45.6	—	89.0	—	7.52	90.6	- 1.6
"	70.6	—	102.5	—	7.40	101.3	+ 1.2
"	141.6	—	120.8	—	7.16	122.7	- 1.9
"	360.3	—	150.7	—	6.86	149.5	+ 1.2

$x = 89.176 \pm 0.5. \quad y = 761.21 \pm 0.5. \quad M = 1.8.$

Saponin haemolysed ox blood C. Curve I. $\Psi = 8$.

19	19.4	—	38.7	20.6	7.54	37.7	+ 1.0
"	60.2	—	56.7	—	7.21	56.8	- 0.1
"	394.1	—	93.6	—	6.62	91.0	+ 2.6
"	166.9	—	73.6	—	6.86	77.1	- 3.5

$x = 57.899 \pm 1.1. \quad y = 474.27 \pm 1.1. \quad M = 3.2.$ Theoretical distance 54.5.

Curve IL

19	44.4	—	91.1	20.3	7.55	91.6	- 0.5
"	398.5	—	135.4	—	6.77	136.6	- 1.2
"	164.9	—	120.1	—	7.10	117.6	+ 2.5
"	82.3	—	102.9	—	7.34	103.7	- 0.8

$x = 57.772 \pm 0.9. \quad y = 527.75 \pm 0.9. \quad M = 2.1.$

Saponin haemolysed horse blood D. Curve I. $\Psi = 11$.

18	18.3	133.9	36.7	25.0	7.53	36.4	+ 0.3
18	18.3	133.9	36.5	24.4	7.53	36.4	+ 0.1
17	52.9	—	57.8	—	7.26	58.3	- 0.5
"	91.9	—	70.3	—	7.11	70.5	- 0.2
18	531.8	—	114.6	—	6.57	114.4	+ 0.2
17	153.4	—	81.5	—	6.95	83.5	- 2.0[1]
"	348.0	—	97.4	—	6.67	106.3	- 8.9[1]

[1] Crystallisation.

$x = 81.260 \pm 0.1. \quad y = 648.25 \pm 0.1. \quad M = 0.5.$ Theoretical distance 54.5.

Table LX (*continued*)

Temp.	mm. Hg		Vols. % combined		$p_{H^{(s)}}$ calculated	Vols. % CO_2 combined calculated	Difference Vols. %
	CO_2	O_2	CO_2	O_2			

Curve II.

18	532·4	—	157·0	—	6·70	157·3	− 0·3
17	16·8	—	66·9	—	7·82	64·0	+ 2·9
,,	52·6	—	91·9	—	7·47	93·2	− 1·3
,,	72·8	—	100·0	—	7·36	102·3	− 2·3
,,	147·5	—	120·4	—	7·14	120·7	− 0·3
,,	515·9	—	158·3	—	6·71	156·5	+ 1·8
,,	320·0	—	141·8	—··	6·87	143·2	− 1·4[1]

[1] Crystallisation.

$$x = 83·322 \pm 0·3. \quad y = 715·57 \pm 0·3. \quad M = 1·6.$$

Saponin haemolysed horse blood E. Curve I. $\Psi = 11$.

18	19·8	—	41·7	25·7	7·55	41·7	0·0
,,	61·7	—	61·6	—	7·23	62·2	− 0·6
17	80·0	—	68·0	—	7·15	67·3	+ 0·7
,,	155·6	—	80·9	—	6·94	80·8	+ 0·1
··	342·3	—	97·1	—	6·68	97·4	− 0·3
,,	545·9	—	107·8	—	6·52	107·7	+ 0·1

$$x = 64·052 \pm 0·1. \quad y = 525·29 \pm 0·1. \quad M = 0·5. \quad \text{Theoretical distance } 54·5.$$

Curve II.

18	22·1	—	78·6	26·0	7·78	77·1	+ 1·5
,,	101·9	—	111·3	—	7·27	113·0	− 1·7
,,	185·9	—	126·1	—	7·06	127·8	− 1·7
17	301·3	—	140·0	—	6·89	139·8	+ 0·2
,,	514·8	—	153·3	—	6·70	153·1	+ 0·2
,,	512·3	—	154·7	26·1	6·70	153·1	+ 1·6

$$x = 70·335 \pm 0·3. \quad y = 624·37 \pm 0·3. \quad M = 1·6.$$

Apart from ox blood haemolysate C, which appears to be subject to a far greater fortuitous error than the other members of the series, the upper curves all exhibit a rather steeper ascent than the lower ones. Although we should be very cautious, as previously explained, in drawing definite conclusions from the size of the mean error about an actual discrepancy between the constants it is extremely probable the upper curve really has a somewhat steeper course than the lower. In accordance with this the equation for the combination of CO_2 in the blood and haemolysate should not be

$$(173) \quad B = (C_I - C_V f (a_{II'})) \; 2226,$$

but $$(176) \quad B = (C_I - C_V f (a_{II'}) - y f_1 (P_{CO_2})) \; 2226,$$

and therefore some CO_2 should be combined in blood in some other form than bicarbonate. There is however an alternative explanation which seems to me more probable. As will be remembered it was explained on p. 280 that the difference between the curves at no place could be less than $2226b$, which is the value entered in the tables as the "theoretical distance." We will now investigate how far this is correct.

			B		Difference		
			At p_H 7·50	At p_H 6·70	At p_H 7·50	At p_H 6·70	Theoretical distance
Horse blood A.		Upper curve	119·5	174·8	56·5	60·5	57·0
		Lower ,,	63·0	114·3			
,, ,, haemolysate B.		Upper ,,	92·4	163·8	52·4	55·2	53·0
		Lower ,,	40·0	108·6			
Ox ,, ,, C.		Upper ,,	94·3	140·7	54·3	54·3	54·5
		Lower ,,	40·0	86·4			
Horse ,, ,, D.		Upper ,,	90·7	157·3	51·9	53·5	54·5
		Lower ,,	38·8	103·8			
,, ,, ,, E.		Upper ,,	96·9	153·1	52·0	57·0	54·5
		Lower ,,	44·9	96·1			

There seems therefore to be an inclination to obtain too small differences in the most alkaline reactions. In the case of the haemolysates D and E this seems to exceed the experimental error as the latter may be estimated as below 2 vols. % of the difference between the curves, determined by five points or more. (The standard deviation for all the determinations with the

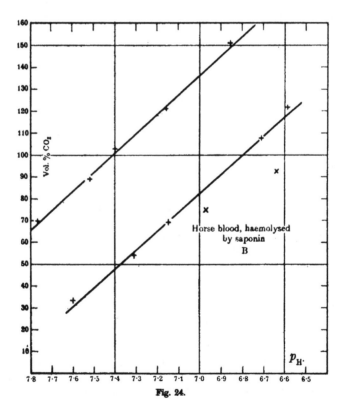

Fig. 24.

exception of the horse blood haemolysate D is 1·2 vols. %.) A possible explanation of the phenomenon is that on the addition of bicarbonate the blood was very little changed so that before the bicarbonate was completely mixed with the blood a slight decomposition of the proteins took place with the formation of new acid radicles.

The result of these experiments is therefore that it is probable that CO_2 is combined in the blood only as carbonic acid, that is to say as bicarbonate ion, but that the possibility cannot be entirely excluded that small quantities may be combined in other ways.

In the haemolysates B and D haemoglobin crystallised out in the acid reactions. It will be noticed that the combination of CO_2 decreases when crystallisation takes place but it must be remembered we cannot be sure of getting a homogeneous sample of the whole system when the crystal phase is in process of formation. Setschenow [1879] has observed a similar decrease of the combined CO_2 of horse blood when the haemoglobin crystallised out.

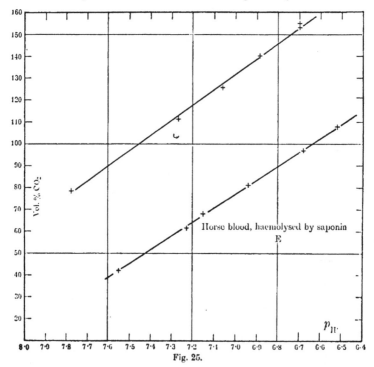

Fig. 25.

In conclusion I will report some experiments carried out in a similar manner with histidine hydrochloride with which Prof. Henriques kindly supplied me, which make it probable the CO_2 combination curve of histidine can be shifted in a p_H·–vols. % CO_2 diagram in a similar way to that of blood. The preparation was impure as it only contained 17·1 % N_2 while the theoretical amount for histidine mono-hydrochloride is 26·5 % and for histidine di-hydrochloride 21·5 %. A 5 % solution was 0·439n as regards chlorine which corresponds to a mixture of mono- and di-hydrochloride. It can however be concluded from the course of the CO_2 combination curve that chloride must also have been present in some other form than histidine hydrochloride. The solutions were prepared in such a way that for the lower curve a 5 % solution of histidine hydrochloride was mixed with an equal part of a sodium bicarbonate solution which combined with 995·9 vols. % just before each experiment, while for the upper curve a bicarbonate solution which combined with 1106·5 vols. % was used. The theoretical distance between the curves should be 55·3 vols. %. The experiments will be found in Table LXI and in Fig. 26. It will be observed the curves are very nearly parallel with an intervening distance of 53·5 to 57·5 vols. % which may be regarded as sufficiently accurate for curves of this type. It should be remarked however that the distance between the curves increases a little with increasing apparent hydrogen ion activity and that the distance in the most alkaline reactions is about 2 vols. % less than the theoretically smallest possible distance so that there must either be small experimental errors or the theory proposed must be incomplete in one way or another.

Table LXI. 2·5 % histidine hydrochloride $p_{K'_I} = 6·300$ at 18°
and 6·305 at 17°. $Y = 8$.

	Curve I.				Curve II.		
Temp.	mm. Hg CO_2	Vols. % CO_2 combined	p_H· calculated	Temp.	mm. Hg CO_2	Vols. % CO_2 combined	p_H· calculated
17°	22·5	35·7	7·41	18°	23·5	76·1	7·72
,,	40·5	44·3	7·24	,,	48·8	86·8	7·46
,,	83·4	57·7	7·05	,,	113·2	101·6	7·17
,,	188·5	78·7	6·83	,,	207·7	119·0	6·97
,,	392·3	103·0	6·63	,,	390·4	141·7	6·77
,,	731·0	127·5	6·45	,,	730·0	167·5	6·50

Theoretical distance 55·3 vols. %.

In the above discussion an assumption has been made which is not really tenable, namely that a complex combination between the CO_2 and proteins or histidine does not drive CO_2 out of the bicarbonate. If such a combination, for example a carbamino-acid, is in itself an acid it will in virtue of its cations be able to drive out CO_2. Equation (176) should therefore—if this possibility be allowed—have the form

$$B = (C_I - C_V f(a_H·) - f_1(P_{CO_2}) - f_2(P_{CO_2}, a_H·)) 2226,$$

and the above experiment would be explained without the assumption of an experimental error. Further experiment can alone determine whether this explanation accords with fact but the above experiment proves at any rate that only very small amounts of carbamino-acid can be formed at the reactions investigated.

As the final result of the investigations reported in chapters VIII, IX, X and XI it is clear that the CO_2 of the blood is exclusively or almost exclusively combined as bicarbonate and that the variable amount combined is due to the presence in the blood of electrolytes the dissociation of which varies with the reaction. Such electrolytes are chiefly proteins. On account of the relatively small concentration of the serum proteins their effect will be subordinate to that of the proteins of the blood corpuscles. Haemoglobin is the most active of the blood corpuscle proteins judging from experiments of Setschenow [1879], Bohr [1905], Hasselbalch [1916, 2] and Campbell and Poulton [1920]. It is impossible to estimate the activity of haemoglobin in this respect but it is not improbable that the other proteins and lecithin (Setschenow) are also active. To a limited extent the phosphates will also function in a similar manner.

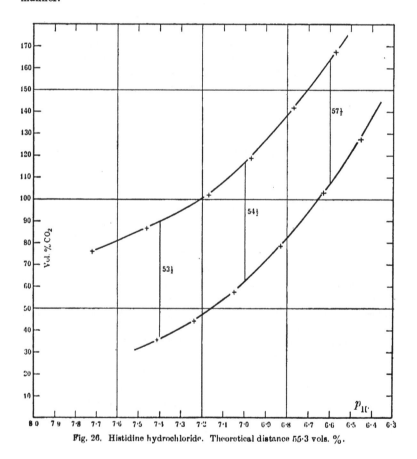

Fig. 26. Histidine hydrochloride. Theoretical distance 55·3 vols. %.

RÉSUMÉ.

I. The theory of the combination of CO_2 in a solution which contains both electrolytes the dissociation of which does and does not vary with the reaction is elaborated.

II. This theory has been tested on phosphate solutions and found to be satisfactory.

III. An empirical relation on a rational basis has been proposed for the combination of CO_2 in serum, haemoglobin solutions and blood.

IV. It has been shown it is only possible for a trifling amount of CO_2 to be adsorbed or bound in a complex manner (as carbamino-acid) in blood or haemolysate.

V. A similar proof has been produced for the combination of CO_2 in a histidine solution.

CHAPTER XII

THE FACTORS WHICH DETERMINE THE PARTITION OF PERMEATING IONS BETWEEN THE BLOOD CORPUSCLES AND THE SERUM, THE VOLUME OF THE BLOOD CORPUSCLES DEPENDENT UPON THE REACTION, AS WELL AS THE POTENTIAL ON THEIR SURFACES

As repeatedly stated A. Schmidt [1867] and N. Zuntz [1867] discovered independently of one another that the amount of alkali in serum increased when the blood was treated with high tensions of CO_2. Zuntz explained this phenomenon by assuming that the effect of CO_2 was to split off sodium from the sodium-protein compounds of the blood cells and that some of the sodium diffused out of the blood cells in the form of sodium bicarbonate until the sodium bicarbonate concentration was the same in blood cells and serum. When we remember the state of physical, and especially physiological, chemistry at that time the absolute genius displayed in Zuntz's hypothesis excites the greatest admiration.

The converse of this bicarbonate diffusion was discovered in 1874 without its importance in this connection being noticed. Hermann Nasse[1], who at that time was an old investigator and who had gained a considerable reputation in haematology, discovered in 1874 that the chlorine in serum decreased when blood was treated with high CO_2 tensions. At the same time he found that the blood cells swelled under CO_2 treatment, taking up water from the serum. Although he again reported his discovery in *Pflüger's Archiv* in 1878 his name has quite disappeared from textbooks on physico-chemical biology because the honour of discovering the phenomena in question was falsely ascribed to Hamburger [1892, 1902] and v. Limbeck [1894]. In the literature

---

[1] Do not confuse with the son O. Nasse, the discoverer of isotonic salt solution, or with the father, the celebrated clinician of Bonn, Christian Friedrich Nasse.

dealing especially with ionic equilibrium in the blood I have only found H. Nasse mentioned by Gürber [1895, 1].

When it was decisively shown by the work of Gryns, Hamburger [1902], Hedin [1915], Overton, Gürber [1895], Koeppe [1897] and many others about the beginning of the century that the membrane of the blood cells was impermeable for cations—Arrhenius' ionic theory was at that time generally accepted by physiologists—it became necessary to seek a new explanation of the diffusion of ions between blood cells and serum. From excellent but wrongly interpreted experiments Gürber [1895] formed a theory that CO_2 split up NaCl with the formation of free HCl and sodium bicarbonate in serum and that the HCl subsequently diffused into the blood cells.

Koeppe [1897] advanced the theory that by CO_2 treatment monocarbonate ions are formed in the blood cells and that these are exchanged with chlorine ions from the serum—a theory which is only a special application of the principle enunciated by W. Ostwald in 1890 for ionic diffusion through a semi-permeable membrane. In his handbook Hamburger [1902] has thoroughly dealt with these theories but considers it is impossible to judge which is right. E. Petry [1903] has partly identified himself with Gürber's theory. But Gürber's theory is undoubtedly fallacious because firstly carbonic acid could not split up an appreciable amount of NaCl, even if the latter was not completely dissociated, and secondly, the same CO_2 tension must be present in the blood cells and serum, so that the CO_2 must have the same effect on the NaCl on both sides of the membrane. Koeppe's theory was falsified by the fact that he assumed monocarbonate ions could be formed in appreciable quantities by CO_2 treatment, but on account of the hydrogen ion activity this is impossible as shown by L. Meyer [1857], Setschenow [1879], Bohr [1905] and others. But if we substitute bicarbonate ions for monocarbonate ions Ostwald and Koeppe's theory is sound. Koeppe was not clear why more carbonate ions should be formed in the blood cells than in serum when the blood is treated with high CO_2 tensions but we can fall back upon Zuntz's [1882] old theory for the explanation, and it is of little importance in this connection whether we imagine the sodium to be originally combined in a complex manner with the proteins of the blood cells or whether we believe it to be present as ions partially balanced by electrolytes the dissociation of which varies with the reaction.

Loewy and Zuntz [1894] have imitated the ionic exchange process by enclosing serum or blood cell fluid in a parchment dialysing bag according to Kühne and dialysing against salt solutions, and they have shown that by treating the inner liquid with CO_2 alkali bicarbonate passed into the outer liquid. Loewy and Zuntz knew that alkali bicarbonate was liberated in this way both in the plasma and in the blood cells, but that by far the greatest amount appeared in the blood cells, and they have drawn attention to the connection between this phenomenon and the great capacity of the blood cells for combining with acid. Gürber [1895, 2] and Spiro and Henderson

[1909] have similarly imitated the exchange of ions with protein and salt solutions separated by parchment membranes with the same results. Spiro and Henderson have put the question very concisely by combining Koeppe's hypothesis (with the revision mentioned) with Loewy and Zuntz's views concerning the different combining properties of blood cells and serum. Hasselbalch and Warburg [1918], without at the time being cognisant of Spiro and Henderson's work, advanced a similar interpretation of the phenomenon, but taking into account the modern view of proteins as ampholytes, which however is not of great moment for the theory. Loewy and Zuntz, Gürber, Rona and György [1913, 2] have made it very probable by dialysis experiments that the greater part of the combined CO_2 in serum is present as bicarbonate, but the explanation of the experiments involves many difficulties. Gürber, Loewy and Zuntz, C. Lehmann [1894] and Petry have shown that the alkali metals do not diffuse on treating the blood with CO_2; the small differences found in the experiments are undoubtedly partly due to slight inaccuracies in the analyses and partly to the fact that the authors, with the exception of Petry, have disregarded volume changes of the blood cells. Hamburger and Bubanovic [1910] alone found a wandering of cations in some briefly reported experiments but this result is in opposition to what is known otherwise of the permeability of the membrane of the blood cells. Reference should also be made to the handbooks and to Ege's important work of 1920.

L. J. Henderson [1920] has quite recently pointed out that O_2 and CO_2 combinations in blood, and Cl and bicarbonate in serum, are so related that we can conclude from the phase rule that there is a common cause for their mutual equilibria, but he did not explain it in detail. A question of considerable interest with regard to the considerations which are to follow is, how many inorganic cations there are in blood cells and serum, but the question cannot be solved at present as the published analyses do not agree with one another. The distribution of inorganic cations between the external and internal phases of the blood cells is unknown but we shall not be far wrong in assuming the concentration in the external phase to be about $0 \cdot 15m$. We can briefly express the Zuntz theory of the distribution of ions between blood cells and serum, taking into account the advance since Zuntz propounded his theory and remembering the work of Koeppe, Zuntz and Loewy, Rona and György, and Spiro and Henderson, as follows.

The blood cells are surrounded by a membrane impermeable to proteins and cations. Both blood cells and serum (plasma) contain electrolytes the dissociation of which varies with the reaction, but in the neighbourhood of the neutral point the content of the blood cells in these substances is greatly in the ascendant. Thus more bicarbonate ions are formed in the blood cells than in the serum (and fewer hydrogen ions in the blood cells than in the serum), when the blood is brought into equilibrium with an atmosphere in which the CO_2 tension is higher than that with which the blood was originally in equilibrium. As the equilibrium is disturbed in this way bicarbonate ions from the blood cells are exchanged with other anions (chiefly chlorine ions) from the serum till equilibrium is again established—until the apparent bicarbonate ion activity in the blood cells and serum is *the same*. It is of no importance whether the bicarbonate ions are balanced by inorganic or by protein cations. The above form of Zuntz's theory is subject to a considerable defect which was foreign to the original theory. While Zuntz thought a part of the alkali of the blood cells was combined undissociated to the blood cell proteins and that these compounds were split up on treatment with CO_2 allowing sodium bicarbonate to diffuse from the blood cells into the serum until the concentration was the same in both, we must assume, since Gryns, Koeppe, Hamburger [1902], Hedin [1915], Gürber [1895 1, 2] and others have demonstrated the impermeability of the membrane of the blood cells to cations, that bicarbonate ions are exchanged for other anions (chiefly Cl') until the concentration is the same in blood cells and serum. This opens up a new problem which has never been adequately discussed. If there was equilibrium before the CO_2 treatment of the blood between the chlorine ions in the blood cells and serum it requires explaining how it is possible there can also be an equilibrium after chlorine ions have wandered from the serum into the blood cells. In the Zuntz theory as modified by Loewy and Zuntz, Koeppe, Spiro and Henderson, and Hasselbalch and Warburg the assumption has however been made that the bicarbonate ion concentration or as it ought to be expressed, the apparent bicarbonate ion activity, is the same in the blood cells and the serum independent of the reaction (and CO_2 tension). If on the other hand we assumed that the exchange of ions only proceeded until the relative decrease in tension of each kind of ion between blood cells and serum was the same we should have a rational explanation.

It will now be shown thermodynamically with the help of Donnan's [1911] distribution law that this explanation is a necessary consequence provided the assumptions relating to the permeability of the membrane of the blood cells and the electrolytes the dissociation of which varies with the reaction are correct.

As early as 1890 W. Ostwald showed that diffusible ions must distribute themselves in a characteristic manner between two solutions separated by a membrane when ions were present which could not pass through it. Ostwald's demonstration of this fact has however not been very fruitful to physiology and its application to the problem of the unequal distribution of ions referred to above was on the whole very difficult to grasp until Donnan published his work. As I fully share Donnan's conviction that a knowledge of his distribution law will be of extreme importance for the appreciation of ionic equilibrium in the organism and a number of other physiological problems, *e.g.* in muscle physiology, I will shortly review the fundamental facts of the theory. A complete exposition is given in W. C. McC. Lewis' *A System of Physical Chemistry* [1916] as well as in Donnan's own paper, to which the reader is referred.

Let us imagine two solutions separated by a membrane which is impermeable to an anion R', while the ions $Na^\cdot$ and Cl' diffuse freely. There are originally only $Na^\cdot$ and R' in the one solution and only $Na^\cdot$ and Cl' in the other according to the scheme

$$\begin{array}{c|c} Na^\cdot & Na^\cdot \\ R' & Cl' \\ (1) & (2) \end{array} \qquad \ldots\ldots\ldots\ldots\ldots\ldots(183)$$

When equilibrium is established $Na^\cdot$ and Cl' have diffused from (2) to (1) so that we now have

$$\begin{array}{c|c} R' & Na^\cdot \\ Na^\cdot & \\ Cl' & Cl' \\ (1) & (2) \end{array} \qquad \ldots\ldots\ldots\ldots\ldots\ldots(184)$$

Diffusion goes on until no work can be done by the simultaneous isothermous transference of a differential mol of $Na^\cdot$ and Cl', so that we can obtain no energy by simultaneously carrying out the two isothermous differential processes

$$\delta n \text{ mol } Na^\cdot (2) \rightarrow (1),$$

$$\delta n \text{ mol } Cl' (2) \rightarrow (1).$$

We therefore have the equilibrium equation

$$\delta n \, R \, T \log_e \frac{C_{Na^\cdot} (2)}{C_{Na^\cdot} (1)} + \delta n \, R \, T \log_e \frac{C_{Cl'} (2)}{C_{Cl'} (1)} = 0. \ldots\ldots\ldots\ldots(185)$$

Therefore
$$\frac{C_{Na^\cdot} (1)}{C_{Na^\cdot} (2)} = \frac{C_{Cl'} (2)}{C_{Cl'} (1)} \qquad \ldots\ldots\ldots\ldots\ldots\ldots(186)$$

on the assumption the solutions are ideal.

Similarly for other ions in general we get

$$\frac{C_{Na^\cdot (1)}}{C_{Na^\cdot (2)}} = \frac{C_{M'_I (1)}}{C_{M'_I (2)}} = \frac{C_{A'_I (2)}}{C_{A'_I (1)}} = \sqrt[2]{\frac{C_{M''_{II} (1)}}{C_{M''_{II} (2)}}} = \sqrt[3]{\frac{C_{M'''_{III} (1)}}{C_{M'''_{III} (2)}}} = \sqrt[2]{\frac{C_{A''_{II} (2)}}{C_{A''_{II} (1)}}} \text{ etc.,} \ldots\ldots(187)$$

where M'_I is any monovalent cation and A'_I any monovalent anion and so on.

Donnan gives the following example: two solutions in equilibrium have originally the constitution

$$\begin{array}{cc|cc} Na^\cdot & R' & Na^\cdot & Cl' \\ C_1 & C_1 & C_2 & C_2 \\ \multicolumn{2}{c|}{(1)} & \multicolumn{2}{c}{(2)} \end{array} \qquad \ldots\ldots\ldots\ldots(188)$$

where C_1 and C_2 are the molar ion concentrations in the solutions. Assuming complete activity and equal volumes of the solutions on either side of the membrane during the whole process, we get, when equilibrium is reached,

$$\begin{array}{ccc|cc} Na^\cdot & R' & Cl' & Na^\cdot & Cl' \\ C_1 + x & C_1 & x & C_2 - x & C_2 - x \\ \multicolumn{3}{c|}{(1)} & \multicolumn{2}{c}{(2)} \end{array} \qquad \ldots\ldots(189)$$

From this it follows that $\frac{x}{C_2} 100$ is the quantity of NaCl which diffuses from

(2) to (1) expressed as a percentage of the original amount of NaCl in (2), and $\frac{C_2 - x}{x}$ is the distribution ratio of NaCl between (2) and (1) corresponding to the equilibrium.

The equation in this case is

$$x = \frac{C_2{}^2}{C_1 + 2C_2}, \quad \dots\dots\dots\dots\dots\dots\dots\dots(I)$$

from which

$$\frac{x}{C_2} = \frac{C_2}{C_1 + 2C_2}, \quad \dots\dots\dots\dots\dots\dots\dots(II)$$

$$\frac{C_2 - x}{x} = \frac{C_1 + C_2}{C_2}. \quad \dots\dots\dots\dots\dots\dots\dots(III)$$

If C_2 is small compared with C_1 we can put

$$\frac{x}{C_2} = \frac{C_2}{C_1}, \quad \dots\dots\dots\dots\dots\dots\dots\dots(IV)$$

and

$$\frac{C_2 - x}{x} = \frac{C_1}{C_2}. \quad \dots\dots\dots\dots\dots\dots\dots\dots(V)$$

If we take for example $C_2 = \frac{1}{100} C_1$, then $\frac{x}{C_2} = \frac{1}{100}$, that is to say only $\frac{1}{100}$th of the original amount of NaCl diffuses from (2) to (1). If on the other hand C_1 is small compared with C_2

$$\frac{x}{C_2} = \frac{1}{2}, \quad \dots\dots\dots\dots\dots\dots\dots\dots(VI)$$

and

$$\frac{C_2 - x}{x} = 1.$$

In the following table Donnan gives some calculations made by equations (I), (II) and (III).

Table LXII.

Original concentration of Na R in (1)	Original concentration of NaCl in (2)	Original ratio of Na R to NaCl	% NaCl diffused from (2) to (1)	Distribution ratio of NaCl between (2) and (1)
C_1	C_2	$\frac{C_1}{C_2}$	$\frac{x}{C_2} 100$	$\frac{C_2 - x}{x}$
0·01	1	0·01	49·7	1·01
0·1	1	0·1	47·6	1·1
1	1	1	33	2
1	0·1	10	3	11
1	0·01	100	1	101

It is clearly seen from the equations and tables that the ions for which the membrane is freely permeable will become very unevenly distributed between the two solutions if the concentration (activity) of the ions for which the membrane is impermeable is large in comparison with the former.

In Fig. 27 we have the conception of Donnan in schematic form. The figure shows the cross-section of a cylinder in which a piston can move. The piston represents the semi-permeable membrane. Solution (1) is on the left of the piston, solution (2) on the right. The piston is situated in the middle of the cylinder and is fixed there, so that the volumes of the two solutions are equal. The ions for which the piston is impermeable will attempt to force the piston to the right (as they are only present in solution (1)), and since the ions of opposite charge (in this case Na·) are held back in the same amount by the electrostatic force, the osmotic pressure (the pressure on a square centimetre of the piston) will be

$$P_0 = 2C_1 R\,T, \qquad\qquad\dots\dots\dots\dots\dots\dots(VII)$$

provided that (2) contains only pure water. But as in the present example (2) also contains Na$^{\cdot}$ and Cl$'$, and as these ions are unevenly distributed in (1) and (2), the sodium chloride will exert a counter-pressure on the piston when equilibrium is established which will be

$$P = 2\,(C_2 - x)\,R\,T - 2x R\,T = 2\,(C_2 - 2x)\,R\,T. \quad\dots\dots(VIII)$$

The force per unit surface which will attempt to drive the piston to the right when equilibrium is established will then be

$$P_1 = P_0 - P = 2R\,T\,(C_1 - C_2 + 2x). \quad\dots\dots\dots\dots(IX)$$

Therefore combining with (I) $\dfrac{P_1}{P_0} = \dfrac{C_1 + C_2}{C_1 + 2C_2}, \qquad\dots\dots\dots\dots\dots\dots\dots\dots\dots(X)$

If C_1 is small compared with C_2, $P_1 = \tfrac{1}{2}P_0$; if C_2 is small compared with C_1, $P_1 = P_0$ as would be expected.

The following table (after Donnan) shows the relations:

$\dfrac{C_2}{C_1}$	$\dfrac{P_1}{P_0}$
0·1	0·92
1	0·67
2	0·60
10	0·52

It will be observed therefore that if the concentration of the permeating ions is large compared with that of the non-permeating ions the counter-pressure of Donnan will have a very important influence on the magnitude of the osmotic pressure.

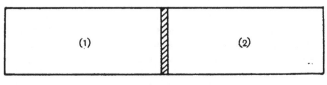

Fig. 27

Donnan and Allmand [1914] have investigated the distribution of KCl between two solutions separated by a copper ferrocyanide membrane, potassium ferrocyanide being also present in one of the solutions, and they have found the theory is supported on the whole. Bayliss [1909, 1911] has demonstrated the effect directly in experiments with congo red, H. R. Procter [1914] and ·Procter and J. A. Wilson [1916] have studied the gelatin-acid systems, and

S. P. L. Sörensen has very thoroughly investigated to what extent such an effect might be expected to influence his measurements of the osmotic pressure of egg albumin.

The unequal distribution is of great importance in the purification of non-permeating substances by dialysis (cf. Donnan [1911] and Sörensen [1915–1917]), and may sometimes be important in experiments to determine the active concentrations (apparent activities) of permeating substances by compensation dialysis and ultrafiltration, as has been shown particularly by Rona and György [1913] and H. J. Hamburger [1918][1].

Donnan has further shown by mathematical treatment analogous to the above that the amount of undissociated NaCl must be the same in the two solutions, which leads to the equation

$$\frac{C_{Na^{.}} C_{Cl'}}{C_{NaCl}} = \text{constant,} \qquad \dots\dots\dots\dots\dots\dots(190)$$

but this relation we know to be wrong (cf. chapter I). Donnan is unable to give an adequate explanation of this but it will be seen the difficulty disappears when we assume with Bjerrum that undissociated NaCl does not exist in watery solutions in the range of concentration in question. This point also brings out the superiority of Bjerrum's dissociation theory over the classical theory of Arrhenius.

In the above example of Donnan it is assumed there are no non-permeating ions in solution (2) and that the volumes of the solutions do not change (the piston in Fig. 27 being fixed). If we assume on the other hand that NaR is present in both solutions and that water can be freely interchanged between them it will be obvious without any special mathematical treatment that the solutions will be of the same constitution so that there will be no uneven distribution of Na' and Cl'. In this case therefore we may imagine that the piston is freely movable without obstruction and that water can freely pass through it.

If there are non-permeating mols without a charge in one or both of the solutions but in a different ratio to that which the non-permeating (in this case negative) ions bear to one another, different concentrations of the non-permeating ions will be found in the two solutions when equilibrium is established and the permeating ions will consequently also be unequally distributed. A point of importance is the possibility that some force or other opposes the change in volume of the solution. In the illustration we can imagine this as happening by the piston doing work against spiral springs. In such a case the volumes, and therefore the concentrations, will not be dependent alone upon the degree of the permeating forces of the mols, and thus unequal distribution of the ions can ensue and conditions dependent thereon. If some of the non-permeating ions have a different valency to the permeating ones, a different number of permeating ions and non-permeating ions will be in

[1] Jaques Loeb has published a series of papers in the *Journal of General Physiol.* 1920-1921 on the influence of the Donnan effect on numerous factors concerned with gelatin solutions.

electrostatic equilibrium mutually. If the volumes of the solutions are only determined by the non-permeating ions and the osmotic activity of the permeating ions electrostatically balanced by the former, unequal distribution of the ions can arise.

The last complicating factor of great significance is that the solutions in question are not ideal. The values in equation (187) are therefore not the concentrations but the apparent osmotic activities so that it becomes

$$\frac{C_{Na^{\cdot}\,(1)} \times F_a(Na^{\cdot})_{(1)}}{C_{Na^{\cdot}\,(2)} \times F_a(Na^{\cdot})_{(2)}} = \frac{C_{M'_I(1)} \times F_a(M'_I)_{(1)}}{C_{M'_I(2)} \times F_a(M'_I)_{(2)}} = \frac{C_{A'_I(2)} \times F_a(A'_I)_{(2)}}{C_{A'_I(1)} \times F_a(A'_I)_{(1)}}. \quad \ldots\ldots(191)$$

Let us briefly collect the results of the above. Donnan's theory shows that the relative decrease in osmotic tension between solutions separated by semi-permeable membranes is the same for all monovalent ions, and that the relative osmotic tension of cations is the reciprocal of that of anions, and also that the relative decrease in osmotic tension of an n-valent ion is the nth root of that of a monovalent ion. If the concentrations of those permeating positive and negative ions which are balanced by non-permeating mols are different in solutions which are in equilibrium through a semi-permeable membrane, the osmotic activity of a permeating ion will be different in the solutions as first discovered by Ostwald [1890].

Let us consider a system of two homogeneous watery solutions separated by a semi-permeable membrane. The solutions contain both permeating and non-permeating ions. The non-permeating ions are distinguished by their symbols being put in parentheses, e.g. $(M'_{I(1)})$, that is, the non-permeating monovalent cation in solution (1). As the following exposition is only developed as far as the present limited problems require and as experiments have not been carried out to make a general expression possible we will assume that in the solutions of permeating ions only chlorine and bicarbonate ions occur in such concentrations as are of significance for the exchange of ions, since the relations of the other ions are analogous with these two.

The following non-permeating ions are also present in the system:

$$(M'_I), \ (M^{\cdot\cdot}_{II}), \ (A'_I) \ \text{and} \ (A''_{II}).$$

According to the law of equal positive and negative charges we have

$$(C_{M'_I(1)}) + (2C_{M^{\cdot\cdot}_{II}(1)}) - (C_{A'_I(1)}) - (2C_{A''_{II}(1)}) - C_{Cl'(1)} - C_{HCO'_3(1)} = 0, \quad (192)$$

and $(C_{M'_I(2)}) + (2C_{M^{\cdot\cdot}_{II}(2)}) - (C_{A'_I(2)}) - (2C_{A''_{II}(2)}) - C_{Cl'(2)} - C_{HCO'_3(2)} = 0, \quad (193)$

therefore $C_{Cl'(1)} + C_{HCO'_3(1)} = (C_{M'_I(1)}) + (2C_{M^{\cdot\cdot}_{II}(1)}) - (C_{A'_I(1)}) - (2C_{A''_{II}(1)}), \quad (194)$

and $C_{Cl'(2)} + C_{HCO'_3(2)} = (C_{M'_I(2)}) + (2C_{M^{\cdot\cdot}_{II}(2)}) - (C_{A'_I(2)}) - (2C_{A''_{II}(2)}). \quad (195)$

From (187) we get $\quad \dfrac{C_{Cl'(2)}}{C_{Cl'(1)}} = \dfrac{C_{HCO_3(2)}}{C_{HCO'_3(1)}} = \dfrac{C_{Cl'(2)} + C_{HCO'_3(2)}}{C_{Cl'(1)} + C_{HCO'_3(1)}}, \quad \ldots\ldots\ldots\ldots\ldots(196)$

and therefore

$$\frac{C_{Cl'(2)}}{C_{Cl'(1)}} = \frac{C_{HCO_3(2)}}{C_{HCO'_3(1)}} = \frac{(C_{M'_I(2)} + (2C_{M^{\cdot\cdot}_{II}(2)}) - (C_{A'_I(2)}) - (2C_{A''_{II}(2)})}{(C_{M'_I(1)} + (2C_{M^{\cdot\cdot}_{II}(1)}) - (C_{A'_I(1)}) + (2C_{A''_{II}(1)})}. \quad \ldots\ldots(197)$$

21—2

If the solutions are not ideal we obtain from (191)

$$\frac{C_{Cl'(2)} F_a(Cl')_{(2)}}{C_{Cl'(1)} F_a(Cl')_{(1)}} = \frac{C_{HCO'_3(2)} F_a(HCO'_3)_{(2)}}{C_{HCO'_3(1)} F_a(HCO'_3)_{(1)}} = \frac{C_{Cl'(2)} F_a(Cl')_{(2)} + C_{HCO'_3(2)} F_a(HCO'_3)_{(2)}}{C_{Cl'(1)} F_a(Cl')_{(1)} + C_{HCO'_3(1)} F_a(HCO'_3)_{(1)}}.$$

$$\dots\dots\dots\dots(198)$$

For the further treatment of the present problem it will be convenient to rearrange equation (198) thus:

$$\frac{\dfrac{C_{Cl'(2}}{C_{Cl'(1)}}}{\dfrac{C_{HCO'_3(2)}}{C_{HCO'_3(1)}}} = \frac{F_a(HCO'_3)_{(2)} F_a(Cl')_{(1)}}{F_a(HCO'_3)_{(1)} F_a(Cl')_{(2)}}. \quad \dots\dots\dots\dots(199)$$

If solutions (1) and (2) are heterogeneous, as is the case with serum and blood cells, (199) will be further complicated as the concentrations in the solutions are then mean concentrations and we therefore have to introduce a correction for this in equation (199) (cf. chapter III). Adding the subscript (c) to the symbols relating to blood cells and (s) to those relating to serum we get

$$D' = \frac{\Phi_a(HCO'_3)_{(c)} \Phi_a(Cl')_{(s)}}{\Phi_a(HCO'_3)_{(s)} \Phi_a(Cl')_{(c)}} = \frac{\dfrac{C_{Cl'(c)}}{C_{Cl'(s)}}}{\dfrac{C_{HCO'_3(c)}}{C_{HCO'_3(s)}}}. \quad \dots\dots\dots\dots(200)$$

The symbol D' is introduced for the sake of convenience.

The value $\frac{\Phi_a(Cl')_{(s)}}{\Phi_a(HCO'_3)_{(s)}}$ can be regarded with very close approximation as being constant on account of the relatively small variations serum undergoes on changing the reaction. This is in the last resort due to the relatively small disperse phase in serum (circ. 8 %).

The value $\frac{\Phi_a(HCO'_3)_{(c)}}{\Phi_a(Cl')_{(c)}}$ is equal to $\dfrac{F_a(HCO'_3)_{(c)} \dfrac{100}{100 - q_{(c)}(1 - d(HCO'_3))}}{F_a(Cl')_{(c)} \dfrac{100}{100 - q_{(c)}(1 - d(Cl'))}}, \dots(201)$

where $d(Cl')$ and $d(HOO'_3)$ are the relations between the concentrations of Cl' and HCO'_3 respectively in the protein phase and external phase of the blood cells, while the activity coefficients refer to the external phase. In accordance with the reasoning in chapter VIII we may undoubtedly regard $\frac{F_a(HCO'_3)_{(c)}}{F_a(Cl')_{(c)}}$ as constant—within the experimental error—for subsequent investigation. The value $\dfrac{\dfrac{100}{100 - q_{(c)}(1 - d(HCO'_3))}}{\dfrac{100}{100 - q_{(c)}(1 - d(Cl'))}}$ $\dots\dots\dots\dots\dots(202)$

must also be nearly constant as will be shown.

$q_{(c)}$ is the volume of the disperse phase expressed in vols. %. Ege [1920, 1921] has shown this is between 35 and 40 % of the blood cells. If we assume that $q_{(c)}$ in horse blood cells at p_{II}. 6·50 is 40 it will be seen from Fig. 9 that $q_{(c)}$ at p_{II}. 7·90 is $40\frac{86·3}{100} = 34·5$. $d(Cl')_{(c)}$ and $d(HCO'_3)_{(c)}$ are the relations between the solubility of the chlorine and bicarbonate ions in the protein and water phases of the blood cells. In chapter IX for the bicarbonate ion we found $d(HCO'_3)_{(c)}$ equal to about $\frac{1}{3}$. Provided that the chlorine ions are not soluble in the protein phase and $d\,(Cl')_{(c)}$ therefore is 0 the variation of the conditions with the reaction will be a maximum. Substituting these values in (202) we obtain at $p_H\cdot$ 6·50 the value 75, and at $p_H\cdot$ 7·90, 79.

It will therefore be appropriate to investigate the value of D' at different reactions with the knowledge that we may expect it to be fairly constant. It is however not improbable it increases a little with decreasing apparent hydrogen ion activity.

In chapter VI there are numerous determinations of $\dfrac{C_{HCO'_3(c)}}{C_{HCO'_3(s)}}$ in defibrinated horse blood at room temperature and the results are given diagrammatically in Fig. 6. I have therefore undertaken some determinations of the distribution of Cl between blood cells and serum, and they are to be found in Table LXIII. The technique employed was as follows: Defibrinated horse blood was saturated in the large saturator with gas containing CO_2 (by

Table LXIII. Horse blood.

Temp.	mm. Hg CO_2	mm. Hg O_2	Vols. % combined CO_2	Vols. % combined O_2	pH'(a) calc.	Haematocrite number		Milligram equivalents of Cl in 1 litre	blood cells	D'
21	34·5	146·4	56·5	22·7?	7·40	38·5	B	83·8	58·6	0·46
							S	106		
						74·5	C	68·3	50·8	0·48
21	616·6	24·2	117·4	9·3	6·46	46·5	B	83·8	71·0	0·75
							S	95·0		
						75·0	C	75·5	69·1	0·73
21	18·9	152·4	46·8	22·2	7·69	37·0	B	83·8	35·7	0·32
							S	112		
						70·5	C	68·7	50·5	0·45
20·5	71·8	137·0	71·1	22·6	7·27	44·0	B	83·8	59·6	0·58
							S	103		
						70·5	C	71·2	57·9	0·56
20[1]	197·5	112·5	91·6	22·0	6·92	45·5	B	83·8	67·5	0·69
							S	97·5		
						77·3	C	71·6	64·0	0·66
20[1]	375·1	75·6	96·7	21·1	6·65	47·8	B	83·8	71·6	0·75
							S	95·1		
						81·8	C	74·1	69·4	0·73
20	15·0	155·9	48·2	21·3	7·81	42·0	B	81·9	47·4	0·44
							S	107		
						75·0	C	66·3	52·8	0·49
20	741·5	0·7	122·0		6·43	47·5	B	81·9	70·6	0·77
				0·6			S	92·1		
						72·5	C	72·6	65·2	0·71
19	96·4	136·6	84·0	20·9	7·20	44·0	B	81·9	59·6	0·60
							S	99·5		
						71·0	C	71·2	59·8	0·60
19	237·3	102·9	107·6	21·9	6·90	46·0	B	81·9	66·6	0·70
							S	95·0		
						76·3	C	74·5	68·2	0·72
19	30·4	150·3	58·2	21·6	7·57	42·5	B	81·9	53·4	0·52
							S	103		
22[2]	19·4	155·4	48·6	20·6		42·5	B	81·9	46·8	0·43
	19·4	155·4	59·4	0·6	7·73		S	108		
	19·4	155·4	37·9	39·1		77·0	C	63·4	50·1	0·46
22[2]	744·9	0·0	122·4	0·4		46·3	B	81·9	70·7	0·77
	744·9	0·0	127·4	0·0	6·47		S	91·6		
	744·9	0·0	116·4	—		78·0	C	74·6	69·9	0·76

[1] The blood was five days old and had a slightly sour smell.   [2] Hirudin blood.

the method described in chapter IV) the constitution of which was determined
by analysis of the spirometer gas at the conclusion of the saturation. Samples
of the blood were taken to determine the amount of combined CO_2, and it
was transferred to a glass cylinder, the CO_2 saturation being maintained, where
it was allowed to sediment in contact with the saturating gas. The chlorine
in the serum and blood cell suspension (and blood) was then estimated by
means of a slightly modified titration according to Mohr and I. Bang's [1916]
method. The volume of the blood cells was estimated in the usual way with
the haematocrite. In estimating the chlorine, serum, blood or blood cell suspension was transferred with a 1 cc. pipette to a 50 cc. measuring flask and
the pipette then washed out three times with distilled water. The flask was
then filled about up to the mark with absolute alcohol with continuous shaking

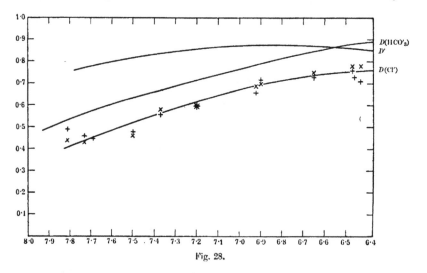

Fig. 28.

which produced a very fine precipitate. After 24 hours' standing, during which
the flask was usually shaken several times, the extraction was complete and
it was filled right up to the mark with absolute alcohol, whereupon the flask
was rotated several times. The solution was then filtered into small Bang-
flasks the funnel being covered with a watch glass to prevent evaporation
and the tube passed through a cork which fitted loosely into the neck of
the flask. 10 cc. of the filtrate was transferred to three flasks and titrated
with $n/100$ silver nitrate with potassium chromate as indicator as described
by Bang [1916].

In explanation of Table LXIII it is only necessary to state that B stands
for blood, S for serum and C for blood cell suspension.

In Fig. 28 all the results are given, those that are obtained from blood
and serum are designated by "recumbent" crosses and those from blood cell

suspensions and serum by "erect" crosses. A curve through the points was drawn freehand. This curve is marked D (Cl'). The curve from Fig. 6 (the curve showing the distribution of combined CO_2) is also given in the figure and is marked D (HCO'$_3$). D' is calculated from the two curves and the result plotted as an independent curve. It will be seen that D' is about 0·85. The experiments do not justify an opinion as to whether the small variation found is a real one or due to experimental error. The result of these experiments very strongly supports the hypothesis advanced for the distribution of ions between blood cells and serum—an hypothesis which may briefly be recapitulated thus: The permeating monovalent anions distribute themselves between blood cells and serum in such a way that the relative decrease in osmotic tension is the same for them all. The relative concentration (mean concentration) of the different anions will be a function of the relative osmotic tension, of the apparent activity coefficients of the ions in the external phase, and the solubility of the ions in the internal phase.

It would have been of great help in the further elucidation of these problems had it been possible to measure the apparent (potentiometric) activity of the chlorine ions in blood cells and serum, but it is not. I shall take this opportunity of explaining why such an estimation must fail because it appears largely to have been overlooked and a knowledge of the facts is of fundamental importance in judging the value of many of the studies of Pauli and his collaborators—as is very evident from Pauli's book which appeared in 1920.

Bugarsky and Liebermann as early as 1898 attempted to measure the chlorine·ion concentration in egg albumin by means of a calomel electrode. Later Manube and Matula [1913] and Blasel and Matula [1914] amongst others used the same electrode for a similar determination in protein solutions. Oryng and Pauli [1915] noticed that serum albumin combined with mercurous ions but they claim to have shown these mercurous ions were split off again at more acid reactions. I cannot see however that the authors have succeeded in demonstrating this or that it has been proved the mercurous combination is independent of the salt concentration. Pauli and his collaborators have themselves made interesting contributions to the problem of the combination by the proteins of various heavy metal ions, a combination which takes place beyond all doubt.

If the chlorine ion concentration (activity) of a solution is to be determined with the help of an electrode of the second class, be it a mercurous or silver electrode, two conditions must be fulfilled.

I. The mercurous or silver ion concentration in chloride-containing solutions must be determined solely by the activity of.the chlorine ions, that is to say calomel or silver chloride must be precipitated as soon as more than a trace of mercurous or silver nitrate is added. If the mercurous or silver ions are combined in a complex manner to any solute thereby diminishing the activity of these ions to an amount smaller than necessary for the precipitation of the calomel or silver chloride, the estimation of the chlorine ion activity is not possible.

II. None of the solutes in the solution (apart from a negligible amount of chlorine ions) must be affected by the presence of mercurous or silver ions in the solution. Several investigators including Pauli and his co-workers have shown that the proteins form complex compounds with mercurous and silver ions which have an extraordinarily low dissociation tension. I have myself some experience from unpublished experiments relating to the effect of the addition of silver nitrate to blood and the following results may be stated:

(1) Considerably more silver nitrate than corresponds to $n/100$ can be added to the blood without silver chloride being precipitated.

(2) Haemolysis gradually takes place which becomes total if sufficient silver nitrate is added.

(3) At body temperature the haemoglobin is relatively slowly destroyed, and quicker in acid than in alkaline reactions.

(4) The combination of CO_2 is rather less than in the original blood at the same reaction.

(5) The O_2 combination curve at body temperature is much steeper than usual.

It will be seen that neither of the two conditions for the determination of the chlorine ion activity is fulfilled. It is appropriate to draw attention here to a fact which in a large degree obscures many of the experiments of Pauli and his collaborators and undoubtedly compromises many of their results. Pauli and his collaborators, as was generally accepted at the time, always took electrically determined apparent activity to be equivalent to the corresponding concentration, which gives rise to an error in the solutions that were poor in salt and moderately rich in protein, which in some cases may be very significant.

The experiments dealing with the distribution of chloride between blood cells and serum are further interesting as they help to throw light upon the conditions of the exchange of bicarbonate and chlorine ions. D. D. van Slyke and G. E. Cullen [1917] have tried to show that only about half the bicarbonate produced by treating serum with CO_2 (measured at roughly constant reaction) is due to exchange with chlorine ions, while L. S. Fridericia [1920] has found that the chlorine diffusion almost completely accounts for the production of bicarbonate. Van Slyke and Cullen as well as Fridericia have however disregarded the fact that the water yield of the serum to the blood cells affects their calculations. Van Slyke and Cullen found that serum from a sample of freshly-drawn blood contained

$$0.1050n \text{ Cl' and } 0.031n \text{ CO}_2.$$

After saturation with CO_2 at atmospheric pressure the serum yielded

$$0.111n \text{ Cl' and acquired } 0.0206n \text{ HCO}_3'.$$

If we assume the serum was concentrated 10 % the chlorine loss would be

$0.0216n$, and the bicarbonate formation $0.0266n$. So that the difference is really much less than van Slyke and Cullen calculated.

Fridericia found that serum separated from blood at 0.08 mm. CO_2 contained $0.0992n$ Cl' and $0.0250n$ HCO'_3, at 163 mm. CO_2.

Serum separated from blood at 162 mm. CO_2 contained

$$0.0814n \ Cl' \text{ and } 0.0463n \ HCO'_3, \text{ at } 163 \text{ mm. } CO_2.$$

If we assume the serum was concentrated 8 % we obtain

	Cl'	HCO'_3
Difference	$- 0.0256n$	$+ 0.0183n$

so that more chlorine has passed over than corresponds to the bicarbonate which presumably is due to experimental error.

In an experiment (Table LXIII) I found that serum separated from blood at 19.4 mm. CO_2 had a chlorine content of $0.108n$, and that the bicarbonate content was $0.027n$ at p_H· 7.73.

The volume of the serum was 57.5 % of the blood.

Serum separated from blood at 744.9 mm. CO_2 had a chlorine content of $0.0916n$, and a bicarbonate content of

$$0.0572n \text{ at } p_H \cdot 6.47.$$

The volume of the serum was 53.7 % of the blood.

In accordance with the determinations in a previous chapter horse serum will combine with about 12.5 vols. % CO_2 in changing from p_H· 7.73 to p_H· 6.47. Subtracting this ($0.006n$) and correcting for the loss of water which is 9.3 % of the original volume of the serum we get

	Cl'	HCO'_3
Difference	$- 0.026n$	$+ 0.022n$

which result like that of Fridericia's experiment gives too large a diffusion of chlorine and is probably likewise due to experimental error.

F. C. McLean, H. A. Murray and L. J. Henderson [1920] have reported experiments according to which only two-thirds of the diffused bicarbonate is accounted for by the chlorine but these experiments are only known to me through a short reference in *Medical Science Abstracts and Reviews*, **5**, p. 88, and it is not clear from this whether the authors corrected for the water loss of the serum[1].

---

[1] In an article which recently appeared in the *J. Biol. Chem.* (1921, **47**, 377) E. A. Doisy and E. P. Eaton have shown by accurate direct analyses that sodium and potassium do not diffuse on changing the CO_2 tension. They further showed that the diffusion of Cl' on varying the CO_2 tension between 35 mm. and 100 mm. accurately corresponded to the diffusion of HCO'_3 (the change of volume was allowed for); on passing to very high CO_2 tensions in agreement with Fridericia and myself they found a too great diffusion of Cl'. Apart from possible experimental errors I am unable, as I have mentioned, to give an explanation of the latter phenomenon. Mukai (*J. Physiol.* 1921, **55**, 356) has likewise shown by direct analysis of the salts that the cations do not diffuse through the surface of the blood cells on changing the CO_2 tension.

Before proceeding further with the investigation of equation (198) it will be well to examine the conditions which determine the change in volume of the blood cells at different CO_2 tensions. It must however be remarked that this investigation deals with the fundamental rather than the quantitative relations of this process because the experimental material in my possession was not performed with the present purpose in view, but is a by-product of the experiments in chapter VI, and the constants used are not very accurately ascertained.

The discovery that the volume of the blood cells is increased by CO_2 treatment of the blood is credited to H. Nasse, 1874, as already stated, but he did not attempt any explanation of the phenomenon. Later it was confirmed by v. Limbeck [1894] and by Hamburger [1902]. Hamburger essayed to give a theoretical explanation of the processes on which it is based by accepting Koeppe's theory of ionic exchange between blood cells and serum. Hamburger's reasoning is as follows:

When blood is treated with CO_2 monocarbonate ions are formed in the blood cells. One of these is exchanged for two chlorine ions from the serum (plasma). In this way the osmotically active mols in the blood cells are increased and they therefore swell up, water from the serum entering them until the osmotic pressure in blood cells and serum is the same. Hamburger's view is erroneous—as is of course generally known—because monocarbonate ions are not present in sufficient quantities in the blood cells at the reactions employed and they do not increase in numbers on treating the blood with CO_2; on the contrary they decrease.

Koeppe's [1897] theory which Hamburger at any rate originally supported is in its earliest form incapable of explaining the problem as Koeppe had no idea as to how the CO_2 (according to Koeppe the monocarbonate ions) was combined.

In Spiro and Henderson's work the old theories of Zuntz [1882], Koeppe [1897] and Loewy and Zuntz [1894] were revived in a modern form. Spiro and Henderson's [1909] views were expressed in a rather different manner in conformity with the theory of ampholytes of Hasselbalch and Warburg [1918], which has been alluded to in more detail above. It is possible from the interpretation given by Loewy and Zuntz, Spiro and Henderson and Hasselbalch and Warburg of the distribution of ions in the blood to explain the change in volume of the blood cells which is determined by the reaction as Spiro and Henderson have briefly indicated.

Rich. Ege [1920], who has made the most important contribution in physiological literature since the beginning of the century to the question of the volume of the blood cells and who from nearly every point of view has decisively shown that it is a simple function of the osmotic pressure of the liquid they are suspended in, failed when it came to explaining the change of volume caused by varying reaction. He has however reported an experiment in which he added HCl to a blood cell suspension in physiological salt solution.

This caused the apparent hydrogen ion activity to rise to p_H· 3·7 and the blood cells swelled 16 %. A freezing point determination of the external liquid and blood cell liquids gave the same osmotic pressure and Ege could therefore conclude the conditions obtaining in the volume change consequent on change of reaction did not differ fundamentally from those which determine the other kinds of volume change of the blood cells[1].

The mols which can exert an osmotic pressure on the membrane of the blood cells must either be incapable of passing through it, that is, be non-permeating, or they must be retained on one side of the membrane by electro-static forces. The non-permeating mols of any importance in this respect are the cations which are all considered to be non-permeating, and protein mols (both charged and uncharged). The permeating mols which are important are consequently the anions. There is over 30 % of haemoglobin in the blood cells and about 8 % of other stroma proteins. If we assume with A. V. Hill that the haemoglobin molecules are aggregated with a mean aggregation number of 2·5, the molecular weight of the aggregated haemoglobin is about 40,000. We shall therefore not incur a large error in assuming the osmolar concen-tration of the blood cell proteins to be $m/100$, provided they ionise without destruction of the molecule which is now accepted by nearly everyone (cf. Pauli's [1920] criticism of T. B. Robertson's hypothesis of protein ionisation); and if the protein aggregates do not further aggregate or split up perceptibly under the given conditions the osmotic pressure exerted by the protein mols will not be altered by varying the reaction. Even if the assumptions con-cerning the aggregation should not turn out to be absolutely correct it will be of no consequence on account of the small osmotic activity of the proteins. On ionisation of the proteins however different amounts of permeating ions can be combined by the blood cells and it is this condition which determines the change in volume of the blood cells at different reactions. We will dwell upon this condition a little and examine it systematically on account of its great theoretical interest.

If we consider the conditions of an apparent hydrogen ion activity where the number of positive and negative charges associated with the blood cell proteins are equal, the amount of permeating anions retained in the blood cells will be equal to the amount of inorganic cations present, which in this connection may be taken to be the same as the potassium and sodium con-centrations. If we now examine the conditions in a more alkaline reaction, protein anions will be formed in the blood cells in greater quantities and protein cations will have disappeared (cf. chapter XI); at the same moment there will be liberated an equivalent amount of bicarbonate ions in the form of CO_2, as protein ions now are balancing the inorganic cations corre-sponding to the expelled CO_2. In this way the osmolar concentration is

---

[1] Mukai (J. Physiol. 1921, **55**, 356), by determining the osmotic pressure of the blood cell fluids and serum by Barger's method, has shown that it is the same in both phases whether we are dealing with defibrinated blood treated with CO_2 or with untreated defibrinated blood.

decreased and the blood cells will shrink as water is transferred from the blood
cells to the serum until the osmotic pressure is equal on either side of the blood
cell membrane.

If we now consider the conditions when the reaction is more acid than
before protein cations will be formed in excess of protein anions, the latter
being simultaneously decreased in amount, and therefore bicarbonate ions
will be held in equilibrium with the protein cations, whereby the osmolar concentration of the blood cells is increased and they will increase in size. It
does not matter whether the bicarbonate ions on reaching equilibrium are
exchanged with the same number of chlorine ions as the osmolar concentration
is not altered in this way. The conditions are actually rather more involved
than indicated here, three complications particularly being active:

I. In serum as well as in blood cells there are proteins (ampholytes), the
dissociation of which depends on the reaction, which will oppose the change
of volume but their action is only about $\frac{1}{40}$th of the blood cells.

II. During the ionic equilibrium an osmotic counter-pressure arises from
the unequal distribution of the permeating ions as previously explained in
accordance with Donnan's views. It may be surmised however that this
pressure must be relatively slight.

III. The apparent osmotic coefficients should change a little on ionisation
of the proteins.

I have made an attempt by a rough approximation to find out the
mathematical relations of the volume change determined by the reaction and
I have compared the result with the experimental one shown in Fig. 9. It
should however be mentioned that the experiments in the first part of
Table LXIII are in poor agreement with my other experiments in this connection but I can give no explanation of the fact. In the following no account
is taken of the Donnan counter-pressure nor of any changes in the osmotic
coefficient which might occur, but it is obvious that the Donnan counter-
pressure will cause the volume change in the alkaline reactions to be smaller
than that calculated.

The osmotic pressure on the membrane of the blood cells exerted by their
fluids is made up of three components, namely the pressures due to the
proteins, the cations, and the anions. If further we assume that approximately no exchange takes place between the mols of the protein phase and
the external phase of the blood cells, we get this expression for the osmotic
pressure in terms of normality exerted by the fluids of the blood cells,

$$(\mathbf{K}^*_{(c)} + C^*_{A'(c)})\, F_0\,(A'_{(c)}),$$

and for that exerted by the serum,

$$(\mathbf{K}^*_{(s)} + C^*_{A'(s)})\, F_0\,(A'_{(s)}),$$

where $C^*_{A'}$ only refers to the anions the concentration of which varies with
the reaction, the phosphates being excluded, and where an asterisk signifies

that the symbol refers to an external phase. Further we assume with Ege that the osmotic pressure in the blood cells is the same as that in serum, which is approximately true.

On this assumption we will now consider a sample of horse blood in which the volume of the blood cells at $p_H \cdot 6 \cdot 50$ is 40 vols. % of the whole blood, therefore at
$$p_H \cdot 6 \cdot 50 \quad Q = 40,$$
and the disperse phase of the blood cells is 35 vols. % of the blood cells, therefore at $p_H \cdot 6 \cdot 50 \; q_{(c)} = 35$. Moreover we put at $p_H \cdot 6 \cdot 50 \; q_{(s)} = 9$.

Fig. 29.

We can represent this blood diagrammatically at $p_H \cdot 6 \cdot 50$ by Fig. 29. The contents of the cylinder are 100 cc. of blood, the piston is the semi-permeable membrane which separates serum from the blood cells, dividing the cylinder into one compartment of 60 cc. which corresponds to the serum and another of 40 cc. which corresponds to the blood cells. From the serum space a portion $5 \cdot 4$ cc. is cut off which is the volume of the protein phase, and from the blood cell space a portion 14 cc. which is the volume of the blood cell proteins. The piston is assumed to be freely movable without resistance and is kept in place by the osmotic forces. We further assume that the osmotic pressure at $p_H \cdot 6 \cdot 50$ both in serum and blood cells corresponds to an osmotic activity of $0 \cdot 3$ mol. We can therefore write
$$K^*_{(c)} = K^*_{(s)} = 0 \cdot 30 \text{ at } p_H \cdot 6 \cdot 50.$$

If during a change of reaction in the water phase of the blood cells a osmotic activities per litre disappear and in the water phase of the serum b osmotic activities, the osmotic activity of the blood cells will be $0 \cdot 30 - a$, and of the serum $0 \cdot 30 - b$.

If there is always to be the same osmotic activity in the blood cells and in the serum, water must be exchanged between them. We will call the quantity of water, expressed in vols. % of the whole blood, which must pass from blood cells to serum x and therefore
$$(0 \cdot 3 - a) \frac{Q - \dfrac{Q q_{(c)}}{100}}{Q - \dfrac{Q q_{(c)}}{100} - x} = (0 \cdot 3 - b) \frac{100 - Q - \dfrac{100 - Q}{100} q_{(s)}}{100 - Q - \dfrac{100 - Q}{100} q_{(s)} + x}, \quad \ldots(203)$$

from which it appears that the volume change of the blood cells is a function of the volume of the blood cells in the blood, which is easy to understand when we remember that in the limiting case where no serum is present the blood cells are quite unable to alter their volume by the appropriation of water.

In an earlier chapter we found that the combination of CO_2 in blood cells and serum can be expressed by

$$\beta = y - x p_{H^{\cdot}}.$$

For haemolysed horse blood with a combined O_2 content of 24·7 vols. % we found $x = 81$, and for the other haemolysate with 26·0 vols. % combined O_2, $x = 87$. In the latter haemolysate there must have been about 50 vols. % serum and since the constant x for horse serum at room temperature is about 10 we obtain $x = 80$ (*circ.*) for the share of the blood cell proteins. In pure blood cell fluids we may therefore expect a constant of about 160. The newly formed ions distribute themselves between the protein phase and the external phase but at the same time their osmotic activity is somewhat depreciated by the Milner effect. I have therefore calculated that the change in the osmotic activity of the bicarbonate ions on passing from $p_{H^{\cdot}(1)}$ to $p_{H^{\cdot}(2)}$—and consequently after exchange of bicarbonate and chlorine ions—can be expressed in terms of activity by

$$\frac{200\,(p_{H^{\cdot}(1)} - p_{H^{\cdot}(2)})}{2226}.$$

In the case of serum I have estimated half of the compound the dissociation of which varies with the reaction comes from the phosphates, and that the change in the osmotic activity of the bicarbonate ions (and the chlorine ions exchanged with them) can be expressed by

$$\frac{5\,(p_{H^{\cdot}(1)} - p_{H^{\cdot}(2)})}{2226}.$$

As the variation in the combined CO_2 is a direct measure of the altered ionisation of the proteins we have, in the case under consideration,

$$a = \frac{200}{2226}\,(p_{H^{\cdot}} - 6 \cdot 50) \quad \text{and} \quad b = \frac{5}{2226}\,(p_{H^{\cdot}} - 6 \cdot 50).$$

Substituting these values in (203) we get

$$x = \frac{125\,(p_{H^{\cdot}} - 6 \cdot 50)}{24 \cdot 2 - 2 \cdot 4\,(p_{H^{\cdot}} - 6 \cdot 50)}, \quad \dots\dots\dots\dots\dots\dots(204)$$

and expressing the volume of the blood cells as a percentage of the volume at $p_{H^{\cdot}}$ 6·50

$$x_1 = 100 - \frac{1250\,(p_{H^{\cdot}} - 6 \cdot 50)}{97 - 9 \cdot 7\,(p_{H^{\cdot}} - 6 \cdot 50)}. \quad \dots\dots\dots\dots\dots(205)$$

In the first column of the following table the $p_{H^{\cdot}}$ values are given; in the second, the volume expressed as a percentage of the volume at $p_{H^{\cdot}}$ 6·50 read directly from curve II, Fig. 9; and in the last column the same amount calculated with (205).

Table LXIV.

$p_{H^{\cdot}}$	Volume of blood cells in % of their volume at $p_{H^{\cdot}}$ 6·50	
	Found	Calculated with (205)
6·50	100	100
6·80	95·2	96
7·00	92·7	93
7·20	90·7	91
7·40	89·1	88
7·60	87·8	85
7·80	86·3	81

It will be seen that the calculation indicates more shrinking of the blood cells than is actually found. In the calculation however the assumption has been made that $p_{H'}$ is the same in blood cells and serum but as shown in chapter VII this is not true. Taking this into account and using for the reaction of the blood cells the value found from Fig. 11, curve II the agreement between the found and calculated results becomes surprisingly good.

By substitution in, and modification of, equation (203) in accordance with the above it becomes

$$x_1 = 100 \; \frac{14200 \; (0\cdot090 \; (p_{H'(e)} - 6\cdot50) - 0\cdot002 \; (p_{H'(e)} - 6\cdot50))}{97 - (9\cdot2 \; (p_{H'(e)} - 6\cdot50) + 0\cdot49 \; (p_{H'(e)} - 6\cdot50))} \; . \quad \ldots\ldots(206)$$

Table LXV below gives the results of calculation with this revised equation compared with the experimentally obtained values.

Table LXV.

$p_{H'(a)}$	$p_{H'(e)}$	Volume of blood cells in % of their volume at p_H 6·50	
		Found	Calculated with (206)
6·	6·49	100	100
6·	6·78	95·2	97
7·	6·96	92·7	94
7·	7·13	90·7	91
7·	7·29	89·1	90
7·50	7·45	87·8	88
7·60	7·60	86·3	86

As stated the agreement is very good, better indeed than we had a right to expect. The result is greatly in favour of the theory being the right one and that complicating factors of much importance do not play a part.

Let us revert to equation (198)

$$\frac{C_{HCO'_3(2)}}{C_{HCO'_3(1)}} \times \frac{F_a \, (HCO'_3)_{(2)}}{F_a \, (HCO'_3)_{(1)}} = \frac{C_{Cl'(2)} \, F_a \, (Cl')_{(2)} + C_{HCO'_3(2)} \, F_a \, (HCO'_3)_{(2)}}{C_{Cl'(1)} \, F_a \, (Cl')_{(1)} + C_{HCO'_3(1)} \, F_a \, (HCO'_3)_{(1)}},$$

from which we approximately assume, at the concentrations employed,

$$\frac{a_{HCO'_3(2)}}{a_{HCO'_3(1)}} = \frac{C_{Cl'(2)} + C_{HCO'_3(2)}}{C_{Cl'(1)} + C_{HCO'_3(1)}}, \quad \ldots\ldots\ldots\ldots\ldots\ldots(207)$$

which in conjunction with (194) and (195) gives

$$\frac{a_{HCO'_3(2)}}{a_{HCO'_3(1)}} = \frac{(C_{M'I(2)}) + (2C_{M''II(2)}) - (C_{A'I(2)}) - (2C_{A''II(2)})}{(C_{M'I(1)}) + (2C_{M''II(1)}) - (C_{A'I(1)}) - (2C_{A''II(1)})} . \quad \ldots\ldots(208)$$

Indicating the concentrations relating to the external phases of blood cells and serum with an asterisk we get

$$\frac{a^*_{HCO'_3(e)}}{a^*_{HCO'_3(s)}} = \frac{(C^*_{M'I(e)}) + (2C^*_{M''II(e)}) - (C^*_{A'I(e)}) - (2C^*_{A''II(e)})}{(C^*_{M'I(s)}) + (2C^*_{M''II(s)}) - (C^*_{A'I(s)}) - (2C^*_{A''II(s)})} . \quad \ldots(209)$$

The numerator in the fraction on the right side of the equation stands for the difference between the non-permeating cations and anions in the external phase of the blood cells, and the denominator for the same difference in the external phase of the serum. At a reaction in the neighbourhood of p_H 6·50 the ionic activity in serum and blood cells is the same as regards the permeating ions. Consequently the difference referred to in the external phase of the blood cells and serum must also be of the same magnitude if (209) is to be satisfied. And since the osmotic pressure of serum is almost identical with that of a $0.15m$ NaCl solution we can put the difference at this reaction without incurring a great error, at 0.15. If we also assume that the variation in the difference can be expressed in the same way as in the case of the volume of the blood cells we obtain

$$\frac{a^*_{HCO'_3\,(c)}}{a^*_{HCO'_3\,(s)}} = \frac{\left(0.15 - \dfrac{200}{2226}(p_{H^·\,(c)} - 6.50)\right)(173 - 0.73x)}{\left(0.15 - \dfrac{5}{2226}(p_{H^·\,(s)} - 6.50)\right)(1.54x_1 - 54)}\,, \qquad \ldots\ldots(210)$$

where x_1 is determined by (206). The results are given in Table LXVI. In the third column the most probable values for $\dfrac{a^*_{HCO'_3\,(c)}}{a^*_{HCO'_3\,(s)}}$ will be found, being the reciprocals of the ratio between the apparent hydrogen ion activities of blood cells and serum.

Table LXVI.

$p_{H^·(s)}$	$p_{H^·(c)}$	$\dfrac{a^*_{HCO'_3(c)}}{a^*_{HCO'_3(s)}}$		$\dfrac{Found}{Calculated}$
		Found	Calculated	
6·49	6·50	(1·02)	1·00	(1·02)
6·80	6·78	0·96	0·90	1·07
7·00	6·96	0·91	0·84	1·08
7·20	7·13	0·85	0·77	1·10
7·40	7·29	0·78	0·66	1·15
7·70	7·45	0·71	0·59	1·20
7·80	7·60	0·63	0·49	1·29

According to calculation the difference in activity should be considerably greater than it is found to be, but it must be remembered the approximation we made in evolving (207) is not very good.

On the whole the results obtained, including the last part of the problem, support the theories advanced although it is quite evident several of the assumptions made are only rough approximations and are undoubtedly susceptible of considerable refinement.

Donnan [1911] has developed a simple relation between the ionic equilibrium of solutions separated by a semi-permeable membrane and the electrical potential difference between the two sides. The following is taken from Donnan and W. C. McC. Lewis.

Let us return again to the system (183) in the equilibrium (184)

R'	$Na^·$
$Na^·$	
Cl'	Cl'

Let π_1 and π_2 be the potentials of the positive electricity in solutions (1) and (2). If we transfer isothermally the very small amount of positive electricity $F\delta n$ from (2) to (1) we have the two following relations:

(a) The increase in free electric energy = $F\delta n\,(\pi_1 - \pi_2)$.

(b) $p\delta n$ mols of the $Na^·$ ions are transferred from (2) to (1) and simultaneously $q\delta n$ mols Cl' ions from (1) to (2), where $p + q = 1$ (p and q represent the fraction of the total current which the respective ions have carried, or in other words they are the transference numbers of the ions). The maximal work for (b) is in an ideal solution given by

$$p\,\delta n\, \mathrm{R\,T} \log \frac{C_{\mathrm{Na}^{\cdot}\,(2)}}{C_{\mathrm{Na}^{\cdot}\,(1)}} + q\,\delta n\, \mathrm{R\,T} \log \frac{C_{\mathrm{Cl}'\,(1)}}{C_{\mathrm{Cl}'\,(2)}},$$

and as the electric forces balance the osmotic we obtain

$$\mathrm{F}\,\delta n\,(\pi_{(1)} - \pi_{(2)}) = p\,\delta n\, \mathrm{R\,T} \log \frac{C_{\mathrm{Na}^{\cdot}\,(2)}}{C_{\mathrm{Na}^{\cdot}\,(1)}} + q\,\delta n\, \mathrm{R\,T} \log \frac{C_{\mathrm{Cl}'\,(1)}}{C_{\mathrm{Cl}'\,(2)}}. \quad \dots\dots(211)$$

It has been shown above that the following relation holds

$$\frac{C_{\mathrm{Na}^{\cdot}\,(2)}}{C_{\mathrm{Na}^{\cdot}\,(1)}} = \frac{C_{\mathrm{Cl}'\,(1)}}{C_{\mathrm{Cl}'\,(2)}},$$

and since
$$p + q = 1, \quad \dots\dots\dots\dots\dots\dots(212)$$

we get
$$\pi_{(1)} - \pi_{(2)} = \frac{\mathrm{R\,T}}{\mathrm{F}} \log \frac{C_{\mathrm{Na}^{\cdot}\,(2)}}{C_{\mathrm{Na}^{\cdot}\,(1)}} = \frac{\mathrm{R\,T}}{\mathrm{F}} \log \frac{C_{\mathrm{Cl}'\,(1)}}{C_{\mathrm{Cl}'\,(2)}}. \quad \dots\dots\dots(213)$$

As it has already been shown that the permeable ions distribute themselves with variable activity between blood cells and serum, a potential difference will exist between the two sides of the membrane of the blood cells. This can easily be calculated by

$$\pi_{(s)} - \pi_{(c)} = \frac{\mathrm{R\,T}}{\mathrm{F}} \log \frac{a_{\mathrm{H}^{\cdot}\,(s)}}{a_{\mathrm{H}^{\cdot}\,(c)}} = \frac{\mathrm{R\,T}}{\mathrm{F}} \log \frac{a^{*}_{\mathrm{HCO}'_3\,(c)}}{a^{*}_{\mathrm{HCO}'_3\,(s)}}. \quad \dots\dots\dots(214)$$

At 18° we therefore have

$$\pi_{(s)} - \pi_{(c)} = 0.058\,(p_{\mathrm{H}^{\cdot}\,(s)} - p_{\mathrm{H}^{\cdot}\,(c)}). \quad \dots\dots\dots\dots(215)$$

If we now calculate the numerical values of (215) with the aid of the most probable corresponding values of $p_{\mathrm{H}^{\cdot}(c)}$ and $p_{\mathrm{H}^{\cdot}(s)}$ we obtain

Table LXVII.

$p_{\mathrm{H}^{\cdot}(s)}$	$p_{\mathrm{H}^{\cdot}(c)}$	$\pi_{(s)} - \pi_{(c)}$ volt
6·49	6·50	− 0·006
6·80	6·78	+ 0·0012
7·00	6·96	+ 0·0023
7·20	7·13	+ 0·0041
7·40	7·29	+ 0·0064
7·60	7·45	+ 0·0087
7·80	7·60	+ 0·0116

The table shows that serum is positively charged against the blood cells, in other words the blood cells wander to the anode. At a reaction in the neighbourhood of $p_{\mathrm{H}^{\cdot}}$ 6·50 the blood cells change the sign of their charge. As far as one can judge, this result is in harmony with the facts but the experimental investigations into the charge of the blood cells will hardly bear a searching criticism. Although it is far from being my purpose to deal thoroughly with the views put forward concerning membrane potentials in the physiological literature (muscle and nerve physiology is particularly rich in them), it will be worth while adding a few remarks about the above calculations. Those especially interested are referred to Zangger's [1908] and Beutner's [1920] monographs and to the textbooks of physico-chemical biology, particularly Höber's [1914] and McClendon's [1917].

The above calculations are distinguished from all earlier calculations and views touching upon the membrane potentials at the surfaces of cells in that they are based on Donnan's theory. The new fact is made use of that it does not matter which permeating ion we credit with the production of the potential as all the permeating ions give rise to the same potential. A condition of the views advanced is here—as elsewhere in this work—that equilibrium is reached in the various systems, or that the equilibria at all events are so nearly approached that any deviations which occur take place very slowly. From this point of view they differ fundamentally from Beutner's ideas as he investigated the potential difference for systems of two aqueous electrolyte solutions separated by non-aqueous phases in which the cations and anions diffused with very different but nevertheless appreciable speed, so that the development of the potential difference between the watery solutions was determined by the different rate of diffusion in the intermediate liquid. Consequently the potential will not have been developed in Beutner's systems when equilibrium is reached.

As early as 1904 Höber [1904, 2; 1914] had experimentally examined the potentials on the surface of the blood cells. He used a chamber the height of which was only 100μ. It seems probable that in such a shallow chamber Höber could not ensure that the movement of the blood cells he observed under the influence of a weak current did not arise from the motion of the water along the surface of the glass since, according to Burton [1916], the stream of water is present for a distance of about 25μ. Höber conjectures from his experiments that the blood cell membrane is originally impermeable for cations and anions, and only becomes permeable for anions after CO_2 treatment. Some of H. Nasse's [1878] and Hamburger's [1902] experiments on the diffusion of chlorine in blood on blowing gas through it had however already made Höber's view improbable. Fridericia [1920] has also recently shown that it must be fallacious. I shall not further discuss Höber's contentions as I feel convinced Höber himself could not wish to retain them in their original form after the appearance of Donnan's work.

A few years ago Fåhraeus [1918] attributed the tendency of the blood cells to aggregate (and consequent rapid sinking) to a diminished potential on their surface. The experimental technique of Fåhraeus however leaves so much to be desired that the hypothesis advanced stands on insecure foundations. In a recently published very exhaustive investigation on "The Suspension-stability of the Blood," Fåhraeus [1921] seems partly to have relinquished his original view as he has shown that the substances which favour the aggregation of blood cells are the fibrinogen and the globulins of the plasma without however having elicited any definite connection between the amount of fibrinogen and the potential. H. C. Gram [1921] independently of Fåhraeus has come to the same conclusion regarding the relation between the fibrinogen of the plasma and the rate of sinking of the blood cells by the examination of a very large number of samples of human blood. Linzenmeyer's

[1920, 1921] work also supports to some extent this view, the relation of which to the haematology of bygone days is so excellently portrayed by Fåhraeus.

I have attempted to carry out new determinations of the charge on the blood cells but I encountered great experimental difficulties. Such determinations would be of considerable interest for the problems before us as we could test the accuracy of (215) with their help.

In concluding this work I would again draw attention to the very great difficulties involved in bringing the blood into equilibrium with a new CO_2 tension, a fact which we time and again encounter in the literature and which vitiates the significance of a large portion of the experiments reported there. It is generally assumed it is the mixing that is the sluggish process in the saturation of the blood cells with CO_2, and that the diffusion of CO_2 through the membrane of the blood cells, the interchanging of the ions and the ionisation of the proteins take place rapidly in comparison with it. This has however not been experimentally proved but the question of the cause of the difficulties of saturation gains physiological importance if we try to argue from the combined CO_2 at equilibrium in the blood *in vitro* to the conditions *in vivo*.

Résumé.

I. Donnan's theory of the distribution of permeating ions in two-phase systems containing non-permeating ions is discussed and developed particularly with regard to its application to blood.

II. The distribution of chloride between the blood cells and serum of horse blood has been determined.

III. It has been proved that the distribution of bicarbonate and chlorine ions between blood cells and serum is in accord with Donnan's theory.

IV. The relation between chlorine ion and bicarbonate ion exchange has been discussed on the basis of experiments by van Slyke and Cullen, Fridericia, and the author.

V. The view that the volume change of the blood cells which is determined by the reaction is an osmotic phenomenon has been strengthened.

VI. The electric potential of the surface of blood cells has been calculated with Donnan's equation.

VII. It has been shown that it is theoretically impossible directly to determine the apparent chlorine ion activity of the blood with an electrode of the second class.

I am greatly indebted to all those who in various ways have helped me while this work was in progress. My thanks are especially due to my teacher, Dr K. A. Hasselbalch, late director of the laboratory of the Finsen Institute in Copenhagen, and to the present director, Dr C. Sonne, for the excellent facilities for working afforded me as well as for their kindness and assistance.

Prof. Niels Bjerrum has on numerous occasions discussed various physico-chemical problems with me and he has kindly read through chapters I, II, III and VIII. Without his aid it would hardly have been possible to write these sections, and I must express my deep gratitude to him.

Lastly I am indebted to Dr E. E. Atkin of the Lister Institute, London, for the great care with which he has translated the work from the Danish.

BIBLIOGRAPHY.

The bibliography has been elaborated essentially with regard to the problem of the nature of the combination of carbon dioxide in the blood.

The older literature dealing with allied subjects is collected in H. J. Hamburger's *Osmotischer Druck und Ionenlehre etc.*, I, Wiesbaden, 1902.

The handbooks and larger monographs in the list are marked with an asterisk. Reference will be found in them to the more recent work on physico-chemical biology. The collection of the literature was practically completed in the first half of 1921.

Abbott, G. A. and Bray, W. C. (1909). The Ionisation Relations of Ortho- and Pyro-phosphoric Acids and their Sodium Salts. *J. Amer. Chem. Soc.* **31**, 729.

Adolph, E. F. (1920). The Liberation of CO_2 from Carbonate by Blood and Serum. *J. Physiol.* **54.** *Proc.* xxxiv.

Arrhenius, S. (1888). Theorie der isohydrischen Lösungen. *Zeitsch. physikal. Chem.* **2**, 284.

* —— (1912). *Theories of Solution.* New Haven.

* —— (1915). *Lehrbuch der Elektrochemie*, 3 Aufl. Leipzig.

Auerbach, F. and Pick, H. (1912). Die Alkalität wasseriger Lösungen kohlensaurer Salze. *Arbeiten aus dem K. Gesundheitsamte*, **38**, 243.

*Bang, I. (1916). *Methoden zur Mikrobestimmung einiger Blutbestandteile.* Wiesbaden.

Barcroft, J. (1913). The Combination of Haemoglobin with Oxygen and with Carbon Monoxide. *Biochem. J.* **7**, 481.

* —— (1914). *The Respiratory Function of the Blood.* Cambridge.

Barr, D. P. and Peters, Jr, J. P. (1920–1921). The carbon dioxide absorption curve and carbon dioxide tensions of the blood in severe anaemia. *J. Biol. Chem.* **45**, 571.

Bayliss, W. M. (1909). The Osmotic Pressure of Congo-red, and some other Dyes. *Proc. Roy. Soc.* B, **81**, 269.

—— (1911). The Osmotic Pressure of electrolytically dissociated Colloids. *Proc. Roy. Soc.* B, **84**, 229.

* —— (1915). *Principles of General Physiology.* London.

—— (1919). The Neutrality of the Blood. *J. Physiol.* **53**, 162.

*Bert, P. (1870). *Leçons sur la physiologie comparée de la respiration.* Paris, 1–30, 66–152.

* —— (1878). *La pression barométrique.* Paris.

—— (1878). Sur l'état daus lequel se trouve l'acide carbonique du sang et des tissues. *Compt. Rend.* **87**.

Beutner, R. (1920). *Die Entstehung elektrischer Ströme in lebenden Geweben u.s.w.* Stuttgart.

Bie, V. and Möller, P. (1913). Undersøgelser af normale Menneskers Blod. Ugeskrift for Læger.

Bjerrum, N. (1909). A new Form for the electrolytic Dissociation Theory. *7th Internat. Congr. Appl. Chem. Sect.* x, 58.

* —— (1914). *Die Theorie der alkalimetrischen und azidimetrischen Titrierungen.* Stuttgart. (Samml. Ahrens.)

—— (1916). De stærke Elektrolyters Dissociation. *Forh. ved 16 skand. Naturforskermøde* 1916.

—— (1918). Die Dissoziation der starken Elektrolyten. *Zeitsch. Elektrochem.* **24**, 321.

—— (1919). On the Activity-coefficients for Ions. *Medd. fråņ Nobelinst.* **5**.

Bjerrum, N. and Gjaldbaek, J. K. (1919). Undersøgelser over de Faktorer, som bestemmer Jordbundens Egenskaber. I. Om Bestemmelser af en Jords sure eller basiske Egenskaber. II. Om Reaktionen af Væsker som er mættede med Calciumcarbonat. Landbohøjskolens Aarsskr. (Deutscher Zusammenfassung: I. Über die Bestimmung der sauren und basischen Eigenschaften des Bodens. II. Über die Reaktion von Flüssigkeiten die mit Calciumkarbonat gesättigt sind.)

Blasel, L. and Matula, J. (1914). Unters. über physik. Zustandsänderungen der Kolloide. XVI. Versuche mit Desaminoglutin. Biochem. Zeitsch. 58, 417.

Bohr, C. (1886). Über die Verbindung des Hämoglobins mit Kohlensäure. Festschrift für Ludwig. Leipzig, 164.

—— (1891). Beiträge zur Lehre von den Kohlensäureverbindungen des Blutes. Skand. Arch. Physiol. 3, 47.

—— (1898). Über die Verbindungen von Methämoglobin mit Kohlensäure. Skand. Arch. Physiol. 8, 363.

—— (1904). Theoretische Behandlung der qualitativen Verhältnisse der Kohlensäureverbindung des Hämoglobins. Centr. Physiol. H. 23.

—— (1905). Absorptionscoefficienten des Blutes und des Blutplasmas für Gase. Skand. Arch. Physiol. 17, 104.

—— (1909). Blutgase und respiratorischer Gaswechsel. Nagel's Handb. d. Physiol. 1, 54–129. Braunschweig.

—— and Bock, J. C. (1891). Bestimmung der Absorption einiger Gase in Wasser bei Temperaturen zwischen 0° und 100°. Wiedemann's Ann. 44.

—— Hasselbalch, K. A. and Krogh, A. (1904). Uber den Einfluss der Kohlensäureverbindung auf die Sauerstoffaufnahme im Blute. Centr. Physiol. H. 23.

*Boruttau, H. (1910). Blut und Lymfe. Nagel's Handb. d. Physiol. Ergänzungsbd. 1–78. Braunschweig.

Botazzi, F. (1911). Physikalisch-chemische Untersuchung des Harns und der anderen Körperflüssigkeiten. C. Neuberg: Der Harn u.s.w. II. Berlin.

Boyle, R., cit. Loewy (1870). Die Gase des Körpers.

Buckmaster, G. A. (1917, 1). The Relation of Carbon Dioxide in the Blood. J. Physiol. 51, 105.

—— (1917, 2). On the Capacity of Blood and Haemoglobin to unite with Carbon Dioxide. J. Physiol. 51, 164.

—— and Gardner, J. A. (1911–1912). The Nitrogen-content of Blood. J. Physiol. 43, 401.

Bugarsky, S. and Liebermann, L. (1898). Über das Bindungsvermögen eiweisshaltiger Körper für Salzsäure, Natriumhydroxyd und Kochsalz. Pflüger's Arch. 72, 51.

*Burton, E. F. (1916). The physical Properties of colloidal Solutions. London.

Campbell, J. M. H. and Poulton, E. P. (1920). The Relation of Oxyhaemoglobin to the CO₂ of the Blood. J. Physiol. 54, 152.

Cannon, W. B., Fraser, J. and Hooper, A. N. (1917). Some Alterations in the Distribution and Character of the Blood in Shock. Medical Research Committee. Wound Shock and Haemorrhage. London, 1919, 49.

Christiansen, J., Douglas, C. G. and Haldane, J. S. (1914). The Absorption and Dissociation of Carbon Dioxide by human blood. J. Physiol. 48, 244.

Clark, W. M. (1915). A Hydrogen Electrode Vessel. J. Biol. Chem. 23, 475.

* —— (1920). The determination of hydrogen ions. Baltimore, reprinted 1921.

—— and Lubs, H. A. (1916). Hydrogen Electrode Potentials of Phthalate, Phosphate and Buffer Mixtures. J. Biol. Chem. 25, 479.

Dale, H. H. and Lovatt Evans, C. (1920). Colorimetric Determination of Reaction of Blood by Dialysis. J. Physiol. 54, 168.

Datta, A. K. and Dhar, N. (1915). A New Method of finding the Second Dissociation Constants of Dibasic Acids. J. Chem. Soc. 107, 824.

Davies, H. W., Haldane, J. B. S. and Kennaway, E. L. (1920). Experiments on the Regulation of the Blood's Alkalinity. I. J. Physiol. 54, 32.

Davy, H. (1803). (Edinburgh Med. and Surg. J. 29), cit. Henry, W. Gilbert's Ann. 591.

Donegan, J. F. and Parsons, T. R. (1919). Some Further Observations on Blood Reaction. J. Physiol. 52.

Donnan, F. G. (1911). Theorie der Membrangleichgewichte und Membranpotentiale bei Vorhandensein von nicht dialysierenden Elektrolyten. Ein Beitrag zur physikalisch-chemische Physiologie. Zeitsch. Elektrochem. 17, 572.

—— and Allmand, A. J. (1914). Ionic Equilibrium across Semi-permeable Membranes. J. Chem. Soc. 105, 1941.

*Edwards, W. F. (1824). De l'influence des agens physiques sur la vie. Paris.

*Ege, R. (1919). *Studier over Glukosens Fordeling mellem Plasmaet og de røde Blodlegemer o.s.v.* København.

—— (1920, 1). Études sur la distribution du glucose entre le plasma et les globules rouges du sang et sur quelques problèmes qui s'y rapportent. *Compt. Rend. Soc. Biol.* 83, 697.

—— (1920, 2). Über die Restreduktion des Blutes. *Biochem. Zeitsch.* 107, 229.

——' (1920, 3). Zur Frage der Permeabilität des Blutkörperchen gegenüber Glucose und Anelektrolyten. *Biochem. Zeitsch.* 107, 246.

—— (1920, 4). Über die Bestimmung des Blutkörperchenvolumen. *Biochem. Zeitsch.* 109, 242.

—— (1921, 1). Die Verteilung der Glucose zwischen Plasma und roten Blutkörper. *Biochem. Zeitsch.* 111, 189.

—— (1921, 2). Wie ist die Verteilung der Glucose zwischen den roten Blutkörperchen und den äusseren Flüssigkeit zu erklären? *Biochem. Zeitsch.* 114, 88.

—— (1921, 3). Studien über das osmotische Verhältniss der Blutkörperchen. I. Untersuchungen über das Volumen der Blutkörperchen in gegenseitig osmotischen Lösungen. *Biochem. Zeitsch.* 115, 109.

—— (1921, 4). II. Der osmotischen Druck in Blutkörperchen und Plasma. *Biochem. Zeitsch.* 115, 175.

*Eichwald, E. and Fodor, A. (1919). *Die physikalisch-kemischen Grundlagen der Biologie.* Berlin.

Enderlin, C. (1844). Physiologisch-chemische Untersuchungen. *Ann. d. Chem. u. Pharm.* 49, 317; 50, 53.

Evans, C. Lovatt (1921, 1). On a probable error in determinations by means of the hydrogen electrode. *J. Physiol.* 54, 353.

—— (1921, 2). The regulation of the reaction of the blood. *J. Physiol.* 55, 159.

Fahr, G. (1912). New Determinations of certain Absorption Coefficients in Blood-gas Investigations. *J. Physiol.* 43, 417.

Fåhraeus, R. (1918). Über die Ursachen der verminderten Suspensionsstabilität der Blutkörperchen während der Schwangerschaft. *Biochem. Zeitsch.* 89, 355.

* —— (1921). The Suspension-stability of the Blood. I–IV. *Acta medica scandinavica*, 55, 1–228.

Falta, W. and Richter-Quittner, M. (1920). Über die Verteilung des Zuckers, der Chloride und der Reststickstoffkörper auf Plasma und Körperchen im strömenden Blute. *Biochem. Zeitsch.* 100, 148.

—— (1921). Über die chemische Zusammensetzung der Blutkörperchen. *Biochem. Zeitsch.* 114, 144.

Farkas (1903). Über die Konzentration der Hydroxylionen im Blutserum. *Pflüger's Arch.* 98, 551.

Fernet, É. (1855). Note sur la solubilité des gaz dans les dissolutions salines, pour servir à la théorie de la respiration. *Compt. Rend.* 41, 1237.

—— (1857). Du rôle des principaux éléments du sang dans l'absorption ou le dégagement des gaz de la respiration. *Ann. Sc. Nat. (Zool.)* 8, 125.

Findlay, A. (1908). Einfluss von Kolloiden auf die Absorption von Gasen, insbesondere von Kohlendioxyd in Wasser. *Kolloid. Zeitsch.* 3, 169.

* —— (1913). *Osmotic Pressure.* London.

* —— (1915). *The Phase Rule and its Applications.* London.

—— and Creighton, H. J. (1909). The Influence of Colloids and Colloid Suspensions on the Solubility of Gases in Water. *7th Internat. Congr. Appl. Chem.* Sect. x.

—— —— (1910). The Influence of Colloids and Fine Suspensions on the Solubility of Gases in Water. I. Solubility of Carbon Dioxide and Nitrous Oxide. *J. Chem. Soc.* 97, 536.

—— —— (1911). Some Experiments on the Solubility of Gases in Ox Blood and Ox Serum. *Biochem. J.* 5, 294.

—— and Shen, B. (1912). II. The Solubility of Carbon Dioxide and Hydrogen. *J. Chem. Soc.* 101, 1459.

—— and Wililams, T. (1913). III. Solubility of Carbon Dioxide at Pressures lower than atmospheric. *J. Chem. Soc.* 103, 637.

—— and Howell, O. R. (1915). IV. The Solubility of Carbon Dioxide in Water in the Presence of Starch. *J. Chem. Soc.* 107, 282.

*Frédéricq, L. (1878). *Recherches sur la constitution du plasma sanguin.* Gand.

*Freundlich, H. (1909). *Kapillarchemie.* Leipzig.

Fridericia, L. S. (1920). Exchange of chloride Ions and of Carbon Dioxide between Blood Corpuscles and Blood Plasma. *J. Biol. Chem.* 42, 245.

*Förster, F. (1915). *Elektrochemie wasseriger Lösungen.* 2. Aufl. Leipzig.

Gad Andresen, K. L. (1920). Über die Verteilung der Reststickstoffkörper auf Plasma und Körperchen im strömenden Blute. *Biochem. Zeitsch.* 107, 250.

*Garoia, J. M. de Corral y (1914). *La reacción actual de la sangre y su determinacion electrometrica.* Valladolid.

Gaule, J. (1878). Die Kohlensäurespannüng im Blut, im Serum und in der Lymphe. *Arch.* (*Anat.*) *Physiol.* 469.

Gjaldbaek, J. K. (1921). Undersøgelser over de Faktorer, som bestemmer Jordbundens Reaktion. III. Magniumkarbonatets forskellige Former, og om Reaktionen af Væsker, som er mættede dermed. *Landbohøjskolens Aarsskr.* (Deutscher Zusammenfassung über die verschiedenen Formen des Magnesium-Karbonats und über die Reaktion von Flüssigkeiten, die damit gesättigt sind.)

Gram, H. C. (1921). *Studier over Fibrinmængden i Menneskets Blod og Plasma.* With an English Summary. (Studies on the percentage of fibrin in human blood and plasma.) København.

Gréhant, N. *Les gaz du sang.* Paris.

Gürber, A. (1895, 1). Die Salze des Blutes. *Verhl. physikal.-chem. Ges. Marburg,* N.F. 28, 129.

—— (1895, 2). Über den Einfluss der Kohlensäure auf die Verteilung von Basen und Säuren zwischen Serum und Blutkörperchen. *Sitzungsber. physikal.-med. Ges. Würzburg,* 28.

Hagedorn, H. C. (1920). Einige Bemerkungen über die Verteilung der Glucose zwischen Blutkörperchen und Plasma. *Biochem. Zeitsch.* 107, 248.

Haggard, H. W. and Henderson, Y. (1919). Hemato-respiratory Functions. I. The CO_2 Diagram of the Blood. II. Laws of Respiration. III. Respiratory Decompensation and Acidosis. IV. The C_H' Scale and the Dissociation Characteristic. J. *Biol. Chem.* 39, 163.

—— —— (1920, 1). VII. The reversible Alteration of the H_2CO_3 : $NaHCO_3$ Equilibrium in Blood and Plasma under Variations in the CO_2 Tension and their Mechanism. *J. Biol. Chem.* 45, 189.

—— —— (1920, 2). VIII. The degree of Saturation of Corpuscles with HCl as a condition underlying the amount of Alkali called into use in the Plasma. *J. Biol. Chem.* 45, 199.

—— —— (1920, 3). IX. An irreversible Alteration of the H_2CO_3 : $NaHCO_3$ Equilibrium of Blood, induced by temporary Exposure to a low Tension of CO_2. *J. Biol. Chem.* 45, 209.

—— —— (1920, 4). X. The Variability of the reciprocal Action of Oxygen and CO_2 in Blood. *J. Biol. Chem.* 45, 215.

Hamburger, H. J. (1892). Über den Einfluss der Athmung auf die Permeabilität der roten Blutkörperchen. *Zeitsch. Biol.* 28, 405.

—— (1893). La constitution du sang veineux et du sang artériel. *Arch. physiol. norm. et pathol.* S. v, 5, 1.

* —— (1902). *Osmotischer Druck und Ionenlehre u.s.w.* I. Wiesbaden.

—— (1918). Anionenwanderungen in Serum und Blut unter dem Einfluss von CO_2, Säure und Alkali. *Biochem. Zeitsch.* 86, 309.

—— and Bubanovic, Fr. (1910). La perméabilité des globules rouges, spécialement vis-à-vis des cations. *Arch. internat. Physiol.* 10, 1.

*Hammarsten, O. (1914). *Lehrbuch der physiol. Chemie.* 8. Aufl., 805. Wiesbaden.

Hardy, W. B. (1900, 1). Über den Mechanismus der Erstarrung in umkehrbaren Kolloidsysteme. *Zeitsch. physikal. Chem.* 33, 326.

—— (1900, 2). Eine vorläufige Untersuchung der Bedingungen welche die Stabilität von nicht umkehrbaren Hydrosolen bestimmen. *Zeitsch. physikal. Chem.* 33, 385.

Hasselbalch, K. A. (1910). Elektrometrische Reaktionsbestimmungen kohlensäure-haltiger Flüssigkeiten. *Biochem. Zeitsch.* 30, 317.

—— (1911, 1). Elektrometrisk Reaktionsbestemmelse af kulsyreholdige Væsker. *Medd. fra Carlsberg Lab.* 10, 64.

—— (1911, 2). Détermination électrométrique de la réaction des liquides renfermant de l'acide carbonique. *Compt. Rend. Carlsberg,* 10, 69.

—— (1916, 1). Die reduzierte und die regulierte Wasserstoffzahl des Blutes. *Biochem. Zeitsch.* 74, 56.

— —— (1916, 2). Die Berechnung der Wasserstoffzahl des Blutes aus der freien und gebundenen Kohlensäure desselben, und die Sauerstoffbindung des Blutes als Funktion der Wasserstoffzahl. *Biochem. Zeitsch.* 78, 112.

—— and Lundsgaard, C. (1912). Elektrometrische Reaktionsbestimmungen des Blutes bei Körpertemperaturen. *Biochem. Zeitsch.* 38, 77.

—— and Gammeltoft, S. A. (1915). Die Neutralitätsregulation des graviden Organismus. *Biochem. Zeitsch.* 68, 206.

—— and Warburg, E. J. (1918). Ist die Kohlensäureverbindung des Blutserums als Mass für die Blutreaktion verwendbar? *Biochem. Zeitsch.* 86, 410.

*Hatschek, E. (1916). *An Introduction to the Physics and Chemistry of Colloids.* London.

*Hedin, S. G. (1915). *Grundzüge der physikalischen Chemie in ihrer Beziehung zur Biologie.* Wiesbaden.

Heidenhain, R. and Meyer, L. (1863). Über das Verhalten der Kohlensäure gegen Lösungen von phosphorsaurem Natron. *Studien des physiol. Inst. zu Breslau*, 2, 103.

Henderson, L. J. (1908). A Note on the Union of the Proteins of Serum with Alkali. *Amer. J. Physiol.* 21, 168.

—— (1908). Concerning the Relationship between the Strength of Acids and their Capacity to preserve Neutrality. *Amer. J. Physiol.* 21, 172.

—— (1908). The Theory of Neutrality Regulation in the animal Organism. *Amer. J. Physiol.* 21, 427.

* —— (1909). Das Gleichgewicht zwischen Basen und Säuren im tierischen Organismus. *Ergebn. Physiol.* 254.

—— (1910). On the Neutrality Equilibrium in Blood and Protoplasm. *J. Biol. Chem.* 7, 29.

—— (1920, 1). The Equilibrium between Oxygen and Carbonic Acid in the Blood. *J. Biol. Chem.* 41, 401.

—— (1920, 2). Blood as a physico-chemical System. *J. Bioch. Chem.* 45, 411.

—— and Black, O. F. (1908). Study of the Equilibrium between Carbonic Acid, Sodium Bicarbonate, Mono-sodium Phosphate and Di-sodium Phosphate at Body Temperature. *Amer. J. Physiol.* 21, 420.

—— and Spiro, K. (1909). Zur Kenntniss des Ionengleichgewichts im Organismus. I. *Biochem. Zeitsch.* 15, 106.

Hermann, R. (1834). Über die saure Beschaffenheit des venösen Menschenbluts und über den Unterschied zwischen arteriellen und venösen Blute. *Poggendorff's Ann.* 1 (2), 311.

Hill, A. V. (1910). The possible Effects of the Aggregation of the Molecules of Haemoglobin on its Dissociation Curves. *J. Physiol.* 40. *Proc. iv.*

—— (1913). The Combination of Haemoglobin with Oxygen and with Carbon Monoxide. *Biochem. J.* 7, 471.

—— (1921). The Combination of Haemoglobin with Oxygen and Carbon Monoxide, and the Effects of Acids and Carbon Dioxide. *Biochem. J.* 15, 577.

Höber, R. (1900). Über die Hydroxylionen des Blutes. I. *Pflüger's Arch.* 81, 522.

—— (1903). II. *Pflüger's Arch.* 99, 572.

—— (1904, 1). Resorption und Kataphorese. *Pflüger's Arch.* 101, 607, 627.

—— (1904, 2). Weitere Mitteilungen über Ionenpermeabilität bei Blutkörperchen. *Pflüger's Arch.* 102, 196.

* —— (1909). Physikalische Chemie des Blutes und der Lymfe. C. *Oppenheimer's Handb. der Bioch.* II. 2. Hälfte, 1. Leipzig.

* —— (1914). *Physikalische Chemie der Zelle und der Gewebe.* 4. Aufl. Leipzig and Berlin.

Hüfner, G. (1901). Neue Versuche über die Dissociation des Oxyhämoglobins. *Arch. (Anat.) Physiol.* 187.

Iversen, P. (1920). Untersuchungen über den "säurelöslichen Phosphor" in Blut und Plasma bei verschiedenen Tieren, sowie einige Studien über die Toxikologie der verschiedenen Phosphate. *Biochem. Zeitsch.* 109, 211.

Jahn, H. (1900). Über den Dissociationsgrad und das Dissociationsgleichgewicht stark dissocierter Elektrolyte. I. *Zeitsch. physikal. Chem.* 33, 544.

Jaquet, A. (1892). Über die Wirkung mässiger Säurezufuhr auf Kohlensäuremenge, Kohlensäurespannung und Alkalescenz des Blutes. Ein Beitrag zur Theorie der Respiration. *Arch. exp. Path. Pharm.* 30, 311.

Jarlöv, E. (1919). Om Syre-Baseligevægten i det menneskelige Blod særlig ved Sygdomme. København.

Joffe, J. and Poulton, E. P. (1920). The Partition of CO_2 between Plasma and Corpuscles in oxygenated and reduced Blood. *J. Physiol.* 54, 129.

Johnston, J. (1916). Determination of carbonic acid, combined and free, in solution, particularly in natural waters. *J. Amer. Chem. Soc.* 38, 947.

Jolyet and Sigalas (1892). Sur l'azote du sang. *Compt. Rend.* 114, 686.

Kendall, J. (1912). The Velocity of the Hydrogen Ion, and a general Dissociation Formula for Acids. *J. Chem. Soc.* 101, 1275.

—— (1916). The specific Conductivity of pure Water in Equilibrium with atmospheric Carbon Dioxide. *J. Amer. Chem. Soc.* 38, 1480.

Koeppe, H. (1897). Der osmotische Druck als Ursache des Stoffaustausches zwischen roten Blutkörperchen und Salzlösungen. *Pflüger's Arch.* 67, 189.

—— (1903). Über das Lackfarbenwerden der roten Blutscheiben. *Pflüger's Arch.* 99, 33.

Konikoff, A. P. (1913). Über die Bestimmung der wahren Blutreaktion mittels der elektrischen Methode. *Biochem. Zeitsch.* 51, 200.

*Korányi, A. v. and Richter, P. F. (1907-1908). *Physikalische Chemie und Medizin.* Leipzig.

*Kraus, F. (1898). Über die Verteilung der Kohlensäure im Blute. *Festsch. Univers. Graz.* Graz.

Krogh, A. and Liljestrand, G. (1921). Eine Mikromethode zur Bestimmung der Kohlensäure des Blutes. *Biochem. Zeitsch.* **104**, 300.

*Lehfeldt, R. A. (1908). *Electro-chemistry.* I. London.

Lehmann, C. (1894). Untersuchung über die Alkalescenz des Blutes und speciel die Einwirkung der Kohlensäure daran. *Pflüger's Arch.* **58**, 428.

Levy, R. L., Rowntree, L. G. and Marriot, W. McK. (1915). A simple method for determining the Hydrogen-ion Concentration of the Blood. *Arch. internal. Med* **16**, 389.

Lewis, G. N. (1908). Umriss eines neuen Systems der chemischen Thermodynamik. *Zeitsch. physikal. Chem.* **61**, 129.

*Lewis, W. C. McC. (1916). *A System of Physical Chemistry.* I-II. London.

Limbeck, R. v. (1894). Über den Einfluss des respiratorischen Gaswechsel auf die roten Blutkörperchen. *Arch. exp. Path. Pharm.* **35**, 309.

Lindhard, J. (1914). *Undersøgelser over Hjærtets Minutvolumen i Hvile og under Muskelarbejde.* København.

—— (1921). Colorimetric determination of the concentration of hydrogen ions in very small quantities of blood by dialysis. *Compt. Rend. Carlsberg,* **14**, No. 13.

Linzenmeyer, G. (1920). Untersuchungen über die Senkungsgeschwindigkeit der roten Blutkörperchen. I. *Pflüger's Arch.* **181**, 169.

—— (1921). II. *Pflüger's Arch.* **186**, 272.

Loeb, J. (1921, 1). Ion series and the properties of proteins. I. *J. General Physiol.* **3**, 85.

—— (1921, 2). II. *J. General Physiol.* **3**, 247.

—— (1921, 3). III. *J. General Physiol.* **3**, 391.

—— (1921, 4). Donnan equilibrium and the physical properties of Proteins. I. Membrane potentials. *J. General Physiol.* **3**, 667.

—— (1921, 5). II. Osmotic Pressure. *J. General Physiol.* **3**, 667.

—— (1921, 6). III. Viscosity. *J. General Physiol.* **3**, 827.

—— (1922). The origin of electrical charges of colloidal particles of living tissues. *J. General Physiol.* **4**, 351.

Loewy, A. (1894). Untersuchungen zur Alkalescenz des Blutes. *Pflüger's Arch.* **58**, 462.

* —— (1911). Die Gase des Körpers. C. *Oppenheimer's Handb. d. Bioch.* IV, 1. Hälfte, 1-79. Leipzig.

—— and Zuntz, N. (1891). Über die Bindung der Alkalien in Serum und Blutkörperchen. *Pflüger's Arch.* **58**, 511.

—— —— (1892). Einige Beobachtungen über die Alkalescenzänderungen des frisch entleerten Blutes. *Pflüger's Arch.* **58**, 507.

Ludwig, C. (1865). Zusammenstellung der Untersuchungen über Blutgase. Separat-Abdruck aus den med. Jahrbüchern. *Zeitsch. Ges. Aerzte Wien.*

*Lundén, H. (1908). *Affinitätsmessungen an schwachen Säuren und Basen.* Stuttgart. (Samml. Ahrens.)

Magnus, G. (1837). Über die im Blute enthaltenen Gase, Sauerstoff, Stickstoff und Kohlensäure. *Poggendorff's Ann.* **40**, 583.

'Manube, K. and Matula, J. (1913). Untersuchung über physik. Zustandsänderungen d. Kolloide. XV. Elektrochem. Unters. am Säureeiweiss. *Biochem. Zeitsch.* **52**, 369.

Marchand, R. F. (1846). Ueber die Anwesenheit der kohlensauren Salze in dem Blute. *J. pr. Chem.* **37**, 321.

*McClendon, J. F. (1917). *Physical Chemistry of Vital Phenomena.* Princeton.

—— Shedlov, A. and Thomson, W. (1917). Tables for finding the alkaline Reserve of Blood Serum in Health and Acidosis, from the total CO_2 or the alveolar CO_2 or the p_H at known CO_2 Tension. *J. Biol. Chem.* **31**, 519.

McCoy, H. N. and Smith, H. J. (1911). Equilibrium between Alkali-earth Carbonates, Carbon Dioxide and Water. *J. Amer. Chem. Soc.* **33**, 468.

—— and Test, C. D. (1911). Equilibrium between Sodium Carbonate, Sodium Bicarbonate and Water. *J. Amer. Chem. Soc.* **33**, 473.

McLean, F. C., Murray, H. A. and Henderson, L. J. (1920). The variable Acidity of the Haemoglobin and the Distribution of Chloride in the Blood. *Proc. Soc. Exper. Biol. and Med.* **17**, 180. Quoted from *Medical Sc.* **5**, 88, 1921.

Meyer, L. (1857). *Die Gase des Blutes. Diss.* Göttingen.

Michaelis, L. (1914). *Die Wasserstoffionenkonzentration.* Berlin.

—— (1920). Über die Analyse der Kohlensäuregleichgewichte im Blute nach H. Straub und K. Meier. *Biochem. Zeitsch.* **103**, 53.

—— and Airila, Y. (1921). Die elektrische Ladung des Hämoglobins. *Biochem. Zeitsch.* **118**, 144.

Michaelis, L. and Bien, Z. (1914). Der isoelektrische Punkt des Kohlenoxydhämoglobins und des reduzierten Hämoglobins. *Biochem. Zeitsch.* 67, 198.

—— and Davidoff, W. (1912). Methodisches und sachliches zur elektrometrischen Bestimmung der Blutalkalescenz. *Biochem. Zeitsch.* 46, 131.

—— and Davidsohn, H. (1911). Der isoelektrische Punkt des genuinen und denaturierten Serumalbumins. *Biochem. Zeitsch.* 33, 456.

—— and Mostynski, B. (1910). Der isoelektrische Punkt des Serumalbumins. *Biochem Zeitsch.* 24, 79.

—— and Rona, P. (1909). Elektrochemische Alkalinitätsmessungen an Blut und Serum. *Biochem. Zeitsch.* 18, 317.

—— —— (1914). Die Dissoziationskonstante der Kohlensäure. *Biochem. Zeitsch.* 67, 182.

—— and Takahashi, D. (1910). Die isoelektrischen Konstanten der Blutkörperchenbestandteile und ihre Beziehungen zur Säurehämolyse. *Biochem. Zeitsch.* 29, 439.

Milroy, T. H. (1917). The Reaction Regulator Mechanism of the Blood before and after Haemorrhage. *J. Physiol.* 51, 259.

Misterlich, E., Gmelin, L. and Tiedemann, F. (1834). Versuche über das Blut. *Poggendorff's Ann.* I (2), 289.

Moscati, P. (1784). Osservazione ed esperienze sul sangue, fluido e rappreso; sopra l'azione dell' arterie; e sui liquori, che bollono poco riscaldati nella machine pneumatica, 1783. Quoted from *Crell's Chem. Ann.* 1, 91.

Müller, F. (1904). Über die Ferricyanidmethode zur Bestimmung des Sauerstoffs im Blute ohne Blutgaspumpe. *Pflüger's Arch.* 103, 541.

Nasse, H. (1878). Untersuchungen über den Austritt und Eintritt von Stoffen (Transudation und Diffusion) durch die Wand der Haargefässe. *Pflüger's Arch.* 16, 604.

*Nernst, W. (1913). *Theoretische Chemie.* 7. Aufl. Stuttgart.

Norgaard, A. and Gram, H. C. (1921). Relation between the Chloride Content of the Blood and its percentic Volume of Cells. *J. Biol. Chem.* 49, 264.

Oryng, T. and Pauli, W. (1915). Untersuchungen über physikal. Zustandsänderungen der Kolloide. XIX. Über Neutralsalzeiweissverbindungen. *Biochem. Zeitsch.* 70, 368.

Ostwald, W. (1890). Elektrische Eigenschaften halbdurchlässiger Scheidewande. *Zeitsch. physikal. Chem.* 6, 71.

* —— (1911). *Grundriss der Kolloidchemie*, I. Dresden.

*Ostwald and Luther. *Hand- und Hülfsbuch zur Ausführung physiko-chemischer Messungen.* 3. Aufl. bei R. Luther und K. Drucker. Leipzig.

Parsons, T. R. (1917). On the Reaction of the Blood in the Body. *J. Physiol.* 51, 440.

—— (1919). The Reaction and Carbon Dioxide carrying Power of Blood—a mathematical Treatment. I. *J. Physiol.* 53, 42.

—— (1920). II. *J. Physiol.* 53, 340.

—— and Parsons, W. (1919–1920). Hydrogen Ion Measurements on Blood in the neighbourhood of the Isoelectric Point of Haemoglobin. *J. Physiol.* 53. Proc. c.

—— —— (1922). The relations of Carbon Dioxide in Acidified Blood. *J. Physiol.* 56, 1.

*Pauli, W. (1920). *Kolloidchemie der Eiweisskörper.* I. Dresden and Leipzig.

—— and Matula, J. (1917). Untersuchungen über physikalische Zustandsänderungen der Kolloide. XXI. Über Silbersalzproteine. *Biochem. Zeitsch.* 80, 187.

Peters, R. A. (1914, 1). A combined Ionometer and Electrode Cell for measuring the H ion concentration of reduced blood at a given tension of CO_2. *J. Physiol.* 48. Proc. vii.

—— (1914, 2). Quoted from Barcroft. *The respiratory Function of the Blood.* Appendix III.

Peters, Jr, J. P. and Barr, D. P. (1921). The Carbon Dioxide Absorption Curve and Carbon Dioxide Tension of the Blood in Cardiac Dyspnoea. *J. Biol. Chem.* 45, 537.

—— —— and Rule, F. D. (1921). The Carbon Dioxide Absorption Curve and Carbon Dioxide Tension of the Blood of normal resting Individuals. I. *J. Biol. Chem.* 45, 489.

Petry, E. (1903). Über die Verteilung der Kohlensäure im Blute. *Hoffmeister's Beiträge*, 3, 1.

Pflüger, E. F. W. (1864). *Über die Kohlensäure des Blutes.* Bonn.

Philip, J. C. and Bramley, A. (1915). Influence of Sucrose and Alkali Chlorides on the solvent Power of Water. *J. Chem. Soc.* 107, 377.

Preyer, W. (1866). Über die Kohlensäure und den Sauerstoff im Blute. *Centr. med. Wiss.* 321.

—— (1867). Beiträge zur Kenntniss des Blutfarbstoffs. *Centr. med. Wiss.* 273.

* —— (1871). *Die Blutkrystalle.* Jena.

Procter, H. R. (1914). The Equilibrium of dilute Hydrochloric Acids and Gelatin. *J. Chem. Soc.* 105, 313.

—— and Wilson, J. A. (1916). The Acid-Gelatin Equilibrium. *J. Chem. Soc.* 109, 307.

Robertson, T. B. (1909). On the Nature of the Chemical Mechanism which maintains the Neutrality of Tissue-fluids. *J. Biol. Chem.* **6**, 313.

—— (1910). Concerning the Relative Magnitude of the Parts played by the Proteins and by the Bicarbonate in the Maintenance of the Neutrality of the Blood. *J. Biol. Chem.* **7**, 351.

—— (1908, 1909, 1910). Note on the Applicability of the Laws of amphoteric Electrolytes to Serum Globulin. *J. Biol. Chem.* **5**, 155; **6**, 313; **7**, 351.

* —— (1912). *Die physikalische Chemie der Proteine.* Dresden.

Rona, P. (1910). Über das Verhalten des Chlors im Serum. *Biochem. Zeitsch.* **29**, 501.

—— and György, P. (1913, 1). Über das Natrium und das Carbonation im Serum. Beitrag zur Frage des nicht diffusiblen Alkalis im Serum. *Biochem. Zeitsch.* **48**, 278.

—— —— (1913, 2). Beitrag zur Frage der Ionenverteilung im Blutserum. *Biochem. Zeitsch.* **56**, 416.

Rose, H. (1835). Über die Verbindungen der Alkalien mit Kohlensäure. *Poggendorff's Ann.* **34**, 149.

*Rothmund, V. (1907). *Löslichkeit und Löslichkeitsbeeinflussung.* Leipzig.

Schmidt, A. (1867). Über die Kohlensäure in der Blutkörperchen. *Ber. k. Sächs. Ges. Wiss. Math.-physikal. Cl.* I. 173.

Schöffer, A. (1861). Die Kohlensäure des Blutes und ihre Ausscheidung mittels der Lunge. *Sitzungsber. k. Akad. Wiss. Wien,* **41**, 589.

—— (1868). Die Kohlensäure des Blutes. *Centr. med. Wiss.* 657.

Scudamore, C. (1826). *Versuche mit dem Blut,* 84–90. Würzburg. Quoted from Zuntz, *Nagel's Handbuch.*

Sertoli, E. (1868). Über die Bindung der Kohlensäure im Blute und ihre Ausscheidung in der Lunge. *Hoppe-Seyler's med.-chem. Unters.* **3**, 350.

Setschenow, I. (1859). Beiträge zur Pneumatologie des Blutes. *Sitzungsber. k. Akad. Wiss. Wien,* **36**, 293.

—— (1879). Über die Konstitution der Salzlösungen auf Grund ihres Verhalten zur Kohlensäure. *Nouv. mém. Soc. Imp. sc. nat. de Moscou,* **15**, 199.

Seyler, C. A. and Lloyd, P. V. (1917, 1). Studies of Carbonates. II. Hydrolysis of Sodium Carbonate and Bicarbonate and the Ionisation Constants of Carbonic Acid. *J. Chem. Soc.* **111**, 138.

—— —— (1917, 2). Studies in Carbonates. III. Lithium, Calcium and Magnesium Carbonates. *J. Chem. Soc.* **111**, 994.

Shields, J. (1893). Über Hydrolyse in wässerigen Salzlösungen. *Zeitsch. physikal. Chem.* **12**, 166.

Siebeck, R. (1909). Über die Aufnahme von StickOxydul im Blut. *Skand. Arch. Physiol.* **21**, 368.

Siegfried, M. (1905, 1). Über die Bindung von Kohlensäure durch amphotere Amidokörper. I. *Zeitsch. physiol. Chem.* **44**, 85.

—— (1905, 2). II. *Zeitsch. physiol. Chem.* **46**, 401.

—— and Neumann, C. (1908). III. *Zeitsch. physiol. Chem.* **54**, 423.

—— and Libermann, H. (1908). IV. *Zeitsch. physiol. Chem.* **54**, 436.

* —— (1910). Der Carbamino- und Hydroxylkohlensäure-reaktion. *Ergebnisse Physiol.* **9**, 334.

Slyke, D. D. van (1921, 1). The normal and abnormal variations of the acid-base balance of the blood. *J. Biol. Chem.* **48**, 153.

—— (1921, 2). The Carbon Dioxide Carriers of the Blood. *Physiol. Rev.* **1**, 141.

—— and Cullen, G. E. (1917). Studies of Acidosis. I. The Bicarbonate Concentration in the Blood Plasma, and its Determination as a Measure of Acidosis. *J. Biol. Chem.* **30**, 289.

Smith, L. W., Means, J. H. and Woodwell, M. N. (1920–1921). Studies on the distribution of carbon dioxide between cells and plasma. *J. Biol. Chem.* **45**, 245.

Sörensen, S. P. L. (1909–1910). Études enzymatiques. II. Sur la mesure et l'impotence de la concentration des ions hydrogènes dans les réactions enzymatiques. *Compt. Rend. Carlsberg,* **8**.

* —— (1912). Über die Messung und Bedeutung der Wasserstoffionenconcentration bei biologischen Prozessen. *Ergebnisse Physiol.* **12**, 393.

—— (1915–1917). Studies on Proteins. *Compt. Rend. Carlsberg,* **15**.

Spiro, K. and Henderson, L. J. (1909). Zur Kenntniss des Ionengleichgewichts im Organismus. II. Einfluss der Kohlensäure auf die Verteilung von Elektrolyten zwischen Blutkörperchen und Plasma. *Biochem. Zeitsch.* **15**, 114.

Stevens, W. (1832). *Observations on the healthy and diseased Properties of the Blood.* London.

—— (1835). Observations on the Theory of Respiration. Communicated by W. T. Brande. *Phil. Trans.* 345.

Stieglitz, J. (1909). The Relation of Equilibrium between the Carbon Dioxide of the Atmosphere and the Calcium Sulphate, Calcium Carbonate and Calcium Bicarbonate in contact with it. *Carnegie Inst. Publ.* **107**, 233.

Straub, H. and Meier, K. (1918, 1). Blutgasanalysen. I. Qualitativer und quantitativer Nachweis von Säuren in kleinen Blutmengen durch Bestimmung von Verteilungsgleichgewichten. *Biochem. Zeitsch.* **89**, 156.

—— —— (1918, 2). II. Hämoglobin als Indicator. Ein Beitrag zur Theorie der Indicatoren. *Biochem. Zeitsch.* **90**, 305.

—— —— (1920, 1). III. Die Chlorionenpermeabilität menschlicher Erytrocyten. *Biochem. Zeitsch.* **98**, 205.

—— —— (1920, 2). IV. Der Einfluss der Alkalikationen auf Hämoglobin und Zellmembran. *Biochem. Zeitsch.* **98**, 228.

—— —— (1920, 3). V. Der Einfluss der Erdalkalien auf Hämoglobin und Zellkolloide. *Biochem. Zeitsch.* **109**, 47.

Stromeyer, E. C. F. (1832). Ist wirklich freie Säure im Blute? *Schweiger's J.* IV, **4**, 95.

Thiel, A. and Strohecker, R. (1914). Über die wahre Stärke der Kohlensäure. *Ber.* **47** (1), 945.

Thorup, S. (1887). *Om Blodets CO₂-Binding.* København.

Tobiesen, F. (1895, 1). *Bidrag til Læren om Blodets Rolle ved Vævenes respiratoriske Stofskifte.* København.

—— (1895, 2). Über den spezifischen Sauerstoffgehalt des Blutes. *Skand. Arch. Physiol.* **6**.

Vogel, A. (1814). Versuche mit Urin. I. Über die Existenz der Kohlensäure im Urin und im Blute. *Schweiger's J.* **9**, 399.

Walbum, L. E. (1914). Die Bedeutung der Wasserstoffionenkonzentration für' die Hämolyse. *Biochem. Zeitsch.* **63**, 221.

Walker, J. and Cormack, W. (1900). The Dissociation of very weak Acids. *J. Chem. Soc.* **77**, 5.

Walpole, G. S. (1913). Gas-electrode for general use. *Biochem. J.* **7**, 410.

Warburg, E. J. (1920). Einige Bemerkungen über die Verteilung von Anionen zwischen Blutkörperchen und Plasma. *Biochem. Zeitsch.* **107**, 252.

*Weimarn, P. P. von (1911). *Grundzüge der Dispersoidchemie.* Dresden.

*Williams, C. J. B. (1834–1835). Observations on the Changes produced in the Blood in the Course of its Circulation. With Experiments. Read before the R. Med. Soc. of Edinburgh 1823, etc. *London Med. Gazette*, **16**, 718, 724, 745–751, 783–788, 807–813, 842–848, 871–876.

*Willstätter, R. and Stoll, A. (1918). *Untersuchungen über die Assimilation der Kohlensäure*, 182–185. Berlin.

*Zangger, H. (1908). Über Membranen und Membranenfunktionen. *Ergebnisse Physiol.* **7**, 99.

*Zigmondy, R. (1918). *Kolloidchemie.* 2. Aufl. Leipzig.

Zuntz, N. (1867). Über den Einfluss des Partialdrucks der Kohlensäure auf die Verteilung dieses Gases im Blute. *Centr. med. Wiss.* 529.

—— (1868). *Beiträge zur Physiologie des Blutes. Diss.* Bonn.

* —— (1882). Physiologie der Blutgase und des respiratorischen Gaswechsels. *Hermann's Handb. Physiol.* Leipzig.

XXIII. THE EFFECT OF COLD STORAGE ON THE CARNOSINE CONTENT OF MUSCLE.

By WINIFRED MARY CLIFFORD.

From the Physiology Department, Household and Social Science Department, Kensington.

(*Received February 1st, 1922.*)

THE flavour of meat has long been believed to be associated with the presence of extractives, although whenever isolated in a pure condition such substances are tasteless. However one extractive—β-alanyl-histidine or carnosine—does disappear in cold storage and this may account for the inferior flavour of imported meat when compared with home-killed.

The idea of such a disappearance was suggested by differences found when estimating carnosine in various samples of beef, by a colorimetric method previously described in this journal [Clifford, 1921]. Research on this extractive was begun just after the war when the beef commonly available was imported. Such samples gave yields of 0·35–0·37 % carnosine. On one occasion some English steak was obtained and gave the unexpectedly large yield of 1·1 %. It has been shown that for a given species the carnosine content of muscle is constant [Clifford, 1922] and the great difference between English and imported beef could not be explained unless cold storage had effected a reduction of the base. In order to test this various samples of English and imported meats were analysed.

CARNOSINE CONTENT OF ENGLISH AND IMPORTED MEAT.

Beef.

Twelve separate samples of English beef have been examined over a period of two years with the following results: 1·1, 1·0, 1·1, 0·98, 0·96, 1·0, 0·97, 0·98, 1·1, 0·98, 0·98, 0·97 %. Average 1·0 %.

Five samples of imported beef gave: 0·37, 0·34, 0·36, 0·35, 0·36 %. Average 0·356 %.

Similar experiments have been carried out on veal, mutton and lamb and in every case the carnosine content of imported meat has been very much lower than that of English meat. Actual figures obtained were:

	English	Imported
Beef	0·96–1·1 % (12 samples)	0·34–0·37 % (5 samples)
Veal	1·05–1·12 (5 ,,)	0·34–0·36 (4 ,,)
Mutton	0·37–0·38 (3 ,,)	0·13–0·16 (2 ,,)
Lamb	0·40–0·42 (2 ,,)	0·15–0·16 (2 ,,)

In all these experiments about one-third of the carnosine value was found in the imported meats as compared with English killed.

EXPERIMENTS ON THE EFFECT OF COLD STORAGE ON MEAT.

Since all the samples of imported meat were perfectly fresh and in an edible condition, the loss of carnosine could not be a putrefactive change, but might have been due to the length of time the meat had been in cold storage.

It was possible to test this idea owing to the kindness of Dr H. H. Dale who allowed a sample of English meat to be placed in a cold room in the Research Institute at Hampstead. Portions were removed at various times for analysis. When first put in, carnosine was present to the normal value of 0·99 %. Samples were taken at intervals of about one month. A fall of carnosine at first great, then slight, followed by a second steep fall was shown (Fig. 1).

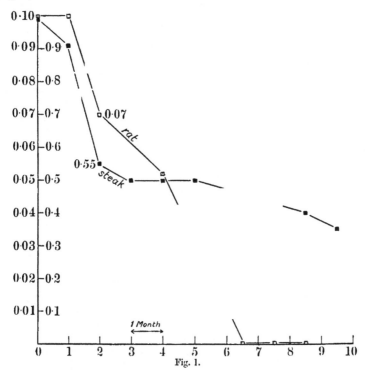

Fig. 1.

The actual figures were:

Original	0·99 % carnosine	4 months	0·50 % carnosine
1 month	0·90 ,,	6 ,,	0·48 ,,
2 months	0·55 ,,	8½ ,,	0·40 ,,
3 ,,	0·50 ,,	9½ ,,	0·35 ,,

EFFECT OF COLD STORAGE ON RAT MUSCLE.

The loss of carnosine in beef might have originated in some process which took place before the meat was put into the cold room. Therefore seven rats were killed with coal gas and put into cold storage immediately after death. The normal value of 0·11 % carnosine was given by one animal, estimated directly, and the others were removed at intervals. The curve of carnosine loss ran parallel with that for beef for a period of four months (Fig. 1). This was followed by a steep fall in the fifth month and by six and a half months all trace of the base had disappeared.

Results were:

Original	0·11 % carnosine	6½ months	Absent
1 month	0·11 ,,	7½ ,,	,,
2 months	0·07 ,,	8½	
4 ,,	0·052 ,,		

The above experiment shows that loss of carnosine occurs in muscle put into a cold room directly after the death of the animal in a similar manner to that shown by bought English meat kept at the same temperature.

The temperature of the cold room was just below 0° C. Spicules of ice were found in every sample analysed and the extractions were made before thawing out had completed.

DISCUSSION OF RESULTS.

From experimental results it may be stated that measurement of the depth of red colour produced on diazotising carnosine in a watery muscle extract gives an easy and rapid test for distinguishing between English fresh killed and cold storage meats. A sample of beef or veal with a carnosine percentage below 0·8 % or of mutton or lamb below 0·3 % would not come from a freshly killed animal, but probably from a carcase which had been chilled in order to preserve it. If the percentage were as low as 0·3 % in the case of beef or 0·15 % with mutton the sample would be 9–12 months old. Further experiments are in progress with the object of obtaining a curve by which the age of cold storage meat may be read off after determining the carnosine content.

The mechanism causing the loss of carnosine is unknown, but experiments are being carried out to investigate the problem. Since the reaction takes place at freezing point it is improbable that bacterial or enzyme action can account for the change, which is more likely to be initiated by a simpler type of catalyst.

Thanks are due to Dr H. H. Dale for allowing the meat to be deposited in the cold room at the Research Institute, Hampstead, and to Prof. V. H. Mottram for suggestions during the course of the work.

The expenses of this research were defrayed by the Medical Research Council.

REFERENCES.

Clifford (1921). *Biochem. J.* 15, 400.
—— (1921). *Biochem. J.* 15, 725.

XXIV. INVESTIGATIONS ON THE NITROGENOUS METABOLISM OF THE HIGHER PLANTS.

PART II.

THE DISTRIBUTION OF NITROGEN IN THE LEAVES OF THE RUNNER BEAN[1].

By ALBERT CHARLES CHIBNALL.

From the Biochemical Department, Imperial College of Science and Technology.

(*Received February 1st, 1922.*)

In the first part of this series [Chibnall and Schryver, 1921] a method of isolating part of the protein complex of fresh leaves was described. Subsequent work carried on with the same object, mentioned in that paper, in view, namely that of throwing light on the protein metabolism in the leaf, showed that it was necessary to investigate two problems in detail. Firstly the nature of the nitrogen retained in the residue of cellular matter after applying the above method remained to be determined, and secondly it was necessary to find out how far it is justifiable to compare leaves picked at different, and even unknown, periods of growth.

The present research on the leaves of the runner bean was primarily undertaken to solve these two problems. A method has been evolved whereby the whole of the nitrogenous material in the leaf is obtained in a state suitable for subsequent analysis. The water-soluble products are freed from all protein substances other than proteoses, about three-fourths of the protein N of the leaf is isolated as an amorphous powder, whilst the residue of the cell matter, containing 5–15 % of the total leaf N, is shown to contain only protein N. This makes it possible to estimate the total protein and water-soluble N in the leaf.

By applying this method at frequent intervals during the life history of the runner bean leaf, an attempt has been made to solve the second problem enumerated above. Not only has the ratio protein N to water-soluble N been found, but by suitable analysis either before or after hydrolysis, the distribution of the protein N in five groups, and of the water-soluble N in seven groups, has been determined.

A study of the protein groups indicates that the protein undergoes little, if any, change during the life history of the leaf, a fact that should be of assistance in future research into the chemistry of the leaf proteins. The water-soluble groups however are continually varying, indicating that comparisons of the composition of leaves, unless regard is paid to exact conditions

---

[1] Thesis approved for the Degree of Doctor of Philosophy in the University of London.

of age and growth, is almost impossible [compare Jodidi, Kellogg and True, 1920].

As the research progressed it was found possible to extend the scope of enquiry to the diurnal variations, and the effects of starvation, the results of which throw considerable light on the protein metabolism in the leaf. The synthesis of protein in the leaf from nitrates appears to take place through the amino acids, whilst the products of protein degradation, in striking analogy to what is found in the animal kingdom, pass into what appear to be urea derivatives.

A certain amount of research has been performed in the past to throw light on most of the problems discussed in this paper [see Czapek, 1920], but it was nearly all carried out before the development of modern methods for estimating the nitrogenous bases and monoamino acids, so that comparison of results is of very little value.

I. GENERAL EXPERIMENTAL METHODS.

Materials used. The plant used was *Phaseolus vulgaris* v. *multifloris* (Scarlet Champion). The seeds were planted in unmanured ground following potatoes on the 28th April, 1921, at the Physic Gardens, Chelsea.

The seedlings appeared above the ground about 20 days later, and all ages of plants stated henceforth are calculated from the 18th May. The growth was normal, flowers appearing when the plant was five to six weeks old. At this time a few aphides appeared, but attention prevented them spreading, and all evidence of the disease had vanished three weeks later. By the middle of July it was found that although the inflorescence was normal, very few of the flowers had set, with the result that the number of pods formed was small, not averaging more than three or four per plant. This failure must be attributed to the abnormal dryness and high temperature of the atmosphere and not to the drought experienced during July, August and September, since the plants received abundance of water daily.

Methods of sampling. The procedure adopted was to pick the leaves from the plants selected, go lightly over the surface with a brush, remove all stems and take the total weight. It is from this weight, obtained before any appreciable evaporation from the surface of the leaves could have taken place, that the "N per fresh weight of the leaves," and hence all the figures given later in Tables XI and XIII, are calculated. The leaves were then counted, sorted out into six pans according to size, the content of each pan weighed, and the sample for grinding (200–250 g.) taken proportionately from each. By this means an average sample, giving a mean weight per leaf similar to that of the whole batch, was obtained. From several of the batches duplicates were taken to test the accuracy of the sampling and experimental methods used. The remainder of the batch was air-dried in an oven at 37° through which a slow stream of air was drawn.

Each batch picked was given a series number, which is retained throughout in all subsequent operations. Details of the pickings, with the weather conditions, are given in Tables I and II. On account of the exceptionally fine weather conditions throughout the experiment the leaves were picked in a normal state without excess of moisture. Furthermore, the absence of high wind and storms kept the surface of the leaves clean and free from soil material. Only in one case, that of series 4, did rain fall in the preceding two or three days. In this rain fell during the night, but when the sample was taken in the morning no excess of moisture, as shown by lightly pressing between filter-papers, was found. Fresh weight determinations then can be considered as exceptionally good.

Dry weight and total nitrogen of the leaves. The air-dried remainder was used for this purpose. The dry weight was taken from a small sample placed for 24 hours in an oven heated to 108°.

The total N was determined by the modified Kjeldahl-Gunning method, whereby the nitrate N is first converted into amino-phenol by salicylic acid and sodium hyposulphite. The practical details followed were those given by Moore [1920].

General methods of manipulation (separation of protein, soluble products and cellular matter). Two methods which aim at this separation have so far been described, that of Osborne and Wakeman [1920] and the one communicated in the first paper on these investigations [Chibnall and Schryver, 1921]. In the former the leaves were passed through a Nixtamel mill three times; the pulp obtained mixed with water and returned twice more to the mill. The result was a green solution containing protein in colloidal suspension, from which the cellular matter was removed by centrifuging. The protein was afterwards flocculated by the addition of alcohol.

Briefly the second method consists of treating the minced leaves with ether-water and pressing out through thick muslin in a tincture press, whereby a green solution, containing most if not all of the water-soluble matter and part of the protein in colloidal suspension, was obtained. The protein was afterwards flocculated by warming. For the purpose of the present research this method, although easy to manipulate and rapid, was of little use unless the nitrogenous matter of the residue, about one-third of the total, could be separated from the cell matter for further investigation.

Osborne's method was certainly more thorough, the residue in this case containing only 18 % of the total N. It was inconvenient for the present purpose however, as it would have required the continuous use of a centrifuge. It was therefore decided to see if the ether-water method could not be improved.

Some cabbage leaves were accordingly treated by this method and the colloidal solution and residue examined microscopically, using a ⅛th objective.

The solution appeared colourless, with two distinct sets of particles. The first, deep green in colour and irregular in shape, were obviously fragments of disintegrated chloroplasts, and were large enough to be free from movement. The second were very numerous, showed no trace of green colour, and were small enough to be in rapid Brownian movement. On warming to coagulate the protein the particles were all removed, and the coagulate appeared as a greyish granular mass in which the fragments of the green chloroplasts could be distinctly seen.

The residue to the naked eye showed large disintegrated lumps of leaf matter. One of the smaller of these was examined under the microscope. The outer layer of cells had their walls broken and the major part of the content expelled; those inside were unruptured but flattened. In addition the vacuole content of the latter had been largely expressed, even though the layer of protoplasm was unbroken.

It seemed clear from this examination that the rupturing of some of the cells was due to the action of the cutter in the mincing machine, and that the function of the ether-water was merely to kill the semi-permeable membrane of protoplasm in the remainder, so that the fluid content could be squeezed out by the press. The large quantity of N remaining in the residue therefore was due to the protoplasmic membrane remaining in the unruptured cells. Clearly then to reduce the N content of the residue it was necessary to increase the number of cells ruptured.

No access could be had to a mill of the type used by Osborne. The only one available was that known technically as an End-runner Mill, which is nothing more than a mechanical pestle and mortar. It was found that if 200 g. of cabbage leaves were broken up and fed slowly into this, then 100 cc. of water added, the whole was reduced to a very fine green pulp in about 20 minutes. On pressing through muslin a solution, deeper coloured than before, was obtained. The grinding and pressing was repeated on the residue, and the N in the two colloidal extracts and final residue determined. The first extract contained 74·2 %, the second 14·8 %, and the residue 11 % of the total N.

Under the microscope the solution appeared much as before, except that the fragments of green chloroplasts were much more numerous and irregular in size, ranging from quite coarse lumps, which appeared however free from cell matter, down to small, but distinctly green, particles in Brownian movement.

The residue appeared different. The large masses of leafy matter had disappeared. Lumps of cell matter however could still be seen under the microscope, though the number of unruptured cells had greatly diminished. In addition numerous shreds of broken cells, containing only infrequent portions of protoplasm, were to be seen.

It is on this property—that the cellular matter is not ground down to a powder (but only ruptured and torn so as to set free the cell content)—that the success of the method, which was used throughout the present research, depends. The thick muslin and the tincture press were found to be quite unnecessary. All that is required is a glass funnel opening to a diameter of about 8 inches and a piece of fine cotton lawn 18 inches square. The whole of the liquid, with its heavy charge of colloidal matter, is expressed by very slight hand pressure in under a minute, leaving behind a pasty mass that can be easily removed from the lawn and returned to the mill for further grinding. The fact that this fine lawn can be used renders the method applicable for quantitative work, since none of the cell matter or protein can be retained in its pores, and the amount of water-soluble N absorbed is negligible.

Reference to Table IV will show how complete a separation was obtained in the present research. Six grindings of the samples (200–250 g.) were given and the colloidal solution, as a safeguard against the pores of the lawn being enlarged by the passage of the liquid under slight pressure, was run through a second piece spread out over another funnel.

In several of the intermediate batches ether-water was used in place of distilled water, but no benefit was experienced and it was later discarded. In case some of the cells should have escaped disruption, and thereby still be retaining their water-soluble contents, the residue from the sixth grinding was thrown into 200 cc. of boiling water. The cellular matter at once swelled up, imbibing the water, and on cooling this was squeezed through the same lawn as before. The operation was then repeated. In the earlier workings, when the total N in the residue was low, very little N passed into these two extracts. In the later however, when the residual N rose to 12 % and over, sufficient to make an appreciable difference was found in the first, though a negligible quantity was always found in the second. The protein was flocculated in the first six series by warming to 45°, but the temperature was afterwards raised to 60°, since it was found that the precipitate was then rendered almost granular, and could be readily filtered in a few minutes, whereas before it had taken about an hour. The filtering was accomplished by means of a Soxhlet extractor thimble standing upright in a glass funnel. After all the liquid had drained through, the retained protein was washed with two extracts from the boiled residue described above. After draining a second time the thimble was then transferred to an extractor described by Schryver [1908] and washed for three or four hours with 95 % alcohol. As first filtered off the protein is a pasty mass, part of which clings to the sides of the thimble, but the majority collects in a large button at the bottom. As the alcohol washes it the surface becomes dehydrated and shrinks, thereby coming away from the wall of the thimble. The whole can then be shaken out and broken, so as to allow the alcohol access to a large surface and thus hasten the washing. It was returned to the thimble and after a further three or four hours in this extractor the thimble was removed and stood overnight in absolute alcohol. In the morning it was placed in a Soxhlet apparatus and extracted with ether until all colouring matter was removed. The protein was then shaken out of the thimble, ground up in a small mortar, returned to the thimble and left 24 hours to dry. The weight of the thimble was determined before the operation, and the increase when it contained the air-dried protein was taken as the weight of the latter.

In the previous communication by Chibnall and Schryver [1921] it was found that after the colloidal protein of cabbage was flocculated by warming, a secondary flocculation could be obtained by raising the temperature to about the boiling point. The same was found to occur in the present case, but the amount was small, and diminished when the temperature of the first flocculation was raised to 60°. Experiment on a moderate quantity of this second flocculent substance prepared from the cabbage showed that it was formed from some water-soluble protein by coagulation, since it was insoluble in all the solvents tried, and the percentage of N in it was about 12. In the present case it was separated by filtering on a Buchner funnel, and the filter paper, with the retained coagulate, transferred to a Kjeldahl flask and the total N estimated.

The aqueous extract, after boiling to remove all flocculent matter, was cooled, its volume taken, and stored in a stoppered bottle with a little chloroform and toluene for future use. As a precaution against any unknown reaction or separation on standing, the distribution of N in this was performed without delay. The alcohol and ether washings of the protein were evaporated down, transferred to a Kjeldahl flask and the total N estimated. The residue was air-dried at 37° in the same oven used for the main bulk of the leaves, as stated above. Total N was estimated by Kjeldahl's method.

It will not be out of place here to give some idea as to the time occupied in the separation. The grinding (10 mins. first, 6–7 mins. the others) and squeezing out take between 1 and 1¼ hours; raising the temperature to 60°, 5 mins., filtering through the Soxhlet thimble 15 mins., boiling the filtrate 10 mins., and the separation of the coagulated protein about 15 mins. About two hours, therefore, after the leaf cells are killed the colloidal protein is in alcohol, the clear aqueous extract has been boiled, cooled and shaken up with antiseptics, whilst the residue, containing none of the water-soluble N, is drying at 37°.

DISTRIBUTION OF N IN THE WATER-SOLUBLE PRODUCTS.

The total N in solution. Determined by the modified Kjeldahl-Gunning method mentioned above.

Nitric nitrogen. Reduction by Devada alloy in alkaline solution in place of the more usual Schulze-Tiemann method was employed, as the same apparatus as that required for free ammonia and amide N could be utilised. This consisted essentially of a two-necked Claisen flask without side tube, connected by means of an ordinary Kjeldahl steam trap to a double-surface condenser leading into the distillate flask of 500 cc. capacity containing $N/100$ sulphuric acid. The apparatus was set up in duplicate, the two larger flasks being immersed in the same water-bath. Both were also connected to the same vacuum pump (Geryk), so that similar conditions as to T and P were maintained in each.

The procedure was as follows: 50 cc. of the extract was run into each flask late in the afternoon, then 10 cc. of 30 % NaOH, and into *one only* a small quantity of powdered Devada alloy. In this latter the evolution of hydrogen proceeded steadily during the night, with reduction of the nitric N to ammonia. Sufficient quantity of $N/100$ acid had been run into the receiver flasks to cover the tube leading from the condenser and the apparatus closed. In the morning 400 cc. of ammonia-free distilled water was run in by means of a tap funnel, the water-bath warmed to about 45° and the duplicate apparatus evacuated. Distillation was allowed to proceed for about two hours, 200 to 300 cc. of liquid coming over. The difference between the ammonia found in the two receivers was taken as that due to reduction of the nitric N. How necessary is the blank distillation with soda will be seen from the fact that the ammonia coming over was about twice that obtained from the same volume of extract using magnesia (as below), indicating a very slight decomposition by the strong alkali of some nitrogenous bodies (proteoses?) in the solution. Duplicate readings of nitric N by this method, without any regard being paid to the exact T or P, agreed to 0·2 cc., an accuracy in the solutions containing the lowest concentration of nitric N of 1 : 100, sufficient for the present research.

Ammonia nitrogen. 50 cc. or, if sufficient solution was available, 100 cc. of the extract were run into the Claisen flasks, then 400 cc. of ammonia-free distilled water followed by sufficient cream of magnesia to make the solution distinctly alkaline. Duplicate determinations were made at the same time. The amount of ammonia measured was equivalent to 1·5–3·5 cc. of $N/100$ acid, a quantity sufficiently small to be affected by errors due to the action of the MgO on the proteoses (see Nash and Benedict [1921] for discussion on methods of estimating free NH_3 content of blood). The figures given in the tables therefore only indicate the very low concentration of ammonium salts.

Amide nitrogen (asparagine and glutamine N). Sachsse's [1873] method was employed, whereby the extract was boiled with 4 % HCl for two hours. Sufficient extract to contain one-fourth to one-third of the total N in solution was used for this purpose. After this mild hydrolysis the solution was concentrated *in vacuo* almost to dryness to remove HCl, transferred with about 400 cc. of ammonia-free water to the Claisen flask described above, then cream of magnesia added to make the solution distinctly alkaline. The total ammonia present was found by distillation *in vacuo*, and from this the free ammonia originally present in the solution, as estimated above, was deducted. When the details of this research were planned, it was not realised that any quantity of proteoses would be present. It is possible that these might give amide N under the conditions stated above [see Osborne and Nolan, 1920], but the fact that no amide N at all can be detected in the samples picked at night shows that the error under this heading is not appreciable.

Humin nitrogen and nitrogen precipitated by phosphotungstic acid. After the distillation for amide nitrogen the procedure follows that of Hausmann for the hydrolysis products of protein, as modified by Osborne and Harris [1903].

Monoamino nitrogen. The filtrate and washings from the phosphotungstic acid precipitation were made just alkaline to litmus with 30 % NaOH, then just acid by a few drops of glacial acetic acid. They were then concentrated *in vacuo* to about 100 cc. and made up to a standard volume. Monoamino N was determined by Van Slyke's method.

"*Other nitrogen.*" Calculated by difference; the value is therefore subject to the sum of the individual errors in the six determinations mentioned above.

Proteose nitrogen. Estimated by saturation with zinc sulphate in acid solution. The precipitate obtained was filtered off, washed with a saturated solution of zinc sulphate until the washings were colourless, and redissolved in water. The solution was then again acidified, saturated with zinc sulphate, the precipitate washed as before, re-dissolved in water and made up to a standard volume. Total N was then estimated in an aliquot part by Kjeldahl's method.

Nitrous nitrogen. This was tested for by both the Griess and meta-phenylenediamine reagents, but none was detected in any of the extracts (before or after heating) prepared in the present research.

DISTRIBUTION OF N IN PROTEINS.

Colloidal protein. Osborne and Harris' modification of Hausmann's method, referred to above, was used, monoamino N being determined by the method of Van Slyke.

Dried residue. When the total N in these (given later in Table IV) was determined they had not been washed in alcohol and ether to remove fats, lecithins, etc., since the extractors were being continuously used for work on the colloidal proteins. Series 7 B and 9 B, which were to be further investigated, were first washed with these solvents, when the N extracted was found to be respectively 0·60 % and 0·78 % of the total leaf N. These two extracted residues were then boiled up with 500 cc. of 1 % HCl for six hours under a reflux condenser, cooled, the undissolved portions filtered off, washed with water, and their N content determined by Kjeldahl's method. In terms of the total leaf N they contained 3·2 % and 4·13 % respectively. About three-fourths of the N left in the dried residue had therefore passed into solution. These extracts were then evaporated to a small bulk *in vacuo*, made up to 100 cc. so that they contained 20 % of HCl, hydrolysed for 16 hours, and the Hausmann numbers determined in a similar way to the colloidal protein above. They are given in Table VI later, and show that the N extracted was of protein origin. It is probable that the 3–4 % of N in the final residues mentioned above was also protein N that had escaped extraction by the dilute acid [compare Osborne and Wakeman, 1920].

Total protein and non-protein N. Previous work on the colloidal leaf proteins [Osborne and Wakeman, 1920; Osborne, Wakeman and Leavenworth, 1921; Chibnall and Schryver, 1921] indicates the possibility of a small quantity of the N being of non-protein origin. In all the analyses therein mentioned, however, either hot or cold alkali was used for extraction, and since the resultant proteins have the properties of alkali-albumins, it is considered here that this "non-protein" N has been set free from the original protein complex by hydrolytic action. The constancy of the distribution of N in the colloidal proteins of the present research (Table V), especially in the starvation experiment to be described later, supports this view. In Table IV therefore the estimated total protein N is taken as the sum of the N in the colloidal precipitate, the dried residue and the protein coagulated by boiling.

There is evidence that more than one protein exists, amongst which may be cited the fact that the Hausmann numbers of the protein in the dried residues differ from those in the colloidal precipitate and also that a small part of the protein, as already mentioned, cannot be flocculated at 60°, but only precipitates when the liquid containing it is boiled. Furthermore, proteoses are present, but separate analysis to determine these could only be performed in the later series.

An analysis of the leaf N (series 9) in terms of protein N, proteose N and non-protein N is given in Table IX, whilst the distribution into seven groups after complete hydrolysis is given in Table X.

II. DESCRIPTION OF EXPERIMENTS.

Three main series of experiments have been carried out to determine:

(a) Seasonal variations.
(b) Diurnal variations.
(c) Results of starvation.

(a) The seasonal variations.

These were studied by a complete analysis of eleven batches of leaves picked at frequent intervals during the life of the plant. Except for the change of temperature at which the colloidal protein was flocculated from 45° to 60°, as stated earlier, the conditions of the individual analyses remained standard throughout.

(b) Diurnal variations (series 5–6 and 9–8).

These were studied at the end of the seventh and eleventh weeks. The day reading was obtained by picking the batch about half-an-hour before sunset, when the leaves had been subjected to 14 hours' light; the night reading by picking half-an-hour before sunrise, when the leaves had been in the dark about six hours.

From a climatic point of view the conditions under which these series were taken were ideal. In each case no rain had fallen for several days previously, nor was there any dew during the night, the relative humidity at the time of the night pickings being 85 % and 75 % respectively. This being so, the large decrease in the dry weight of the leaves during the night (10 % in both cases) can be taken as due entirely to translocation away from the leaf.

(c) Starvation experiments (series 10).

These were designed so to alter the conditions in the leaf that light would be thrown on the mechanism of either protein synthesis or degradation.

At the time that the batch for series 10 was picked two samples, 10 c and 10 D, were cut with long stems, placed in cups containing ordinary tap water so that their stems only were immersed, and left in a subdued light, 10 c for 100 hours and 10 D for 114 hours. In this way the leaves were kept quite fresh, but starved of nitrate and of sugar (other than that due to photosynthesis).

Table I. *Giving details of samples picked.*

Series number	Age above ground in days	Time of day when picked, G.M.T.	Date when picked	Condition of weather	Condition of plants
*1	14	8 A.M.	June 14	Sunny	Height 6–8 ins.
*2	21	,,	,, 21	Dull, no sun	Not very different from above
3	26	,,	,, 13	Sunny	Average height 12 ins. Just starting to climb
*	33	,,	,, 20	Dull, rainy night	Starting to flower
5	46	8.30 P.M.	July 3	Very fine day	Height 6–8 ft., flowers just developing
6	47	2.15 A.M.	,, 4	Fine night	As above
7	60	3 P.M.	,, 17	Very fine day	An appreciable increase of new shoots and leaves
†7N	66	2 A.M.	,, 23	Fine night	Pods just starting to form
8	77	,,	Aug. 3	,, ,,	Fully grown. Fertilisation had not taken place on account of the heat. A few pods still forming
9	81	3 P.M.	,, 7	Very fine day	Ditto
10	97	9 A.M.	,, 23	Fine day	Only a few pods fully formed. Numerous new shoots and leaves still forming
11	143	,,	Oct. 8	,, ,,	Most of the leaves old and beginning to show signs of chlorophyll degeneration. A few new shoots and leaves

* Planted 14 days later than the rest of the plants.
† No analyses were made with this batch.

Table II. *Giving further details of samples picked.*

Series number	Number of plants used	Average number of leaves per plant	Average weight of leaves per plant in g.	Average weight of leaf in g.
1	68	5·7	5·3	0·935
2	50	7·6	5·4	0·715
3	25	15·3	15·6	1·027
4	13	30·5	29·7	0·975
5 } day	12	56·3	50·9	0·904
6 } night	12	53·6	55·6	1·086
7	12	109·0	90·0	0·827
7N	12	—	126·0	—
8 } night	12	218·0	172·0	0·791
9 } day	12	259·0	202·0	0·781
10	6	372·0	245·0	0·683

Table III. (Compare Fig. 1.) *Showing the dry weight and total nitrogen.*

Series number		Fresh weight of sample in g.	Percentage dry weight	Percentage of N per fresh weight	Percentage of N per dry weight
1		160	11·43	0·623	5·44
2		88	13·06	0·671	5·13
3		177	13·43	0·658	4·90
4		191	13·16	0·588	4·48
5	day	193	15·85	0·678	4·28
6	night	236	14·46	0·653	4·51
7		590	15·90	0·762	4·76
8	night	1365	14·42	0·656	4·53
9	day	1925	15·51	0·675	4·35
10		921	16·31	0·724	4·44
11		164	16·13	0·622	3·86

Table IV. *Distribution of protein and water-soluble nitrogen in the leaves.*
(In percentages of total leaf N.)

Series number	N of colloidal protein isolated	N of protein coagulated by boiling	N of residue	Estimated protein N *	N of alcohol and ether washings of colloidal protein	N remaining in solution after boiling
1	59·55	2·84	6·88	69·27	2·53	28·20
2	62·56	1·76	7·15	71·47	2·78	25·75
3	64·90	3·30	5·44	73·64	2·81	23·55
4	60·18	5·29	7·86	73·33	2·93	23·74
5A	56·32	3·61	13·34	73·27	2·44	24·29
5B	50·70	3·88	19·00	73·58	2·52	23·90
6A	56·02	2·97	13·37	72·36	2·75	24·89
6B	48·52	4·36	19·62	72·50	2·80	24·70
7A	59·61	1·34	12·25	73·20	2·46	24·34
7B	58·87	0·74	14·11	73·72	2·30	23·98
8A	61·28	0·68	14·92	76·88	2·02	21·10
8B	59·37	0·91	16·62	76·90	2·06	21·04
9A	58·11	0·68	16·81	75·60	2·72	21·68
9B	59·70	0·76	15·09	75·55	2·88	21·57
10A	56·82	0·68	18·17	75·67	2·17	22·16
10B	52·47	0·82	21·99	75·28	1·07	22·75
11	58·43	1·43	15·83	75·69	1·90	22·41
Starvation experiments:						
10c	49·55	1·79	12·30	63·64	2·22	34·14
10D	51·85	0·72	8·75	61·32	2·50	36·18

* The sum of the three columns on the immediate left.

Table V. *Percentage distribution of nitrogen in the colloidal protein.*

Series number	Amide N	Humin N	Basic N	Monoamino N	"Other" N
1	7·44	3·52	21·65	58·90	8·49
2	6·56	3·50	22·40	60·25	7·29
3	6·68	3·30	21·56	58·60	9·86
4	6·59	3·63	21·08	58·20	10·50
5A	day	6·68	4·18	20·55	59·67	8·78
6B	night	6·36	3·64	19·82	59·54	10·64
7B		6·37	3·57	20·55	59·10	10·41
8B	night	6·14	3·55	19·33	59·25	11·73
9B	day	6·34	3·64	22·00	58·88	9·14
10B		6·50	3·82	21·78	59·22	8·68
*10c		6·60	3·56	21·63	58·50	9·71
11		6·40	3·82	21·32	58·50	9·96

* Starvation experiment.

Table VI. *Showing the distribution of N in the products extracted from the residues by 1 % HCl (after complete hydrolysis).*

Series number	Amide N	Humin N	Basic N	Monoamino N	"Other" N
7ʙ	8·38	5·73	25·18	48·20	12·53
9ʙ	8·19	5·82	24·66	48·78	12·64

Table VII. *Distribution of the water-soluble nitrogen (in percentages of total leaf nitrogen). After hydrolysis with 4 % HCl for two hours.*

Series number	Ammonia N	Asparagine N	Nitric N	Humin N	Proteose + basic N	Mono-amino N	"Other" N
1	0·67	0·79	7·78	0·78	7·63	6·43	4·64
2	0·22	0·62	7·35	0·79	6·77	6·41	3·59
3	0·27	0·94	3·93	0·72	6·29	5·54	5·87
4	0·27	0·51	2·45	0·74	6·67	5·52	7·58
5ᴀ } day	0·47	0·49	2·71	1·32	6·81	9·57	2·92
5ʙ }	0·36	0·43	2·96	0·66	6·25	9·25	3·99
6ᴀ } night	0·45	nil	3·89	1·19	6·60	10·37	2·39
6ʙ }	0·34	nil	3·66	1·07	6·79	10·16	2·77
7ᴀ }	0·39	0·21	3·77	0·83	7·90	9·33	1·91
7ʙ }	0·32	0·24	4·25	1·06	7·49	8·60	2·02
8ᴀ } night	0·34	trace	3·38	0·94	6·87	7·08	2·49
8ʙ }	0·38	trace	3·74	1·07	6·62	7·05	2·18
9ᴀ } day	0·41	0·48	3·04	1·03	6·67	7·72	2·33
9ʙ }	0·44	0·53	2·75	0·99	6·35	7·99	2·45
10ʙ	0·37	0·47	3·53	1·06	6·89	6·81	3·62
11	0·61	0·92	1·28	0·96	6·97	6·61	5·05

Table VIII. *Showing the percentage of nitrogen found in the colloidal precipitates.*

Series number	Percentage of nitrogen	Series number	Percentage of nitrogen	Series number	Percentage of nitrogen
1	11·42	6ᴀ }	11·84	9ᴀ }	11·66
2	11·44	6ʙ }	10·75	9ʙ }	11·71
3	11·76	7ᴀ }	11·75	10ᴀ }	11·54
4	9·73	7ʙ }	11·81	10ʙ }	11·36
5ᴀ }	10·88	8ᴀ }	11·84	10ᴄ } *	11·35
5ʙ }	10·68	8ʙ }	11·75	10ᴅ }	11·42

* Starvation experiment.

Table IX. *Showing the amounts of protein, proteose and non-protein nitrogen in the leaves of series 9 (age 14 weeks).*

	Percentage of total N
Colloidal protein	59·70
Protein in residue extracted by 1 % HCl	10·24
Undetermined in residue	4·13
Coagulated by boiling	0·76
Total protein N	74·83
Proteose N	4·35
Non-protein N, soluble in water... ...	17·22
Total	96·40
Soluble in alcohol and ether	3·60

Table X. *Showing the final analysis of the leaf nitrogen in series 9 B (in terms of total leaf N). After complete hydrolysis.*

	Ammonia N	Nitric N	Amide N	Humin N	Basic N	Mono-amino N	"Other" N
Colloidal protein	—	—	3·78	2·17	13·13	35·16	5·46
Residue	—	—	1·18	0·82	3·54	7·01	1·82
Proteoses, water-soluble N	0·44*	2·73*	1·26	1·08	2·78	9·01	4·25
Coagulate	—	—	—	—	—	Total N {0·76	
Soluble in alcohol and ether	—	—	—	—	—	{3·60	
Total	0·44	2·73	6·22	4·07	19·45	51·18	15·89

* Determined before hydrolysis.

Table XI. (Compare Fig. 1.) *Showing the amounts of protein N, water-soluble N and th components of the water-soluble N in percentages of the fresh weight of the leaves.*

Series number	Protein N	Water-soluble N	Ammonia N	Nitric N	Aspara-gine N	Humin N	Proteose +basic N	Mono-amino N	"Other" N	
1	0·4315	0·1757	0·0042	0·0485	0·0049	0·0048	0·0475	0·0401	0·0289	
2	0·4795	0·1728	0·0045	0·0493	0·0041	0·0053	0·0454	0·0430	0·1241	
3	0·4846	0·1550	0·0017	0·0259	0·0062	0·0047	0·0414	0·0358	0·0386	
4	0·4322	0·1396	0·0016	0·0144	0·0030	0·0043	0·0392	0·0325	0·0446	
5A day	0·4967	0·1647	0·0027	0·0192	0·0031	0·0067	0·0444	0·0637	0·0237	
5B	0·4989	0·1620								
6A night	0·4725	0·1626	0·0026	0·0247	0·0	0·0072	0·0438	0·0670	0·0168	The last seven columns ar the mean o the respectiv A and B value
6B	0·4732	0·1613								
7A day	0·5575	0·1855	0·0026	0·0291	0·0016	0·0072	0·0567	0·0683	0·0149	
7B	0·5620	0·1828								
8A night	0·5045	0·1382	0·0024	0·0234	0·0	0·0066	0·0443	0·0463	0·0152	
8B	0·5045	0·1380								
9A day	0·5102	0·1463	0·0028	0·0196	0·0034	0·0067	0·0439	0·0523	0·0172	
9B	0·5100	0·1458								
10A	0·5477	0·1605	—	—	—	—	—	—	—	
10B	0·5450	0·1647	0·0027	0·0256	0·0034	0·0077	0·0499	0·0493	0·0262	
11	0·4708	0·1394	0·0038	0·0080	0·0058	0·0059	0·0423	0·0411	0·0314	

Table XII. *Showing variations introduced in the distribution of the water-soluble N of series 10 by starvation.*

(Figures given are in percentages of the total leaf N)

Series number	Hydrolysis Strength of HCl	Hydrolysis Time in hours	Nitric N	Ammonia N	Aspara-gine N	Amide N other than asparagine N	Humin N	Proteose N	Basic N	Mono-amino N	"Other" N
10A & B	4%	2	3·53	0·37	0·47	—	1·06	4·59	2·30	6·81	3·62
10c (1)	4	2	3·42	0·46	0·49	—	0·88	(3·60)[1]	0·67	6·95	17·66
10c (2)	20	16	3·42	0·46	0·49	5·69	1·71	—	5·03	8·88	8·46
10D (1)	—	—	—	—	—	—	—	3·52	—	—	—
10D (2)	20	16	(3·42)[2]	(0·46)[2]	(0·49)[2]	6·57	1·72	—	5·61	9·64	8·27
9B (1)[3]	4	2	2·73	0·44	0·53	—	0·99	4·35	2·00	7·76	2·77
9B (2)[3]	20	16	2·73	0·44	0·53	0·73	1·08	—	2·78	9·01	4·25

[1] Estimated by comparison with 10D (1).
[2] Not determined; assumed, by comparison with 10c, to be unchanged.
[3] Extract 9B was used as none of 10A or 10B was available.

Table XIII. *Showing the diurnal variation in the N of series 5–6 and 9–8, in percentages of the fresh weight of the leaves.*

Series number		Total N	Protein N	Water-soluble N	Ammonia N	Nitric N	Aspara-gine N	Humin N	Proteose +basic N	Mono-amino N	"Other" N
5A 5B	day. Mean	0·6780	0·4980	0·1634	0·0027	0·0192	0·0031	0·0068	0·0444	0·0637	0·0237
6A 6B	night. Mean	0·6530	0·4730	0·1620	0·0026	0·0247	0·0	0·0072	0·0438	0·0670	0·0168
Diurnal difference		−0·0250	−0·0250	−0·0014	−0·0001	+0·0055	−0·0031	+0·0004	−0·0006	+0·0033	−0·0069
9A 9B	day. Mean	0·6750	0·5101	0·1460	0·0028	0·0196	0·0034	0·0067	0·0439	0·0523	0·0172
8A 8B	night. Mean	0·6560	0·5045	0·1392	0·0024	0·0234	0·0	0·0066	0·0443	0·0463	0·0152
Diurnal difference		−0·0190	−0·0056	−0·0068	−0·0004	+0·0038	−0·0034	−0·0001	+0·0004	−0·0060	−0·0020

+ Indicates that there is a diurnal increase.   − Indicates that there is a diurnal decrease.

Table XIV. *Showing the difference between certain duplicate readings in Tables IV and VII. (Expressed as a percentage of the mean value of the two readings.)*

Series number	Protein N	Water-soluble N	Nitric N	Monoamino N	"Other" N
5A 5B	0·44	1·69	8·44	3·37	25·0
6A 6B	0·15	0·83	5·91	2·05	14·0
7A 7B	0·75	1·48	11·30	8·49	5·5
8A 8B	0·0	0·15	9·60	0·43	12·0
9A 9B	0·0	0·50	9·54	3·50	6·0
10A 10B	0·50	0·77	—	—	—

The results of the above experiments are given in Tables III to XI. All the figures in the hydrolyses are the means of duplicates. Since the fresh weight of the sample taken for grinding might be too low on account of evaporation from the surface of the leaves whilst sorting, the distribution of N was calculated in terms of the total N. To convert this distribution into one in terms of the fresh weight of the leaves, as is given in Tables XI and XIII, the appropriate value for "N per fresh weight" given in Table III was used. In connection with these tables the following definitions must be noted:

(1) *"Amide N"* in Tables V and VI. This is used in the standard way to denote the ammonia set free on the complete hydrolysis of proteins, etc.

(2) *"Asparagine N."* Ammonia obtained by hydrolysis with 4 % HCl for two hours (after deduction of free ammonia nitrogen). (See reservation *re* Osborne and Nolan given on p. 350.)

(3) *"Amide N other than asparagine N."* In Table XII this is the ammonia set free on complete hydrolysis minus that estimated as asparagine in (2), and as free ammonia.

(4) *"Monoamino N"* refers only to the amino N that is estimated by Van Slyke's method.

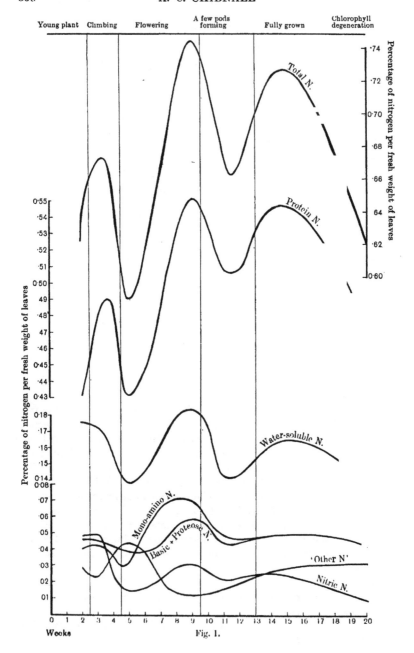

Fig. 1.

(5) *"Other N"* is the N. not estimated in any of the other groups given. Calculated by difference.

(6) *"Protein N"* is the sum of the N in the colloidal precipitate, dried residue and coagulated protein found by boiling the extract (see p. 351).

Fig. 1 illustrates Table XI. The curves are drawn to show a value for about 9 a.m. to 11 a.m. (after 6–8 hours daylight). Whenever only sunrise and sundown values have been determined, the value shown on the curves (Fig. 1) is the mean of the two.

Sampling and experimental errors in the distribution of N^1. In no case was it possible to take more than duplicate samples of the same batch of leaves, so that probable errors, based on Peter's formula, cannot be obtained. Table XIV however shows the differences between duplicate samples of the more important groups given in Tables IV and VII calculated as a percentage of the mean reading. The protein N shows a maximum difference of 0·75 % only, so that all the changes indicated in Fig. 1 can be taken as significant. The water-soluble N shows a maximum difference of 1·69 %. The nitric N shows a maximum difference of 11·3 % and a minimum of 5·91 %. This is much greater than the experimental error, and must be due to variation of the concentration in different leaves. As none of the changes indicated by Fig. 1 is less than 20 % they can be regarded as significant.

Monoamino N shows a maximum difference of 8·49 %, although that in the other four readings does not exceed 3·5 %. The experimental error here is undoubtedly high, due to the small volume of N actually measured, but a variation of 12 % or over should be fairly significant.

"Other N" is calculated by difference, and shows a maximum of 25 %. Considering that it is subject to all the errors in the rest of the N groups, a variation of 50 % is required to be at all significant.

III. DISCUSSION OF RESULTS.

The main results, as indicated in the tables and figure, may be summarised as follows:

(a) *Seasonal variations.* (Tables I, II, III, V; Fig. 1.)

(1) Total N and protein N, calculated as percentages of fresh weight, vary with growth, decreasing when this is very rapid or pods are forming.

(2) The distribution of N in the colloidal protein remains practically unchanged throughout the series.

(3) Asparagine N and free ammonia N remain low throughout the series.

(4) Nitric N and monoamino N vary directly with the protein N, indicating that they may be connected with protein synthesis. Similarly the "other N" varies inversely as the protein N, indicating that it may be connected with protein degradation.

---

[1] Sampling errors for dry weight and total N were not obtained in the present research. Earlier work on the leaves of the broad bean showed, for a batch of 250 g., the following probable errors calculated by Peter's formula: dry weight, 0·48 %; N per dry weight, 0·63 %; N per fresh weight, 0·91 %.

(5) The water-soluble products show considerable variation throughout the series, indicating that comparison of leaves from plants of different age is impossible [see Jodidi and colleagues, 1920].

(b) *Diurnal variations.* (Table XIII.)

(1) There is a diminution of total solids at night.

(2) This is accompanied by a diminution of total N and protein N.

(3) There is a distinct rise in the nitric N at night.

(4) The asparagine N disappears at night.

(5) The other water-soluble products remain more or less unchanged in amount during the night.

(6) The most significant diminution of nitrogenous products at night must be assigned to the proteins, indicating translocation of unchanged protein or its decomposition products.

The results for series 9–8 are not so reliable, since three days elapsed between the pickings. The conclusions to be drawn from them are similar to those stated above, except that there is a loss of both water-soluble N and monoamino N. This difference may be connected with the fact that pods were forming at the time.

(c) *Starvation experiments* (series 10). Tables IV, V and XII.

(1) There is a considerable decrease of protein N, with increase of water-soluble N.

(2) The distribution of N in the colloidal protein remains unaltered.

(3) Nitric N and ammonia N remain unaltered.

After mild hydrolysis:

(4) Asparagine N and monoamino N remain unchanged.

(5) Proteose N and basic N have decreased.

(6) There is a large increase of "other N," equivalent to the decrease in the protein N.

After complete hydrolysis:

(7) The "other N" is greatly reduced, and allowing for proteoses in solution, half of the loss reappears as ammonia (amide N other than asparagine N in Table XII) and one-third as bases.

Perhaps the most interesting of these results is that the protein disappears, and is replaced, not as one might expect by amide N and amino N, which remain more or less constant, but by N products of undetermined composition, given under the heading "other N." Light is thrown on the nature of these products by the different results of mild and complete hydrolysis. It would appear that they consist, in large part, of substances that fulfil the following conditions:

After mild hydrolysis:

(1) Yield no free ammonia, *i.e.* contain no acid-amide linkage R—CO—NH$_2$.

(2) Give no precipitate with phosphotungstic acid.

(3) Give no increase of free amino groups capable of reacting with nitrous acid.

After complete hydrolysis:

(4) Break down giving about one-half their N as ammonia and one-third as bases.

This great increase of free ammonia on complete hydrolysis suggests a urea derivative in which the amino groups are coupled to form compounds capable of resisting mild hydrolysis. The ureides are substances of this type; further they are, as a class, strongly acidic and capable of forming salts. As an example take barbituric acid,

$$
\begin{array}{c}
NH-CO \\
/ \qquad \backslash \\
CO \qquad \qquad CH_2. \\
\backslash \qquad \quad / \\
NH-CO
\end{array}
$$

This substance contains both replaceable imino and methylene hydrogen atoms, and only on prolonged hydrolysis breaks down giving ammonia from the urea nucleus. A condensation product of such an acid with one or more of the nitrogenous bases might possibly resist disruption on mild hydrolysis, and so give no precipitate with phosphotungstic acid due to liberated bases. In the absence of further chemical evidence this view can only be regarded as a speculation, but it is worth noting that these products appear to be present in small quantities in the unstarved leaf extract. A complete hydrolysis (see 9 B (1) and (2), Table XII) of this extract gives an increase of ammonia equivalent to 17 % of the proteose N, a value three times greater than one would expect from published data concerning the amide N content of proteoses. Furthermore, Fosse [1912, 1913, 1914] has demonstrated the presence of traces of urea in the leaves of higher plants.

Summarising the results under (a), (b) and (c) above, it would appear that the nitrogenous metabolism of the leaf runs on the following lines:

$$
\begin{array}{c}
\text{Protein} \\
\nearrow \qquad \searrow \\
\text{monoamino N} \qquad \text{"other N"} \longrightarrow \text{translocated} \\
\nearrow \\
\text{nitric N}
\end{array}
$$

This will explain the following points brought out in the above discussions:

(1) The graphs of the nitric and monoamino N in Fig. 1 vary with that of the protein N.

(2) In spite of rapid protein degradation at night, there is very little change in most of the constituents of the water-soluble N.

(3) There is an accumulation of "other N" during the fourth and fifth weeks when protein degradation appears to be rapid, also later in the season when the plant is becoming aged, and translocation less active.

(4) In the starvation experiment, though protein degradation is rapid, there is no increase in amide N or monoamino N, and, since translocation away from the leaf has stopped, there is a large increase in "other N."

Protein metabolism, however, does not depend on N alone, and due regard must be paid to the supply of carbohydrate and other substances when interpreting some of the results outlined above. Thus in the starvation experiments, since nitric N is present, it might be asserted that protein synthesis would go on until this was used up. But in this case the leaves were starved of sugar, and since they were placed in a subdued light photosynthesis would be weak. In all probability then, protein synthesis had stopped for lack of carbohydrate [compare Suzuki, 1898]. Similarly the accumulation of nitric N at night is probably due to the fact that starch, stored during the day, has been partially or entirely used up, so that the supply of carbohydrate for protein synthesis is reduced or is stopped.

The low concentration of free ammonia and asparagine N observed throughout the series confirms the work of Emmerling [1887] on the broad bean, and of Kosutany [1897] on the vine. The researches of Wasilieff [1901] and Prianischnikoff [1899] have shown that the growth of seedlings depends in part on asparagine derived from the decomposition of the reserve proteins in the cotyledons. This may indicate that the function of asparagine in the general metabolism of the plant is connected with growth, so that the total disappearance at night, an observation confirmed by Kosutany, may be due to translocation away from the leaf to the growing points or roots. The starvation experiments, where the asparagine N remains unchanged, support this view, and show that this substance is not derived from the decomposition of the leaf proteins. No indication as to its origin can be deduced from the results of the present research, but they lend support to the theory of Butkewitsch [1909] that its formation is to remove excess of free ammonia.

This research was carried out at the instigation of Prof. S. B. Schryver, whom the author thanks, together with Prof. V. H. Blackman, F.R.S., for continued advice and criticism. His thanks are also due to Mr Hales, of the Chelsea Physic Gardens, for the care bestowed on the plants used.

This research was made possible by a grant from the Department of Scientific and Industrial Research.

REFERENCES.

Butkewitsch (1909). *Biochem. Zeitsch.* **16**, 411.
Chibnall and Schryver (1921). *Biochem. J.* **15**, 60.
Czapek (1920). Biochemie der Pflanzen (Jena), II, pp. 291–307.
Emmerling (1887). *Land. Vers. Stat.* **34**, 1.
Fosse (1912). *Compt. Rend.* 155, 851.
—— (1913). *Compt. Rend.* 156, 1934.
—— (1914). *Compt. Rend.* 158, 1374; 159, 253.
Jodidi, Kellogg and True (1920). *J. Amer. Chem. Soc.* **32**, 1061.
Kosutany (1897). *Land. Vers. Stat.* **48**, 13.
Moore (1920). *J. Ind. Eng. Chem.* **12**, 669.
Nash and Benedict (1921). *J. Biol. Chem.* **48**, 2.
Osborne and Harris (1903). *J. Amer. Chem. Soc.* **25**, 323.
—— and Nolan (1920). *J. Biol. Chem.* **43**, 2.
—— and Wakeman (1920). *J. Biol. Chem.* **42**, 1.
—— Wakeman and Leavenworth (1921). *J. Biol. Chem.* **49**, 1.
Prianischnikoff (1899). *Land. Vers. Stat.* **52**, 137, 347.
Sachsse (1873). *J. pr. Chem.* n. F. **6**, 118.
Schryver (1908). *J. Physiol.* **37**. *Proc.* xxiii.
Suzuki (1898). *Bull. Coll. Agr. Tokyo*, **2**, 409.
Wasilieff (1901). *Land. Vers. Stat.* **55**, 45.

XXV. MILK AS A SOURCE OF WATER-SOLUBLE VITAMIN. III[1].

By THOMAS BURR OSBORNE and LAFAYETTE BENEDICT MENDEL

WITH THE COOPERATION OF

HELEN C. CANNON.

From the Laboratory of the Connecticut Agricultural Experiment Station and the Sheffield Laboratory of Physiological Chemistry, Yale University, New Haven.

(*Received February 14th, 1922.*)

THE question of the vitamin content of milk has, very properly, attained considerable prominence not only in its scientific aspects in relation to nutrition but also from the standpoint of public health. The classic experiments of F. G. Hopkins [1912] indicating an unexpectedly favourable effect upon health and growth secured by the addition of as little as 2 cc. of cow's milk per day to a "synthetic diet" fed to young rats have given the occasion for the current emphasis upon the richness of milk in that nutritive factor now commonly designated as vitamin B. Our own early experiments [1911] published in 1911 had indicated that what we originally termed "protein-free milk"[2] furnishes something without which rats cannot grow satisfactorily when they are kept upon "synthetic" diets consisting of mixtures of more or less isolated food substances. However, the quantities of "protein-free milk" which we have found necessary to supply the essential food factor to rats in numerous trials at various times have been decidedly larger than would correspond to 2 cc. of milk. This might have been due to a variety of obvious possibilities including the loss of vitamin in the process of preparation of the milk product. Consequently we [1918] undertook a re-investigation of the subject, using milk as the added source of vitamin B in the ration. The experiments indicated the necessity of feeding as much as 16 cc. of fresh cow's milk to secure the desired growth. Again [1920] we made further attempts to demonstrate a possible greater vitamin B potency by feeding unpasteurised milk of known origin. In some of these trials the diets used by Hopkins were imitated as closely as the materials at our disposal would permit. Again "the outcome

---

[1] The expenses of this investigation were shared by the Connecticut Agricultural Experiment Station and the Carnegie Institution of Washington, Washington, D.C.

[2] This product consists of the dried solids of skim milk from which the casein has been removed by precipitation with acid and the coagulable proteins removed by heat.

was in harmony with all our experience in showing that even additions of 10 cc. of fresh milk per day were insufficient to effect a food intake adequate for growth at a normal rate" [1920].

Neither seasonal variations, differences in the rations fed to the lactating cows, nor manipulations of the milk prior to marketing appeared to offer a satisfactory explanation of the differences between the results recorded by Hopkins and ourselves. He [1920][1] has lately reported the outcome of a new series of tests of 2 cc. of milk per day as a source of vitamin B for rats. His experiences are related as follows:

"On receipt of Osborne and Mendel's private communication, I made a few experiments in the winter of 1919. At this time the results were frankly disappointing. When the animals receiving the 2 cc. of milk in addition to the synthetic diet were compared with others which had the latter alone, the favourable effect upon health and upon the survival periods of the animals was unmistakable. Growth, however, was very slow, and the death rate of the rats was higher than that of animals normally fed.

"In April of this year fresh experiments were begun and continued during the summer. The results now became such as to confirm entirely my earlier experiments. Out of 20 animals, each receiving daily 2 cc. of fresh cows' milk as an addition to a highly purified synthetic diet, not one failed to grow almost normally throughout the period of experiment. The observations were not extended beyond 60 days, as they were meant only for comparison with those described in my original paper, and not to determine the amount of milk necessary for continued growth. The technique was exactly that originally used by me. In each experiment control animals were fed upon the synthetic diet alone and the contrast between them and a corresponding set receiving 2 cc. of milk was just as marked as in my original experiments. There has been no selection of results."

It will be noted in most of Hopkins' experiments that he began the feedings when the rats were still quite young, often weighing not more than 40 g. We concluded to repeat our tests upon rats somewhat younger than those with which our previous experiments were begun. The standard diet consisted of

Casein	...	...	18 %	Butter fat	...	9 %
Salt mixture[2]	...	4	Lard	...	...	20
Starch	...	...	54			

Each day 2 cc. of milk, supplied in a separate dish, were promptly consumed by the rats. The tests began October 19, 1921. The milk was obtained unpasteurised from a nearby supply. A bacterial count made in November on samples from the same source showed 230,000 bacteria per cc.

---

[1] In this paper a reference (p. 724) to previous experiments by Osborne and Mendel is erroneously cited as *Biochem. J.* **41**, 515. It should be *J. Biol. Chem.* **41**, 515.

[2] [Osborne and Mendel, 1919.]

The changes in body weight are shown graphically in the appended chart, the weekly food intakes (exclusive of milk solids) being indicated, where available, on the curve of growth. During the periods represented by the broken lines (- - - - - -) 40 mg. per day of a yeast fraction prepared by Osborne and Wakeman's method [1919] replaced the milk in the case of rats 7462, 7485, 7488, 7489. Rat 7506 received 0·1 g. dried brewery yeast per day. Rat 7460 died with symptoms characteristic of lack of vitamin *B*.

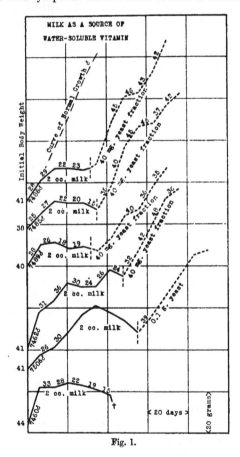

Fig. 1.

If variations in the vitamin *B* content of cow's milk from different sources actually reach the magnitude represented by the discrepancies between the English reports and our own they can scarcely be due to variations in the diet of the cows. The differences seem too large to be accounted for by such a probability. Hopkins himself states:

"The incomplete observation just mentioned, and the experience of Osborne and Mendel, suggest, at any rate, that the apparent seasonal variation in the results of my experiments was not likely to be due to differences in the milk. That there is a seasonal factor in the growth energy of rats is I think sure, but it is doubtful if it could account for the large difference in the experimental results now described. I am endeavouring to obtain further light on the matter."

In harmony with our own experiences alike with fresh cow's milk and "protein-free milk" prepared therefrom are reports from other investigators. Thus Johnson [1921], who studied the "growth-promoting properties" of milk and dried milk preparations used in the form of "reconstructed" milk in the Hygienic Laboratory of the U.S. Public Health Service, found that his animals required, as an adequate source of vitamin B, an amount of the milk equivalent to that reported by ourselves. Thus he states that "a food mixture consisting of purified foodstuffs plus milk as the sole source of water-soluble vitamine must contain at least $2\frac{1}{2}$ parts of milk to 1 part of the basal ration in order to produce normal growth. Since a full-grown male rat weighs about 280–300 grams, and a full-grown female about 180 grams, a rat upon such a diet consumes just about 16–18 cubic centimeters of milk daily. These figures agree with those of Osborne and Mendel. It is also seen that the rate of growth of rats receiving less than $2\frac{1}{2}$ parts of milk in their diet was accelerated after increasing the amount of milk in the diet."

According to Johnson and Hooper [1921] "Pigeons fed upon mixtures of spray process skim milk powder with polished rice require 30 per cent. of the food in skim milk powder in order to get full protection from polyneuritis. This corresponds to about 75 cc. daily liquid milk."

Dutcher [1921], Dutcher, Kennedy and Eckles [1920], Dutcher and colleagues [1920, 2], have found spring milk obtained after the cows were placed on green grass to be superior to winter milk in "antiscorbutic and nutritive properties." However, when the milk was at its best it required 10 cc. to furnish "sufficient water-soluble and fat-soluble vitamine for normal growth in the albino rat." Even the greatest reported increases in antiscorbutic potency in the milk of cows fed upon especially suitable rations [Hart, Steenbock and Ellis 1920; Dutcher and colleagues, 1920, 1; Hess, Unger and Supplee, 1920] do not parallel the increment which the differences represented by 2 cc. (Hopkins) and 10 cc. respectively per day denote. If the variations in the vitamin-content of the diet of cattle in different places will account for the marked differences in the vitamin B content of cow's milk they must be far more pronounced than has been suspected hitherto.

Through the cooperation of the Department of Obstetrics in the School of Medicine, Yale University, we have been enabled to test mixed samples of human milk for its content of vitamin B in the same way as has been reported for cow's milk. The product, which represented the secretion from several persons, was not essentially richer than we have found cow's milk to be. 5 cc. of human milk per day added to the standard diet devoid of vitamin *B* was insufficient to secure continued increments of body weight. When the milk was desiccated and fed in the form of tablets the equivalent of 10 cc. sufficed, in addition to the standard food eaten, to secure growth at a fairly good rate. In no case was the human milk equivalent in its potency as a source of vitamin *B* to the cow's milk used in Hopkins' most successful tests.

REFERENCES.

Dutcher (1921). *J. Ind. and Eng. Chem.* **13**, 1102.
Dutcher, Eckles, Dahle, Mead and Schaefer (1920, 1). *J. Biol. Chem.* **45**, 119.
Dutcher, Eckles, Dahle, Mead and Schaefer (1920, 2). *Science*, **52**, 589.
Dutcher, Kennedy and Eckles (1920). *Science*, **52**, 588.
Hart, Steenbock and Ellis (1920). *J. Biol. Chem.* **42**, 383.
Hess, Unger and Supplee (1920). *J. Biol. Chem.* **45**, 229.
Hopkins (1912). *J. Physiol.* **44**, 425.
—— (1920). *Biochem. J.* **14**, 721.
Johnson (1921). *Public Health Reports*, **36**, 2044; *Chem. Abstracts*, **15**, 3664.
Johnson and Hooper (1921). *Public Health Reports*, **36**, 2037; *Chem. Abstracts*, **15**, 3664.
Osborne and Mendel (1911). *Carnegie Institution of Washington, Publication* 156, pt. ii, 83.
—— —— (1918). *J. Biol. Chem.* **34**, 537.
—— —— (1919). *J. Biol. Chem.* **37**, 572.
—— —— (1920). *J. Biol. Chem.* **41**, 515.
Osborne and Wakeman (1919). *J. Biol. Chem.* **40**, 383.

XXVI. THE ESTIMATION OF NON-PROTEIN NITROGEN IN BLOOD.

By ERIC PONDER.

From the Department of Physiology, Edinburgh University.

(*Received February 16th, 1922.*)

THE object of this paper is to describe a rapid micro-method for the estimation of non-protein nitrogen in blood. Several methods have hitherto been described, but, for various reasons, are unsatisfactory. The method described by Cole [1919] is very lengthy, and requires 5 cc. of blood—an amount which cannot be obtained from small animals, and only with some inconvenience from man. The method of Folin and Denis [1916] is even more unsatisfactory. Several cc. of blood are required. The digestion mixture does not appear to be a suitable one, the contents of the incineration tube turning solid before the incineration is complete, and being afterwards quite insoluble in water: this defect is apparently due to an excess of phosphoric acid. Even if this occurrence be avoided, Nesslerisation is very difficult: Cole has observed this, and notes that a cloud rapidly appears. Frequently a brick-red precipitate appears also.

The following method, which may be applied to small quantities of blood, obviates these difficulties, and may be used with success when estimating the non-protein nitrogen in the blood of small animals.

Special apparatus etc. required:

1. The solutions required for the preparation of blood filtrates, as described by Folin [1919].

2. A digestion mixture. To 50 cc. of a 5 % copper sulphate solution add 100 cc. of 85 % phosphoric acid, and 300 cc. of pure concentrated sulphuric acid.

3. A boiling tube, 10 × 1 cm. The tube need not be of special glass. It is graduated at 3·5 cc. It is provided with a small loose stopper, shaped like a specific gravity bulb, which is introduced neck downwards into the boiling tube, so as to close the orifice.

4. A micro-filtration apparatus, such as is described in the paper dealing with the author's method for the estimation of blood-sugar [Ponder and Howie, 1921]. This is essentially a small filter which filters under slight suction.

5. Small calibrated pipettes, to deliver 0·2, 0·15, 0·5 and 1·5 cc.

6. Standard ammonium sulphate solution, containing 0·4716 g. of the purified salt per litre; Nessler's reagent, as described by Folin or Cole.

The estimation is performed as follows:

1. Blood is drawn from the finger, or from an animal's vein, into a 0·2 cc. pipette. The contents of the pipette are added to 1 cc. of distilled water:

the pipette is twice filled with water, and these volumes, carrying with them all traces of blood from the walls of the pipette, are added to the tube containing the blood and distilled water. 0·2 cc. of sodium tungstate and 0·2 cc. of 2/3 N sulphuric acid are added: the tube is allowed to stand for five minutes at least, its contents being occasionally shaken.

2. A filtrate is produced from the contents of the tube by about two minutes filtration in the micro-filter.

3. 0·5 cc. of filtrate is placed in the boiling tube, and to it is added 0·2 cc. of the digestion mixture diluted 1 in 4. The contents are boiled over a micro-burner, very gently. As soon as boiling occurs, the stopper is placed in the mouth of the tube. The boiling is continued for two minutes. The flame should be very low; bumping is thus prevented. At the end of the incineration, the flame is removed, and a few drops of water are added. The tube is allowed to cool, and is then filled to the mark with distilled water.

4. A standard is prepared. Place 3·15 cc. of distilled water in a tube, add 0·15 cc. of standard ammonium sulphate solution, and 0·2 cc. of the diluted digestion mixture.

5. Nesslerise the contents of the boiling tube and the standard simultaneously, by adding 1·5 cc. of Nessler's reagent. There is no difficulty in obtaining a clear yellow solution.

6. The solutions are matched in the colorimeter in the usual way. If the standard be set at 15, the non-protein nitrogen in 100 cc. of blood is 15 divided by the reading, and multiplied by 30.

It has not been possible, for the reasons given above, to compare this method with that of Folin and Denis. The method has however been applied to nitrogenous substances, of known nitrogen content, which require incineration before producing a colour with Nessler's reagent, and has been found accurate and satisfactory, the results obtained by the method and the calculated results agreeing closely.

Some results obtained in the examination of blood of various animals are given in the following table:

Animal	mg. in 100 cc.	Animal	mg. in 100 cc.
Man	36	Rabbit	43
,,	25	Frog	93
Cat	67	Lizard	76

In the case of the frog and lizard, the blood was obtained from the heart.

It is claimed for this method that (1) a very small quantity of blood is required, (2) that the technique is simple and rapid, (3) that the final Nesslerisation presents no difficulty, and (4) that the results given are accurate.

REFERENCES.

Cole, S. W. (1919). Practical Physiological Chemistry, 260.
Folin, O. (1919). A Laboratory Manual of Biological Chemistry, 179.
Folin and Denis (1916). J. Biol. Chem. 26, 491.
Ponder and Howie (1921). Biochem. J. 15, 171.

XXVII. THE CONDITIONS INFLUENCING THE FORMATION OF FAT BY THE YEAST CELL.

By IDA SMEDLEY MACLEAN.

(*Received February 22nd, 1922.*)

VERY little is known of the story of fat metabolism in the lower organisms although a considerable amount of work has been carried out on the conditions attending the formation of fat in yeast. Yeast offers a particularly favourable field of study as it can readily be grown in large quantity and considerable variations can be effected in the conditions of its growth. In spite of the work that has been done but little progress has been made since the work of Naegeli and Loew [1878, 1879] carried out more than forty years ago.

The conclusions arrived at by Naegeli and Loew were briefly as follows:

(1) The fatty acid of the yeast cell consisted chiefly of oleic acid.

(2) The amount of fat obtainable from yeast was about 1 to 2 % if the dried yeast was directly extracted by ether; this figure might be raised to about 5 % by first evaporating the yeast with concentrated HCl several times on a water-bath. The acid destroyed the cell wall and the ether then was no longer prevented from extracting the fat by the impermeability of the cell wall.

(3) The more vigorous the growth of the yeast cell, the greater was the amount of fat formed. Both the total amount of dry substance formed and the percentage of fat it contained were raised.

(4) Other conditions being similar, the percentage of fat formed increased with the supply of oxygen. They found that the fat content of a yeast grown in a solution of sugar containing ammonium tartrate (2 %) which was aerated throughout the time of growth was 12·5 %, whereas a yeast grown on peptone and sugar at a low temperature with scanty respiration contained only 5 % of fat.

(5) Naegeli discussed the question of the origin of the fat formed and concluded that under different conditions both carbohydrate and protein might act as sources of fat. He drew attention to the rape seed which, before maturity, is filled with starch grains and from which, when ripe, the oil is pressed and to the case of fungi in which, when put into water, the plasma diminishes with the appearance of fat, the cellulose membrane also increasing during fat formation.

(6) Naegeli recognised also that the fat contained a sterol which he termed cholesterol and further that the conditions which led to an increase of fat led also to an increase of sterol.

The nature of yeast fat.

This has since been elucidated by the work of Hinsberg and Roos [1903, 1904], Neville [1913], Gérard [1895], and Smedley MacLean and Thomas [1920]. As the result of these investigations it has been shown that the fatty acids present are palmitic, oleic and linoleic with a small quantity of lauric. The fat is chiefly remarkable for the large proportion of sterol which it contains, the sterol being apparently identical with ergosterol, a characteristic constituent of the fats of all the lower plant world.

The amount of fat present in yeast.

In the normal yeast cell the percentage of fat present has been described as from 1 to 5 % of the dried yeast. Naegeli pointed out that the extraction of the air-dried yeast with ether only removed part of the fat, and that if the cells were first treated with concentrated HCl from two to three times as much fat could be extracted, this being however hydrolysed to the free fatty acids. Naegeli's results have been criticised by later observers who consider that prolonged treatment with the strong acid may give rise to ether-soluble decomposition products of the yeast cell and that the higher figures obtained do not represent the true percentage of fat.

Hinsberg and Roos [1903] and Bokorny [1916, 3] therefore retain the ether extraction method in determining the fat percentage of dried yeast. Bokorny [1916, 2] indeed did not confirm Naegeli's results; he treated the air-dried yeast for 24 hours with concentrated HCl and found that the percentage of fat was only 0·66 % of the weight of dried yeast compared with 2·66 % of fat obtained by ether extraction without the preliminary treatment with acid: the sticky mess obtained by treating the yeast with acid, he found unsuitable for ether extraction.

In a series of experiments carried out with the object of determining the proportion of fat to carbohydrate in the yeast cell under different conditions, I adopted the method of hydrolysing the yeast by boiling with N HCl for two hours, filtering and washing the residue with water until the washings no longer reduced Fehling's solution; the residue was then air-dried overnight at the laboratory temperature and extracted with ether in a Soxhlet apparatus and the filtrate and washings used for the estimation of carbohydrate. After evaporating off the ether, the residual fat was taken up with dry ether and dried to constant weight in a vacuum desiccator at the laboratory temperature.

I found that the amount of fat found in this way might be several times as great as the amount obtained by the direct extraction of the dried yeast with ether. In the latter method the yeast was dried by treating it with a large volume of absolute alcohol, the alcoholic extract was evaporated and the residue added to the dried yeast before extracting it with ether.

A comparison of the fat obtained by the two methods showed that the two specimens were similar in appearance and from both sterol separated on standing; the iodine values of both varied considerably in different experiments but no consistent difference could be detected: the difference between the Wijs and Hubl numbers which may be taken as an indication of the amount of sterol present [Smedley MacLean and Thomas, 1921] also showed no constant variation between the two series of experiments.

Table I. *Showing the amount of fat extracted by ether before and after hydrolysis of the yeast.*

| | | Amount of fat | | | |
| | Weight of dried yeast g. | (a) Without hydrolysis | | (b) After hydrolysis | |
Sample of yeast		Weight g.	% on dry yeast	Weight g.	% on dry yeast
Pressed yeast (12·5 g.)	2·99	0·1118	3·74	—	—
	3·07	—	—	0·1948	6·35
	3·05	—	—	0·1892	6·20
12·5 g. above sample incubated 48 hours at 26° in glucose solution					
Oxygenated	4·17	0·0960	2·29	—	—
„	4·80	—	—	0·5402	11·24
Not oxygenated	4·18	0·1114	2·67	—	—
„ „	4·80	—	—	0·2434	5·07
A pure culture of yeast grown on wort containing lactic acid (N/10) at 30°					
Oxygenated	8·05	0·1076	1·34	—	—
„	8·05	—	—	0·1975	2·45
Not oxygenated	6·49	0·0996	1·53	—	—
„ „	6·49	—	—	0·1524	2·30

It must be admitted therefore that ether extraction of the yeast dried either in air or by means of alcohol, does not remove all the fat from the cells. The criticisms brought forward against Naegeli and Loew's method of repeated evaporation of the yeast with concentrated HCl cannot be urged against the much less drastic treatment of boiling for two hours with 3·6 % or even with 1·8 % HCl, a method by which the fat is not hydrolysed, its acid value being barely affected.

The state in which the fat occurs in the yeast cell.

Naegeli and Loew apparently regarded the hydrochloric acid as acting by impairing the cell membrane and thus permitting the entrance of the solvent into the cell containing the fat. They considered it probable that continued treatment with alcohol or ether would completely remove the fat, an expectation which does not appear to be realised even when the extraction is continued for a very long time. Two views present themselves: (1) a proportion of the fat may exist in the free state in the cell, being probably formed as a decomposition product of the cell-plasma. The remainder of the fat may be in combination in the plasma of the cell and may only be liberated on hydrolysis when some complex substance in the plasma is itself decomposed. (2) The sub-microscopic fat particles may be retained in a protein meshwork, only the larger fat particles being extracted by the ether. The smaller fat particles would then only be liberated by the breaking down of the protein when they would be extracted by the ether.

All the known facts as to the extraction of fat from yeast are in agreement with the hypothesis that the fat is closely associated with the sterol and protein and possibly with the carbohydrate of the cell; this association may be of the nature of a chemical combination.

The evidence upon which this view is based, is as follows:

(a) Extraction with alcohol and ether removed readily from 1 to 3 % of yeast fat, calculated on the dried yeast, after which only traces of fat were obtained by long continued extraction.

(b) As stated above about twice as much fat may be removed from the yeast cell after boiling with normal or semi-normal acid as is obtained by direct ether extraction of the dried yeast. The greater part of the fat which is obtainable from dried yeast by direct extraction with ether is removed in a comparatively short time; further prolonged extraction only results in the separation of traces of fat.

Thus in an experiment carried out by Miss D. Hoffert, 12·5 g. yeast were soaked overnight in alcohol, the alcohol evaporated, the residue added to the yeast, the whole dried overnight and extracted with ether in 14 hours; the amounts of fat extracted varied from 0·0822 to 0·1060 g. 12·5 g. of the same sample of yeast were hydrolysed with N HCl for two hours, the solid residue dried overnight and extracted for 14 hours with ether; 0·2111 g. fat was obtained.

The amounts of fat obtained per hour by direct extraction with ether are shown in the following table.

Table II.

Time in hours...	1st	2nd	3rd	4th	5th	6th	7th	7th to 14th	Total in 14 hours
(1) Weight of fat from 12·5 g. yeast	0·0334	0·0216	0·0146	0·0090	0·0082	0·0079	0·0055	0·0058	0·1060
(2) „ „	0·0652	0·0077	0·0023	0·0013	0·0014	0·0013	0·0011	0·0019	0·0822
(3) „ „	—	—	—	—	—	—	—	—	0·0888

(c) Old yeast cells or cells grown under unfavourable conditions, e.g. an abnormally low or high temperature, give considerably higher fat percentages than normal cells when extracted directly with alcohol and ether. Such cells when examined microscopically show small globules of fat staining with osmic acid. In experiments where the period of incubation is long, partial autolysis of the yeast probably takes place—the amount of yeast formed is much reduced and the proportion of fat extracted by ether is greater. The total amount of fat obtained is decreased; but since the total amount of yeast is proportionately still less, the percentage of fat is raised.

Tubes containing 10 cc. wort were inoculated with a pure culture of brewer's yeast and after 48 hours the contents added to 1500 cc. sterilised wort and incubated. The yeast was centrifuged, filtered and dried with alcohol; the residue from the alcohol was added to the dried yeast and the product extracted with ether in a Soxhlet apparatus for 14 hours. From the figures given below it will be noted that (1) the quantity of yeast formed is in inverse ratio to the fat percentage, (2) the percentage of fat tends to be higher when the time of incubation is long and (3) incubation in the presence of 1 % lactic acid at 35° is particularly unfavourable and leads to the production of the smallest amount of yeast and the highest percentage of fat.

Table III. *Showing that a higher percentage of fat is extracted by ether from yeast which has been grown under unfavourable conditions.*

Temperature 25–26°

| | (a) Reaction neutral | | | | (b) 1 % lactic acid added | | |
No. days	Weight fat. g.	Weight dry yeast. g.	Fat %	No. days	Weight fat. g.	Weight dry yeast. g.	Fat %
2	0·105	7·4	1·4	2	0·131	6·5	2·0
2	0·205	8·9	2·3	2	0·117	5·3	2·2
3	0·143	7·7	1·8	2	0·110	5·0	2·2
3	0·152	7·4	2·0	3	0·168	5·8	2·8
3	0·155	6·5	2·4	3	0·130	5·8	2·2
6	0·155	4·6	3·3	5	0·149	4·2	3·6
				6	0·153	3·7	3·3
Mean...	0·153	7·1	2·2		0·134	5·2	2·6

Temperature 35–36°

4	0·127	4·1	3·1	4	0·148	3·1	4·6
4	0·101	2·1	4·7	4	0·063	2·6	2·4
4	0·105	2·9	3·5	4	0·076	1·3	5·2
6	0·083	2·4	3·4	8	0·087	0·96	9·0
8	0·051	2·9	1·7	12	0·032	0·49	6·5
12	0·058	1·5	3·8	12	0·028	0·69	4·0
Mean...	0·087	2·65	3·4		0·072	1·52	5·3

(d) Bokorny [1916, 3] found that when yeast is submitted to the action of protoplasmic poisons the amount of fat obtained by extraction was very considerably increased. He soaked pressed yeast for some hours in such solutions as phenol (5 %), formaldehyde, mercuric chloride, etc. While it seems unlikely that living processes would continue to be carried on by yeast soaked in 5 % phenol solution, it is certainly conceivable that such treatment might decompose the complex substance in the plasma and liberate fat from it, if fat be indeed one of its constituents.

Bokorny's experiments were carried out on small amounts of yeast and the quantities of fat weighed were small. It is known that protein readily absorbs phenol [Cooper, 1912] and it is possible that when the phenol-treated yeast was extracted with ether, the small amount of fat present may have been augmented by traces of phenol which contributed to the 12 % of fat obtained. In repeating this work it was found very difficult completely to remove the phenol, but I think there is no doubt that after the soaking with phenol the proportion of fat extracted by ether is appreciably increased. Bokorny regarded the increase as being caused by an abnormal secretion of fat deposited as a protection against unfavourable conditions of growth. More probably it is to be regarded as fat liberated from combination in the cell contents by the action of the poison. It is interesting to note that a German patent (D.R.P. 309,266) recommends the auto-digestion of the yeast to ensure the liberation of the fat globules from the cells before extracting with solvents.

Conditions affecting the amount of fat in the cell.

During the Great War the question of using the lower organisms as sources of fat became one of practical importance especially in Germany where a good deal of work was carried out on these lines. Lindner's [1916, 1919] "mineral yeast" (*Endomyces vernalis*) was cultivated as a source of both fat and protein and in this organism a fat percentage of 18 % was claimed. The work of Bokorny and other observers was directed to producing a similar result with yeast; Bokorny [1916, 1] found that by using peptone as his nitrogenous food and repeatedly adding sugar, the fat content of yeast could be raised. In one German patent (D.R.P. 320,560) it is claimed that by applying the methods used by Lindner to increase the fat content of mineral yeast, the fat content of beer yeast and of pressed yeast may be raised to from 20 to 50 % viz. by growing a surface culture on a non-nitrogenous medium. The method of estimating the fat is not given and only the microscopic appearance is described. Another patent (D.R.P. 307,789) describes the application of hydrogen peroxide and of violent aeration of the glucose solution to increase the fat content of yeasts which do not form surface cultures. Here the increase of fat is not stated nor is the method of extraction indicated.

In the first series of experiments I carried out, the air-dried yeast was extracted with alcohol and ether. Pure cultures of yeast were grown on wort with and without aeration of the medium; pressed yeast was added to glucose solutions and to wort and incubated with and without aeration of the medium. No marked differences were produced in the amount of fat extracted and the oxygenation of the medium appeared to be without result. If however the yeast was first hydrolysed with normal HCl in the manner already described and the solid residue extracted with ether, very marked variations were observed and the fat content appeared to be largely raised by the aeration of the medium.

Pressed yeast incubated for 44 hours in glucose solution without oxygenation showed a marked increase in the total weight of the dried yeast, and the total weight of fat present was increased although the percentage of fat calculated on the dry weight was decreased. But in the oxygenated glucose solution, not only was the total dry weight of the yeast increased, but the percentage of fat was sometimes more than doubled. Thus in one experiment the fat percentage rose from 6·0 to 11·6 %; but while 11·6 % of fat was extracted after hydrolysis, part of the same material treated with alcohol and then extracted with ether showed only 3 % of fat. The increased amount of fat formed when yeast is grown in a glucose solution which is oxygenated throughout the experiment appears to be held in combination in the cell plasma and is not extracted by ether until by hydrolysis it is set free from the cell complex. The ether-soluble material which was weighed as fat contained both fat and sterol.

Yeast incubated in a solution of glucose to which nitrogenous material has not been added is characterised by a high percentage of carbohydrate. If a given quantity of yeast be incubated for 44 hours in (1) a solution of glucose and in (2) wort containing the same percentage of carbohydrate, the total amount of yeast formed in the wort is from two to three times as much as in the carbohydrate solution. The total amounts of carbohydrate contained in the yeasts after incubation in (1) and in (2) respectively are approximately the same, though, since much more yeast is formed during the incubation in the wort, the *percentage* of carbohydrate in the latter specimen is much less. The total amount of fat as well as the percentage are considerably higher in the yeast incubated in the glucose solution.

It is not clear whether the decreased amount of fat in the yeast incubated in the medium rich in nitrogen is due to a lessened synthesis of fat or to an increased breaking down of the fat after it has been formed. The amount of fat is however markedly greater in the yeast from the oxygenated wort than in that from the wort which has not been aerated.

The part played by the oxygen requires further elucidation and is at present being further studied; it is uncertain whether, as Slator[1921] suggested in his study of the conditions affecting the growth of yeast, it acts by removing the detrimental influence of the carbon dioxide or by some specific action of its own. In all the experiments which I have so far carried out an increase in the total amount of fat has been associated with a high percentage of carbohydrate in the cell.

The observation of Naegeli that the more vigorous the growth of the yeast cell (as is the case in the aerated medium) the greater the amount of fat formed is therefore confirmed. It will be remembered that Naegeli extracted the fat after previously warming the yeast with concentrated HCl. Bokorny's observation that in strongly growing yeast cells there is a diminished fat content probably depends on the fact that he extracted the dried yeast directly with ether, for when this method of extraction is used his results are in agreement with those quoted above [Bokorny, 1916, 2].

In the following experiments 12·5 g. of pressed yeast were incubated for 44 hours at 26° in 1500 cc. of the sterilised medium, in some of the experiments a current of oxygen being passed through the medium during the experiment. The yeast was then filtered and the total reducing substance determined after hydrolysis by Bertrand's method, the result being calculated as glucose. The figures given for fat refer to the total amount of material soluble in dry ether.

The nature of the substance from which fat is formed.

Naegeli and Loew [1879] appear to have recognised clearly that fat is formed in moulds and yeasts from substances existing in the plasma of the cell; they noticed that as the fat globules appeared the plasma diminished, and argued therefore that the fat could not be derived from the nitrogen-free carbon compounds since these were only present in very small quantity in the cell contents. In this case therefore they claimed that the fat must have been formed from protein. When peptone was used as the source of nitrogen, they regarded the sugar or tartaric acid in the medium as the source of fat.

Table IV. *Showing the effect of oxygenation.*

Medium	Oxygen	Yeast dry weight. g.	Fat weight. g.	Fat %	Carbohydrate weight. g.	Carbohydrate %	Nitrogen %
Original yeast		{3·17	0·1643	5·2	0·9	28·2	—
		{3·19	0·1812	5·7	—	—	—
In water	−	2·64	0·1734	6·2	0·55	20·8	—
„ „	+	2·65	0·1343	5·1	0·55	20·6	—
„ glucose	−	3·72	0·1892	5·1	2·0	53·8	—
„ „	+	4·17	0·4532	10·8	1·85	44·1	—
Original yeast		{3·40	0·2277	6·7	0·92	27·2	—
		{3·48	0·2158	6·4	0·88	26·0	—
In water	−	2·87	0·1542	5·4	0·66	23·0	—
„ „	+	3·03	0·1812	6·0	0·64	21·2	—
„ glucose	−	4·98	0·2824	5·7	2·50	50·1	—
„ „	+	5·69	0·6623	11·6	2·64	46·3	—
Original yeast		{3·13	0·1892	6·05	0·49	15·7	7·87
		{3·07	0·1948	6·37	0·55	17·9	7·67
In glucose	−	4·80	0·2440	5·07	2·40	50·0	5·10
„ „	+	4·80	0·5402	11·24	2·21	46·1	4·73
Original yeast	−	{2·98	0·2183	7·32	0·51	17·2	8·78
		{2·90	0·2193	7·56	0·52	18·1	8·69
In glucose	+	4·65	0·6267	12·43	2·00	23·3	4·77
Original yeast		3·93	0·2481	6·30	0·84	21·5	7·96
In wort	−	11·26	0·1637	1·45	2·42	21·5	8·55
„ „	−	12·27	0·2153	1·76	2·56	20·9	8·72
„ „	+	11·73	0·2846	2·43	2·63	21·1	8·51
Original yeast							
(1) In wort	−	6·44	0·1487	2·31	1·72	26·7	8·70
(2) „ „	−	3·29	0·1169	3·51	1·05	28·4	8·97
(1) „ „	+	10·04	0·4165	4·15	2·61	26·0	8·67
(2) „ „	+	9·27	0·3535	3·65	2·30	23·72	—

(1) Instead of 12·5 g., 2·5 g. of yeast containing 0·0505 g. fat was added to the solution.
(2) The inoculating dose was here 0·25 g. containing 0·005 g. fat.

They thought it probable that all materials for fat formation were first built into the protoplasm protein and in this way utilised for the formation of fat, the fat being subsequently split off and stored as reserve material.

The figures already quoted (Table IV) show conclusively that when yeast is grown in oxygenated glucose solution, where no external supply of nitrogen is available, the loss in protein even if the protein were wholly converted to fat could account only for a small portion of the fat formed. As it is extremely unlikely that all the amino-groups present would be converted into fatty chains, it is clear that some other source must be found for the manufacture of the fat. When yeast is grown in glucose solution there is a large accumulation of carbohydrate in the cell and since the protein which has disappeared is insufficient to account for the increase in fat, either the carbohydrate of the external medium or that of the yeast cell itself must have acted as the starting material for the formation of fat. The following figures will make this apparent.

	Dry yeast g.	Carbohydrate g.	Fat g.	Nitrogen g.	Protein g.
1. 12·5 g. of the original sample of yeast contained	3·1	0·52	0·192	0·2409	1·50
12·5 g. yeast incubated 44 hours in oxygenated glucose solution at 26° and filtered. The residue contained	4·80	2·21	0·540	0·2270	1·42
Gain or loss	+1·7	+1·69	+0·348		−0·08
2. 12·5 g. of the original sample of yeast contained	2·94	0·52	0·219	0·257	1·61
After incubating 44 hours at 26° in glucose solution and filtering. The residue contained ...	4·65	2·00	0·627	0·222	1·39
Gain or loss	+1·71	+1·48	+0·408		−0·22

I endeavoured to find out whether the carbohydrate stored in the yeast cell was converted in the presence of a free supply of oxygen into fat; a sample of yeast which had been grown in glucose solution for 44 hours and contained a high carbohydrate percentage (46·7 %) was then transferred to water and a rapid current of oxygen passed through for about 20 hours. The carbohydrate and fat were determined in the yeast before and after it was submitted to the action of the oxygen.

3·84 g. dry yeast contained before being submitted to the action of oxygen 0·1932 g. fat and 1·79 g. carbohydrate; afterwards 0·2146 g. fat and 0·96 g. carbohydrate were detected. In another experiment 3·32 g. dry yeast contained 0·1477 g. fat and 1·446 g. carbohydrate, after growing in the glucose solution. After oxygenation in a solution containing 1 % of potassium and magnesium phosphates 0·1736 g. fat and 0·88 g. carbohydrate were found. Though the differences in the amounts of fat are not very large the increase suggests that the fat is formed from carbohydrate material stored inside the yeast cell.

Conclusions.

Free fat exists in the normal yeast cell in small amount; the percentage of fat is increased in old and degenerating cells, in cells grown under unfavourable conditions and in cells exposed to the action of protoplasmic poisons. A large part of the fat normally present in yeast and of the sterol associated with it, is in some form of combination in the plasma of the cell and is not extracted from it by treatment with alcohol and ether. This complex is decomposed by boiling with dilute mineral acids and the fat can then be readily extracted from it by ether. It is possible that part of the protein and carbohydrate which are hydrolysed by the dilute acid are also combined with the sterol and fat.

A free supply of oxygen and a non-nitrogenous medium rich in carbohydrate are conditions producing an increased amount of fat in yeast but the fat formed in this way appears to be entirely held in combination in the yeast cell, and is only set free by hydrolysis.

The fat and sterol formed in this way are derived from carbohydrate.

This investigation has been carried out for the Food Investigation Board of the Department of Scientific and Industrial Research, and I desire to express my thanks for the grants which have been received for this purpose. I should like also to thank Miss Florence Banks for the assistance she has given in the experimental part of the work.

REFERENCES.

Bokorny (1916, 1). *Allgemein. Brauer- und Hopfen-Zeitung*, **56**, 603.
—— (1916, 2). *Allgemein. Brauer- und Hopfen-Zeitung*, **56**, 1479.
—— (1916, 3). *Biochem. Zeitsch.* **75**, 346.
Cooper (1912). *Biochem. J.* **6**, 362.
Gérard (1895). *Compt. Rend.* **126**, 909.
Hinsberg and Roos (1903). *Zeitsch. physiol. Chem.* **38**, 1.
—— —— (1904). *Zeitsch. physiol. Chem.* **42**, 189.
Lindner (1916). *Wochsch. Brauerei*, **33**, 193.
—— (1919). *Zeitsch. techn. Biol.* **7**, 79.
Naegeli and Loew (1878). *Annalen*, **193**, 322.
—— —— (1879). *München Akad. Sitzber.* 289.
Neville (1913). *Biochem. J.* **7**, 347.
Slator (1921). *J. Chem. Soc.* **119**, 115.
Smedley MacLean and Thomas (1920). *Biochem. J.* **14**, 483.
—— —— (1921). *Biochem. J.* **15**, 319.

XXVIII. THE ACTION OF WHOLE BLOOD UPON ACIDS.

By ERNEST LAURENCE KENNAWAY AND JAMES McINTOSH.

*From the Bland-Sutton Institute of Pathology, Middlesex Hospital,
and the Cancer Hospital Research Institute.*

(*Received February 23rd, 1922.*)

ALTHOUGH very many investigations have been made upon the alkali reserve of serum and plasma, and upon the movements of ions which take place when red corpuscles are exposed to different tensions of CO_2, it seems that no quantitative study of the acid-neutralising power of whole blood has been made. It is commonly assumed and stated that whole blood is more perfectly buffered than plasma. There is no difficulty in investigating this matter if the blood is placed in an isotonic medium.

METHOD.

Whole blood was shaken with $0.01N$ H_2SO_4 containing 0.9% NaCl. The mixture was then centrifuged and the supernatant fluid titrated to ascertain how much of the acid had been neutralised, and examined by the electrode for H-ion concentration. The number of volumes of $0.01N$ acid added to one volume of blood varied from 2.28 to 9.66; the actual amounts used are given in the tables. The mixture of blood and acid was in all cases made up with 0.9% NaCl to 8.25 cc. as this volume gave a convenient amount for duplicate titrations and H-ion determination. It was then shaken vigorously for five minutes in a wide centrifuge tube in a strong current of CO_2-free air to remove CO_2, and centrifuged in corked tubes. Of the supernatant fluid duplicate portions of 2.5 cc. were taken, mixed in test tubes with 5 cc. freshly boiled distilled water and nine drops of 0.1% phenol red in alcohol, and titrated with $0.01N$ NaOH in a comparator, a phosphate solution of p_{II} 7.4 being used as colour standard. The remainder of the supernatant fluid was used for H-ion determination. Plasma was treated in exactly the same way as the whole blood except that the centrifuge was not used. Two series of observations were made, namely upon human blood derived from one healthy subject over a period of about three months, and upon guinea-pig's blood (Tables I and II, and Figs. 1 and 2).

It is at once evident in Fig. 1 that the behaviour of whole blood is quite different from that of plasma. However much acid is added to the human plasma, roughly the same amount is neutralised; in fact plasma behaves, though with much wider variations, like a solution of caustic soda. With whole blood on the other hand the amount of acid neutralised increases with the amount added in such a way that round about 80 % of the acid is neutralised in every case. The exact percentages can be seen in Table I; they range from 89 to 78 %, and tend to diminish as the amount of acid increases.

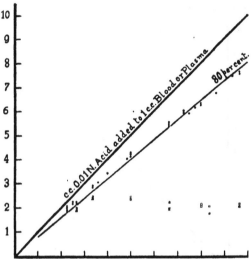

Fig. 1. NEUTRALISATION OF ACID BY HUMAN WHOLE BLOOD AND PLASMA. The points of intersection of the diagonal line with the ordinates represent amounts of acid added to 1 cc. blood or plasma; the points marked ● and □ on these ordinates represent the fractions of these amounts which are neutralised by the blood (●) or plasma (□). A second oblique line has been drawn at a height which is 80 % of the height of the diagonal.

The results obtained with the whole blood of guinea-pigs (Fig. 2) are confirmatory; the amounts neutralised keep rather closer to the 80 % line. With the plasma, however, extremely wide variations were found; in some cases the plasma had scarcely any neutralising power. This is probably due to the disturbances of circulation and respiration brought about in the taking of the blood, which are quite absent when blood is drawn from the human subject. It is remarkable that there is no indication in the results obtained from whole blood of these variations in the plasma. The maximum amount of $0.01N$ acid neutralised by 1 cc. plasma is about the same (2·3 to 2·4 cc.) in both cases (man and guinea-pig) and indicates that the "alkali reserve" of the plasma is equivalent to that of a $0.024N$ solution.

The addition of increasing amounts of acid to whole blood cannot of course be carried on indefinitely, because haemolysis supervenes. This occurs in human blood when more than about 11 cc. $0.01N$ acid is added to 1 cc. blood, whereas 1 cc. guinea-pig's blood is lysed by 7 or 8 cc. $0.01N$ acid. The amounts of acid used in these experiments are not so large as to be altogether beyond the range of what may occur in the body; thus in diabetic coma the concentration of compounds of aceto-acetic and β-hydroxybutyric acids in the blood may be that of a 0·03N solution [Kennaway, 1918]. This would be in some degree comparable with an experiment in which 3 cc. 0·01N acid was added to 1 cc. blood (see Figs. 1 and 2). No methhaemoglobin was detected with the spectroscope in these mixtures even when more than enough acid to produce lysis was added.

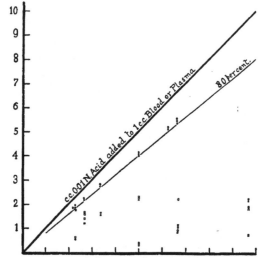

Fig. 2. NEUTRALISATION OF ACID BY GUINEA-PIG'S WHOLE BLOOD AND PLASMA.
For description, see Fig. 1.

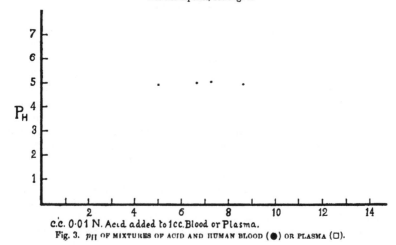

c.c. 0·01 N. Acid added to 1cc. Blood or Plasma.

Fig. 3. p_H OF MIXTURES OF ACID AND HUMAN BLOOD (●) OR PLASMA (□).

Table I (see Figs. 1 and 2). *Neutralisation of acid by whole blood.*

All mixtures of blood and acid were made up to 8·25 cc. with 0·9 % NaCl.

Mixture cc.			Ratio cc. 0·01N acid to 1 cc. blood	Human blood			Guinea-pig's blood		
Blood	+ Acid			1 cc. blood neutralised cc. 0·01N acid	Percentage of acid neutralised	Final p_H	1 cc. blood neutralised cc. 0·01N acid	Percentage of acid neutralised	Final p_H
	0·01N	0·005N							
1·75	4·0	—	2·286	1·83	80·2	7·05	1·906	83·4	—
,,	,,	—	,,	1·83	80·2	—	—	—	—
,,	,,	—	,,	1·94	85·0	7·1	—	—	—
,,	,,	—	,,	2·03	88·8	6·6	—	—	—
1·0	—	5·0	2·5	2·20	88·0	—	—	—	—
1·50	4·0	—	2·66	2·22	83·4	—	2·20	82·7	—
1·50	5·0	—	3·33	2·90	87·1	6·5	2·76	82·9	6·15
,,	,,	—	,,	2·84	85·3	—	2·84	85·4	—
,,	,,	—	,,	2·87	86·0	6·35	—	—	—
0·75	—	5·0	,,	2·83	85·0	—	—	—	—
1·0	—	7·25	3·6	3·04	84·4	—	—	—	—
0·75	—	6·0	4·0	3·39	84·8	—	—	—	—
0·75	—	7·25	4·8	4·05	84·3	—	—	—	—
1·0	5·0	—	5·0	4·15	83·0	5·2	4·01	80·2	5·3
,,	,,	—	,,	4·08	81·6	5·65	4·13	82·6	5·8
0·75	—	7·5	,,	4·25	85·0	—	—	—	—
0·8	5·0	—	6·25	—	—	—	5·14	82·2	5·55
0·75	5·0	—	6·66	5·35	80·3	5·4	5·52	82·9	5·05
,,	,,	—	,,	5·53	83·0	5·3	5·34	80·2	5·8
1·0	7·25	—	7·25	5·93	81·8	5·05	lysis	—	—
,,	,,	—	,,	6·01	82·5	4·8	—	—	—
0·5	—	7·75	7·75	6·16	79·5	—	—	—	—
0·75	6·0	—	8·0	6·28	78·5	5·65	—	—	—
0·75	6·5	—	8·66	6·79	78·4	4·95	—	—	—
0·75	7·0	—	9·33	7·48	80·1	—	—	—	—
0·75	7·25	—	9·66	7·55	78·2	4·5	—	—	—
,,	,,	—	,,	7·58	78·5	—	—	—	—
0·5	5·5	—	11·0	lysis	—	—	—	—	—

Table II (see Figs. 1 and 2). *Neutralisation of acid by plasma.*

Acid and plasma were mixed in the same amounts as those given in the first column of Table I for acid and whole blood.

Ratio cc. 0·01N acid to 1 cc. plasma	Human plasma			Guinea-pig's plasma		
	1 cc. plasma neutralised cc. 0·01N acid	Percentage of acid neutralised	Final p_H	1 cc. plasma neutralised cc. 0·01N acid	Percentage of acid neutralised	Final p_H
2·286	—	—	—	0·579	25·3	5·05
,,	—	—	—	1·797	78·6	6·4
,,	—	—	—	1·825	80·0	6·7
2·66	1·93	72·5	6·5	1·308	50·0	5·55
,,	2·25	84·6	—	1·59	60·0	6·6
3·33	2·41	72·4	6·15	1·595	47·7	5·8
5·0	2·38	47·6	—	0·347	7·0	4·2
,,	2·40	48·0	—	2·18	43·6	4·5
,,	—	—	—	2·31	46·2	4·35
6·66	1·915	28·8	4·0	0·84	12·6	4·0
,,	2·22	33·3	4·25	2·18	32·7	3·95
,,	—	—	—	1·10	16·5	4·1
8·0	2·06	25·8	—	—	—	—
8·33	1·88	22·6	—	0·96	11·6	—
9·66	2·09	21·6	3·65	0·73	7·6	3·45
,,	2·06	21·3	—	1·86	19·3	3·05
,,	2·11	21·8	—	2·16	22·4	3·4

The measurements of H-ion concentration (Fig. 3 and Tables I and II) show that the whole blood does not bring this concentration to any constant level when different amounts of acid are added. With the addition of increasing amounts the H-ion concentration increases at a diminishing rate, roughly from p_H 7 to 4. In striking contrast to their effects upon titratable acidity, the whole blood and the plasma have very much the same effect upon p_H; the plasma shows a rather greater acidity, as one would expect. The rather wide variations in p_H are no doubt due to the errors of pipette measurement, and they would be wider still if the mixtures were not buffered; thus an error of only 0·01 cc. in the pipetting of 0·01N acid in these experiments would, in an unbuffered solution, be nearly sufficient to produce the difference between p_H 5 and 6.

This quite unexpected behaviour of whole blood towards acids requires further investigation. The word "neutralisation" has been used above, for want of any less committal term, to denote simply a disappearance of titratable acid, which is all that is actually observed; this disappearance might be due to combination with inorganic or organic bases, or to some form of adsorption, or to both of these. The inorganic bases of the red corpuscles are insufficient to account for the neutralisation of these amounts of acid. 1 cc. of blood can neutralise 7·5 cc. 0·01N acid, of which 1·2 cc. is to be assigned to the plasma and 6·3 cc. to the corpuscles (see Tables I and II). The amounts of K and Na present, according to Schmidt's analyses, in 0·5 cc. of red corpuscles, are equivalent to about 5 cc. 0·01N. Hence these bases would be unable to neutralise such an amount of acid even if they were wholly combined with carbonic acid, whereas the analyses show that they must be combined chiefly as chlorides, phosphates, and sulphates.

The following points suggested themselves for investigation:

(a) Is the amount of acid neutralised dependent upon its dilution?

Fig. 4 and Table III show the amounts of acid neutralised when constant amounts of human blood (0·75 cc.) from the same subject and of 0·01N acid (5 cc.) were allowed to react in the presence of varying amounts of saline, the acid being diluted before the blood was added, and the total volume ranging from 6·25 to 45·75 cc.; the concentration of acid thus varied seven-fold. The differences observed in the case of any one of the four samples of blood examined thus are mostly within the range of experimental error. Hence, under the conditions studied, dilution does not have any distinct influence upon this neutralisation of acid.

(b) Is the acid taken up by the corpuscles liberated if they are re-suspended in a fresh neutral medium? This was found not to be the case. 1 cc. blood was shaken with from 5 to 7·5 cc. 0·01N acid in saline in the manner described under "Method" above, and centrifuged, and the supernatant fluid drawn off as completely as possible; the corpuscles were then re-suspended in neutral saline, shaken for five minutes, let stand for a time, the mixture then centrifuged and the supernatant fluid titrated. Only traces of acid (0·05 cc. 0·01N)

were recovered; such an amount might well have remained enclosed between
the packed-down corpuscles when the first supernatant fluid was removed.

(c) Is the neutralisation of acid by whole blood dependent upon the
structure of the red corpuscles, or does lysed blood act in the same way?

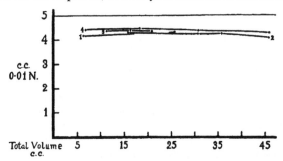

Fig. 4. NEUTRALISATION BY WHOLE BLOOD OF ACID IN DIFFERENT CONCENTRATIONS.
0·75 cc. blood was mixed with 5 cc. 0·01N acid diluted with saline. Points plotted
represent the amount of acid neutralised. The numbers 1, 2, 3, 4 refer to four
different samples of blood.

Table III (see Fig. 4). *Neutralisation by whole blood of acid
in different concentrations.*

0·75 cc. blood + 5 cc. 0·01N acid.

Sample of blood	Total volume cc.	cc. 0·01N acid neutralised	Sample of blood	Total volume cc.	cc. 0·01N acid neutralised
1	6·25	4·17	3	11·25	4·37
	25·75	4·31		16·25	4·40
	15·75	4·32		20·75	4·37
	30·75	4·24		6·75	4·41
	35·75	4·25		18·25	4·43
	45·75	4·05		30·75	4·37
				45·75	4·27

The difficulty in settling this question is due to the very deep colour of
lysed blood; this necessitates considerable dilution before titration can be
carried out. The small amount of acid to be titrated is in very dilute form
and it is difficult to prevent the presence of appreciable quantities of CO_2 in
the large volume of water required. It might be possible to employ dialysis,
but it seemed that this might introduce various other sources of error. It
may be best to describe one of the experiments in detail. The lysed blood
was prepared by freezing and thawing four times. The following mixtures
were made:

Blood +	$\begin{cases} 0·01N\ H_2SO_4\ in \\ normal\ saline \end{cases}$	+ Normal saline	Total volume	Burette reading cc. 0·01N NaOH
Unlysed $\begin{cases} 0·5 \\ 0·5 \end{cases}$	4 0	4 8	8·5 ,,	0·20, 0·21 0·11, 0·12
Lysed $\begin{cases} 0·5 \\ 0·5 \end{cases}$	4 0	4 8	,, ,,	0·50, 0·50 0·115, 0·11

The mixtures were shaken for five minutes and centrifuged, in the ordinary
way; the two with 8 cc. saline only and no acid served as "blanks." Duplicate
portions of 1 cc. of each supernatant fluid were taken, diluted with 36 cc.

freshly boiled water and 2 cc. 0·1 % phenol red and titrated in the comparator; the colour in the lysed samples was very slight at this dilution (1 of blood in 663) but was matched by diluted blood placed in front of the colour standard. The results of three such experiments were as follows:

							% neutralised
I.	{1 cc. unlysed blood neutralises	6·21 cc.	out of 8·0 cc. 0·01N				77·7
	{1 cc. lysed	,,	,,	1·54	,,	,,	19·2
II.	{1 cc. unlysed	,,	,,	6·28	,,	,,	78·5
	{1 cc. lysed	,,	,,	1·37	,,	,,	17·1
III.	{1 cc. unlysed	,,	,,	6·93	,,	,,	86·9
	{1 cc. lysed	,, ·	,,	-2·31	,,	,,	29·0

As has been explained above, this form of experiment is not adapted to a high degree of accuracy; but it does show beyond doubt that lysed blood has only from one-quarter to one-third of the acid-neutralising power of un-lysed blood. The actual difference measured in this comparison of lysed and unlysed blood is that between two burette readings; with the burette employed the difference of level was about $2\frac{1}{2}$ inches, a sufficiently obvious amount. The acid-neutralising power of lysed blood is not greater than that of plasma (see Table II). The remarkable property of whole blood which is illustrated in Figs. 1 and 2 seems then to be due to the structure or physical condition of the red corpuscles. Probably there are many such forms of adsorption or pseudo-adsorption which have not yet been described.

SUMMARY.

The action of whole blood upon acids is quite different from that of plasma. If a series of amounts of sulphuric acid is added to plasma, and to whole blood, so as to produce p_H between 7 and 4, it is found that the amount of titratable acid removed by whole blood increases with the amount added in such a way that a roughly constant percentage, namely about 80 %, of the acid disappears; whereas the amount neutralised by plasma does not vary with the amount added.

It was not found possible to wash out the acid taken up by the red corpuscles. Within the range investigated, the amount of acid removed was independent of the dilution. The inorganic bases of the corpuscles are in-sufficient to neutralise the acid. Since lysed blood does not take up acid in this way it seems that the action of whole blood upon acids must be due to some form of adsorption which is dependent upon the structure or physical condition of the red corpuscles.

. REFERENCE.

Kennaway (1918). Biochem. J. 12, 120.

XXIX. SALIVARY SECRETION IN INFANTS.

By CLEMENT NICORY.

(*Received February 28th, 1922.*)

THERE seems to be great difference of opinion as to the age of onset of the secretion of ptyalin in the saliva and its relation to the food of the child.

Schiffer [1891] says that the ptyalogenous function is present in the new-born child in the submaxillary gland, but is not established in the parotid under the age of two months.

Moll [1905] obtained saliva containing ptyalin from a parotid fistula in a seven months foetus.

Ibrahim [1908] confirmed the work of Moll, and found that ptyalin appeared earlier in the parotid than in the submaxillary gland. This seems natural when one remembers that the parotid is a purely serous gland, and most of the ptyalin in the mixed saliva of the adult comes from it as shown by Pavlov.

Berger [1899] examined the dry residue of the saliva in new born babies, and found that the foetal and adult salivary secretions do not differ very considerably in their consistency. The dry residue in new-born babies being 0·82 % and in adults about 0·79 %.

According to most investigators the secretion of saliva is very slight until after the sixth week of life when the amount increases considerably. Korowin succeeded in collecting as much as 1¼ cc. in five minutes in a child of four months.

Finzi [*Rev. Hyg.*] found that up to the age of six months the amylolytic power of the saliva was the same, whether the food contained starch or not, but after that age the power increased if the food contained starch.

Histologically the infant's salivary glands are relatively rich in connective tissue and in blood vessels, and the mucous cells are few in number. At two years of age however it is impossible to distinguish microscopically an infant's salivary gland from that of an adult.

In an extensive investigation of the subject I have recently obtained the following results at St Thomas's Hospital and the Evelina Hospital for Children, London.

The saliva was collected by giving infants plugs of gauze weighing 500 mg. to suck, the amount of saliva being estimated by the increase in weight of the plugs. The latter were extracted with 5 cc. of distilled water for one hour and then two series of tests were done.

(a) Equal parts of salivary extract and 1 % starch solution were mixed.

(b) To 1 cc. of the salivary extract 2 cc. of 1 % starch solution were added.

The test tubes were kept at a temperature of 40° C.

The following table shows some of the results obtained:

Age	Condition of infant ·	Nature of diet	Equal parts of starch solution and salivary extract	Two parts starch solution to one part salivary extract
34 weeks (premature)	—	Whey	98 minutes	130 minutes
36　,,　　,,	—	,,	30　,,	104　,,
36　,,　　,,	Not as good as others	,,	No change after two hours	
5 hours (after birth)	Fair	Breast	23 minutes	30 minutes
24　,,	Good	,,	16　,,	30　,,
24　..	Poor		29　,,	41　,,
48　,,	Good	,,	17　,,	29　,,
48　,,	Very poor	,,　　·	No change after two hours	
12 days	Poor (one of twins)	Cows' milk and water	17 minutes	65 minutes
9 weeks	Fair	Breast	8	11　,,
10　,,	Good	,,	4	5
11　,,	Fair	,,	7	8
3 months	Poor	Cows' milk	6　,,	8　　..
5　,,	Very poor	Breast and bread	9　,,	13　,,
8　,,	Fair	Breast	1 minute	3½　,,
1 year	Good	—	42 seconds	1 minute

The time indicated in the table was taken at the moment when the mixture of enzyme and starch solution had reached the achromic point as indicated by the iodine test. At this point the starch has reached the stage of convertion into achroodextrin and some maltose. The presence of reducing sugar was demonstrated with Fehling's solution.

An analysis of the results obtained from the saliva of 80 different infants supported the following conclusions:

(1) Ptyalin is present in the saliva of infants at least 1½ months before term although only in small quantities.

(2) The amount of enzyme increases gradually to the age of one year, when the composition of the child's saliva becomes identical with that of the adult.

(3) At all ages the quantity of ptyalin present varies with the general condition of the child, being larger in strong than in weak infants.

Inquiries were made among the mothers attending the out-patient departments as to the nature of the diet administered to their infants, and one was surprised to find that a large number of them fed their babies on bread soaked in milk and ground rice at the early age of 11 weeks and even younger, and barley water was a diluent used by most of them. Quite a number of these children were in fair condition and suffering from no digestive trouble.

According to Halliburton [1913] and others no diastatic enzyme is present in the pancreatic juice of infants.

In view of the fact that an amylolytic enzyme is present in the saliva of babies in quite appreciable quantities, and that a number of infants fed on milk plus starchy foods manage to utilise a large quantity of the starch, there seems every reason for assuming that diastase is present in the pancreatic secretion of infants.

At the commencement of my investigations it appeared as if children who had taken starch in their diet secreted more ptyalin than those who had had none, indicating a capacity of adaptation as postulated by Pavlov, but after examining a larger number it seemed that the presence of starch in the diet makes no appreciable difference in the quantity of ptyalin present in the saliva.

REFERENCES.

Berger (1899). *Inaug. Dissert.* St Petersburg.
Finzi. *Rev. Hyg. et Méd. Inf.* 8, No. 3, 224.
Halliburton (1913). *Handbook of Physiology,* p. 520.
Ibrahim (1908). *Naturforsch. Vers. Köln.*
Moll (1905). *Monatschr. f. Kinderheilk.*
Schiffer, Korowin and Sweifel (1891). Cited in Halliburton's *Chemical Physiology.*

XXX. THE RELATION OF SALIVARY TO GASTRIC SECRETION.

By TOMOICHI NAKAGAWA (Osaka).

From the Institute of Physiology, University College, London.

(*Received March 1st, 1922.*)

SALIVA, besides its main functions in assisting deglutition and exercising some amylolytic action, has other functions of secondary importance which also play a certain rôle in digestive processes. Maxwell [1915] found that boiled starch has a marked inhibitory action on peptic digestion, but that the products of amylolytic hydrolysis have not this effect. Thus the amylolytic ferment of the saliva acquires a new importance by removing the inhibitory factor and thus indirectly assisting in the gastric digestion. Maxwell, in all his experiments, used commercial pepsin. The object of the experiments described in this communication was to repeat Maxwell's experiments with natural gastric juice and to extend them to the rennet action of gastric juice.

A. SALIVA AND PEPTIC DIGESTION.

Pure gastric juice was collected by "sham feeding" from a dog which had a gastric fistula and an oesophagostomy. Fresh samples were used for each experiment. Mett's method was found to be sufficiently accurate for the purpose of these experiments and was used throughout the research. The Mett's tubes were left to digest for ten hours at 40°. All the experiments were repeated a considerable number of times, and those given in this communication serve only as examples.

Exp. 1. The seven small flasks containing the solutions named in Table I were kept at 40° for one hour; at the end of this time 5 cc. of gastric juice were added to each flask, and after another half hour the Mett's tubes were placed in each sample. In ten hours the readings of the digested tubes given in Table I were observed.

Table I.

	Digestion of Mett's tubes in mm.
1. 3 cc. of 2 % boiled potato starch + 1 cc. fresh human saliva	5·3
2. „ „ „ „ „ + „ boiled „ „	3·9
3. „ „ raw „ „ + „ „ „ „	5·4
4. „ „ soluble starch „ + „ „ „ „	5·3
5. „ „ erythrodextrin „ + „ „ „ „	5·2
6. „ „ maltose „ + „ „ „ „	5·4
7. „ „ glucose „ + „ „ „ „	5·5

The figure for flask 2 is the only one that shows a deviation from the rest. This flask was also the only one that contained the starch in a colloidal form. The hydrolysis of the starch in flask 1 was carried to the achromic point.

Exp. 2. In this experiment the actions of hydrolysed and unhydrolysed starches of different origin were compared. The different pairs given in Table II must not be compared with each other, because, although the hydrolysis was carried on for the same length of time, it reached a different stage with each starch as determined by the iodine reaction.

Table II.

			Digestion of Mett's tubes in mm.
1. Potato starch	hydrolysed		5·3
,, ,,	unhydrolysed		3·9
2. Maize ,,	hydrolysed		3·0
,, ,,	unhydrolysed		2·2
3. Wheat ,,	hydrolysed		3·3
,, ,,	unhydrolysed		2·6
4. Rice ,,	hydrolysed		5·2
,, ,,	unhydrolysed		4·2

In each case 1 cc. of saliva, boiled or unboiled, was added to 3 cc. of the 2 % solution, kept in the thermostat for one hour, and then mixed with 5 cc. of gastric juice. The readings were taken at the end of ten hours.

Maxwell's observations made with commercial pepsin are extended in these experiments to normal gastric juice.

The theory of adsorption of the pepsin by the colloidal starch advanced by Maxwell seems to be the most probable explanation of the experiments described. There is one factor of general interest which should be mentioned. As has been described by several observers, the gastric juice secreted on a carbohydrate food is always the richest in pepsin. On the other hand, carbohydrates, when in a colloidal form, hinder peptic digestion. This seems to be another case of a delicate adaptation of the organism in providing more ferment when its action meets with difficulties.

B. SALIVA AND MILK CLOTTING.

We have seen that the carbohydrates in a colloidal form behave towards normal gastric juice and the commerical preparation of pepsin in the same manner. We now proceeded to determine whether the carbohydrates behave in the same way towards the rennet of gastric juice.

The experiments with milk clotting were performed in the same way as those with peptic digestion, but instead of the Mett's tubes, 5 cc. of cow's milk were added to each test tube and the time required for clotting noted.

The action of the boiled starch on milk clotting was found to be the opposite to its action on peptic digestion. In each experiment the addition of starch in a colloidal form produced a considerable acceleration of the clotting. Exp. 3 serves as an example.

Exp. 3. The seven test tubes containing the solutions were placed in a thermostat at 40°. After one hour 0·3 cc. of gastric juice was added to each test tube and after a further 30 minutes 5 cc. of cow's milk. The time of clotting of each sample is given in Table III.

Table III.

Time of clotting in mins.

1.	3 cc. boiled potato starch	+ 1 cc.	fresh	human	saliva	15
2.	„ „ „ „	+ „	boiled	„	„	3
3.	„ raw „ „	+ „	„	„	„	15
4.	„ soluble starch	+ „	„	„	„	14
5.	„ erythrodextrin	+ „	„	„	„	14
6.	„ maltose	+ „	„	„	„	15
7.	„ glucose	+ „	„	„	„	16

As in the first experiment, the only figure which deviates from the others is the one corresponding to the test tube containing the starch in a colloidal form. But in this experiment the difference is in the opposite direction, showing a marked acceleration.

Exp. 4. Some other starches (beans, sago, barley, oats, etc.) were tried, and always with the result that the starch in a colloidal form produced an acceleration of milk clotting. Table IV gives some of the figures obtained.

Table IV.

		1 cc. saliva	Gastric juice	Milk cc.	Time of clotting mins.
3 cc. boiled potato starch		fresh	0·3	5	9
„ „ „ „		boiled	„	„	3
„ „ bean „		fresh	..		8
„ „ „ „		boiled	„		3
„ „ rice „		fresh			8
„ „ „ „		boiled	„		5
„ „ maize „		fresh	„		6
„ „ „ „		boiled	„	„	4

In all the above experiments human filtered saliva was used. The filtered saliva was quite watery and contained hardly any mucin as determined by precipitation with acetic acid. The result of the experiments with milk clotting indicates that the amylolytic action of saliva indirectly slows the clotting of the milk in the presence of colloidal starch. The experiments provide one more example of the difference in behaviour of pepsin and rennin.

Under normal conditions it is rather exceptional for saliva to have this effect on milk clotting, and it certainly never occurs in animals deprived of the salivary amylase. It is of some interest, however, that even in such animals (dog) a fair amount of saliva is secreted on milk although it is of no use for its deglutition and has no direct relation to its digestion. In the dog the submaxillary saliva secreted on milk is richer in mucin than the saliva secreted on any other food. In sucklings there is a profuse secretion of viscid saliva. These facts suggest that saliva itself, besides its indirect action described above, has some importance in the digestion of milk. Borisov [1914] has shown that in dogs with gastric fistula milk gives finer clots when mixed with saliva, and ascribes it to the physical action of the viscous saliva. Allaria [1912] found occasionally a slight slowing in milk clotting when mixed saliva from an infant was added.

The following experiments were performed in order to determine whether saliva has any direct effect on milk clotting.

Parotid and submaxillary saliva were collected from a dog having salivary fistulae of the two glands. The saliva was diluted ten times with thrice distilled water. To each 5 cc. of the diluted saliva 5 cc. of milk were added and then mixed with 0·5 cc. of gastric juice. The time required for milk clotting in the thermostat at 40° is given below.

Table V.

	Time of clotting in mins.
1. { 5 cc. tap water + 5 cc. milk	2
„ ordinary distilled water + 5 cc. milk	2
„ parotid saliva + 5 cc. milk	3
„ submaxillary saliva + 5 cc. milk	6
2. { „ submaxillary saliva (10 times diluted) + 5 cc. milk	5
„ tap water + 5 cc. milk	2
3. { „ pure submaxillary saliva (undiluted) + 5 cc. milk	∞
„ tap water + 5 cc. milk	2

The sample of parotid saliva used in this experiment did not contain any mucin, which is often present in dog's parotid secretion.

Brennemann [1917] showed that in rapid clotting larger curds are formed than when the clotting of milk is slow. As the presence of the viscid saliva delays the clotting of the milk, this explains the observations of Borisov quoted above. It is certain that this effect of the submaxillary saliva does not depend on a change in the reaction. The addition of colloidal starch as used in the above experiments to acid solutions did not change their p_H as determined colorimetrically. Parotid saliva, being not less alkaline than submaxillary, has no such delaying effect on milk clotting. The action of the viscid saliva seems to be better explained by ascribing a protective action to the mucin.

Conclusions.

1. Maxwell's experiments on the inhibitory effect of colloidal starch on peptic digestion are confirmed and extended to natural gastric juice.

2. The amylase of saliva hydrolysing the starch removes its inhibitory effect on gastric digestion.

3. Pepsin and rennin behave towards colloidal starch differently—the action of the pepsin is inhibited by colloidal starch and the action of the rennin accelerated.

4. The amylase of saliva hydrolysing the starch removes its accelerating action on rennin.

5. The saliva has a delaying action of its own on milk clotting; this action was found to be due to the protective action of mucin.

6. The secretion of saliva in sucklings is important because it delays the milk clotting and thus produces finer and more easily digested curds.

I wish to express my thanks to Dr G. V. Anrep for his kind help and criticism during my experiments. The gastric juice and saliva were collected from dogs operated on by Dr Anrep.

REFERENCES.

Allaria (1912). Zentr. Biochem. Biophys. 13, 536.
Borisov (1914). Quoted after Babkin's Die äussere Sekretion der Verdauungsdrüsen, 8, 19.
Brennemann (1917). Arch. Pediatrics, 34, 81.
Maxwell (1915). Biochem. J. 9, 323.

XXXI. THE RELATION OF THE FAT SOLUBLE FACTOR TO RICKETS AND GROWTH IN PIGS. II.

By JOHN GOLDING, SYLVESTER SOLOMON ZILVA, JACK CECIL DRUMMOND AND KATHARINE HOPE COWARD (*Beit Memorial Research Fellow*).

From the National Institute for Research in Dairying, Reading, the Biochemical Department, Lister Institute, the Institute of Physiology, University College, London.

(*Received March 14th, 1922.*)

THE experiments described in this communication form a part of an inquiry now in progress, the main purpose of which is to study the part played by the accessory factors in the nutrition of agricultural stock and in the etiological factors concerned in the causation of rickets in pigs. In a previous communication [Zilva, Golding, Drummond and Coward, 1921] in which the etiology of rickets in pigs was investigated we have shown that a rigorous elimination of vitamin A from the diet of young pigs did not conduce to the production of well-defined rickets. We were, however, able to demonstrate in those experiments that such a dietetic deficiency has a very marked effect on the development of these animals. Indications were also obtained that the deprivation of this dietetic principle possibly has some bearing on the production of healthy young.

Having been unsuccessful in producing rickets in pigs experimentally by depriving them of vitamin A alone we next attempted to ascertain whether a dietetic deprivation of calcium and vitamin A would lead to the production of the disease. The results obtained by us form the subject of this communication.

EXPERIMENTAL.

Our animals were divided into four groups. Group I received a diet deficient in vitamin A and in calcium ($- A - Ca$), group II received a diet deficient in vitamin A ($- A + Ca$), group III received a diet deficient in calcium only ($+ A - Ca$), group IV received a diet containing calcium and vitamin A ($+ A + Ca$). Two pigs were placed in each group. Group I consisted of two boars, group II of one boar and one sow, group III of two sows and group IV of two boars. The young pigs, which belonged to the same litter, were started on their special diets when they were 53 days old. They were young Berkshires, born of a sow 14 months old. The sow was kept for some time on a diet of toppings and whey deficient in vitamin A and manifested this deprivation by retarded growth [Drummond, Golding, Zilva and Coward, 1920]. On supplementing the above diet with lucerne the animal resumed growth and continued an apparently normal existence. After 74 days of correct feeding she was served by a pedigree boar and soon after was placed again on a diet of toppings, whey and swedes. This diet which was poor in vitamin A was purposely planned as we did not desire the young pigs to be born with a store of the vitamin. After 116 days nine pigs were born and at the age of 65 hours were divided into two sections. The chart below shows the history and subsequent allocation of the pigs. One lot received its supply of vitamin A entirely from the mother, the other received additional vitamin A in the form of cream ($\frac{1}{8}$ oz. per day) and eventually in the form of cod liver oil ($\frac{1}{2}$ rising to 1 oz. per day). Two of the animals in the former section received an equal supply of additional oil in the form of inactive olive oil, made into an artificial cream at first and later given to balance in nutrients the cod liver oil given to Section II.

The two remaining pigs in the first section received only the sow's milk; the sow being fed all the time on a diet shown to be poor in the vitamin A.

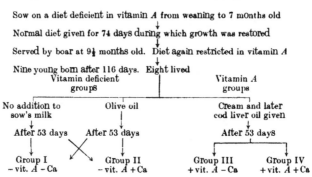

As soon as it was convenient the young pigs were given additional toppings which at the end of this preliminary period of 53 days (period I) reached a ration as high as 3 lbs. per day for the litter of eight pigs. The actual intake of food and the corresponding increase in weight for the entire experiment is summarised in Table I. The growth of the animals is graphically represented in Fig. 1. The average daily gains in pounds for period I were for pigs in section I receiving no oil 0·533 and 0·514, for those receiving olive oil 0·377 and 0·5, while the daily gains of the pigs in section II receiving cod liver oil were 0·481, 0·453, 0·344 and 0·509.

The agreement between the rates of growth of the pigs in the two sections is made more evident by employing the formula advised by R. A. Fisher,

Table showing feeding-experiment data for four groups of pigs.

Treatments: **Group I** — −Vitamin A, Low calcium; **Group II** — −Vitamin A, +calcium and phosphate; **Group III** — +Vitamin A, Low calcium; **Group IV** — +Vitamin A, +calcium and phosphate.

	Group I Boar lbs.	Group I Per 2 pigs lbs.	Group II Sow lbs.	Group II Per 2 pigs lbs.	Group II Boar lbs.	Group III Sow lbs.	Group III Per 2 pigs lbs.	Group III Sow lbs.	Group IV Boar lbs.	Group IV Per 2 pigs lbs.	Group IV Boar lbs.
Period I — With the sow, April 11th to June 3rd											
Weight on June 3rd	29·5	30	22·5		30·5	28·5		26·5	30		21
„ April 11th	3·0	2·75	2·5		2·25	3·0		2·5	3		2·75
Gain in 53 days	26·5	27·25	20·0		28·25	25·5		24·0	27		18·25
Gain in lbs. per day	0·5	0·514	0·377		0·533	0·481		0·453	0·509		0·344
Relative growth rate	4·31	4·50	4·14		4·91	4·25		4·45	4·34		3·83
Oil given	Olive	None	Olive		None	Cod liver		Cod liver	Cod liver		Cod liver
Period II — Of equal intake, June 3rd to July 27th											
Weight on July 27th	50·38	51	46		56	63·125		54·56	55·625		70
Gain in 54 days	20·88	21	23·5		25·5	34·625		28·06	34·625		40
Relative growth rate	0·99	0·98	1·32		1·13	1·47		1·34	1·80		1·57
Toppings, dry matter		170		170			170			170	
Caseinogen		5-7 H		5-7 H			5-7 U			5-7 U	
Oil		6·06 O		6·06 O			6·06 C			6·06 C	
Chalk		—		2·0			—			2·0	
Charcoal		4 W		5·36 A			4 W			5·36 A	
Pounds dry matter per 1 lb. increase in live weight		4·4		3·8			3·0			2·5	
Period III — July 27th to Aug. 31st											
Weight on Aug. 31st	55·0	52·25*	59		71·75	84·75		75	85		99
Gain in 35 days	4·62	1·25	13		15·75	21·6		20·4	29·4		29
Relative growth rate	0·25	0·08	0·71		0·71	0·84		0·91	1·21		0·99
Toppings, dry matter		Per 1 pig 55·2		111·5			155			157·28	
Caseinogen		2·85 H		5-7 H			6·2 U			6·2 U	
Oil		2·8 O		5·6 O			5·6 C			5·6 C	
Charcoal		1·12 W		4·75 A			2·25 W			4·75 A	
Chalk		—		2·2			—			2·2	
Pounds dry matter per 1 lb. increase in live weight		14·6		4·5			4·0			3·02	
Period IV — Unheated caseinogen for all groups, Aug. 31st to Sept. 22nd											
Weight on Sept. 22nd	65·25		68·25		81·5	108·25		93·25	116·5		129·25
Gain in 22 days	10·25		9·25		9·75	23·5		18·25	31·5		30·25
Relative growth rate	0·78		0·66		0·58	1·11		0·99	1·43		1·22
Toppings, dry matter		29·69		42·5			146			147·8	
Caseinogen, "White Light"		1·8		3·9			4·12			4·12	
Oil		1·52 O		3·28 O			3·52 O			3·52 O	
Charcoal		1·75 W		3 A			2·6 W			2·3 A	
Chalk		—		1·1			—			1·37	

Period V
Groups I and II

(a) Heated caseinogen

Weight on Sept. 29th	64·75		70		82	114·25		99·5	125		1·42
Gain in 7 days	− 0·5		1·75		0·5	6·0		6·25	8·5		12·75
Relative growth rate	Loss		0·36		0·09	0·77		0·92	1·0		1·34
Toppings, dry matter		10·2		13·72			57·08			58·41	
Caseinogen		0·62 H		1·3 H			1·3 U			1·3 U	
Oil		0·52 o		1·12 o			1·12 c			1·12 c	
Charcoal		0·44 w		0·87 A			0·87 w			0·87 A	
Chalk		—		0·42			—			0·43	
Pounds dry matter in food per 1 lb. increase in live weight		—		7·75			4·92			2·92	

(b) Unheated caseinogen

Weight on Oct. 2nd	69		71·5		86·5	116		101	130		145·5
Gain in 3 days	4·25		1·5		4·5	1·75		1·5	5		3·5
Relative growth rate	2·11		0·7		1·78	0·51		0·50	1·31		0·81
Toppings, dry matter		4·64		7·96			25·32			26·55	
Caseinogen		0·28 U		0·56 U			0·56			0·56 U	
Oil		0·24 o		0·48 o			0·48 c			0·48 c	
Charcoal		0·2 w		0·37 A			0·37 w			0·37 A	
Chalk		—		0·18			—			0·18	
Pounds dry matter per 1 lb. increase in live weight		1·26		1·59			8·23			3·31	

(c) Heated caseinogen

Weight on Oct. 22nd	76·25		80·75		93	135·5		120·25	162		167
Gain in 20 days	7·25		9·25		6·5	19·5		18·25	32		21·5
Relative growth rate	0·50		0·61		0·36	0·776		0·87	1·10		0·69
Topping's dry matter		25·2		57			180			197	
Caseinogen		1·95 H		3·9 H			3·9 U			3·9 U	
Oil		—		—			—			—	
Charcoal		1·3 w		2·6 A			2·6 w			2·6 A	
Chalk		—		1·3			—			1·3	
Pounds dry matter per 1 lb. increase in live weight		3·9		4·1			4·9			3·8	

*= Weight of Pig, group I on Aug. 25th. *Caseinogen* U = Unheated, H = Heated. *Oil* c = Cod Liver Oil, o = Inactive Olive Oil. *Charcoal* A = Animal Charcoal, w = Wood Charcoal.

NOTE. The total dry matter consumed is calculated on the ingredients given and expressed in pounds and decimals of a pound. The small daily dose of lemon juice and the occasional doses of marmite are set against loss of charcoal and chalk which were not always cleaned up.

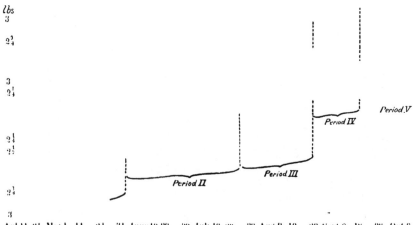

Fig. 1.

Chief Statistician, Rothamsted Experimental Station, for calculating the relative growth rate per day per cent. using natural logarithms, viz.:

$$\frac{\log_e M_2 - \log_e M_1}{K_2 - K_1} \, 100$$

where $M_2 =$ the weight at the end of the period,

$M_1 =$ „   „   commencement of the period,

$K_2 - K_1 =$ duration of period in days.

The following figures are obtained for Section I $(-A)$ without oil 4·91 and 4·50, with olive oil 4·14 and 4·31, for Section II $(+A)$ 4·25, 4·45, 3·83 and 4·34.

It is evident from the above figures that a sow, even when fed on a diet deficient in the fat-soluble factor and having undergone a previous deprivation in this factor, is capable of rearing her young satisfactorily. This confirms further our earlier observation that the requirements of the pig for the fat-soluble factor are not of a high order.

At the end of this period, namely 53 days after birth, the animals were weaned and placed on their respective experimental diets. Each group was placed in a separate sty. The sties faced south and were divided by wooden partitions, the floor being partly wood and partly concrete. The bedding consisted of wood shavings or sawdust.

The basal diet for all groups consisted of toppings or wheat offal having the following average composition:

Moisture	11·51 %	Mucilage	59·60 %
Oil	4·29	Woody fibre	5·86
Protein	14·81	Ash	3·93

The dry matter of the above contained 0·338 % calcium. Besides the basal diet the pigs received supplementary protein in the form of caseinogen, which in the case of the vitamin-free diets was previously inactivated by being heated for 24 hours at 120° C. This inactivation was carried out for us by Dr R. T. Colgate in Messrs. Huntley and Palmer's Laboratory, for which we wish to express our thanks.

The food was weighed out three times a day at 8 a.m., 12 noon and 4.30 p.m.; it was mixed with cold water just before feeding and given in the form of a thin cream. The food was given on a live weight basis, being regulated to the quantity which the pigs would clean up.

The other two accessory factors, namely the antiscorbutic and the antineuritic factors, were supplied in the form of freshly prepared lemon juice and marmite (Commerical Yeast extract), about 7 cc. of the former being the daily dose, whilst $\frac{1}{3}$ oz. of the latter was given at intervals. Groups III $(+A - Ca)$ and IV $(+A + Ca)$ received a daily dose of cod liver oil as a source of the fat-soluble factor whilst the other two groups received an equivalent dose of inactive olive oil. Groups II and IV also received a daily dose of 1 oz. each of precipitated chalk and 1 oz. animal charcoal containing

67·3 per cent. of calcium phosphate as a source of additional calcium and phosphate; the other two groups received only the small amounts of calcium from the basal diet.

During the following 54 days (period II) the intake in all the four groups was the same. The increase in weight in the animals of the various groups however showed a marked disparity. The two animals in group I gained 41·88 lbs., in group II 49 lbs., in group III 62·7 lbs., and group IV 74·6 lbs. The relative growth rates, as calculated from the formula given above, were group I 0·99 and 0·98, group II 1·32 and 1·13, group III 1·47 and 1·34, and group IV 1·80 and 1·57. The average dry matter in the food consumed per pig for each pound of gain in live weight was: group I 4·4 lbs., group II 3·8 lbs., group III 3·0 lbs. and group IV, 2·5 lbs. By this time very marked differences in the general appearance of the animals in the respective groups could be discerned. The animals in group I developed a scurfy skin and saddle back, weak legs and joints painful to pressure. They were easily tired and were not playful. Those in group II also showed lack of vitality and a saddle back, those in group III were in good condition possessing glossy coat, whilst the animals in group IV were decidedly in the best condition. One animal in group I received an injury from a fall and died 87 days after the commencement of the experiment. At the post mortem examination it was found that the vertebral column was broken.

During period III, i.e. 35 days following period II, the increase in weight was as follows: group I 4·62 lbs. (one animal), group II 28·75 lbs., group III 42 lbs., and group IV 58·4 lbs., and the average dry matter in the food consumed per pig for each pound of gain in live weight was: group I (one animal) 14·6 lbs., group II 4·5 lbs., group III 4·0 lbs. and group IV 3·02 lbs.

Owing to the low condition of the remaining animal in group I it was decided to administer a small amount of the fat-soluble factor in order to save the pig. This was done by introducing caseinogen, which had not been previously inactivated by heating, in the diet during period IV. This addendum had its desired effect and the animal responded after about five days by resuming growth.

During this period of 22 days (period IV) the animals gained in weight as follows: group I (one animal) 10·25 lbs., group II 19 lbs., group III 41·7 lbs., and group IV 61·7 lbs. In the last period (period V) the inactivated caseinogen was alternated with crude caseinogen in order to keep the weight of the animals in group I and group II in check.

The experiment was terminated 145 days after its commencement. The animals were slaughtered with the exception of one sow in each of the groups II and III. The condition of the animals before slaughter was as follows: groups I and II wrinkled skin, ears carried forward, down on hind legs; group III skin rather rough, lack of size and bloom, flesh not quite firm, otherwise normal; group IV skin healthy and animals perfectly normal.

The photographs of the pigs before slaughtering are shown in Plate II, two photographs having been taken of the pig in group I. The photographs were taken to scale.

The post mortem examination revealed no abnormalities of the organs beyond that the ribs of the animals in group I were easily fractured. It is also to be reported that the fat of the pigs of groups I and II was, in the butcher's opinion, softer and not of such good quality as the fat of the other pigs. This was confirmed by the estimation of the melting and solidifying points of the fats.

		Melting points		Solidifying points
Group	I	21°	22·9°	19°
„	II	21°	23·8°	19·3°
„	III	25°	25·20°	22°
„	IV	31·5°	32·8°	22·5°

Table II gives the size, weight, breaking points, and calcium content of the bones.

Table II.

	Weight of humerus in g.	Distance of bearing points in inches	Breaking weight in tons	Same corrected to 3¼″ length	CaO percentage of dry matter
Group I	103	3½	0·210	0·210	29·52
„ II	127	3½	0·345	0·345	29·05
„ III	117	4	0·292	0·336	33·52
„ IV	152	4	0·466	0·532	38·08

No abnormal flavour or taste was reported by a number of people who consumed the joints derived from the animals fed on the cod liver oil.

The fifth, sixth and seventh ribs of one animal in each group were examined for us histologically by Prof. V. Korenchevsky to whom we are also indebted for the interpretation of the results. The following is a summary of the observations made:

(1) Group IV ($+ A + Ca$) showed a somewhat abnormal picture with very slight osteoporosis and a belated deposition of lime salts in the newly formed bone.

(2) In all cases the bone marrow especially in the region of the secondary spongiosa consisted of a fine fibrous reticulum with but few bone marrow cells.

(3) The histological condition in animals belonging to groups II ($- A + Ca$) and III ($+ A - Ca$) was not very different from that of group IV ($+ A + Ca$). Only a higher degree of osteoporosis resulting from a diminished activity of osteoblasts could be seen.

(4) In group I ($- A - Ca$) the condition of osteoporosis was more pronounced. Moreover a more frequent incursion of the proliferating cartilage into the bone marrow was in evidence. In these places was also noticed defective calcification in the zone of provisional calcification.

It is quite evident that in spite of the very marked changes which have been effected by our restricted diets, no rickets in the pathological sense of the word has been induced. The animals in groups I and II have on various occasions during the experiment displayed a condition which would have been described by the practical man as the pigs being "off their feet." No doubt such a condition has been before now loosely referred to as "rickets." Although defective calcification was found in the zone of provisional calcification in the case of group I $(-A - Ca)$ no increase in the amount of osteoid tissue could be established and therefore no faulty deposition of calcium in the newly formed bone in the sense of rickets can be asserted.

A remarkable feature in our experiments is that even in the case of group IV $(+A + Ca)$ which acted as our control, a normal histological picture was not obtained. This requires further investigation. Possibly the restricted diet of the mother may be responsible for this. Another point to be considered is that the animals consuming the calcium-deficient diet received 0·338 % of calcium in their food. With animals, such as pigs, which consume large bulks of food it was difficult from a technical point of view to reduce this cálcium intake. However the calcium deficiency was definite enough to impair the growth and diminish the calcium content in the bones of the animals in group III $(- Ca + A)$ and it is very doubtful whether pigs which develop rickets under farm conditions, receive a diet much poorer in this element. We refrain from reviewing the literature which has appeared during the last few months in connection with the etiology of rickets. Most of the experimental work was done on rats and the results and conclusions of the various investigations are conspicuously contradictory. We cannot however conclude without briefly referring to the results obtained by Korenchevsky [1921]. This investigator, working with rats, obtained a definite condition of rickets on a diet free from the fat-soluble factor and calcium. Whether our apparently different results were due to the higher content of calcium in our experimental diets, or whether it was due to the different character of the experimental animal will most probably be decided by future investigation. We are continuing our experiments and although the dietetic hypothesis of the etiology of rickets forms our main line of work, we are not excluding such a hypothetical factor as light, especially in view of the latest work of Hess and Unger [1921], of Powers, Park, Shipley, McCollum and Simmonds [1922].

The expenses of this research were defrayed from a grant made by the Medical Research Council, to whom our thanks are due.

REFERENCES.

Drummond, Golding, Zilva and Coward (1920). Biochem. J. 14, 742.
Hess and Unger (1921). Soc. Exp. Biol. Med. 18, 298.
Korenchevsky (1921). Brit. Med. J. 547.
Powers, Park, Shipley, McCollum and Simmonds (1922). J. Amer. Med. Ass. 78, 159.
Zilva, Golding, Drummond and Coward (1921). Biochem. J. 15, 427.

Group I   – Vitamin *A*.

Group II   – Vitamin *A* + Calcium and Phosphate.

Group III   + Vitamin *A*.

Group IV   + Vitamin *A* + Calcium and Phosphate.

Pigs from Groups IV, III, II, I.

Pigs from Groups IV, III, II, I.

Cross sections of pigs from Groups IV, III, II, I.
Group I − Vitamin A.
Group II − Vitamin A + Additional Calcium and Phosphate.
Group III + Vitamin A.
Group IV + Vitamin A + Additional Calcium and Phosphate.

XXXII. THE ESTIMATION OF CALCIUM IN BLOOD[1].

BY ARTHUR ROBERT LING AND JOHN HERBERT BUSHILL.

*From the Department of Biochemistry of Fermentation,
University of Birmingham.*

(*Received March 17th, 1922.*)

INTRODUCTORY.

THE estimation of calcium in blood, serum etc., is a matter of extreme delicacy seeing that, in the case of human blood at least, it is not often easy to obtain more than 1 to 5 cc. of the sample. Working with such small quantities therefore it is necessary to employ a method capable of measuring calcium to a limit of accuracy of at least 0·006 mgm. It is obvious also that in these circumstances the reagents employed must be of the highest purity, especially so far as their freedom from calcium is concerned.

Among published methods are some based on the use of the haemacytometer (counting the calcium oxalate crystals) and on principles, such as (1) the determination of the concentration of calcium ions necessary to produce clotting of the blood, (2) mixing the blood with ammonium oxalate solutions made isotonic with sodium chloride, and calculating the calcium from that concentration of oxalate which just prevents clotting, (3) mixing the blood with sufficient ammonium oxalate to prevent coagulation and subsequently determining the quantity of calcium chloride necessary to produce clotting. We agree with de Waard [1919] that these give comparative rather than absolute results. Recourse must therefore be had in our opinion to chemical methods.

McCrudden [1909] suggests a gravimetric method. The calcium is precipitated as oxalate in a solution in which the hydrogen ion concentration is so adjusted that the iron and phosphates remain in solution.

Halverson and Bergeim [1917] propose to estimate calcium in blood which has not been incinerated, but from which the proteins have been precipitated by picric acid. The precipitation of the calcium is carried out under McCrudden's conditions, the hydrogen ion concentration being adjusted to the point at which alizarin gives a violet coloration. The calcium oxalate is washed by centrifuging and after dissolving it in dilute sulphuric acid the solution is titrated with $N/100$ potassium permanganate. This method, according to the examples shown in the paper, gives results agreeing closely,

---

[1] This investigation was undertaken at the request of my colleague, Prof. Carlier. (A. R. L.)

i.e. within the limit of experimental error, with similar estimations made on the ash of the blood, etc.

Marriott and Howland [1917] have devised an ingenious method for the estimation of calcium and magnesium in blood serum. It depends on the fact that solutions of ferric thiocyanate are decolorised by oxalates and by phosphates. Calcium is precipitated as oxalate and magnesium as ammonium magnesium phosphate. The precipitates are dissolved in acid, and added to solutions of ferric thiocyanate, the degree of decolorisation being determined by comparison in small Nessler tubes. It is said to be possible by this method to work on 2 cc. samples of serum with a maximum error of less than 5 %.

Jansen [1918] describes a gravimetric method of estimating calcium. He incinerates the blood; and he considers the removal of iron and phosphoric acid necessary before precipitating the calcium as oxalate.

De Waard [1919] has devised a method for the estimation of calcium in organic substances for which he claims an accuracy of 4 %. The substance is incinerated, the ash dissolved in hydrochloric acid in a special tube, drawn out to a capillary end, which is subsequently used for centrifuging the liquid when the calcium has been precipitated. The tube containing the hydrochloric acid solution of the ash is placed in a bath of boiling water, 0·5 cc. of saturated ammonium oxalate solution added, then a slight excess of ammonia, and finally a slight excess of acetic acid. In these circumstances he finds that the iron, magnesium and phosphates remain dissolved, whilst the calcium is precipitated quantitatively.

EXPERIMENTAL.

The method we have adopted is based on the methods of McCrudden, Halverson and Bergeim, and de Waard.

The blood is incinerated in a platinum dish, the ash dissolved in concentrated hydrochloric acid, and the solution washed into a centrifuge tube of special shape. In this tube the calcium is precipitated as oxalate as described later. When the liquid is centrifuged, the precipitate collects in the narrow tube leaving the supernatant liquid clear. This liquid can then be removed by a tube drawn out to a capillary without disturbing the precipitate. The calcium oxalate is washed twice in this way, dissolved in dilute sulphuric acid and the solution titrated with $N/100$ potassium permanganate.

The shape of the centrifuge tube is shown in Fig. 1. It has the following dimensions: diameter 1·5 cm., total length 12 cm., length of narrow tube 1·3 cm., bore of capillary tube 0·5 cm. The tube is graduated for 2 cc. and for 25 cc. A glass rod capable of entering the narrow tube (as shown) is used for stirring the precipitate after the addition of sulphuric acid before titrating. The advantage of this centrifuge tube is that it is of such dimensions that all the operations can be performed in it. It is large enough for the precipitation to be carried out, the narrow tube at the base enables the precipitate to be washed easily without being disturbed, and finally the end point can be seen easily when titrating as the tube does not taper gradually to a point. The solutions may be titrated to 0·01 cc.

The procedure is as follows:

Incineration. About 2–5 cc. of blood is measured into a platinum dish. This is heated on a piece of wire gauze with a coating of asbestos, by a very small flame until all the water has evaporated and a spongy mass of carbonaceous matter is left

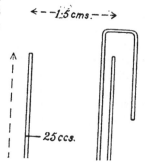

4ꝺ

ERRATUM

THE ESTIMATION OF CALCIUM IN BLOOD (LING AND BUSHILL)

The diameter of the centrifuge tube should be 2·6 cm. and not 1·5 as stated in the text and diagram on p. 405.

McCrudden's method as modified by Hal- in the estimation of calcium in blood. verson and Bergeim. The solution in the centrifuge tube is neutralised with 0·880 ammonia using alizarin as indicator. The neutral point is overshot and then titrated back with $N/2$ HCl until the colour just commences to change. Then 2·5 cc. of $N/2$ HCl and 2·5 cc. of 2·5 % oxalic acid solution are added and water to make to the 25 cc. mark. The tube is then placed in a water-bath and brought to the boiling point. To the hot solution 5 cc. of 3 % ammonium oxalate is added and the tube kept at the boiling point for 15 minutes when it is removed, placed in ice water and 5 cc. of 20 % sodium acetate solution added (or until the colour commences to change). When the calcium oxalate has precipitated which takes about 6–8 hours (it is usually allowed to remain over night), the solution is centrifuged. The clear super-natant liquid is removed by means of a tube (drawn out to a capillary) leading to a bottle connected with a filter pump. By this means the liquid is removed to the 2 cc. mark on the tube. The outside of the capillary tube is washed with water to prevent the loss of any precipitate which may adhere to it.

i.e. within the limit of experimental error, with similar estimations made on the ash of the blood, etc.

Marriott and Howland [1917] have devised an ingenious method for the estimation of calcium and magnesium in blood serum. It depends on the fact that solutions of ferric thiocyanate are decolorised by oxalates and by phosphates. Calcium is precipitated as oxalate and magnesium as ammonium magnesium phosphate. The precipitates are dissolved in acid, and added to solutions of ferric thiocyanate, the degree of decolorisation being determined by comparison in small Nessler tubes. It is said to be possible by this method to work on 2 cc. samples of serum with a maximum error of less than 5 %.

Jansen [1918] describes a gravimetric method of estimating calcium. He incinerates the blood; and he considers the removal of iron and phosphoric acid necessary before precipitating the calcium as oxalate

------ .... ..rgein, and de Waard.

The blood is incinerated in a platinum dish, the ash dissolved in concentrated hydrochloric acid, and the solution washed into a centrifuge tube of special shape. In this tube the calcium is precipitated as oxalate as described later. When the liquid is centrifuged, the precipitate collects in the narrow tube leaving the supernatant liquid clear. This liquid can then be removed by a tube drawn out to a capillary without disturbing the precipitate. The calcium oxalate is washed twice in this way, dissolved in dilute sulphuric acid and the solution titrated with $N/100$ potassium permanganate.

The shape of the centrifuge tube is shown in Fig. 1. It has the following dimensions: diameter 1·5 cm., total length 12 cm., length of narrow tube 1·3 cm., bore of capillary tube 0·5 cm. The tube is graduated for 2 cc. and for 25 cc. A glass rod capable of entering the narrow tube (as shown) is used for stirring the precipitate after the addition of sulphuric acid before titrating. The advantage of this centrifuge tube is that it is of such dimensions that all the operations can be performed in it. It is large enough for the precipitation to be carried out, the narrow tube at the base enables the precipitate to be washed easily without being disturbed, and finally the end point can be seen easily when titrating as the tube does not taper gradually to a point. The solutions may be titrated to 0·01 cc.

The procedure is as follows:

Incineration. About 2–5 cc. of blood is measured into a platinum dish. This is heated on a piece of wire gauze with a coating of asbestos, by a very small flame until all the water has evaporated and a spongy mass of carbonaceous matter is left, when it is heated more strongly. During the final heating the dish is placed on a pipe-clay triangle and brought to redness. To prevent extraneous matter from gaining access to the dish during the first part of the incineration, an inverted funnel is fixed over it by a clamp. The residue is treated with 1·5 cc. concentrated hydrochloric acid 0·5 cc. at a time—each time the solution is gently warmed and washed into the centrifuge tube. The dish is then washed three times with water. Care must be taken in the operation to keep the volume to a minimum.

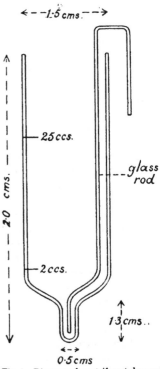

Fig. 1. Diagram of centrifuge tube used in the estimation of calcium in blood.

Precipitation. This was carried out by McCrudden's method as modified by Halverson and Bergeim. The solution in the centrifuge tube is neutralised with 0·880 ammonia using alizarin as indicator. The neutral point is overshot and then titrated back with $N/2$ HCl until the colour just commences to change. Then 2·5 cc. of $N/2$ HCl and 2·5 cc. of 2·5 % oxalic acid solution are added and water to make to the 25 cc. mark. The tube is then placed in a water-bath and brought to the boiling point. To the hot solution 5 cc. of 3 % ammonium oxalate is added and the tube kept at the boiling point for 15 minutes when it is removed, placed in ice water and 5 cc. of 20 % sodium acetate solution added (or until the colour commences to change). When the calcium oxalate has precipitated which takes about 6–8 hours (it is usually allowed to remain over night), the solution is centrifuged. The clear supernatant liquid is removed by means of a tube (drawn out to a capillary) leading to a bottle connected with a filter pump. By this means the liquid is removed to the 2 cc. mark on the tube. The outside of the capillary tube is washed with water to prevent the loss of any precipitate which may adhere to it.

Water is then added to the 25 cc. mark, the centrifuging repeated, and the supernatant liquid removed as before to the 2 cc. mark, etc. It is washed twice in this way, when 4 cc. of 5 % sulphuric acid are added and the precipitate stirred with the glass rod. The tube is then placed in a water-bath at 65° C. for 1–2 minutes, when the oxalic acid is titrated with $N/100$ potassium permanganate by means of a burette capable of reading to 0·01 cc. 1 cc. of $N/100$ potassium permanganate is equivalent to 0·020 mgm. of calcium.

A "blank" experiment must be performed on the materials. Coagulation of the blood was prevented by the addition of sodium citrate to the extent of 2·5 %.

Solutions used. All materials used were recrystallised twice from water which had been redistilled twice from a Jena glass flask, and such water was used throughout the operations. It was found necessary to have the end of the burette drawn out to a fine point and vaseline was placed on the outside so that it was possible to reduce the volume of one drop to 0·01 cc.

The permanganate solution. About 0·5 g. of potassium permanganate crystals is dissolved in 1 litre of redistilled water in a litre flask. The flask is placed on the water-bath for 2–3 days, a funnel and watch glass being placed in the neck to prevent extraneous matter from entering. The solution is then carefully filtered through an ashless filter paper on a Buchner funnel to remove the deposit of manganese dioxide. The solution should be kept in the dark. It is standardised against an oxalic acid solution prepared by dissolving about 0·25 g. of recrystallised oxalic acid in 250 cc. of water. When standardising allowance must be made for the titre of the sulphuric acid used. It is as well to standardise the permanganate about once every fortnight.

Alizarin solution. An aqueous solution of 0·2 % strength was used.

The following are some of the results obtained:

							Calcium per 100 cc. g.
1. Standard calcium solution (containing 0·0076 g. per 100 cc.)							0·0077
2. „ „ „ „ „							0·0077
3. „ „ „ .. „							0·0077
4. „ „ „							0·0077
1. Human blood ...	...	...	...	...	...	...	0·0082
2. „ „ ...	...	...	...	...	...	...	0·0080
1. Ox blood ...	...	...	...	...	...	...	0·0070
2. „ „	...	...	...	...	...	...	0·0070
1. „ „ (clotted)	...	...	...	...	...	...	0·0055
2. „ „ „	...	...	...	...	...	...	0·0055
3. „ „ „	...	...	...	...	...	...	0·0056

In all cases 5 cc. were used for the determination with the exception of the human blood when 3 cc. and 2 cc. respectively were used.

Duplicate experiments show an agreement equal to two drops of $N/100$ permanganate or in other words to 0·006 mgm. of calcium.

REFERENCES.

Halverson and Bergeim (1917). *J. Biol. Chem.* **32**, 159.
Jansen (1918). *Zeitsch. physiol. Chem.* **101**, 176.
McCrudden (1909). *J. Biol. Chem.* **7**, 83.
Marriott and Howland (1917). *J. Biol. Chem.* **32**, 223.
De Waard (1919). *Biochem. Zeitsch.* **97**, 176.

XXXIII. THE MINIMUM NITROGEN EXPENDITURE OF MAN AND THE BIOLOGICAL VALUE OF VARIOUS PROTEINS FOR HUMAN NUTRITION.

By CHARLES JAMES MARTIN AND ROBERT ROBISON.

From the Lister Institute, London.

(*Received March 22nd, 1922.*)

HISTORICAL.

UNTIL comparatively recently, the search for the minimum protein requirements of the human body has been made on the assumption that protein is an entity, little regard being paid to whether the proteins were derived from meat, milk, cereals, etc.

The more important of the numerous investigations undertaken with this object are those of Hirschfeld [1887], Kumagawa [1889], Klemperer [1889], Peschel [1891], Lapicque and Marrette [1894], Sivén [1900] and Albu [1901][1].

In all of them the protein fed was derived from more than one source and often from several. The observations, which vary in precision, were made under conditions which were not uniform, particularly in regard to the total calories taken. Nevertheless, each experimenter succeeded in establishing that nitrogenous equilibrium could be maintained over short periods with one-third to two-thirds the standard laid down by Voit of 118 g. protein, equal to 0·39 g. N per kilo.

The minimum arrived at by the above experimenters varied between 0·08 g. and 0·18 g. N per kilo, most of the results being round about 0·1 g. The lowest value is that of Sivén, who considers that he ultimately attained nitrogenous equilibrium on a mixed diet containing 0·08 g. N per kilo, only 0·03 g. of which he regards as true protein, but as the only evidence that this small amount was sufficient is a positive balance of ·04 g. N on the last day of a four days experiment, decided negative balances occurring on the first three days, this conclusion would appear questionable. In another series in which the nitrogen intake was 37 % higher the evidence of nitrogenous equilibrium is satisfactory.

[1] The earlier observations have been collected by Atwater and Langsworthy in their "Digest of Metabolism Experiments," *Bulletin* 45, U.S. Dept. of Agriculture, 1898. An excellent review of the work previous to his paper is given by Sivén. Most of the literature on the subject to date is referred to in Mendel's "Theorien des Eiweissstoffwechsels," *Ergebnisse der Physiologie,* 11 Jahrgang, 1911; Caspari's article "Eiweissstoffwechsel" in Oppenheimer's *Handbuch der Biochemie,* 1911, and Cathcart's *Physiology of Protein Metabolism* (1921). All of these contain good bibliographies.

The investigations of Neumann [1902] and Chittenden [1904] were undertaken with a somewhat different object, namely, to ascertain whether health and activity could be maintained over prolonged periods on a mixed diet of low content in protein. Neumann made three experiments on himself, each lasting four to ten months. The total calories of the diet amounted to 30–40 per kilo. Chittenden's observations were made upon 26 individuals and the duration varied between six and nine months in different cases. The total calories varied between 35 and 45 per kilo body weight. Nitrogenous equilibrium was obtained by Neumann on an intake of 0·15 g. N per kilo and in Chittenden's experiments with 0·1 g. to 0·17 g. per kilo in the different individuals. There is no reason to suppose that these figures represent minima. The conclusion drawn is that a nitrogen intake of one- to two-thirds of the amount laid down by Voit is sufficient to maintain health and efficiency over the periods during which the observations endured.

During the last decade of the 19th century, physiologists were becoming increasingly alive to the possible significance of differences discovered in the elementary composition and chemical properties of the proteins, and the uniform value hitherto attributed to them in nutrition was coming under suspicion. Rubner [1897] appears to have been the first to formulate the view that proteins had different biological values. He used this conception to interpret some early experiments of his on the utilisation of various foodstuffs [1879] in which he had observed that less nitrogen was excreted in the urine on potato diets than on bread diets although the adverse N-balance was smaller on the former. In the same article Rubner expressed the opinion that the old search for a protein minimum must be fruitless since there will be, not one, but many minima according to the nature of the foodstuff used. He does not appear to have attributed such variation to any difference in chemical constitution although doubtless this possibility was in his mind. At this date the chemical constitution of the proteins was obscure although a number of amino acids had been isolated from the decomposition products of proteins and differences in the amounts of these had been observed. A little later Kossel and Kutscher [1900] determined the histidine, arginine, lysine and ammonia derived from the hydrolysis of a number of proteins and Kossel [1901] as a result of his own and others' work came to the conclusion that the habit of regarding protein as a physiological unit was unsound and that, since proteins possess different chemical compositions, they will also have different values for the organism.

About the same time discoveries were being made in another direction. The researches of Cohnheim [1901, 1906], Kutscher and Seemann [1901], Abderhalden and his co-workers and others, proved that a much more exhaustive break up of the protein molecule than had previously been supposed takes place in the small intestine prior to absorption, and that what the body really receives is a mixture of amino acids and simple polypeptides. Loewi [1902], and later, Abderhalden and Rona [1904] showed by means of feeding experiments with previously digested proteins that the abiuret products of such digestion were capable of maintaining animals in nitrogen equilibrium. Abderhalden believed at first that such amino acids were at once utilised for building up of blood and tissue proteins and in conjunction with Samuely [1905] attempted to ascertain whether the composition of the body proteins varied with the character of food proteins. A horse was fed for three days on gliadin containing 36·5 % of glutamic acid, but no increase in the very low content of this amino acid in the serum proteins could be detected. From these results Abderhalden and Rona [1906] drew the conclusion that in the renewal of body proteins a proportion of the amino acids arising from the food will be left over unless the body is capable of synthesising one amino acid from another. Since this proportion will depend on the relative composition of the food and body proteins the protein minimum must also be variable.

These discoveries provided a theoretical basis for Rubner's empirical conclusions and a stimulus for the further investigation of the subject.

Meanwhile the rôle play'd by protein in satisfying the energy requirements of the body, and the effect on the protein minimum of insufficient as against abundant provision for these needs from non-protein sources, was becoming more clearly realised.

The discovery of the variable composition of proteins and of the fact that certain of them are almost entirely deficient in one or more of the amino acids was followed by interesting researches to determine to what extent the animal body could, by the practice of economy or synthesis, dispense with the missing complexes. The deficiency of gelatin in tyrosine was ascertained early and the absence from it of cystine and tryptophan was discovered when these amino acids became known as protein constituents. The inability of gelatin to preserve the body in nitrogenous equilibrium has been shown by many investigations, but as this aspect of the subject has recently been dealt with by one of us in this journal [Robison, 1922, 1] it is unnecessary to review it here.

The discovery of tryptophan by Hopkins and Cole [1901] and of the deficiency of this amino acid in zein by Osborne and Harris [1903] was followed by an experimental enquiry by Willcock and Hopkins [1907] to ascertain whether zein would serve as an exclusive source of nitrogen for mice and if not whether the addition of tryptophan would enhance the nutritive value of this protein. Zein alone failed to maintain the animals but this was achieved by supplementing with tryptophan.

Following this pioneer work Osborne and Mendel [1911] planned a lengthy investigation of the biological value of different proteins in the light of the new knowledge of their chemical structure. Their earlier work was carried out before the importance of accessory factors was recognised but they became aware from their experiments with individual proteins that some factor other than the supply of protein, salts and energy was complicating their observations. In the continuation of their researches, the results of which have appeared in some 30 papers in the *Journal of Biological Chemistry* from 1912 up to the present, the error due to absence of vitamins was obviated. Osborne and Mendel [1912–1920] confirmed and extended the observations of Willcock and Hopkins and showed that the addition of 3 % tryptophan to the zein given was sufficient to maintain rats over a period of 182 days, but that they failed to grow. When 2 % of lysine, in which zein is also deficient, was added, growth occurred. The problem of maintainance is therefore distinct from that of growth. The observations of Osborne and Mendel are some of the most important contributions to our knowledge of nutrition. The choice of rats enabled great numbers of experiments to be carried out and the extension of the periods of observation to cover a large fraction of the normal life of the animal. They prove that rats cannot supply some of the missing amino acids and that the minimum requirements and relative nutritive value of any particular protein depend upon the proportion of essential amino acids it contains. Their work also shows how a knowledge of the composition of particular proteins may be used for the economical adjustment of the nitrogenous portion of an animal's dietary by arranging that one protein shall compensate for the deficiencies of another.

The supplementary value of proteins from different sources has also been investigated by McCollum, Simmonds and Pitz, and McCollum, Simmonds and Parsons [1917 to 1921], whose observations, like those of Osborne and Mendel, were carried out upon rats. They found that cereal proteins could be satisfactorily supplemented by the proteins of milk, meat, kidney and casein and gelatin. Proteins of various leguminous seeds also usefully supplemented cereal proteins, *e.g.* wheat together with navy beans or peas.

The ultimate test of the nutritive adequacy of a protein is its capacity to nourish a young animal and provide for its complete growth and development and this, as far as the rat is concerned, is the criterion of the American investigators to which we have briefly referred. It may be surmised that, broadly speaking, conclusions arrived at from experiments on rats will be applicable in general to human nutrition. On the other hand the human mechanism may differ in detail. It will, for obvious reasons, be long before information as to the complete adequacy of individual proteins and quantitative data as to their biological values is forthcoming for human nutrition. In the meantime the findings of Osborne and Mendel, and McCollum and his co-workers have been applied with advantage to the feeding of stock.

From this more general survey of the subject we will now return to consider observations upon the minimum requirements for equilibrium when nitrogen is supplied in different forms. We have already referred to the observations of Rubner which led him to the conclusion that a different nitrogen minimum would be discovered for different proteins. This surmise was subsequently investigated in his laboratory by Karl Thomas [1909] who introduced the term "biological value." The expression "physiological value" had been previously suggested by Voit and Korkunoff [1895] for a similar conception. Karl Thomas defined biological value as the number of parts of body nitrogen replaceable by 100 parts of the nitrogen of the foodstuff. Thomas's definition is not concerned with the relative digestibility of the protein. The replacement of the "Wear and Tear" quota was recognised as the only proper basis for comparison and in order to determine this value he fed himself on a carbohydrate diet (starch, sucrose, lactose) of high calorie value for periods of several days, during which the daily output of nitrogen in faeces and urine was determined. The figure to which this output fell was taken as his minimum requirements for the time being. During succeeding periods varying from one to four days a similar carbohydrate diet supplemented by a certain amount of the foodstuff under examination was taken and the N-intake and output determined as before. The N-intake was not as a rule kept constant and sometimes varied considerably on the different days. In most cases a negative N-balance was obtained. From the results Thomas calculated his biological value by three formulae based on the above definition, but differing from one another according to the way in which the nitrogen of the faeces is dealt with.

Some of Thomas's experiments lasted four to five weeks though no individual foodstuff was taken for longer than four days at a time. A period of one or two days on nitrogen-free diet was usually interposed between the experiments. Sixteen foodstuffs were investigated and their biological values recorded. These varied from 100 % in the case of milk to 30 % in the case of maize. We shall have occasion to discuss some of his results after dealing with our own experiments.

Shortly before this work of Thomas appeared the results of experiments upon dogs with a similar object were published by Michaud [1909]. The output of nitrogen on diets of dog-flesh, sugar and fat was compared with that on diets of horse-flesh, caseinogen, gliadin, and edestin and on the carbohydrate and fat alone. Nitrogenous equilibrium was attained with an amount of nitrogen in the form of dog-flesh equivalent to the nitrogen output on the nitrogen-free diet. Negative balances were obtained with the other proteins, the greatest being in the case of gliadin and edestin.

Zisterer [1910] found differences between caseinogen, flesh and gluten. These were however, in his opinion, too small to have practical significance.

Observations upon pigs were made by McCollum [1911]. These animals lend themselves to metabolism experiments of this kind as they will consume sufficient of a diet free from nitrogen to obtain the necessary calories over a considerable period. Their minimum nitrogen expenditure can therefore be determined with reasonable accuracy.

After a period of a week upon a diet of starch alone, the animals were fed for several days with the same ration to which was added a small amount of gelatin, zein, caseinogen or other protein. This was followed by the starch ration for a further period of some days. An amount of nitrogen in the form of gelatin equal to that of the urine upon the starch diet was found to cover 39 % and in the form of zein 73 % of the animal's expenditure. The same amount of nitrogen in the form of cereal protein did not cause any rise in the nitrogen of the urine and with caseinogen the rise was small. McCollum's experiments seem to avoid all the obvious pitfalls and his results indicate a much higher biological value for cereal proteins when fed to the pig than those arrived at by Thomas's experiments upon himself.

Hindhede [1913, 2] concludes that nitrogen equilibrium may be attained on a diet of potatoes and margarine containing only 20 g. of digestible protein. The figure is, however, arrived at by deducting the nitrogen of the faeces from the intake. This method of calculation is not in accordance with our knowledge of the origin of a considerable portion of the faecal nitrogen and will furnish a too favourable balance sheet.

Hindhede [1914] vigorously contests the findings of Rubner and Thomas and claims to have attained nitrogenous equilibrium on as small an amount of protein in the form of bread as of potatoes. He declares as a result of his lengthy experiments that the proteins of potatoes, bread and meat can replace those of the body gram for gram. With Hindhede's criticisms of some of Thomas' experiments and treatment of his data, we are, for the most part, in agreement but must at the same time admit the justice of a great part of Rubner's equally severe criticisms of Hindhede's evidence, in particular, as regards the justification for assuming that all the nitrogen of the faeces represents undigested food proteins.

Abderhalden, Fodor and Röse [1915] carried out some experiments to determine the minimum requirement of nitrogen in the form of different kinds of bread and potatoes. The subject of the experiments was Hofrat Röse who possessed some peculiarly advantageous characteristics. Röse was accustomed to a monotonous diet, neither smoked nor drank alcohol and was in the habit of chewing his bolus 120 times before swallowing it. Experiments of three to eight days' duration were made on diets of potatoes, white wheaten bread, Swedish bread and kommiss brot, the last two being made from rye and containing bran. The experimental facts seem to us to warrant the conclusion that a gross intake of 4·5 g. of potato nitrogen, equal to 0·074 g. N per kilo, were adequate in the case of this Hofrat who chewed so long and so well. 9 g. N in the form of white wheaten bread was not quite sufficient and 10·8 g. N as supplied in the rye bread was only just enough to reach equilibrium. This is not, however, the interpretation placed upon the results by Abderhalden and his co-workers, who conclude that bread nitrogen is as good as potato nitrogen and that for both of them the minimum nitrogen requirement is round about 4 g. for a man of 60 kilos.

Rubner [1919] in the course of some studies of the capacity of certain vegetable nutriments to satisfy nitrogen needs, undertaken during the war, investigated different sorts of bread and the effect of milling to varying extent on the value of the product as a source of nitrogen. The paper covers a good deal of ground and contains some particularly useful experiments with white wheat bread which can be compared with our own upon whole wheat. The bread was made, in one series, from white flour, 30 % milled, in the other of the same flour mixed with rye-bran to the extent of 30 %, so-called "Finkler brot." 10 g. of the N as contained in the fine flour and between 10 and 11 g. of that in the Finkler bread were adequate to maintain equilibrium.

Recently Sherman and his co-workers [1918, 1 and 2, 1919, 1920] have obtained results which are difficult to harmonise with those of Thomas. In experiments upon men and women, nitrogenous equilibrium was attained with an intake of 0·08 g. N per kilo, nine-tenths of this being supplied by cereal proteins and one-tenth by those of milk or apple. Wheat, maize and oats were found of equal value as a source of nitrogen and the view is taken that these cereal proteins possess a higher biological value than Thomas found. The effect of the supplementary action of the small quantity of milk may, in the light of the observations of Osborne and Mendel [1917] and of McCollum, Simmonds and Parsons [1921] be considerable.

Boruttau [1915] believes that the low value of cereals as a source of nitrogen is greatly improved when these are consumed without the removal of the bran, etc. The biological value of 145 % he obtained for the nitrogen of bran, is, in our opinion, an instance of the misuse of arithmetical formulae.

R. O. Neumann [1919] made an excellent experiment upon himself in 1917 in which he lived *exclusively* on rye bread, cane sugar and water for 40 days. Nitrogenous equilibrium was attained with 1000 g. bread and 300 g. cane sugar (= 9·9 g. N). The total calories of this diet amounted to 3630 or 63·8 per kilo. On raising the calorie value of the intake to 4434 (or 73 per kilo) the nitrogen excreted steadily fell to 7·3· This indicates that Neumann in a long continued experiment could maintain nitrogenous equilibrium on less than 7 g. of nitrogen in the form of rye-proteins if excess of calories were furnished by sugar. The experiment is also interesting as indicating the sensitiveness of the nitrogen balance to the addition of carbohydrate. This aspect of the experiment will be discussed later.

From a survey of the literature it is clear that certain of the proteins possess very different biological values both for growth and maintainance. There is, however, much uncertainty as to the degree to which the admixed proteins occurring in individual foodstuffs, where one protein to some extent complements the deficiencies of another, vary in value as a source of nitrogen.

The divergence of opinion is most marked when it is based upon metabolism experiments on man over limited periods.

OUR OWN OBSERVATIONS.

INTRODUCTORY.

We commenced our investigation lightheartedly with the comparatively modest object of re-determining the relative values of certain cereal proteins in human nutrition, in particular that of maize, in view of the significance given by Goldberger and others [1915, 1920] and Wilson [1921] to the low biological value of maize in the causation of pellagra. The difficulties in arriving at values which could justifiably be compared were soon, however, apparent and it became essential to investigate thoroughly the conditions under which valid results might be obtained. In so far as the problem can be solved by metabolism experiments on adult animals the one unexceptional way to determine the relative biological values of proteins would be to ascertain the minimum intake on which nitrogen equilibrium can be maintained in each case. This sounds simple but unfortunately a positive balance only tells one that the intake is sufficient but not how much it is in excess and a number of experiments have to be performed to ascertain the minimum quantity.

We were ourselves the subjects of the experiments. This is inconvenient but advantageous, for the experiments are exacting and necessitate constant supervision of one's actions if sources of error are to be avoided. The partial abandonment of the joys of life is to some extent compensated by interest in the results.

Nevertheless, the unnecessary multiplication of irksome experiments on one's self, each extending over many days, is a thing to be avoided and it would be very desirable if a couple of observations could be made and the minimum requirements calculated from these with sufficient accuracy. This is what Thomas attempted to do. But in adopting such a method an assumption is made, the truth of which is by no means self-evident, namely, that the value of any protein for biological purposes remains uniform whatever the amount taken. The assumption would be justified if the nitrogen were utilised in the first instance to form some complex, such as " Vorratseiweiss."

In this case the biological value, as pointed out by Abderhalden, would be determined by the ratio of the percentages, in food protein and body complex respectively, of that amino acid for which this ratio has the lowest value, unless the body has the capacity to synthesise that particular amino acid from others.

It might also be true, if the nitrogen requirements are of varying nature so long as they are also indivisible, that is that no single requirement can be satisfied unless at the same time all the others are satisfied.

The former of these two conceptions would appear to have been accepted without question by Rubner and Thomas though the case of gelatin obviously could not be treated in this way. Gelatin was considered to be capable of sparing body protein to the extent of 30–40 % when fed in relatively small

amounts but unable to do more than this however much was taken. Its
biological value, if calculated by any of Thomas's formulae would therefore
appear quite appreciable when the intake was small but almost zero if the
intake was very large. Yet, Boruttau [1919] has actually made use of these
formulae to calculate the biological value of gelatin and has obtained a result
of 58·2 %.

Another possible disturbing factor (which we have reason to suppose
occurs) is the varying economy with which the body deals with the amino
acids supplied to it, according to their abundance.

The various possibilities stated above may be made clearer by a diagram
in which abscissae represent real nitrogen intake and ordinates the real
nitrogen output.

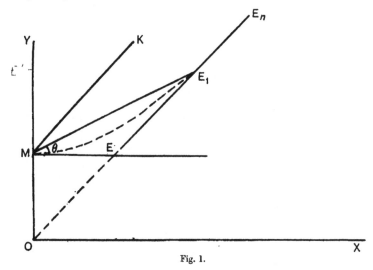

Fig. 1.

Let OM (= m) be the output on a N-free diet of adequate fuel value.
Then m is equal to the nitrogen minimum.

Suppose that an ideal protein (B.V. = 100) is fed in gradually increasing
amounts and is utilised without waste. So long as the intake remains lower
than m the output will remain constant and equal to m since the food protein
saves an equal amount of body protein. The graph of intake and output will
therefore be a line parallel to the x-axis and at E where $ME = MO$ the body
will be in nitrogen equilibrium.

If now the intake be further increased, equilibrium will again result (unless
the body is in a growing condition or has been previously starved of N) and
the graph will now follow the line EE_n at an angle of 45° to the axis.

In the case of a protein of value less than that of the ideal protein just
considered, equilibrium will not be attained on an intake equal to m but on

some greater amount e_1 (at the point E_1). On all amounts less than this, the output will exceed the intake and the graph will follow some line joining ME_1. Whether this line is straight or curved will depend on the conditions set out above, viz.:

(1) indivisibility of the nitrogen requirements of the body;

(2) uniform economy with varying nitrogen intake.

If these conditions are obtained the line ME_1 will be straight and its equation will be $y = m + x \tan \theta$ where y is the real output corresponding with any real intake x less than e_1.

For higher values of x the graph will follow the line E_1E_n.

Thomas's formulae can be very simply expressed in terms of θ; thus formula B

$$\text{B.V.} = 100 \, \frac{\text{Urine N in N-free diet} + \text{faeces N} + \text{balance}}{\text{N-intake}}$$

becomes

$$\text{B.V.} = 100 \, \frac{m + (x - y)}{x}$$

$$= 100 \, \frac{m + x - (m + x \tan \theta)}{x}$$

$$= 100 \, (1 - \tan \theta).$$

If the above conditions do not obtain, e.g. if a number of different amino acids are required for specific purposes, which are distinct and can be separately satisfied, the graph of a protein rich in certain of these acids but poor in others would be a curved line such as the dotted line joining M and E_1 in the diagram. This curvature would express the fact that a certain fraction of the body's needs could be satisfied by a smaller amount of this protein than would correspond with the amount required to obtain equilibrium. The angle θ and the biological value would then vary for different values of x.

The graph of a protein, unable by itself to satisfy any portion of the body's nitrogen requirements, would be a straight line MK parallel to OE_n, since the nitrogen output would always be equal to the intake $+ m$. For this line $\theta = 45°$ and the equation $y = m + x \tan \theta$ becomes $y = m + x$ while the biological value $= 100 \, (1 - \tan 45°) = 0$.

In our opinion there was very little reason for assuming that these graphs would necessarily prove to be straight lines. It is true that Thomas calculates his values from individual daily balances and takes the average of the results, but such daily balances are too variable and are subject to too great experimental errors to offer any satisfactory proof of the uniformity of the value. We therefore set out to obtain evidence on this question by determining as accurately as possible a number of points for the same protein but for different values of x. Our results will be considered later but we may here state that in the case of bread proteins the points do lie on, or close to, a straight line. In the experiments with nitrogen in the form of milk results were at first obtained indicating pronounced curvature of the line, and nitrogen equilibrium was not obtained until more than 11 g. of milk nitrogen was taken per day. By increasing the amount of carbohydrate however, so that the fuel value of the diet was greatly in excess of requirements and the respiratory quotient greater than 1, equilibrium was finally reached with half this amount, and bearing in mind that when x, the intake, is very small, the physiological errors of experiment become relatively great, the observations could now perhaps be expressed by a straight line. As long as any doubt exists of the rectilinear character of the line ME it will obviously be prudent to place reliance only upon observations in which x and y are as large as possible short of equality.

FACTORS WHICH MUST BE CONSIDERED IF VALID RESULTS ARE TO BE OBTAINED.

1. *The time required to reach a uniform N-output on a constant intake.*

The effect of the previous diet upon the N-output and the length of time required to reach a constant output on a constant intake which is either greater or less than that of the preceding period, was clearly demonstrated by the old experiments of C. Voit [1866, 1867]. The N-output of a dog during the first days of starvation varied with the amount of protein in the previous diet but fell gradually until a relatively constant figure was reached on the fifth or sixth day. When the dog was given a constant meat diet for some days and then a considerably greater (or less) amount daily during a further period a similar gradual increase (or decrease) in the N-output was observed during five or six days before equilibrium again set in at the new level. These observations have been repeatedly confirmed by Grubner [1901], Landergren [1903], Kinberg [1911] and many others.

The rapidity with which the nitrogen excretion diminishes obviously depends on the difference between the N-intake during the experimental and the preceding periods and will be greatest when a period of nitrogen starvation follows one of high protein intake or *vice versa*. There is no reason to suppose, however, that this gradual change is ever replaced by an immediate jump to the new level even when the difference between the two planes of N-intake is but small, though naturally the absolute amounts of the variations will be correspondingly less.

Whether the N-output is also influenced by the nature as well as the amount of the protein taken in the foregoing period is more difficult to decide. If part of the nitrogen of the previous diet is stored up in any form that can be utilised by the body (*e.g.* amino acids) and not merely in the form of unexcreted end products, we should expect the N-output during the first few days of the succeeding period to be influenced thereby—unless during both periods the body is in N-equilibrium. For example if the diet during the first period contains 10 g. of N from caseinogen, and during the second period a negative balance occurs on 10 g. of N from zein, the amount of this negative balance might very well be less during the first few days of the zein diet than on the latter days of the same period owing to the supplementary action of amino acids stored up during the caseinogen diet. The results of experiments [Robison, 1922] in which a diet containing gelatin as the sole protein followed a diet of mixed proteins suggest that this does occur, and that therefore when the diet is changed in any way—either in amount or nature of the protein—the N-output cannot be considered to represent that of the second diet until some days have elapsed.

It follows that metabolism experiments are subject to error from these causes and that the error diminishes as the duration of the experiment increases. As a compromise, we have exlcuded from calculation the figures for the first three or four days in arriving at the N-output. It is also advisable that the N-intake should not vary during the experiment. Most of Thomas's experiments are subject to both these sources of error.

2. The time required for the elimination of errors due to the fluctuation in the N-output.

The very considerable fluctuations that may occur in the amount of nitrogen excreted in the urine by men receiving an absolutely constant diet have been noted by Bornstein [1898], Atwater and Benedict [1902] and by Falta [1906], by all of whom the cause was considered to be psychical. Atwater and Benedict also noted the increased diuresis which often accompanied a high N-output and thought it probable that the diuresis was the direct result of the psychical stimulus and the cause of the increased N-elimination. Falta observed variations of 4–5 g. in the N-output on individual days although in equilibrium over the period as a whole.

Neumann [1899] studied the influence of variations of the urinary flow upon the daily output of nitrogen, the intake remaining constant. When diuresis was produced by increasing the water drunk from one to three litres, the N-output increased from 10·5 to 14·3 g. and did not reach the original level until the third day.

In almost all our experiments such fluctuations, amounting frequently to 25 % of the mean output, have occurred and not least in those experiments in which the most rigorous attention has been paid to constancy of diet and fluid intake, and to regularity in the mode of life.

At first we also were disposed to attribute these fluctuations to increased diuresis but the frequent lack of any correlation between the two compelled us to modify our opinion. Some factors (e.g. mental strain or excitement) may possibly affect both diuresis and N-output, but the latter does not always coincide with the increased volume of urine and may even vary in the opposite sense. In one experiment, for example, the minimum N-output corresponded with the maximum volume of urine.

Judging from the experiments of Voit and other workers on dogs and of McCollum on pigs, it would appear that in the case of these animals the fluctuations in N-output on a constant diet are usually less considerable than with man. It is clear that calculations based upon the nitrogen balance sheet for single days, a procedure frequently resorted to, are subject to very large errors and that these can only be eliminated by taking the average results over a number of days.

3. *The necessity for abundant energy supply.*

It has been universally recognised that proteins will be used as fuel unless an adequate supply of fat and carbohydrate is provided, but there have been different opinions as to what constitutes adequacy in this respect. In their search for the protein minimum some of the earlier workers considered it necessary to supply a diet of fuel value very greatly in excess of the energy requirements of the organism. Thus in Klemperer's [1889] experiments on two young men, a diet of 5020 calories, equal to about 75 calories per kilo was given. Sivén, however, was of the opinion that such excess was unnecessary and that the minimum could be reached without increasing the calorie value of the diet above the normal. The fuel value of Sivén's diets was equal to about 40 calories per kilo body weight.

Hindhede's [1913, 2] attitude is somewhat difficult to understand. He considers that an abundant calorie intake is necessary if the protein minimum is to be attained, but that this minimum will vary with the calorie value of the diet. He does not believe that a quiet old man, for whom a diet of 1500–2000 calories is sufficient, can have the same minimum as an active young man who requires a diet of 3000–5000 calories. From the results of his experiments, he calculates by simple proportion, the minimum for a standard diet of 3000 calories. As on a diet of 3900 calories F. Madsen's minimum was equal to 25 g. of digestible protein so, according to Hindhede, for 3000 calories the minimum would be 19 g. of digestible protein.

Rubner [1919] has criticised this procedure, and in our opinion justly, on the ground that it is unwarranted by the facts, and considers that the values so calculated to 3000 calories possess no scientific basis.

Rubner's own conclusions are that the N-minimum may sometimes be reached when no more than a third of the total energy requirements are satisfied, in other cases only when they are fully met, while in others a diet considerably in excess of these requirements will be necessary. These differences he considers are due to the varying nutritional condition of the body cells. The minimum is, however, most easily reached on an abundant carbohydrate diet.

In our own experiments on a diet nearly N-free we appeared to reach our N-minimum with an intake of 45–50 calories per kilo body weight of which about one-third was taken in the form of fat, whereas on a diet of milk (with additional carbohydrate), equilibrium was not readily obtained until the fuel value was increased to about 55 calories per kilo of which only 10 % was in the form of fat.

The effect of diets containing varying proportions of fat and carbohydrate on the protein minimum has been studied by Zeller [1914], who found that the mixed diet was just as efficacious in reducing the consumption of body protein as one of carbohydrate alone, so long as the proportion of carbohydrate to fat did not fall below 1 : 4.

Neumann's [1919] experiment upon himself, with a diet composed of bread and sugar, affords a striking demonstration of the effect of excess of carbohydrate in lowering the protein minimum. It is obvious that Prof. Neumann readily stores fat. Otherwise, he could not consume 73 calories per kilo over a period of three weeks, unless doing hard work. In his case, presumably, a greater excess of carbohydrate would be required to maintain the blood sugar at a high level than in our own, owing to the greater greed of his connective tissue cells.

It would seem, therefore, that the most certain way of determining the protein minimum would be to take a diet consisting mainly of carbohydrate and so much in excess of the energy requirements that the blood sugar is maintained high, the liver and muscles are kept well stocked with glycogen and the surplus is being stored as fat, as indicated by a respiratory quotient above unity. This was accomplished in the latter part of our experiment on milk. It is by no means easy for one of the meagre habit of the subject of the experiment (C.J.M.) as it increases the distaste for the sufficiently unappetising ration of starch and lactose and if persisted in too enthusiastically it produces unpleasant symptoms.

4. *Reduction of the nitrogen in the basal diet to a minimum.*

The carbohydrate (starch, lactose, etc.) and fat, etc. which form the basal diet for these experiments are nominally but not absolutely nitrogen-free. The amount of nitrogen taken in this form can of course be estimated but its biological value is unknown and this complicates the results. It is therefore important to reduce the nitrogen in the basal diet to a minimum by careful selection of the most suitable forms of such foods.

Most observers have neglected to take account of the nitrogen in the starch, etc. fed. As large quantities of such basal ration are consumed it may not be negligible in the case of experiments in which small quantities of some protein are being given.

5. *Accessory food factors and inorganic salts.*

The duration of metabolism experiments is limited by the difficulty of providing an adequate supply of the accessory food factors. The ill effects of long continued low-protein diets observed in some of the older animal experiments was no doubt sometimes due to the deficiency of one or other of these factors.

Fat soluble *A* can be introduced in the form of rendered butter fat or cod liver oil, and water soluble C in the form of lemon juice, the nitrogen content of which is very low, but we have found no method of introducing the water soluble *B* without an undue amount of possibly very valuable nitrogen.

When the diet is deficient in inorganic salts or is such as to afford an acid ash, adjustment by suitable amounts of a salt mixture is essential. McCollum and Hoagland [1913] found that the endogenous metabolism of the pig reached its lowest level when the animal was given an abundant carbohydrate diet together with a salt mixture of an alkaline character. When an acid salt mixture was given the urinary output rose, the increase occurring in the amount of ammonia. They concluded that this animal is not able to use the nitrogen of the urea fraction for the neutralisation of acid.

6. *The apportioning of the nitrogen in the faeces.*

The difficulty of correctly apportioning the nitrogen in the faeces to unabsorbed food nitrogen and excretion from the alimentary tract respectively, is the limiting factor in most experiments in which the total intake of nitrogen is small, and the way in which different observers treat the faecal nitrogen has given rise to much controversy and recrimination.

The daily nitrogen output in the faeces on a protein-free diet usually amounts to about 1 g. On other diets the amount may be considerably greater than this and the question arises—what is the significance of this excess? Does it represent unabsorbed food residues or increased loss of body nitrogen? This difficulty was recognised and discussed by Karl Thomas, whose three formulae for the calculation of the biological value of protein differ only in the assumptions that are made with regard to this point. In formula A the whole of the nitrogen of the faeces is assumed to represent unabsorbed food; in formula B it is assumed to arise entirely from the body, while in formula C 1 g. N (which is taken as the average output on protein-free diet) is assumed to be body nitrogen, any excess over this amount being ascribed to unabsorbed food.

Rubner [1915] has investigated this problem in connection with his researches on the digestibility of various foodstuffs. He has devised a method for estimating the amount of nitrogen in the faeces present in the form of undigested food residues (vegetable cell membranes) by making use of the insolubility of the latter in acid alcohol and in a concentrated solution of chloral hydrate in which he states bacteria, epithelial cells, etc. are dissolved. He considers that the rest of the nitrogen comes from the body and represents metabolic products, and he concludes that such body nitrogen forms a very considerable proportion of the increase in the total nitrogen of the faeces that commonly occurs when the diet consists largely of whole cereals, vegetables or fruit.

Rubner found that on a vegetable diet nearly half the total nitrogen of the faeces was soluble in acid alcohol whilst on a diet consisting chiefly of animal foodstuffs the proportion of soluble nitrogen was still greater. We have obtained similar results with faeces resulting from a mixed diet, but do not know in what form the whole of this soluble nitrogen is present, and are therefore not able to draw definite conclusions as to its origin.

So long as this uncertainty remains, the nitrogen of the faeces will be the limiting factor for the accuracy of such experiments as those here described. We have therefore followed Thomas's plan and have calculated our results in two ways; the first, assuming that the whole of the nitrogen of the faeces comes from the body and that this amount plus the urine N on N-free diet represents the body's minimum requirements under the conditions of the experiment; the second, assuming that the amount by which the nitrogen of the faeces exceeds the average amount excreted on a N-free diet represents unabsorbed food and is therefore to be subtracted from the total intake in order to arrive at the true nett intake, i.e. the absorbed nitrogen. The truth will probably lie somewhere between these two extremes.

EXPERIMENTAL.

Firstly as to general procedure. The N of all foodstuffs and beverages used was determined. The food was weighed in the same condition as that in which the N was determined and the total intake of N recorded in the tables can be relied upon to plus or minus 0·01 g. The urinary excretion of each 24 hours was collected in the presence of toluene, weighed, and duplicate determinations of the N content made. Care was exercised to see that invalid sampling from the deposition of either urates or ammonium magnesium phosphate did not occur. The faeces, after collection each day, were mixed with some H_2SO_4 to prevent decomposition and possible loss of NH_3 and great care was taken at the end of each period to ensure the validity of the samples taken for analysis. N determinations were made upon about 10 g. in duplicate, only closely concordant results being accepted.

The evacuation of the intestines is not usually so regular and complete that any great value can be placed on the figures for the nitrogen excretion on individual days. The best that can be done is to determine the nitrogen in the faeces for the whole experimental period and from this to calculate the average daily output. The latter may vary within rather wide limits even when the diet contains little or no nitrogen, and appears to depend to some extent on the bulk and character of the faeces. In order that these might be kept as uniform as possible agar-agar was taken when the diet consisted wholly or largely of completely digestible foodstuffs.

Food was taken in approximately equal amounts three times a day at the customary hours, and we led our usual life. In the experiments upon milk we abandoned all attempts to make our basal ration of fat and carbohydrate resemble a repast and drank a suspension of uncooked corn-starch in a saturated solution of lactose. This was followed by an alkaline salt mixture and 2 g. of agar-agar. In this way 600 calories were contained in about a tumbler full. Microscopical examination of the faeces showed that the starch was completely digested. When the starch-lactose mixture exceeded 250 g. at one meal some glycosuria occurred temporarily. On the N-minimum experiment the faeces were olive green as the bile pigment was unreduced. Little gas was formed.

The following are the essential data regarding the subjects of the experiments, ourselves:

C.J.M. Age 56. Weight 61 kilo. Height 183 c. Very thin, stores fat with difficulty. Mode of life: laboratory work. Exercise: lawn tennis before breakfast for three-quarters òf an hour and about three miles walk during the day.

R.R. Age 37. Weight 59 kilo. Height 173·5 c. Spare, does not store fat readily. Usual mode of life consists chiefly in laboratory work. Very little regular exercise beyond daily walk of two to four miles.

Minimum nitrogen expenditure on carbohydrate-fat diet.

Our minimum nitrogen expenditure was determined in two experiments, during which our diet was as nearly as possible nitrogen-free. In the vain attempt to make this diet appetising much labour was expended in endeavouring to prepare the food in a varied and attractive manner. Biscuits made of corn starch with 20 % of fat proved quite palatable when taken in small quantities but nauseating in bulk. A biscuit, whose chief defect was its hardness, made from starch, dextrin and a little fat was finally adopted and formed, with butter and honey, the chief article of diet. A starch mould flavoured with lemon juice was also taken. Weak tea with lemon and sometimes black coffee and a little vermouth was drunk. After the first few days of this diet no desire was felt either for this or for any other food, nor did the sight of our first normal meal at the close of the experiment arouse any appetite. The drinking of a glass of hot milk, however, excited in a few minutes a very keen appetite and desire for food. The quantities of the individual constituents of the diet varied somewhat from day to day but were always accurately measured and noted. Those for a single typical day are set out in Table I while the total nitrogen intake and fuel value of the diet for each day are shown in Table II in which are also set out the daily output of nitrogen in the urine and faeces and the nitrogen balance. In this and in all other experiments the nitrogen in the tea and coffee has been assumed to consist chiefly of caffeine and to be excreted unchanged in the urine. It has therefore been subtracted in all cases from the total intake and from the urinary nitrogen output. Any error involved in this method of treatment must be of negligible dimensions.

The nitrogen in the urine fell steadily until the last day of the experiment when a rather considerable rise occurred in the case of both C.J.M. and R.R. This is probably to be explained by the fact that, owing to the difficulty of consuming the food when all appetite was in abeyance, the fuel value of the diet was reduced during the last day or two.

Table I.

	N-minimum 2.		Diet R.R. 4. xii. 20		
Foodstuff	N per 100 g.	Calories per 100 g.	Weight g.	N g.	Calories
Corn starch	0·042	360	280	0·118	1008
Dextrin	0·065	360	50	0·033	180
Butter	0·080	775	90	0·072	698
Margarine (rendered)	0·010	900	15	0·002	135
Honey	0·023	327	55	0·013	180
Sucrose	—	395	25	—	99
Lactose	0·013	370	65	0·008	241
Lemon juice	0·067	40	30 cc.	0·020	12
Vermouth	0·005	140	25 „	0·001	35
Tea infusion	0·006	—	1200 „	0·072	—
Agar-agar	0·242	-·	15 g.	0·036	—
Salt mixture	—	—	5 „	—.	—
			Total	0·375	2588
			Excluding tea N	0·303	45 per kilo

Table II.

N-minimum 1. Subject: R.R. Fluid Intake: 2000–2500 cc.

		Daily Intake			Output			
Date 1920	Body weight k	N (excluding tea and coffee) g.	N tea and coffee g.	Calories per kilo	N urine g.	N faeces g.	N total g.	Balance N g.
Nov. 15	58·5	0·30	0·16	56	8·64	—	—	—
„ 16		0·28	0·06	54	5·31	1·05	6·36	− 6·08
, 17		0·30	0·15	55	3·63	„	4·68	− 4·38
, 18		0·25	0·10	51	2·66	„	3·71	− 3·46
, 19		0·22	0·06	43	2·25	„	3·30	− 3·08
„ 20	57·7 (21. xi)	0·23	0·15	49	2·82	„	3·87	− 3·64

N-minimum 2. Subject: R.R. Fluid Intake: 4000 cc.

„ 29	58·6	0·27	0·07	49	8·79	1·13	9·92	− 9·65
„ 30	(23. xi)	0·30	0·07	50	4·71	„	5·84	− 5·54
Dec. 1		0·30	0·08	51	3·46	„	4·59	− 4·29
„ 2		0·35	0·08	51	2·71	„	3·84	− 3·49
3		0·32	0·09	52	1·99	„	3·12	− 2·80
„ 4		0·30	0·07	45	2·17	„	3·30	− 3·00
„ 5	57·7	0·27	0·06	44	2·01	„	3·14	− 2·87
Average of last 3 days		0·30	0·07	—	2·06	1·13	3·19	− 2·89

N-minimum 1. Subject: C.J.M. Fluid Intake: 2000 cc.

Nov. 14	61·7	0·33	0·20	56	8·61	—	—	—
„ 15		0·30	0·20	47	4·89	1·24	6·13	− 5·83
, 16		0·30	0·20	47	3·28	„	4·52	− 4·22
, 17		0·38	0·21	52	3·28	„	4·52	− 4·14
, 18		0·32	0·21	46	2·48	„	3·72	− 3·40
, 19		0·30	0·21	45	2·25	„	3·49	− 3·19
„ 20	60·9 (21. xi)	0·21	0·18	35	2·49	„	3·73	− 3·52

N-minimum 2. Subject: C.J.M. Fluid Intake: 4000 cc.

„ 29	61·9	0·34	0·31	59	8·89	—	—	—
„ 30		0·34	0·27	55	4·78	•—	—	—
Dec. 1		0·34	0·33	55	3·40	1·17	4·57	− 4·23
„ 2		0·33	0·32	53	2·41	„	3·58	− 3·25
3		0·35	0·32	52	2·51	„	3·68	− 3·33
„ 4		0·35	0·29	52	2·40	„	3·57	− 3·22
„ 5	60·5	0·33	0·26	48	2·13	„	3·30	− 2·97
Average of last 3 days		0·34	0·29	—	2·34	1·17	3·51	− 3·17

On plotting the amounts of urine nitrogen as ordinates against the time in days as abscissae it became apparent that all the points except the last would lie on, or close to a curve for which a simple logarithmic expression was found (Figs. 2 and 3). Among several possible interpretations of this curve is the simple one that the falling nitrogen output represents washing out of some metabolic products from the tissues, the amount washed out each day being proportional to that still present.

A second similar experiment was therefore carried out in order to confirm this result and also to discover whether the steepness of the curve could be altered by drinking large quantities of water and thus greatly increasing the volume of the urine. Apart from the volume of total fluid, which was doubled, the diet did not differ from that taken in the first experiment. The results are set out in Table II.

The regularity in the fall of the nitrogen output was again observed but the change in the rate of this fall was very small.

The average outputs in urine and faeces on the last three days of this experiment have been taken as representing the minimum nitrogen expenditure on such a diet. Whether this amount also represents the minimum expenditure on a diet that is absolutely nitrogen-free will depend on the biological value of the small amount of nitrogen in the food consumed during ᴠthe above experiment. If this has a value of 100 %, *i.e.* if it can replace and therefore spare an equal amount of body nitrogen, the output thus determined will be equal to the real minimum expenditure. If the value of the food nitrogen is zero, *i.e.* if it is unable to satisfy any fraction of the body's requirements, it must be excreted in addition to the amount representing the latter and the minimum expenditure will therefore be equal to the observed output less the full amount of the nitrogen intake. Probably this nitrogen, which was present chiefly in the corn starch and butter, has a value intermediate between 100 and zero. The minimum nitrogen expenditure will therefore be some amount between those shown in the last two columns below.

Minimum nitrogen expenditure.

Subject	Urine g.	Faeces g.	Total output g.	Total output – intake g.
C.J.M.	2·34	1·17	3·51	3·17
R.R.	2·06	1·13	3·19	2·89

THE BIOLOGICAL VALUE OF THE PROTEINS OF WHOLE WHEAT.

A series of experiments was carried out with whole wheat flour with two objects in view:

(1) to determine the Biological Value of the wheat proteins from the minimum amount with which nitrogen equilibrium can be attained;

(2) to discover whether the Biological Value is uniform for varying amounts of wheat nitrogen.

The first of these has been investigated upon man by other workers, whose results will be considered with our own. The great practical importance of this question and the astonishing divergence between the results of previous investigations were sufficient reasons for further study.

Method of experiment.

The large variations in the percentage of water, and consequently of nitrogen, in different parts of a loaf of bread and the difficulty of obtaining a satisfactory sample, render it impossible to estimate the total nitrogen content of the loaf with sufficient accuracy for these experiments. We decided therefore, to base our calculations on the flour and to bake the bread ourselves. By suitable manipulation it was found possible to prepare loaves from 500 g. of flour with a maximum loss of less than 0·1 %. The other materials used were butter (5 %), salt, baking powder (prepared from tartaric acid, sodium bicarbonate and corn starch) and water. The loaves were baked in the laboratory for about one hour at a temperature of 240°–250°, and were only very slightly browned so that no appreciable loss of nitrogen can have occurred during the baking. For the first period, a somewhat coarsely ground whole wheat flour containing 1·85 % N was used but on increasing the daily ration from 300 g. to 450 g., considerable discomfort was experienced from the large particles of bran, and the bread was poorly absorbed. For the second and remaining periods a very finely ground flour prepared from a mixture of English and foreign whole wheat was employed. We found it very palatable and well absorbed, as the figures for the nitrogen in the faeces indicate. Only when the daily consumption had been raised to 550 g. and the total calories to 63 per kilo did we experience any discomfort.

The experiment was carried out in duplicate on ourselves and commenced with a total nitrogen intake of nearly twice the amount of our minimum requirements. On this diet a considerable negative balance occurred and the amount of wheat nitrogen was therefore increased during successive periods until equilibrium was finally attained.

The daily ration of flour was kept constant during each separate period of the experiment, but some latitude was permitted in the amounts of the remaining constituents of the diet. The actual quantities taken were, however, measured and the variations in any one period were not such as to affect appreciably the total nitrogen intake or greatly alter the fuel value of the diet. The diet set out in Table III shows the average amounts consumed by R.R. from the 24th to the 31st of October but except for the quantity of flour it would with slight variations serve for the whole experiment.

The experimental results are set out in Tables IV and V.

As in the previous experiments it has been assumed that the nitrogen consumed in tea and coffee would be excreted unchanged and the amount has therefore been subtracted from both intake and output (urine N).

During certain periods indicated by the letter (A) in the first column of

Table III.

Foodstuff	Whole wheat. N per 100 g.	Calories per 100 g.	R.R. Diet during period 5, 24–31 Oct. 1920 Weight g.	N g.	Calories
Flour, whole wheat	2·162	360	450	9·730	1620
Corn starch	0·042	360	36	0·015	130
Butter	0·080	775	56	0·045	434
Dripping	0·016	885	80	0·013	708
Honey	0·023	327	40	0·009	131
Marmalade	0·052	341	20	0·010	68
Cane sugar	—	395	56	—	221
Lemon juice	0·067	40	20 cc.	0·013	8
Tea	0·006	—	960 ,,	0·058	—
Coffee	0·041	—	100 ,,	0·041	—
Vermouth	0·005	140	25 ,,	0·001	35

Total 9·935   3355

Excluding tea and coffee N 9·836   58 per kilo

Table IV.

Diet: whole wheat. Subject: C.J.M.

Period	Date 1920	Body weight k	Daily Intake N bread g.	N total g.	Calories per kilo	Daily Output N urine g.	N faeces g.	N total g.	Balance N g.
1. (A)	Sept. 30–Oct. 2 (26. ix)	61·80	5·55	5·81	46	—	—	—	—
(O)	Oct. 3–5 (3. x)	61·50	,,	5·70	44	5·62	2·14	7·76	−2·06
2. (A)	,, 10–14 (10. x)	60·45	8·32	8·63	56	—	—	—	—
	,, 15		,,	,,	,,	8·26	2·17	10·43	−1·80
	,, 16		,,	,,	,,	7·62	,,	9·79	−1·16
	,, 17	61·44	,,	,,	,,	6·98	,,	9·15	−0·52
	Average		8·32	8·63	56	7·62	2·17	9·79	−1·16
3. (A)	,, 18		8·65	8·97	57	7·86	2·35	10·21	−1·24
	,, 19		,,	,,	,,	7·25	,,	9·60	−0·63
	,, 20		,,	,,	,,	7·15	,,	9·50	−0·53
	Average		8·65	8·97	57	7·42	2·35	9·77	−0·80
4. (A)	,, 21	60·65	9·73	9·98	53	7·25	1·74	8·99	+0·99
	,, 22		,,	,,	,,	7·92	,,	9·66	+0·32
	,, 23	61·03	,,	,,	,,	7·30	,,	9·04	+0·94
	Average		9·73	9·98	53	7·49	1·74	9·23	+0·75
5. (O)	,, 24	60·95	9·73	9·85	60	8·65	1·94	10·59	−0·74
	,, 25		,,	,,	,,	9·23	,,	11·17	−1·32
	,, 26		,,	,,	,,	8·77	,,	10·71	−0·86
	,, 27		,,	,,	,,	7·84	,,	9·78	+0·07
	,, 28		,,	,,	,,	8·93	,,	10·87	−1·02
	,, 29		,,	,,	,,	8·63	,,	10·57	−0·73
	,, 30		,,	,,	,,	7·68	,,	9·62	+0·23
	,, 31	61·60	,,	,,	,,	7·67	,,	9·61	+0·24
	Average		9·73	9·85	60	8·43	1·94	10·36	−0·51
6. (O)	Nov. 1		11·89	12·00	61	9·13	2·53	11·66	+0·34
	,, 2		,,	,,	,,	8·33	,,	10·86	+1·14
	.. 3		,,	,,	,,	9·83	,,	12·36	−0·36
	,, 4 (5. xi)	61·40	,,	,,	,,	9·63	,,	12·16	−0·16
	Average		11·89	12·00	61	9·23	2·53	11·76	+0·2

the table, a quantity of stewed apples (225 g. raw fruit containing 0·11 g. N) was included in the diet. The letter (O) indicates that this fruit was omitted.

For several days prior to Sept. 30, 1920 (period 1) a bread diet containing about 7·3 g. N had been taken in order to eliminate the disturbing effect of the previous high protein dietary. For the same reason the first three days of this period and the first five days of period 2 have been excluded in calculating the average nitrogen balance.

Table V.

Diet: whole wheat bread. Subject: R.R.

Period	Date 1920	Body weight k	Daily Intake			Daily Output			Balance N g.
			N bread g.	N total g.	Calories per kilo	N urine g.	N faeces g.	N total g.	
1. (A)	Sept. 30–Oct. 2	58·65 (27. ix)	5·55	5·82	48	—	—	—	—
(O)	Oct. 3–5	57·70 (6. ix)	,,	5·71	49	5·62	1·47	7·09	− 1·38
2. (A)	,, 10–14	58·20 (13. x)	8·32	8·57	54	—	—	—	—
	,, 15		,,	,,	,,	7·21	1·98	9·19	− 0·62
	,, 16		,,	,,	,,	8·30	,,	10·28	− 1·71
	,, 17	58·13	,,	,,	,,	6·41	,,	8·39	+ 0·18
		Average	8·32	8·57	54	7·31	1·98	10·29	− 0·72
3. (A)	,, 18		8·65	8·91	56	8·06	1·25	9·31	− 0·40
	,, 19		,,	,,	,,	8·70	,,	9·95	− 1·04
	,, 20		,,	,,	,,	7·79	,,	9·04	− 0·13
		Average	8·65	8·91	56	8·18	1·25	9·43	− 0·52
4. (A)	,, 21–23	57·89 (21. x)	9·73	10·00	58	8·32	1·70	10·02	− 0·02
5. (O)	,, 24		9·73	9·84	58	8·41	1·81	10·22	− 0·38
	,, 25		,,	,,	,,	8·25	,,	10·06	− 0·22
	,, 26		,,	,,	,,	10·51	,,	12·32	− 2·48
	,, 27	57·75	,,	,,	,,	8·83	,,	10·64	− 0·80
	,, 28		,,	,,	,,	7·67	,,	9·48	+ 0·36
	,, 29		,,	,,	,,	9·74	,,	11·55	− 1·71
	,, 30		,,	,,	,,	9·50	,,	11·31	− 1·47
	,, 31	57·44	,,	,,	,,	8·54	,,	10·35	− 0·51
		Average	9·73	9·84	58	8·93	1·81	10·74	− 0·90
6. (O)	Nov. 1		11·89	11·98	63	9·30	2·68	11·98	0
	,, 2		,,	,,	,,	9·75	,,	12·43	− 0·45
	.. 3		,,	,,	,,	8·86	,,	11·54	+ 0·44
	,, 4	57·70 (5. xi)	,,	,,	,,	9·46	,,	12·14	− 0·16
		Average	11·89	11·98	63	9·34	2·68	12·02	− 0·04

A marked rise in the nitrogen output occurred with both C.J.M. and R.R. at the beginning of period 5 and the coincidence of the omission of the apples that had been included in the diet during the preceding period led us to consider whether there was here any relation of cause and effect. Two possibilities suggest themselves: (1) the small amount of apple protein (0·11 g. N) might possess high value to supplement the wheat protein; (2) the alkaline ash of the fruit would partially neutralise the acid ash of the bread and this might affect the nitrogen expenditure. Neither of the above appears adequate to explain the facts and further, the results obtained during period 4 do not agree any better with those for periods 2 and 3, in which apples were included in the diet, than with those for period 5 in which they were omitted. The increased output must therefore, like the variations which occur from day to day, be left for the present unexplained.

From the results of these experiments we have calculated the biological value of the wheat proteins by two formulae, which differ only in the assumption made with regard to the nitrogen of the faeces. Both are based on Thomas's definition of this value as the number of parts of body nitrogen spared by 100 parts of the nitrogen of the food, and, when reduced to their simplest form, can be thus expressed

$$\text{B.V.} = 100 \frac{\text{Body N spared}}{\text{Food N absorbed}} = 100 \frac{\text{Balance } [P] - \text{Balance } [M]}{\text{Intake } [P] - \text{Intake } [M]}$$

where P signifies the experiment with the protein under investigation and M the nitrogen minimum experiment.

This correction for the small amount of nitrogen in the diet of the N-minimum experiment is strictly accurate only if certain provisos hold, viz. (1) that the same amount of nitrogen in the same form, or of the same biological value, enters also into the second diet (P); (2) that the value of this nitrogen is not increased by supplementary action with the other proteins of the second diet.

In our experiments with wheat and milk proteins the first proviso is partly but not entirely satisfied. Whether or not the second is also satisfied cannot be decided. The method, however, certainly involves a less error than if the N-intake in the N-minimum experiment is altogether ignored.

When the different assumptions as to the faeces are made the two formulae become:

I. $\text{B.V.} = 100 \dfrac{\text{Balance } [P] - \{\text{Intake } [M] - (\text{Urine N } [M] + \text{Faeces N } [P])\}}{\text{Intake } [P] - \text{Intake } [M]}$.

II. $\text{B.V.} = 100 \dfrac{\text{Balance } [P] - \text{Balance } [M]}{\text{Intake } [P] - (\text{Faeces N } [P] - \text{Faeces N } [M]) - \text{Intake } [M]}$.

In I the whole of the food nitrogen is assumed to have been absorbed so that the total intake is also the real intake. In calculating the minimum expenditure corresponding with the period in question the faeces N for this period is added to the urine N $[M]$ and the intake $[M]$ subtracted from the sum. This formula corresponds with Thomas's formula B.

In II when the nitrogen of the faeces is in excess of that occurring in the N-minimum experiment this excess has been assumed to represent unabsorbed food and has been subtracted from the total intake to obtain the real intake. The minimum expenditure has been taken as the actual balance on N-minimum diet, i.e. Urine N $[M]$ + Faeces N $[M]$ − Intake $[M]$. This formula corresponds with Thomas's formula C except that in the latter an average figure of 1 g. N, has been taken to represent the faeces N on a N-free diet. Both procedures have obvious disadvantages but there is very little difference in the results whichever is adopted.

A summary of the results for the separate periods with the biological values calculated from both the above formulae is given in Table VI.

Table VI.

Diet: whole wheat bread. Subject: C.J.M.

Period	Intake			Balance N g.	Biological value	
	Total N g.	Absorbed N g.	Calories per kilo		(1)	(2)
1. (O)	5·70	4·73	44	− 2·06	38·8	25·3
2. (A)	8·63	7·63	56	− 1·16	36·3	27·6
3. (A)	8·97	7·79	57	− 0·80	41·1	31·8
4. (A)	9·98	9·41	53	+ 0·75	41·0	43·2
5. (O)	9·85	9·08	60	− 0·51	36·0	30·4
6. (O)	12·00	10·64	61	+ 0·24	40·9	33·1
				Average	39·0	31·9

Diet: whole wheat bread. Subject: R.R.

Period	Total N g.	Absorbed N g.	Calories per kilo	Balance N g.	(1)	(2)
1. (O)	5·71	5·37	49	− 1·38	34·2	29·8
2. (A)	8·57	7·72	54	− 0·72	36·8	29·3
3. (A)	8·91	8·79	56	− 0·52	28·7	27·9
4. (A)	10·00	9·43	58	− 0·02	35·5	31·4
5. (O)	9·84	9·16	58	− 0·90	28·0	22·5
6. (O)	11·98	10·40	63	− 0·04	37·7	28·1
				Average	33·5	28·2

THE BIOLOGICAL VALUE OF THE PROTEINS OF COW'S MILK.

The first experiment with nitrogen in the form of milk protein followed immediately after the second period on low nitrogen diet and was carried out in duplicate on C.J.M. and R.R. The total nitrogen intake was approximately equal to the minimum nitrogen expenditure and though equilibrium was not obtained the negative balance was not very large, from which we concluded that milk proteins would be found to possess a relatively high value. When the experiments were repeated with larger amounts of milk we were surprised to find the nitrogen balance still negative and equilibrium was only reached with diets containing over 11 g. of milk nitrogen. This result appeared so extraordinary as to lead us to suspect that it might be due to a deficiency in the energy value of the diet, although this was amply sufficient to cover our normal requirements. The unusually rapid loss in weight which occurred in some of these experiments pointed in the same direction, as did also the coincidence of the fall in the nitrogen output occurring in one period (C.J.M. 6–R.R. 3) with a pronounced rise in the temperature of the air.

A further series of experiments was therefore carried out on one of us (C.J.M. periods 7, 8, 9), the fuel value of the diet being increased to the maximum that could be consumed. In the first of these, nitrogen equilibrium was practically attained with a diet containing 6·84 g. N and furnishing 57 calories per kilo. After an interval of one day, on which the basal ration together with a very little milk (3·0 g. N) was consumed, period 8 was begun. The diet contained 5·28 g. N and furnished 55 calories per kilo, but the effects of the continued excessive diet made themselves unpleasantly obvious in the form of a bilious attack, which threatened to terminate the experiment. By reducing the amount of carbohydrate consumed, so that the fuel value fell to 36 calories per kilo it was however just possible to carry on and after two days the condition was so far improved that the full diet was resumed. The nitrogen intake was not altered at all but the effect of the reduced diet was very marked in the increased nitrogen output, which persisted for several days after the calories were again increased. The period was extended for six days after the output had reached a fairly constant level and only the results for these days have been considered in calculating the average output. A decided though small negative balance occurred.

Table VII.

Milk. C.J.M.

Foodstuff	Nitrogen per 100 g.	Calories per 100 g.	Diet during period 9 20–25. viii. 21		
			Weight g.	Nitrogen g.	Calories
Milk	0·508	65	830	4·216	539
Starch	0·027	360	280	0·076	1008
Lactose	0·013	370	210	0·027	777
Honey	0·023	327	250	0·057	818
Margarine (rendered)	0·010	900	18	0·002	162
Sucrose	—	395	10	—	40
Tea	0·008	—	750 cc.	0·060	—
Agar-agar	0·242	—	6	0·014	—
Salts	—	—	4	—	—
			Total	4·452	3344
			Excluding tea N	4·392	54 per kilo

In period 9 the quantity of milk was again reduced but the high calorie value of the diet was maintained. The usual daily game of tennis was discontinued and no exercise was taken so that the excess of energy supplied was even greater than before. The weather also was very hot. The average of the last six days of this period showed a considerable negative balance and the biological value calculated from this agrees fairly well with that obtained from the results of periods 7 and 8 and also of period 2 (C.J.M.). This point is interesting because the calorie value of the diet in period 2 was lower than in any other, only 44 per kilo. The composition of the diet during period 9 (C.J.M.) is given in Table VII and the results of the experiments are set out in Tables VIII (R.R.) and IX (C.J.M.), while Table X is a summary of these showing the biological values calculated from the two formulae. It is obvious from the amounts of nitrogen in the faeces that the milk proteins were very completely absorbed so that formula 1 probably gives the closest approximation to the truth in this case.

Table VIII.

Diet: milk. Subject: R.R.

			Daily Intake			Daily Output			
Period	Date 1920	Body weight k	N milk g.	N total g.	Calories per kilo	N urine g.	N faeces g.	N total g.	Balance N g.
"N-free" 1.	Dec. 5	57·75	0	0·27	44	2·01	1·13	3·14	− 2·87
	„ 6		2·73	2·98	48	2·49	0·66	3·15	− 0·17
	„ 7		2·69	2·94	„	2·95	„	3·61	− 0·67
	„ 8		2·76	3·09	„	3·22	„	3·88	− 0·79
	„ 9		2·59	2·91	„	2·86	„	3·52	− 0·61
	„ 10	57·50 (11. xii)	2·67	2·97	„	2·83	„	3·49	− 0·52
	Average of last 4 days		2·68	2·98	48	2·97	0·66	3·63	− 0·65
2.	1921 May 28	59·65	4·10	4·30	40	7·12	—	—	—
	„ 29		6·14	6·29	48	5·97	1·18	7·15	− 0·86
	„ 30		„	„	„	5·91	„	7·09	− 0·80
	„ 31		„	„	„	6·34	„	7·52	− 1·23
	June 1		„	„	„	7·09	„	8·27	− 1·98
	„ 2		„	„	„	6·55	„	7·73	− 1·44
	„ 3	57·90 (4. vi)	„	„	„	6·21	„	7·39	− 1·10
	Average of last 4 days		6·14	6·29	48	6·55	1·18	7·73	− 1·44
3.	June 16		Mixed diet			12·29	—	—	—
	„ 17	59·60	13·61	13·69	47	13·48	1·37	14·85	− 1·16
	„ 18		„	„	„	13·13	„	14·50	− 0·81
	„ 19		„	„	„	12·29	„	13·66	+ 0·03
	„ 20		„	„	„	13·76	„	15·13	− 1·44
	„ 21		„	„	„	12·39	„	13·76	− 0·07
	„ 22		„	„	„	12·60	„	13·97	− 0·28
	„ 23	58·50	„	„	„	11·10	„	12·47	+ 1·22
	„ 24		„	„	„	11·19	„	12·56	+ 1·13
	Average of last 5 days		13·61	13·69	47	12·21	1·37	13·58	+ 0·11

Table IX.

Diet: milk. Subject: C.J.M.

			Daily Intake			Daily Output			
Period	Date 1920	Body weight k	N milk g.	N total g.	Calories per kilo	N urine g.	N faeces g.	N total g.	Balance N g.
"N-free" 1.	Dec. 5	60·45	0	0·33	48	2·13	1·17	3·30	− 2·97
	„ 6		3·15	3·37	49	2·42	1·29	3·71	− 0·34
	„ 7		„	3·36	„	2·57	„	3·86	− 0·50
	„ 8		„	3·42	„	3·01	„	4·30	− 0·88
	„ 9		„	3·40	„	3·12	„	4·41	− 1·01
	„ 10	59·30 (11. xii)	„	3·45	„	3·22	„	4·51	− 1·06
	Average of last 3 days		3·15	3·42	49	3·12	1·29	4·41	− 0·99
2.	1921 April 30	62·60	4·02	4·17	44	4·74	1·22	5·97	− 1·80
	May 1		„	„	„	4·58	„	5·80	− 1·63
	„ 2		„	„	„	4·09	„	5·32	− 1·15
	„ 3		„	„	„	4·25	„	5·47	− 1·30
	„ 4	62·20	„	„	„	4·53	„	5·75	− 1·58
	Average of last 4 days		4·02	4·17	44	4·36	1·22	5·58	− 1·41
3.	May 10		Mixed diet			8·00	—	—	—
	„ 11	62·20	6·20	6·35	48	6·13	1·18	7·31	− 0·96
	„ 12		„	„	„	5·71	„	6·89	− 0·54
	„ 13		„	„	„	5·26	„	6·44	− 0·09
	„ 14		„	„	„	5·80	„	6·98	− 0·63
	„ 15		„	„	„	6·08	„	7·26	− 0·91
	„ 16		„	„	„	6·07	„	7·25	− 0·90
	„ 17	62·15 (18. v)	„	„	„	6·21	„	7·39	− 1·04
	Average of last 4 days		6·20	6·35	48	6·04	1·18	7·22	− 0·87

Table IX (continued)

Diet: milk. Subject: C.J.M.

Period	Date 1921	Body weight k	N milk g.	N total g.	Calories per kilo	N urine g.	N faeces g.	N total g.	Balance N g.
"N-free"	May 27	—	—	—	—	8·38	—	—	—
4.	„ 28	62·40	7·63	7·78	47	7·05	0·96	8·01	− 0·23
	„ 29		„	„	„	7·11	„	8·07	− 0·29
	„ 30		„	„	„	7·55	„	8·51	− 0·73
5.	„ 31		8·62	8·77	47	7·48	0·98	8·46	+ 0·31
	June 1		„	„	„	8·58	„	9·56	− 0·79
	„ 2	61·30	„	„	„	8·41	„	9·39	− 0·62
	„ 3		„	„	„	8·67	„	9·65	− 0·88
	Average of last 3 days		8·62	8·77	47	8·55	0·98	9·53	− 0·76
	June 16			Mixed diet		10·12	—	—	—
6.	„ 17	61·80	11·32	11·44	46	9·71	1·45	11·16	+ 0·28
	„ 18		„	„	„	10·46	„	11·91	− 0·47
	„ 19		„	„	„	10·21	„	11·66	− 0·22
	„ 20		„	„	„	10·19	„	11·64	− 0·20
	„ 21		„	„	„	10·82	„	12·27	− 0·83
	„ 22		„	„	„	8·09	„	9·54	+ 1·90
	„ 23		„	„	„	9·60	„	11·05	+ 0·39
	„ 24		„	„	„	9·71	„	11·16	+ 0·28
	„ 25	61·70	„	„	„	10·28	„	11·73	− 0·29
	(26. vi)								
	Average of last 6 days		11·32	11·44	46	9·78	1·45	11·23	+ 0·21
	July 2			Mixed diet		6·33	—	—	—
7.	„ 3	61·10	6·65	6·84	57	5·66	1·28	6·94	− 0·10
	„ 4		„	„	„	5·55	„	6·83	+ 0·01
	„ 5		„	„	„	5·35	„	6·63	+ 0·21
	„ 6		„	„	„	5·70	„	6·98	− 0·14
	„ 7		„	„	„	6·35	„	7·63	− 0·79
	„ 8	61·80	„	„	„	5·04	„	6·32	+ 0·52
	Average of last 4 days		6·65	6·84	57	5·61	1·28	6·89	− 0·05
	July 9	62·10	—	3·00	54	5·16	—	—	—
8.	„ 10		5·16	5·28	55	4·54	1·09	5·63	− 0·35
	„ 11		„	„	36	3·99	„	5·08	+ 0·20
	„ 12	61·00	„	„	36	5·57	„	6·66	− 1·38
	„ 13	60·80	„	„	53	5·28	„	6·37	− 1·09
	„ 14	61·10	„	„	57	5·29	„	6·38	− 1·10
	„ 15	61·30	„	„	55	4·71	„	5·80	− 0·52
	„ 16	61·60	„	„	54	4·63	„	5·72	− 0·44
	„ 17	61·90	„	„	53	4·83	„	5·92	− 0·64
	„ 18	61·85	„	„	52	4·06	„	5·15	+ 0·13
	„ 19	61·85	5·04	5·18	52	4·65	„	5·74	− 0·56
	„ 20	62·00	„	„	52	4·24	„	5·33	− 0·15
	Average of last 6 days		5·12	5·25	53	4·52	1·09	5·61	− 0·36
9.	Aug. 16	61·60	4·24	4·58	53	15·79	—	—	—
	„ 17	61·85	„	„	„	10·78	—	—	—
	„ 18	61·85	„	„	„	6·46	—	—	—
	„ 19	61·90	„	„	„	5·83	—	—	—
	„ 20		„	4·39	„	4·69	1·00	5·69	− 1·30
	„ 21	61·80	„	„	„	4·00	„	5·00	− 0·61
	„ 22	61·80	„	„	„	4·10	„	5·10	− 0·71
	„ 23	61·60	„	„	„	5·04	„	6·04	− 1·65
	„ 24	61·60	„	„	58	4·78	„	5·78	− 1·39
	„ 25	61·75	„	„	55	4·72	„	5·72	− 1·33
	Average of last 6 days		4·24	4·39	55	4·56	1·00	5·56	− 1·17

Table X.

Diet: milk. Subject: C.J.M.

	Intake			Balance	Biological value	
Period	Total N g.	Absorbed N g.	Calories per kilo	N g.	(1)	(2)
1.	3·42	3·30	49	−0·99	74·7	73·6
2.	4·17	4·12	44	−1·41	47·3	46·6
3.	6·35	6·34	48	−0·87	38·4	38·3
5.	8·77	8·77	47	−0·76	26·3	28·6
6.	11·44	11·16	46	+0·21	33·0	31·2
7.	6·84	6·73	57	−0·05	49·7	48·8
8.	5·25	5·25	53	−0·36	55·6	57·2
9.	4·39	4·39	55	−1·17	45·2	49·4
				Average for last 3 periods	50·2	51·8

Diet: milk. Subject: R.R.

1.	2·98	2·98	48	−0·65	66·0	83·6
2.	6·29	6·24	48	−1·44	25·0	24·4
3.	13·69	13·45	47	+0·11	24·2	22·9

Table XI. *Basal metabolism of C.J.M. on normal diet and on carbohydrate and fat (= 55 cals. per kilo) with 4·4 g. milk N.*

Date	Diet during previous 24 hrs.	Oxygen consumed per min. cc.	R.Q.	Calories per 24 hrs.
Aug. 10	Normal	225·6	0·711	1557
,, 12	,,	236·3	0·928	1699
,, 13		227·1	0·684	1566
,, 14		220·9	0·820	1557
			Average	1595
,, 20	Milk	211·5	0·974	1529
,, 21	(4·39 g. N, 55 cals.	178·0	1·159	1359
,, 22	per kilo)	197·3	1·093	1478
,, 23	,,	185·7	1·152	1415
,, 24	,,	195·0	1·044	1440
,, 25		196·1	1·055	1453
,, 26		210·4	1·017	1540
			Average	1459
,, 27	Normal	221·9	0·915	1590
,, 28	,,	216·8	0·889	1550
,, 31		228·0	0·811	1604
			Average	1581

During period 9 the subject's basal metabolism was determined on waking and the results were compared with similar determinations carried out during previous and succeeding periods when the diet was normal. These results are set out in Table XI and show a decrease of about 8 % in the basal metabolism on the milk diet.

DISCUSSION OF RESULTS.

THE MINIMUM NITROGEN EXPENDITURE.

During the second experiment to determine our nitrogen minimum the output of nitrogen in the urine fell to an amount equal to 0·035 g. per kilo body weight in the case of C.J.M. and 0·034 g. per kilo for R.R. The average amounts for the last three days of this period were slightly above the minima, being 0·038 g. and 0·035 g. respectively. If the nitrogen of the faeces is included the average output for the same period was 0·057 g. per kilo for C.J.M., and 0·055 g. for R.R. The nitrogen intake amounted to 0·005 g. per kilo.

These figures are somewhat lower than most of those recorded by other workers [Landergren 1903; Folin, 1905; Kinberg, 1911; Graham and Poulton, 1912], but this may be explained by the fact that their diets unavoidably contained more nitrogen than ours. Karl Thomas [1910] determined his minimum expenditure on a purely carbohydrate diet in seven experiments carried out over a period of two and a half years and found that this minimum fell from experiment to experiment, the final amount for the output in the urine alone being 2·2 g. or 0·029 g. per kilo, while that for urine and faeces combined was 2·9 g. or 0·039 g. per kilo. At this period Thomas weighed 75 kilos and had put on a good deal of body fat. In some of his earlier experiments, however, it seems probable that the minimum output was never reached as the diet was continued for too short a period.

In McCollum's [1911] experiments on pigs, which were fed on a diet of starch, a salt mixture and water, the minimum nitrogen output fell to a level corresponding closely with that reached by us. Thus a pig weighing 68·4 kilos excreted 0·039 g. N per kilo in the urine. For smaller animals the output per kilo was somewhat greater.

The determination of the nitrogen in urine and faeces does not, of course, give a complete account of the loss of nitrogen from the body. To these must be added the loss through hair, beard and nails, through loss of epidermis and in sweat. Except for the last named these losses are all very small in amount but the loss through the sweat may be considerable. Benedict [1906] has shown that a resting man may excrete 0·071 g. N per day in this way while with moderate work the loss may amount to 0·13 g. N per hour. McCollum considers that the nitrogen of the faeces should also be classed with these as representing losses that may be termed accidental in character and that the nitrogen of the urine alone is to be taken as representing the essential tissue metabolism of the body.

The amount of nitrogen excreted in the urine on the successive days of the experiments in which the diet was nearly free from nitrogen, is plotted in Figs. 2, 3, and 5. The amount diminishes in a fairly regular manner, at first quickly and then more slowly until the minimum is reached. The points lie on or near the graph of a simple logarithmic equation

$$\log (y - \lambda) = a - kx$$

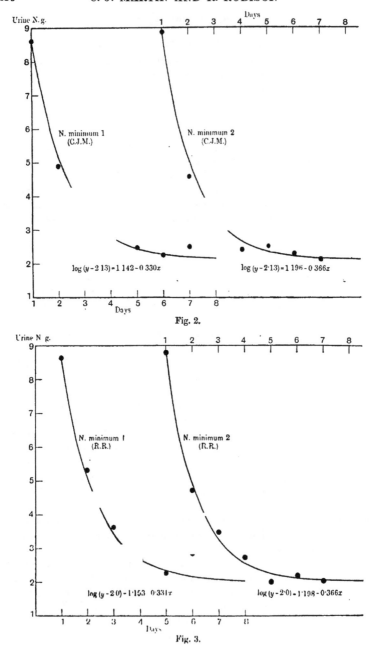

Fig. 2.

Fig. 3.

in which x is the number of days, y is the nitrogen output in the urine, and λ the minimum value of y. A closer agreement is obtained if λ is given a value slightly lower than the minimum actually reached.

Thus in Fig. 3 the curve R.R. 1 is drawn from the equation

$$\log (y - 2 \cdot 0) = 1 \cdot 153 - 0 \cdot 331x$$

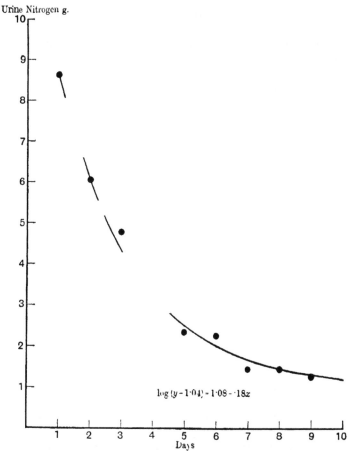

Fig. 4. E. V. McCollum's experiment on a pig. Daily output of nitrogen in urine on starch diet after ingestion of zein.

and fits the points reasonably well, but a still closer approximation is obtained with the equation

$$\log (y - 1 \cdot 73) = 1 \cdot 1209 - 0 \cdot 2832x$$

the agreement between the observed and calculated values of y being extraordinarily good for all points except the last.

Day of experiment	y calculated from equation $\log (y - 1 \cdot 73) = 1 \cdot 1209 - 0 \cdot 2832x$	y obtained
1	8·61	8·64
2	5·31	5·31
3	3·60	3·63
4	2·70	2·66
5	2·24	2·25
6	1·99	2·82
20	1·73	—

A question is thus raised: Does this amount 1·73 g. represent the real minimum expenditure, which, from some cause was not realised in either experiment? At present this must be left unanswered, but further experiments may give some information on the point.

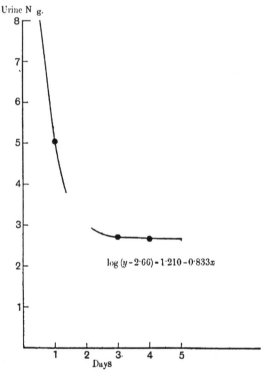

Urine N g.

$$\log (y - 2 \cdot 66) = 1 \cdot 210 - 0 \cdot 833x$$

Days

Fig. 5. E. V. McCollum's experiment on a pig. Daily output of nitrogen in urine on starch diet after ingestion of urea.

The difference between experiments 1 and 2 lies only in the amount of total fluid taken which was about 2000 cc. daily in the first and 4000 cc. in the second. This makes but slight difference in the value of k, that is in the rate at which the nitrogen output falls. Kinberg [1911] has previously drawn attention to this regularity but has not attempted to deduce any mathematical expression from his results. Thomas [1910] also recognised that on a protein-free diet the nitrogen of the "Vorratseiweiss" leaves the body with varying rapidity according to the amount present, but was unable to find any exact relationship either in his own results or in those of Landergren. He calculated the amount of "Vorratseiweiss" excreted, by subtracting the minimum output ("Abnutzungsquota") from the urine nitrogen, but the value of this minimum was taken from experiments of four days' duration and was probably too high. Landergren's results show a fair agreement with the graph of an equation of the type given above if λ is taken as 2·5 instead of 3·0. Thomas's results are more irregular. In considering the agreement of the results of such experiments

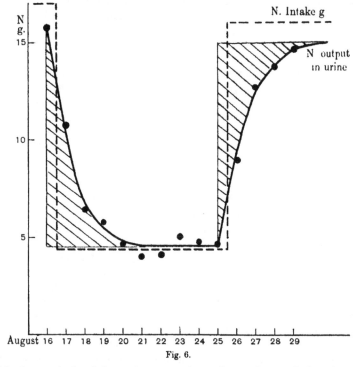

Fig. 6.

with those calculated from these equations, the tendency of the nitrogen output to oscillate even when the nitrogen intake is constant must be borne in mind. Such oscillations are present in most of the experiments considered, but do not alter the general character of the curve. In McCollum's experiments on pigs a similar regularity also appears as may be seen from Figs. 4 and 5. Fig. 4 shows the nitrogen output on a starch diet after a diet containing zein and Fig. 5 shows the output after the ingestion of urea. The curves are of the same type but "k," i.e. the rate at which the output falls, is very much greater after urea than after zein. This point is of interest as it seems to indicate that the store of nitrogen which exists in the body after a protein diet, and which is rapidly given up on a diet free from nitrogen is not present entirely in the form of urea. As to the form in which this nitrogen occurs, very little can be definitely stated. It may be as resynthesised protein (Vorratseiweiss), or as amino acids adsorbed in the tissues, or compounds of intermediate complexity. It is certainly present partly as urea and other end products of metabolism. If our conclusions as to the rate at which this storage nitrogen is excreted are correct the interpretation is, that the amount removed from the body on any day is proportional to the amount still present. This might hold whether the reaction involved was the hydrolysis of protein, the deaminisation of amino acids or simply the washing out from the tissues of the end products of nitrogen metabolism.

On reversing the process, and after a minimal N-intake for ten days, suddenly increasing the nitrogenous food consumed, nitrogen at first remains in the body and equilibrium between intake and output does not occur for several days. Fig. 6 is a graph of the results of an experiment designed to show this. The broken line represents intake of N, the solid line output. The shaded area on the left hand represents the stored N gradually removed on dropping the intake from 17 g. to 4·4 g. and the right-hand area the amount again stored on resuming a diet containing 16 g. N. The curves are reciprocal and the two shaded areas approximately correspond.

The Nature of the Minimum Nitrogen Requirements of the Body.

The low level to which the nitrogen output falls on a protein free diet is evidence of the smallness of the body's daily requirements in this respect. We know from Folin's [1905] researches that the reduction in the nitrogen on such a diet occurs mainly at the expense of the urea fraction, the ammonia and uric acid being reduced to a relatively much smaller extent while the creatinine remains constant. These facts led Folin to conclude that protein metabolism is of two types, (1) "tissue" or "endogenous," which tends to be constant, and is represented largely by such products as creatinine, neutral sulphur, and to a less extent by uric acid, (2) "exogenous" which varies with the amount of protein consumed and is represented chiefly by urea. The nitrogen required for processes of the first type is alone essential, but Folin recognised that equilibrium at this low level may not be possible since a certain amount of protein may always fall prey to the exogenous metabolism. The distribution of the nitrogenous constituents of the urine was determined during our second minimum experiment and the milk diet immediately succeeding this.

The results [Robison, 1922, 2] are similar to those of Folin. The minimum nitrogen output was lower than any recorded by him, and the percentage of urea was correspondingly reduced, the minimum figure being 37·2 %, while the sum of the urea and ammonia amounted to 54·8 % of the total nitrogen.

The multifarious transactions involved in endogenous metabolism are not likely to be conducted with perfect economy. When much protein is hydrolysed and the products mobilised and used for the synthesis of proteins of another composition such as those of the blood or for the manufacture of thyroxin or adrenaline it is unlikely that the whole balance of unwanted amino acids escapes deaminisation and conversion. Leakage of this kind may account for no inconsiderable fraction of the minimum nitrogen expenditure and it is perhaps in this direction that, by adaptation, the body may effect some saving. The experiments of Thomas and Hindhede would appear to show that this does in fact occur.

The amount of carbohydrate eaten may also influence the degree of this waste since the process of deaminisation is reversible and will be affected by the rapidity with which the non-nitrogenous products, hydroxy or ketonic acids, are removed by oxidation or conversion into carbohydrate. It is probable that an excess of carbohydrate in the blood would retard either action and in consequence deaminisation.

The Question of Uniformity of the Biological Value.

We have seen that the validity of the method adopted by Thomas for the determination of the biological values of proteins depends in the first place on the uniformity of this value when varying amounts of the same protein are consumed. The investigation of this question was one of the objects of our experiments and the results must now be considered from this point of view. The three proteins so far studied were chosen on account of the wide difference in the values attributed to them. The biological values of milk and wheat proteins as given by Thomas are 99·71 and 39·56 respectively, while gelatin has long been known to be deficient in several essential amino acids, so that no amount, however great, can completely satisfy the body's nitrogen requirements.

The results of the experiments described in this paper together with those on gelatin, previously recorded by one of us [Robison, 1922, 1] are plotted in Figs. 7 and 8, in which the values of the nitrogen intake for different periods are the abscissae and the corresponding amounts for the total output in urine and faeces are the ordinates. For this purpose the nitrogen of the faeces in excess of the amount on the "protein-free" diet has been taken as representing unabsorbed food and has been subtracted from both intake and output.

In the case of whole wheat proteins, the results are strikingly similar with both individuals and the points obviously lie on or close to a straight line passing through the points representing the minimum expenditure on the low nitrogen diet. The agreement is remarkably good considering the errors inseparable from such experiments. The results for periods 4 and 5 show some discrepancy but these periods were continuous and should probably be considered as a whole, the only difference in the diets being the inclusion of

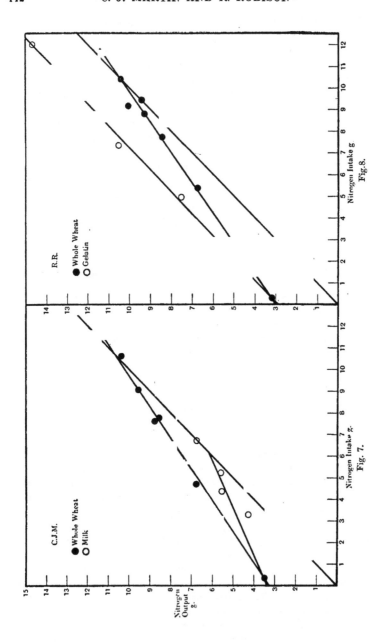

Fig. 7.

Fig. 8.

a small quantity of apples in the former and its omission from the latter. The average for the two periods gives a point lying close to the line. The uniformity of the biological value is therefore satisfactorily proved for the proteins of whole wheat.

In the case of gelatin also the points fit a straight line reasonably well, but if this line passes through the point representing the nitrogen minimum (as drawn in the figure), it will at some distant point intersect the "equilibrium line" (at 45° to the axis), i.e. nitrogenous equilibrium will be attained at this point, which we believe to be impossible. On the other hand if it is drawn parallel to the "equilibrium line" it will cut the axis of y at a point slightly below the observed minimum. We must conclude, therefore, either that the line is slightly curved near the minimum or that the observed value of the latter is somewhat higher than the real one. If the capacity of the gelatin is limited to the reduction of "leakage" these two alternatives have practically the same significance.

Milk presents a more difficult case. Only the results for those periods in which a diet of abundant fuel value was taken (C.J.M. 1, 7, 8, 9) have been plotted but even these do not agree well with any straight line or other regular curve. The most that can be said is that in view of the possible errors in such experiments where the intake is small, the results are not inconsistent with the uniformity of the biological value of this protein.

From the consideration of these three cases we may conclude that the general assumption of this principle made by Thomas, occasioned no serious errors in his conclusions. Caution must be used in extending this principle to all cases without investigation and reliance should not be placed upon results of experiments in which the negative balance is large.

The Biological Values of the Proteins of Whole Wheat and Milk.

Whole Wheat.

The biological values calculated from the different periods of our experiments with whole wheat proteins agree very well amongst themselves and give an average of 35 for those obtained on C.J.M. and 31 for those obtained on R.R. Thomas's value is somewhat higher, 39·56. The latter figure was based on the result of two experiments each of three days' duration and one of four days. During none of these experiments was the nitrogen intake constant; in one, the amounts for the separate days were 4·0 g., 7·3 g., 9·0 g. respectively. The minimum requirements corresponding with these three periods were taken as 4·63 g., 3·991 g. and 3·316 g. N (urine only), these amounts being determined in experiments of four and three days' duration. In the first of these the nitrogen output on the four days was 18·32 g., 10·17 g., 7·39 g., 4·63 g. but there is no evidence that the last figure represented Thomas's minimum requirements. The biological value of wheat proteins was calculated from each separate day's balance and out of these widely varying results those for certain days were selected in a somewhat arbitrary fashion. The experimental results of his third period, in which the intake was nearly constant, indicate a similar biological value for wheat proteins to that found by ourselves.

The result Hindhede [1913, 1] obtained upon F. Madsen with white bread, namely a positive balance of about 1 g. per day over a 28 day experiment on 13·73 g. N was certainly not a minimum quantity and the amount necessary for N equilibrium may be considerably less. His further experiments [1914] were made with rye bread ("Schwarzbrot") and the diets contained considerable amounts of fruit or vegetables, which accounted for 0·5 g. to 1·9 g. N. These experiments were carried out in duplicate on F. and H. Madsen and were continued for four months so that they possess great value. In the final period of six days the fruit was omitted and positive balances were obtained on 13·52 g. (F.M.) and 10·68 g. (H.M.) bread nitrogen respectively. By subtracting the whole of the nitrogen in the faeces Hindhede calculates that the amounts absorbed were 8·49 g. and 6·28 g. respectively. We do not agree with this method of treating the faecal nitrogen and cannot accept his conclusion that bread proteins possess equal value with those of potatoes, of meat and of the body; but there is no doubt that in his experiments, equilibrium was obtained on a lower intake of nitrogen in the form of rye bread (and still lower when supplemented by fruit) than our minimum for whole wheat. This is confirmed by the experiments of Neumann [1919], who, by long continued diet of very high calorie value, was ultimately able to retain nearly 3 g. N daily with an intake of 9·9 g. N in the form of rye proteins. Neumann's experiment was upon himself, and was in every way unexceptionable, but it may perhaps be significant that it was preceded by a prolonged period of semistarvation during which he was investigating the German civilian ration of 1916–17. This is a further indication that, apart from the influence of loss of body weight, the organism can gradually accommodate itself to a lower ration of nitrogen, perhaps by the exercise of greater economy.

Abderhalden's [1915] observations on Röse, although as pointed out earlier in this paper, not susceptible of the interpretation he places upon them, do indicate that the latter could get into equilibrium with about a gram less N in the form of white bread than we could with bread made from the whole grain. Röse's nitrogen expenditure on a nitrogen-free diet was not ascertained so we cannot estimate the biological value for this diet.

The recent observations of Rubner [1919] upon the proteins of white flour are interesting in relation to our own, because they show, we think, that from the point of view of biological value, the proteins of the endosperm are equal, if not superior to, those of the whole seed.

In most of Sherman's experiments on the value of cereal proteins these are supplemented by a certain amount of milk and the results are not directly comparable with ours, but in some experiments on white bread [1920] the diet consumed would not appear to be greatly different from that taken by us and we are therefore the less able to explain the difference between his results and ours. Sherman's subject, a man weighing 80 kilos, attained equilibrium on a diet containing 6·0 g. N over 95 % of which was derived from white bread and the remainder from apples and butter. The energy value was only 34 calories per kilo. The bread was purchased from a bakery and probably contained a small amount of milk but how much was not known.

Milk.

The results of our experiments with milk proteins do not agree so closely as those for the whole wheat. The very low values calculated from the results of periods 3, 5, 6 (C.J.M.) and 2, 3, (R.R.) are not easy to explain. That they are in some way due to the lower calorie value of the diet seems clear but this was in no case below that of our normal diet and amply covered our energy requirements. The high values obtained with both subjects in period 1 are perhaps connected with the previous nitrogen starvation and a consequent increase in the economy with which the body may deal with the protein supplied to it.

If we consider only periods 7, 8, 9 (C.J.M.) in which the conditions were the same and the energy supply abundant the biological value for milk proteins is equal to 51 %. This value is only half of that found by Thomas, but the criticisms we have made in discussing his experiments with bread apply with still greater force in this case. His value (100 %) was calculated from the nitrogen balance on a single day of an experiment lasting only two days on which the intake was 6·24 g. and 7·28 g. respectively. The minimum requirements were taken as 3·99 g. which is probably much too high. If the value of milk protein were as high as Thomas makes out it is difficult to see how he could explain the large negative balances occurring in his experiment with "Frauenmilch" (cow's milk with extra sugar and cream). During the first two days of this experiment the fuel value of his diet was obviously too low, but in the last three days it was equal to 40–45 calories per kilo, the N-intake being 15·3 g.–17·3 g. yet the negative balance was never less than 1·0 g. The results of this experiment appear to agree with our experience, and to suggest that a high intake of milk nitrogen tends to result in increased expenditure of body nitrogen, unless the fuel value of the diet is raised very much above the normal. So far as we are aware the value of milk protein has not been the subject of any other investigation on man.

Conclusions.

From our observations upon ourselves we conclude:

(1) That our minimum nitrogen expenditure by the urine is somewhat less than 0·038 g. and 0·035 g. per kilo in C.J.M. and R.R. respectively.

(2) That on taking a diet of carbohydrate and fat of adequate calorie-value the nitrogen excreted in the urine falls in a regular and orderly manner,

capable of simple mathematical expression, approaching a minimum in five to seven days. On resuming an ordinary nitrogenous diet the reciprocal phenomenon occurs.

(3) Bearing in mind the considerable experimental errors, the ratio $\frac{\text{Body N saved}}{\text{Food N absorbed}}$ appears to remain constant. whatever amount of nitrogen is taken in the form of whole wheat bread, until equilibrium is reached.

(4) In the case of milk the experimental errors are proportionately greater and the most we can say is that this ratio may remain constant.

(5) In the case of gelatin the ratio certainly does not remain constant and there is no indication that the amount of body nitrogen saved increases beyond that effected by the smallest quantity of gelatin fed.

(6) The application of Thomas's method of determining biological values is justified in the case of bread, doubtful with milk and impossible with gelatin.

(7) Until Thomas's procedure has been ascertained to be justifiable for the particular proteins concerned, the ratio $\frac{\text{Body N saved}}{\text{Food N absorbed}}$ should be determined close to, but below, the point of equilibrium.

(8) The mean biological value of the nitrogen contained in the whole wheat grain as determined by six experiments on each of two adults was 35 % (C.J.M.) and 31 % (R.R.).

(9) The mean biological value of the nitrogen in cow's milk, derived from three experiments upon C.J.M. in which an excess of calories (55 per kilo) was taken, was 51 %.

(10) Biological values arrived at from experiments of comparatively short duration, however well justified, have a limited significance.

REFERENCES.

Abderhalden and Rona (1904). *Zeitsch. physiol. Chem.* **42**, 528.
—— —— (1906). *Zeitsch. physiol. Chem.* **47**, 397.
Abderhalden and Samuely (1905). *Zeitsch. physiol. Chem.* **46**, 193.
Abderhalden, Fodor and Röse (1915). *Pflüger's Archiv.* **160**, 511.
Albu (1901). *Zeitsch. klin. Med.* **43**, 75.
Atwater and Benedict (1902). *Mem. Nat. Acad. Science*, **8**, 233.
Benedict (1906). *J. Biol. Chem.* **1**, 263.
Bornstein (1898). *Berl. klin. Woch. Nr.* 36.
Boruttau (1915). *Biochem. Zeitsch.* **69**, 225.
—— (1919). *Biochem. Zeitsch.* **94**, 194.
Chittenden (1904). Physiological Economy in Nutrition, New York.
Cohnheim (1901). *Zeitsch. physiol. Chem.* **33**, 451.
—— (1906). *Zeitsch. physiol. Chem.* **49**, 64.
Falta (1906). *Arch. klin. Med.* **86**, 517.
Folin (1905). *Amer. J. Physiol.* **13**, 66 and 117.
Goldberger, Waring and Willets (1915). *U.S., P.H. Reports.* Reprint 307, 5.
Goldberger, Wheeler and Sydenstricker (1920). *U.S., P.H. Service Reports*, **35**, 648.
Graham and Poulton (1912). *Quart. J. Med.* **6**, 82.
Grubner (1901). *Zeitsch. Biol.* **42**, 407.

Hindhede (1913, 1). *Skand. Arch. Physiol.* **28**, 165.
—— (1913, 2). *Skand. Arch. Physiol.* **30**, 97.
—— (1914). *Skand. Arch. Physiol.* **31**, 259.
Hirschfeld (1887). *Pflüger's Archiv.* **41**, 533.
—— (1889). *Virch. Archiv.* **114**, 301.
Hopkins and Cole (1901). *J. Physiol.* **27**, 418.
Kinberg (1911). *Skand. Arch. Physiol.* **25**, 291.
Klemperer (1889). *Zeitsch. klin. Med.* **16**, 550.
Kossel (1901). *Ber.* **34**, 3214.
Kossel and Kutscher (1900). *Zeitsch. physiol. Chem.* **31**, 165.
Kumagawa (1889). *Virch. Archiv.* **116**, 370.
Kutscher and Seemann (1901). *Zeitsch. physiol. Chem.* **34**, 527
Landergren (1903). *Skand. Arch. Physiol.* **14**, 112.
Lapicque (1894). *Arch. Physiol. Norm. et Path.* **26**, 596.
Lapicque and Marrette (1894). *C.R. Soc. Biol.* 10 ser. **1**, 274.
Loewi (1902). *Arch. Exp. Path. Pharm.* **48**, 303.
McCollum (1911). *Amer. J. Physiol.* **29**, 210.
McCollum and Hoagland (1913). *J. Biol. Chem.* **16**, 317.
McCollum, Simmonds and Pitz (1917). *J. Biol. Chem.* **29**, 341.
McCollum, Simmonds and Parsons (1921). *J. Biol. Chem.* **47**, 111, 139, 175, 207, 235.
Michaud (1909). *Zeitsch. physiol. Chem.* **59**, 405.
Neumann (1899). *Arch. Hyg.* **36**, 248.
—— (1902). *Arch. Hyg.* **45**, 1.
—— (1919). *Vierteljahrschrift gericht. Med.* 52.
Osborne and Harris (1903). *J. Amer. Chem. Soc.* **25**, 853.
Osborne and Mendel (1911). Pub. no. 156. Carnegie Inst. Washington.
—— —— (1912–1920). *J. Biol. Chem.* 12 to 45.
—— —— (1917). *J. Biol. Chem.* **29**, 69.
Peschel (1891). *Inaug. Dissert.* Berlin.
Robison (1922, 1). *Biochem. J.* **16**, 111.
—— (1922, 2). *Biochem. J.* **16**, 131.
Rubner (1879). *Zeitsch. Biol.* **15**, 115.
—— (1897). V. Leyden "Handbuch der Ernährungstherapie."
—— (1915). *Arch. Physiol.* 145.
—— (1919). *Arch. Physiol.* 81.
Sherman (1920). *J. Biol. Chem.* **41**, 97.
Sherman, Wheeler and Yates (1918, 1). *J. Biol. Chem.* **34**, 383.
Sherman and Winters (1918, 2). *J. Biol. Chem.* **35**, 301.
Sherman, Winters and Phillips (1919). *J. Biol. Chem.* **39**, 53.
Sivén (1900). *Skand. Arch. Physiol.* **10**, 91.
Thomas (1909). *Arch. Physiol.* 219.
—— (1910). *Arch. Physiol.* Suppl. 249.
Voit, C. (1866). *Zeitsch. Biol.* **2**, 307.
—— (1867). *Zeitsch. Biol.* **3**, 1.
Voit, E. and Korkunoff (1895). *Zeitsch. Biol.* **32**, 58.
Willcock and Hopkins (1907). *J. Physiol.* **35**, 88.
Wilson (1921). *J. Hygiene*, **20**, 1.
Zeller (1914). *Arch. Physiol.* 213.
Zisterer (1910). *Zeitsch. Biol.* **53**, 157.

XXXIV. ON THE CHANGE OF THE OSMOTIC PRESSURE OF SOLUTIONS OF CERTAIN COLLOIDS UNDER THE INFLUENCE OF SALT SOLUTIONS.

BY DAIZO OGATA (*Fukuoka*).

From the Department of Biochemistry, Oxford University.

(*Received March 17th, 1922.*)

IN spite of certain controversies among investigators on the osmotic pressure exerted by colloidal solutions the positive proof that a definite, though small, osmotic pressure is set up was given by Starling [1896], Moore and Parker [1902], Lillie [1907], Moore and Roaf [1907] and others.

This osmotic pressure is affected by several factors, such as the mode of the formation of the colloidal solution [Lillie, 1907], the change of temperature [Moore and Roaf, 1907], the acidity or alkalinity of solution [Adamson and Roaf, 1908]. Further that the added salts also increase or diminish the size of the colloidal particles before any visible precipitation takes place, *i.e.* lower or raise the osmotic pressure, has been proved by Biltz and Vegesack [1909], Lillie [1907], Moore and Roaf [1907] and Loeb [1918].

The present research was started, by suggestion of Prof. Moore and under his direction, to prove by means of osmotic pressure measurements whether, first, salt solutions affect the "solution aggregate" of different colloidal solutions always in the same way, and, secondly, what is the influence, if any, of certain anaesthetics upon colloidal solution. While definite and regular results were obtained in the former experiments, those in the latter were not at all decisive.

METHODS.

As colloidal solutions gum arabic (1 %), egg albumin (5 %) and sheep serum (half diluted) were used. The membrane (dialyser) was of collodion and made on the outside of a large test tube. This was fixed to a well-fitting rubber cork with a strong string. The cork had two holes, one for a glass manometer tube and the other for an inlet and outlet tube. The length of the membrane was so adjusted as to limit its content to approximately 15 cc.

In the earlier part of the experiments the colloidal solution only was introduced into the dialyser, *i.e.* without mixing previously with any salt solution, thus letting the latter penetrate gradually from the outside. But the procedure became more elaborate with the progress of the experiments, and the final method consisted of mixing the colloidal solution with the salt solution just before setting up the apparatus and then introducing into the membrane. The membrane was immersed up to the upper surface of the cork in a definite volume (120 cc. as a rule) of salt solution contained in a beaker. The mano-meter tube, together with a millimetre scale, was clamped in a vertical position during the experiments. The salts employed were copper sulphate, calcium chloride and alkaline sodium phosphate respectively.

The experiments were performed in series, e.g., 1/10 N, 1/100 N, 1/1000 N solution and solvent (distilled water) only, the last as a control. The same salt solution of the same concentration was always placed inside and outside the membrane.

Usually the osmotic pressure was not very high, therefore the consideration of the distension of the membrane was of little importance.

As the salt solution could easily pass through the membrane the column of solution in the manometer increased in length, at first, very rapidly. After a certain time it began to fall, hardly staying at the highest point any measurable length of time. Then the column fell very slowly to a certain level at which it remained for a considerable time. When almost the same level was kept for at least two days, it was understood, in my experiments, that the expected equilibrium of osmotic pressure was attained. In most cases when this equilibrium was reached the outside liquids were replaced with distilled water, four or five times in succession with an interval of one day or more. The reversibility of pressure was fairly satisfactory.

Results of Experiments.

The following are the examples of the abridged records of experiments carried out during five months. Similar results were always obtained when due precautions were taken.

Experiments with gum arabic solution. (See Fig. 1.)

Inside the membrane: 1 % gum arabic solution, 15 cc.
Outside the membrane: salt solutions of different concentrations, 120 cc.

Table I. SO_4Cu.

Day	Hour	Temp.	Osmotic pressure			
			1/10 N SO_4Cu	1/100 N SO_4Cu	1/1000 N SO_4Cu	distilled water
Sept. 1	afternoon (set up)					
,, 2	4.00 p.m.	18·9° C.	2·5 mm.	13·2 mm.	48·0 mm.	163·3 mm.
,, 3	2.30	16·9	7·5	14·0	45·1	150·5
,, 5	4.00	18·9	8·5	12·8	42·0	135·5
,, 6	3.20	18·3	9·0	13·0	42·2	130·0
p_H of the outside solution, about			9·0	7·2	6·5	6·5
The reversibility was tested:						
Sept. 13	10.00 a.m.	17·3	92·0	100·0	106·0	120·0

The curves in Fig. 1 indicate the rate at which equilibrium was reached both in direct and "reversed" experiments.

Table II. Cl_2Ca.

Day	Hour	Temp.	Osmotic pressure			
			1/10 N Cl_2Ca	1/100 N Cl_2Ca	1/1000 N Cl_2Ca	distilled water
Sept. 2	about noon (set up)					
„ 6	3.30 p.m.	17·5° C.	12·3 mm.	17·0 mm.	50·0 mm.	108·5 mm.
p_H of the outside solution, about			6·0	6·5	6·0	6·5
The reversibility was tested:						
Sept. 13	10.00 a.m.	16·9	93·0	94·0	113·0	117·5

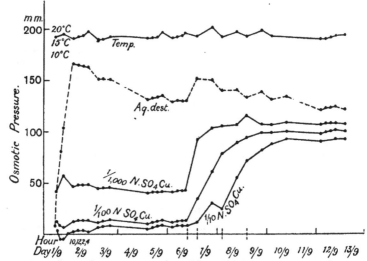

Fig. 1. Exp. with gum arabic solution. Effect of SO_4Cu. At the ⋮ line on the abscissa, the outside liquid was replaced by dist. water in succession.

Table III. PO_4HNa_2.

Day	Hour	Temp.	Osmotic pressure			
			1/10 N PO_4HNa_2	1/100 N PO_4HNa_2	1/1000 N PO_4HNa_2	distilled water
Sept. 3	morning (set up)					
„ 7	12.00	17·2° C.	15·2 mm.	27·0 mm.	110·5 mm.	149·0 mm.
p_H of the outside solution, about			9·0	7·2	6·5	6·5
The reversibility was tested:						
Sept. 13	10.00 a.m.	17·2	251·0	220·0	129·0	130·0

Experiments with egg albumin.

Inside the membrane: 5 % egg albumin solution, 15 cc.
Outside the membrane: salt solutions of different concentrations, 120 cc.

Table IV. SO_4Cu.

Day	Hour	Temp.	Osmotic pressure			
			1/2000 N SO_4Cu	1/20,000 N SO_4Cu	1/200,000 N SO_4Cu	distilled water
Sept. 16	morning (set up)					
„ 20	12.00	16·1° C.	12·0 mm.	53·9 mm.	55·1 mm.	52·2 mm.
p_H of the outside solution, about			6·0	6·0	6·0	6·0
The reversibility was tested:						
Sept. 27	4.00 p.m.	16·0	36·0	64·0	62·0	59·0

Table V. Cl_2Ca.

Day	Hour	Temp.	Osmotic pressure			
			(1/20 N Cl_2Ca	1/200 N Cl_2Ca	1/2000 N Cl_2Ca	distilled water
Sept. 16 afternoon (set up)						
,, 20	12.00	15·7° C.	14·0 mm.	17·0 mm.	36·0 mm.	50·0 mm.
p_H of the outside solution, about			6·75	6·0	6·0	6·0
The reversibility was tested:						
Sept. 27	4.00 p.m.	15·5	40·0	41·0	49·0	58·0

Table VI. PO_4HNa_2.

Day	Hour	Temp.	Osmotic pressure			
			(1/20 N PO_4HNa_2	1/200 N PO_4HNa_2	1/2000 N PO_4HNa_2	distilled water
Sept. 16 afternoon (set up)						
,, 21	2.00 p.m.	16·5° C.	22·6 mm.	30·0 mm.	44·0 mm.	47·0 mm.
p_H of the outside solution, about			9·0	7·2	6·75	6·5
The reversibility was tested:						
Sept. 27	4.00 p m.	15·9	40·0	48·0	51·0	59·0

Experiments with blood serum.

Inside the membrane: sheep serum half diluted, 15 cc.
Outside the membrane: salt solutions of different concentrations, 120 cc.

Table VII. SO_4Cu.

Day	Hour	Temp.	Osmotic pressure			
			(1/2000 N SO_4Cu	1/20,000 N SO_4Cu	1/200,000 N SO_4Cu	distilled water
Sept. 30 afternoon (set up)						
Oct. 5	12.00	16·6° C.	161·0 mm.	189·0 mm.	192·0 mm.	184·0 mm.
p_H of the outside solution, about			6·75	7·2	7·2	7·2
Sulphate in 100 cc. of the outside solution as SO_3			5·36 mg.	2·86 mg.	0·06 mg.	0·00 mg.
The reversibility was tested:						
Oct. 17	10.00 a.m.	13·4	115·0	168·0	163·0	170·0

Table VIII. Cl_2Ca.

Day	Hour	Temp.	Osmotic pressure			
			(1/20 N Cl_2Ca	1/200 N Cl_2Ca	1/2000 N Cl_2Ca	distilled water
Sept. 30 afternoon (set up)						
Oct. 5	12.00	16·0° C.	105·0 mm.	120·0 mm.	169·0 mm.	176·0 mm.
p_H of the outside solution, about			7·2	7·2	7·2	7·2
Chloride in 100 cc. of the outside solution as ClNa			37·7 mg.	11·5 mg.	9·25 mg.	8·0 mg.
The reversibility was tested:						
Oct. 17	10.00 a.m.	13·1° C.	178·0 mm.	168·5 mm.	155·0 mm.	163·0 mm.

Table IX. PO_4HNa_2.

Day	Hour	Temp.	Osmotic pressure			
			(1/20 N PO_4HNa_2	1/200 N PO_4HNa_2	1/2000 N PO_4HNa_2	distilled water
Oct. 1	morning (set up)					
,, 5	12.00	16·5° C.	133·5 mm.	163·0 mm.	166·0 mm.	176·0 mm.
p_H of the outside solution, about			9·0	7·2	6·75	6·75
Phosphate in 100 cc. of the outside solution as P_2O_5			204·0 mg.	27·2 mg.	8·0 mg.	3·0 mg.
The reversibility was tested:						
Oct. 17	10.00 a.m.	13·2° C.	69·0 mm.	101·0 mm.	137·0 mm.	158·0 mm.

Unless specially mentioned the protein did not appear in the outside solution through the membrane (acetic acid and potassium ferrocyanide test).

As clearly seen in the above tables the salts used have all a more or less prominent influence upon the osmotic pressure exerted by the colloidal solutions such as gum arabic, egg albumin or blood serum. The salt solutions always reduced the osmotic pressure of the colloidal solution, and the more dilute the salt solution, the less decrease was brought about, thus gradually approaching the control with distilled water only. This fact suggests, at once, the important influence of electrolytes, because within the limits of concentration of the salt solutions employed here it may safely be concluded that the more concentrated the solution, the greater will be the number of electrically dissociated molecules. The trouble is the appearance of precipitation in the long run of experiments of this kind, and, therefore, all the statements here are based necessarily on the comparison with the control otherwise under the same conditions. Of copper sulphate an extremely dilute solution was made use of simply to avoid the unnecessary precipitation as far as possible.

The temperature, no doubt, affects the height of the osmotic pressure, but this effect is rather slight compared with that due to the electrolytes themselves. Notwithstanding the fluctuation of osmotic pressure parallel to that of temperature, the main course of change of osmotic pressure cannot be ignored.

The hydrogen-ion is neither the sole nor the principal cause in these cases, since the osmotic pressure does not always keep pace with its concentration. To quote a few examples, the hydrogen-ion concentration was about the same through the series in Tables II, IV, VII and VIII respectively. Nevertheless the pressure difference is very notable.

The more or less prominent reversibility of the pressure leads us to seek the probable cause in the physical reaction rather than the chemical change, as was propounded by several previous investigators. Mayer, Schaeffer and Terroine [1907] observed under the microscope that the submicroscopical granules of colloidal solutions (metal colloid, starch, albumin and others) disappeared on alkalisation and appeared on acidification. This is one of the convincing proofs that colloid aggregation can change at different reactions. It, moreover, strongly suggests that the decrease of the osmotic pressure of colloidal solutions under the influence of electrolytes may be most conveniently ascribed to a similar phenomenon.

So far as the results of our experiments were concerned the effects of three salts on the three colloidal solutions showed no qualitative difference among themselves.

Briefly speaking I am inclined to think, with previous investigators, that the diminished osmotic pressure under the influence of salt solutions depends upon an increased aggregation of colloidal particles.

The effects of anaesthetics (chloroform, ether, alcohol) were tested in the same way. The only difference of the method lay in the use of a flask, instead of a beaker, to prevent the evaporation of these anaesthetics, connection with the manometer tube being made by a side tube. The solution inside the membrane was 15 cc., outside about 500 cc. The concentrations of anaesthetics were not always the same. Saturated solutions of chloroform in distilled water, 10 % ether, and 5 % alcohol were the most concentrated solutions in the series of experiments in each case. Roaf and Alderson [1907] found that anaesthetics such as chloroform, ether and carbon dioxide detached no salts from blood serum, which differed thus from other tissues (red blood corpuscles, brain, liver, muscle, kidney). Moore and Roaf [1907] observed that the presence of anaesthetics (chloroform) or organic solvents (benzene) did not alter the osmotic pressure of the serum protein (pig's serum). In their experiments the serum saturated with chloroform was placed against saline (0·175 %) also saturated with chloroform.

Despite the imperfections of the present osmotic pressure determinations the results of experiments with anaesthetics seem to favour somewhat positive effects though they are not very decisive.

SUMMARY.

1. Salt solutions (SO_4Cu, Cl_2Ca, PO_4HNa_2) added to colloidal solutions (gum arabic, egg albumin, blood serum) decreased the osmotic pressure exerted by the latter. No qualitative difference was confirmed among the salts or among the colloidal solutions within the concentrations here employed. The remarkable reversibility of the osmotic pressure suggests, at once, that it is a physical phenomenon, *i.e.* increased aggregation of colloidal particles by the electrolytes.

2. The effects of anaesthetics (chloroform, ether, alcohol) on the osmotic pressure of colloidal solutions were studied. The results, although not quite definite, seem to show that there is some effect.

I wish to express my great indebtedness for Prof. B. Moore's constant direction in this work.

REFERENCES.

Adamson and Roaf (1908). *Biochem. J.* **3**, 422.
Biltz and Vegesack (1909). *Zeitsch. physikal. Chem.* **68**, 357.
Lillie, R. S. (1907). *Amer. J. Physiol.* **20**, 121.
Loeb, J. (1918). *J. Biol. Chem.* **35**, 497.
Mayer, Schaeffer and Terroine (1907). *Compt. rend.* **145**, 91⁹
Moore and Parker (1902). *Amer. J. Physiol.* **7**, 261.
Moore and Roaf (1907). *Biochem. J.* **2**, 34.
Roaf and Alderson (1907). *Biochem. J.* **2**, 412.
Starling (1896). *J. Physiol.* **19**, 312.

XXXV. FURTHER OBSERVATIONS ON THE NATURE OF THE REDUCING SUBSTANCE IN HUMAN BLOOD.

By EVELYN ASHLEY COOPER and HILDA WALKER.

From the University of Birmingham.

(*Received April 3rd, 1922.*)

In the previous communication [1921] it was shown that the reducing power of human blood was sometimes increased by acid hydrolysis. This pointed to the presence in addition to glucose of a more complex substance, which was shown not to be glycogen, and thus appeared to be a disaccharide. The quantitative determination of this accessory substance was found to be complicated by two factors:

(1) the destructive effect of the HCl used for hydrolysis upon the reducing substance;

(2) the inhibitory effect of NaCl (formed by neutralisation after hydrolysis) upon the reduction of the copper carbonate by sugar.

Further work on the subject has since been carried out, and the results are recorded in the present paper.

First of all, it was necessary to select an acid, which did not itself destroy sugar, and the salts of which had no effect on the reducing process.

In the previous work sodium sulphate had been found to have no inhibitory effect, but the use of sulphuric acid for hydrolysis is inadmissible, because sodium sulphate is used for removing the proteins from the blood, and its presence would diminish the ionisation of the acid.

Further experiments showed that potassium chloride, like sodium chloride, had a retarding action on the reducing process, while the sulphate had no effect. The use of other chlorides for comparative purposes was not possible for various reasons. For example, ammonium chloride is decomposed by the alkali in the copper solution, and the liberated ammonia destroys the glucose, while the salts of several metals, *e.g.* Ca, Ba, Mg, give precipitates with the copper carbonate. The results, however, show that the chloride ion has a specific retarding effect.

Experiments with varying concentrations of sodium and potassium chloride showed that the retarding action first appeared when the concentration was approximately $N/1$. This inhibitory effect was found to be irreversible, *i.e.* when a solution of glucose containing NaCl was diluted with a salt-free sugar solution, the reducing action of the sugar upon the copper solution was still (proportionately) retarded. The reduction was also retarded when the salt was added to the boiling mixture of sugar and copper solutions. This suggests that the compounds of glucose with the chlorides are much more stable than is commonly supposed. Bromides and iodides of the alkali metals also inhibited the reducing process.

Sodium citrate was moreover also found to exert an inhibitory action when present in concentration equivalent to 2 % citric acid, which is employed for hydrolysis in sugar analysis.

Di-sodium phosphate, on the other hand, had no inhibitory effect, and it thus seemed that phosphoric acid might be a suitable acid for hydrolysing the blood-sugar extracts. Separate portions of an extract (prepared by MacLean's method, as in the previous communication) were hydrolysed with $N/10$ HCl and $N/10$ H_3PO_4, and it was found that the reducing substance was destroyed to the extent of about 10 % by both acids.

Pure glucose was next dissolved in sodium sulphate solution, or in blood-extract, and the mixture hydrolysed with $N/10$ or $N/100$ HCl. The results obtained were similar to those obtained with blood-extract itself [Cooper and Walker, 1921]; sometimes there was no destruction of reducing substance at all, but occasionally a destruction occurred amounting to about 10 %.

These results afford an explanation of our previous work. Evidently the hydrolysable substance, which may be a disaccharide, is only occasionally present in blood, so that on hydrolysis there is often no increase in the reducing power of the blood-extract, or there may actually be a loss in reducing power, owing to destruction of some of the glucose. In fact, out of 18 samples of normal human blood examined, only six cases showed an increased reducing power after acid hydrolysis, and as a general rule the chief reducing substance present is glucose.

Two blood-extracts were subjected to dialysis. In one case the hydrolysable substance dialysed completely. In the other case, however, it was only partially dialysable, and this suggests that occasionally substances of intermediate complexity may be present in blood, possibly related to malto-dextrins.

THE ESTIMATION OF BLOOD-SUGAR.

Up to the present time nearly all the methods for estimating sugar in physiological work are modifications of Fehling's method. Although fairly reliable, this is an empirical method, and the interaction of sugar with alkali copper solution is of great complexity. Recently, the iodometric method for estimating sugars has come to the forefront [see Baker and Hulton, 1920]. This is a simple method, consisting in the quantitative oxidation of aldoses to the corresponding carboxylic acid by means of iodine in alkaline solution, and it can be carried out in a few minutes at room temperatures. We have therefore attempted to adapt it to physiological work.

We have found, however, that in estimating pure glucose in concentration approximating to the average content of the blood-sugar extracts by the iodine method, slightly low results are obtained. Varying the proportion of alkali, and extending the period of reaction did not affect the results. In estimating the sugar in blood by this process, however, the results were twice as high as those obtained by MacLean's method. The alkali copper reagent is thus more selective than iodine in its action, and the iodometric method does not appear to be suitable for physiological work.

INFLUENCE OF FATIGUE ON THE BLOOD-SUGAR CONCENTRATION.

Estimations of the blood-sugar by MacLean's method were carried out on normal persons immediately before, and directly after exercise. The persons examined were students, both men and women, of ages ranging from about 18 years to 26. Estimations were made on one man before and after half-an-hour of strenuous boxing, the other estimations on men were made before and after Rugby football, and on the women, before and after hockey. Of twelve experiments, there was a rise in the blood-sugar concentration in nine cases. The increase, as shown in the appended table, varied from 10 to 50 %, but in one case the amount was nearly trebled after 1¼ hours' play.

			% before	% after
Boxing. ½ hour.		A	0·04	0·05
Rugby. 1 hr. 10 mins.		B (1)	0·052	0·077
		(2)	0·072	0·130
		C	0·097	0·090
		D	0·093	0·10
		E	0·115	0·10
		F	0·087	0·094
		G	0·118	0·094
Hockey. 1 hr. 15 mins.		H	0·087	0·253
		I	0·065	0·075
		J	0·118	0·150
		K	0·083	0·092

The player B is a first class forward, and in good training; he is of a very nervous and imaginative temperament, and before the match is always in a state of suppressed excitement. C and D are also fast players; C is of a quiet type, not easily roused to excitement. E, F and G were not in training, and did not take the game so seriously as the others.

Of the hockey players, H was playing back; she is of a highly nervous temperament, but well controlled. I and K were half-backs, and J played forward.

It will be seen that the players most easily roused to excitement over the game (B and H) are those whose blood-sugar has increased the most, although H, playing back, probably underwent less physical fatigue than most of the others. Psychic influences, therefore, probably play a considerable part.

CONDITION OF THE SUGAR IN THE BLOOD.

We next considered the question of the structural condition of the sugar in the blood, and its bearing on physiological and pathological problems. It is well known that sugars do not merely exist as aldoses and ketoses, but may also pass into cyclic forms known as oxides. The ethylene oxide form:

$$O \begin{array}{c} \diagup CH.OH \\ \diagdown CH \\ | \\ [CH(OH)]_3 \\ | \\ CH_2OH \end{array}$$

is characterised by being extremely chemically reactive, and it decolorises permanganate rapidly.

Hewitt and Pryde [1920] showed that this active form of sugar could be actually formed by contact of an aqueous solution of glucose with the intestinal wall *in vivo*.

It is possible that in the animal organism sugar is normally metabolised in this active condition, and that an enzyme exists for transforming ordinary sugar as ingested into this form. Now it is known that fructose is more readily converted into the ethylene oxide structure than glucose, and that in diabetes the organism may still be able to metabolise fructose, although it has lost the power to deal with glucose. This suggests that the causation of diabetes is associated with some disturbance in the enzyme mechanism, which normally converts inactive sugar into the reactive ethylene oxide form.

We therefore proceeded to ascertain whether normal human blood can cause ordinary sugar to pass into this reactive state. A few cc. of fresh blood were placed in a small dialyser, immersed in $\frac{1}{2}$–1 % solutions of glucose and fructose. At varying intervals of time, samples of the dialysate were withdrawn and examined, either polarimetrically or with a solution of permanganate. No evidence, however, was obtained that blood, under the above conditions, can produce the ethylene oxide form from ordinary glucose or fructose.

Since these experiments were carried out, Hewitt and Souza [1922] have found that even *in vivo* normal blood is unable to induce this structural change.

SUMMARY.

1. Chlorides, bromides, iodides and citrates inhibit the reduction of copper carbonate by glucose, as carried out by MacLean's method. Sulphates and phosphates have no effect.

2. Glucose is slightly destroyed by boiling with $N/10$ HCl and $N/10$ H_3PO_4. The reducing substance present in blood is also destroyed by acid to about the same extent. This affects the determination of the hydrolysable reducing substances in blood.

3. Glucose is the chief sugar occurring in human blood, and reducing substances of a more complex nature are only occasionally present.

4. The estimation of blood-sugar by the iodine method gives results much higher than those obtained by MacLean's process, and the iodine method does not seem suitable for physiological work.

5. The blood-sugar concentration may rise considerably as the result of muscular exertion.

6. There is no evidence that human blood can transform ordinary glucose or fructose into the reactive ethylene oxide form.

7. A theory as to the causation of diabetes is put forward.

REFERENCES.

Baker and Hulton (1920). *Biochem. J.* **14**, 756.
Cooper and Walker (1921). *Biochem. J.* **15**, 415.
Hewitt and Pryde (1920). *Biochem. J.* **14**, 395.
Hewitt and Souza (1922). *Biochem. J.* **15**, 667.

XXXVI. BLOOD ENZYMES. II. THE INFLUENCE OF TEMPERATURE ON THE ACTION OF THE MALTASE OF DOG'S SERUM.

By ARTHUR COMPTON.

From Laboratoire de Chimie Biologique, Institut Pasteur, Paris.

(*Received April 4th, 1922.*)

IN a recent communication [Compton, 1921, 2], we have called attention to the fact that it is possible to divide mammals into two groups, depending upon whether the enzyme maltase is present, or not, in their blood-stream. For convenience, these groups have been named the "dog group" and the "rabbit group." The former group is represented in our experiments by such mammals as the dog, the pig, the ox, the horse, the goat, the sheep and the rat; while the latter group is represented by the rabbit, the guinea-pig, the cat and man.

In view of the general interest of the question, and the fact that it offers a starting-point for various researches, we have set out in the present investigation to determine, under certain well-defined conditions, the effect of heat on the enzyme as met with in the blood of the dog—an animal typical of the maltase group.

EXPERIMENTAL.

For this investigation three dogs were utilised (2-*P*, 3-*P*, 4-*P*). Blood was withdrawn aseptically by venous puncture from the jugular, the animals having been in general in the fasting state for 24 hours previously[1]. Dog 2-*P* was examined three times, an interval of three weeks elapsing between the examinations (1) and (2), and one week between the examinations (2) and (3); dog 3-*P* twice, 18 days elapsing between the examinations; while blood was only once examined from dog 4-*P*.

As soon as withdrawn, the blood was transferred to a sterile centrifuge tube, centrifuged, and the supernatant serum pipetted off from the clot. The serum thus collected was conserved in presence of a few drops of toluene in a closed test-tube, and utilised as required in the studies which follow.

I. *Activity in maltase of various specimens of dog's serum.*

The activity in maltase of the different specimens of serum was determined at the temperature of 47°, in an action of 16 hours' duration and concentration

[1] I desire to express here my indebtedness and thanks to Monsieur A. Frouin, for kindly supplying me with specimens of dog's blood for these experiments.

in maltose $M/20$: that temperature having previously been found under like circumstances to be the optimum temperature of a specimen of vegetable maltase prepared from *Aspergillus* or*yzae* [Compton, 1914].

For the determination, 90 mg. of pure hydrated maltose were weighed out into each of four clean test-tubes, to which were added 2 cc. of pure water (redistilled) to make a solution; then doses of the serum under investigation were added to the tubes as follows: 0·1, 0·2, 0·4, 0·8 cc. The contents of the tubes were then completed rapidly to 5 cc. with a further addition of water respectively as follows: 2·9, 2·8, 2·6 and 2·2 cc. Next, three drops of toluene were added to each tube, the latter shaken, closed with clean corks, and plunged into a water-thermostat regulated at or about 47°. After 16 hours' incubation, the tubes were withdrawn, the corks removed and the tubes heated for seven minutes in boiling water to stop the enzyme action. When cold, the contents of each tube were diluted to 50 cc., and 20 cc. of the diluted mixture employed to determine, by Bertrand's method, the proportion of maltose hydrolysed. The numbers obtained are set out in Table I.

Table I.

Lab. No. of the animal	Dose of serum employed cc.	Temperature of experiment ° C.	Percentage of maltose hydrolysed	Information as to the age of serum at the time of testing
2-P (1)	0·05	47·0–46·8	25·4	Fresh
	0·1		39·1	
	0·2		50·9	
	0·4		63·5	
	0·8		78·5	
2-P (2)	0·1	47·4–46·8	34·3	Two days
	0·2		46·4	
	0·4		61·9	
	0·8		71·7	
2-P (3)	0·1	47·5–46·0	31·2	Fresh
	0·2		49·4	
	0·4		60·3	
	0·8		71·7	
3-P (1)	0·1	47·0–46·9	18·3	Two days
	0·2		32·8	
	0·4		49·4	
	0·8		58·7	
3-P (2)	0·1	46·8	26·8	Fresh
	0·2		46·4	
	0·4		49·4	
	0·8		70·0	
4-P	0·1	47·2	47·9	One day
	0·2		68·4	
	0·4		78·5	
	0·8		80·3	

On plotting the percentage of maltose hydrolysed against the quantities of serum in action, these numbers give the curves indicated in Fig. 1.

An outstanding feature of this figure is the close relationship existing between the individual examinations of a given animal. Thus, we find dog 2-*P* giving a series of curves: 2-*P* (1), 2-*P* (2) and 2-*P* (3), crowded together in the same region of the figure. Indeed, the curves 2-*P* (2) and 2-*P* (3) are practically identical. Again, the same is true, although to a lesser degree, of the dog 3-P: the curves 3-P (1) and 3-P (2) presenting a very similar allure, and being situated lowest in the diagram. The animal 4-P, on the other hand, represented by only a single examination, gives a curve which stands well apart, in a quite different region from the curves of the other dogs, being situated highest.

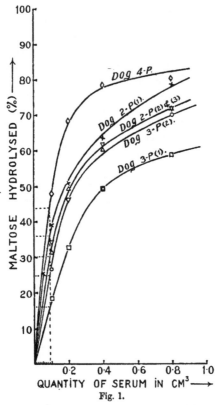

Fig. 1.

The conclusion therefore, would seem justified, that while the content of the blood serum of the dog in the enzyme maltase may vary considerably from animal to animal, still for the individual it is remarkably constant, varying only between comparatively narrow limits.

II. *Optimum temperature of the maltase of dog's serum.*

For reasons relating to another investigation, the quantity of serum utilised for the determinations of optimum temperature described in this paper is 0·09 cc. To employ conveniently this dose we operated as follows: 1·35 cc. of serum were measured out into a small 15 cc. measuring flask and the content made up to the 15 cc. mark with pure water. After shaking, the mixture was distributed in portions of 1 cc. to a series of seven or eight test-tubes which had been prepared in advance, containing 90 mg. of maltose dissolved in 4 cc. of water and five drops of toluene, each tube being already placed in a water-thermostat regulated to a constant temperature. The object of this latter procedure was that the substrate solution in the tubes should already be in temperature equilibrium with the thermostat at the moment of adding the enzyme solution. The tubes were closed with clean sterile corks and incubated for 16 hours, after which the corks were removed, and the enzyme action stopped as previously. The proportion of maltose hydrolysed in each tube was determined as already described. The numbers obtained are set forth in Table II.

Table II.

Temperature at beginning and end of each experiment °C.	Maltose hydrolysed %					
	Dog 2·P			Dog 3·P		Dog 4·P
	(1)	(2)	(3)	(1)	(2)	
30·3–29·1	—	—	—	—	—	12·8
36·8–37·2	—	—	—	—	—	24·0
37·0–37·1	—	12·8	—	—	12·8	—
37·3–37·0	—	—	—	10·0	—	—
37·5	19·7	—	—	—	—	—
41·9–42·0	—	—	15·5	—	24·0	—
44·5–45·2	26·8	—	—	—	—	—
45·8–46·0	—	—	—	—	—	46·4
46·1–46·0	—	—	—	24·0	—	—
46·8	—	—	—	—	—	43·5
46·9–46·8	—	25·4	—	—	—	—
46·9	—	—	—	—	34·3	—
47·0–46·8	—	—	—	24·0	—	—
47·0–47·2	31·2	—	—	—	—	—
47·2	—	—	21·1	—		—
49·6	34·3	—	—			—
49 8	—	—	—	—	—	52·5
49·8–49·	—	28·2	—	—	—	—
49·8–50·9	—	—	—	—	40·6	—
50·0–49·6	—	—	—	28·2	—	—
54·5	43·5	—	—	—	—	—
54·5–54·8	—	—	—	32·8	—	—
54·6–54·9	—	—	—	—	45·0	—
54·8–55·0	—	—	31·2	—	—	63·5
	—	37·5	—	—	—	—
55·0	—	—	—	—	43·5	—
58·0–57·6	—	29·7	—	—	—	—
58·0–58·5	31·2	—	—	—	—	—
58·1–58·5	—	—	—	—	—	46·4
58·2–58·0	—	—	—	28·2		—
58·4–58·3	—	—	25·4	—		—
60·0–60·1	—	28·2	—	—	—	—
61·0–62·0	11·4	—	—	—	—	—
61·5–61·8	—	—	—	—	24·0	—
61 8–62·0	—	—	—	12·8	—	37·5

On plotting the percentage of maltose hydrolysed against the mean temperature of the experiment, these numbers give the series of optimum temperature curves indicated in Fig. 2.

Fig. 2 reveals for the three animals studied the same optimum temperature: 55°, in an action of 16 hours' duration and reaction of medium that of the serum diluted in pure water. It is of interest that this same optimum temperature is found, whether successive studies are made at different times of the same animal, or whether different animals are studied.

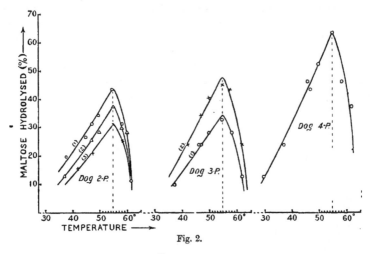

Fig. 2.

Conclusion.

The enzyme maltase, met with in the blood of the dog, exhibits from animal to animal a constant optimum temperature. This stability of optimum temperature constitutes an experimental fact which cannot but prove to be˙ of considerable practical value as a reference point, in connection with any subsequent study of the enzyme. It is in contrast with the variability which characterises the actual amount of enzyme present at any time in the blood of different animals.

From what we know of the remarkable sensitiveness of the optimum temperature of an enzyme to variations of p_H of the medium [Compton, 1915; 1921, 1], there can be little doubt that the theoretical interpretation of the above effect lies in a certain constancy of the chemical reaction, or hydrogen-ion concentration, of the animal's blood serum. The same delicate mechanism which maintains constancy of the one, will obviously maintain constancy of the other.

From the point of view of general physiology, this stability of optimum temperature of a blood enzyme raises problems of even wider interest. For instance, one may ask how far it may be possible to define a species by humoral reactions of enzymic nature. That is a question to which we hope to return in the course of subsequent studies in this series.

REFERENCES.

Compton (1914). *Proc. Roy. Soc.* B, **88**, 258.
—— (1915). *Proc. Roy. Soc.* B, **88**, 408.
—— (1921, 1). *Proc. Roy. Soc.* B, **92**, 1.
—— (1921, 2). *Biochem. J.* **15**, 681.

XXXVII. THE MODE OF OXIDATION OF FATTY ACIDS WITH BRANCHED CHAINS. II. THE FATE IN THE BODY OF HYDRATROPIC, TROPIC, ATRO-LACTIC AND ATROPIC ACIDS TOGETHER WITH PHENYLACETALDEHYDE.

By HERBERT DAVENPORT KAY AND HENRY STANLEY RAPER.

From the Department of Physiology and Biochemistry, the University of Leeds.

(*Received April 11th, 1922.*)

IN a previous communication, Raper [1914] advanced an hypothesis to explain the mode of oxidation in the body of fatty acids substituted at the α-carbon atom by a methyl group. The experimental evidence available at that time indicated that these acids underwent oxidation in the same way as the straight chain fatty acids from which they were derived. It was suggested that this behaviour might be accounted for if oxidation of the α-methyl group took place first, giving rise to a derivative of the semi-aldehyde of malonic acid, which being very unstable would lose carbon dioxide to produce a normal fatty aldehyde. This aldehyde, in turn, would give rise to its corresponding fatty acid by further oxidation. The series of reactions would thus be as follows:

$$\underset{\text{R--CH--COOH}}{\overset{\text{CH}_3}{|}} \longrightarrow \underset{\text{R--CH--COOH}}{\overset{\text{CHO}}{|}} \longrightarrow \text{R--CH}_2\text{--CHO} + CO_2 \longrightarrow \text{R--CH}_2\text{--COOH}.$$

In this way an α-methylated acid would give rise to a fatty acid with a straight chain structure but containing one carbon atom less. In support of this hypothesis it was shown that isobutyric acid and α-methylbutyric acid, on oxidation with hydrogen peroxide, yielded propaldehyde and normal butaldehyde respectively.

Owing to the relative scarcity of data concerning the fate of branched chain fatty acids in the animal organism it was decided to investigate the fate of phenyl derivatives of the α-methylated fatty acids. The present communication is concerned with the simplest member of this series, namely, hydratropic acid (α-phenylpropionic acid), and some of its derivatives. Experiments with other members of the α-substituted series are in progress and will be communicated in due course. The results have been disappointing so far as the isolation of intermediate products of oxidation is concerned but the fate of the various acids investigated, (I) to (IV), has enabled conclusions to be drawn which indicate the probable manner in which hydratropic acid is oxidised.

$$\text{(I)} \quad C_6H_5-\underset{\overset{|}{CH_3}}{CH}-COOH, \qquad \text{(II)} \quad C_6H_5-\underset{\overset{|}{CH_2OH}}{CH}-COOH,$$

$$\text{(III)} \quad C_6H_5-\underset{\overset{|}{CH_3}}{C}(OH)-COOH, \qquad \text{(IV)} \quad C_6H_5-\underset{\overset{||}{CH_2}}{C}-COOH.$$

Hydratropic acid (I) administered to dogs produced slight toxic effects in doses of 0·25 g. per kilo and was oxidised to the extent of about two-thirds. The remaining third was excreted in the urine, partly combined with glycine, and was dextrorotatory. A resolution of hydratropic acid, not yet complete, has shown that this was not the pure dextro form but a mixture of the dextro and laevo modifications in which the dextro component was in excess. No intermediate products of oxidation were detected. Inactive tropic acid (II) was found to be very resistant to oxidation, over 90 % of the acid given being recovered from the urine. The acid excreted, as expected, was optically inactive. Inactive atrolactic acid (III) was also very resistant to oxidation and over 80 % of this acid was excreted unchanged. It also was optically inactive. Atropic acid (IV), on the other hand, was found to be the most toxic of the four acids investigated, but in doses of 0·13 g. per kilo, which were tolerated, it was completely oxidised. No intermediate products were detected in the urine but a small but definite trace of succinic acid was excreted on two occasions and was not found on careful examination of the normal urine of the same dog. It is difficult however to connect this directly with the oxidation of atropic acid.

It is of great general interest to find that tropic acid is much more resistant to oxidation in the body than hydratropic acid, a result which the greater *in vitro* oxidisability of the hydroxylated acids (when acted on by the usual oxidising agents) would hardly lead one to expect. Dakin [1909] has observed the same phenomenon with β-phenylpropionic acid and β-phenyl-β-hydroxypropionic acid. The former is much more easily oxidised than the latter. These results show that the introduction of a hydroxyl group in the β-position in phenyl derivatives of propionic acid does not facilitate their oxidation in the body but rather hinders it. The fact that tropic and atrolactic acids are much less easily oxidised than hydratropic acid excludes them as possible intermediate products in its oxidation. It is also very improbable that hydratropic acid undergoes α-oxidation since either atrolactic acid, phenylglyoxylic acid, mandelic acid or benzoic acid would be expected as intermediate products. A careful search failed to reveal any of these, so that it may be assumed that hydratropic acid is oxidised in the β-position. Further, since the methyl group containing the β-carbon atom is unsubstituted both in atrolactic and hydratropic acids, it would be expected, that the former would be oxidised to a greater extent than was found to be the case. This makes it probable that some other factor besides the non-substitution of the methyl group is important in facilitating the oxidation of hydratropic acid. We believe this factor to be the ability to form an unsaturated linkage between the α- and β-carbon atoms by the loss of two atoms of hydrogen. This is not possible in atrolactic acid but it is in hydratropic acid and would result in the formation of atropic acid. Atropic acid is not resistant to oxidation and although it may be urged that it is more toxic than hydratropic acid and therefore is not a likely intermediate product in the oxidation of the latter, it has to be borne in mind that the toxicity of a substance when produced slowly by metabolic processes in the tissues may be much less marked than when a comparatively large dose is introduced quickly by subcutaneous injection. Hydratropic acid, also, is distinctly more toxic than atrolactic and tropic acids. In the first case, doses of 0·25 g. per kilo give toxic effects whereas 1 g. per kilo of either of the other two acids gives no toxic symptoms. For these reasons therefore we are led to believe that atropic acid is the most likely primary oxidation product of hydratropic acid. It follows from this that the change from hydratropic acid to atropic acid is brought about by the direct oxidative removal of two hydrogen atoms and not by the loss of water from either tropic or atrolactic acids, since these are apparently not capable of conversion to atropic acid in the body. If they were they would be oxidised to a greater extent than was found to be the case. It is also possible that atropic acid may be produced from hydratropic acid by a process of dehydrogenation as pictured by Wieland [1912].

As to the further stages in the oxidation of hydratropic and atropic acids we can only speculate. From analogy with cinnamic acid β-oxidation might be expected with the formation of formylphenylacetic acid (V):

$$\underset{C_6H_5-\overset{||}{C}-COOH}{\overset{CH_2}{}} \longrightarrow \underset{C_6H_5-\overset{|}{C}-COOH}{\overset{CH(OH)}{}} \longrightarrow (V)\ \underset{C_6H_5-\overset{|}{C}H-COOH}{\overset{CHO}{}}.$$

This would be unstable and lose carbon dioxide to give phenylacetaldehyde, or by further oxidation it might give phenylmalonic acid. The fate of the latter substance in the body is not known, but Dakin [1909] has described experiments in which phenylacetaldehyde in dilute alcoholic solution was given subcutaneously to dogs. About 84 % was completely oxidised and the remaining 16 % converted into phenylacetic acid which was excreted in the urine as phenaceturic acid. We have repeated and confirmed Dakin's experiment and found that by oral administration also a considerable part of the phenylacetaldehyde administered undergoes complete oxidation. Since neither hydratropic nor atropic acid gave rise to the excretion of phenylacetic acid it is not certain that phenylacetaldehyde represents an intermediate stage in their oxidation, although it would be expected to be such if the original hypothesis put forward previously [Raper, 1914] were correct. It is however not improbable that phenylacetaldehyde produced *in vivo* may undergo complete oxidation in spite of the fact that when introduced orally or subcutaneously it is partially oxidised to phenylacetic acid.

In a recent communication, Hanke and Koessler [1922] have called attention to the phenomena of the oxidation of fatty acids in the body in relation to their electronic structure. The arguments put forward in the case of butyric acid, for which the following electronic structure is developed, will serve as an example:

$$-\overset{-}{C}- +\overset{-}{C}+ -\overset{-}{C}- +\overset{+}{C}+$$
$$\quad -\quad -\quad -\quad +$$
$$\gamma\quad \beta\quad \alpha$$

In this formula the α- and γ-carbon atoms are represented as quadruply negative and the carboxylic carbon atom as quadruply positive, whereas the β-carbon atom is doubly positive. Evidence is brought forward by Hanke and Koessler which indicates that quadruply positive or negative carbon, and especially the former, is resistant to oxidation. This being so, it would be expected that butyric acid when oxidised would be attacked most readily at the β-carbon atom which is only doubly positive. It is therefore suggested that in this way the predominance of β-oxidation of fatty acids in the body may be explained. On our part, we feel that there is, as yet, insufficient knowledge, either qualitative or quantitative, and especially the latter, of the mechanism of oxidation processes on which an explanation in relation to their electronic structure can be based. The idea that β-oxidation is predominant is supported essentially by qualitative evidence for there is little of a quantitative nature. There are no animal experiments on record which enable one to postulate that butyric acid is oxidised only or chiefly in the β-position. Further, Cahen and Hurtley [1917] have shown that when butyric acid is oxidised with hydrogen peroxide, a process which simulates to some extent oxidation in the body, the main product is succinic acid and not acetone, so that oxidation in this case is most marked at the γ-carbon atom. There is also evidence which indicates that in certain cases, β-oxidation, when it occurs, may be preceded by "desaturation" between the α- and β-carbon atoms. The oxidation of hydratropic acid to produce atropic acid, as suggested above, is a case in point, but others are: the production of cinnamic acid from phenylpropionic acid [Dakin, 1909] and the formation of furfurylacrylic acid from furfurylpropionic acid [Sasaki, 1910; Friedmann, 1911]. Again, the electronic formulae of β-phenyl-β-hydroxypropionic acid and β-phenyl-propionic acid, developed by Hanke and Koessler, show that in both compounds the β-carbon atom is doubly positive and the α- and carboxylic-carbon atoms quadruply negative and quadruply positive respectively. If the ease of oxidation in the body of a particular carbon atom of a substance is sufficiently explained by the electronic formula then there would be no reason to expect that these two substances should not undergo oxidation equally readily. As a matter of fact the non-hydroxylated acid is much more readily oxidised. The experiments with hydratropic and tropic acids, described in the present communication, show equally marked differences and do not seem to be capable of adequate explanation on an electronic basis alone. It would .seem that until we have a more definite knowledge of the mechanism of the oxidation process, its explanation, based solely on views of polarity in the substance oxidised, or on its electronic structure, cannot be entirely successful. These views, developed largely by a study of the phenomena of addition and substitution in organic compounds, have given a satisfactory general explanation of previously inexplicable phenomena; but with the meagre quantitative data at our disposal and our lack of knowledge as to how oxidation of very simple substances takes place either inside or outside the body, the polarity alone of the atoms in the substance oxidised is insufficient as a basis of explanation. As an instance of the probable complexity of an apparently simple process such as the oxidation of carbon monoxide to carbon dioxide when the former is burnt in moist air, the work of von Wartenberg and Sieg [1920] may be referred to. It is suggested that in this oxidation four intermediate reactions are concerned. Firstly the production of formic acid by the addition of water, secondly its dehydrogenation to produce carbon dioxide and hydrogen, thirdly, oxidation of the hydrogen to produce hydrogen peroxide and lastly decomposition of the peroxide with the formation of water and oxygen. If the oxidation of carbon monoxide is as complicated as this scheme represents, then it is at least probable that the processes of oxidation of fatty acids are not less complex.

EXPERIMENTAL.

Methods. Unless stated otherwise, the acids used in this investigation were administered hypodermically as sodium salts in aqueous solution. The urine was collected from the time of the first dose until 36 hours after the last dose was given. Urines were preserved on ice. Volatile products were sought for by distilling all, or part of the urine in steam and examining the distillate. The urine was then concentrated under reduced pressure to about one-fifth of its original volume and acidified strongly with syrupy phosphoric acid. Subsequently it was extracted with ether in a continuous extractor for 30 to 48 hours. The ether extract was then examined for unchanged acid or products of oxidation.

Hydratropic acid. The acid was obtained in two ways. Firstly, by the reduction of atropic acid with sodium amalgam and secondly, by the oxidation of hydratropic aldehyde with silver oxide in the presence of a slight excess of sodium hydroxide. The hydratropic aldehyde was prepared by the method of Tiffeneau [1907]. The former method was found to be the more expeditious. The acid was purified by distillation under reduced pressure and had B.P. 148–9° (14 mm.). In doses of 0·25 g. per kilo a slight lassitude was developed but no other ill effects were noted. Dogs, only, were used in this experiment and larger doses were not tried. No volatile products of oxidation were detected when the urine was distilled with steam. The ether extract from the acidified urine was obtained as a light brown syrup which refused to crystallise. Extraction with warm light petroleum removed some unchanged hydratropic acid and traces of fatty acids. The latter were removed as insoluble barium salts and the hydratropic acid recovered was weighed. In this way, in two experiments, 3 and 22 % of the acid originally given was recovered. Its equivalent was 151 (calculated 150), and it was dextrorotatory. In absolute alcohol (concentration 3 %), it had $[\alpha]_D^{20} + 21 \cdot 2°$ to $+ 26 \cdot 9°$. The ether extract, after removal of the hydratropic acid as just outlined, was treated in many ways to induce it to crystallise but all were fruitless. It was dextrorotatory, reduced Fehling's solution on boiling and gave the orcinol reaction, indicating the presence of a glycuronic acid derivative. The original urine also gave these reactions. In addition, the ether extract contained some nitrogenous substance. Having failed to obtain any crystalline compound from this extract, and suspecting the presence of a glycine derivative because of the marked nitrogen content, it was hydrolysed by boiling with concentrated hydrochloric acid for an hour. After hydrolysis, the solution was diluted and extracted with light petroleum and then with ether. The latter yielded only a trace of oily substance which was neglected, but the former was found to contain dextrorotatory hydratropic acid (equivalent 150·8). In 3 % solution in absolute alcohol it gave $[\alpha]_D^{20} + 27 \cdot 2°$. The hydrolysate after ether extraction was evaporated to small bulk under reduced pressure and esterified with absolute alcohol saturated with hydrochloric acid. It was seeded with a crystal of glycine ester hydrochloride and placed on ice for two days, whereby a mass of crystals of glycine ester hydrochloride was obtained. The identity of the glycine was confirmed by benzoylation, when hippuric acid, M.P. 186–7°, was obtained. Careful search failed to reveal any other definite products of hydrolysis, phenylacetic, tropic and atropic acids being specially sought for. The chief substance present in the ether extract of the urine thus appeared to be hydratropylglycine. A sample of this substance was made by the Schotten-Baumann method from hydratropyl chloride and glycine and was found to be more difficult to crystallise than the commoner acyl derivatives of glycine such as phenaceturic and hippuric acids, and its contamination with some oily impurity, possibly a glycuronate of hydratropic acid, with about the same solubilities, explains its non-crystallisation from the ether extract. In a later experiment it was found possible to obtain crystalline hydratropylglycine from this extract with considerable loss, by the following method. The acids extracted from the urine by ether, after removal of unchanged hydratropic acid, were converted in turn into sodium salts and ferric salts. The acids obtained from the latter by decomposition with dilute sulphuric acid and extraction with ether were mixed with half their equivalent of sodium hydroxide and the free acids present in the solution shaken out with ether. The remaining sodium salts were now decomposed with dilute sulphuric acid and the liberated acid extracted with ether. This extract crystallised readily on inoculation with hydratropyl glycine and after two recrystallisations from water had M.P. 103°, which agreed with that of the synthetic product made from the dextrorotatory hydratropic acid recovered from the urine, and glycine. The pure inactive glycine derivative has M.P. 105°. In one experiment carried out as quantitatively as possible for the estimation of the hydratropic acid excreted, either free or combined, three doses, each of 1·5 g., of the acid were given on three successive days. It was found that 0·156 g. (= 3·5 %) of the acid administered was present in the free state and 1·225 g. (= 27 %) in combination with glycine, or possibly partly with glycuronic acid. In order to determine whether any derivative was present in the urine as a glycuronate the concentrated urine after the usual ether extraction was freed from phosphoric acid by baryta and the reducing substance precipitated by basic lead acetate. This precipitate was decomposed with hydrogen sulphide and to the resulting solution, after removal of hydrogen sulphide in a current of air, dilute sulphuric acid was added to make the concentration about 5 %. It was boiled for eight hours and the resulting solution, after clarification with charcoal, extracted continuously with ether for eight hours. The ether extract contained only a trace of resinous substance. The reducing substance was probably, therefore, free glycuronic acid which had originally been combined with hydratropic acid and which had been split off during the ether extraction of the concentrated and strongly acidified urine. It is well known that many glycuronates are easily hydrolysed under such conditions.

Tropic acid. Inactive tropic acid was used in these experiments and its fate in both cats and dogs was determined. Doses of 1 g. per kilo caused no noticeable ill effects. The urine on steam distillation yielded no volatile oxidation products. The ether extract of the concentrated urine, on removal of the ether, gave a brown mass which crystallised at once on cooling and after one recrystallisation gave pure tropic acid, M.P. 118–119°. The acid was optically inactive. In two preliminary experiments, one on a cat and the other on a dog, the urine was extracted with ether for only 12 hours. In the first, 2·8 g. of the acid were given and 2·25 g. (80 %) of the pure acid were recovered from the urine. In the second, 5·9 g. of the acid yielded 4·4 g. (74 %) in the urine. A third experiment was carried out and the extraction with ether continued for 30 hours, when all the acid had been extracted. Of 2·94 g. of the acid administered, 2·77 g. were recovered, *i.e.* 94 %. Examination of the mother liquors after the separation of the tropic acid failed to reveal the presence of any oxidation products.

Atrolactic acid. The inactive acid was prepared from acetophenone by Spiegel's method using the modification described by McKenzie and Clough [1912]. Both dogs and cats were used for the experiments and doses of 0·9 g. per kilo were well tolerated, no ill effects being noted. No volatile oxidation products were detected in the steam distillate. The residue from the ether extract crystallised on cooling and was extracted with hot benzene to remove unchanged atrolactic acid. Most of the residue was soluble in benzene but a small amount of brown tarry substance remained after the extraction. This was dissolved in water and re-extracted with ether. The ether residue was then repeatedly extracted with moist light petroleum whereby a further quantity of atrolactic acid was obtained. A small amount of brown oily substance remained which was insoluble in cold water, benzene or light petroleum and would not crystallise. The only definite substance obtained from the urine, therefore, was atrolactic acid. It was optically inactive and melted at 92–3° after one recrystallisation from water. In one experiment on a dog, 5·8 g. of the acid were administered and 4·9 g. (84 %) recovered from the urine. In another case two cats were used and of 4·5 g. of the acid administered, 3·42 g. (76 %) were recovered from the urine.

Atropic acid. The acid used in these experiments was prepared from tropic acid by boiling with 10 % aqueous baryta for 18 hours. It was obtained quite pure after one crystallisation from alcohol and had M.P. 106–7°. The yield was 67 % of the theoretical. Doses of 0·4 g. per kilo were invariably fatal both in cats and dogs within 30 hours and no oxidation products or unchanged acid were found in the urine. A dose of 0·2 g. per kilo, given to a dog, produced only slight lassitude, but when repeated after 24 hours caused death at the end of 60 hours. Doses of 0·13 g. per kilo were well tolerated. In one instance a dose of 0·18 g. per kilo, given to a dog, caused haematuria but the animal recovered. The urine on steam distillation yielded no detectable oxidation products. The residue of the ether extract was very small in amount. It was dissolved in hot water, clarified with charcoal, and on standing deposited a small amount of a colourless, crystalline substance. In one experiment in which 1·9 g. of the acid were given, 0·04 g. of this crystalline substance was obtained, and in another, 2·8 g. yielded 0·05 g. It was recrystallised from a mixture of nine parts benzene and one part alcohol and had M.P. 178–9°. On heating above its melting point it gave a crystalline sublimate. It was soluble in water and alcohol, slightly in ether and almost insoluble in chloroform and benzene. It gave insoluble ferric and lead salts but a soluble calcium salt. When recrystallised from dilute nitric acid it melted at 183°. The ammonium salt heated with zinc dust gave the pyrrole reaction. These properties agree with those of succinic acid but enough material was not obtained for analytical identification. No other substance was isolated from the urine. A control experiment with a 48 hours sample of urine from the same dog to which atropic acid had been given yielded none of the crystalline acid on ether extraction.

Phenylacetaldehyde. In a mixture of four parts alcohol to six parts water (by volume), 0·93 g. of phenylacetaldehyde was dissolved and given subcutaneously to a cat weighing 2 kilos. The cat appeared to be affected by the alcohol and was drowsy for two days, but recovered. The urine was collected for four days and on steam distillation yielded no detectable oxidation products. The ether extract was dissolved in water, distilled with steam to remove acids, again acidified with phosphoric acid and extracted with ether. The ether extract yielded a small amount of oily substance which crystallised after standing for two days. The mass was rubbed up with a little water and filtered. In this way 0·26 g. of a crystalline acid was obtained. After re-crystallisation from water it gave a very unsharp M.P., the main part melting at 135–140° and the rest at 170–180°. It appeared therefore to be phenaceturic acid mixed with some higher melting substance. By boiling with 20 % sulphuric acid for a short time the phenaceturic acid was hydrolysed and the resulting phenylacetic acid was extracted with benzene. It weighed 0·102 g. and was not quite pure. The higher melting substance was extracted by ether from the acid liquid and proved to be hippuric acid, which is more resistant to hydrolysis with hot dilute acids than phenaceturic acid. In this experiment therefore, not more than 10 % of the phenylacetaldehyde given appeared as phenaceturic acid. A second experiment was carried out on a dog to test the effect of oral administration, the use of alcohol as a solvent for subcutaneous injection being objectionable. 1·9 g. of phenylacetaldehyde in its own volume of alcohol were administered to a dog by the mouth in gelatin capsules. No abnormal symptoms were observed. The urine was collected for the ensuing 48 hours and was treated in the same way as that from the cat. A fairly pure crop of crystals of phenaceturic acid was obtained (M.P. 141–3°). It weighed 0·85 g. The mother liquors on hydrolysis yielded 0·1 g. of impure phenylacetic acid and again a little hippuric acid was isolated. The phenaceturic and phenylacetic acids thus obtained correspond to 0·62 g. of phenylacetaldehyde, which represents 32 % of that administered. Nearly 70 % had therefore been completely oxidised. These experiments thus confirm those of Dakin [1909].

Summary and Conclusions.

1. The experiments described in this paper were carried out to determine the mode of oxidation in the body of phenyl derivatives of fatty acids with a branched chain structure.

2. The acids investigated were, hydratropic, tropic, atrolactic and atropic acids. In addition previous observations by Dakin on the fate of phenylacet-aldehyde were confirmed.

3. Hydratropic acid is moderately well oxidised and undergoes a partial resolution in the body, the unoxidised acid found in the urine being dextro-rotatory. Atropic acid is also well oxidised and is more toxic than the other three acids. Tropic and atrolactic acids are comparatively resistant to oxidation.

4. From the observations made, it is concluded that hydratropic acid on oxidation is converted directly into atropic acid (though this was not directly proved), and that this then undergoes complete oxidation.

5. Phenylacetaldehyde largely undergoes complete oxidation in the body and may therefore be an intermediate product in the oxidation of hydratropic and tropic acids.

We desire to express our thanks to Messrs Boots Pure Drug Co. Ltd., for a gift of residues from the saponification of atropine. From these the tropic, atropic, and part of the hydratropic acids used in our experiments were obtained.

REFERENCES.

Cahen and Hurtley (1917). *Biochem. J.* **11**, 164.
Dakin (1909). *J. Biol. Chem.* **7**, 203, 242.
Friedmann (1911). *Biochem. Zeitsch.* **35**, 40.
Hanke and Koessler (1922). *J. Biol. Chem.* **50**, 193.
McKenzie and Clough (1912). *J. Chem. Soc.* **101**, 393.
Raper (1914). *Biochem. J.* **8**, 320.
Sasaki (1910). *Biochem. Zeitsch.* **25**, 272.
Tiffeneau (1907). *Ann. Chim. Phys.* (8) **10**, 176.
von Wartenberg and Sieg (1920). *Ber.* **53**, 2192.
Wieland (1912). *Ber.* **45**, 484, 679, 2606.

XXXVIII. STRUCTURES IN ELASTIC GELS CAUSED BY THE FORMATION OF SEMIPERMEABLE MEMBRANES.

By EMIL HATSCHEK.

(*Received April 20th, 1922.*)

WHEN a reaction is produced in a gel, the latter may undergo slight changes which, however, are not striking or even obvious, and have received little attention compared with that devoted to the characteristics of insoluble precipitates obtained in the gel. These changes are, generally speaking, alterations in the water content of the gel, such as dehydration by contact with very concentrated salt solutions, *e.g.* the 20 or 25 % silver nitrate solution used in the classical Liesegang experiment; or increased hydration caused by the diffusion of acid or of one of the strongly lyotropic salts such as iodide or thiocyanate. If the experiments are carried out in test-tubes, these changes show themselves most readily in alterations of the gel meniscus.

Much more complicated conditions arise, and the gel itself develops a complicated structure, when the reaction leads to the formation of a semipermeable membrane. The most convenient and striking instance is the formation of copper ferrocyanide in gelatin gel, and the most suitable arrangement is as follows: clean test-tubes are filled to about half their height with a gelatin sol containing 10 g. of air dry gelatin in 100 cc. of a 2 % solution of crystallised potassium ferrocyanide ($K_4Fe(CN)_6.3H_2O$). After setting completely—which is somewhat retarded by the ferrocyanide—the gel is covered with a 5 % solution of crystallised cupric chloride or cupric sulphate ($CuCl_2.2H_2O$ or $CuSO_4.5H_2O$ respectively), or, as will be explained below, with a mixed solution of cupric and sodium sulphate. The molar concentrations are accordingly $M/21$ $K_4Fe(CN)_6$ in the gel, and $M/3.4$ $CuCl_2$ or $M/5$ $CuSO_4$ in aqueous solution. The molar concentration of the latter is thus considerably greater than the molar concentration of the salt in the gel, the necessary condition when a reaction producing an insoluble compound is to proceed into the gel.

A few minutes after putting on the copper salt solution a pale brown and perfectly transparent membrane of copper ferrocyanide can be noticed. The next change to be noticed is a gradual sinking of the surface of the gel. In the first stages there is no great alteration in the curvature of the meniscus, but it appears to travel down bodily, leaving a thin film of gel adhering to the glass. The constant formation or exposure of new surface entails fresh formation of copper ferrocyanide, and the layer of gel on the glass as well as the meniscus turns dark brown and opaque. If the preparation is observed superficially only and at long intervals, this shifting of the meniscus and the formation of a dark brown zone on the glass seem to be all that happens. Careful observation however shows the process to be discontinuous: the curvature of the meniscus increases and finally the membrane composing it *bursts* somewhere near the centre, the copper salt solution penetrates through the crack and a fresh membrane is formed.

What happens can be readily deduced *a priori* from known properties of the semipermeable membrane and of elastic gels. As soon as the membrane is formed, water flows through it from ferrocyanide solution in the gel to the strongly hypertonic copper salt solution above it: in other words, the gel "dries." In doing so it behaves exactly as if drying proceeded in the usual way by evaporation into the atmosphere, *i.e.* the meniscus gradually sinks and becomes more and more curved. The surface film now consists of very concentrated gel and—provided the gelatin adheres well to the glass—is in tension. At the same time, owing to the withdrawal of water and the low diffusion velocity of the potassium ferrocyanide, with a molecular weight of 368·5, the concentration of the latter in the zone immediately below the membrane increases markedly; this can easily be seen by the colour, which is a clear yellow, although the rest of the gel, with the initial concentration of 2 %, is barely coloured. The ferrocyanide has, like the other cyanides, a marked softening effect on gelatin, which shows itself in concentrations of about 10 %, so that the layer below the tough surface film is considerably softer than the body of the gel. The surface film accordingly is torn off when the tension has increased sufficiently, and is at the same time ruptured by the atmospheric pressure, the copper salt solution penetrating under it. The whole process then repeats itself, and this continues until the copper salt solution, which is continually diluted, becomes approximately isotonic with the potassium ferrocyanide solution in the gel.

If the gelatin does not adhere to the glass wall sufficiently, it begins to contract below the reaction zone and copper salt solution flows into the space thus formed, covering the exposed gelatin surface with a semipermeable membrane. This leads to further shrinkage and finally a thin cylinder of very tough gel only may be left.

The analysis of the process, set forth above, was confirmed by a detailed examination of a large number of preparations. In the first instance the test-tubes were cut through about 1 cm. below the bottom of the reaction zone. Immediately the cylinder of gel is cut across, it contracts sharply and develops a deep constriction below the last membrane. The dehydrated gel containing the precipitate of copper ferrocyanide no longer adheres to the glass, and the reaction zone can, with a little care, be removed intact from the tube and cut through longitudinally. Two halves of such preparations, made in large boiling tubes 35 mm. in diameter, are shown in Fig. 1 (Plate IV).

An intact test-tube, in which the reaction has proceeded for some days, is illustrated in Fig. 2. The rupture in the last membrane but one is clearly visible, and the membrane in course of formation, which is still almost transparent, can be seen faintly. Fig. 3 shows a preparation in which the gel has detached itself from the glass and has become covered completely with copper ferrocyanide; the consequent shrinkage is very marked.

If the reaction proceeds throughout the gel, without the latter detaching itself, the final aspect of the preparation is extremely remarkable. It then consists of a thin tube of tough gel with a large number of septa, convex towards the lower end of the cylinder and always ruptured somewhere near the centre. A cross section through a specimen of this description is shown in Fig. 4. It may be mentioned that this preparation was obtained by using a mixed solution of copper and sodium sulphate, to which reference has been made, instead of copper sulphate alone. The function of the copper salt is a twofold one: it provides material for the formation of the semipermeable membrane, and it maintains the osmotic pressure so that dehydration of the gel takes place. The latter function can of course be performed equally well by a salt which does not enter into the reaction, and sodium sulphate has the advantage of not tanning the gelatin to the same extent as copper and other heavy metal sulphates. It is the latter action which seems to lead, or to contribute largely, to the detaching of the gel from the glass, which makes it difficult to obtain specimens showing more than two or three septa with copper salt alone. Incidentally the condition of the glass surface is of considerable importance: test-tubes cleaned carefully with bichromate-sulphuric acid mixture etc. consistently gave bad results, while new test-tubes simply dusted with a dry cloth gave the best.

If the explanation of the formation of septa, which has been given above, is correct, it follows that they should be formed in the course of any reaction which satisfies the following three conditions: (1) a semipermeable membrane is produced; (2) the solution above the gel is, and remains, hypertonic, and (3) the gel contains a lyotropic agent which, in moderately high concentration, reduces its modulus and tensile strength. Only one such combination—apart from the copper ferrocyanide reaction—readily suggests itself: gelatin gel containing potassium iodide as lyotropic agent, and on it a concentrated sugar solution containing tannin. The general results were the same as with copper ferrocyanide: depression of the meniscus, followed by the detachment and rupture of membranes. The latter were very irregular, probably because one set only of concentrations, chosen at random, was tried; a detailed investigation appeared unnecessary in view of the somewhat artificial nature of the arrangement.

The results described show that the possibilities of periodic structures in gels, as the result of continuous and uncontrolled diffusion, are by no means exhausted by the normal Liesegang phenomenon. In the latter the gel, while of course influencing the character of the results very considerably, primarily serves as the support for periodic deposits of insoluble reaction product, and does not undergo any profound changes. If a semipermeable membrane is formed by the reaction, the gel itself becomes segmented and the final structure is one which it would be extremely difficult to explain *a posteriori*—quite as difficult as it was to explain other periodic structures before the study of the Liesegang rings provided a clue. While the author does not know any natural structure resembling those described and while he is, generally speaking, by no means inclined to overrate the cogency of the conclusions drawn from artificial imitations of natural forms, he ventures to think that the highly complicated consequences which follow the formation of a semipermeable membrane in a suitable elastic gel may have some biological or histological interest.

47

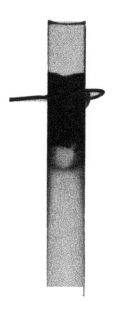

Fig. 2

Fig. 3

Fig. I

Fig 4

XXXIX. A MODIFICATION OF BASAL DIET FOR RAT FEEDING EXPERIMENTS.

BY MURIEL BOND.

*From the Physiological Laboratory, London (Royal Free Hospital)
School of Medicine for Women.*

(*Received April 21st, 1922.*)

CASEINOGEN, the usual protein constituent of a basal diet for rat feeding experiments, possesses several disadvantages. Owing to its natural association with milk fat it is difficult to render it free of fat-soluble *A*—if the purification be done by alcohol and ether extractions the cost of the material is greatly increased, and if the caseinogen is treated by the Drummond method [Coward and Drummond, 1921], careful supervision is required to ensure a proper oxygenation and a constant temperature. In addition it often seems impossible when using caseinogen to obtain a deficiency curve, although apparently no fat-soluble *A* is present in the diet. The rats continue to grow steadily and seem apparently healthy.

In view of these difficulties which were experienced while investigating the fat-soluble *A* content of certain foods, some rat feeding experiments were started, in which for the caseinogen dried egg white was substituted. Dried egg white, unlike caseinogen, contains very little associated fat and if it is an adequate protein when used in a percentage similar to that normally used for caseinogen, would be considerably cheaper.

The first series of experiments was done with a diet containing 30 % protein—instead of the usual 20 %—as analyses of egg white have shown it to be somewhat deficient in tyrosine.

When the diet was first used the dried egg white was crushed and the dry protein mixed with the other ingredients of the diet. It was found, however, that on this diet many of the rats developed diarrhoea and little or no growth was obtained. A diet was then tried in which the dried egg white was soaked overnight in tap water and the solution gently heated to coagulate the proteins. While heating the solution it was constantly stirred and the broken clot was then mixed with the other constituents. A fairly solid mass was thus obtained, which was readily taken by the rats and was satisfactorily digested[1].

The experiments were carried out on young healthy albino rats, weighing about 40–60 g. at the beginning of the experiments (see Fig. 1). Usually the

---

[1] Subsequently I read Bateman's paper [1916] on the digestibility of egg proteins and find that his results and mine are completely in accord.

young rats were fed on bread and milk for about a week previously and any animal not showing a decided gain in weight during that time was discarded. The rats were fed twice daily—usually at 9 a.m. and 3 p.m.— and were weighed twice a week. The amount of food given was so arranged that there was still a little left uneaten when a fresh quantity was due.

Each litter was divided into two groups. One group received diet A and the other group diet B.

	Diet A	Diet B
Soaked and cooked egg white	30 g.	30 g.
Salts (Hopkins' mixture) ...	5 g.	5 g.
Marmite	5 g.	5 g.
Strained lemon juice	5 cc.	5 cc.
Potato starch	40 g.	40 g.
Fat { butter	15 g.	—
{ hardened cotton seed oil	—	15 g.
Water to drink	ad lib.	ad lib.

The rats on the B diet showed a growth curve, which in a few weeks began to flatten out. These rats also frequently suffered from eye inflammations. The rats on the A diet grew well, appeared to be healthy and when they were mated reproduction was obtained (Fig. 1). In this series of experiments 20 rats were used.

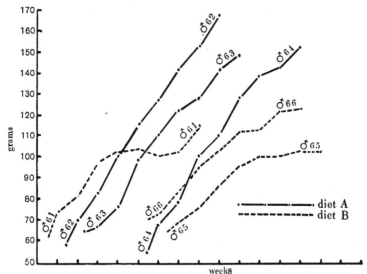

Fig. 1. Litter of 6. Feeding extending over 8 weeks, A and B diets.

The results of feeding the lactating rat and her growing young on the egg white diet are not very conclusive. One litter has been successfully reared on this diet, but as a rule it was found that the young died about the 20th–30th day after birth, and the mother became ill unless the diet was changed to one of bread and milk as soon as the young rats began to move about the cage.

This is probably due to the large proportion of protein in the diet [Hartwell, 1921, 1922]. Hartwell has shown 30–40 % protein to be excessive during lactation, although the experiments described by her are somewhat different from those described here.

1. In the experiments described in this paper the diet used for the lactating rats always contained all the necessary food factors and had been given to the mother rats during growth and gestation.

2. The egg white used has been soaked and cooked, whereas in the cases quoted by Hartwell, the protein was used crushed, dry and uncooked. This condition of the egg white in itself frequently leads to diarrhoea and death in non-lactating and non-sucking rats.

3. The unhealthy symptoms were not observed as early as in her experiments—the young rats always having begun to move about the cage and get food for themselves before they began to show signs of weakness. The characteristic spasms and screaming fits observed by Hartwell were never noticed, but the young rats became less active, were cold and died after a period of coma. Weighing of the young rats was not carried out so that it is not possible to say when growth ceased.

A second series of experiments is now in progress, in which the rats are receiving a diet containing only 20 % egg white. This is a diet comparable to that normally used, but with the substitution of egg white for caseinogen. The growth of rats on the 20 % diet is being compared with that of rats on the 30 % diet and so far no signs of inadequacy can be detected among the rats on the lower protein diet. There has not yet been time to ascertain whether this 20 % diet is adequate for reproduction nor whether it is more satisfactory than the 30 % diet during lactation.

A point in favour of egg white as a source of protein seems to be the constancy with which, when this material is used, the characteristic flattening of the growth curve is obtained if the diet be deficient in fat soluble A—a constancy not always obtained when caseinogen is used.

REFERENCES.

Bateman (1916). *J. Biol. Chem.* **26**, 263.
Coward and Drummond (1921). *Biochem. J.* **15**, 531.
Hartwell (1921). *Biochem. J.* **15**, 140, 563.
—— (1922). *Biochem. J.* **16**, 78.

XL. SYNTHESIS OF VITAMIN *A* BY A MARINE DIATOM (*NITZSCHIA CLOSTERIUM* W.Sm.) GROWING IN PURE CULTURE.

By HENRY LYSTER JAMESON, JACK CECIL DRUMMOND AND KATHARINE HOPE COWARD (*Beit Memorial Research Fellow*).

From the Biochemical Laboratories, Institute of Physiology,
University of London.

(*Received April 28th, 1922.*)

DURING the early stages of an investigation of the nutrition of certain species of marine molluscs it became necessary to gain some information concerning the food value of a number of marine plants upon which the shell-fish directly or indirectly feed[1].

Amongst other dietary factors to be studied that known as vitamin *A* claimed our early attention. By now it is generally understood that the higher land animals draw their supplies of this indispensable factor directly or otherwise from green plants, since their own tissues do not appear to possess the power to synthesise this organic complex. That a parallel condition would exist in the sea appeared to us to be highly probable from the fact that certain marine algae had been found to share with land plants the power to synthesise vitamin *A* [Coward and Drummond, 1921]. The algae examined in the former work were Fucus, Ulva, Cladophora, Polysiphonia and Enteromorpha, but these are not truly representative of the ultimate food supply of the species of molluscs with which we were working. We therefore decided to test whether a typical unicellular marine alga can synthesise vitamin *A*.

To obtain the necessary material it was decided to grow the organisms in pure culture following the technique described by Allen and Nelson [1910]. Pure cultures of three common marine diatoms, *Nitzschia closterium* W. Sm. *forma minutissima, Thalassiosira gravida,* and *Skeletonema* were very kindly placed at our disposal by Dr E. J. Allen, F.R.S., Director of the Marine Biological Laboratory, Plymouth.

To grow the relatively large quantities of the organisms required for the feeding tests was not a simple matter, and in the case of two of the three species named our attempts failed owing to their slow rate of growth. With Nitzschia, however, we found it a relatively simple matter to prepare ample

---

[1] Unfortunately owing to the sudden and untimely death of our colleague Dr H. Lyster Jameson the main line of this work has been discontinued. The present communication records the chief result which we had obtained up to the time of his death, by which science and particularly marine biology suffers a grievous loss. J. C. D.

quantities of the diatom. The cultures were prepared by carefully isolating a few of the organisms from the original Plymouth sample and allowing them to grow in small flasks containing Miquel's culture fluid or sterilised sea water (see reference to Allen and Nelson's work). At the end of a month the bottom of the flask was covered with a thick brown growth of the diatom, but, as Allen and Nelson also observed, the cultures do not thrive well after more than a month so that at that stage the cultures were transferred to a number of large Kilner jars of 4-litre capacity containing the sterile culture fluid. The covered jars were allowed to stand in diffuse daylight in front of a window facing north, and the growth was accompanied by quite an appreciable evolution of oxygen on the brighter days. Frequent microscopical examination of the cultures demonstrated their freedom from contamination. By repeated sub-culturing in this manner large amounts of the organism were prepared without much trouble.

We were at first doubtful whether the administration of the fresh diatoms to rats in the usual manner of testing for the presence of vitamin *A* would be of any value but decided to test this method before going to the trouble of preparing by chemical means a special fraction. Somewhat to our surprise the organisms were digested during their passage through the alimentary canal of the rats as was indicated by the recovery of growth and by the presence of empty frustules in the faeces. The usual method of testing was followed [Drummond and Coward, 1920], the supplement being administered before the daily ration of basal diet was given. To prepare the organisms for administration the fluid was decanted off from a month-old culture and the thick flocculent growth of diatoms on the bottom filtered on to a thin layer of starch at the bottom of a Gooch crucible. After washing with a very small quantity of distilled water the almost dry mixture of starch and organisms was thoroughly mixed and an aliquot part given to the animals who invariably consumed it with readiness.

. The organisms although possessing a deep brown colour whilst in culture assumed a dark green tint on drying. The total fat content (ether extract) was 5·25 % of the dry weight, and in addition to the characteristic pigment of the brown algae, fucoxanthin, there were also present chlorophyll and relatively large amounts of the lipochrome pigments carotene and xanthophyll.

Although it was impossible accurately to measure the dose of Nitzschia given daily an approximation was made on the basis of the average dry weight of a month-old culture which was 0·32 g.

The curves shown in Fig. 1 show that this diatom may serve as a source of vitamin *A*, the resumption of growth being especially marked at the higher dosage of approximately 0·08 g. (dry weight) per day.

The extraordinary potency of these diatoms as a source of the vitamin *A* may be judged from the fact that this ration of diatom represents approximately 4 mg. of fat; which recalls the observations of Zilva and Miura on the potency of cod liver oils [1921].

It is significant that this organism promoted growth very much more effectively than did the common seaweeds we previously tested. This difference may possibly be accounted for by taking into account the much greater relative surface of the diatoms.

The enormous stores, relatively speaking, of vitamin A which are found in the tissues of certain marine animals lead one to conclude that they are derived from some highly potent source of that factor such as is represented by the diatom we have been studying.

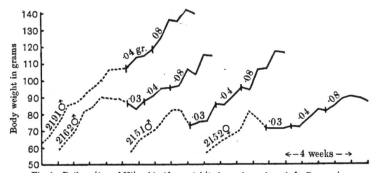

Fig. 1.   Daily ration of Nitzschia (dry weight) shown in each period. Preparatory period shown by dotted line.

We essayed to prove the transference of the vitamin from the diatoms to the molluscs we were investigating but owing to the sad death of our colleague the experiments were carried no further than the demonstration that many species of molluscs are valuable sources of the A factor. The production of vitamin A in green land plants capable of carbon assimilation [Coward and Drummond, 1921] is therefore paralleled by a similar synthesis in marine plants containing photocatalytic pigments. Further, the fundamental dependence of terrestrial animal life on the fresh green leaf for its supplies of vitamin A would appear to be paralleled by a similar dependence of marine animals on the marine flora, particularly the microscopic plants.

SUMMARY.

1. Pure cultures of a common marine diatom, *Nitzschia closterium*, grown in Miquel's solution or sterilised sea water synthesise large amounts of vitamin A.

2. A parallel is drawn between the dependence of land animals on fresh green leaves and that of marine animals on the synthetic activity of the marine flora for their supplies of vitamin A.

3. A number of molluscs were found to contain considerable amounts of vitamin A.

The greater part of the expense incurred in this research was defrayed out of a grant made by the Medical Research Council to whom our thanks are due. We also wish to express our appreciation of the assistance extended to us by the Ministry of Agriculture and Fisheries.

REFERENCES.

Allen and Nelson (1910). *Quart. J. Micros. Sci.* **55**, 361.
Coward and Drummond (1921). *Biochem. J.* **15**, 530.
Drummond and Coward (1920). *Biochem. J.* **14**, 661.
Zilva and Miura (1921). *Biochem. J.* **15**.

XLI. THE PHOSPHOLIPIN OF THE BLOOD AND LIVER IN EXPERIMENTAL RICKETS IN DOGS.

By JOHN SMITH SHARPE.

From the Institute of Physiology, University of Glasgow.

(Received May 1st, 1922.)

IN a paper in which he advances a theory of the metabolism in rickets, Prof. Noël Paton [1922] points out that recent work seems to indicate that the supply of phosphorus rather than of calcium may be the limiting factor in ossification and develops the idea that a disturbance in the metabolism of the phospholipins may be involved.

Since the phospholipins and especially lecithin play so important a part in the metabolism of phosphorus an attempt has been made to discover whether there is any modification in the amount in the blood and liver in rickets.

Unfortunately, the material available has not been of such a nature as to enable a conclusive answer to be given, for in each of the experiments in which the blood and liver were preserved for analysis, either all the pups remained free of rickets or all developed the condition. The results obtained are however sufficient to justify their being recorded and to allow a tentative conclusion to be drawn.

A considerable amount of work has been recorded upon the phospholipin of the blood but little of it has a bearing upon the present investigation. Bloor [1915] showed that in feeding with fat the lecithin content of the blood is raised and [1918] that this is due to an increase of phospholipin in the corpuscles.

METHOD.

The blood and the livers were taken at the time of death by chloroform and bleeding, from experiments 5, 6, 7 and 8 of the series described by Noël Paton and Watson [1921]. The details of experiment 12 have not yet been published, but all the pups remained free from rickets. Unfortunately the non-phospholipin phosphoric acid was determined only in lots 7 and 8.

METHODS OF ANALYSIS.

Livers. The livers were preserved in formalin. A weighed amount was repeatedly extracted with alcohol-ether (10-1) by rubbing up in a mortar. The extract was dried in a steam-bath and in a vacuum desiccator and weighed and from this total fats and alcohol-ether soluble P_2O_5 were determined. The alcohol-ether-insoluble P_2O_5 was estimated in the residue. Each was ignited

NORMAL.

I. Pups of VII killed at 16 weeks old, those of VIII at 18 weeks old.

Experiments	VII. B 7 (Blood)	VII. B 7 (Liver)	VIII. B 5 (Blood)	VIII. B 5 (Liver)	VIII. B 4 (Blood)	VIII. B 4 (Liver)	VII. D 4 (Blood)	VII. D 4 (Liver)	VIII. D 2 (Blood)	VIII. D 2 (Liver)	VII. D 5 (Blood)	VII. D 5 (Liver)
Diet per diem	Separated milk (dried) 22·5 g. Bread 40 g.		Separated milk (dried) 22·5 g.		Separated milk (dried) 22·5 g. Bread 30 g.		Full cream milk (dried) 22·5 g. Lard 4 g.				Full cream milk (dried) 22·5 g. Bread 30 g.	
Gain in weight %	620		280		240		600		300		366	
Total fats %	2·94	2·19	1·60	1·2	1·25	2·45	2·12	3·4	2·04	2·45	2·34	2·9
Total P_2O_5 %	0·140	0·462	0·098	0·309	0·085	0·310	0·085	0·435	0·113	0·286	0·108	0·324
P_2O_5 in phospholipins %	0·049	0·098	0·035	0·096	0·027	0·161	0·026	0·167	0·038	0·109	0·042	0·115
P_2O_5 insoluble in alcohol %	0·091	0·364	0·063	0·213	0·058	0·149	0·059	0·268	0·075	0·177	0·066	0·209
P_2O_5 in phospholipins as % of total fats	1·7	4·5	2·1	8·0	2·1	6·5	1·2	4·9	1·9	4·4	1·8	3·9

II. Pups killed at 15 weeks old.

XII

Diet per diem: Full cream milk (dried) 22·5 g. increased to 34 g. Bread 30 g. increased to 50 g.

	1 (Blood)	1 (Liver)	2 (Blood)	2 (Liver)	3 (Blood)	3 (Liver)	4 (Blood)	4 (Liver)	(Blood)	(Liver)
Gain in weight %	200		120		83		200		50	
Total fats %	1·86	5·5	1·80	5·8	1·32	0·7	2·15	3·4	1·90	—
P_2O_5 in phospholipins	0·052	0·209	0·059	0·186	0·048	0·320	0·052	—	0·042	—
P_2O_5 in phospholipins as % of total fats	2·8	3·8	3·2	3·2	3·6	4·7	2·4	—	2·2	—

RICKETS.

Pups of V killed at 17 weeks old, those of VI at 13 weeks old.

Experiments	VI. B 5	V. B 1	V. B 2	VI. D 2	VI. B 6	V. D 5	VI. D 3
Diet per diem	Whole milk 250 cc. Oatmeal 45 g. after 5th week	Oatmeal	Separated milk 250 cc. Oatmeal 45 g. after 5th week		Whole milk 250 cc. reduced to 150 cc. Oatmeal 45 g. after 5th week	Whole milk (dried) 30 g. Oatmeal 45 g. after 5th week	Separated milk 250 cc. reduced to 150 cc. Oatmeal 45 g. after 5th week
Gain in weight %	156	66	22	116	161	135	65
Condition	Slight rickets	Rickets (condition between V. B 2 and V. D 5)	Marked rickets	Rickets	Rickets	Slight rickets	Rickets

with an excess of phosphorus-free calcium carbonate, the residue dissolved
in hydrochloric acid with the addition of a little nitric acid and the phosphorus
estimated by the molybdate method in the usual way.

Blood. The blood was drawn into a measured quantity. of methylated
spirit, well shaken and allowed to stand several days. This was filtered into
a weighed flask and the residue extracted with alcohol-ether (10–1) many
times, the washings being added to the filtrate. This was evaporated and the
total fat and the P_2O_5 determined as in the liver.

SUMMARY OF TABLES.

P_2O_5 in Phospholipin.

	Non-rachitic			Rachitic		
	Maximum	Average	Minimum	Maximum	Average	Minimum
Blood	0·059	0·043	0·026	0·033	0·025	0·019
Liver	0·320	0·162	0·096	0·100	0·082	0·068

CONCLUSIONS.

The tables show:

1. That the percentage of phospholipin in the blood and in the liver bears
no relation to the gain in weight, to the amount of fat in the diet, to the total
fat in blood and liver or to the age of the animal.

2. That in the rachitic pups the phospholipins of the blood and of the
liver are lower than in the non-rachitic.

3. That the amount of phospholipins of the liver is smaller in proportion
to that of the blood in rachitic than in non-rachitic animals.

I have to thank Prof. Noël Paton for the help and advice he gave me
during the investigation. The expenses for the work were defrayed by a
grant from the Medical Research Council to which body I tender my thanks.

REFERENCES.

Bloor (1915). *J. Biol. Chem.* **23**, 317.
—— (1918). *J. Biol. Chem.* **36**, 33.
Paton, D. Noël (1922). *Brit. Med. J.* i, 379.
Paton, D. Noël and Watson, A. (1921). *Brit. J. Exp. Path.* **2**, 75.

XLII. THE CONSTITUENTS OF THE FLOWERING TOPS OF *ARTEMISIA AFRA*, JACQ.

By JOHN AUGUSTUS GOODSON.

From the Wellcome Chemical Research Laboratories.

(*Received May 2nd, 1922.*)

THE genus Artemisia comprises some 350 species, of which only about 30 have been chemically examined. In most cases the investigation has been confined to the essential oil, but a few of the more important species such as wormwood, *Artemisia Absinthium* Linn., used in the preparation of absinthe, and wormseed, *A. maritima* var. *Stechmanniana* Bess. which grows in Turkestan and is the sole source of the anthelmintic, santonin, have been more thoroughly examined. In consequence of the difficulty of obtaining santonin since 1914, other sources of this indispensable drug have been sought and in this connection attention has been given to other species of Artemisia, of which the author has examined four. *A. brevifolia* Wall. (which according to the Index Kewensis is a form of *A. maritima* Linn.) from India was found to contain santonin, as previously recorded by Greenish and Pearson [1921] and Simonsen [1921]; specimens of *A. afra* Jacq. from South Africa, *A. mexicana* Willd. from Mexico, and *A. monosperma* Delile (*A. deliliana* Bess.) from Egypt were found to contain no santonin. The closely related Egyptian plant *Santolina chamaecyparissa* was also found to be free from santonin.

Judging from these results and others previously recorded it would seem that the occurrence of santonin is restricted to species of Artemisia indigenous to East Europe and Asia, the only exception so far recorded being *A. gallica* Willd., which occurs in France, and according to Heckel and Schlagdenhauffen [1884] contains santonin. These authors, however, give no evidence for this statement and their observation has not been confirmed[1].

Of the four species examined by the author only one, *A. afra*, was available in quantity and this was fully investigated to see if it contained anything that could be regarded as a precursor or a derivative of santonin. The results given below show that this plant contains camphor, a wax-ester probably ceryl cerotate, triacontane, scopoletin and quebrachitol, none of which can be regarded as connected with santonin.

l-Camphor has been recorded in several species of Artemisia, *e.g.* in *A. Herba-alba* Ass. [Grimal, 1904], *A. cana* Pursh [Whittelsey, 1909] and *A. annua* Linn. [Yoshitomi, 1917]. In *A. afra* the camphor is dextrorotatory,

---

[1] Since this paper was submitted for publication a preliminary notice of a communication by Viehoever and Capen [1922] has appeared, in which it is stated that santonin has been isolated from *A. mexicana* and *A. neo-mexicana*. J.A.G.

but the rotation is much lower than that of normal d-camphor. Two specimens were found to have $[\alpha]_D^{20} + 9\cdot7°$ and $[\alpha]_D^{20} + 9\cdot3°$ instead of $[\alpha]_D^{20} + 42\cdot4°$. As it is usually considered that only two optically active isomerides can exist notwithstanding the fact that the camphor molecule contains two asymmetric carbon atoms, and as camphor is not easily racemised, it would appear that both d- and l-camphor are present in A. *afra*. It may be noted that inactive camphor has been found in *Chrysanthemum sinense* var. *japonicum* [Keimatsu, 1909].

Scopoletin and quebrachitol have not been recorded previously in the genus Artemisia or even in the natural order Compositae. The former has been found in *Atropa Belladonna* Linn. [Kunz, 1885], *Fabiana imbricata* Ruiz et Pav. [Kunz-Krause, 1899] and *Scopolia japonica* Maxim. [Eykman, 1884] of the natural order Solanaceae; *Ipomoea purga* Hayne (Convolvulaceae) [Power and Rogerson, 1910]; *Gelsemium sempervirens* Ait. (Loganiaceae) [Moore, 1910] and *Prunus serotina* Ehrh. (Rosaceae) [Power and Moore, 1909]. Quebrachitol has only been found in four other plants, viz. *Aspidosperma quebracho* Schlect (Apocyanaceae) [Tanret, 1889]; *Grevillea robusta* A. Cunn (Proteaceae) [Bourquelot and Fichtenholz, 1912]; *Heterodendrum oleaefolium* Desf. (Sapindaceae) [Petrie, 1918] and *Hevea brasiliensis* Muell. (Euphorbiaceae) [Pickles and Whitfield, 1911], all of which belong to different natural orders. The other components ceryl cerotate and triacontane are fairly widely distributed in plants.

EXPERIMENTAL.

For the material used in this investigation the author is indebted to Mr I. B. Pole Evans, Chief of the Division of Botany, Union of South Africa. It consisted of the flowering tops of the plant including stems, leaves and florets.

When extracted with Prollius's fluid it yielded a mere trace of alkaloid and furnished the following percentages of extract on exhaustion in a Soxhlet apparatus with solvents in the order named: petroleum (b.p. 35–60°), 7·0; ether, 9·5; chloroform, 2·1; ethyl acetate, 3·4; alcohol, 15·5. Special search was made for santonin with negative results. For the purpose of a more complete examination a quantity (11·1 kilograms) was extracted in succession with hot solvents, petroleum (b.p. 35–60°), ether and alcohol. The alcoholic extract was further fractionated by drying on a quantity of the flowering tops previously exhausted with petroleum and ether and re-extracting hot with chloroform, ethyl acetate and alcohol in succession. The petroleum extract on distillation in steam and extraction of the distillate with ether yielded 55·2 g. of essential oil; a further 28·6 g. was subsequently obtained from the ether extract, equivalent in all to 0·75 % by weight of the flowering tops.

The material left behind in the distillation flask on boiling with petroleum (b.p. 35 to 60°), deposited on cooling about 81 g. of crude wax-ester melting at 74–76°. The residue left on removal of the petroleum was boiled with ether and this solution on cooling gave 17 g. of crude hydrocarbon, melting at 62 to 66°. The ethereal solution was then extracted with dilute hydrochloric acid, but yielded no alkaloid; it still contained some free and combined fatty acids, which were not investigated, and some unsaponifiable matter, which appeared to contain a sterol giving a red coloration with acetic anhydride and sulphuric acid.

Examination of the Essential Oil. The oil possessed the odour of camphor, had a specific gravity 0·9453 at 15°/15° and specific rotation $[a]_D^{15} + 5·8°$. On washing with solutions of sodium carbonate, sodium hydroxide and sodium bisulphite (saturated), it lost to each quantities of material too small to be investigated in detail. The acids extracted by sodium carbonate were fractionally precipitated as silver salts, the fractions containing 24·7, 34·9, 38·6; 39·9 and 41·3 % of silver respectively. Silver pelargonate, $C_8H_{17}COOAg$ requires Ag = 40·4. The saponification value of the oil before acetylation was 33·9 and after acetylation 73·9.

On distillation the following fractions per cent were obtained (1) below 180°/760 mm., 31·8; (2) below 100°/25 mm., 27·5; (3) at 100–120°/25 mm., 12·1; (4) at 120–140°/25 mm., 12·3; (5) at 140–180°/25 mm., 8·0; (6) and above 180°/25 mm., 7·7.

Isolation of Camphor. Fractions (2) and (3) on standing deposited camphor, a further quantity of which separated on redistilling the fractions and freezing the distillates with solid carbon dioxide, the total quantity obtained amounting to 13·5 % of the crude oil, but this does not represent the total amount of camphor in the oil as much still remained in the various fractions. The crude camphor was recrystallised from dilute alcohol until its melting point was constant at 178° (corr.) and showed no depression of melting point on admixture with natural d-camphor melting at 180° (corr.). The specific rotation was low, being only $[a]_D^{20} + 9·7°$ in 95 % alcohol ($a_D^{20} = + 0·98°$, $l = 1$ dcm. C = 10·148), that of natural d-camphor used for comparison being $[a]_D^{20} + 42·4°$ ($a_D^{20} = + 4·35°$, $l = 1$ dcm., C = 10·261). Found C = 78·9; H = 10·5. Calculated for camphor, $C_{10}H_{16}O$, C = 78·9; H = 10·6.

The oxime was prepared and after recrystallisation melted at 118–119° (corr.), the melting point remaining unchanged on admixture with camphoroxime prepared from natural d-camphor. It was laevo-rotatory $[a]_D - 3·4°$ ($a_D = - 0·16°$, $l = 0·5$ dcm., C = 9·296). The semicarbazone after recrystallisation from alcohol melted with decomposition at 241° (246° corr.) and when mixed with camphor-semicarbazone prepared from natural d-camphor, m.p. 246° (corr.) with decomposition, it melted at 242° (247° corr.). The melting point of the semicarbazone of d-camphor is usually given in the literature as 236–238°, but this is undoubtedly too low. (Found C = 63·4; H = 9·4. Calculated for camphor-semicarbazone, $C_{11}H_{19}N_3O$, C = 63·1; H = 9·2.)

Examination of wax-ester. The wax-ester was recrystallised many times from ethyl acetate or petroleum and then melted constantly at 79°, although it appeared still to contain a small quantity of hydrocarbon. From the analyses of the ester and its hydrolytic products, it would appear to be ceryl cerotate or a closely related ester. (Found C = 82·5, 82·5; H = 13·5, 13·7. Calculated for ceryl cerotate, $C_{26}H_{53}COOC_{25}H_{51}$, C = 82·0; H = 13·8.)

The alcohol obtained on hydrolysis melted at 74°, and gave an acetyl derivative melting at 64–66°. Ceryl alcohol melts at 81°, and ceryl acetate at 64·5°. (Found C = 81·6, 81·5; H = 14·2, 13·9. Calculated for ceryl alcohol, $C_{26}H_{54}O$, C = 81·6; H = 14·2.)

(Found C = 79·0; H = 13·3. Calculated for ceryl acetate, $CH_3COOC_{26}H_{53}$, C = 79·2; H = 13·3.)

The fatty acid produced on hydrolysis melted at 76°. (Found C = 78·7, 78·8; H = 13·3, 13·5. Calculated for cerotic acid, $C_{26}H_{52}O_2$, C = 78·7; H = 13·2; m.p. 82°.)

Isolation of Triacontane. The crude hydrocarbon was distilled under reduced pressure and recrystallised several times from ethyl acetate and petroleum until it melted constantly at 66°. (Found C = 84·9; H = 14·9. Calculated for triacontane, $C_{30}H_{62}$, C = 85·2; H = 14·8; m.p. 65·5°.)

After removal of the essential oil, wax-ester and hydrocarbon from the ether extract, the latter was extracted with dilute hydrochloric acid, which removed no alkaloid or other basic material and then with sodium carbonate solution, followed by potassium hydroxide solution.

Isolation of Scopoletin. The sodium carbonate extract was acidified with hydrochloric acid, and extracted with ether. The ethereal solution on concentration deposited a crystalline substance, a further quantity of which was obtained by extracting the dilute hydrochloric acid extract of the original ether extract with ether. The substance was purified by recrystallisation from ethyl acetate, yielding 2 g. of pure material, which formed pale yellow needles and dissolved in sodium carbonate solution giving a solution possessing a striking blue fluorescence. It melted at 203° (208° corr.) the melting point being uninfluenced on admixture with scopoletin. (Found C = 62·4, 62·4; H = 4·2, 4·5. Calculated for scopoletin (4-hydroxy-5-methoxycoumarin), $C_{10}H_8O_4$, C = 62·5, H = 4·2.)

The chloroform extract yielded no crystalline substance and dilute hydrochloric acid removed from it only a trace of material giving alkaloid reactions.

Isolation of Quebrachitol (Methyl l-inositol). The residues of the ethyl acetate, and alcoholic extracts, after concentration, deposited a mixture of resinous matter and crystals, which was extracted with water, the solution was boiled with animal charcoal, filtered, and concentrated and finally alcohol was added when a quantity of crude methyl-*l*-inositol crystallised out corresponding to 0·43 % of the flowering tops used. After several recrystallisations it melted constantly at 194° (corr.), and had a specific rotation $[\alpha]_D - 81·6°$ ($\alpha_D = - 8·16°$, $l = 1$ dcm., C = 10·0). Found C = 43·2, 43·4; H = 7·0, 7·3. Calculated for quebrachitol, $CH_3OC_6H_6(OH)_5$, C = 43·3; H = 7·3.

The acetyl derivative, prepared by the action of acetic anhydride in presence of pyridine, was recrystallised from ethyl acetate and obtained in rosettes of needles, melting at 94–95°. (Found $C = 50.8$; $H = 6.1$. Calculated for penta-acetylmethylinositol, $CH_3OC_6H_6(OOC.CH_3)_5$, $C = 50.5$; $H = 5.9$.)

There is no published record of previous chemical work on *Artemisia afra* but the author desires to state that Messrs H. W. B. Clewer, and R. R. Baxter have made preliminary examinations of the plant in these laboratories, and their records, which include the isolation of camphor and of the crystalline substance now shown to be quebrachitol, have been available to him.

In conclusion, the author desires to express his warmest thanks to Dr T. A. Henry for his advice and criticism throughout the course of the work.

REFERENCES.

Bourquelot and Fichtenholz (1912). *J. Pharm. Chim.* 6, 346.
Eykman (1884). *Rec. trav. chim. Pays-Bas*, 3, 171.
Greenish and Pearson (1921). *Pharm. J.* (4), 52, 2.
Grimal (1904). *Bull. Soc. Chim.* 31, 694.
Heckel and Schlagdenhauffen (1884). *Compt. Rend.* 100, 804.
Keimatsu (1909). *J. Pharm. Soc. Japan*, 326, 1.
Kunz (1885). *Arch. Pharm.* 223, 701.
Kunz-Krause (1899). *Arch. Pharm.* 237, 1.
Moore (1910). *J. Chem. Soc.* 97, 2223.
Petrie (1918). *Proc. Linnean Soc.* N.S.W. 43, Part 4.
Pickles and Whitfield (1911). *Proc. Chem. Soc.* 54.
Power and Moore (1909). *J. Chem. Soc.* 95, 243.
Power and Rogerson (1910). *J. Amer. Chem. Soc.* 32, 95.
Simonsen (1921). *J. Indian Industries and Labour*, Nov. 539.
Tanret (1889). *Compt. Rend.* 109, 908.
Viehoever and Capen (1922). *J. Amer. Pharm. Assoc.* 11, 393.
Whittelsey (1909). *Wallach Festschrift*, 668.
Yoshitomi (1917). *J. Pharm. Soc. Japan*, 424, 1.

XLIII. A METHOD FOR THE ESTIMATION OF SMALL QUANTITIES OF CALCIUM.

By PATRICK PLAYFAIR LAIDLAW AND
WILFRED WALTER PAYNE.

From the Pathological Department, Guy's Hospital.

(*Received May 5th, 1922.*)

OUR knowledge of calcium metabolism is limited, and probably hampered, by the lack of a good and accurate method of estimating small quantities of calcium. In quite a number of biological problems the estimation of amounts of calcium of the order of 0·1 mg. is frequently desired. The method about to be described is applicable to even smaller quantities than this; and is, we believe, remarkably accurate if used. with reasonable care, and superior to those in use at the present time.

The modern methods are of two types (*a*) biological, (*b*) chemical. The biological methods depend on the fact that calcium is necessary for blood coagulation. Wright [1912] and later Vines [1921] have developed methods with this fundamental fact as basis. Inasmuch as the nature of the coagulation process is only imperfectly known these methods suffer from being built on an insecure foundation and are liable to break down for unknown reasons. They are, however, sensitive and within limits not clearly defined useful. If a method equally sensitive with a secure basis were available they would rarely be employed. The more recent chemical methods such as those of Howland and Kramer [1920], Kramer and Tisdall [1921] are difficult to use, and consisting as they do of microtitrations of oxalate by $N/100$ potassium permanganate are liable to error. Thus small amounts of organic matter such as carbon from imperfect incineration of a biological product, dust, or even the organic material in poor distilled water, may each and all cause significant error. The weak solution does not keep well, and the end point is not sharp. With elaborate precautions and in the hands of the select few they give good results; but we obtained very disappointing results on trying them out.

The principle of the method. The calcium is separated in the first place as oxalate and the oxalate is converted into calcium alizarinate under defined conditions. The crystalline calcium alizarinate is collected and washed on a Gooch filter and decomposed by oxalic acid. The alizarin is dissolved in alcohol made up to known volume with a relatively large volume of dilute ammonia, and the resultant solution of ammonium alizarinate is compared in a Dubosc colorimeter with a standard dilution of ammonium alizarinate. The primary separation as oxalate is necessary since in spite of many trials no clean separation of magnesium and calcium alizarinates proved feasible. Since the calcium in the alizarinate is only one-sixth the weight of the alizarin, the initial weight of calcium to be estimated is multiplied by six; and since the colorimetric comparison is carried out against a 0·001 % solution of alizarin, the method is very sensitive. Given cleanliness, reagents free from calcium, pure alizarin, and avoiding undue hurry, 0·1 mg. calcium may be estimated, to 0·002 or 0·005 mg. without much trouble.

Precipitation of calcium oxalate. This is done along the usual lines and may be carried out on the ash of a biological product, or sometimes, as in the case of blood serum, directly. After careful incineration the ash is dissolved in 0·5 cc. of N HCl and transferred with about 2 cc. wash water to a centrifuge tube. One cc. of a saturated solution of ammonium oxalate is added, followed by 2 cc. of a saturated solution of sodium acetate, to which some ammonium oxalate has previously been added to exclude calcium contamination. The whole is mixed and allowed to stand for three hours, or if more convenient till next day. In our experience three hours' standing, especially in the case of direct estimations in blood serum, gives more uniform results than shorter periods, though most of the oxalate separates as a rule in a much shorter time. The oxalate precipitate is centrifugalised to the bottom of the tube (two minutes at 4000 per min. is sufficient) and the supernatant fluid sucked off as completely as possible without disturbing the oxalate precipitate at the bottom of the tube. Three cc. of a 0·1 % solution of ammonium oxalate are then added, the precipitate stirred up and the centrifugalisation repeated at once. The supernatant fluid is once more sucked off as completely as possible. In this way all the calcium is obtained as oxalate free from all but traces of magnesium, and with only a *little* excess of oxalate.

Conversion of oxalate into alizarinate. The oxalate is dissolved in 0·5 cc. of N HCl, and transferred to a clean test-tube. The centrifuge tube is washed out four times with water. The resultant dilute solution in the test-tube should measure 8 to 10 cc. A small volume is disadvantageous, as oxalate may separate at a later stage if the volume is kept too small. Excess of an alcoholic solution of alizarin is then added (1 cc. of a 1 in 1000 solution is sufficient for quantities of calcium of about 0·1 mg.). The test-tube is then warmed on a water bath to about 80° C., and five drops of strong ammonia solution added. The contents of the tube turn deep purple on mixing. It is best to add at this stage a minute amount of crystalline calcium alizarinate though it is not essential. The step ensures the full and ready separation of the calcium alizarinate in a crystalline form which facilitates filtration later. The tube is kept warm for about one hour then set aside to cool and stand till the following day. The whole of the calcium should then have separated in clumps of blue black microcrystalline needles at the bottom of the tube and the supernatant fluid should be pale in colour.

The calcium alizarinate is filtered off on a Gooch crucible with a fine asbestos layer at a water pump, and the precipitate and filter washed with dilute ammonia. At this point care must be taken not to suck much air through the filter since acids, even the carbonic acid of the air, will decompose the calcium salt. Paper filters are not so good as asbestos. The ordinary papers contain significant amounts of salts which interfere; and extracted papers have their own difficulties.

The delivery from the Gooch crucible is now arranged to return fluid to the test-tube in which crystallisation of the alizarinate took place, and 1 cc. of a strong solution of oxalic acid in 50 % alcohol allowed to run all over the asbestos and precipitate. The calcium alizarinate decomposes at once and the blue black colour becomes orange. The alizarin set free is washed through with warm 95 % alcohol along with excess of oxalic acid into the test-tube. The asbestos should now be pink or very pale purple. The nature of this colour is unknown, but it does not appear to be due to any calcium compound and is usually greater when the oxalate is precipitated directly from blood serum, and less in the case of ash estimations. The alcoholic solution of alizarin is transferred to a 50 cc. standard flask, the test-tube washed out with dilute ammonia, the contents of the flask made just alkaline with ammonia, and the volume made up to the mark with water.

The standard solution for comparison is made up as follows. Two cubic centimetres of an accurate $M/1000$ solution of alizarin in alcohol are run into a 50 cc. standard flask, followed by 1 cc. of the oxalic acid solution in 50 % alcohol, and a volume of alcohol approximately equal to that used in extracting the alizarin from the Gooch crucible when dealing with the unknown. Water is added, the contents of the flask are made just alkaline with ammonia, and the volume made up to the mark. In this way a perfect colour match is readily obtained as the reactions of the test and standard solutions are nearly the same. A large excess of ammonia in one flask yields a purple tint as compared with a red one. A small difference in ammonia content is insignificant owing to the buffer effect of the oxalate.

The unknown and the standard solutions are then compared in a Dubosc colorimeter. 20 to 25 mm. depth of the standard solution yields a suitable tint for colour match.

The alizarin in the unknown solution is calculated from the depth of the column of fluid which gives a perfect match with the standard, and this figure divided by six gives the weight of the calcium.

The method is simple and though time must be spent over the whole process owing to the waits for complete precipitation at two stages, the actual time spent on manipulation is not great, and if a series of estimations is being done the time spent on any one is still smaller. It may be stated that there are many indications that if still smaller amounts than 0·1 mg. are in question, the use of small scale apparatus would enable these to be estimated with fair accuracy.

EXPERIMENTAL.

The composition of the calcium alizarinate as prepared from the oxalate, under the defined conditions of excess alizarin, in the presence of ammonia, and in weak alcoholic solutions, was determined by incineration of three separate specimens of the crystalline solid; conversion of the calcium carbonate so obtained to sulphate; and weighing after heating and cooling:

Alizarinate taken	Sulphate	Calcium	Theory
0·2424 g.	0·1023	0·0301	0·0307
0·4855 g.	0·2053	0·0604	0·0614
0·2535 g.	0·1111	0·0327	0·0321

0·2535 g. calcium alizarinate dried over sulphuric acid heated to

110° C. lost 0·0158 g.			$1H_2O = 0.0144$ g.
120° C. ,, 0·0213 g.			
160° C. ,, 0·0276 g.			$2H_2O = 0.0288$ g.
190° C. began to decompose.			

The weighing of the heated alizarinate was difficult since there was gain in weight during the weighings. The data show that under the conditions of experiment a salt of the composition $Ca_1 Aliz_1 + 2H_2O$ separates. We also have evidence of the formation of another calcium salt of a different composition when there is not excess of alizarin present.

The method was tried out on an artificial salt mixture, made up to imitate the ash of blood serum but with the magnesium in excess, of the following composition:

NaCl	...	...	...	4·6340
KH_2PO_4		...	...	0·0395
$MgSO_4$ (dry)	...		...	0·7577
$CaCO_3$		...	...	0·2410
HCl	...	...	...	8 cc. Water to 500 cc.

A trace of iron was present probably due to impurity in the magnesium sulphate. This was no disadvantage since this metal may be present in blood serum ash from haemoglobin.

The following table shows some of the results we obtained when estimating various amounts of the salt mixture:

Solution taken cc.	Ca found mg.	Calculated mg.	Difference mg.
0·9	0·171	0·1737	−0·0027
0·7	0·140	0·135	+0·005
0·6	0·117	0·115	+0·002
0·5	0·102	0·096	+0·006
0·4	0·0815	0·077	+0·0045
0·3	0·056	0·058	−0·002

Repeated estimations of quantities of about the amount normally present in blood serum gave the following figures:

Calcium taken mg.	Found mg.	Difference mg.
0·116	0·113	−0·003
0·116	0·118	+0·002
0·116	0·1196	+0·0036
0·096	0·098	+0·002
0·096	0·098	+0·002
0·096	0·096	0·000

Estimations on mixed samples of human blood serum are given below. It will be observed that the figures for the ash of the serum and the direct determinations are practically the same. Other figures show that the agreement is even closer if 24 hours be allowed for the separation of the oxalate in the case of the direct estimation. Other experiments on heated serum indicate that after heating some of the calcium becomes bound to the serum proteins and cannot be estimated directly:

	Ca in ash, mg.	Ca directly without ashing, mg.
Sample 1	{ 0·098	0·096
	{ 0·100	0·095
Sample 2	{ 0·097	0·091
	{ 0·094	0·093

SUMMARY.

A method for estimating small quantities of calcium is described. The method is colorimetric and depends on the insolubility of calcium alizarinate. It is accurate on 0·1 mg. to 0·002 mg.

REFERENCES.

Howland and Kramer (1920). *J. Biol. Chem.* **43**, 35.
Kramer and Tisdall (1921). *J. Biol. Chem.* **48**, 223. *Bull. Johns Hopkins Hospital*, **32**, 44.
Vines (1921). *J. Physiol.* **55**, 86.
Wright (1912). The Technique of the Teat and Capillary Tube, 85.

XLIV. PRODUCTION OF HYDROGEN PEROXIDE BY BACTERIA.

By JAMES WALTER McLEOD and JOHN GORDON.

From the Department of Pathology, University of Leeds.

(Received May 8th, 1922.)

A STUDY of phenomena of auto-inhibition in bacterial cultures was published by McLeod and Govenlock [1921] in 1921 showing that a series of investigations on this subject had led to the demonstration of such phenomena in connection with the growth of many bacteria. C. Eijkman's results [1904 and 1906, 1, 2], which have been more questioned than accepted, were fully confirmed.

It was shown further that the Pneumococcus, a bacterium, which did not appear to have been investigated previously in this respect, produced in the course of growth a substance inhibitory to its own growth and to that of most bacteria. This inhibitory effect was more powerful than that observed in connection with the growth of any other bacterium examined and the inhibitory substance was found to be thermolabile (85°). The labile character of the substance produced suggested an analogy with toxins and ferments and for this reason the name "bactericidins" was suggested for such substances. Subsequent work has shown however that the substance produced by the Pneumococcus, at all events, is simpler in character and is in all probability hydrogen peroxide: but the question whether other bacteria produce specific inhibitory substances analogous to ferments remains an open one.

A short summary of the work done to show that the Pneumococcus produces hydrogen peroxide in culture has already been published as an abstract of a communication to the Pathological Society [McLeod and Gordon, 1922].

It is the purpose of this paper to give in some detail the experimental evidence for this conclusion together with some more recent observations on the subject.

Conditions necessary for the production of pneumococcal inhibitory substance in fluid media.

The original observations establishing the existence of a thermolabile bactericidal product in pneumococcal cultures had been made in solid media—serum agar—and such were obviously unsuitable for attempting to concentrate the bactericidal substance.

33—2

In trying to obtain evidence of the production of these substances by the Pneumococcus in fluid media, the Staphylococcus was used as a test micro-organism because it grows freely and with great constancy in all ordinary bacteriological media and because it had proved to be particularly sensitive to the inhibitory body produced by the Pneumococcus.

The experiment consisted in preparing a series of dilutions in peptone water of a 10 % serum bouillon culture of Pneumococcus, which had been heated for 30 minutes at 60° to kill off the Pneumococcus, and inoculating these with Staphylococcus. The inhibitory potency of the pneumococcal culture was judged by the extent to which it had to be diluted in order to permit of growth of the Staphylococcus.

By using such methods it was soon determined that aeration of the culture was of importance; thus little or no inhibitory effect would be produced by growing a Pneumococcus in 10 cc. of 10 % serum broth in a narrow test-tube and the Pneumococcus was always alive at the end of 48 hours, whereas the same organism grown in 10 cc. of the same medium spread out over the base of an Erlenmeyer flask gave rise to a markedly inhibitory culture in which the Pneumococcus had usually died at the end of 48 hours. It was found further that inhibitory cultures could be obtained more constantly if a current of air was kept bubbling through the culture during incubation, by connecting the flask containing the medium with a water suction pump. Control observations in which air was passed through sterile media kept for similar periods of time in the incubator gave negative results—the medium did not develop any inhibitory quality.

All the earlier observations were made with a bouillon medium prepared from meat extract and Parke Davis peptone, 1 %, to which 10 % of horse serum was added. The horse serum had been heated for an hour at 56° and was several months old. It contained very little catalase and subsequent observations have shown that it is important for constant production of the inhibitory body that the medium should contain little or no catalase. Some of our observations also tended to show that the amount of inhibitory substance produced was influenced by the kind of peptone used but the point has not yet been investigated carefully enough to permit of a definite statement.

Methods of concentrating the inhibitory substance.

Under the impression that the substance under investigation was probably a labile protein similar to toxin an attempt was made to obtain concentration by precipitating out the proteins of the culture with absolute alcohol at 0°, and redissolving the precipitate in a small quantity of water. The solutions so obtained showed little or no power of inhibiting bacterial growth. Further experiment showed however, that the bactericidal substance dissolved in alcohol and passed into the filtrate. By concentrating such filtrates *in vacuo* at 35°–45° fluids of considerably enhanced bactericidal potency were obtained.

The average culture before concentration had an antiseptic potency sufficient to prevent development of *Staphylococcus aureus* in peptone water containing from 1 part in 3 to 1 part in 8 of pneumococcal culture, and by concentrating this culture to one-sixth or one-eighth of its original volume in the manner outlined above a residue was obtained at least four times as active as the original culture, *i.e.* capable of completely inhibiting growth of Staphylococcus in peptone water if present in the proportion of 1 part to 20 of peptone water. That is to say the fluid obtained was about equal to 4 % carbolic acid as regards its antiseptic effect on *Staphylococcus aureus*: it differed from the latter however in the marked instability of the antiseptic substance contained.

By reprecipitating the first concentrate with an excess of alcohol and reducing to a considerably smaller bulk it was possible to obtain fluids of higher antiseptic potency.

Chemical evidence of the presence of peroxide in the inhibitory cultures and concentrates.

The antiseptic concentrates from pneumococcal cultures had not been long under investigation before the observation was made that the addition of such fluids to emulsions of blood was followed by evolution of gas. We are indebted to Mr H. D. Kay of the Physiology Department, Leeds University, for suggesting that the effervescence was probably due to the presence of peroxide. This appears to be the correct explanation. Both cultures and concentrates give a strong blue colour when mixed with starch iodide paste containing traces of $FeSO_4$ (Schönbein's reagent). Wolff [1912] has pointed out that under certain conditions nitrites may give this reaction and cause confusion in regard to presence of peroxides in plant juices; but the absence of nitrites from the cultures is proved by failure to obtain any colour reaction on adding diphenylamine. Cultures so concentrated also give quite definitely a yellow or orange colour with titanium sulphate solution. Lastly a compound blue in colour and soluble in ether but disappearing rapidly if not extracted with ether immediately was obtained on adding 10 % chromic acid to a concentrate.

The chief evidences from the chemical standpoint for considering this substance to be hydrogen peroxide rather than some organic peroxide are:

(1) Its sensitiveness to decomposition by the catalase of blood and serum or by special catalase preparations from the liver [Morgulis, 1921]; Novy and Freer [1902] state that the various organic peroxides with which they worked differed from H_2O_2 in being relatively insensitive to the action of catalase.

(2) The fact that it tends to decompose on heating, but only does so very rapidly when the temperature is raised to 85°, in which respect it closely resembles H_2O_2, which distils at 84° under reduced pressure when relatively pure and concentrated but is rapidly destroyed about the same temperature in dilute and impure solutions [Baur, 1908].

(3) The gas evolved on adding catalase to a concentrate was proved to be oxygen since it was not absorbed by KOH solution but was absorbed to the extent of 90–95 % by alkaline pyrogallol solution and was capable of rekindling a glowing splint placed in it. Further the oxygen evolved corresponded very nearly to the amount calculated as available when the concentrate was titrated with potassium iodide, sodium thiosulphate and starch with a view to determining its H_2O_2 content.

Bacteriological evidence of identity of inhibitory substance in pneumococcal cultures with H_2O_2.

Two lines of investigation have been pursued.

First of all a number of bacteria have been compared as regards their varying sensitiveness to the inhibitory effect of (a) small concentrations of H_2O_2 in agar plates (b) the substances diffusing from deep plate cultures of Pneumococcus to superimposed layers of agar [McLeod and Govenlock, 1921]. It has been found that if bacteria are graded according to the greatest concentration of H_2O_2 in an agar plate which is compatible with their growth the order of their resistance to the H_2O_2 is as follows: B. subtilis, B. prodigiosus, some Streptococci, B. coli, other Streptococci, Staphylococcus aureus, Anthrax, Cholera, Shiga, Typhoid.

Among the bacteria which were observed often enough on inhibitory pneumococcal plates to give a reliable average result the order of resistances was: Streptococcus, B. coli, Staphylococci and Shiga. Although complicating factors such as favouring effect of other products of pneumococcal growth on certain bacteria may be present, the results are remarkably alike.

The second line was to determine how far the antiseptic effect of the concentrate of a pneumococcal culture corresponded to that of a dilute solution of H_2O_2 of similar strength, i.e. one which would give a similar figure for available oxygen when that was determined by titration with potassium iodide and thiosulphate solution in presence of starch. Four experiments of this kind were performed and the results are given in Table I.

Table I.

		Antiseptic potencies expressed as highest dilution in peptone water which inhibited growth of *Staphylococcus aureus* completely			
Experiment	H_2O_2 of concentrated culture calculated in "volumes" by KI, $Na_2S_2O_3$ + starch titration	(a) Concentrated culture	(b) Solution of H_2O_2 in saline. Strength as calculated for concentrate	(c) Concentrated culture inactivated by heat: + H_2O_2 of same strength as in (b)	(d) Concentrated culture inactivated by heat
1	0·5	1–34	1–34	1–24	—
2	0·8	1–24	1–50	1–50	—
3	0·4	1–12	1–32	1–18	—
4	0·38	1–25	1–50	1–25	nil

It will be seen that the figures for antiseptic potency are nearly identical for concentrates and for heat inactivated concentrates which had received addition of H_2O_2 sufficient to restore them to the same titration figure for available oxygen as that of the concentrates.

Further in the fourth experiment, where figures for concentrate and heat inactivated concentrate + calculated amount of H_2O_2 exactly correspond, a determination of volumes of gas liberated by catalase from 2 cc. of H_2O_2 of calculated strength was 0·7 cc. and that from 2 cc. of concentrate 0·67 cc.

That the antiseptic potencies of the concentrates should be usually less than those of solutions in 0·85 % saline of H_2O_2 of similar strengths is not surprising since in the strong solution of amino-acids etc. and suspended fatty bodies which constituted the concentrate the tendency to decomposition of the H_2O_2 is likely to be greater.

Changes in blood pigment effected by growth of Pneumococcus.

It has long been recognised that certain streptococci produce a green colour when grown on a blood agar plate, hence the name "viridans" [Schottmüller, 1903]. The Pneumococcus resembles this micro-organism in the appearance of its growth on blood agar plates. It has also been recognised for some time that pneumococci and certain streptococci produce methaemoglobin [Stadie, 1921].

Methaemoglobin is not however a green pigment and so far as we know no full investigation of the pigmentary changes associated with the development of the green colour has been made.

In recent years a medium called "chocolate agar" [Crowe, 1915, 1921; Neurin and Gurley, 1921], consisting of agar and heated blood mixed in varying proportions, has been much used. The blood here is no longer present in the form of haemoglobin or methaemoglobin and on such media the pigmentary changes produced by pneumococci etc. are much more striking than on unheated blood media. All gradations in colour between dark olive green and light yellow may be seen around pneumococcal colonies. If, however, small drops of dilute solutions of H_2O_2—$\frac{1}{4}$ to 1 "vol."—are repeatedly applied to the surface of such media a very similar range of colours is obtained. If the application of H_2O_2 is sufficiently powerful the medium is completely bleached. It would seem therefore that the varying colour changes which develop around pneumococcal colonies on such media are due to H_2O_2 formation in varying degrees of intensity. Also that the heated blood agar plate may reasonably be used to detect bacteria capable of forming H_2O_2.

Production of H_2O_2 by bacteria other than the Pneumococcus.

The following conclusions have been reached by using the "chocolate agar" plate as a primary indicator of H_2O_2 formation and then proceeding to confirm the fact that peroxide is produced by the bacteria which form green colonies, by growing them in bouillon or serum bouillon media in presence of abundant oxygen supply and testing their cultures with hydrogen peroxide reagents such as Schönbein's or titanium sulphate solution. No bacteria have been met in the course of routine work or on going through a large series of stock cultures which produce a green colour on "chocolate agar,' with the exception of various forms of cocci. Amongst the streptococci green-forming strains predominate but both amongst the haemolytic and the non-haemolytic forms strains are met which do not produce green on chocolate agar, and all such strains which have been examined have also failed to produce fluid cultures giving H_2O_2 reactions.

As a general rule, although with some distinct exceptions, the streptococci are differentiated from the pneumococci by producing a green colour later and with lesser intensity, similarly the presence of H_2O_2 is usually detected with the appropriate reagents at a later period in streptococcal than in pneumococcal culture, i.e. after 36 hours rather than after 18 hours.

In addition to the streptococci, bacteria from the urethra of sarcinal type and a coarse coccus unclassified have been found to produce a green coloration on "chocolate agar" and also to give H_2O_2 reactions in fluid media.

Amongst the many bacteria which do not produce green coloration on "chocolate agar" plates three were also tested for production of H_2O_2 in flasks of fluid media in which an air current was maintained. These were B. coli, Paratyphoid B. and Staphylococcus. All gave negative results and a number of different bacteria occurring from time to time as contaminations in such experiments have also given negative results.

Importance of catalase in connection with these phenomena.

It is an old finding in bacteriological work that most bacteria produce catalase [Gottstein, 1893].

If the production of H_2O_2 by bacteria as described above is a fact, certain observations may be reasonably expected to follow; amongst these are

(1) that bacteria producing H_2O_2 will not produce catalase;

(2) that such bacteria will grow much better in media containing catalase;

(3) that a medium containing some substance producing the catalase effect will be the best for maintaining stock cultures of such bacteria.

These observations have all been made; no evolution of oxygen over Pneumococcus colonies has ever been observed on pouring dilute solutions of H_2O_2 over cultures of these bacteria. The same holds for most streptococci. A more marked turbidity is developed in a serum bouillon culture of Pneumococcus than in a parallel culture where the serum bouillon has been heated to 65° for 30 minutes so as to destroy the catalase; that is if the cultures are incubated under conditions allowing of moderate access of oxygen. In fact it is very probable that the "vitamin" effects of fresh tissue fluids in promoting growth of Pneumococcus which Kligler [1919] describes are partly due to catalase.

Lastly it has been found that peptone bouillon containing 10 % of washed blood, a medium which retains to a considerable extent the power of decomposing H_2O_2 even after heating for 20 minutes at 115°, is the most convenient for maintaining stock cultures of the Pneumococcus; no difficulty has been met in preserving various strains over long periods when subcultures have been made at 3–4 week intervals.

Possible importance of H_2O_2 formation by bacteria.

The most interesting possibility that arises in this connection is that H_2O_2-forming bacteria may under certain circumstances tend to kill themselves in the body by peroxide formation, just as they do in the test-tube. The most concrete case in which this possibility arises is the crisis in pneumonia. In the explanation of this there is admittedly a good deal of difficulty at the present moment, since the observed phenomena of phagocytosis, development of bactericidal quality in the serum, antitoxin formation etc. are inadequate. The chief difficulty in accepting such a theory of the crisis in pneumonia is that the pneumonic exudate is rich in catalase and it is most unlikely that any concentration of H_2O_2 similar to that which kills off the pneumococcus in culture can occur. There are two possible ways out of this difficulty. One is to suppose that something occurs to paralyse the catalase of the pneumonic exudate. Changes of reaction and enzyme reactions have been demonstrated in the pneumonic exudate which do not occur in the healthy tissues of the body [Lord, 1919, 1, 2; Lord and Nye, 1921, 1, 2]. Another is to suppose that a much smaller concentration of H_2O_2 may be effective in killing off the Pneumococcus in the lung than is required in cultural experiment on account of the oxidising ferments present in the exudate. Both however are subjects for extended experimental investigation.

Explanation of the formation of H_2O_2 by bacteria.

Wieland's [1912, 1, 2; 1913] theory of oxidation supposes that the essential phenomenon is the liberation of hydrogen and that the rôle of oxygen is that of an acceptor for the hydrogen liberated, while Wartenberg and Sieg [1920] show that, under certain conditions at all events, the first stage in the union of H and O is H_2O_2. Such a sequence of events would fit in well enough with the observed phenomena in pneumococcal cultures in which H_2O_2 is only formed as a bye-product where there is sufficient access of oxygen and little or no catalase.

The reason why H_2O_2 does not appear in the cultures of other bacteria may be either that they are catalase formers—the great majority of known bacteria; or that although not catalase formers they are too sensitive to the antiseptic action of low concentrations of H_2O_2 to grow sufficiently in the presence of oxygen to produce any recognisable traces of that substance— the anaerobes. When however certain strains of streptococci are considered which we have found to be incapable of producing catalase and relatively insensitive to H_2O_2, but which do not produce H_2O_2 in their cultures, it is obvious that other factors which await investigation must enter into the problem.

Conclusions and Summary.

1. The inhibitory substance developed in pneumococcal cultures to which there is abundant access of oxygen is H_2O_2.

2. This is proved both by chemical reactions and by the comparison of the antiseptic effects of the pneumococcal cultures and those of dilute solutions of H_2O_2 reckoned by titration etc. to contain a similar amount of available oxygen.

3. The early death of Pneumococcus in culture is usually due to accumulation of excess of H_2O_2.

4. The green or yellow colorations produced on heated blood media by certain bacteria are due to H_2O_2 formation.

5. In addition to pneumococci the only bacteria which have been shown to produce H_2O_2 are many streptococci, both haemolytic and non-haemolytic, and a few other coccal forms.

6. These findings can be utilised practically in putting bacteriological technique on a more definitely scientific basis in several respects.

7. The substance in fresh tissue fluids which specially promotes the growth of the Pneumococcus and which has been supposed to be of the nature of vitamin is most probably catalase.

In conclusion we have pleasure in expressing our indebtedness to Prof. M. J. Stewart in whose department this work was carried out and to Prof. H. S. Raper of Leeds University for valuable advice.

REFERENCES

Baur (1908). *Abegg's Handb. Anorg. Chem.* 2, 1te Abth. 87.
Crowe (1915). *Lancet*, ii, 1127.
—— (1921). *J. Path. Bact.* 24, 361.
Eijkman (1904). *Centralbl. Bakt. Par.* O.I. 37, 436.
—— (1906, 1). *Centralbl. Bakt. Par.* O.I. 41, 367.
—— (1906, 2). *Centralbl. Bakt. Par.* O.I. 41, 471.
Gottstein (1893). *Virchow's Archiv*, 133, 295.
Kligler (1919). *J. Exp. Med.* 30, 31.
Lord (1919, 1). *J. Amer. Med. Assoc.* 72, 1364.
—— (1919, 2). *J. Amer. Med. Assoc.* 73, 1420.
Lord and Nye (1921, 1). *J. Exp. Med.* 34, 199.
—— (1921, 2). *J. Exp. Med.* 34, 201.
McLeod and Govenlock (1921). *Lancet*, i, 900.
McLeod and Gordon (1922). *J. Path. Bact.* 25, 139.
Morgulis (1921). *J. Biol. Chem.* 47, 341.
Neurin and Gurley (1921). *J. Immunology*, 6, 5.
Novy and Freer (1902). *Amer. Chem. J.* 27, 161.
Schottmüller (1903). *Münchener Med. Woch.* 50, 849.
Stadie (1921). *J. Exp. Med.* 33, 627.
Wartenberg and Sieg (1920). *Ber.* 53, 2192.
Wieland (1912, 1). *Ber.* 45, 484.
—— (1912, 2). *Ber.* 45, 2600.
—— (1913). *Ber.* 46, 3327.
Wolff (1912). *Ann. Inst. Past.* 27, 501.

XLV. NOTE ON URINARY TIDES AND EXCRETORY RHYTHM.

By JAMES ARGYLL CAMPBELL AND
THOMAS ARTHUR WEBSTER.

*From the Department of Applied Physiology, National Institute
for Medical Research.*

(*Received May 10th, 1922.*)

INTRODUCTION.

IN previous papers [1921, 1922] we gave the results of observations on the composition of the day and night urine of the same subject under four different routines, each of five days duration. We showed that the differences between day and night urine were constant under a routine of complete rest in bed, a laboratory routine, a light muscular work routine and a severe muscular work routine. We found, in each case, that the total N, urea N, water and chloride were excreted in much greater amounts during the day than at night; that the amino-acids and uric acid were excreted in slightly greater amounts during the day than at night; that the acid, ammonia N and phosphate were excreted in much greater amounts during the night than during the day; and that creatinine and sulphate were more or less evenly distributed between day and night. In other words, there was an excretory rhythm which was not altered by any of the conditions of the four routines examined. In this rhythm there were definite tides during the day of water, chloride, total N and urea N, whilst at night there were phosphate, ammonia N and acid tides. It was considered that the greater excretion of acid, ammonia N and phosphate at night was due to delayed excretion of "certain fixed acids" formed in the cells during the day and that the phosphate tide at night was connected with the greater acidity at night.

To investigate further these various tides and this excretory rhythm we employed the same subject under further changes of routine. In the present paper we refer to differences between day and night urine, (1) during a routine of day starvation for 48 hours with food only at night, (2) during a routine of complete starvation for 24 hours and (3) during a reversed routine, for 96 hours, in which work was performed and food was taken during the night whilst the subject slept during the day.

METHODS.

We have kept the methods constant throughout all routines.

Our subject took his usual diet but it was controlled to exclude articles which are known to influence greatly the composition of the urine and also to keep the average daily diet for each routine as similar in substance as possible. We employed his usual diet as we were undertaking a prolonged research in which the subject would at times be exposed to somewhat strenuous conditions. By making each routine cover several—and the same—consecutive days and by keeping the quality of the average daily diet constant, we thought that errors would be lessened. Moreover, preliminary observations under his normal daily diet showed that the differences between day and night urines were constant.

We collected the urine between 7 a.m. and 5 p.m. for the day and between 5 p.m. and 7 a.m. for the night. It was not considered that these periods were the best. They were chosen as being most convenient for our research. For a large part of this research the subject lived at the laboratory.

There is no accurate way to decide whether a waste product formed in a cell is excreted as soon as it is formed by the cell, or for how long the cell retains it. All periods for collecting urine have their advantages and disadvantages. The best results will probably be attained by a comparison of figures obtained during various periods of collection from each of many subjects under various conditions.

In estimating the acidity of the urine, we employed four methods—p_H, Folin's titration method, titration to p_H 7·4, and Leathes' acidity percentage method [1919]. As other observers have pointed out we found that the agreement between these methods as regards detail was not marked, but nevertheless the broad significance of the conclusion as regards acidity was similar with each method. Thus we were able to demonstrate the presence of the so-called alkaline tide in our subjects and to determine the difference in acidity between the day and night urines by any of these methods.

Starvation Routines.

Our subject starved all day taking food only between 5 p.m. and 10 p.m., the object of this experiment being to determine whether taking food at night would alter any of the tides, particularly the acidity and total N tides. For two days this routine was followed, and on the next day the subject starved for 24 hours. Table I shows the results for both these routines, namely, "day starvation" and "complete starvation." On comparing these results with our previous work [1921, 1922] it will be noticed that practically no change from our previous figures occurred as regards the differences between day and night urine. That is total N, urea N, water and chloride were still excreted in greater quantities during the day than at night, and acidity, phosphate and ammonia N were still higher at night than during the day.

Lusk [1917] records that the total N excretion during starvation was higher during the day than at night. From our results during starvation, it was obvious that the urine, under normal conditions, was not more acid during the night than during the day because taking of food during the day produced so-called alkaline tides. We have carried out many observations on our subject with regard to the so-called alkaline tide and found that any normal meal was followed by such a tide, in many cases the tide being well marked.

During the "day starvation" routine, our subject carried out sedentary laboratory work, whilst during the "complete starvation" routine he rested in bed reading most of the day.

We found about 20 mg. of creatinine N per hour in our subject, as seen in Table I. In our previous papers [1921, 1922] owing to an error we gave about 30 mg. per hour as the figure for this subject; all the figures in these previous papers for creatinine N should be reduced by one-third and the figures for undetermined N correspondingly increased. The general conclusions, however, are not disturbed by this correction, as the error was constant.

Reversed Routine.

In the "reversed routine" our subject worked and had most meals and fluid at night whilst he slept during the day from about 10.30 a.m. to 5.30 p.m. Meals were taken at 8 p.m., 2 a.m., 5 a.m. and 8 a.m. instead of the usual times 7.30 a.m., 1 p.m., 4 p.m. and 7.30 p.m. respectively. Four consecutive days of this "reversed routine" were spent by our subject C.P. and one of us (J.A.C.). Neither subject had become accustomed to the new routine by the fourth day; although both subjects had slept well during the day they felt very sleepy at night. Table I shows C.P.'s average results for the last 72 hours of this routine and Table II A gives J.A.C.'s average results. Average figures are given because the results for any one day were very similar to these. Subject C.P.'s results show that no change from the normal rhythm was obtained except that the phosphate for the day was much increased, the night phosphate remaining about the usual figure.

In the case of J.A.C. a similar increase of phosphate during the day was noted but also there were other changes; now the acidity and ammonia N were higher during the day when sleep was taken, so that the reversal of routine produced some reversal in composition in J.A.C.'s urine (see Table II A).

It is interesting to note that after four days of complete reversal of routine and meals (including fluid) the day and night partition of water and nitrogen and of practically all other excretions remained unchanged. This points towards a fixed physiological rhythm of the cells which cannot be altered easily. It is not likely that the kidney was responsible for this rhythm. All the body cells were probably responsible. The rhythm may be due to following closely a fixed daily routine—of work, meals and sleep—for a long period. Allowance for this phenomenon must, therefore, be made where necessary when carrying out experiments of this nature.

Table I. *Average hourly results. Subject C.P.*

	48 hours' day starvation (food between 5 p.m. and 10 p.m.)		24 hours' complete starvation		72 hours' reversed routine	
	DAY	NIGHT	DAY	NIGHT	DAY	NIGHT
	(Sedentary work, no food)	(Food and sleep)	(Rest in bed)	(Sleep)	(Sleep)	(Work and food)
Amount cc.	56·0	25·0	88·0	17·9	71·3	34·5
Acidity %	46·4	78·6	33·0	58·3	42·4	70·4
Titratable acidity (Folin) cc. N/10	6·0	15·7	6·0	10·0	8·0	12·4
Total acidity, cc. N/10...	15·0	30·7	14·0	24·3	17·3	29·5
Total N g.	·379 (100)*	·307 (100)*	·392 (100)*	·195 (100)*	·373 (100)*	·337 (100)*
Urea N g.	·315 (83·00)	·248 (80·80)	·305 (77·80)	·142 (72·80)	·313 (84·00)	·250 (74·20)
Ammonia N (A) g. ...	·010 (2·64)	·019 (6·19)	·010 (2·55)	·018 (9·23)	·010 (2·68)	·018 (5·34)
Ammonia N (B) g. ...	·012	·021	·011	·020	·013	·024
Amino-acid N g. ...	·002 (0·52)	·002 (0·65)	·001 (0·25)	·002 (1·02)	·003 (0·80)	·006 (1·78)
Creatinine N g.	·020 (5·28)	·019 (6·19)	·019 (4·85)	·020 (10·25)	·021 (5·63)	·022 (6·53)
Uric acid N g.	·006 (1·58)	·004 (1·30)	·006 (1·53)	·003 (1·54)	·006 (1·61)	·007 (2·07)
Undetermined N g. ...	·026 (6·98)	·015 (4·87)	·050 (13·02)	·010 (5·16)	·020 (5·28)	·036 (10·08)
Chloride (NaCl) g. ...	·472	·213	·497	·134	·335	·261
Phosphate (P₂O₅) g. ...	·068	·082	·068	·078	·093	·081
Total S (SO₃) g. ...	·055 (100)	·068 (100)	·056 (100)	·041 (100)	·063 (100)	·072 (100)
Inorganic S (SO₃) g. ...	·039 (70·9)	·052 (76·5)	·044 (78·6)	·035 (85·4)	·054 (85·7)	·057 (79·2)
Ethereal S (SO₃) g. ...	·006 (10·9)	·009 (13·2)	·005 (8·9)	·001 (2·4)	·002 (3·2)	·006 (8·3)
Neutral S (SO₃) g. ...	·010 (18·2)	·008 (10·3)	·007 (12·5)	·005 (12·2)	·007 (11·1)	·009 (12·5)
Purine N g.	·0015	·0026	—	—	—	—
Calcium (CaO) g. ...	—	—	—	—	·011	·014

(A) Van Slyke's method. (B) Malfatti's method. Total acidity is ammonia (B) + titratable acidity.

* Figures in brackets are percentages.

Table II A. *Hourly averages. Subject J.A.C.*

	Reversed routine		Ordinary routine	
	Night Work and food	Day Sleep	Day Work and food	Night Sleep
Amount cc.	43·1	76·0	60·7	31·5
Acidity %	53·8	57·3	42·8	72·3
Titratable acidity, Folin cc. N/10	11·2	13·6	8·2	10·7
Total acidity cc. N/10	28·8	35·3	26·5	30·6
Total N g.	·408	·596	·533	·451
Ammonia N (B) g.	·025	·030	·026	·028
Phosphate (P₄O₅) g.	·074	·097	·055	·075

Table II B. *Hourly averages. Subject C.P.*

Experiment no.	Acidity ⁰/₀	Phosphate (P₂O₅) g.	Titratable acidity cc. N/10	Total acidity cc. N/10	Ammonia N (A) g.	Total N g.	Time	Sleep	Routine
1	20·6	·045	5·4	18·8	·014	·386	day	−	Rest in bed
2	70·2	·080	18·2	36·7	·022	·296	night	+	
3	44·8	·065	10·4	25·4	·015	·428	day	−	Ordinary
4	64·6	·080	16·8	33·0	·019	·332	night	+	
5	50·0	·066	11·6	28·1	·011	·447	day	−	Light muscular work
6	64·0	·075	15·4	31·6	·010	·357	night	+	
7	90·0	·070	14·0	29·0	·015	·446	day	−	Severe muscular work
8	70·0	·080	14·4	30·7	·019	·414	night	+	
9	46·4	·068	6·0	15·0	·010	·379	day	−	Day starvation
10	78·6	·082	15·7	30·7	·019	·307	night	+	
11	33·3	·068	6·0	14·0	·010	·392	day	−	Complete starvation
12	58·3	·078	10·0	24·3	·018	·195	night	+	
13	42·4	·093	8·0	17·3	·010	·373	day	+	Reversed
14	70·4	·081	12·4	29·5	·018	·337	night	−	

In both subjects the phosphate for the day and therefore for the whole 24 hours was distinctly increased. This increase was probably due to some special nervous metabolism connected either with sleep itself or with the desire to sleep. In Table II B, we have drawn up some results from all our experiments including those previously published [1921, 1922]. These show that a high phosphate excretion always accompanied sleep and that the higher of the day and night figures for phosphate excretion in each routine could be separated from every other factor in our experiments but sleep (see Table II B).

SLEEP AND ACIDITY OF URINE.

In our previous papers we found that sleep was accompanied by a definite increase of ammonia, acidity and phosphate and a relative increase of sulphate. We suggested that there was a delayed excretion of "certain fixed acids" formed in the cells during the day, and that when formed in certain quantities, they were responsible for sleep and fatigue. In our present paper the higher figures for phosphate whether found by day or by night always accompanied sleep and sleep was independent of every other factor including acidity (see Table II B). Therefore, although these "certain fixed acids" might have been responsible for sleep and although higher acidity usually accompanied sleep it was not a necessary accompaniment of sleep (see experiment 13, Table II B). This is contrary to the finding of Leathes [1919], who considered that the CO_2 tension of alveolar air was increased at night above that for the day, as a result of depression of the respiratory centre during sleep. Collip [1920] also considered that an increase of C_H of the blood at night was due to sleep.

We see no reason to abandon the suggestion that sleep might be due to "certain fixed acids" since there was in the case of subject C.P. in the "reversed routine" an increased acidity of the urine previous to sleep although not accompanying it. He was obviously sleepy during the time of excretion of the larger amount of acid, but kept awake purposely. If the high acidity was due to "certain fixed acids" formed by the activity of the cells and capable of causing sleep, the cells would only recover after sleep itself and not after excretion of the "certain fixed acids." However, it was obvious that the subject had not become accustomed to the "reversed routine," so that our results really belonged to a transitional period and it is at present difficult to draw conclusions. We have carried out some observations on a subject who was somewhat more accustomed to work and food at night, with sleep during the day. He worked on a night shift every third week. The results we have obtained from him are in complete agreement with the results here described. Results from subjects accustomed to long periods of night work are required.

PHOSPHATE EXCRETION.

We conclude from all our results regarding the higher figure for day and night phosphate excretion in each routine that it was intimately related to metabolism either connected with sleep or occurring at the same time as sleep. We were able to separate the higher phosphate excretion in each routine from every factor, but sleep (see Table II b). Although, as a rule, the higher phosphate excretion accompanied the higher acid and the higher ammonia N excretion there was a marked exception to this rule (see experiment 13, Table II b).

Broadhurst and Leathes [1920] have suggested that the phosphate tide at night may be connected with some special muscular or nervous metabolism. Our previous results indicated that there was no connection with muscular metabolism and less clearly that there was none with nervous metabolism. Broadhurst and Leathes also found that the phosphate tide was not dependent upon food. Our results during the starvation routines confirm this finding. Fiske [1921], who considered that the phosphate tide was due in part to "retention of phosphorus," also showed that the phosphate curve was not affected by the taking of sodium bicarbonate so that the phosphate curve was not connected with excretion of acid. This may be the case, but there is much evidence from other observers that phosphate may have, as one of its functions, the removal of acid. Haldane [1921] found that acid was removed by phosphate if there was any phosphate available. In our subject C.P. a high acidity, as interpreted by a high titratable acidity and a high ammonia N excretion, was always accompanied by a high phosphate excretion.

We estimated the calcium in the urine and found that it was not connected with the increase of phosphate, the calcium being low when the phosphate was high and *vice versa*. When in the "reversed routine" the phosphate was absolutely increased, no such marked change was noted in the amount of calcium excreted.

According to text books [Hawk, 1919], some investigators hold that during extensive decomposition of nervous tissue the phosphate is increased. Mendel found that phosphate was increased after sleep produced by potassium bromide or chloral hydrate, so that it is possible that special metabolism connected with sleep was the main factor for the higher of the day and night phosphate excretions in our subject. Nervous metabolism may have been concerned.

It is interesting to note that in children we have found no such correlation; in them the phosphate followed closely the total N. In children, the condition is rendered more complex by the presence of growth metabolism together with the maintenance metabolism. Children did not show the same rhythm, probably because of different conditions with regard to sleep.

Summary.

1. Observations on day and night urine during a routine of day starvation, a routine of complete starvation and a reversed routine—that is with work and meals at night and with sleep during the day—are recorded.

2. Evidence was obtained that the urine was more acid at night, neither because the taking of food during the day produced alkaline tides, nor because the respiratory centre was depressed during sleep.

3. After four days of complete reversal of habit most of the differences between day and night urine still remained as before, so that there must be a fixed physiological rhythm connected with excretion by the body cells. Thus, although in the "reversed routine," most of the fluid was swallowed at night, the greater amount of urine was still excreted during the day which now included the sleep period. Also, although most of the food taken was eaten at night, the total N excretion still remained higher during the day than at night.

4. In the excretory rhythm referred to, the total N, urea N, water and chloride were excreted in greater amount during the day whilst the ammonia N, acidity and phosphate were higher during the night than during the day.

5. The higher figures for phosphate whether found by day or by night always accompanied sleep and could be separated from every other factor except sleep; and, with one exception, the higher phosphate excretion accompanied both the higher acidity and the higher ammonia N excretion in all routines examined. An absolute increase in phosphate occurred in the " reversed routine."

REFERENCES.

Broadhurst and Leathes (1920). *J. Physiol.* **54.** *Proc.* xxviii.
Campbell and Webster (1921). *Biochem. J.* **15,** 660.
—— (1922). *Biochem. J.* **16,** 106.
Collip (1920). *J. Biol. Chem.* **41,** 473.
Fiske (1921). *J. Biol. Chem.* **49,** 171.
Haldane (1921). *J. Physiol.* **55,** 272.
Hawk (1919). Practical Physiological Chemistry, p. 435.
Leathes (1919). *Brit. Med. J.* ii, 165.
Lusk (1917). The Science of Nutrition, p. 110.

XLVI. NOTE ON A NEW TANNASE FROM *ASPERGILLUS LUCHUENSIS*, INUI.

By MAXIMILIAN NIERENSTEIN.

*From the Biochemical Laboratory, Chemical Department,
University of Bristol.*

(*Received May 2nd, 1922.*)

PAULLINIA tannin[1] is hydrolysed by emulsin, dextrose and β-gambier-catechincarboxylic acid being produced [Nierenstein, 1922]. It was therefore surprising to find that tannase, which had been prepared by growing *Aspergillus Luchuensis*, Inui in a solution of gallotannin and which is known to hydrolyse gallotannin into dextrose and gallic acid [Rhind and Smith, 1922], had no effect on paullinia tannin. This suggested the possibility that the hydrolysing properties of tannase depended on the medium in which it had been produced by the fungus and that a catechutannin, such as paullinia tannin, required a tannase, which had been formed in a solution of a catechu- and not gallotannin. This has actually been found to be the case, since *Aspergillus Luchuensis* yields in a medium in which gallotannin had been replaced by catechutannin from cube-gambier (*Ourouporia Gambier*, Baill), a new tannase, which hydrolyses paullinia tannin, but which has no action on gallotannin. We have therefore to distinguish between these two kinds of tannase and it is proposed to refer to them as *gallotannase* and *catechutannase* respectively. In this connection reference must be made to what may in time prove a dangerous practice, which has recently been introduced by Freudenberg and Vollbrecht [1921]. These chemists prepare tannase by growing an unknown species of *Aspergillus* in the aqueous extract of ground myrobalams, the fruit of *Terminalia chebula*, Retz. Such a medium must obviously affect the general properties of the tannase and ought to be avoided.

Freudenberg's method [1920], in which gallotannin is replaced by an aqueous solution of the residues of cube-gambier, from which the catechin is removed by extraction with ethyl acetate, is used for the preparation of catechutannase. The catechutannase thus obtained resembles gallotannase in

---

[1] This catechutannin is present in the Seeds of *Paullinia Cupana*, H.B. and K., which is a synonym of *P. sorbilis*, Mart. These Seeds are used for the preparation of the paste known as "pasta guaraná" and the tannin is consequently also referred to as guarana tannin [compare, for example, Perkin and Everest, 1918]. Harvey [1921] and also Procter [1922] have wrongly assigned the name *P. sorbilis* to "guara," which contains a gallotannin and not a catechutannin, as "guara" signifies *Cupania americana*, L. For this information I am indebted to the authorities of Kew Gardens, who also inform me that in Cuba the names "guaraná" and "guara" are apparently used for *Cupania macrophylla*, A. Rich. and *Guarea trichiliodes*, L. respectively.

appearance and produces when added to a solution of paullinia tannin (0·5 g. catechutannase are used for 2 g. paullinia tannin in 150 cc. of water to which 5 cc. of chloroform are added and the mixture kept in a dark incubator at 23° for ten days), a bulky precipitate, which when filtered off crystallises from water in small, pointed needles, which melt at 252–253°, carbon dioxide being evolved. These are in every respect identical with the inactive β-gambier-catechincarboxylic acid, previously obtained by the action of emulsin. (Found: C = 57·2; H = 4·6. Calculated: C = 57·5; H = 4·2 %.) For further identification a small quantity of the acid was methylated with diazomethane, when the corresponding pentamethoxy methyl ester was obtained. It crystallises from light petroleum in long needles, which melt at 74°. This melting point is not depressed on admixture with the same substance (m.p. 74°) previously obtained by the action of diazomethane on the acid.

The filtrate from the crude β-gambier-catechincarboxylic acid, freed from unchanged paullinia tannin with the aid of lead acetate and hydrogen sulphide, contains dextrose, which is identified as the dextrosazone (m.p. and mixed m.p. 200–203°).

In conclusion it is interesting to note that catechutannase produces the identical β-gambier-catechincarboxylic acid when added to the aqueous extract of fat-free cocoa-beans or to a solution of catechutannin from cube-gambier. This indicates a very close relationship between these tannins which has so far not been suspected.

REFERENCES.

Freudenberg (1920). *Ber.* **53**, 958.
Freudenberg and Vollbrecht (1921). *Zeitsch. physiol. Chem.* **116**, 277.
Harvey (1921). Tanning Materials, p. 23.
Nierenstein (1922). *J. Chem. Soc.* **121**, 23.
Perkin and Everest (1918). The Natural Organic Colouring Matters, p. 442.
Procter (1922). The Principles of Leather Manufacture, p. 325.
Rhind and Smith (1922). *Biochem. J.* **16**, 1.

XLVII. A QUALITATIVE TANNIN TEST.

By ETHEL ATKINSON and EDITH OLIVE HAZLETON.

*From the Biochemical Laboratory, Chemical Department,
University of Bristol.*

(*Received May 15th, 1922.*)

TANNING consists in the fixation of the tannins by animal fibre. A reliable tannin test must therefore demonstrate this specific property and it is remarkable that no such test has so far been devised. The reactions which are generally used for the identification of the tannins, such as the colorations produced by iron salts or the precipitations which are given by potassium dichromate and gelatin [compare, for example, Onslow, 1920], obviously only indicate some incidental properties of the tannins. These reactions are also shared by other organic substances, thus phenols and hydroxybenzoic acids are known to give colorations with iron salts, whilst both gallic acid and gallotannin are precipitated by potassium dichromate [Drabble and Nierenstein, 1907]. Similarly gelatin is not only precipitated by tannins, but also by gum arabic [Pelletier, 1813], starch [Tollens, 1914], inulin [Tollens, 1914], methyl gallate [Nierenstein, 1905, 1912; Fischer, 1919; Freudenberg, 1920] and other substances. The want of a specific tannin test was therefore felt for many years in this laboratory, with the result that the following method has been devised by us at the suggestion of Dr Nierenstein.

A small piece of gold-beater's skin[1], about $\frac{1}{2}$ inch long and $\frac{3}{4}$ inch wide, is pinned on a flat surface of paraffin-wax, which is prepared by pouring melted paraffin-wax into a watch-glass. The skin is covered with a few cc. of water, in which it is left soaking for five minutes so as to make it better permeable to the tannins. The water is then poured off and the skin tanned by covering it with 1 cc. of a tannin solution, which is prepared by extracting for half-an-hour on a boiling-water bath 1 g. of the material to be tested with 50 cc. of water. After many experiments we have come to the conclusion that half-an hour's tanning suffices even for the weakest tannin solution, but that solutions of usual strength (about 1–2 %) tan in a few minutes. It is, however, advisable to tan not less than 15 minutes. The tanned skin is washed for two minutes at a constant drip of two drops per second and then stained for five

---

[1] Gold-beater's skin is the outside membrane of the large intestine of the ox. It has many advantages over other animal tissues. It is easily obtainable (the material used by us was supplied by the British Drug Houses) and it is very thin (1 yard × 6 inches weighs only about 3 g.). It consequently tans rapidly and requires only little tannin, which should make this test of use not only for chemical, but also for botanical work.

minutes with 1 cc. of a 1 % solution of ferric chloride. It is then again washed, dried and mounted for reference.

We find that only tanned gold-beater's skin is stained by ferric chloride. By this method we have been able to demonstrate the presence of tannins in the following plant products[1], all of which are known to contain tannins:

Aleppo galls, Chinese galls, Chinese plum galls, blue Basra galls, white Basra galls, Knopper galls, Sumac, Mangrove bark, Mimosa bark, Myrobalams, Valonia, Oak bark, Quebracho wood, Hemlock bark, Divi-Divi, Algarobilla, Canaigre, *Pistacia Lentiscus*, Golden Wattle, *Acacia Arabica*, Larch bark, Tea, Coffee, and Cocoa.

Gallotannin behaves exactly in the same manner, whereas gallic acid and pyrogallol, both of which are known not to possess tanning properties, are not fixed by gold-beater's skin, which is consequently not stained by ferric chloride. Identical results were also given by the following substances, all of which are known not to possess tanning properties:

Phenol, catechol, quinol, resorcinol, phloroglucinol, salicylic acid, protocatechuic acid and β-resorcylic acid.

The combination between the tannins and the gold-beater's skin is of a very permanent character. This is evident from the fact that the tanned and subsequently stained gold-beater's skin may be decolorised with dilute hydrochloric acid and again re-stained with ferric chloride. This process of decolorising and re-staining may be repeated several times without the slightest effect on the colour of the re-stained material.

In conclusion we would like to suggest that this test be referred to as the "gold-beater's skin test for tannins." We also wish to thank Prof. McCandlish of the University of Leeds and Mr M. C. Lamb of the Leathersellers' Technical College at London for the different tanning materials used in this investigation. Our thanks are also due to the Research Fund Committee of the Chemical Society for a grant allocated to Dr Nierenstein for his work on gallotannin, from which the expenses of the present investigation were defrayed.

REFERENCES.

Dekker (1913). Die Gerbstoffe.
Drabble and Nierenstein (1907). *Biochem. J.* 2, 96.
Fischer (1919). *Ber.* 52, 821.
Freudenberg (1920). *Ber.* 53, 236.
Nierenstein (1905). *Collegium*, p. 307.
—— (1912). *Ber.* 45, 837.
Onslow (1920). Practical Plant Biochemistry, p. 91.
Pelletier (1813). *Ann. Chim.* 87, 106.
Tollens (1914). Kurzes Handbuch der Kohlenhydrate, pp. 525 and 551.

[1] We refrain from giving the Latin names of the plants we have used, since they are well-known tanning materials, which can be found in any book which deals with them [compare for example, Dekker, 1913].

XLVIII. THE ORIGIN OF THE VITAMIN A IN FISH OILS AND FISH LIVER OILS.

By JACK CECIL DRUMMOND and SYLVESTER SOLOMON ZILVA,

WITH THE CO-OPERATION OF

KATHARINE HOPE COWARD (*Beit Memorial Research Fellow*).

From the Biochemical Laboratories Institute of Physiology, University College, London, and Biochemical Department, Lister Institute, London,

(*Received May 18th, 1922.*)

DURING the last two years the authors have been collaborating in an exhaustive investigation of the nutritive value of the edible oils and fats with particular reference to the dietary unit referred to as vitamin A. Of the time spent in this work a very large proportion has been devoted to the study of the oils of marine origin and in particular the fish liver oils, chiefly because it has been shown by Zilva and Miura [1921] that these oils represent by far the most valuable sources of the vitamin of all the groups into which the edible oils and fats are divided. Our study of the liver oils has been naturally divided into several phases each concerned with a distinct aspect of the subject and the present communication deals with the question of the origin of the relatively enormous stores of vitamin A which are found in the livers of many species of fish. The results set out in this paper supplement some recorded by Hjort who at our suggestion has studied a number of marine organisms in connection with this enquiry [1922].

It is now generally agreed that the higher land animals derive their supplies of vitamin A either directly or indirectly from green plants and the probability of a similar relationship existing in the sea has been suggested by Jameson, Drummond and Coward [1922]. The fundamental dependence of all marine animal life on the marine flora, particularly the microscopic flora, has been recognised for many years past, and the studies of Jameson, Drummond and Coward have shown that a typical marine diatom grown in pure culture can synthesise relatively large quantities of vitamin A, whilst Coward and Drummond [1921] showed earlier that higher forms of marine algae, whilst far less potent sources of this factor than diatoms, may contain a concentration approximately as high as that of typical land plants such as cabbage.

There appear to be no adequate grounds for doubting that the ultimate origin of the vitamin A in marine oils and liver oils is represented by the marine plants possessing photocatalytic pigments, particularly those which are included in the term plankton. Our studies of fish eggs show us that the young fish normally begins its life with a considerable store of vitamin *A* derived from the yolk-sac. We have reasons for believing that there may be a transference of vitamin from the liver or tissues of the spawning fish to the eggs although as yet our experiments on this subject are incomplete.

Apart from our experimental work the existence of such a transference would appear probable from our knowledge concerning the transport of vitamin from tissues to the eggs or milk of certain terrestrial species.

Probably the vitamin-*A* content of the yolk-sacs is sufficient for the requirements of the developing larvae for some time, but it is well known that the period between the stage at which the contents of the yolk-sac are nearly absorbed and that at which the young fish are entirely dependent on external food is a very critical one.

The careful studies of Lebour [1918, 1919, 1, 2; 1920, 1921] and others have shown us that post larval fish feed very largely on copepods. Her work may for our purpose be summarised by quoting two paragraphs from one of her most valuable papers [1920].

"One finds that certain copepods and other entomostraca constitute by far the larger part of the food of nearly all the very young fish, and that usually each species of fish selects its own favourite food to which it keeps, indiscriminate feeding seldom or never taking place, and one can usually assign to each fish its own particular food." "Few fish are vegetarians, and it is unusual for any but the youngest fish to eat diatoms." "Very young herring or sprat and a few others often contain green remains which probably belong to some algae, and occasionally diatoms can be recognised in this but even before the yolk sac is absorbed the gut may contain larval molluscs and small crustaceans."

But whilst there is little or no direct feeding of fish on diatoms these and other microscopic organisms form the diet of the copepods and larval molluscs which play so important a rôle in the nutrition of young fish. The relationship between these two groups of organisms has been traced by a number of investigators with some care, but it is only necessary here to call attention to the very rapid multiplication of the plant life of the sea which takes place in the spring and which is intimately connected, as Moore has shown [1921], with the increase in light intensity during that period. Some short time after the rapid rise in diatoms and other plant organisms there is an associated rise in the numbers of microscopic animals, particularly copepods, and the occurrence of these two related phenomena has a very important bearing on the survival and development of the young fish hatched out at that season [Hjort, 1914].

By the kindness of Dr O. Borley of the Fisheries Laboratory, Lowestoft, we were enabled to study the vitamin-*A* content of a number of plankton samples collected during the spring of 1922.

The samples after being dredged, sifted through muslin and pressed, were preserved in an equal bulk of absolute alcohol and sent directly to us in London. On arrival they were evaporated to dryness *in vacuo* at 40° and ground well to mix the organisms thoroughly with the material extracted by the alcohol. The dry material was administered to rats in the usual manner for testing for vitamin A, and all four samples were found to be highly potent as sources of that factor when administered in daily supplements of 0·1 g. to the rats.

The four samples studied were representative of the staple food of a large majority of the small fish. The analysis of the samples for which we are indebted to Dr Borley is given below.

Table I. *Plankton samples.*

	1	2	3	4
Locality	52·2 N. x 3·8 E.	Sandetti Lightship	Pas de Calais	Pas de Calais
		51·15 N. x 1·52 E.	50·58 N. x 1·30 E.	50·58 N. x 1·30 E.
Principal contents .				
(1) Copepods	*Temora longicornis** Pseudocalanus elongatus Calanus finmarchicus Acartia longiremus	*Temora Pseudocalanus* Calanus Acartia Centrophages typicus	*Temora* Pseudocalanus Calanus	
(2) Mysids	Schistomysis spiritus Gastrosaccus spinifer		Macropsis Gastrosaccus	
(3) Amphipods	Paratylus swammerdami			
(4) Larvae	Decapod	Decapod	Decapod	Decapod
(5) Miscellaneous	Sagitta hexaptera Pleurobrachia	Oikiopleura Polynoid polychaetes	Sagitta	
(6) Post-larval fish	Herring			

* The species in italics were the most numerous in each sample.

Samples 1 and 2 were particularly interesting in that they consisted almost entirely of the very widespread genera of copepods Temora, Pseudocalanus and Calanus. These organisms directly or indirectly represent a very important stage in the food supply of the cod and thrive on diatoms such as *Nitzschia closterium* [Allen and Nelson, 1910], which has been shown to be a very rich source of vitamin A [cf. Jameson, Drummond and Coward, 1922].

Small fish. Having shown that the food of the majority of the small fish and certain other marine organisms is relatively rich in vitamin A, the next step was to study whether this factor is transferred to the tissues of the species which feed upon them. For this purpose we have taken one or two representative species from different groups without attempting to study the matter exhaustively, and have where possible selected those which most frequently form the food of the cod and the related species which yield large quantities of marketable medicinal oils.

The food of the cod is not at all times the same. In Norway the early spring rise in diatoms leads to the vast multiplication of copepods, which form the food of enormous shoals of small fish which come in to the northern coast (Finmarken) to spawn. These shoals are chiefly composed of the caplin or capelan (*Mallotus vilosus*) and they are followed by great numbers of cod, coal-fish, haddock and other species which devour them in enormous quantities [Hjort, 1914]. Other species of young fish are frequently consumed by these larger species, but their diet also includes certain shell fish, larger crustacea, salps, and especially in the case of cod on the Newfoundland banks, squid.

The results of our few tests are given in Table II, but they only represent the testing of a few miscellaneous products we had on hand. No attempt has been made to examine these and similar products exhaustively.

Table II.

Species	Approximate daily dose to rats	Results
Caplin, *Mallotus vilosus*	0·1 g.	Rapid growth
Sprat, *Clupea sprattus*	1·0 g.	,, ,,
Young herring, *Clupea harengus*	0·4 g.	,, ..
Mussel, *Mytilus edulis*	4·0 g.	Good
Periwinkle, *Littorina littorea*	1·0 g.	,,
Shrimp, *Crangon allmanni*	0·01 g.*	

* Of unsaponifiable matter.

Of these species the periwinkle is interesting as being an example of a direct transference of the vitamin from the algae on which it feeds (Enteromorpha, etc.). The fat extracted from these animals after feeding is deeply coloured with chlorophyll and other pigments derived from these green weeds.

It would be an almost insuperable task to demonstrate an actual transference of vitamin *A* right through from a diatom such as Nitzschia to its final location in the liver of the cod, but it would appear that we are justified. in assuming that such transference does indeed take place by relying on the fact that as yet synthesis of this dietary factor by an animal organism has not been demonstrated, whereas such synthesis is readily carried out by certain plants both marine and terrestrial. The dependence of marine animals on certain products of the synthetic powers of marine plants may indeed be as fundamental as that which exists between these two great groups of organisms on land, for many examples may be recalled such as that given by Fabre-Domergue and Bietrix [1900] who showed that a supply of microscopic plant life would save the lives of larval fish (soles) which were failing from malnutrition in spite of the fact that their yolk was not completely absorbed.

We have included in this paper a few of the unpublished results obtained with molluscs by the late Dr H. Lyster Jameson as well as the examination of certain material he had collected for other purposes since they have a certain bearing on the main subject.

SUMMARY.

1. The ultimate origin of the vitamin A found in the oils derived from fish, and particularly the fish liver oils, would appear to be chiefly the unicellular marine plants. Except very occasionally these organisms are not consumed directly by the fish.

2. The extraordinary rise in the number of marine plants which begins as soon as the intensity and duration of sunlight increase early in the year is followed by a rapid rise in the organisms, largely copepods, and larval decapods and molluscs, whose growth and development are dependent on their food supply which consists of minute plants. These minute animals, which form a large proportion of plankton, contain relatively large quantities of vitamin A presumably derived from the diatoms on which they have thriven.

3. The plankton forms the staple food of innumerable species of marine animals from small fish to some whales. This no doubt accounts for the presence of vitamin A in the tissues or fat depots of these animals.

4. The fish which yield the large bulk of the liver oils, cod, haddock, coal-fish, etc., feed on many species. At one season in northern Norway and Newfoundland they feed extensively on a small fish, the capelan or caplin, which has been found to be very rich in vitamin A, doubtless derived from its food.

5. The origin of the vitamin A in fish oils and fish liver oils has therefore been traced back to the synthetic powers of the marine algae which form the fundamental food supply of all marine animals.

It is a pleasure to acknowledge the valuable advice and assistance given to us by Dr Allen and Dr Lebour of the Marine Biological Laboratory, Plymouth, Dr Borley of the Fisheries Laboratory, Lowestoft, Professor Hjort and our late colleague Dr H. Lyster Jameson.

We also beg gratefully to acknowledge a financial grant from the Medical Research Council which defrayed the cost of this investigation.

REFERENCES.

Allen and Nelson (1910). *Quart. J. Micros. Sci.* 55, 361.
Coward and Drummond (1921). *Biochem. J.* 15, 530.
Fabre-Domergue and Bietrix (1900). *Bull. de la marine marchande*, Paris.
Hjort (1914). *Rapports, Conseil permanent international pour l'Exploration de la Mer*, Copenhagen, 20.
—— (1922). *Proc. Roy. Soc.* B. 93, 440
Jameson, Drummond and Coward (1922). *Biochem. J.* 16.
Lebour (1918). *J. Mar. Biol. Ass.* 11, 433.
—— (1919, 1). *J. Mar. Biol. Ass.* 12, 9.
—— (1919, 2). *J. Mar. Biol. Ass.* 12, 21.
—— (1920). *J. Mar. Biol. Ass.* 12, 262.
—— (1921). *J. Mar. Biol. Ass.* 12, 458.
Moore (1921). *J. Chem. Soc.* 149, 1555.
Zilva and Miura (1921). *Lancet*, i, 323.

XLIX. THE RESPIRATORY EXCHANGE IN FRESH WATER FISH. III. GOLD FISH.

By JOHN ADDYMAN GARDNER, GEORGE KING
AND EDWIN BOOTH POWERS.

*From the Physiological Laboratory, University of London,
South Kensington.*

(*Received May 22nd, 1922.*)

IN part I of this series of papers [Gardner and Leatham, 1914, 1] a detailed account was given of apparatus for measuring the respiratory exchange in fish, and measurements were recorded showing the influence of temperature, size of the animal, etc. on the respiratory exchange in the case of brown trout.

In this paper we give an account of experiments by the same method on gold fish.

METHOD.

Preliminary experiments showed that gold fish live at a much lower plane of metabolism than trout, and it was found necessary to modify slightly the procedure adopted with the latter animals.

In the case of trout the duration of the experiments varied from four to six hours. Gold fish, however, both at low and at ordinary temperatures use much smaller volumes of oxygen per hour, and it was found necessary to continue the experiments over periods of 24 hours in order to obtain accurate measurements. At higher temperatures more oxygen was used and satisfactory results could be obtained in experiments of six hours' duration.

In most of the 24 hour experiments it was not found practicable to continue the pumping of the air through the water during the whole period, owing to irregularities during the night of the electric power used in working the pumps, though in one or two experiments the pumping was continuous. Usually the experiment was commenced about 11 a.m. and the pumping of the air through the water bottle was continued until as late as convenient in the evening. The pump was started again early in the morning and continued until the end of the experiment. Experience showed that the absorption of oxygen was so slow that, with the large bulk of water used, the tension of the dissolved oxygen was not reduced sufficiently during the time the pumps were out of action to affect the fish. Oxygen, equivalent to that used during the 24 hours, was gradually added to the air above the water in the experimental bottle during the period before the cessation of pumping.

To obtain accurate results by this method, it is essential to take the greatest care in measuring the temperature and pressure of the air above the water at the beginning and end of the experiment.

Some minor improvements were made in the thermostatic arrangements of the apparatus used for the trout, and the pressure gauges were fitted with spirit levels. Care was also taken to paint over all joints, and particularly the line of contact of the rubber bung and the neck of the bottle with a rapidly drying acid proof enamel paint. The gold fish used were about five to six inches long and averaged about 50 g. in weight. They were kept in a large tank outside in slowly running water. The tank had a gravel bottom and contained suitable water plants. The fish were fed in the usual way and remained perfectly healthy during the long period—more than a year—over which the experiments extended.

EXPERIMENTAL RESULTS.

A detailed example of the method of calculation was given in the paper referred to above [1914, 1, pp. 379–380].

In the following protocols we give only details of the number and weight of fish used, the temperatures, and the *total* initial and final volumes of *free and combined carbon dioxide* of *oxygen* and of *nitrogen* in cc. reduced to 0° and 760. The volume of nitrogen should theoretically remain constant and the small positive and negative differences actually found afford a ready indication of the experimental errors.

Experiments at low temperature.

(1) Seven fish; weight 362 g.; individual variations 45–60 g. The fish prior to experiment had lived outside in cold weather. Initial temperature of water in experimental bottle 5·8°, final temperature 6·1°. Duration of experiment 22·43 hours.

CO_2	Initial	1352 cc.	Final	1459·4 cc.	Difference	+107·4 cc.
Oxygen	„	5646·8	„	5454·4	„	−192·4
Nitrogen	„	20356·4	„	20349·3	„	−7·1

Nitrogen error −0·034 %.

(2) Seven fish; 348 g.; variations 42–59 g. I.T. 7·4°, F.T. 6·2°. These fish had been kept at from 3° to 6° for several days prior to the experiment. Duration 23·17 hours.

CO_2	Initial	1571·8 cc.	Final	1689·1 cc.	Difference	+117·3 cc.
Oxygen	„	5162·2	„	5069·1	„	−93·3
Nitrogen	„	1868·8	„	18705·4	„	+18·6

Nitrogen error +0·099 %.

(3) Twelve fish; 601 g.; variations 38–65 g. These fish had had a long spell of mild winter weather prior to experiment. I.T. 6·15°, F.T. 1·4°. Duration 23·75 hours.

CO_2	Initial	1475 cc.	Final	1667 cc.	Difference	+192 cc
Oxygen	„	5261·7	„	5048·8	„	−212·9
Nitrogen	„	18991·9	„	19036·7	„	+44·8

Nitrogen error +0·23 %.

small. Seven fish; 382 g.; variations 40–65 g. I.T. 6·2°, F.T. 6·3°. The fish had mild winter temperature prior to experiment. Duration 23·78 hours.

CO₂	Initial 2167 cc.	Final 2230·3 cc.	Difference +63·3 cc.
Oxygen	„ 4321·9	„ 4181·8	„ −140·1
Nitrogen	„ 15469·4	„ 15426·4	− 43·0

Nitrogen error −0·28 %.

(5) Seven fish; 368 g.; variations 40–60 g. These fish had been in water covered with a film of ice for the previous three days. I.T. 5·7°, F.T. 6·05°. Duration 23 hours.

CO₂	Initial 1648·6 cc.	Final 1781·5 cc.	Difference +132·9 cc.
Oxygen	„ 5216·4	„ 5091·2	„ −125·2
Nitrogen	„ 18894·4	„. 18884·1	− 10·3

Nitrogen error −0·054 %.

At medium temperatures.

(6) Seven fish; 355 g.; variations 35–60 g. I.T. 13·55°, F.T. 16·7°. Duration 24·25 hours.

CO₂	Initial 2066·9 cc.	Final 2459·2 cc.	Difference +392·3 cc.
Oxygen	„ 4427·2	„ 4020·7	„ −406·5
Nitrogen	„ 16044·3	„ 16017·6	− 26·7

Nitrogen error −0·17 %.

(7) Seven fish; 355 g.; variations 35–60 g. I.T. 13·55°, F.T. 15·9°. Duration 23·25 hours.

CO₂	Initial 2123·9 cc.	Final 2392·0 cc.	Difference +268·1 cc.
Oxygen	„ 4413·4	„. 3976·3	„ −437·1
Nitrogen	„ 15966·4	„ 15931·4	− 35·0

Nitrogen error −0·22 %.

At higher temperatures.

(8) Seven fish; 370 g.; variations 45–62 g. I.T. 21·4°, F.T. 20·6°. Duration 23·18 hours. In this experiment the valves of the pumping apparatus were out of order and the air could not be pumped through the water. At the end the water in the experimental bottle had a strong faecal smell.

CO₂	Initial 1518·4 cc.	Final 2256·5 cc.	Difference +738·1 cc.
Oxygen	„ 5098·7	„ 4521·2	. „ −577·5
Nitrogen	„ 18551·2	„ 18519·1	− 32·1

Nitrogen error 0·17 %.

(9) Seven fish; weight 361 g.; variations 45–63 g. I.T. 20·7°, F.T. 20·4°. Duration seven hours. At the end no objectionable smell was noted.

CO₂	Initial 1649·1 cc.	Final 1855·2 cc.	Difference +206·1 cc.
Oxygen	„ 4640·8	„ 4354·0	„ −286·8
Nitrogen	„. 16805·0	„ 16819·9	+ 14·9

Nitrogen error +0·088 %.

The results of these experiments are given in Table I.

Table I.   ↄ thↄ

No. of exp.	Average weight of fish used in g.	Volume of oxygen consumed per fish per hour in cc. at 0° and 760	Volume of CO₂ produced per fish per hour in cc at 0° and 760	Volume of oxygen consumed per kilo of fish per hour in cc. at 0° and 760	Volume of CO₂ produced per kilo of fish per hour in cc. at 0° and 760	Respiratory quotient
At low temperature:						
(1)	51·7,	1·225	0·684	23·70	13·23	0·56
(2)	49·7	0·574	0·723	11·56	14·55	1·26
(3)	50·0	0·747	0·674	14·91	13·45	0·98
(4)	54·5	0·842	0·380	15·42	6·97	0·45
(5)	52·5	0·778	0·825	14·79	15·70	1·06
	Mean 0·833		0·657	16·07	12·78	
At medium temperature:						
(6)	50·7	2·395	2·309	47·22	45·54	0·97
(7)	50·7	2·686	1·647	52·96	32·48	0·61
	Mean 2·541		1·978	50·09	39·01	
At higher temperature:						
(8)	52·8	3·560	4·550	67·35	86·08	1·28
(9)	51·6	5·853	4·207	113·50	81·55	0·72
	Mean 4·706		4·379	90·43	83·81	

INFLUENCE OF TEMPERATURE.

It will be seen from Table I that the volume of oxygen consumed per fish, or per kilo of fish, increases with the temperature and very roughly in proportion to the temperature.

We should scarcely expect to find any exact proportion in experiments on the living animal, more especially as, owing to the long period over which the work extended, different sets of fish were used, and some latitude must be allowed for individual idiosyncrasy. Further in most of the experiments it was necessary to keep the animals in the experimental bottle for about 24 hours and during some hours the pumping of air was discontinued, so that during some part of the experiment the oxygen tension in the water would be below saturation. Experiments 8 and 9 at higher temperatures need some comment, as they show marked difference in the oxygen absorption, though most of the fish used were the same in each.

In experiment 8, which was of 24 hours' duration, the valves of the apparatus went wrong at the beginning, so that it was impossible to pump the air of the bottle through the water. Consequently the oxygen tension in the water progressively decreased, and at the end the partial pressure of the oxygen in the water was only 0·583 % of an atmosphere. At the end it was also noticed that the water had an objectionable faecal smell, a condition not found in any of the other experiments. It seemed probable therefore that the oxygen consumption figure was below normal. The experiment was therefore repeated (No. 9) and the duration was reduced to seven hours. Pumping was continuous. The oxygen figure is markedly higher than in No 8, and the respiratory quotient approximately normal for an unfed animal.

In the experiments at low and at normal temperatures, owing to the relatively slow oxygen absorption of the fish, it did not seem likely that the small reduction of the oxygen tension in the water during the non-pumping stage of the experiment would have much effect.

To test this, however, we made the following experiment at a moderate temperature keeping the oxygen tension below normal during the whole time.

(10) Seven fish; weight 329 g.; variations 22–65 g. I.T. of water 16·8°, F.T. 16°. The water had been previously partially denuded of oxygen and the gas over the water was air and nitrogen. Duration six hours. The partial pressure of the dissolved oxygen was found to be at the beginning 6·17 % and at the end 2·69 % of an atmosphere.

CO_2	Initial	436·0 cc.	Final	510·1 cc.	Difference	+ 74·1 cc.
Oxygen	„	2301·9	„	2168·2	„	− 133·7
Nitrogen	„	23818·5	„	23847·6		+ 29·1

Nitrogen error +0·12 %.

Oxygen absorbed per fish per hour 3·114 cc. or per kilo of fish per hour 66·26 cc. Carbon dioxide evolved per fish per hour 1·726 cc., or per kilo per hour 36·72 cc. The respiratory quotient is 0·55·

At the temperature used the oxygen absorption should be a little higher than the mean of experiments 6 and 7. This is evidently the case, and we may conclude that the non-pumping interval in the experiments at low and at medium temperatures did not appreciably affect the result.

COMPARISON OF THE OXYGEN REQUIREMENTS OF TROUT AND GOLD FISH.

In Table II the oxygen requirements of trout and gold fish, at various temperatures, are compared. The gold fish were 5 to 6 inches in length, and the sets of trout 4 inches and 8 inches respectively, but comparisons in terms of length are somewhat imperfect, owing to the difference in shape of the two kinds of fish. We were unable to obtain gold fish of the same average weight as the larger trout, in sufficient number. A glance at the table will

Table II. *Comparison of oxygen consumption of gold fish and trout.*

	4 inch trout, average weight 23 g.		8 inch trout, average weight 102 g.		5 to 6 inch gold fish, average weight 51 g.	
	Per fish per hour	Per kilo of fish	Per fish	Per kilo	Per fish	Per kilo
4°–7°	2·05	89·04	10·48	102·53	0·83	16·07
13°–16°	2·82	117·40	17·84	192·36	2·54	50·09
20°–22°	4·63	213·80	19·57	204·24	4·71	90·42
25°	5·64	258·87	—	—	—	—

show that the gold fish live at a much lower plane of metabolism than trout, and require much less oxygen. They also appear to react to temperature over a wider range than trout. At temperatures round 16° gold fish use about three times as much oxygen as at 4° to 7°, whereas the 8 inch trout use about double. A further rise in temperature from about 16° to between 20° and 22° again doubled the oxygen used by the gold fish, but caused only a small increase in the case of trout. It would appear at first sight that in the case of the large trout the organism ceases to react to rise of temperature about the limit of 16° to 17°, but that in the case of gold fish this limit is higher, for the ratio of oxygen consumption of gold fish to trout at 4° to 7° is about 1:6, at about 16° 1:4, and at 20° to 23° nearly 1:2. It is probable, however, that this difference in reactivity to temperature is more apparent than real, for it has been shown [Gardner and Leatham, 1914, 2] that trout can live and keep well for long periods at a lower level of oxygen partial pressure than that corresponding to full saturation, and it is well known that gold fish flourish at very much lower levels of saturation; in addition the large trout were somewhat restricted in their movements in the experimental bottle compared with the smaller animals, and were consequently less favourably situated for assisting the pumping of water through the gills by swimming movements. If we make the assumption that a fish absorbs the whole of the oxygen from the water passing through its gills, then calculating on the basis of Winkler's figures [1905] for the oxygen content of water saturated with air at various temperatures, a 4 inch trout at 7° to 8° would need to pump about 4 cc. per minute through its gills, while at 21° he would need about 12·4 cc. per minute and at 25°, 16 cc. per minute. On the same basis an 8 inch trout would need to pump at 6°, 20 cc. per minute, and 16°, 44 to 45 cc. The above assumption is perhaps scarcely warrantable, so that these figures must be regarded as minimum values. It is therefore perhaps not surprising that the large fish should apparently cease to react to increases of temperature at a lower temperature than the small fish.

Another factor which appears to bear on the difference between trout and gold fish in this respect is that gold fish are able to bear with impunity higher temperatures than trout. As was shown in the paper referred to above [1914, 1] temperatures round about 25° appear to be dangerous or fatal to trout. Measurements of respiratory exchange were successfully carried out on 4 inch trout at 25°, but on attempting to determine the oxygen absorption for 8 inch trout at this temperature, of the three fish used in the experiment, one died in about two minutes, though the water was fully oxygenated, and the other two turned over on their backs in about ten minutes, and would no doubt have died, but on adding cold water they recovered completely and rapidly, and appeared to be quite well the next day. This result was not due to any sudden change of temperature, as the fish were carefully and gradually warmed up beforehand during the course of one hour. In another experiment four 8 inch fish were accidentally plunged into well-oxygenated water at 33°. The fish gave a few violent leaps and collapsed as one might perhaps imagine a warm-blooded animal doing on falling into boiling water, and were all quite dead in under one minute. The water was afterwards found to be quite frothy, as though it contained some saponaceous material. Some experiments were made with gold fish to ascertain the maximum temperature compatible with life. A strong healthy gold fish (about 50 g.) was placed in a large earthenware basin of water at 20°. After 15 minutes warm water was carefully added, with good stirring, until the temperature rose to 25°. At intervals of 10–15 minutes more hot water was gradually added, and the temperature very gradually increased. At 30° the fish was very active, swimming about more rapidly and continuously than at lower temperatures, but beyond this showed no apparent sign of distress, though the breathing, as one would expect, was rather rapid. At 35° the fish apparently died without any struggle, remaining on his side at the surface, the mouth being closed. No sign of life being observed after ten minutes at this temperature, the water was then gradually cooled down to 17°. After about one hour the fish was observed to be breathing feebly. It then began to recover more rapidly; it was put into the store tank outside, and the next day appeared to have completely recovered.

It would seem from these experiments that trout can live in water up to a temperature of from 20–25°, but gold fish can exist without apparent harm up to at least 30° or even higher. This result is, we believe, in accordance with the geographical distribution of these fish. As far as we are aware trout are never found in tropical rivers, except such as have a snow mountain source.

RESPIRATORY QUOTIENT.

We are not yet able to explain the variations noted in the respiratory quotient, particularly the very low values observed with trout at low temperatures, but a careful examination of our experimental results has convinced us that the variations cannot be attributed to experimental error.

Experiments are in progress which we hope will throw light on this point.

We take this opportunity of expressing our thanks to the Government Grant Committee of the Royal Society for aid in carrying out this work.

REFERENCES.

Gardner and Leatham (1914, 1). *Biochem. J.* **8**, 374.
——— ——— (1914, 2). *Biochem.* **8**, 591.
Winkler (1905). Landolt Bórnstein Meyerhoffer, Tabellen, 3 Auflage, 184.

L. NOTES ON SOME PROPERTIES OF DIALYSED GELATIN.

By DOROTHY JORDAN LLOYD.

From the Biochemical Laboratory, Cambridge, and the Laboratory of the British Leather Manufacturer's Association.

(*Received May 29th, 1922.*)

I. *Material.*

THE gelatin on which the following observations was made was purified by dialysis at the iso-electric point ($p_H = 4·6$) and subsequent precipitation in strong alcohol. Details have been published in a previous paper [Jordan Lloyd, 1920]. The dry gelatin contained 0·06 to 0·00 % of ash. It was obtained as a snow white, fibrous and brittle solid, with a low solubility. In one specimen the maximum solubility was 1·5 % in boiling distilled water. In others 2–2·5 % could be dissolved under the same conditions, but in no case was it possible to dissolve more than 2·5 % of the pure fibrous gelatin in pure boiling water. This agrees with the observations of Dheré [1910] and Dheré and Gorgolewski [1913] that prolonged dialysis is accompanied by diminished ash content and decreased solubility. The hot water solutions were clear and colourless. On cooling to 15°, if the concentration were above 0·9 % the sol set to a turbid white gel in 24 hours, below 0·9 % but above 0·6 %, gelation took from 2–5 days. At a concentration of 0·6 % or less, the clear hot sols cooled to form turbid white sols at 15°, the turbidity decreasing with decreasing gelatin content. The gelatin in these sols is not precipitated by ferric, mercuric or lead salts, nor by ferrocyanides, dichromates or tannic acid. Gels formed under these conditions are not stable, the gelatin separating from the water [see Jordan Lloyd, 1920; Smith, 1921]. The solubility of dialysed gelatin is, as would be anticipated, immediately increased by the addition of acid or alkali. Gels formed in the presence of free acid or base are glassy clear, colourless and stable.

II. *The influence of hydrochloric acid, sodium hydroxide and sodium chloride on the gelling power.*

"Gelling power" is a difficult phrase to define precisely, and the property which it purports to describe is equally difficult to measure accurately. The following notes do not claim to set out any satisfactory measurement or definition of gelling power, the differences described between the experimental fluids being purely comparative, and observed under conditions kept as uniform as possible. The solutions of gelatin used for this comparative work were made by dissolving dialysed gelatin in hot water, raising to boiling point, and then allowing to stand for 48 hours at 15°. The minimum concentration of gelatin required to produce a gel under these conditions was assumed to give an inverse measure of the gelling power. A fluid was taken as gelled when it remained in its place in an inverted test-tube. The influence of hydrochloric acid and sodium hydroxide was examined in the absence of sodium chloride and in the presence of 0·1 N, 0·4 N and N sodium chloride. The reactions of the 1 % gelatin solutions were measured electrometrically and also by adding a few drops of indicator at the time of mixing, and comparing the colours with suitable standards at the end of the experiment. It is well known that the addition of sodium chloride to a solution containing hydrochloric acid increases the concentration of hydrogen ion but Harned [1915] has· shown that the increase is not great. In the present case it will be well within the experimental error. The measurements of reaction are accurate to $\pm$ ·2 p_H. The results of the experiment are summarised in Tables II, III, IV and V, and a diagram, in which the minimum concentration that will cause gelling is plotted against reaction, is given in **Fig. 1**. Areas above the curves are sols, those below are gels.

Conc. %

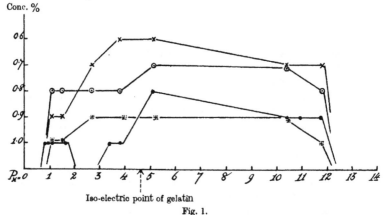

Iso-electric point of gelatin

Fig. 1.

Abscissae = p_H.
Ordinates = Minimum concentration per cent. at which gel forms. 15° and 48 hours.
●———● = 0·0 N NaCl   ◉———◉ = 0·4 N NaCl
×———× = 0·1 N NaCl   ✳———✳ = 1·0 N NaCl

Pure gelatin under the given conditions (48 hours' standing at 15°) requires a minimum concentration of 0·8 % to form a gel, and will only form gels at 1 % concentration from p_H = 12–3, and again rather surprisingly between p_H 2 and 0·7. The disappearance of the power of gelation in acid solutions and its temporary reappearance in stronger acids is a phenomenon which has not previously been described. The conditions necessary to observe it are rather restricted, temperature being an important factor as is shown in Table I.

Table I.

p_H .	Concentration of HCl or NaOH	1 % gelatin after 48 hours		
		0° C.	15° C.	20° C.
0·7	·25 N HCl	—	Clear fluid	—
0·8	·20	Transparent fluid	„ very soft gel	Clear fluid
0·9	·15	„ „	„ soft gel	„ „
1·1	·10	„ „	„ gel	„ „
1·5	·05	„ soft gel	„ „	„ „
1·6	·04	—	„ soft gel	—
1·7	·03	—	„ „	—
2·0	·02	—	„ fluid	—
2·8	·01	Transparent fluid	„ „	Clear fluid
3·8	·005	„ soft gel	„ soft gel	„ „
	·002	Gel with ice crystals	—	„ gel
	·0025	—	Clear gel	—
	·0015	Very faintly turbid gel	—	Clear gel
	·0010	Faintly turbid gel	—	Faintly turbid gel
	·0005	Turbid gel; hysteresis	—	Turbid gel
5·1	Distilled water	Opaque white gel	Opaque white gel	„ „
	·0005 N NaOH	Faintly turbid gel	—	Very faintly turbid gel
	·0010	Very faintly turbid gel	—	Clear gel
	·0015	„ „	—	„ „
	·002	„ „	—	„ „
10·4	·005	Clear gel	Clear gel	„ fluid
11·7	·01	„ viscous fluid	„ „	„ „
12·15	·02	—	„ very soft gel	—
12·4	·03	—	„ viscous fluid	—
12·5	·04	—	„ fluid	—
12·6	·05	Clear fluid	„ „	—
12·9	·10	„ „		Clear fluid
13·1	·20	„ „	„ „	„ „

At concentrations of 0·1 N, and 0·4 N, sodium chloride lowers the minimum concentration of gelatin required to produce gelation, except at reactions more alkaline than $p_H = 12·2$ where in no instance was a gel obtained. At normal concentration sodium chloride may either increase or decrease the gelling power, according to the value of the hydrogen ion concentration of the solution under investigation. Its influence is most marked from $p_H = 5$ to $p_H = 1·5$, *i.e.* at the range where pure gelatin gels least readily. It is interesting that this range is immediately on the acid side of the iso-electric point. The relationship between sodium chloride content and gelling power is obviously not a simple one and requires further study. One clear point however emerges from the study of Fig. 1, *i.e.* the greater the concentration of neutral salt, the less the influence of the hydrogen ion.

If gelation is regarded as a function of viscosity, the influence of sodium chloride as found by the experiments described here, does not agree with previous work by Mörner [1889] and Freundlich [1909].

The inhibition of setting in strongly acid solutions both in presence and absence of salt can be shown to be due to the hydrogen ion since on neutralisation with sodium hydroxide the gelling power is restored. Neutralisation does not restore gelling power to the strongly alkaline sols. The gelling power in this case is permanently destroyed.

III. *The influence of hydrochloric acid, sodium hydroxide and sodium chloride on the turbidity of sols and gels.*

The observations on turbidity were made in all cases on solutions of gelatin of 1 % or weaker, and refer only to their condition at 15°. 1 % gels of gelatin in water are turbid under these conditions; in 0·1 N sodium chloride they are less turbid, in 0·4 N sodium chloride they are again slightly less turbid than in 0·1 N salt and in normal sodium chloride they are only faintly opalescent. In acid solutions the influence of sodium chloride is reversed. In the absence of sodium chloride the 1 % gelatin solutions in hydrochloric acid are generally clear. A slight turbidity develops on standing if the acid concentration does not exceed 0·005 N [Jordan Lloyd, 1920]. The addition of sodium chloride in concentrations of 0·1 N, 0·4 N and 1·0 N causes the clear gels to become turbid. In the last case the gels containing 0·01–0·008 N HCl are actually cloudier than those with salt alone. In dilute alkaline solutions (0·002 N or less) the addition of sodium chloride causes the clear gels to become turbid. In stronger alkaline solutions (0·05 N or more) the addition is without effect, but evidence from other sources shows that the gelatin is actually destroyed at this reaction.

The influence of acid, alkali and salt is summarised in Tables II, III, IV and V. Miss Laing has very kindly examined the turbid water gels for me by means of the ultra-microscope and finds that they contain long fibres similar to the fibres in soap curds.

The relations of the turbidity or "degree of dispersion" of the gelatin to acid, alkali and salt form a close parallel to those for globulin described by Hardy [1905].

Table II.

48 hours at 15°. Salt concentration = 0·00.

Reaction			1·0 %	0·9 %	0·8 %	0·7 %	0·6 %
				Concentration of gelatin			
·25 N HCl	p_H =	0·7	Clear fluid	—	—	—	—
·20 N		0·8	,, soft gel	Clear gelatinous fluid	—		
·15 N		0·9	,, ,,	,, ,,	—		
·10 N		1·1	,, gel	,, ,,	—		
·05 N		1·5	,, ,,	,, ,,	—		
·04 N		1·6	,, soft gel	,, ,,	—		
·03 N		1·7	,, ,,	,, ,,	—		
·02 N		2·0	,, fluid	Clear fluid	—		
·01 N		2·5	,, ,,	,, ,,	—		
·005 N		3·8	,, soft gel	,, ,,			
·0025 N		4·4	,, ,,	Clear gelatinous fluid	—	—	—
Water		5·0	Firm opaque gel	Turbid soft gel	Turbid soft gel	Turbid gelatinous fluid	Turbid fluid
·005 N NaOH		10·4	Clear gel	Clear gel	Clear gelatinous fluid	Clear fluid	—
·01 N		11·7	,, ,,	,, ,,	,, ,,	,, ,,	
·02 N		12·1	Clear very soft gel	,, fluid	,, — ,,	,, — ,,	
·03 N		12·4	Clear fluid	—			
·04 N		12·5 ⎫					
·05 N		12·6 ⎬					
·10 N		12·9 ⎭					

Table III.

48 hours at 15°. Salt concentration $= 0.10$ N.

Concentration of gelatin

Reaction	1.0 %	0.9 %	0.8 %	0.7 %	0.6 %	0.5 %
Hydrochloric acid						
$p_H = 0.8$	Faintly turbid gelatinous fluid					
$p_H = 1.1$	Faintly turbid gel	Faintly turbid soft gel	Clear fluid	—		
$p_H = 1.5$	,, ,,	,, ,,	,, ,,	—	—	—
$p_H = 2.7$	,, ,,	Faintly turbid gel	Faintly turbid soft gel	Clear gel	Clear gelatinous fluid	Clear fluid
$p_H = 3.8$	,, ,,	,, ,,	Faintly turbid gel	,, ,,	Clear soft gel	,, ,,
Water						
$p_H = 5.1$	Turbid gel	Turbid gel	,, ,,	Faintly turbid gel	Faintly turbid soft gel	Faintly turbid fluid
Sodium hydroxide						
$p_H = 10.4$	Faintly turbid gel	,, ,,	,, ,,	Faintly turbid soft gel	Faintly turbid fluid	
$p_H = 11.8$	,, ,,	,, ,,	,, ,,	Faintly turbid gel	,, ,,	—
$p_H = 12.6$	Clear fluid	—		—		
$p_H = 13.0$	,, ,,	—	—	—		

Table IV.

48 hours at 15°. Salt concentration $= 0.40$ N.

Concentration of gelatin

Reaction	1.0 %	0.9 %	0.8 %	0.7 %	0.6 %	0.5 %
Hydrochloric acid						
$p_H = 0.8$	Faintly turbid fluid					
$p_H = 1.1$	Turbid gel	Turbid soft gel	Faintly turbid soft gel	Clear fluid	—	
$p_H = 1.5$	,, ,,	,, ,,	,, ,,	Faintly turbid gelatinous fluid	Faintly turbid fluid	
$p_H = 2.7$	,, ,,	Turbid gel	Faintly turbid gel	,, ,,	Faintly turbid gelatinous fluid	Clear fluid
$p_H = 3.8$	,, ,,	,, ,,	,, ,,	,, ,,	,, ,,	,, ,,
Water						
$p_H = 5.1$	,, ,,	,, ,,	,, ,,	Faintly turbid gel	,, .	Faintly turbid fluid
Sodium hydroxide						
$p_H = 10.4$	Turbid gel	,, ,,	,, ,,	Clear soft gel	,, ,,	Clear fluid
$p_H = 11.8$	,, ,,	,, ,,	Faintly turbid soft gel	Clear gelatinous fluid	Clear fluid	—
$p_H = 12.6$	Clear fluid	—	—			
$p_H = 13.0$	,, ,,	—				

Table V.

48 hours at 15°. Salt concentration = 1·0 N.

Reaction	Concentration of gelatin		
	1·0 %	0·9 %	0·8 %
Hydrochloric acid			
$p_H = 0.8$	Faintly turbid fluid	—	
$p_H = 0.9$	Faintly turbid gelatinous fluid	Faintly turbid fluid	
$p_H = 1.1$	Faintly turbid soft gel	„ „	—
$p_H = 1.5$	Faintly turbid gel	Faintly turbid gelatinous fluid	Faintly turbid fluid
$p_H = 2.5$	Turbid gel	Faintly turbid gel	Faintly turbid gelatinous fluid
$p_H = 3.8$	„ „	.. ..	„ „
Water $p_H = 5.1$	Faintly turbid gel		
Sodium hydroxide			
$p_H = 10.4$	Clear gel	Clear gel	„
$p_H = 11.7$	„ „	„ fluid	—
$p_H = 12.1$	„ fluid	—	—
$p_H = 12.5$	„ „	—	

IV. Optical rotation.

The rotatory power of a solution of gelatin is not a constant one but varies with temperature. This property has been examined by Trunkel [1910] and Smith [1919].

Smith considers that above 35° gelatin is converted to a form which he calls gelatin A which has not the property of forming gels. Below 15° Smith considers gelatin exists in a form which he calls gelatin B, which is the substance capable of gel formation. Between these two temperatures, a solution of gelatin is considered to be a mixture of the two isomers. The conversion of A to B is reversible. Smith, using the sodium line, gives the following values:

$$[\alpha]_D \text{ at } 35° \text{ and above} = -141°,$$

$$[\alpha]_D \text{ at } 15° \text{ and below} = -313°.$$

The rotatory power of the purified gelatin used throughout these experiments was examined in solutions of distilled water, dilute acid, or dilute alkali. The reaction of the solution in distilled water is approximately $p_H = 5$. It was found that all three solutions showed mutarotation, and that the rotatory power differed in acid, alkaline and distilled water solution. The mercury green line was used as the source of light in all experiments. $[\alpha]_{Hg\ green}$ is the rotation caused by a column of fluid 10 cm. in length when each cc. of solution contained 1 g. of pure gelatin. Constant values were obtained for the higher temperature by keeping the tubes for three hours at 37°: at the lower temperature by keeping the tubes for 48 hours at 10° before taking a reading. Intermediate values were not taken. All three solutions set to gels at the lower temperature, the acid and alkaline gels are clear, the water gel cloudy.

For this reason a short tube had to be used in the last case. The following experimental figures were obtained:

Solvent	0·017 NHCl	Water $p_H = 5$	0·017 N NaOH
Concentration of gelatin	1·67 %	2·00 %	1·67 %
Length of polarimeter tube	18·9 cm.	9·9 cm.	20 cm.
Observed rotation at 37°	− 5·03°	− 2·81°	− 4·97°
Observed rotation at 10°	− 9·55°	− 6·08°	− 9·34°
$[\alpha]_{Hg}$ at 37°	− 160°	− 142°	− 150°
$[\alpha]_{Hg}$ at 10°	− 302°	− 307°	− 280°

These values for $[\alpha]_{Hg}$ at 10° and 37° are reversible. After several heatings and coolings, the value of the alkaline, and to a less extent of the acid tube, tends to sink. The difference in the values of the acid, alkaline and neutral solutions at 37° is a considerable one. A further investigation of the influence of temperature and reaction on the value of $[\alpha]_{Hg\ green}$ is being carried out.

V. *Note on the combination of gelatin with hydrochloric acid and sodium hydroxide and on molecular structure in sols and gels.*

If the gelling power of a gelatin solution be plotted against the reaction, in a system free from inorganic salts, as in Fig. 1, it can be seen that the former is reduced over a range of reaction immediately on the acid side of the iso-electric point, Michaelis and Grineff's [1912] figure being taken for the last point. Over this same range, the gelatin is combining rapidly with hydrochloric acid to form soluble hydrochlorides [Procter, 1914; Loeb, 1919; Lloyd and Mayes, 1922]. On the theory that gels are formed of a framework of uncombined gelatin, penetrated with a solution of soluble gelatin salts, it is possible to suppose that the loss of gelling power at these reactions is due to the reduction of the uncombined gelatin to below the minimum necessary for the formation of a rigid framework. The temporary reappearance of the gel state at higher reactions is probably due to the reduction of the active mass of the water owing to the hydration of the protein ions. At higher p_H values the gel state is again inhibited. Immediately on the alkaline side of the iso-electric point, the presence of free hydroxyl ions has very little effect. Yet it is known that the combination of gelatin with sodium hydroxide begins immediately on the alkaline side of $p_H = 4·6$, though the rate of combination is slower by about a third for bases up to a concentration of 0·01 N of free hydrogen or hydroxyl ions. At this point on the curve of "base fixed" the curve inflects and this point corresponds to a p_H value of 12 beyond which gelation cannot exist [see Lloyd and Mayes, 1922, Fig. 5]. Smith [1921] finds that the rise of osmotic pressure to a maximum is more rapid in alkaline solutions than in acid. Evidence has been given elsewhere [Lloyd and Mayes, 1922] that acids and bases combine with gelatin at different positions in the molecule and this is supported by the observations given above on the gelling power. It even seems possible that the sodium salts of gelatin may be able to build themselves into the jelly framework in the undissociated state. The loss of gelling power at $p_H = 12$ is to be attributed to actual destruction of thé gelatin molecule in the presence of the strong alkali. Patten and Johnson [1919] also found that liquefaction accompanied high alkalinity ($p_H = 8$, 9 or 10 in their experiments) but as their observations were all made in the presence of buffer salts they are not comparable with those given in this paper. The effect of sodium chloride in diminishing the minimum concentration of gelatin required to form a gel is obscure, but a possibility suggests itself that it may be due to withdrawal of water from the system by the hydration of its ions. This hypothesis is consistent with the fact that sodium chloride increases the turbidity of gelatin in acid or alkaline solution but is difficult to harmonise with the observation that sodium chloride decreases the turbidity of gels near the iso-electric point, and when in sufficient concentration also decreases the setting power. (It may be remarked in parenthesis that Dr Holker has examined the influence of sodium chloride on some of the gelatin used in this work, at a temperature of 40°. At this temperature, according to Smith, the gelatin would be entirely transformed into the non-gelling form. Dr Holker finds that at this temperature the curve of turbidity against sodium chloride content shows the characteristic sine form [Holker, 1921]. This fine variation of turbidity at 40° is however different from the coarser variations described in this paper which are more of the nature of an actual separating out of gelatin from its solutions.)

In a recent paper [Jordan Lloyd, 1920] the theory was put forward that gelatin gels consist structurally of a framework of solid iso-electric or neutral gelatin together with water, through the open meshes of which lies an aqueous solution of ionised gelatin salts, either acid or basic. The solid phase was assumed to be the seat of the elastic forces of the gels, and the liquid phase of the osmotic forces. The necessity for gelatin in both the uncombined, uncharged, and the ionised, charged condition for the formation of a stable gel, was postulated. This theory has been criticised adversely by both Procter [1920] and Laing and McBain [1920] on the grounds that the structure postulated was supposed to be of microscopic dimensions. This, however, is a misinterpretation of the position taken up by the writer. There is no need to assume that the framework, even if solid, is necessarily of an order of magnitude that would be visible under a microscope, and the existence of glass-like gels of gelatin, showing that they are free from particles of an order of magnitude greater than 10^{-6} cm. in diameter, is against this. Adam [1921] has shown that films of palmitic acid on water only one molecule in thickness may either be in the solid or the liquid state. This proves that the solid state is not incompatible with structure of molecular dimensions. The writer does not consider that there is any fundamental difference between the "molecular network" of Procter, the "exceedingly fine filamentous structure" of Laing and McBain, or the "solid framework" postulated by herself, though originally suggested by Hardy. (It is noteworthy that though Hardy made his gelatin frameworks of microscopic dimensions by the arbitrary use of alcohol, he explicitly states in his original paper [1900] that the droplets may be any size from macroscopic to sub-microscopic according to the conditions under which they separate.) The existence of structure in both gels and sols would seem to have been proved by work on the viscosity of melted gels [see P. van Schroeder, 1903; Garrett, 1903; Levites, 1908; Gokum, 1908; Schorr, 1911; and more recently Hess, 1922], and further the anisotropic properties demonstrated by Hatschek [1921] are evidence that a part at any rate of a gel exhibits the characteristics of a solid. Bearing in mind the work of Adam already referred to, it does not seem impossible that the liquefaction and solidification of gels may be associated with a liquefaction and solidification of the framework similar to the liquefication and solidification of a film of palmitic acid one molecule in thickness.

Procter considers gelatin gels to be of the nature of solid solutions, presumably similar to glass. Laing and McBain state that "a gel is identical with a sol except for its mechanical properties." It is perhaps helpful to consider the matter in the light of the kinetic theory. According to this theory it is possible to visualise the solid state as one in which the molecules composing it are able to vibrate only round a mean position which is fixed, and the fluid state as one in which the molecules (or ions) composing it are free to migrate in any direction. If the molecules of solid are displaced from their mean position by the application of an external force, they will return towards it on the removal of the force, i.e. provided that the body has not been strained beyond its elastic limit. Molecules in a liquid displaced by an external force will not return to their previous position on the removal of the force. Gelatin gels are elastic solids with a modulus that has been determined [Leick, 1904], but the classical researches of Graham [1850, 1851] and Dumánski [1907] show that ions can migrate through them as rapidly as through water provided the gel be not too concentrated; i.e. experiments on diffusion show that part of the gel at least is fluid and that in gels not too concentrated the fluid phase is continuous. Hardy's observations on the reversal of continuity of phases in the alcohol-water-gelatin system, with increasing concentration of gelatin, have a bearing on this point. Laing and McBain have shown the similarity of gels and sols for other physical properties. The simplest theory harmonising all these facts does seem to be that in sols there is a meshwork of protein in the fluid state through which is an aqueous solution non-miscible with it; in gels the framework is not fluid but solid. The justification for this is to be found in the work of Adam on palmitic acid films. The extension of the theory to suppose that the framework was iso-electric gelatin, and the solution gelatin salts, led to the deduction that perfectly pure gelatin would be unable to form stable gels. This supposition has since been confirmed by the observations of the writer [Jordan Lloyd, 1920]; Field [1921] and Smith [1921].

SUMMARY.

1. Gelatin purified by dialysis and precipitation in alcohol is only soluble with difficulty in boiling distilled water. On cooling, if the concentration be 0·8 % or above, the solution sets to an opaque white gel; if the concentration be 0·7 % or less, the solutions on cooling become turbid fluids. The gels are not stable but show syneresis. The addition of sodium chloride in small quantities (0·4 N) *decreases* turbidity and *increases* gelling power, 0·6 % gelatin setting to a soft almost clear gel at 15° in 48 hours in the presence of 0·1 N sodium chloride: at 1·0 N concentration sodium chloride *decreases* turbidity and *decreases* gelling power.

2. In the presence of hydrochloric acid, gelatin becomes much more readily soluble. Sols and gels are transparent to the naked eye. Hydrochloric acid diminishes the gelling power, the degree of diminution passing through a maximum between the reactions $p_H = 2$ and $p_H = 3$, and again beyond $p_H = 0·7$ at 15°. The addition of sodium chloride in small quantities (0·1 to 1·0 N) *increases* turbidity and *increases* gelling power up to a p_H of 0·9. In stronger acid solutions the effect of salt is to decrease gelling power. The presence of the sodium chloride does not appreciably affect the reaction of the system.

3. In the presence of caustic soda, gelatin is more soluble than in water. Sols and gels are transparent to the naked eye. Caustic soda slightly decreases the gelling power between $p_H = 10$ and $p_H = 12$ and completely prevents gelation at concentrations above $p_H = 12$. At concentrations $p_H = 12$ and above slow hydrolysis occurs. The addition of sodium chloride in small quantities (0·4 N) *increases* turbidity and *increases* gelling power. In greater concentrations (1·0 N) gelling power is decreased. The presence of the sodium chloride is without appreciable effect on the reaction of the system.

4. Gelatin therefore can exist in the forms clear sol, clear gel, turbid sol, turbid gel, at the same temperature and the same concentration.

5. Aqueous solutions of pure gelatin are not precipitated by the salts of the heavy metals.

6. The rotatory power of gelatin is not a constant but varies (*a*) with temperature, (*b*) in acid, neutral or alkaline solutions.

7. The bearing of the above properties on the theory of gelation is discussed.

REFERENCES.

Adam (1921). *Proc. Roy. Soc.* A, **99**, 336.
Dheré (1910). *J. Physiol. Path. Gen.* **13**, 158, 167.
Dheré and Gorgolewski (1913). *Compt. Rend.* **150**, 934. *J. Physiol. Path. Gen.* **12**, 646.
Dumanski (1907). *Zeitsch. physikal. Chem.* **60**, 553.
Field (1921). *J. Amer. Chem. Soc.* **43**, 667.
Freundlich (1909). Kapillarchemie (Leipzig), 418.
Garrett (1903). *Phil. Mag.* (6), **6**, 374.
Gokum (1908). *Koll. Zeitsch.* **3**, 84.

Graham (1850). *Phil. Trans.* **140**, 1, 805.
—— (1851). *Phil. Trans.* **141**, 483.
Hardy (1900). *Proc. Roy. Soc.* **66**, 95.
—— (1905). *J. Physiol.* **33**, 251.
Hamed (1915). *J. Amer. Chem. Soc.* **37**, 2460.
Hatschek (1921). *Trans. Faraday Soc.* **16**, Appendix I, 35.
Hess (1922). *Koll. Zeitsch.* **27**, 154.
Holker (1921). *Biochem. J.* **15**, 216.
Leick (1904). *Ann. Phys.* **14**, 139.
Levites (1908). *Koll. Zeitsch.* **2**, 210, 239.
Laing and McBain (1920). *J. Chem. Soc.* **117**, 1506.
Lloyd (1920). *Biochem. J.* **14**, 147, 584.
Lloyd and Mayes (1922). *Proc. Roy. Soc.* B, **93**, 69.
Loeb (1919). *J. Gen. Physiol.* **1**, 379, 487.
Michaelis and Grineff (1912). *Biochem. Zeitsch.* **41**, 373.
Mörner (1889). *Zeitsch. physiol Chem.* **28**, 471.
Patten and Johnson (1919). *J. Biol. Chem.* **38**, 178.
Procter (1914). *J. Chem. Soc.* **105**, 313.
—— (1920). *J. Soc. Leather Trades Chem.* **4**, 187.
Schorr (1911). *Biochem. Zeitsch.* **37**, 1911.
Schroeder, P. v. (1903). *Zeitsch. physikal. Chem.* **45**, 75.
Smith (1919). *J. Amer. Chem. Soc.* **41**, 135.
—— (1921). *J. Amer. Chem. Soc.* **43**, 1350.
Trunkel (1910). *Biochem. Zeitsch.* **26**, 493.

LI. THE MECHANISM OF THE REVERSAL IN REACTION OF A MEDIUM WHICH TAKES PLACE DURING GROWTH OF *B. DIPHTHERIAE*.

By CHARLES GEORGE LEWIS WOLF.

From the John Bonnett Memorial Laboratory, Addenbrooke's Hospital, Cambridge.

(*Received June 1st, 1922.*)

A Report to the Medical Research Council.

IN the preparation of diphtheria toxin for immunising purposes it is known that the initial and final reactions of the medium are factors of prime importance in harvesting a toxin of high potency. The more recent investigations of L. Davis [1918], Bunker [1918, 1919] and Hartley [1922] have placed this beyond a doubt. Bunker, indeed, goes so far as to say that, outside certain well-defined final hydrogen ion concentrations, potent toxins are never harvested, and for this reason has suggested that a determination of the hydrogen ion concentration by modern methods is a criterion for ascertaining when the organism has elaborated the greatest amount of toxin. Hartley questions this and has been able to show that while Bunker's statement is approximately correct, there are cases in which a very high grade antigen is produced outside the limits set by Bunker.

For the purposes of the present paper it is only necessary to refer to the course which the medium takes when reaching this final reaction.

If one starts with a bouillon peptone medium of $p_H = 7.8$, during the earlier period of growth the reaction becomes more acid, often exceeding the neutral point of $p_H = 7.0$. A sudden change in reaction occurs when this acid point is reached and the medium becomes alkaline. The alkalinity may reach the high value of $p_H = 8.6$ or even 8.8. It is during the period of alkalinity that the toxin is formed.

The chemical reactions which occasion the change to the acid side of neutrality and then to the alkaline side with this bacillus have been little studied and not much is known about them. The acid production early received the attention of Theobald Smith [1895, 1899] who believed that it was closely connected with the presence of glucose in the bouillon. Later on he seems to have modified his view and to have attached more importance to the presence of inositol, the so-called muscle sugar, as a producer of acid, and in certain experiments on toxin production he sought to remove all fermentable carbohydrates by the preliminary treatment of the medium with *B. coli* or yeast. It is noteworthy that he suggests that after this treatment a small amount of glucose should be added to the medium for the more efficient production of toxin. Smith's experiments have had a very considerable influence on the technique of toxin production. Even as late as 1921, Dernby and David [1921] have recommended the preliminary fermentation of media by yeast before proceeding to the inoculation with the diphtheria bacillus.

Smith's statements regarding glucose in the bouillon were based on amounts of gas obtained in a fermentation tube after sowing the bouillon with *B. coli*, and there does not seem to be any real information available as to the amount of glucose in meats as ordinarily purchased. In a subsequent paper I hope to be able to give some data on this point. For the present one may say that, assuming the amount of glucose to be 0·06 %, this would yield, if completely converted into carbon dioxide by fermentation, 0·088 % by weight of this gas. This expressed as gas per 100 g. of meat at room temperature and normal pressure would be about 47 cc. It is improbable that such a reaction occurs. One might also assume that a complete transformation of glucose to lactic acid took place. In this event, the amount of lactic acid produced would be equivalent to 0·66 cc. of normal lactic acid per 100 cc. of nutrient broth. Wolf and Harris (1917, 2) have shown that by adding this amount of lactic acid to broth the reaction shifted to the acid side to about the same extent as is seen when *B. diphtheriae* acts upon peptone bouillon. As the amount of meat corresponding to the bouillon is at least half the quantity reckoned as above, it seems inevitable to conclude that the acids produced are not derived solely from the fermentation of the sugar contained in the medium.

The only statement regarding the amount of inositol to be found in meat is an early one of O. Jacobsen [1871] who estimated that fresh muscle contained 0·035 %. In any case, so far as I am aware, no controls have ever been made to ascertain the effect of inositol added to a medium which has been fermented with yeast or with *B. coli*.

That there are other views on the origin of the acid appears from the discussion which took place between Madsen [1897] and Lubenan [1908]. The former author contended that the reversal of the medium was determined by the initial reaction. Lubenau asserted that with the absence of fermentable carbohydrates there was never any acid reaction, provided that the growth took place in the presence of air. Jacobsen [1911] attempted to reconcile these two points of view and tried to show that peptone also has a capacity for forming acids. Petruschky [1891] also believed that peptone furnishes material for the production of acid substances.

The probability is that the truth lay rather with Madsen. With a high initial alkalinity and enough efficient buffer substances in the medium, despite the formation of acids, the reaction of the medium would never reach an acid point as determined with litmus.

When it came to the question of reversal, the matter was on less certain ground. The obvious product in fermentation which can lead to an alkaline reaction is ammonia, and to this was attributed the change to an alkaline reaction.

There were certain indications that this could not be the case. For example, Teruuchi and Hida [1912] examined the daily production of ammonia in 2 % bouillon peptone, with and without the addition of small quantities of glucose. While they obtained a small increase of ammonia, commencing on the fourth day, they did not find that the alkalinity bore any relation to the amount of ammonia produced.

The more recent work of Ayers and Rupp [1918] gives a strong clue to the mechanism whereby the medium becomes alkaline. In their paper on the well-known reversion produced by organisms of the colon-aerogenes group it is shown that, with *B. coli*, acid and alkaline fermentations proceed simultaneously. Glucose is converted to organic acids and these, in turn, are transformed to carbonates. They suggest that the reversion of reaction is probably due to bicarbonates. A similar simultaneous acid and alkaline fermentation has been analysed by Harris and the writer [1917, 1] in the case of strongly proteolytic anaerobic organisms. Here the stabilisation of the reaction, in spite of the production of large amounts of organic acids, is due in great part to the production of ammonia. It seemed, therefore, of some importance to ascertain to what the reversal in reaction with *B. diphtheriae* was due.

The first experiment was designed to follow out the changes in hydrogen ion concentration, and volatile acid and ammonia production. For this purpose portions of 100 cc. of a standard bouillon-peptone medium were sterilised in 450 cc. rectangular medicine bottles. These were incubated lying on their sides, so that the surface of medium exposed to the air was about 115 sq. cm. At stated times a bottle was taken from the incubator and the hydrogen ion concentration, volatile acids and ammonia determined. Table I gives the results of the experiment.

Table I.

Hours	p_H	Mgm. ammonia nitrogen per 100 cc.	Volatile acids, cc. $N/10$ alkali per 100 cc.
Control	7·5	6·13	4·4
24	7·0	7·85	21·2
48	7·1	8·40	24·8
72	7·9	8·98	8·8
96	8·4	8·52	7·2
120	8·5	8·52	6·8
144	8·5	8·42	7·2
168	8·5	8·98	7·8
192	8·4	9·25	8·0
216	8·5	11·20	8·0
240	8·5	10·07	8·0
264	8·5	8·12	6·0

Certain quite clear results emerge from the analyses. Starting with an initial hydrogen ion concentration of $p_H = 7·5$ the reaction follows the usual course, becoming as acid as $p_H = 6·9$. When reversion took place, a final

36—2

value of $p_H = 8.5$ was reached which persisted as long as the experiment was continued. The amount of ammonia produced is definite but very small. Starting from an initial value of 6·13 mgm. ammonia N per 100 cc. of medium it does, on one occasion, reach a concentration of 11·2 mgm., but the average is somewhat less than this. This experiment confirms what Teruuchi and Hida found, viz. that the amount of ammonia formed during the metabolism of the bacillus is very small indeed. At the same time a very remarkable change takes place in the volatile acid production. Starting with an amount of acid equivalent to 4·4 cc. of $N/10$ NaOH per 100 cc. of medium, in the first 24 hours the acid had risen to 21·2 cc. with a further rise to 24·8 cc. at the end of 48 hours. With the sudden reversal of reaction from $p_H = 7·1$ to 7·9 a large proportion of the acid disappears and in the fourth sample only 8·8 cc. are found. From this time on the amount of acid rises only slightly with a final alkalinity of the medium of $p_H = 8.5$.

It is thus obvious that a large part of the volatile organic acids formed in the first stages of fermentation are utilised in the sense of Ayers and Rupp to form basic substances.

A second experiment was undertaken to ascertain to what extent the fixed acids took part in the changes above referred to. The results are embodied in Table II.

Table II.

Hours	p_H	CO_2, cc. per 100 cc.	Volatile acids $N/10$ alkali per 100 cc.	Fixed acids $N/10$ alkali per 100 cc.
Control	7·8	3·0	3·6	27·2
24	7·5	—	4·4	20·5
48	7·3	—	14·8	6·6
72	7·4	—	15·4	11·1
96	7·5	—	15·8	7·4
120	8·1	18·0	7·2	4·5
144	8·2	—	6·8	4·1
192	8·5	27·0	7·2	2·9
240	8·5	27·0	7·4	3·5

The initial alkalinity of the medium was greater than in Exp. 1 and possibly for this reason the medium did not become so acid. The final reaction was, however, the same as in the previous test.

As the amount of medium was not sufficient, no analyses for ammonia were performed. The general trend of the volatile acid curve is the same as in the preceding experiment; one finds a sudden increase in the concentration of these compounds corresponding with a decrease in p_H. The increase in alkalinity is coincident with the sudden lowering of the volatile acids. The course of the fixed acid curve is less easy to explain. From the time of inoculation there is a steady fall, only one value—that obtained at 72 hours—being outside the curve.

The main non-volatile organic acid in bouillon is undoubtedly lactic acid, and this appears to have been metabolised, apparently during the earlier hours of the fermentation, to substances still having an acid reaction and it may well be that some of these are represented in the increase to be found in the volatile acids. Finally, however, the fixed acids reach a level represented by about 4·0 mgm. of $N/10$ alkali per 100 cc. of medium.

During the course of this experiment I was led to ascertain whether the available carbon dioxide in an old culture was different from that in a medium which had not been inoculated. For this work I employed the van Slyke apparatus, using 1·0 cc. of medium for a determination. It appeared at once that the amount in the former was over five times that of the latter.

Using the present series of cultures it was found that at the 120th hour an amount of carbon dioxide was disengaged by dilute sulphuric acid equivalent to 20·0 cc. of gas per 100 cc. of medium. This was at a time when the reaction of the medium was $p_H = 8·1$. When the fluid reached an alkalinity of $p_H = 8·5$ the amount of carbon dioxide disengaged was 29·0 cc. per 100 cc. of medium. The amount of carbon dioxide disengaged from the control medium was only 3·0 cc., and of this 2·0 cc. were due to gases dissolved in the reagents used for the determination.

It would thus appear that the reversal of reaction was closely connected with the formation of carbonates, confirming what Ayers and Rupp had found with *B. aerogenes*. The matter was therefore studied from the commencement of fermentation.

At the same time it was decided to obtain some information regarding the action of *B. diphtheriae* on certain well-known organic salts. Preliminary experiments performed some time previously made it appear that certain organic acids added to the medium would be available as a source of energy, and a detailed examination of this point was made. Not a great deal of information is available on the latter point.

Table III.

Hours	Control			Sodium succinate			Sodium tartrate		
	Pellicle	p_H	CO_2	Pellicle	p_H	CO_2	Pellicle	p_H	CO_2
0		7·8	1		7·8	1		7·8	1
48	+ + +	7·6	10	+ + +	8·0	23	+ + +	7·3	9
72	+ + +	8·1	9	+ + +	8·1	21	+ + +	8·0	17
120	+ + +	8·3	23	+ + +	8·6	57	+ + +	8·2	19
168		8·3	34		8·6	65		7·9	12
240		7·3?	5?	+ + +	8·5	71	+ + +	8·0	11
336		8·3	27		8·6	75		8·2	22
456		8·5	27		8·8	90		8·5	28
624		8·4	21		8·6	56		8·3	13

Hours	Sodium malate			Sodium citrate			Sodium formate		
	Pellicle	p_H	CO_2	Pellicle	p_H	CO_2	Pellicle	p_H	CO_2
0		7·8	1		7·8	1		7·8	1
48	+ + +	7·7	29	0	7·3	4	+ + +	7·5	10
72	+ + +	8·3	50	0	7·2	5	+ + +	7·5	16
120	+ + +	8·6	75	0	7·3	7		7·9	20
168		8·6	66		7·5	7		8·3	17
240	+ + +	8·1?	20?	0	7·5	6	+ + +	8·3	20
336		8·8	90		8·3	21		8·4	26
456		8·8	92		8·3	23		8·4	26
624		8·6	71		7·4	3		8·3	19

Maassen [1896] many years ago in an elaborate investigation of the action of bacteria on organic salts found in a series of 21 organic acids that malic acid was the only one attacked by *B. diphtheriae* and that very feebly.

Altobelli [1914] in a somewhat recent paper also states that this acid is the only one attacked by *B. diphtheriae*.

In the present experiments the fermentations were carried out in test-tubes, the quantity of medium used being about 10 cc.

The salts were added in a concentration of about 1·0 %. After the addition of the salts the medium was adjusted to $p_H = 7·8$ and sterilised at 120° for 30 minutes. The tubes were all inoculated from the same culture and kept at 37°. The culture used (Park and Williams, No. 8) had been transferred at 24-hour intervals in bouillon peptone and was growing very actively (Table III).

It is clear that various organic acids are utilised in quite a special way by *B. diphtheriae*. The two which are most strongly attacked are malic and succinic acids. Judged by the amount of carbonate formed, succinic acid is nearly as completely used as malic acid and it is difficult to understand how Maassen overlooked its effect on metabolism. Tartaric acid and formic acid do not seem to have been utilised at all, for the amounts of carbonates formed are very close to those given by the control medium. Citric acid appears to have had an inhibiting effect on the growth of the bacillus. Reversal was very much delayed and the alkalinity did not reach the height obtained with other cultures. The carbonate fermentation was below that of the control medium.

A special interest attaches to formic acid, as this compound has been shown to be one of the steps through which the higher acids are finally oxidised to carbon dioxide. In the katabolism of glucose by *B. coli* the intermediary between this compound and carbon dioxide is said to be formic acid [Harden and Penfold, 1912]. This does not appear to be the case with the diphtheria bacillus, for the addition of formic acid to the medium is not associated with an increase in carbon dioxide above that found in the control medium.

SUMMARY.

The reversal of reaction in carbohydrate-free media caused by the growth of *B. diphtheriae* is due, in its acid phase, in part to the production of volatile acids. When the medium reverses to a more alkaline reaction, these acids are converted into carbonates.

Certain organic acids, such as malic and succinic acids, are also utilised to produce carbonates.

The formation of carbonates from the acids examined does not appear to take place with formic acid as an intermediary product.

REFERENCES.

Altobelli (1914). *Atti Soc. Toscan. d'Igiene,* quoted from *Centr. Bakt.* 1916, **64**, 188.
Ayers and Rupp (1918). *J. Inf. Dis.* **23**, 188.
Bunker (1918). *Abstr. Bacteriol.* **2**, 10.
—— (1919). *J. Bacteriol.* **4**, 379.
Davis, L. (1918). *J. Lab. Clin. Med.* **3**, 358.
Dernby and David (1921). *J. Pathol. Bact.* **24**, 180.
Harden and Penfold (1912). *Proc. Roy. Soc.* B, **85**, 417.
Hartley, P. (1922). In press.
Jacobsen, O. (1871). *Annalen,* **157**, 227.
Jacobsen, K. A. (1911). *Centr. Bakt.,* Orig. I, **57**, 17.
Lubenau (1908). *Arch. Hyg.* **66**, 305.
Maassen, A. (1896). *Arb. kaiserl. Gesundhmt.* **12**, 340.
Madsen (1897). *Zeitsch. Hyg.* **26**, 157.
Petruschky, J. (1891). *Baumgarten's Jahresb.* **6**, 488 (footnote).
Smith (1895). *Centr. Bakt.* **18**, 1.
—— (1899). *J. Expt. Med.* **4**, 373.
Teruuchi and Hida (1912). *Saikingashi,* Nos. 205 and 768, quoted from *Biochem. Zentr.* 1913, **14**, 748, 749.
Wolf and Harris (1917, 1). *J. Pathol. Bacteriol.* **21**, 386.
Wolf and Harris (1917, 2). *Biochem. J.* **11**, 218.

LII. THE PROPERTIES OF DIBENZOYLCYSTINE.

By CHARLES GEORGE LEWIS WOLF[1]
AND ERIC KEIGHTLEY RIDEAL.

*From the John Bonnett Memorial Laboratory, Addenbrooke's Hospital,
Cambridge, and the Chemical Laboratory, Cambridge University.*

(*Received June 20th, 1922.*)

NEARLY all the organic substances exhibiting gel formation in water are compounds of complex structure and high molecular weight.

In order to elucidate the mechanism of gel formation, in recent years search has been made for substances of relatively low molecular weight and simple constitution, and the dye-stuffs, benzo-purpurin, chrysophenone, thionin blue, the soaps, the organic sulphonic acids and camphorylphenylthio-semicarbazide have been used. As Gortner and Hoffman have recently pointed out [1921], benzoylcystine is the simplest known type of organic substance which exhibits gelation with water and it therefore appeared to present a suitable material with which to study some of the elementary properties of gels in greater detail.

The preparation of dibenzoylcystine used in this work was essentially that of Brenzinger [1892] and Gortner and Hoffman [1921], except that the substance, after recrystallisation from dilute alcohol, was extracted for several hours with benzene in a Soxhlet apparatus to remove completely all traces of benzoic acid and benzoyl chloride. The melting-point of our product was 189° (uncorrected) which is some eight degrees higher than that given by the older authors.

While the method of dispersion in water from an alcoholic solution is well known, yet for many purposes the presence of a second component is undesirable, and accordingly attempts were made to obtain direct dispersion of the material in water. It was found that aquagels may be prepared by triturating the solid in small quantities at a time with warm water and completing peptonisation by stirring the solution for long periods at a temperature of 80°–90°. The rigidity of the gels obtained by this method was essentially of the same order as that of those described by Gortner and Hoffman using alcohol for dispersion.

The sol-gel transformation was found to be perfectly thermally reversible, thus differing from that of silicic acid, and decomposition does not appear to occur as in the case of gelatin. Dilute gels are perfectly transparent; but at higher concentrations, 0·2 % and over, an opalescence makes its appearance, increasing with the amount present in the liquid. Dilute gels, on standing,

[1] Working under a grant from the Medical Research Council.

become somewhat cloudy. On prolonged standing syneresis occurs with contraction of the gel to a core and the exudation of water. Simultaneously small nuclei make their appearance in the gel, which continue to grow to form spherites. Microscopic examination of the spherites shows them to consist of minute elongated radiating crystals. The appearance of these crystalline spherites in the gel, differing from the shot-like spheroids formed in the gelation of gelatin [Bradford, 1921], lends additional support to the two-phase theory of gel structure. The thread or fibrillar nature of the spherite group, in addition, supports the assumption that the natural crystal growth of dibenzoylcystine in water is in the form of poly-molecular crystalline rods, the gel consisting of a number of intersecting fibrils built up of these rods in which the water is enclosed. The fibrils in the gel are evidently relatively coarse, since the gel is both moist and elastic and exhibits syneresis.

The examination of the spherites thus classifies dibenzoylcystine as belonging to the fibrillary crystalline gels, which include nearly all the gels which have been examined, with the possible exception of dilute gelatin, agar and silica. These latter, although generally regarded as a class apart, may be brought into the same category, if one assumes, in the disperse phase, the existence of highly elastic or mobile fibrils undergoing a process analogous to crystal twinning.

Dibenzoylcystine is a relatively strong acid, a gel containing 2·66 g. per litre showing a p_H of 3·05 at 20°, the value being obtained both by indicators and the hydrogen electrode. This value, if the acid be considered as monobasic, corresponds to a dissociation constant of 0.3×10^{-3}. The degree of dissociation was also determined by the conductivity method. The molecular conductivities of a 0·1 % solution and solutions more dilute were determined at 25° in a conductivity cell of the usual type, with the aid of a potentiometer, telephone and coil. The value of $\lambda \infty$ was found to be 250, a limit to which the molecular conductivity had almost attained at a dilution of 0·00014 molar. The value of K, the dissociation constant, varied from 1.4×10^{-3} to 1.5×10^{-3} with a mean value of 1.48×10^{-3}. The acid is thus comparable to monochloroacetic acid and monobromosuccinic acid. No discontinuity in conductivity in the 0·1 % solution, which exhibits a feeble tendency to set, was observed, whether the liquid was agitated or kept at rest prior to the determination. This is a phenomenon which has been observed by McBain [1919] and his co-workers in their studies of soap solutions. The discrepancy between the dissociation constant value as calculated from the hydrogen ion concentration and determined by the conductivity method is probably to be attributed to a liquid junction potential in the former due to the non-diffusibility of the anion in the dibenzoylcystine. This discrepancy is of the order of 10 millivolts.

On the alteration of the hydrogen ion concentration of the medium by the addition of sodium hydroxide or hydrochloric acid several interesting points were noted. On the addition of increasing amounts of alkali the gel formation gradually became less, until the contents of the tube were entirely liquid.

On the addition of acid, the rate of setting increased and the gels were distinctly more opalescent in character, while on the addition of relatively large quantities of acid the gels became distinctly granular and less resistant to penetration. The gel possessed the maximum stability at the isoelectric point. The marked change in the physical characteristics on the alteration of the p_H of the solution was made evident by the following experiments.

Water Retention. One gram of dibenzoylcystine was dissolved in 6·0 cc. of hot alcohol and poured into 250 cc. of boiling water as quickly as possible with constant stirring. From this stock solution 20 cc. were run into boiling tubes which contained 10 cc. of solutions of alkali and acid of varying concentration as given in Table I. 20 cc. of the mixtures were then pipetted into Petri dishes for determinations of their setting power. The remaining 10 cc. were used for hydrogen determinations. At the end of 72 hours the jellies contained in the Petri dishes were removed and allowed to drain under their own gravity on filter papers in funnels. The liquid running through was measured. The results of these determinations are given in Table I.

Table I. 0·30 % *dibenzoylcystine.*

No.	$N/20$ H_2SO_4 cc.	$N/10$ NaOH cc.	Water cc	Dibenzoyl-cystine solution cc.	p_H 20° C.	Water drained in 2 hours from 20 cc. of mixture after setting for 3 days cc.	Observations made on mixing and cooling
1	0	2	8	20	3·65	16·5	Quite fluid
2	0	1	8	20	3·48	16·5	„ „
3	0	0·5	9·5	20	3·36	15·5	Moderately set, not very opalescent
4	0	0·25	9·75	20	3·26	13·0	Not quite stiff, pours
5	0	0	10	20	3·05	13·0	Stiff, does not pour
6	0·5	0	9·5	20	2·65	13·0	Very stiff
7	1·0	0	9·0	20	2·59	14·0	Full of air bubbles, set at once
8	1·5	0	8·5	20	2·52	13·0	„ „
9	2·0	0	8·0	20	2·41	13·0	Opalescent, set at once „
10	3·0	0	7·0	20	2·29	11·0	„ „
11	4·0	0	6·0	20	2·19	11·5	Very opalescent
12	8·0	0	2·0	20	1·93	11·5	Solid had separated out

It is evident as the hydrogen ion concentration decreases below p_H 3·0 the water-retaining power of the gel is very seriously diminished. In the more acid solutions a slight alteration takes place but the curve of drainage is not so regular.

In the determination of the p_H of the alkaline solutions it was noticed that the internal resistance of the cell became extremely high, possibly owing to the formation of a pellicle at the interface of the solution and the saturated potassium chloride junction fluid. It was not, however, sufficient to interfere with the readings.

Effect of Electrolytes. In common with other gels, the water-retaining power of the fibrils is much affected by the presence of neutral electrolytes. The effect of alteration of the osmotic pressure and the reduction in the free water content and of ion hydration naturally varies with the nature of the added electrolyte. Quantitative investigations on the effect of the addition of ammonium thiocyanate to a typical emulsoid albumin to gelatin and to silicon hydroxide gels are to be found in the literature.

Ammonium thiocyanate exerts a similar lyotropic effect on the dibenzoyl-cystine gels. Syneresis is more pronounced in the presence of small quantities of this salt and relative instability of the gel structure may be obtained in fairly dilute solutions as is indicated by the following draining experiments performed in a manner similar to those previously described.

Table II. *Effect of Ammonium Thiocyanate on the setting of Dibenzoylcystine.*
0·4 % dispersion of dibenzoylcystine.
5·0 % solution of ammonium thiocyanate.

Number	Ammonium thiocyanate cc.	Water cc.	Dibenzoylcystine solution cc.	Filtrate cc. in 1 hour
1	1	9	20	14·5
2	2	8	20	17·5
3	3	7	20	18·0
4	4	6	20	19·0
5	6	4	20	21·0
6	8	2	20	21·0
7	10	0	20	21·0
8	0	10	20	14·0

Nos. 1 and 8 went solid in about 10 minutes. The others some time afterwards.

Resistance to Penetration. In order to obtain some information on the tenacity of the gels as affected by change in the p_H of the dispersion medium a simple penetration test was devised. This consisted essentially in measuring the rate of fall of a hemispherically ended plunger through a column of the gel maintained at room temperature. It was noted that a uniform velocity was attained after a fall of two or three centimetres. The time in seconds for a penetration of 1 cm. is plotted in the following curve.

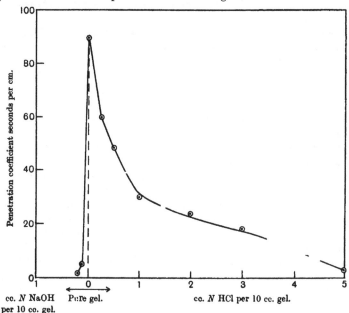

It will be noticed that the influence of the alkali is extremely marked, whilst a distinct reduction in the internal viscosity is produced in relatively acid solutions. The conclusion that one may draw from these experiments is that on the addition of alkali the number of fibrils decreases, presumably because the sodium salt has no gelatinising qualities and is entirely soluble, while, on the addition of acid, the fibrils become relatively much coarser and consequently fewer in number.

Protective Action. An attempt was made to determine the protective power of dibenzoylcystine, using red colloidal gold prepared according to Zsigmondy's method. This gold gave correct values when tested against gelatin. Owing to the sparing solubility of the compound, it was impossible to carry out the test in exactly the way which Zsigmondy [1920] prescribes.

With a 0·2 % solution of the compound partial protection of 10 cc. of gold suspension occurred with 4·0 cc. of the gel. Complete protection was afforded by 6·0 cc. The gold number should therefore be about 10, which is close to that of dextrin and is nearly a negligible quantity.

Diffusion Experiments. In dilute concentrations gels such as gelatin and agar offer no abnormal resistance to the diffusion of electrolytes. With more concentrated gels the coefficient of diffusion is somewhat reduced. Attempts to estimate the fibril mesh of such gels by observations on the rate of diffusion of electrolytes, semi-colloids and colloids of varying particle diameter have frequently been made. The results, however, have been extremely variable. The same abnormalities have been noticed with dibenzoylcystine which being of much simpler chemical constitution permitted partial interpretation of the results obtained.

The experiments were made by taking a gel of 0·2 % and allowing it to set in tubes of uniform size. As soon as the gel was fairly stiff, solutions of dyes were carefully pipetted on top of it and the level marked. At intervals the entrance of the dye into the gels was estimated with calipers. All the dyes used were of Grübler's make. The results are given in Table III.

It will be noted that the gel offers very slight resistance to an electrolyte such as potassium dichromate, a distance of 100 mm. being traversed in a few hours.

Dibenzoylcystine is a negative colloid, being itself a relatively strong acid, and although the ion, as evidenced by the highly dissociated sodium salt, possesses feeble if any gelatinising properties, yet there is no doubt that the fibrils are partially ionised. In the presence of positive crystalloid or colloid dyes adsorption or mutual precipitation is to be anticipated. For such basic dyes, diffusion far into the interior is not to be looked for, and the fibrils should be markedly stained, the partition coefficient between fibril and surrounding liquid being very high. The low apparent diffusibility is for these dyes to be ascribed to adsorption and not to a small fibrillary mesh. The zone of precipitation naturally varies with the colloidal character of the dye and its diffusibility.

Table III.

Diffusion of	Nature	Adsorption	Precipitation	Diffusion rate mm. in 24 hrs.
Potassium dichromate	Crystalloid	–	–	135
Basic dyes, cationic:				
Night blue	Colloid	× ×	× ×	0
Neutral red	Semi-colloid	× × ×	×	0·2
Methyl violet 5B	,, ,,	× ×	× ×	1·0
Bismarck brown	Crystalloid	× ×	–	1·6
Toluidine blue	Colloid	× ×	⌣	2·0
Methylene blue	,,	×		2·0
Malachite green		×		2·0
Thionin		× ×		2·4
Safranin	Crystalloid	× ×		3·0
Fuchsin	Semi-colloid	–		3·0
Chemically reacting dyes:				
Eosin	Colloid	× ×	× ×	0·2
Rose bengale	,,	× ×	× ×	1·0
Acid dyes, anionic:				
Azo blue		× –		1·1
Congo red		× –		1·6
Benzopurpurin		× –		1·6
Alkali blue		–		2·0
Nigrosin				2·4
Carminic acid				3·3
Water blue				4·2
Light green F.S.				4·4
Acid fuchsin	Semi-colloid			5·0

POSSIBLE STRUCTURE OF THE GEL FIBRIL.

Evidence is collecting to show that the union of molecules to form molecular complexes is brought about by forces identical with, though generally more feeble in character than, those uniting atom to atom within the molecule; further that these residual force fields, the product of these internal molecular asymmetries, are localised within certain definite areas in the molecule itself. These forces are electrical in character, and for convenience the various residual affinities may be designated as positive or negative relative to other such fields. The active areas on the molecules are associated with the presence of specific chemical groups which are electrically unsaturated.

The results on the diffusion of the dyes indicate that mutual precipitations of the cationic dyes and the anionic gel do actually take place, and that with the more colloidal dyes the precipitation zone is more limited and the colour more intense, confirming the work of Biltz and Teague and Buxton.

The halide-containing dyes, eosin and rose bengale, appear to react chemically with the dibenzoylcystine. But feeble, if any, adsorption was observed in the case of the anionic dyes. Those that were distinctly colloidal in character were feebly adsorbed. Benzopurpurin, a dye very similar to electrolytes, was apparently relatively easily precipitated. In general, the order of diffusibility observed with dyes and benzoylcystine was very similar to that observed by Bechold and others [1906, 1919].

Some idea of the nature of the linkages involved in the gel fibril may be obtained from a survey of the effect of substitution in dibenzoylcystine of various groups on the gel formation.

These may be:

1. Replacement of the carboxylic hydrogen by a more electropositive element. For example, Na destroys the gel-forming properties.

2. Replacement of the electronegative benzoyl group attached to the amino nitrogen by an electropositive group destroys the gel-forming power of the substance. Attempts to prepare the cinnamyl derivative, which should be more electronegative than the hydrogen or the acetyl, have not succeeded. The electronegative di-*m*-nitrobenzoylcystine has been prepared by Dr F. W. Dootson and exhibits the characteristic gelatinising qualities of the simple benzoyl compound.

3. Replacement of the —S—S— linkage by groups such as —CH$_2$—CH$_2$— or —CH = CH— results in a loss of gelatinising properties. Gel formation thus appears to be dependent on the presence of an electronegative group attached to the amino nitrogen, which must not be too polar in character, on the presence of a relatively negative carboxyl group and on the presence of the electropositive —S—S— grouping. Evidence for the unsaturated and electropositive character of the —S—S— group is furnished by the interaction of cystine with halide aromatic substances in the body, *e.g.* the formation of bromophenylmercapturic acid from bromobenzene. Linkages of this double character would result in the formation of molecular aggregates arranged in echelon permitting the growth of needle-like or fibrous crystals, as shown in the following suggested formula.

Formula.

SUMMARY.

Gels of dibenzoylcystine may be prepared by peptisation with hot water. The substance is a relatively strong acid with a dissociation constant of $1 \cdot 49 \times 10^{-3}$. The gel structure appears to be fibrillary and relatively coarse. The sodium salt of the acid exhibits no gelatinising properties. The presence of acids greatly reduces the solubility of the compound. This was confirmed by observations on the water-retaining capacity and internal viscosity of the gel. Lyotropic salts, such as ammonium thiocyanate, reduce the water-retaining power of the gel, producing ultimate liquefaction. Cationic dyes are adsorbed and precipitated by the gel. Anionic dyes diffuse in a normal manner, whilst halogen dyes apparently react chemically with the sulphur group. Di-*m*-nitrobenzoylcystine has similar properties to the simple compound. A possible structure for the gel fibril based upon the effect of chemical substitution on the gel formation is suggested.

REFERENCES.

Bechold (1919). Colloids in Biology and Medicine, 427.

Bechold and Ziegler (1906). *Zeitsch. physikal. Chem.* **56**, 105.

Bradford (1921). *Koll. Zeitsch.* **28**, 214 and *Biochem. J.* **15**, 553.

Brenzinger (1892). *Zeitsch. physiol. Chem.* **16**, 573.

Gortner and Hoffman (1921). *J. Amer. Chem. Soc.* **43**, 2199.

McBain, Laing and Titley (1919). *J. Chem. Soc.* **115**, 1280.

Zsigmondy (1920). Kolloidchemie, 174.

LIII. THE NITROGEN-DISTRIBUTION IN BENCE-JONES' PROTEIN, WITH A NOTE UPON A NEW COLORIMETRIC METHOD FOR TRYPTOPHAN-ESTIMATION IN PROTEIN.

By ERY LÜSCHER.

From the Biochemical Laboratory, Cambridge.

(Received June 6th, 1922.)

THE physical properties of Bence-Jones' protein have been more or less fully described in all investigated cases. On the other hand, there is but little reference to the exact chemical constitution and especially to the amount of individual amino-acids. Magnus-Levy [1900] determined the N-balance by Haussmann's method [1899]. Abderhalden and Rostoski [1905] estimated the monoamino-acids by esterification and Hopkins and Savory [1911] all the amino-acids by a similar method. This last investigation was carried out in two different cases and it is to one of these that the protein belongs that I have used for the analysis described in this paper. I determined the nitrogen distribution by van Slyke's method and the tryptophan content by the colorimetric estimation of von Fürth and Lieben [1920]. The details of the analysis are described later. For the clinical symptoms of this case and the physical properties of the protein the original paper by Hopkins and Savory [1911] must be consulted in which the case is described as Case C.

Table I gives the results of the different analyses carried out on Bence-Jones' protein.

Table I.

The figures are % of total nitrogen.

	Magnus-Levy	Abder-halden	Hopkins Case A	Hopkins Case C	Present analysis
Amide-N	8·6	.	8·02	8·00	9·43
Melanin-N			.	.	.
Total-N of bases	24·7		.	.	23·11
Cystine-N	.	.	.	0·41	1·25
Arginine-N		+	11·9	12·0	9·27
Histidine-N	.	+	1·44	1·33	4·54
Lysine-N	.	+	.	4·34	8·04
Total-N of filtrate	64·2	.		.	66·84
Amino-N of filtrate	.	.		.	61·69
Non-amino-N of filtrate	.	.	.	+	5·15
Glycine-N	.	1·95	+	+	.
Alanine-N	.	4·37	+	+	
Valine-N	.	.	.	4·22	
Leucine-N	.	6·98	.	4·29	
Aspartic acid-N	.	2·89	1·34	1·41	
Glutamic acid-N	.	3·52	4·42	4·73	
Phenylalanine-N	.	0·78	2·47	2·57	
Tyrosine-N	.	0·81	1·97	2·02	
Proline-N	.	1·40	1·93	2·00	.
Oxyproline-N	.	−	−	−	.
Tryptophan-N	.	0·69	.	.	2·44

In the papers of Abderhalden [1905] and Hopkins [1911] the amounts of amino-acids are given in percentage of the protein. Assuming the total nitrogen of the protein to be 16·21 %, a value found by Hopkins, it is possible to calculate the nitrogen of each amino-acid in percentage of the total nitrogen. The above table is expressed according to the latter.

It is a matter of course that only those estimations are directly comparable which have been performed by the same method. This holds good in the two cases examined by Hopkins [1911], which show almost identical values. Moreover Hausmann's estimation of the nitrogen distribution and that by van Slyke are based on the same principle, so that those cases may also be compared. They also show a close agreement. This indicates that in the two cases examined by Hopkins [1911] and in the case of Magnus-Levy [1900] the protein shows the same relative composition of mono- and diamino-acids. The assumption that these three cases have so far the same chemical constitution by mere chance is not probable. It seems more likely that in all cases of Bence-Jones' protein-uria there appears one and the same protein or mixture of proteins in the urine. Abderhalden [1905] found in his case somewhat different values from those of Hopkins in spite of a similar method. But esterification is a very complicated method and does not give quantitative results.

Table II shows the relations between Bence-Jones' protein and those of serum and tissue.

Table II.

The figures are % of total nitrogen.

	Bence-Jones' protein	Serum-globulin[1]	Serum-albumin[2]	Normal tissues[3]	Tumors[4]
Amide-N	9·43	7·95	6·65	5·82	4·73
Melanin-N	0·90	2·30	0·95	3·09	2·42
Total-N of bases	23·11	25·25	34·75	30·71	30·54
Cystine-N	1·25	1·65	3·10	0·93	1·12
Arginine-N	9·27	8·00	9·90	11·13	10·09
Histidine-N	4·54	4·80	5·85	7·86	7·09
Lysine-N	8·04	10·80	15·90	9·83	12·04
Total-N of filtrate	66·84	64·80	58·15	60·82	62·14
Amino-N of filtrate	61·69	62·65	56·55	55·88	54·24
Non-amino-N of filtrate	5·15	2·15	1·60	4·94	7·90
N-recovered	100·28	100·50	100·30	100·4	99·8
Tryptophan-N	2·44	2·5–3·5[5]	—	—	—

[1] and [2] carried out by Hartley [1914].
[3] and [4] carried out by Drummond [1916].
[5] carried out by von Fürth and Lieben [1920].

Hartley's [1914] analysis and that described in the present paper show the same nitrogen distribution for mono- and diamino-acids in Bence-Jones' protein and serum-globulin. The serum-albumin is characterised by a higher proportion of diamino-acids. Moreover, the tryptophan content is almost identical in serum-globulin and Bence-Jones' protein. On the other hand, the latter as also the serum-albumin yields less melanin. But this fraction depends much on the hydrolysis and is therefore the most uncertain. The only real difference between serum-globulin and Bence-Jones' protein has been found in the non-amino-nitrogen of the filtrate, i.e. the proline and oxyproline

fraction. It is much greater in Bence-Jones' protein than in serum-globulin
and approaches the amount in tissue proteins. But these, on the other hand,
contain the diamino-acids in a greater proportion than Bence-Jones' protein.
Analyses of the tumor-proteins in the bone-marrow have not yet been carried
out. Those registered in the above table were prepared from breast-cancers.
They do not show any close resemblance in their amino-acid balance with
Bence-Jones' protein. But it must be understood that Drummond coagulated
the proteins in tissues and tumors by heat in acid solution, which also precipitates the nucleo-proteins. It is not known in which fraction the nitrogen
of the purine and pyrimidine-ring appears and in what manner the relations
are changed. This may make the comparison between his analyses and those
in which no nucleo-proteins are present somewhat uncertain. On the whole,
Bence-Jones' protein seems to be a substance, not only characterised by its
physical behaviour, but also by its distribution of amino-acids, which differs
from all the proteins analysed up to the present time.

EXPERIMENTAL DETAILS.

A. Preparation of Bence-Jones' protein: it was precipitated by heat not
exceeding 60°, washed with water, alcohol and ether, and dried at 110° till of
constant weight. [See Hopkins and Savory, 1911.]

B. Analysis by van Slyke: 3–4 g. of protein were hydrolysed with a volume
of 20 % HCl 35 times its weight for 48 hours at 100°. For each of the two
analyses the hydrolysis was carried out separately.

The following modifications mentioned by Crowther and Raistrick [1916]
were applied:

(a) To the solution of $N/10$ sulphuric acid were added some crystals of
KI and 1–2 cc. of saturated solution of KIO_3. The iodine produced was then
titrated with sodium thiosulphate using soluble starch as indicator.

(b) The melanin-precipitate was filtered off under slight suction.

(c) The amino-nitrogen was determined by the micro-apparatus. Triplicates
done on 2·5 cc. varied within 2 %.

(d) I found it easier to estimate the total nitrogen of the bases in a different
portion from arginine as Plimmer [1916] has pointed out. I carried out the
arginine estimation itself in the original way described by van Slyke. Bumping
was almost entirely avoided by using a rose-burner.

After concentrating the hydrolysed protein the solution was made up to
100 cc. and I proceeded in the following manner:

1. Total-N in 10 cc.: duplicate.
2. Ammonia-N in 75 cc.
3. Melanin-N in 75 cc.

The solution of the bases was made up to 100 cc. In it were estimated:

1. Total-N in 10 cc.
2. Cystine-N in 20 cc.
3. Arginine-N in 50 cc.

4. Amino-N in 2·5 cc.: triplicate.

The filtrate of the bases was made up to 200 cc. and in this were estimated:

1. Total-N in 50 cc.: duplicate.

2. Amino-N: 10 cc. were diluted to 25 cc. and 2·5 cc. taken for estimation of amino-N: triplicate.

Table III gives the results in percentage of total nitrogen of the two analyses carried out, the values corrected for solubility of the bases and according to blank tests.

Table III.

	I	II	Average
Amide-N	9·36	9·50	9·43
Melanin-N	0·92	0·82	0·90
Total-N of bases	23·05	23·18	23·11
Cystine-N	1·26	1·24	1·25
Arginine-N	9·15	9·40	9·27
Histidine-N	4·82	4·27	4·54
Lysine-N	7·82	8·27	8·04
Total-N of filtrate	66·88	66·81	66·84
Amino-N of filtrate	61·98	61·40	61·69
Non-amino-N of filtrate	4·90	5·41	5·15
Total-N recovered	100·26	100·31	100·28

C. Colorimetric estimation of tryptophan by von Fürth and Lieben. For details the original paper must be consulted [1920].

The ordinary cuvettes of the Dubosque colorimeter can be used, putting a rubber ring between the bottom and the upper part and screwing the latter tightly down. Experiments show that the fluid does not come in contact with the metal and the colour remains unchanged.

The colour reaction is better produced in two measuring flasks of 25 cc. than in two test-tubes. Thus the volume of the standard solution and that of the unknown solution can be made equal with greater accuracy.

Each time to 2 cc. of a standard solution of 0·1 % pure tryptophan (prepared in the Biochemical Laboratory of Cambridge) and to 2 cc. of a solution of Bence-Jones' protein of about 5 % were added 1 drop of formaldehyde $2\frac{1}{2}$ % and 20 cc. of saturated hydrochloric acid solution. Ten minutes afterwards sodium nitrite was added, the amount varying from 10–50 drops of a 0·05 solution in order to get the maximum depth of colour.

The following estimations were carried out:

1. 4·80 % solution of Bence-Jones' protein in 0·6 % NaOH. A 20 minutes' heating at 100° dissolved the protein. 20 drops of sodium nitrite produced the maximum colour: triplicate.

2. 6·29 % solution of Bence-Jones' protein in 15 % NaOH. Dissolved in 20 minutes at 100°. 50 drops of sodium nitrite produced the maximum colour.

3. 4·57 % solution of Bence-Jones' protein in 20 % NaOH. Dissolved in 15 minutes at 100°. 50 drops of sodium nitrite produced the maximum colour.

Table IV gives the calculated tryptophan content of Bence-Jones' protein in percentage:

Table IV.

1					
a	b	c	2	3	Average
2·74	2·88	2·95	3·14	2·75	2·89

The difference between the extreme values is 14 % of the highest. The difference in the amount of alkali used in dissolving the protein and therefore the difference in the acidity of the colour solution has no influence on the colour depth. This fact ought to be noticed since von Fürth and Lieben [1920] pointed out that the colour depth in pure tryptophan solutions varies with the acidity.

In the course of the experiments I found two difficulties:

1. The standard solution often changes the character of the colour in a short time and matching becomes uncertain even with a colour-screen. This is probably due to the beginning of clouding. On the other hand, the protein solution does not cloud but reaches its maximum colour more slowly than the pure tryptophan.

2. The colour in pure tryptophan is blue, that in the protein is more or less red, so that matching without a colour-screen is impossible. It is not difficult to find a screen making the two colours identical and a solution of $CuSO_4$ proved to be best. But the following must be considered. Fearon [1920] has shown that formaldehyde and glyoxylic acid give with tryptophan two coloured compounds, which he calls tryptophan-red and tryptophan-blue respectively. The colour obtained under ordinary conditions is a mixture of both and in pure tryptophan the blue prevails, in protein the red. Since both colours are due to tryptophan it must be considered what influence a colour-screen may have on the estimation. If the screen takes away the red rays, then one part of the tryptophan molecules is eliminated and it is uncertain if the calculation only based on the remaining blue would give right values for the whole amount of tryptophan. As Fearon remarks, a reaction producing two different colours is not suitable for a colorimetric method.

The blue colour of pure tryptophan turns more red by cooling, but at 10° the colour of pure tryptophan and that of protein are not yet matchable.

There is one aldehyde which gives nearly the same blue colour in pure tryptophan and in protein solutions, namely benzaldehyde. Salicylaldehyde gives also red and blue. As to colour, benzaldehyde would therefore be the most suitable reagent for a colorimetric method. The reaction is specific for tryptophan, no other known amino-acid producing it. Pure gelatin does not give any blue colour.

I therefore tried with benzaldehyde and used the following solutions:

1. Solution of pure tryptophan of 0·05 % in distilled water.

2. Solution of pure benzaldehyde of 1·9 % in pure hydrochloric acid of about 38 %.

3. Solution of sodium nitrite of 0·05 % in distilled water.

4. Pure hydrochloric acid of about 38 %. It is most essential to use only

very pure hydrochloric acid without or only with a very slight yellow colour. Otherwise the colour reaction does not reach its maximum. On the other hand, I found it unnecessary to saturate with hydrochloric acid gas if the acid is 38 % or more.

After a good many trials the following technique proved to be best: I mixed 2 cc. of the tryptophan solution, 1 cc. of the benzaldehyde solution and 10 cc. of hydrochloric acid in a measuring flask of 25 cc. After 5 minutes I added 5 drops of the sodium nitrite solution and let it stand together for 15–60 minutes. Then I filled up to the mark with hydrochloric acid.

The experiments revealed several advantages in using benzaldehyde instead of formaldehyde.

1. The colour solution does not cloud and even after several days no precipitate is produced. There is a slight diminution of the colour depth in 24 hours, but it does not change appreciably in the course of 1 or 2 hours.

2. The maximum colour reached with benzaldehyde is about double as deep as with formaldehyde, using in both cases the same concentration of tryptophan. This is, of course, welcome in the case of more or less insoluble proteins. The solution of 0·05 % tryptophan gives a satisfactory colour depth. The reaction goes on as quickly or even more quickly with benzaldehyde.

3. Both pure tryptophan and protein give blue colours. There is a slight difference in shade and sometimes they are not matchable without a colour-screen. On page 560 I discussed the error possibly arising from this source by using formaldehyde and this holds good for benzaldehyde too. But in the latter case the two solutions are nearly the same, whereas in the former they are very different. Therefore one would assume that the error with formaldehyde might be much greater than with benzaldehyde. The estimation is uncertain in both cases with coloured protein solutions or very impure proteins. Traces of iron, for instance, hinder the colour reaction.

In order to decide if the colour depth varies proportionally to the concentration, I estimated solutions of varying concentrations of tryptophan by comparing them with a standard solution and obtained the following results:

Table V.

Number	Calculated tryptophan %	Estimated tryptophan %	Error %
1	0·1092	—	—
2	0·0546	0·0510	− 6·6
3	0·0546	0·0574	+ 5·0
4	0·0546	0·0510	− 6·6
5	0·0273	0·0256	− 6·0
6	0·0546	0·0549	+ 0·5
7	0·0546	0·0535	− 1·7
8	0·0273	0·0273	± 0·0
9	0·0546	0·0549	+ 0·5
10	0·0546	0·0560	+ 2·5
11	0·0273	0·0278	+ 2·0
12	0·0546	0·0546	± 0·0
13	0·0546	0·0568	+ 4·0
14	0·0273	0·0282	+ 5·0
Average	—	—	− 0·2

There is a maximum error of 6·6 %. Von Fürth had one of 20 %, therefore benzaldehyde seems to give more accurate results. I can not give any explanation of the error. The colour reaction is capricious and an unknown influence may sometimes come into play.

I tried the same with caseinogen and got a maximum error of 5 %.

In all cases I used the same amount of benzaldehyde and of sodium nitrite. Within the above limits it is not necessary to vary them proportionally to the amount of tryptophan, since the maximum colour is reached without such a complication of the method. In comparing protein and pure tryptophan the concentration of tryptophan in both solutions ought to be about the same.

In the case of water-insoluble proteins von Fürth dissolved them in more or less alkaline solutions. Since I did the same in my estimations I had to determine the influence of alkali on the colour reaction. Instead of a neutral solution of pure tryptophan different concentrations of NaOH were used as indicated in the following table:

Table VI.

Number	% NaOH	Calculated %	Estimated %	Error %
1	0	0·1092	0·1092	—
2	10	0·1092	0·0729	−50
3	5	0·1092	0·0926	−18
4	2·5	0·1092	0·1051	−4
5	1	0·1092	0·1081	−1
6	0·5	0·1092	0·1050	−4

Up to a concentration of about 2·5 % NaOH there is practically no influence on the maximum colour. Higher concentrations diminish the colour depth.

It is essential to know that things are different in protein solutions. The following table gives the tryptophan content of Witte-Pepton dissolved in different concentrations of NaOH:

Table VII.

Number	% NaOH	Tryptophan %	Error %
1	0·6	3·0	—
2	15	2·85	−5
3	20	3·0	±0
4	30	2·9	−3

After filtering off the precipitate of NaCl the colour depth is the same in all cases. Therefore insoluble proteins may be dissolved in strong alkali of 20–30 % without changing the colour reaction. It is not sure that this holds good for all proteins and it ought to be proved in every special case by using different concentrations of alkali.

The above data show that the colour reaction is not so easily disturbed in proteins as in pure tryptophan. Therefore in making comparisons one is apt to get too high values of tryptophan in proteins.

The following table gives the tryptophan content of several proteins estimated by formaldehyde and benzaldehyde:

Table VIII.

Protein	Tryptophan-content % estimated by		Difference %
	Benzaldehyde	Formaldehyde	
Caseinogen	1·1	1·39	27
Serum-protein	1·65	2·2	33
Ovalbumin	0·97	1·5	53
Witte-Pepton	3·0	4·8	60

In all cases formaldehyde gives values 30–60 % higher than benzaldehyde. Most probably the estimation by benzaldehyde is the more correct, since it avoids several disadvantages of the formaldehyde method. No other method for exact tryptophan estimation is known, so that a direct test is impossible. Therefore the method described in this paper seems to be the best at present available for tryptophan estimation in proteins.

SUMMARY.

1. The nitrogen distribution of Bence-Jones' protein is determined by van Slyke's method.

2. Bence-Jones' protein is a substance, not only characterised by its physical behaviour, but also by the distribution of nitrogen, which differs from that of all the proteins analysed up to the present time.

3. There is some evidence that the same protein appears in the urine in all cases of Bence-Jones' protein uria.

4. The tryptophan content of Bence-Jones' protein is estimated by von Fürth's colorimetric method. Objections to this method are discussed.

5. The writer proposes to use benzaldehyde instead of formaldehyde. There are several advantages in doing so and the method becomes more correct. The values of tryptophan estimated by formaldehyde are probably 30–60 % too high.

My very best thanks are due to Prof. F. G. Hopkins for his kind hospitality in his laboratory and his kind help and advice during the progress of this work.

REFERENCES.

Abderhalden and Rostoski (1905). *Zeitsch. physiol. Chem.* **46**, 125.
Crowther and Raistrick (1916). *Biochem. J.* **10**, 434.
Drummond (1916). *Biochem. J.* **10**, 473.
Fearon (1920). *Biochem. J.* **14**, 548.
von Fürth and Lieben (1920). *Biochem. Zeitsch.* **109**, 124, 153.
Hartley (1914). *Biochem. J.* **8**, 541.
Hausmann (1899). *Zeitsch. physiol. Chem.* **29**, 136.
Hopkins and Savory (1911). *J. Physiol.* **42**, 189.
Magnus-Levy (1900). *Zeitsch. physiol. Chem.* **30**, 200.
Plimmer (1916). *Biochem. J.* **10**, 115.

LIV. THE AUTOLYSIS OF BEEF AND MUTTON.

By WILLIAM ROBERT FEARON and DOROTHY LILIAN FOSTER.

*Report to the Food Investigation Board from the Biochemical Laboratory,
Cambridge.*

(Received June 13th, 1922.)

I. THE AUTOLYSIS OF BEEF AND MUTTON COMPARED.

THE study of the course of autolysis in beef and mutton described in the
following experiments was undertaken in the hope that it would reveal some
explanation of the inherent differences observed in the effect of cold storage
on these two kinds of muscle.

It is a fact of common knowledge that beef cannot be frozen and thawed
again without marked changes taking place in the appearance, palatability,
and general physical state of the meat, while nothing of the sort happens when
mutton is similarly treated. Frozen beef on thawing becomes bluish in colour, ·
flabby to the touch, and, most serious of all, there is an exudation of juice
which contains valuable nutrient material. "Frozen" beef must be carefully
distinguished from "chilled" beef. The latter is beef which has been cooled
to a temperature low enough to prevent bacterial action, but not so low as to
cause the tissue fluids to become solid. Meat so treated is not immune from
attack from moulds, and can only be preserved for a limited length of time.

The fundamental differences in the two kinds of flesh underlying the
differences in their behaviour on freezing and subsequent thawing may be
chemical or physical. Autolysis is a term used to describe a series of post-
mortem chemical changes in tissues, and it was hoped that any radical
difference between the chemical constitution of beef and that of mutton would
reveal itself in the course of the autolysis of the two kinds of meat.

Autodigestion was the term used by early workers on this subject, and it
implies that, under certain conditions, tissues digest themselves. The digestion
of an organ by its own self-contained enzymes must clearly be distinguished
from that produced by external agencies, as, for instance, the bacteria of
putrefaction. If an excised organ is kept aseptically *in vitro*, it undergoes
characteristic autolytic changes. Since certain antiseptics inhibit bacterial
action, while they are without effect on enzymes, autolysis pursues a normal,
or nearly normal, course *in vitro* in the presence of antiseptics such as toluene,
chloroform, or acriflavine. In some pathological conditions autolysis can occur
in the living animal.

As in the case of digestion by the ordinary enzymes of the alimentary canal,
autodigestion is a series of processes by which the complex compounds of the
tissues are broken down into simpler ones. It is certain that there are changes
in the carbohydrates and fats, but the course of the protein autolysis is that
which has received most study. Proteolysis implies a degradation of coagulable

protein material into simpler substances, which are not coagulated by heat, nor precipitated by certain protein reagents, and hence autolysis is accompanied by a relative increase in the non-coagulable, or soluble, nitrogen. As this is capable of easy estimation, the course of autolysis is usually followed by making successive determinations of the soluble nitrogen present.

All animal tissues are subject to post-mortem autolysis, but the process proceeds at very different rates in different organs, being most rapid in glandular tissue, such as the liver, and slowest in striated muscle. In order to develop the details of technique, a preliminary examination and comparison was made of the course of autolysis in sheep and ox liver, and essentially the same methods were later applied to the study of autolysis in muscular tissue.

Experimental details.

About one and a half pounds of the liver were transferred immediately after the death of the animal, and with as little handling as possible, to a sterile, covered dish. On arrival at the laboratory it was cut into small slices and scraped to a fine pulp with sharp scalpels on a glass plate. In this way the glandular material could be almost entirely freed from connective tissue. The pulp was well mixed, weighed in a glass dish, and made up into a 20 % suspension in toluene water. For example, in one case (Exp. 12), to exactly 400 g. of well-mixed liver pulp, 50 cc. of toluene were added and sufficient distilled water to make a total volume of 2 litres. The suspension was poured into a sterile flask with a well-fitting cork which had previously been soaked in toluene. The mixture was incubated, generally at 37°, the flasks being shaken thoroughly twice a day. The course of the autolysis was followed by successive estimations of the soluble nitrogen in an aliquot part of the whole suspension. Fractions were obtained by withdrawing a known volume by means of a blunt-ended pipette. The tip was cut off an ordinary, rather wide, 20 cc. pipette, which was then found to deliver 19·4 cc., and this was used throughout all the experiments. Provided that the suspension had been carefully made up, no difficulty was found in obtaining representative samples by this means. Samples were withdrawn immediately after making up the suspension for determination of the initial total and soluble nitrogen, and then at regular intervals, generally once a day for the first four days, and later every two or three days.

The total and soluble nitrogen were estimated by Kjeldahl's method, and the amino-nitrogen by Sörensen's formaldehyde method. The value for the total nitrogen was obtained by incinerating 20 cc. of the suspension, corresponding to a known weight of muscle, and that of the soluble nitrogen by incinerating an aliquot part of the filtrate after precipitating the coagulable protein. In all cases the ammonia was steam-distilled into the standard acid, which was N for the total, and $N/10$ for the soluble nitrogen. The caustic soda was used in equivalent strengths, and was CO_2 free. The indicator used was methyl red.

The coagulable protein was precipitated either by 25 % metaphosphoric acid, or 2·5 % trichloroacetic acid, according as the filtrate was to be used for the determination of amino-nitrogen or not. It was found impossible to do a formaldehyde titration in the presence of so large an excess of phosphate, while trichloroacetic acid showed no such "buffer" action.

The following is a detailed account of the procedure. 20 cc. of the suspension were run into a stoppered graduated cylinder, and 10 cc. of a 25 % solution of glacial phosphoric acid (freshly made) added, and the volume made up to 100 cc. After thorough shaking, the cylinder was allowed to stand over night. The precipitate was then removed by filtration through a dry, fluted paper, when a crystal clear filtrate was obtained. The nitrogen was estimated in 10 cc. of this filtrate. Duplicate estimations were always made. When trichloroacetic acid was used, 50 cc. of a 2·5 % solution were added instead of the phosphoric acid, but the procedure was otherwise the same.

Expression of results.

In reviewing the literature on this subject, a great diversity in the method of expressing results has been found. The most general practice has been to give the number of cc. of acid neutralised in the Kjeldahl estimation of the nitrogen; sometimes the amount of alkali equivalent to the excess of acid. As each observer chooses his own standards, this method is far from satisfactory, and it is suggested that a uniform method of expressing results should be adopted. We have found that the clearest picture of the course of autolysis is presented by expressing the amount of non-coagulable nitrogen present as a percentage of the original total nitrogen. In this way the progressive degradation of the nitrogenous complexes is most easily followed and graphically represented.

THE AUTOLYSIS OF OX AND SHEEP LIVER.

It was found that the autolysis of the livers of both these animals follows a normal course, and they are strictly comparable. There is nothing in the breakdown of the chemical complexes after death to suggest any fundamental difference during life. The curves representing the increase in soluble nitrogen follow the same course and are similar to those obtained by other workers. Estimations were not made sufficiently soon after death to show the latent period, but both show a rapid initial increase followed after above five days by equilibrium. The curves obtained from amino-acid titrations show the same characteristics. Curves from a typical experiment are appended (Figs. 1 and 2).

THE AUTOLYSIS OF SHEEP AND OX MUSCLE.

In the main the methods used for the examination of the autolysis of liver were applicable to muscle tissue, but slight modifications were necessary in preparing the suspension. The meat was separated from fat and connective tissue, and then shredded on a glass plate. The resulting pulp was passed

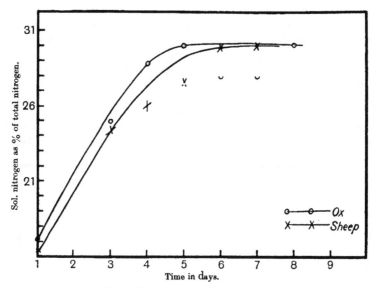

Fig. 1. The autolysis of sheep and ox liver.

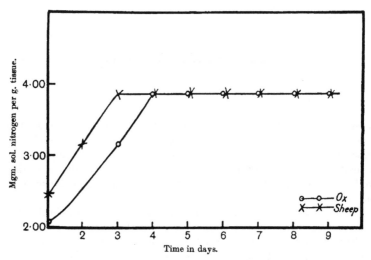

Fig. 2. The amino-nitrogen curve of liver autolysis.

twice through a fine mincer, and then rubbed to a paste in a mortar. A definite amount (*e.g.* 100 g.) was then weighed on a tared clock glass and transferred quantitatively to the mortar, and well mixed with water and toluene. The suspension was washed into a cylinder, and made up to a known volume; a fixed volume of water was then used to wash this into a sterile flask, so that the total volume of suspension was accurately known. The flask was closed by a cork soaked in toluene. In general, 20 % solutions in 5 % toluene were used, and 20 cc. fractions withdrawn as already described. The technique of precipitation etc. was exactly the same as for liver autolysis.

As in the case of the liver, the study of post-mortem autolysis in these two species of muscle failed to reveal any intrinsic differences. It was found,

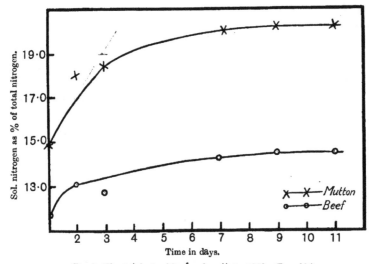

Fig. 3. The autolysis of beef and mutton at 37°. (Exp. 13.)

however, that in mutton the soluble nitrogen forms a larger percentage of the total nitrogen, so that the curves are parallel, but not superposable. The course of both, however, is perfectly regular and typical, and there is nothing to point to any important differences in the chemical constitution of these two kinds of muscle sufficient to account for their different behaviour in the cold store.

The progress of the autolysis is markedly slower in muscle than in glandular tissue, and equilibrium is not reached till about the eighth day. Moreover, the degradation stops short at a smaller percentage of soluble nitrogen.

Curves are given of a typical autolysis at 37°, and also of one at a much lower temperature. It is to be noted that though autolysis is retarded at the lower temperature, the two curves remain parallel (Figs. 3, 4 and 5).

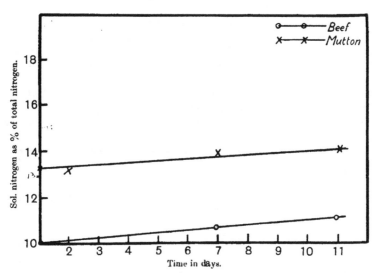

Fig. 4. Autolysis of beef and mutton at 0°. (Exp. 13.)

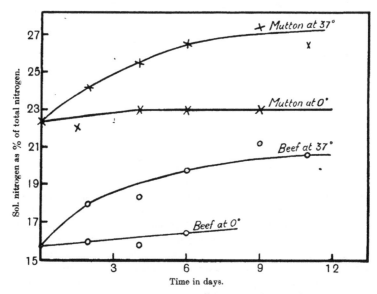

Fig. 5. The autolysis of beef and mutton at 37° and at 0°. (Exp. 14.)

GENERAL CONCLUSIONS AND SUMMARY.

An examination of the post-mortem autolysis of beef and mutton has thrown no light on the cause of their different behaviour after being frozen. The processes in both are exactly parallel, both at incubator temperature and at low temperatures. In the case of mutton, equilibrium is reached at a higher percentage of soluble nitrogen than in beef; but the initial non-coagulable nitrogen is higher, so that the curves are comparable. It is probable, therefore, that the differences in the two kinds of muscle in this respect lie not in their chemical constitution, but in the structure and physical properties of the fibres.

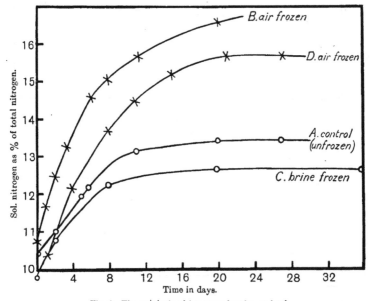

Fig. 6. The autolysis of frozen and unfrozen beef.

II. THE AUTOLYSIS OF FROZEN BEEF.

While working on the autolysis of beef it seemed of interest to determine whether previous freezing had any influence on the course of autolysis after thawing. It was found that not only was the course and degree of autolysis profoundly altered by such treatment, but that the method of freezing, that is the rate of freezing, is of great significance. This is made clear by the curves given in Fig. 6. It will be seen that previous freezing of the muscle results in a much greater degradation of the nitrogenous constituents during autolysis, and also that equilibrium is greatly delayed. The rate of freezing of carcasses is known commercially to be most important, and recent work has shown that, if frozen sufficiently rapidly, the meat on thawing approaches more nearly in appearance and palatability to fresh beef. On the large scale, rapid cooling is achieved by immersing the carcass in brine at about − 20° C. instead of freezing in air chambers. The beneficial effect of rapid freezing is borne out by the autolysis curve of beef frozen in brine. This curve approaches much more closely to that of the control from fresh beef, than that of air-frozen meat.

The following are the details of one such experiment. The animal was killed at 2.30 p.m. on June 29th, and immediately, before skinning, about 2½ lbs. were cut off the shin, and placed in a sterile pan in ice. On arrival at the laboratory, the skin and fat were removed, and the muscle divided into four portions.

"A" was used as control, and a 20 % suspension was made by the method described above, and placed in the incubator at once.

"B" was suspended in a glass jar, and placed in the cold store at 18° F. at 3.20 p.m. It was removed the next day, and allowed to thaw for six hours at room temperature, and then the autolysis suspension made up as "A."

"C" was treated in the same way as "B" except that, in the cold store, it was sunk in saturated brine at 18° F.

"D" was kept at 0° for 20 hours, and then placed in cold store on June 30th, after which it received the same treatment as "B."

In each case the autolytic mixtures consisted of 100 g. muscle in 5 % toluene. Samples were removed at the times indicated on the curve, and the subsequent procedure was exactly as previously described.

Curves are given for one experiment only, but similar relationships were found to hold in other experiments.

It is very difficult to explain this effect on autolysis produced by freezing. Obviously the equilibria of the cell have been disturbed by the rupture of the membranes which occurs during slow freezing. As there is an absence of "drip" after brine freezing it would seem that by this method any extensive damage to the cell is avoided, and this is borne out by the autolytic changes in the brine-frozen beef.

Our sincere thanks are due to Prof. Hopkins for his interest and help throughout this work.

LV. NOTE ON THE OXIDATION OF CARBO-HYDRATES WITH NITRIC ACID.

By PAUL HAAS and BARBARA RUSSELL-WELLS.·

(*Received June 16th, 1922.*)

DURING the course of an examination of the carbohydrate constituent of *Chondrus crispus* we had occasion to oxidise the material with nitric acid, with a view to determining the amount of mucic acid produced. On examining the mother liquor from the mucic acid it was found to have a powerful reducing action upon Fehling's solution in the cold. On repeating the oxidation under rather more drastic conditions and evaporating the mixture to a syrup the same result was obtained. This seemed sufficiently remarkable to be worth investigating and samples of cane sugar, lactose, glucose and levulose were accordingly oxidised in the same way, and in every case it was found that the resulting product readily reduced Fehling's solution in the cold. Experiments were thereupon started upon pure glucose with the intention of isolating the material responsible for this strong reducing action. While these experiments were in progress a paper appeared by Kiliani [1922] in which he described the production of certain new reducing acids formed by carefully oxidising various carbohydrates at room temperature in the absence of air. As this author expressed the wish to reserve this field for himself we felt bound to discontinue our investigation, but thought it desirable to put on record our observations so far as they go in order to draw attention to the fact that powerful reducing substances are produced, not only under the rather peculiar conditions described by Kiliani, but even by oxidation of carbohydrates with hot nitric acid and without special precautions, although the yields obtained may possibly be smaller than those obtained by Kiliani. 25 g. of glucose were dissolved in 100 cc. of nitric acid (sp. gr. 1·15) and warmed over a boiling water-bath; after a short time a violent action set in which gradually subsided and the heating was continued until the solution had been reduced to one-third of its original volume. In order to avoid undue dilution the solution was neutralised by the addition of powdered crystallised sodium carbonate. The resulting liquid, which was of a light yellow colour, was found to darken slowly even when kept in a stoppered bottle protected from the light; it reduced Fehling's solution at once in the cold and likewise ammoniacal silver oxide.

The solution was treated with 20 % lead nitrate until no further precipitate was formed. This precipitate, *A*, was filtered off and when washed, suspended in water and decomposed with hydrogen sulphide yielded a solution which reduced ammoniacal silver nitrate but not Fehling's solution. The filtrate from precipitate *A*, though containing an excess of lead nitrate, gave no further precipitate on standing; the fact that the solution still reduced Fehling's solution at once in the cold was taken to indicate that the reducing substance did not form an insoluble lead salt and was possibly not an acid, but on investigation it was found that the solution had acquired a markedly acid reaction, and on addition of a little alkali until the reaction was just on the acid side of neutrality a further heavy precipitate B was formed, which, after washing and decomposing with hydrogen sulphide, yielded an acid which reduced ammoniacal silver nitrate and Fehling's solution on heating but not in the cold. The filtrate from precipitate B which still reduced Fehling's solution in the cold was next treated with basic lead acetate, when it yielded a third precipitate C; the latter, when decomposed, gave the unknown acid X for whose isolation these experiments were undertaken; it was found to reduce both ammoniacal silver nitrate and Fehling's solution in the cold. The filtrate from precipitate C no longer had any reducing properties.

Subsequently it was found that the same acid X could be more rapidly obtained by replacing lead nitrate by lead acetate, since in using this reagent there was no change in the reaction of the solution; the addition of an excess of lead acetate caused the simultaneous precipitation of both the precipitates A and B of the previous experiment, leaving a filtrate containing only the acid X which could then be precipitated by the addition of basic lead acetate; after adding an excess of this reagent the precipitate was filtered and washed repeatedly with hot water until the washings were free from nitrate, an operation which resulted in considerable loss owing to the tendency of the precipitate to go through the filter paper. The washed precipitate decomposed with hydrogen sulphide yielded a powerfully reducing acid solution which, evaporated *in vacuo*, left a pale brown, viscous material which would not crystallise and which gradually darkened on keeping, especially if heated. This substance, dissolved in water and heated with concentrated hydrochloric acid, evolved a considerable quantity of furfural, thus showing it to contain at least five carbon atoms. The aqueous acid solution, treated with baryta solution, precipitated a small quantity of a sparingly soluble salt, the filtrate from which, however, still contained barium and reduced Fehling's solution in the cold. The neutral solution now also gave an immediate precipitate with lead acetate although this reagent would not precipitate it from the original oxidation mixture; this precipitate dissolved in caustic soda still reduced Fehling's solution in the cold.

Attempts to prepare a crystalline oxime, phenylhydrazone, p-bromophenylhydrazone and a cinchonine salt were unsuccessful. From the above facts it would appear that this substance, though in some respects similar to glycuronic acid, is not identical with it, but in view of the circumstances set forth above (with regard to the claim by Kiliani to reserve the field) and the possible identity of this substance with one of the compounds obtained by this author we have decided not to pursue the investigation any further.

REFERENCE.

Kiliani (1922). *Ber.* 55, B. 75.

LVI. OPSONINS AND DIETS DEFICIENT IN VITAMINS.

By GEORGE MARSHALL FINDLAY and RONALD MACKENZIE.

From the Royal College of Physicians' Laboratory, Edinburgh.

(Received June 19th, 1922.)

ALTHOUGH it is well known that animals fed on diets deficient in vitamins frequently die from intercurrent bacterial infections, the reason for this apparent failure in the protective mechanism of the body is at present unknown. Hektoen [1914] found that the serum reactions were quite normal in animals fed for considerable periods on artificial diets, while later, Zilva [1919] could not determine any decrease in the power to form agglutinins, complement and immune body in animals fed on diets deficient in vitamins.

In the present investigation the opsonic activities of the serum were studied in animals fed on diets deficient in vitamins.

DIETS DEFICIENT IN VITAMIN *A*.

Rats were employed for this experiment. Four partially grown rats varying from 106 to 120 g. were placed on a diet consisting of caseinogen, starch, cotton seed oil, marmite and an inorganic salt mixture. After ten weeks the animals were killed. One of the animals was suffering from keratomalacia at the time of its death. The sera from the four animals were collected in the usual manner and pooled before use.

Four normal animals fed on a diet of caseinogen, starch, cod-liver oil, marmite and inorganic salts were also killed and their sera collected to serve as a control. Leucocytes were obtained from a normal rat. The technique employed was that recommended by Wright [1921].

The opsonic activity of the two groups of sera was determined for two types of organisms, *Bacillus coli* and *Staphylococcus aureus*. One hundred polymorphonuclear leucocytes were counted on each slide. Notes were taken of (1) the number of leucocytes containing bacteria, (2) the total number of bacteria phagocytosed. The average number of bacteria per polymorphonuclear leucocyte was termed the phagocytic index. The data are recorded in Table I.

Table I.

Date of examination	Organism	Condition of animal	No. of polymorphonuclear leucocytes counted	No. of cells with bacteria	Total no. of organisms ingested	Phagocytic index
After 10 weeks on the diet	Staphylococcus	– Vitamin A	100	79	128	1·3
	B. coli		100	40	46	0·5
	Staphylococcus	+ Vitamin A	100	84	136	1·4
	B. coli		100	· 42	69	0·7

It will be seen that there is no significant difference in the readings obtained with the two sera and thus little evidence of any decrease in the opsonic activity as the result of a diet deficient in vitamin *A*.

DIETS DEFICIENT IN VITAMIN *B*.

Rats were also used for this experiment. Six rats were fed on a diet of caseinogen, starch, cod-liver oil and an inorganic salt mixture. Three rats were killed after 21 days, the remainder after 6 weeks on the diet. Rats fed on a complete diet were used as controls. The technique employed was the same as in the first experiment. The data are given in Table II.

Table II.

Date of examination	Organism	Condition of animal	No. of polymorphonuclear leucocytes counted	No. of cells with bacteria	Total no. of organisms ingested	Phagocytic index
After 3 weeks on the diet	Staphylococcus	– Vitamin B	100	73	96	1·0
	B. coli		100	74	98	1·0
	Staphylococcus	+ Vitamin B	100	70	129	1·3
	B. coli		100	82	126	1·3
After 6 weeks on the diet	Staphylococcus	– Vitamin B	100	77	135	1·3
	B. coli		100	59	79	0·8
	Staphylococcus	+ Vitamin B	100	67	141	1·4
	B. coli		100	76	88	0·9

DIETS DEFICIENT IN VITAMIN C.

Guinea-pigs were used for this experiment. As is well known, when fed on a diet containing only small quantities of vitamin C these animals develop chronic scurvy, a condition characterised by no very definite clinical symptoms except a failure of normal growth. Six guinea-pigs, varying in weight from 250 to 300 g., were fed on an *ad libitum* diet of oats and bran, and 70 cc. of autoclaved milk a day, with the addition of 2 cc. of orange juice every third day. Six controls were placed on the same basal diet but with the addition of 10 cc. of orange juice every day. Experiments, as before, were carried out with two organisms, *Staphylococcus aureus* and *Bacillus coli*. Readings were made after four weeks and after eight weeks on the dietary, three scorbutic and three control animals being killed on each occasion. The technique was the same as in the experiments with rats except that the leucocytes in this instance were those of a normal guinea-pig. The results are summarised in Table III.

Table III.

Date of examination	Organism	Condition of animal	No. of polymorphonuclear leucocytes counted	No. of cells with bacteria	Total no. of organisms ingested	Phagocytic index
After 4 weeks on the diet	Staphylococcus	− Vitamin C	100	94	315	3·1
	B. coli		100	87	115	1·1
	Staphylococcus	+ Vitamin C	100	91	330	3·3
	B. coli		100	84	148	1·5
After 8 weeks on the diet	Staphylococcus	− Vitamin C	100	87	261	2·6
	B. coli		100	82	104	1·0
	Staphylococcus	+ Vitamin C	100	92	282	2·8
	B. coli		100	76	95	0·9

THE PHAGOCYTIC ACTIVITY OF THE POLYMORPHONUCLEAR LEUCOCYTES IN SCURVY.

Since the opsonic activity of the serum was practically the same both in scorbutic and in control animals, an attempt was next made to determine whether the liability to infection in scurvy was in any way associated with a decreased phagocytic power of the polymorphonuclear leucocytes themselves. For this purpose two guinea-pigs of equal weight were selected, one being quite healthy, while the other had been fed on a diet deficient in vitamin C for the previous eight weeks. 1 cc. of a suspension of S. aureus heated for an hour at 55° and containing 200 million organisms was injected intraperitoneally into each animal. At intervals some of the peritoneal exudate was removed by means of a capillary tube; smears were made, stained with "Leishman" and examined microscopically. Two hundred polymorphonuclear leucocytes were counted on each film, a note being taken of the number of polymorphonuclear leucocytes containing 0, 1, 2, 3, and more than 3 staphylococci. The results are shown in Table IV.

Table IV.

Time	Condition of animal	Percentage no. of polymorphonuclear leucocytes containing				
		No Staphylococci	1 Staphylococcus	2 Staphylococci	3 Staphylococci	More than 3 Staphylococci
3 hours after injection	Scorbutic	50	27	13	8	2
	Control	44	26	16	10	4
6 hours after injection	Scorbutic	16	32	22	16	14
	Control	33	18	25	12	12
18 hours after injection	Scorbutic	36	24	26	6	8
	Control	36	18	24	16	6
30 hours after injection	Scorbutic	48	16	21	9	6
	Control	37	20	21	13	9

An examination of the pleural exudate 54 hours after the injection showed that mononuclear leucocytes were now the predominant type of cell in both animals while the polymorphonuclear leucocytes still present were undergoing degeneration.

There is no evidence to suggest any decrease in the phagocytic activity of the polymorphonuclear leucocytes as the result of a diet deficient in vitamin C.

In many ways the bacterial infections so frequently found in association with the deficiency diseases are analogous to the terminal infections met with in other chronic diseases. It is therefore of some interest to note that quite recently Cross [1921] has investigated the question of the opsonic index in relation to terminal infections associated with chronic bacterial infections. He was unable to determine any decrease in activity against any bacteria not concerned in the primary infection even in the very last stages of the disease.

As Bordet [1909] was the first to point out, the phagocytic power of the body would appear to be a relatively stable function and one not easily influenced by conditions which profoundly affect other vital activities.

Conclusions.

1. Rats fed on diets deficient in vitamins A and B do not show any decrease in the opsonic activity of the serum.

2. Guinea-pigs fed on a diet deficient in vitamin C do not show any decrease in the opsonic activity of the serum.

3. The phagocytic activity of the polymorphonuclear leucocytes of guinea-pigs with chronic scurvy is not decreased.

REFERENCES.

Bordet (1909). Studies in Immunity. New York.
Cross (1921). *Johns Hopkins Hosp. Bull.* **32**, 350.
Hektoen (1914). *J. Infect. Dis.* **15**, 278.
Wright (1921). The Technique of the Teat and Capillary Glass Tube. 2nd edition. London.
Zilva (1919). *Biochem. J.* **13**, 172.

LVII. ON CARRAGEEN (*CHONDRUS CRISPUS*). III.
THE CONSTITUTION OF THE CELL WALL[1].

By BARBARA RUSSELL-WELLS.

(*Received June 21st, 1922.*)

HISTORICAL.

MANY workers have from time to time investigated *Chondrus crispus*, but of recent years the most exhaustive work has been done by Sebor. The presence of galactose had been established by Flückiger and Obermayer [1868], and Bénte [1875, 1876] had found among the products of the action of acid upon carrageen levulinic acid and a sugar, which showed little or no optical activity. Haedicke, Bauer and Tollens [1867] established in carrageen mucilage the presence of fructose and calculated that it contained 28 % of galactose. They suggested that these sugars might be present under the form of raffinose. Sebor [1900] determined to identify the remaining sugars. He confirmed the presence of galactose and fructose, found small quantities of pentosan and showed that raffinose is not present. He also stated that he had found evidence of glucose. He oxidised the carrageen mucilage, but, after filtering off the mucic acid, failed to obtain any acid potassium saccharate on treating with potassium carbonate and acetic acid. As an analysis of the silver salt gives the same result for both mucate and saccharate he tested the potassium solution for optical activity, but found it inactive, whereas the saccharate should be active. He therefore abandoned the oxidation products and tested for glucose in a hydrolysed solution of mucilage. He hydrolysed with $6N$ HCl to destroy the fructose, and when the solution no longer gave a Selivanoff reaction, added benzylphenylhydrazine. He separated off the benzylphenylhydrazone of galactose and found the remaining compound had a melting point of 163°, that of glucose benzylphenylhydrazone being 165°. He declared that the hydrazone was a mixture and his paper gives the impression that he was not satisfied with the agreement of melting points referred to above. This appears to be all the evidence he had in favour of the presence of glucose among the constituents of carrageen. Müther and Tollens [1904] state that glucose is among the constituents of carrageen, but all the evidence they offer is an estimation of the amount of silver present in the silver salt of an oxidation product; this is obviously useless since the amount of silver is the same in the salts of both saccharic and mucic acids.

Abderhalden's *Biochemisches Handlexikon* [1911] contains the statement that the pure mucilage of carrageen corresponds to the formula $C_6H_{10}O_5$, but

---

[1] Thesis approved for the Degree of Doctor of Philosophy in the University of London.

it is not stated on what evidence this formula is based; it must be borne in mind that the mucilage is known to contain both nitrogen and considerable quantities of ash. Stocks and White [1903], on the other hand, have shown by analysis that the mucilage does not possess this formula.

	Found for ash free substance	Calculated for $C_6H_{10}O_5$
Carbon	37·94 %	44·44 %
Hydrogen	5·92	6·17
Oxygen	54·95	49·39
Nitrogen	1·19	—

These figures conclusively demonstrate that the formula $C_6H_{10}O_5$ does not agree with the results of analysis.

Since, as will appear later, it has now been proved that carrageen mucilage contains at least two substances which are carbohydrate esters of sulphuric acid, and various metallic radicles, any calculations of the above nature are futile.

Previous investigators have commented on the high ash content of carrageen and reported that it is largely composed of calcium sulphate, but most of them have ignored the significance of this and proceeded to examine the organic radicles obtained by various means from the plant. They also appear to have worked either on the seaweed itself or on a direct hot water extract. In 1921 Haas and T. G. Hill [1921] showed that this extract consists in reality of two distinct fractions, and in the same year Haas [1921] published a method for separating the two, together with a systematic investigation of the inorganic constituents of one of them. He found that while one was readily soluble in both hot and cold water, the other was readily soluble in hot, but only sparingly in cold water. The method of extraction and separation was based on this difference of solubility, and was, with slight modification, the method adopted in the preparation of larger quantities of the two substances required for the present investigation.

The distinctions between the hot extract (H.E.) and the cold extract (C.E.) described by Haas and Hill were confined to physical characteristics such as different solubility in cold water and different gelatinising properties.

It was also shown by Haas that the H.E. contained 17·6 % of ash which could not be reduced by dialysis and was found to be due to the presence of the calcium salt of an ethereal sulphate, in which the calcium is fully ionised and can be precipitated by the ordinary reagents, while the sulphate radicle is not ionised and can only be precipitated by barium chloride after hydrolysis of the compound.

Haas and Hill also state that they found nitrogen in both fractions. They also got positive reactions for protein in the seaweed itself and in the water-soluble product obtained from it.

From the foregoing it will be seen that carrageen can be separated into three constituent fractions, the cold water and hot water soluble portions and the residue. The physical characteristics of the two extracts have been described by these previous workers, and the presence of an ethereal sulphate complex definitely established in the H.E.

PRESENT INVESTIGATION.

The present investigation was undertaken with a view to instituting a chemical comparison between the two extracts of carrageen, both as to their inorganic and organic constituents, to examine the residue of the weed, remaining after the extraction of both fractions, for cellulose and to discover if other examples could be found among the seaweeds of ethereal sulphate grouping.

The hand picked carrageen was first washed free from dust by holding it for a few seconds in running water, and then soaked for about an hour in distilled water. The aqueous extract thus obtained was poured off, filtered and evaporated, the resulting scales constituting the "cold extract." This process was then repeated and thus a further quantity of C.E. obtained. After only two hours' soaking much C.E. still remained in the weed but it was decided to sacrifice this quantity lest a more prolonged extraction should result in contamination with some of the H.E. which is also slightly soluble in cold water. The weed was therefore washed for several days by soaking in water until it was no longer slimy, all the wash water being rejected. The weed thus deprived of its C.E and some of its H.E. was then washed several times in distilled water and air-dried. In order to obtain the H.E. the dried carrageen was heated on a water-bath with fresh distilled water and the liquid from this yielded, on filtration and evaporation, the "hot extract."

EXAMINATION OF THE COLD EXTRACT.

The crude C.E., just as it is taken from the evaporating pans, has a distinctly saline taste and on incineration was found to contain about 32 % of ash which showed a marked tendency to fuse. After dialysis the salt taste had disappeared and the ash content had fallen to 21·86 %. No amount of dialysis would reduce the ash further. The salts which dialysed out were, no doubt, mainly impurities arising from contact of the seaweed with sea-water. During dialysis it was observed that the water outside the dialyser turned slightly yellow as though some organic substance had come through the parchment. This water was therefore collected and evaporated down over a water-bath on tin-lined copper trays. A small quantity of a brown, sticky, amorphous material was obtained; this did not dry in scales like the hot and cold extracts, and although composed very largely of the salts which had dialysed out from the C.E., gave a strong reaction with Molisch's reagent, thus showing the presence of a diffusible carbohydrate; as it did not reduce Fehling's solution to any appreciable extent it was thought possible that it might be trehalose, which Kylin [1915] claims to have found in carrageen. Accordingly a dilute solution was tested in a polarimeter, but not the slightest trace of rotation could be observed.

A qualitative analysis of the ash of the C.E. showed the presence of sulphate, calcium, magnesium and small quantities of sodium and potassium with traces of iron. These were all found in the ash of the H.E. as well, but here the amount of alkali metals was so small that the ash showed no tendency to fuse during the process of incineration.

The calcium in the C.E. as in the H.E. is fully ionised and can be precipitated quantitatively from a 0·35 % solution by means of ammonium oxalate. The actual figures obtained were as follow:

1. Calcium by incineration 3·98 %.
2. Calcium by direct precipitation 4·03 %.

The sulphate however is not ionised, no precipitate being formed by the addition of barium chloride. But, after hydrolysis with hydrochloric acid, barium chloride gives a copious precipitate of barium sulphate, showing that hydrolysis liberates the sulphate radicle in an ionised condition, while previously it was bound to some organic radicle. The method of hydrolysis employed was as follows:

About 0·5 g. of C.E. was dissolved in 200 cc. of water and 5 cc. of concentrated hydrochloric acid added. This solution was boiled over a Bunsen flame for at least six hours to ensure complete hydrolysis. On estimation the percentage of sulphate obtained after hydrolysis was found to be rather more than twice that obtained on incineration, the actual figures being:

	I	II	Mean
1. SO_4 by incineration	13·75 %	14·16 %	13·95 %
2. SO_4 by hydrolysis	30·14	30·32	30·23

As in the case of the H.E. it is extremely difficult to get rid of all moisture, the substance tending to char when heated in a steam oven owing to the liberation of sulphuric acid.

If the C.E. contained only a calcium salt of the ethereal sulphate corresponding to the formula which has been given for H.E., viz.

$$O.SO_2.O$$
$$R \diagup \diagdown Ca$$
$$O.SO_2.O$$

the sulphate ratio would be exactly two to one. Actually, however, it is more than two to one. This discrepancy could be accounted for on the assumption that some of the ethereal sulphate was combined with ammonium instead of with calcium. Such a combination might be present in a mixed calcium and ammonium salt, or there might be slight traces of the pure ammonium salt mixed with the calcium salt. There is not, however, enough evidence to show which combination is the more probable. It is obvious that in a salt of this nature the sulphate moiety combined with (NH_4) would leave no ash on incineration, although such sulphate could be estimated after hydrolysis.

In order to discover if any ammonium radicle were in fact present, some of the C.E. was distilled with magnesium oxide, when ammonia was given off. This ammonia was estimated and the results, calculated as N, were as follow:

	I	II	Mean
N by magnesium oxide	0·24 %	0·22 %	0·23 %

The quantity of nitrogen found would, in the form of ammonium, satisfy 0·80 % of sulphate calculated as SO_4. Thus assuming the nitrogen actually present is contained in such an ammonium salt as suggested above, the experimental value for incinerated sulphate is raised by 0·80 % from 13·95 % to 14·75 %, which is only 0·36 % below the value required by theory, a difference well within the limits of experimental error. In addition to the nitrogen which can be driven off by distillation with magnesium oxide, C.E. contains other nitrogen not liberated in this manner.

	I	II
N by Kjeldahl	0·55 %	0·57 %

As shown above, nearly half this nitrogen can be liberated by boiling with magnesium oxide and is therefore probably present in an ammonium radicle. The rest is almost certainly there as protein nitrogen. That protein is present in both H.E. and C.E. is indicated by the fact that both give a positive reaction with Millon's reagent and with nitric acid. Whether this protein is combined or not is at present impossible to say, but it is constantly present though it may be removed by prolonged heating with alkali.

Ammonium salt of extracts.

An attempt was made to prepare the ammonium salt of the C.E. in the hope of removing all the calcium and magnesium and so obtaining an ash free material. For this purpose some C.E. was dissolved in water, ammonia and ammonium chloride were added and then enough ammonium phosphate to precipitate all the calcium and magnesium. The solution was left to cool and settle over night and was filtered next morning. The liquid was filtered with the aid of the filter pump first through a Chardin and then through a quantitative filter-paper on a Buchner funnel. The filtrate, which was much clearer than a solution of ordinary C.E., was dialysed for five days to remove excess of ammonium phosphate and other impurities. Since it charred when evaporated to dryness over a water-bath, the filtrate was evaporated in a vacuum-desiccator and clear, almost colourless scales were obtained. These were dried to constant weight in the desiccator, and it was then found that the ash content had fallen from 21·26 % to 5·87 %. A qualitative analysis of the ash showed it to consist very largely of magnesia, from which it was concluded that the magnesium present in the C.E. is not in an ionised condition, or else it would have been precipitated by the phosphate.

In addition to magnesium the ash of the ammonium salt of the C.E. contained potassium and sodium but no calcium.

The same procedure was adopted with the H.E. The filtrate, after precipitation with ammonium phosphate, was dialysed for six days, and then evaporated over a water-bath in a platinum basin. Like the C.E. it showed a tendency to char, and when evaporated to small bulk turned acid and reduced Fehling's solution. It was therefore kept alkaline with ammonia all the time it was being evaporated, and when reduced to a small volume it was taken from the water-bath and set aside to cool. After cooling it set to a stiff jelly, proving that the loss of calcium does not destroy the gelatinising properties of the H.E. On complete evaporation in a vacuum the substance was found to be very like the ammonium salt of the C.E. in appearance, and consisted of clear, nearly colourless scales which readily absorbed moisture from the atmosphere. On incineration a considerable amount of ash was left, which like that of the C.E. contained magnesium. So here again the magnesium is present in an un-ionised condition.

Another sample of the ammonium salt of the H.E. was prepared by precipitating with ammonium oxalate instead of with ammonium phosphate and the ash in this was estimated and found to have been reduced from the original 17·6 % to 4·38 %.

Comparison of organic substances in hot and cold extract.

The organic complexes in the two extracts of carrageen were also investigated with a view to discovering further differences of composition between the two fractions.

Both H.E. and C.E. consist largely of carbohydrates. Neither of them will reduce Fehling's solution without previous hydrolysis, but both will do so after such treatment.

Since galactose has been found in carrageen mucilage by Flückiger and Obermayer [1868] and in Chondrus elatus by Takahashi [1920] it was decided to test for it in the two extracts. They were therefore oxidised with nitric acid according to the Creydt modification of the Kent and Tollens method. Mucic acid was obtained from both, thus proving galactose to be present in both fractions. The amount of mucic acid obtained was then estimated, according to the method of van der Haar [1920], the average values obtained being 21·12 % from the H.E. and 24·82 % from the C.E. These figures correspond to 29·47 % and 33·72 % of galactose respectively in the two extracts.

The mother liquors from the mucic acid were examined qualitatively for saccharic acid, but in no case was any found; on the other hand, both tartaric and oxalic acid were shown to be present, the latter in larger quantity in the case of the H.E.

ESTIMATION OF PENTOSES.

Sebor [1900] expressed the view that carrageen mucilage contained a small quantity of pentosan or methyl-pentosan. The H.E. and C.E. were accordingly tested by the Kröber-Tollens method to discover whether pentoses could be discovered in either or both of these fractions. The actual results obtained were: H.E. 1·89 % of pentosan (mean of two estimations); C.E. 1·38 % of pentosan.

Since concentrated solutions of carrageen mucilage will set to a jelly on cooling it was thought that this mucilage might contain pectic bodies. Accordingly a direct hot water extract of Chondrus was tested according to the method of von Fellenberg [1914], which consists in steam distilling an alkaline solution of the substance under consideration, and testing the distillate for methyl alcohol and acetone [Tutin, 1921]. Two grams of the extract of carrageen were therefore gently boiled for $6\frac{3}{4}$ hours under·a reflux condenser with 100 cc. of $N/10$ NaOH, and the alkaline solution, which had darkened slightly, was then steam distilled. The distillate was practically neutral and slightly yellow in colour. It was tested for alcohol by the acid and bichromate test and for acetone by the iodoform and nitroprusside tests, but the result was negative in each case. This was therefore taken to prove the absence of pectic bodies from the mucilage of carrageen.

RESIDUE LEFT AFTER EXTRACTION BY COLD AND HOT WATER.

The residue of the Chondrus left after the removal of the C.E. and H.E. showed a great reluctance to part with its last traces of water-soluble material. It was boiled for several hours on a water-bath, the water being changed periodically, but each time this water on being filtered off and tested with Molisch's reagent gave a strong positive reaction. A fresh portion of residue was therefore heated over a Bunsen burner with three successive changes of water, but this treatment also failed to remove the last traces of water-soluble material. Recourse was then had first to a steriliser and finally to an autoclave. After several hours in each of these, however, water-soluble material was still coming out, and was found to contain calcium.

Kylin [1915] claims to have found an insoluble calcium compound in Chondrus crispus. He apparently cut sections of fresh weed, boiled them in water and then left them to soak over night, concluding that this treatment removed all water-soluble substances present. He then found calcium to be present in the sections and assumed it to be in an insoluble form. In view of the extreme difficulty experienced during the present investigation in getting rid of the last traces of water-soluble material from the residue of carrageen which had already been soaked for several days in water, it seems unlikely that all was removed by Kylin's treatment. Hence the calcium-containing substance he found was in all probability merely the last traces of the H.E.

Since it was found so difficult to extract the residue completely merely by heating with water, a portion was boiled under a reflux condenser with

3 % NaOH. After two days of this treatment the contents of the flask were filtered, and the residue on the filter-paper washed free from alkali. It was then soaked in dilute hydrochloric acid to remove the carbonate found to have been formed during the alkali decomposition. When washed free from acid the residue was found to be protein- and ash-free, and completely soluble in cuprammonia, showing it to be pure cellulose. Cellulose was also obtained from the residue by acid decomposition. This was effected by boiling for some hours with hydrochloric acid. The solution was filtered and the solid matter left on the filter-paper was washed and tested with cuprammonia. The total amount of cellulose in carrageen, estimated by alkali decomposition, was found to be 1·3 %.

CERAMIUM RUBRUM.

After it had been definitely established that the sulphate in both the C.E. and H.E. of carrageen is present as an ethereal sulphate linked to an organic radicle it seemed probable that this grouping might be found to occur in other seaweeds. Accordingly it was decided to examine another alga which was known to yield an ash with a high sulphate content, and for this purpose *Ceramium rubrum* was selected.

Three grams of dried *Ceramium* were heated with distilled water on the water-bath for an hour. The seaweed was then removed and the liquid filtered and evaporated. Scales of a transparent, horny material were obtained. Some of this material was dissolved in distilled water and the solution was divided into two portions *A* and *B*, barium chloride being added to each. To solution *A* a small quantity of concentrated hydrochloric acid was added while solution *B* was left neutral; both were then boiled for one and a half hours. At the end of this time a precipitate of barium sulphate had been formed in the acid solution *A* only, thus demonstrating the presence of the ethereal sulphate grouping in *Ceramium rubrum*. The investigation is being continued.

GENERAL.

Microscopic examination of sections of the thallus of *Chondrus* show that the cell walls become very thick and swollen in distilled water. This is what one would expect if they were largely composed of H.E. and C.E., since on being placed in water dried scales of these materials absorb much moisture and swell up enormously. While only 1·3 % of the dry weight of carrageen consists of cellulose the cell walls appear under the microscope to constitute the greater part of the section, and over 80 % of its weight consists of H.E. and C.E. Since these contain complex polysaccharides they are allied to cellulose and are possibly degradation products of it.

The exact nature of the organic radicle or radicles present in the ethereal sulphates has not been determined. There may be one bi-valent or two uni-valent groups in each molecule (see formula on p. 581).

The present investigation has shown that the two extracts, though differing chemically as well as physically, are nevertheless closely allied. It may be that each extract contains a mixture of various closely related bodies and that different ones predominate in each extract. The similarity of the oxidation products points to both extracts consisting of substances belonging to the same group of compounds, but there is not enough evidence to show whether the differences are due to different bodies or mixtures in different proportions of the same substances. On the whole the former of these alternatives appears the more likely.

SUMMARY.

1. Dialysable organic matter can be separated from the C.E. of carrageen.

2. The C.E. contains calcium and ammonium ethereal sulphates and its ash contains, besides sulphate and calcium, magnesium, sodium, potassium and traces of iron. The ash of H.E. also contains these radicles, but it has less sodium and potassium and more calcium than that of C.E.

3. The ammonium salts of both extracts have been prepared by replacing the ionised calcium by the ammonium radicle.

4. Un-ionised magnesium is present in both extracts.

5. The main oxidation products of both extracts consist of mucic, oxalic and tartaric acids. More mucic acid, but less oxalic acid, is obtained from C.E. than from H.E.

6. Pentose radicles are present in both extracts, but more in cold than in hot.

7. Pectic bodies are absent from both extracts.

8. Cellulose is present in the residue of carrageen left after extraction with both cold and hot water.

9. There are indications that the ethereal sulphate grouping is present in *Ceramium rubrum*.

In conclusion the writer wishes to express her thanks to Dr Haas for his help and advice during the progress of the work, and his kind interest throughout.

REFERENCES.

Abderhalden (1911). *Biochemisches Handlexikon*, 2, 75.
Bente (1875). *Ber.* 8, 417.
—— (1876). *Ber.* 9, 1158.
von Fellenberg (1914). *Chem. Zentr.* 2, 942.
Flückiger and Obermayer (1868). *Repert. Pharm.* 380.
van der Haar (1920). *Monosaccharide u. Aldehydsäuren*
Haas (1921). *Biochem. J.* 15, 469.
Haas and Hill (1921). *Ann. Appl. Biol.* 7, 352.
Haedicke, Bauer and Tollens (1867). *Annalen*, 238, 302.
Kylin (1915). *Zeitsch. physiol. Chem.* 94, 337.
Müther and Tollens (1904). *Ber.* 37, 298.
Sebor (1900). *Oesterr. Chem. Zeit.* 3, 441.
Stocks and White (1903). *J. Soc. Chem. Ind.* 22, 4.
Takahashi (1920). *J. Coll. Agr. Hokkaido Imp. Univ. Japan*, 8, 183.
Tutin (1921). *Biochem. J.* 15, 494.

LVIII. STUDIES OF THE COAGULATION OF THE BLOOD.

PART II. THROMBIN AND ANTITHROMBINS.

By JOHN WILLIAM PICKERING and JAMES ARTHUR HEWITT.

From the Physiological Department, the University of London (King's College).

(*Received June 26th, 1922.*)

IN a recent paper in this *Journal* the present writers questioned the current view that antithrombin is secreted by the liver. It was suggested that the action of that organ in promoting the fluidity of the blood after the injection of "peptone" depends on variation of carbon dioxide content of the blood [Pickering and Hewitt, 1921]. Experiments were devised to test this suggestion with the somewhat unexpected result that in pithed cats respiring air, typical inhibition of the coagulation of the blood followed the rapid injection of "peptone" into animals with the liver out of the circulation. Further, the anticoagulant action of "peptone" was annulled by partial asphyxia and was restored by administration of air. It was found also, when suitable precautions were taken to preserve the surface conditions of the blood, that the addition of "peptone" to blood *in vitro* inhibited clotting when the concentration of "peptone" was no greater than was required to produce a like effect *in vivo*.

Immunity to the anticoagulant action of "peptone" was also found to follow the slow injection of that substance into cats with the liver out of circulation [Pickering and Hewitt, 1922].

Having thus demonstrated that the typical action of "peptone" on the blood of the cat can be obtained without invoking the aid of the liver, it became of interest to examine the action of thrombin on the circulating blood of animals under similar conditions, and also to re-examine the evidence for the extraction of antithrombin from the liver, blood and other tissues. Observations on these latter points are presented in this communication.

THE INTRAVASCULAR INJECTION OF THROMBIN INTO ANIMALS DEPRIVED OF HEPATIC CIRCULATION.

The thrombin was dissolved in mammalian Ringer's solution and was extremely active as a coagulant *in vitro*; thus the addition of 0·25 cc. of a solution, prepared by dissolving 0·0048 g. of solid thrombin in 100 cc. Ringer, to 0·75 cc. of cat's blood caused coagulation to commence[1] in 45″ on glass, completion[1] of this process being one minute later. On paraffined surfaces these figures were 45″ and 2′ 30″ respectively.

Experiment 1. Black cat, 2060 g. Pithed. Artificial respiration was maintained for 20′ before the aorta and inferior vena cava were ligatured. The following table shows the coagulation times of blood shed on to glass before and after the injection of thrombin into the circulation.

Table I.

Time	Notes	Commencement of coagulation	Completion of coagulation
0′ 0″	Intact animal ... ... _	7′ 40″	12′ 0″
1′ 0″	Animal pithed, vessels ligatured	—	—
36′ 30″	— — —	7′ 50″	11′ 48″
36′ 45″	7·5 cc. of 0·0048 % thrombin in Ringer injected	—	—
39′ 0″	— — —	1′ 0″	7′ 10″
44′ 0″	— — —	1′ 10″	7′ 50″
71′ 0″	— — —	2′ 50″	9′ 30″
86′ 0″	— — —	1′ 45″	8′ 50″

Notes: (1) Between 44′ and 86′ several observations were made. The coagulation times given at 71′ show the greatest variations.

(2) *Post-mortem* examination revealed no intravascular clots.

(3) Other experiments of the same nature yielded similar results.

Experiment 2. In this observation 0·01 g. of thrombin was injected into a 2¼ kilo. cat under precisely similar conditions to those of Exp. 1. Similar results were obtained. The subsequent injection of 7·5 cc. of 10 % calcium chloride in distilled water caused no material alteration in the coagulation times of shed blood. A further injection of 5 cc. however killed the animal. *Post-mortem* examination did not show any clots in those portions of the vascular system where blood had been circulating.

These experiments show: (1) That relatively large doses of thrombin can be injected into cats respiring air and with the liver out of circulation without causing intravascular coagulation. (2) That subsequent injection of a lethal amount of calcium chloride in distilled water fails to induce thrombosis. (3) That blood shed after the injection of thrombin into animals deprived of hepatic activity coagulates much more rapidly than the normal blood of the same animal under otherwise similar conditions.

It may be remarked that the present authors have found, in the course of a large number of observations on pithed cats, that if the air supply of the animals is regulated and kept constant, then constant times of commencement and of completion of clotting of shed blood can be maintained for considerable periods.

The great rapidity of clotting of shed blood after the injection of thrombin may in part be explained by the fact that in the experiments recorded above no excretion of injected thrombin by the kidneys was possible; thus, directly the blood was shed, the surface conditions of the plasma were altered and the thrombin was able to exert its coagulant action.

[1] "Commencement" of clotting indicates the first visible departure from normal fluidity: "completion" that coagulation had advanced so far that the vessel could be inverted without spilling.

The somewhat delayed coagulation reported by Davis [1912] following the injection of thrombin into intact animals under ether anaesthesia and preceding the typical hypercoagulability has not been observed in this series of experiments. Unfortunately Davis records only the times of completion of clotting of shed blood after injection of thrombin and does not give the coagulation times before, recording instead the "bleeding times," estimated by the rate of decrease of haemorrhage from a small skin wound, such as in the lobe of the ear. Davis employed the method of Duke [1910]. It should be noted that Weil [1920] has shown that prolonged bleeding time may coexist with normal coagulation time. Moreover Mendenham [1915] found that coagulation is hastened by ether anaesthesia. Unless precautions were taken to maintain a constant depth of anaesthesia and a constant carbon dioxide and oxygen content in the blood slight variations of coagulability, like those observed by Davis, might well occur. These would be masked later by the coagulant action of the thrombin.

The bearing of the foregoing experiments on the "thrombin theories" of coagulation.

The experiments reported in this paper are directly opposed to a commonly accepted view, advocated by Howell [1918], that the injection of thrombin stimulates the liver to secrete an excess of antithrombin, which latter neutralises the injected thrombin and so maintains the fluidity of the blood. They are also dissonant with the suggestions of Howell [1912] "that prothrombin or thrombin itself constitutes a hormone which excites the secretion of antithrombin," and are concordant with the fact that the typical action of "peptone" on blood can be obtained in cats deprived of hepatic activity. Support is thus given to the conclusion [Pickering and Hewitt, 1922] that the retarded coagulability of shed blood after the rapid injection of "peptone" into cats is due neither to the secretion of antithrombin nor of excess of alkali by the liver. This latter view, advocated originally by Dastre and Flouresco [1897], and later by Mellanby [1909] is also dissonant with the recent work of Gratia [1921] who, using Marriott's colorimetric method [Marriott, 1916], was unable to find any evidence of the slightest difference in the alkalinity of normal and "peptone" plasmas.

In the hypotheses of Morawitz [1904, 1] and of Fuld and Spiro [1904], the formation of thrombin from thrombogen (prothrombin) is said to be the prelude of the actual coagulation of the blood. Thrombin, it is postulated, acts directly on the fibrinogen of the plasma and is the immediate cause of clotting.

So long as it was reasonable to assume that the liver secreted sufficient antithrombin to neutralise the effect of massive doses of thrombin these theories remained tenable. As large amounts of thrombin can be injected without thrombosis into the circulation when such hepatic secretion is impossible, the view that thrombin acts directly on the fibrinogen of *unaltered* plasmas must be discarded and the views of Morawitz and of Fuld and Spiro become untenable. The theory of Howell [1912] explains the fluidity of normal blood by the presence of antithrombin which prevents the calcium of the blood from activating prothrombin to thrombin. The coagulation of shed blood is attributed to the neutralisation of antithrombin by thromboplastin, derived from platelets and tissue cells. Thrombin, so soon as formed, is again regarded as the immediate cause of clotting, a conclusion which is opposed to the evidence brought forward in this communication.

The most recent development of the thrombin theories, that propounded by Bordet [1920], accepts the belief in the existence of antithrombin and other anticoagulants in normal blood, but dissents from the view of Howell as to the rôle played by cytozyme (thromboplastin) in coagulation, citing the work of Gratia [1920] on the neutralisation of hirudin by cytozyme in support of his contention. There is also another important difference between the views of Howell and of Bordet. The former author accords primacy to antithrombin in the maintenance of fluidity, the latter, together with his co-workers [Bordet and Gengou, 1903; Bordet and Delange, 1913], assigns equal importance to surface conditions. Bordet maintains that contact of the blood with foreign bodies causes the interaction of lipoidal cytozyme (thromboplastin), formed largely from platelets but also from leucocytes and other sources, with serozyme, a product of the plasma formed immediately prior to coagulation. Serozyme and cytozyme are then said to combine, in the presence of calcium, to yield thrombin which in turn acts on fibrinogen to give clots of fibrin. Thus apart from the preliminary processes involving the production of thrombin, the first act of coagulation is believed to be the action of thrombin on fibrinogen. The evidence in this paper shows that if the physical conditions of the blood are preserved by remaining in the living vessels, then thrombin, in amounts sufficient to cause coagulation *in vitro*, does not cause thrombosis *in vivo*. It thus appears that although Bordet is correct in assigning importance to the surface conditions of the blood in the maintenance of fluidity, it is not to the disintegration of formed elements yielding thrombin, under the influence of physical change, that the *inception* of coagulation is due, but rather to physical change destroying the stability of the clotting complexes of the plasma. It is only after such change that thrombin can act as a coagulant. In this view, thrombin would be regarded as an accelerator of coagulation rather than as an initiator of that process. The work of Vines [1921] indicates that the complex associated with the inauguration of clotting exhibits distinctive differences from the prothrombin of Howell.

This view of the action of thrombin is supported by the observation of the present writers [Pickering and Hewitt, 1921] that in blood surrounded by oil, coagulation commences as a reversible gel. Further the work of Howell [1916] indicates that the change from a gel to a typical clot is a physical process due to ageing and condensation.

The addition of thrombin to blood in the gel stage immediately transforms che gel into typical fibrin, the speed of coagulation being identical with the speed of clotting when thrombin, in the same concentration, is added to blood shed on to clean glass by means of a paraffined cannula. This latter observation is concordant with the fact discovered by Gratia [1918], that the coagulation of blood induced by thrombin proceeds at the same rate on paraffined as on glass surfaces.

Bürker [1904] found platelets to remain intact in blood on a paraffined surface, yet under these conditions the first stage in coagulation—the formation of a reversible gel—sets in. This falls into line with the observations of Achard and Aynaud [1908] that if coagulation takes place in a moist vaselined chamber at 16–18°, then the platelets remain independent of the fibrin filaments. These authors maintain that physical changes of platelets are the result, rather than the cause of coagulation. Morawitz [1904, 2] found that thrombokinase (cytozyme) could be obtained from platelets. Cramer and Pringle [1913] showed that thrombokinase is not present in oxalate plasma and concluded that recalcification "induces coagulation primarily by causing the breaking up of platelets and only secondarily by influencing the chemical reaction which takes place in the presence of the substance (thrombokinase) liberated by the platelets," also the "filtered oxalate plasma...is readily coagulated by both filtered (through a Berkefeld filter) and unfiltered platelet extract in the presence of soluble calcium salts." On the other hand Gratia [1918] concluded that contact with foreign substances capable of being wetted by blood, does not act solely on platelets and other formed elements but also on the colloids of the plasma. The same writer also states that calcium chloride, if dissolved in distilled water, disintegrates platelets, liberating, presumably, cytozyme (thromboplastin or thrombokinase). Attention is drawn to experiment 2 of this paper in which massive doses of calcium chloride in distilled water were injected into the circulation of an animal deprived of hepatic activity and subsequent to a massive dose of thrombin. In this experiment, if Gratia is correct, platelets were disintegrated; it appears therefore that thrombin plus cytozyme from platelets is unable to cause clotting when the surface conditions of the blood are maintained by contact with the living walls of the blood vessels. It also indicates that platelets in concentrations existing in circulating blood, play only a secondary part in provoking coagulation.

It has been shown that the rapid addition of a tissue extract (a source of cytozyme) to bird's blood in vitro has a coagulant action, while the slow addition of similar amounts has an anticoagulant action [Pickering and Hewitt, 1921]. Demonstration has also been given that the intravascular injection of either calcium chloride [Löwit, 1892] or of inactive nucleoprotein [Halliburton and Brodie, 1894], each of which causes leucolysis and therefore liberates cytozyme into the circulation, is not followed by thrombosis.

These facts indicate that the relatively slow liberation of cytozyme by the disintegration of formed elements in the blood is insufficient to induce thrombosis.

Bloch [1920] suggested that in circulating blood the calcium ions exist in an "inactive and latent state" and that on issuing from the vessels, contact with the *air*, dust and so-called thromboplastic material may transform "inactive" calcium to an active electrolytic precipitant of the colloids of the plasma. The work of Chio [1917] has indicated that the tension of carbon dioxide in the blood may affect the equilibrium of the calcium salts which in turn regulates the physical state of the blood, while Vines [1921] has shown that a change from combined calcium to ionised calcium takes place during clotting.

In Cramer and Pringle's experiments [1913] blood was shed through the air directly into oxalate[1]. If Bloch's view is correct change towards clotting had already taken place. Caution should, it is thought, be exercised in deducing any theory of coagulation from the behaviour of blood after exposure to the air, oxalation and replacement of ionised calcium.

In the experiments of Cramer and Pringle referred to, platelets, leucocytes and some erythrocytes were removed from oxalate plasma, and from these was prepared a "platelet extract" whose coagulant action was tested *in vitro*.

The work of Spring [1900], Höber and Gordon [1904], Paine [1911] and Galecki [1912] shows that the speed of addition of a coagulant electrolyte is an important factor in coagulation of inorganic sols, while the present writers have demonstrated that similar conditions determine the coagulation of bird's blood by tissue extract *in vitro* [Pickering and Hewitt, 1921]. If experiments are to show that lysis of platelets is the initial factor in *normal* coagulation, it would be necessary to prove: (*a*) that the "platelet extract," or the aggregates of platelets, was free from products extracted from leucocytes and erythrocytes; (*b*) that the addition of the "platelet extract" was made at the same rate as the liberation of the coagulant material occurring normally in shed blood; (*c*) that the concentration of either "platelet extract" or of aggregates of platelets added, was not greater than the concentration of extract resulting from the disintegration of platelets in normal shed blood; (*d*) that the lysis of platelets added to oxalate plasma in the presence of ionised calcium took place at the same rate as the disintegration of platelets in normal shed blood.

[1] In a private communication to one of the present authors (*J. A. H.*) Dr Cramer expressed the opinion that our remarks on p. 719 of Vol. xv of this *Journal* regarding the technique of Cramer and Pringle were liable to misinterpretation; he suggested that we might refer to it in our next communication on the subject in question. We stated "*that blood was not withdrawn through a paraffined or similarly treated cannula.*" The method actually was [Cramer and Pringle, 1913] "*to cut the artery so that the blood flowed directly into the receiving vessel.*" It will be seen that the statement of fact is strictly correct.

THE SIGNIFICANCE OF THE METHODS OF PREPARING ANTITHROMBINS
AND ALLIED SUBSTANCES.

Doyon and his colleagues Morel, Policard, Dubrulle and Sarvonat [1910–1919] describe the following ways of preparing antithrombin. The liver of the dog is frozen and thawed three times over a period of 48 hours; during the intervals between freezing and thawing it is exposed to the air at room temperature. The liver is then perfused with 0·9 % sodium chloride and dilute alkali carbonate. The perfusate has no anticoagulant action until heated to 100° when it yields antithrombin. Alternative methods involve such processes as autoclaving liver at 120° or exposing minced liver to chloroform vapour or allowing intestine to undergo autolysis.

By the use of one or other of these methods Doyon has prepared antithrombins from the majority of the organs of the body, from the muscles of the cray-fish and from macerated earthworms. Recently however Doyon [1921, 1, 2] finds that antithrombin can only be obtained from the liver of graminivorous birds by freezing and thawing, a process which involves breaking up the colloidal complexes of the cells. In this connection it is noteworthy that Howell ascribes the fluidity of shed bird's blood on glass to excess of antithrombin secreted by the liver. Doyon also found that even freezing and thawing, in the case of the rabbit's liver, failed to produce an anticoagulant; yet Davis [1912] has shown that thrombin can be injected into the circulation of the rabbit, in large doses, without causing thrombosis.

The method adopted by Howell [1914] to demonstrate the presence of antithrombin in the blood was by heating oxalate plasma or peptone plasma to 60°. The filtrate delays the action of thrombin on fibrinogen and is consequently said to contain antithrombin. This antithrombin, Howell states, is destroyed by heating at 80°–85°. Collingwood and Macmahon [1914] found that serum, particularly when kept for two days, exhibits an inhibitory influence on thrombin; this antithrombin is destroyed by heating to 60 or 65°. The retarded coagulability of "peptone" plasma is commonly ascribed [Howell, 1918] to excess of antithrombin secreted by the liver. This antithrombin is completely destroyed only at 100° [Nolf, 1908].

Even allowing a wide margin for differences in stability due to the presence of various protective colloids in the respective solutions and also to individual experimental error, it is difficult to believe that liver antithrombin stable in solution at 100°, or even after autoclaving at 120°, is the same substance as the antithrombin prepared from plasma which is destroyed at 80–85°, or that which appears in disintegrating serum and is destroyed at 60°, or the anticoagulant of "peptone" plasma which is active up to 100°. Yet if the current view is accepted that antithrombin obtained from the liver is actually secreted into the circulation, these anomalies must be ignored.

The most recent method for the preparation of antithrombin is the extraction of rapidly dried and hashed lungs by benzene; the extract is dissolved in

0·9 % sodium chloride and purified by precipitation at its iso-electric point. The anticoagulant substance produced is stated by Mills, Raap and Jackson [1921] to be antithrombin.

The earlier methods of manufacturing antithrombin demand drastic dis-integration of the colloidal complexes from which the anticoagulant is pre-pared. The term "manufacturing" is used as such methods cannot imply unaltered extraction. Moreover, Mills, whose technique is less drastic, admits his method involves removal of a phospholipin from a protein-phospholipin complex, so that again there is only a product of cellular break-down. Like-wise in serum, in self-digested viscera and in the exudation of decaying liver exposed to chloroform, the anticoagulant may well be considered as a product of autolysis.

When more physiological methods were employed negative results were obtained. Menten [1920] found that perfusion of the liver with normal saline failed to yield any anticoagulant substance. Minot [1916] applied Doyon's chloroform method to solutions of Howell's antithrombin (prepared from oxalate plasma by heating to 60°), and also to plasma and serum both un-heated and heated to 60°. In each case no anticoagulant was found. Loeb [1904] was unable to obtain any antithrombin-like material by extraction of linings of blood vessels, a result which throws doubt on the view, recently revived, that antithrombin is secreted by the endothelium of the vascular wall. Rettger [1909] found no evidence of the production of antithrombin after the serial injection of minimal doses of thrombin into the dog and into the rabbit.

Attempts to cause the liver to secrete antithrombin by means of bile, bile salts, secretin and by electrical stimuli gave negative results [Denny and Minot, 1915]. It is noteworthy that Bulger [1918] found an increased coagula-tion time in cats, rabbits and guinea-pigs suffering from anaphylactic shock, associated with a decrease of antithrombin estimated by Howell's method. Thus the altered coagulability of anaphylactic shock, which closely resembles that of "peptone" shock, also affords no evidence of the secretion of anti-thrombin.

In short, no direct evidence has been obtained of the existence of anti-thrombin in living animals.

Two other organic anticoagulants merit notice. Schikele [1912] stated that the extract obtained from the uterus by pressure retarded coagulation. Bell [1914] repeated these experiments but failed to find any anticoagulant. Howell and Holt [1918] extracted dried powdered liver with ether. The solution thus obtained was precipitated by acetone, redissolved in ether and reprecipitated by absolute alcohol at 50°. This last process was repeated from 12 to 20 times and yielded a substance (or mixture) termed by them *heparin*, which is a powerful anticoagulant *in vivo* and *in vitro*; it differs from antithrombin in that it does not neutralise the action of thrombin. Here again the product obtained appears to be the result of more or less drastic disintegration of cells. Howell and Holt admit that there is no direct evidence of the existence of heparin in the blood. Nevertheless they believe it is an important factor in the maintenance of the fluidity of the blood, and suggest that it inhibits clotting by preventing the activation of prothrombin to thrombin and also by activating pro-antithrombin with the production of antithrombin.

THE PREPARATION OF ANTITHROMBIN FROM VEGETABLE CELLS.

In view of the evidence indicating that antithrombins are artificial products formed by the dissolution of complex substances of animal origin, it became of interest to enquire if the employment of Doyon's technique would produce an anticoagulant from vegetable sources.

The raw materials selected were baker's yeast and a very pure crystalline specimen of edestin. It may at once be stated that from both these vegetable products active anticoagulants were prepared.

In each case the materials were suspended in water and heated to 120° under pressure for 4 hours on three separate occasions; when not being heated they were freely exposed to the air. The product was centrifuged and to the clear supernatant liquid sodium carbonate was added until the solution was neutral to litmus. In the case of the yeast the neutralised solution was dried at room temperature at low pressure and was dissolved in water when required.

The action of the products obtained from yeast will be evident from the following experiment.

Blood was shed from a pithed cat respiring air and with the liver out of circulation, through a paraffined cannula into similarly treated vessels. Thrombin was dissolved in 0·9 % sodium chloride. The yeast extract was dissolved in distilled water. 5 cc. of blood were employed in each experiment.

	Blood on paraffined surface	Blood plus 2 cc. distilled water	Blood plus 0·75 cc. 0·5 % thrombin	Blood plus 2 cc. yeast extract	Blood plus 2 cc. yeast extract plus 0·5 cc. thrombin	Blood plus 2 cc. yeast extract plus 0·75 cc. thrombin
Commencement of clotting	26' 10"	4' 45"	0' 45"	Unclotted 2 hours later	24' 10"	18' 40"
Completion of clotting	29' 50"	8' 45"	2' 45"	Small clots next morning	43' 36"	41' 10"

Note: Similar results were obtained when the yeast extract was dissolved in 0·9 % sodium chloride.

In other experiments of this nature in which non-paraffined vessels have been used, clotting has been observed to occur in fractions. In this, the first clot formed contracted rapidly, while a portion of the fluid remained uncoagulated. Later a second small clot appeared and contracted normally and so on. In clotting induced however by addition of thrombin the clot invariably assumed the typical gel form and later contracted in the normal manner.

Similar results were obtained with rat's blood and with human blood. In the latter case the shed blood had been in contact with cut tissues. With dog's blood shed through a paraffined cannula into glass vessels, from animals in which the liver was not acting, the anticoagulant action, though evident, was not so pronounced. Employing the same materials as in the experiment just given, delays of 6 and 10½ minutes were obtained in the times of commencement and completion of coagulation.

The action of the yeast extract on the coagulation of recalcified oxalate plasma was investigated.

Fresh oxalate plasma was centrifuged for 30′ at 4000 r.p.m. On recalcification with calcium chloride the times of clotting were: commencement 8′, completion 9′ 15″.

To the same volume of plasma was added 0·192 g. of yeast extract; after solution and recalcification the times were 28′ and 31′ respectively.

Employing the usual animal technique previously described, 0·1 g. of the desiccated material, dissolved in 0·9 % sodium chloride, was injected into the heart. Blood withdrawn 10 minutes after injection yielded a minute clot in 8 minutes, but the great bulk of the blood was fluid after 22 hours. After 26 hours however large gelatinous clots were present. The normal coagulation times of the blood of the animal in question were 8′ 15″ and 9′ 50″ for commencement and completion. Other experiments gave similar results.

The crude material employed in the above experiments was extracted with boiling ether and the ether removed at room temperature. The resultant product was also found to possess well marked anticoagulant properties when tested on recalcified oxalate plasma and on unsalted blood; it did not however neutralise the coagulant action of thrombin added to shed blood. It thus resembles in its action the substance extracted by Howell and Holt from the liver and termed heparin.

Treatment of edestin as mentioned above gave a product which behaved like that obtained from yeast except that the coagulant action of added thrombin was not neutralised.

The preceding experiments show that extract of yeast exhibits the typical characteristics of the antithrombin prepared by Doyon. It not merely acts as an anticoagulant *in vivo* and *in vitro* but antagonises the action of thrombin; this and the preparation of a material resembling in its action heparin, when considered in relation to the methods of obtaining antithrombins from the colloids of cells and of serum, indicate that these substances are not secreted by organisms possessing blood streams but are *post-mortem* products of the break-down of colloidal complexes.

Doyon [1921, 1, 2] has recently shown that the nucleic acids prepared by Neumann's method from thymus and from ganglia are anticoagulants *in vivo* and *in vitro*; he assigns the former action to the secretion of a nucleo-protein by the liver. Experiments to be reported in a subsequent paper have demonstrated that in an animal deprived of hepatic activity the anticoagulant action is equally well marked.

Summary.

1. The massive intravascular injection of thrombin into cats deprived of hepatic activity does not produce thrombosis but accelerates the coagulation of shed blood. A phase of retarded coagulability of shed blood preceding hyper-coagulability was not observed when the above conditions were adhered to.

2. The view that the continuance of the fluidity of circulating blood after the injection of thrombin is due to the secretion by the liver of an excess of antithrombin is dissonant with the results of the foregoing experiments.

3. In the theories of Morawitz, of Fuld and Spiro and of Bordet, apart from the preliminary processes involving the formation of thrombin, the first act of coagulation is held to be the action of thrombin on fibrinogen. The evidence in this paper indicates that the inception of coagulation is due to physical change destroying the stability of the colloidal complexes of the plasma. Thrombin appears to be an accelerator of clotting rather than an initiator of that process.

4. The addition of thrombin to blood in the state of a reversible gel causes immediate coagulation.

5. Accepting the statement of Gratia that platelets are destroyed by calcium chloride, it is shown that detritus of platelets, in concentrations not greater than can be formed by lysis in circulating blood, plays only a secondary part in provoking coagulation.

6. Attention is drawn to certain inherent difficulties in interpreting experimental results obtained by the use of "platelet extracts."

7. An analysis of the methods of obtaining antithrombins from liver, sundry tissues and from serum, indicates that the anticoagulant substances so formed are *post-mortem* products.

8. The employment of Doyon's technique for preparing antithrombin from liver, yielded from yeast an extract which exhibited the properties of antithrombin. Further extraction with ether yielded a material similar in its action to heparin. The hydrolysis of edestin also gave an anticoagulant.

9. It is concluded that antithrombins are not phylogenetic products of the animal kingdom arising late in evolution as a protection against thrombosis, but are products resulting from the hydrolysis of protein.

The authors would express their thanks to Prof. M. Doyon and to Dr P. A. Levene for placing at their disposal preparations of nucleic acids, and to Prof. B. J. Collingwood for an exceedingly active sample of thrombin.

REFERENCES.

Achard and Aynaud (1908). *Compt. Rend. Soc. Biol.* **64**, 716; **65**, 459.
Bell (1914). *J. Path. Bact.* **18**, 462.
Bloch (1920). *Lancet*, ii, 301.
Bordet (1920). *Ann. Inst. Pasteur*, **34**, 561.
Bordet and Gengou (1903). *Ann. Inst. Pasteur*, **17**, 822.
Bordet and Delange (1913). *Ann. Inst. Pasteur*, **26**, 657; **27**, 341.
Bulger (1918). *J. Infect. Diseases*, **23**, 522.
Bürker (1904). *Pflüger's Arch.* **102**, 36.
Chio (1917). *Arch. Farm. Sperim.* **23**, 206.
Collingwood and Macmahon (1914). *J. Physiol.* **47**, 53.
Cramer and Pringle (1913). *Quart. J. Exp. Physiol.* **6**, 1.
Dastre and Flouresco (1897). *Arch. de Physiol.* 5th series, **9**, 216.
Davis (1912). *Amer. J. Physiol.* **29**, 161.
Denny and Minot (1915). *Amer. J. Physiol.* **38**, 246.
Doyon (1910, 1). *Compt. Rend. Soc. Biol.* **69**, 340.
—— (1910, 2). *Compt. Rend.* **151**, 1074.
—— (1912, 1). *Compt. Rend. Soc. Biol.* **72**, 26, 485, 727 and 766.
—— (1912, 2). *J. Physiol. et Pathol.* **14**, 229.
—— (1919). *Compt. Rend. Soc. Biol.* **82**, 570.
—— (1921, 1). *Arch. Internat. de Physiol.* **16**, 343.
—— (1921, 2). *Compt. Rend.* **172**, 1212; **173**, 1120.
Doyon and Dubrulle (1912). *Compt. Rend. Soc. Biol.* **73**, 546.
Doyon, Morel and Policard (1911). *Compt. Rend. Soc. Biol.* **70**, 175, 232, 433 and 797.
Doyon and Sarvonat (1913). *Compt. Rend. Soc. Biol.* **74**, 312.
Duke (1910). *J. Amer. Med. Assoc.* **55**, 1185.
Fuld and Spiro (1904). *Beiträge*, **5**, 174.
Galecki (1912). *Zeitsch. anorgan. Chem.* **74**, 179.
Gratia (1918). *J. Physiol Pathol. Gén.* **17**, 772.
—— (1920). *Compt. Rend. Soc. Biol.* **83**, 313.
—— (1921). *Ann. Inst. Pasteur*, **35**, 529.
Halliburton and Brodie (1894). *J. Physiol.* **17**, 172.
Höber and Gordon (1904). *Beiträge*, **5**, 432.
Howell (1912). *Amer. J. Physiol.* **29**, 208 and 209.
—— (1914). *Arch. Int. Med.* **13**, 76.
—— (1916). *Amer. J. Physiol.* **40**, 526.
—— (1918). *A Text-Book of Physiology*, 466, 469.
Howell and Holt (1918). *Amer. J. Physiol.* **47**, 328.
Loeb (1904). *Virchow's Arch.* **176**, 10.
Löwit (1892). *Studien z. Physiol. u. Pathol. d. Blutes u. d. Lymphe.* Jena.
Marriott (1916). *Arch. Int. Med.* **17**, 840.
Mellanby (1909). *J. Physiol.* **38**, 502.
Mendenham (1915). *Amer. J. Physiol.* **38**, 51.
Menten (1920). *J. Biol. Chem.* **43**, 383.
Mills, Raap and Jackson (1921). *J. Lab. Clin. Med.* **6**, 379
Minot (1916). *Amer. J. Physiol.* **39**, 135.
Morawitz (1904, 1). *Beiträge*, **5**, 133.
—— (1904, 2). *Arch. klin. Med.* **79**, 215.
Nolf (1908). *Arch. internat. Physiol.* **7**, 42.
Paine (1911). *Proc. Camb. Phil. Soc.* **16**, 430.
Pickering and Hewitt (1921). *Biochem. J.* **15**, 710.
—— —— (1922). *Proc. Roy. Soc.* B, **93**, 367.
Rottger (1909). *Amer. J. Physiol.* **24**, 431.
Schikele (1912). *Biochem. Zeitsch.* **38**, 169.
Spring (1900). *Rec. trav. chim. Pays-bas* (2), **4**, 204.
Vines (1921). *J. Physiol.* **55**, 294.
Weil (1920). *Rev Méd.* **37**, 81.

LIX. INVESTIGATIONS ON THE NITROGENOUS METABOLISM OF THE HIGHER PLANTS.

PART III. THE EFFECT OF LOW-TEMPERATURE DRYING ON THE DISTRIBUTION OF NITROGEN IN THE LEAVES OF THE RUNNER BEAN.

By ALBERT CHARLES CHIBNALL.

*From the Biochemical Department, Imperial College
of Science and Technology.*

(*Received June 27th, 1922.*)

IN any research into the metabolic processes taking place in plants, where fresh material is used, the period during which the research can be carried out must necessarily be restricted to the appropriate season during which the plants can be grown. To overcome the difficulty, thus entailed, some workers have used leaves that had been previously dried, either at low temperatures or from 50–70°.

As far as is known, however—and this statement applies especially to those who have investigated the N and protein metabolism in plants—no one has first investigated the possible changes taking place during drying. Both protein synthesis and degradation are continually going on in the leaf cell; consequently, proteolytic enzymes must be present, and it is strange that so many workers have assumed that these would remain inactive during the process of drying, often at temperatures of 30–50°, the most favourable for enzyme action.

The present research shows that in as far as the leaves of the runner bean are concerned, considerable proteolysis takes place, with consequent increase in the simpler water-soluble nitrogenous products, chiefly ammonia (in the form of ammonium salts or as the amide N of asparagine) and monoamino acids, the former being increased from 1 % to 5 % or 6 % of the total leaf N, the amount of increase depending on the conditions of drying.

On reviewing, therefore, the work done in the past to throw light on the N metabolism in the leaf, it would appear that the predominance given to asparagine as a final product of protein metabolism is based, to a certain extent, on an erroneous conception as to the amount of it actually present in the living leaf.

Boussingault [1860] was one of the first systematically to study the asparagine content of leaves, and he emphasised that it could only be isolated when the plants were young. His results were confirmed by Meunier [1880] and Müller [1887], who extended their researches over a wide range of plants. Emmerling [1887], who studied the N content of the fresh leaves of *Vicia Faba maj.* throughout the life history of the plant, never found that either ammonia N or asparagine N exceeded 1·25 % of the total leaf N, a result the author of the present paper confirmed in an unpublished series of experiments in 1921. The latter also [Chibnall, 1922] found less than 1 % for each of these substances during the life of *Phaseolus vulgaris* v. *multifloris.* Kosutany [1897] found a similar result for *Vitis vinifera.* Yet side by side with these results others were being published which indicated that the asparagine content of leaves could be quite high. On investigation it would appear that in all these cases the leaves had been dried before analysis. Thus Suzuki [1897] gives 5 % of asparagine in leaves of *Phaseolus vulgaris* dried in an air oven at an unstated temperature. Wassilieff [1901] found 21·46 % of the total N as asparagine in young leaves of *Lupinus alba* dried at 70°. Jodidi, Kellogg and True [1920] found in spinach leaves (age unstated) dried at atmospheric temperature for 7 days, 4·8 % of the total N as ammonia and about 12 % as "amide N of asparagine." Yet the present author found in the fresh leaves from plants three weeks old only 0·6 % as ammonia and 2·05 % as "amide N of asparagine."

From the results given later, combined with those stated above, it seems fairly certain that these high values must be attributed, at any rate in part, to proteolysis during the drying. The same remark must apply to the results obtained by Prianischnikoff [1899], who investigated the seedlings of several legumes, though he analysed the plants as a whole, and not the leaves alone.

Whilst proteolysis thus affects the water-soluble products of the leaf, it is found that the general character of the leaf proteins, as shown by Hausmann's method of analysis, undergoes but little change, if any, during the drying.

EXPERIMENTAL DETAILS.

The plant used was *Phaseolus vulgaris* v. *multifloris* (Scarlet Champion). The samples of leaves investigated were from series 8, 9, and 10, of a former research [Chibnall, 1922]. In these the total weight of leaves picked was—in series 8, 2064 g. from 12 plants; in series 9, 2424 g. from 12 plants; in series 10, 1860 g. from 8 plants. For the distribution of N in the fresh leaves only about 500 g. were required, the remainder, in each case, being air-dried in a closed oven through which a slow stream of air was drawn. Series 8 was dried at 50° in approximately 100 hours, series 9 at 40° in approximately 120 hours, and series 10 at 40° in approximately 50 hours. The dried leaves were afterwards ground to a moderately fine powder and stored in a stoppered bottle. About six months elapsed before the present research was commenced.

Table I gives further details of the air-dried samples. As the drying took place in the dark, the leaves of series 8, which had been picked one hour before dawn, received no opportunity for photosynthetic action before death.

Table I. *Showing details of the air-dried samples used.*

Series number	Approximate age of plant above ground in weeks	Time of day when picked (G. M. T.)	Fresh weight of sample in g.	Air-dried weight of sample in g.	% of moisture in air-dried sample
8	11	2 a.m.	1365	213·0	7·65
9	11	3 p.m.	1925	335·1	10·92
10	14	9 a.m.	921	161·8	7·17

General methods of manipulation (separation of protein, soluble products and cellular matter). The method of grinding in an end-runner mill, as used with the fresh leaves [Chibnall, 1922], was found to be applicable in the case of the dried leaves. The ground up particles of cellular matter imbibe the water and swell up, so that only the very fine ones pass through the pores of the lawn when gentle pressure is applied. By using lawn of very close texture, as in the case of series 8, this can be reduced to a minimum, the percentage of N per dry-weight of the colloidal protein being but very little lower than that from the fresh leaves. Even when moderately coarse lawn is used, so as to allow the separation to be conducted in the minimum of time, the amount of carbohydrate mixed with the protein is not sufficient to interfere with the chemical analysis of the latter (*vide* series 9 in Table IV).

In some preliminary experiments the ground leaves were first shaken up with a small quantity of water and allowed to stand for some hours, with the object of facilitating the extraction of the less soluble products, but subsequent analysis of the extracts, as will be shown later, indicates the presence of proteolytic enzymes, which are activated by the addition of water. It is therefore necessary to carry out the extraction in the minimum of time.

In the preparations from 50 g. of the dried leaves of series 8 and 9, upon which the subsequent analyses given herein are based, 6 extractions by grinding, followed by six extractions with boiling water were given. Extraction with boiling water without the preliminary grinding in the cold was also tried. The protein- and proteose-free water-soluble N agreed with that given by the other method, but only a part of the proteoses passed into solution and, of course, no colloidal protein was obtained. On account of lack of material, the analysis given for series 10 in Table IV was of an extract prepared by this method. Table II illustrates the difference between the extracts prepared by the two methods.

Table II. *Showing the water-soluble N extracted from the dried leaves by cold and hot extraction.*

(In percentages of total leaf N.)

Series number	Number of extractions		Total water-sol. N	Precipitated by ammonium sulphate		Protein- and proteose-free water-soluble N
	By cold grinding	With hot water		Not resoluble in water (protein?)	Proteose N	
{ 8	6	6	30·78	0·51	3·62	26·65
{ 8	0	10	27·65	0·48	1·45	25·71
{ 9	6	6	33·68	0·39	3·74	29·55
{ 9	0	10	30·94	0·28	0·69	29·97

Analytical methods. The details given in the former paper were followed, with the following exceptions:

Proteose N. This was estimated by ammonium sulphate instead of zinc sulphate in acid solution [Baumann and Bömer, 1898]. No difficulty had been experienced with the latter method when using the extracts from the fresh leaves, but in those from the dried leaves the zinc sulphate separates a quantity of non-nitrogenous material of a slimy nature, which renders the subsequent filtering slow, and the washing of the separated proteoses almost impossible. No difficulty was experienced with the ammonium sulphate, the proteoses filtering readily at the pump. They were then washed with a saturated solution of ammonium sulphate until the washings were colourless and redissolved in water. The operation was repeated, and the N of the aqueous solution of the proteoses, after expulsion of the free ammonia by distillation with magnesia *in vacuo*, was determined by Kjeldahl's method. The N of the small insoluble residue on the filter paper, probably coagulated protein, was also determined, after thorough washing, by Kjeldahl's method.

Residue after extraction of the colloidal protein and water-soluble products. As this still contained about half the protein N of the leaf, three extractions with 700 cc. of 1 % HCl were given, the extracts being united and analysed as before.

RESULTS.

Effect of drying on the proteins. Table III gives a detailed account of the proteins extracted from the dried leaves of series 9, with the corresponding data for the fresh leaves. The amount of protein passing into colloidal solution is proportionately much smaller than with the fresh leaves, but could no doubt be increased by further grindings, since the 6th extract appeared as heavily charged with colloidal matter as the earlier ones. The bulk of the protein not passing into colloidal solution, however, is extracted by the HCl. Assuming, as in the case of the fresh leaves, that the small amount of N in the final residue of cellular matter is due to unextracted protein, it will be seen that the total protein N has fallen from 74·83 % to 61·15 % of the total leaf N.

Table III. *Comparison of the proteins and water-soluble N in the fresh and dried leaves of series 9.*

(In percentages of total leaf N.)

	Fresh leaves	Dried leaves
Colloidal protein extracted	59·70	29·44
Protein extracted by 1 % HCl	10·24	25·00
Remaining in residue	4·13	5·90
Coagulated by boiling	1·76	0·42
Coagulated by saturation with ammonium sulphate (?)	0·00	0·39
Total protein N	74·83	61·15
Proteose N	4·35	3·74
Protein- and proteose-free water-soluble N	17·22	29·55
Extracted by alcohol and ether	3·60	2·96

Table IV gives the distribution of N in the colloidal proteins from the fresh and dried leaves of series 8 and 9. It will be observed that no apparent change has taken place in the protein molecule.

Table IV. *Comparison of the distribution of N in the colloidal proteins from the fresh and dried leaves.*

(In percentages of total protein N.)

Series number	% of N per dry weight	Amide N	Humin N	Basic N	Monoamino N	Other N
8 fresh	11·75	6·14	3·55	19·33	59·25	11·73
8 dry	10·11	6·26	4·63	21·20	57·93	9·98
9 fresh	11·71	6·34	3·64	22·00	58·88	9·14
9 dry	8·64	6·73	4·71	21·67	56·30	10·51

Table V gives the distribution of N in the proteins extracted by the HCl from the fresh and dried leaves of series 9. In comparing them it must be remembered that they represent very different proportions of the total leaf protein. The HCl extract from the fresh residues represents only about ⅙ of the total protein, that from the dried leaves nearly ½. Since the leaf cell undoubtedly contains more than one protein, it follows that if the dried leaves, like the fresh, contain a small amount of protein rich in amide and basic N, these characteristics, on analysis, may be masked by the predominance of unextracted protein, of the type represented by the colloidal protein. If allowance be made for soluble carbohydrates due to the mild hydrolytic action of the HCl on the pentosans etc. of the cell wall material, which will increase the humin N at the expense of the monoamino N, the distribution of N in the HCl extract from the dried leaves does not differ from that of the corresponding colloidal protein, indicating that the major part of it is of this type. Interpreting the various distributions of N in terms of the amounts of the total leaf protein that they represent, it would appear that the leaf proteins do not undergo any appreciable change in character during the process of drying.

Table V. *Comparison of the distribution of N in the proteins extracted by 1 % HCl.*

(In percentages of total N in the extract.)

Series number	Amide N	Humin N	Basic N	Monoamino N	Other N
9 fresh	8·19	5·82	24·66	48·78	12·64
9 dry	6·96	6·16	21·57	54·01	11·30

Effect of drying on the proteoses. Table VI shows that in the cold extracts prepared from the dried leaves the proteoses have diminished. Since Table II shows that they are but slowly soluble in hot water, it may well be that after drying they are not so readily soluble in cold water, in which case the apparent loss may be due to incomplete extraction.

Table VI. *Comparison of the distribution of N in the aqueous extracts from the fresh and dried leaves.*

(In percentages of total leaf N.)

Series number	Total water-soluble N	Protein- and proteose-free water-soluble N	Ammonia N	Amide N of Asparagine	Nitric N	Humin N	Basic N	Mono-amino N	Other N
8 fresh	21·07	*	0·36	0·0	3·56	1·00	*	7·06	2·34
8 dry	30·78	26·65	6·02	0·0	3·57	1·15	1·81	10·55	3·55
Diff.	9·71	—	5·66	—	—	—	—	3·49	1·21
9 fresh	21·57	17·22	0·44	0·53	2·73	0·99	2·00	7·76	2·77
9 dry	33·68	29·55	5·32	1·51	1·62	0·82	2·84	13·22	4·22
Diff.	12·11	12·33	4·88	0·98	1·11	—	0·84	5·46	1·45
10 fresh	22·75	18·16	0·37	0·47	3·53	1·06	2·30	6·81	3·62
10 dry	(1)	27·07	2·68	2·26	3·01	0·89	3·51	10·60	4·12
Diff.	—	8·91	2·31	1·79	0·52	—	1·21	3·79	0·50

* Proteoses not determined.
(1) Determined from an extract made with hot water only, which gives a low value for the proteoses.

Effect of drying on the protein- and proteose-free water-soluble products. Table VI shows that in all three cases this has increased, the amount being roughly proportional to the time of drying. The products of the protein autolysis fall chiefly into ammonia (as ammonium salts or "amide N of asparagine") and monoamino acids. This is illustrated better by Table VII. Leaving for a moment the relation between the ammonium salts and asparagine, which is discussed later under the heading of enzymes, it will be seen that in the two samples picked after some hours' daylight, series 9 and 10, the amount of ammonia N is nearly equal to the monoamino N, whereas in series 8, picked after 6 hours' dark, the ammonia is in considerable excess. In the latter case this may possibly be due to the higher temperature of drying (50° instead of 40°) or, more probably, to the fact that the leaves, before drying, were deprived of most of their reserve carbohydrate by translocation away from the leaf at night.

Table VII. *Showing the ammonia and amino acids formed by proteolysis during drying.*

(In percentages of total autolysed protein N.)

Series number	Ammonia N + amide N of asparagine	Monoamino N	Total: Ammonia N + Amide N of asparagine / Monoamino N
8	58·3	35·9	94·2
9	47·5	44·3	91·8
10	46·0	42·5	88·5

In series 9 and 10 again, there is a loss of nitric N, probably due to the necessary carbohydrate for the synthesis of organic nitrogenous products being present, since in series 8 there is no loss. The higher temperature of drying may have accounted for this, however, since Couperot [1909] has shown that leaves dried at room temperature lost 25–50 % of their nitric N as hydrocyanic acid whereas those dried rapidly at 60° suffered no such loss.

Table VIII. *Showing the enzyme action produced by allowing the dried leaves to stand with cold water.*

(In percentages of total leaf N.)

Series No.	Hours standing with water	Total water-soluble N	Protein and proteose N removed by ammonium sulphate	Protein- and proteose-free water-soluble N	Ammonia N	Amide N of asparagine	Nitric N	Humin N	Basic N	Monoamino N	Other N
8	0	30·78	4·13	26·65	6·02	0·0	3·57	1·15	1·81	10·55	3·55
8	16	35·90	3·16	31·74	2·30	2·96	3·59	0·89	3·76	13·56	5·68
9	0	33·68	4·13	29·55	5·32	1·51	1·62	0·82	2·84	13·22	4·22
9	16	39·35	3·39	35·94	1·69	4·58	1·67	0·90	5·63	15·80	5·67
9	116	39·82	2·86	36·96	1·20	4·10	1·56	1·04	5·67	17·48	5·91
10	0	*	*	27·07	2·68	2·26	3·01	0·89	3·51	10·60	4·12
10	16	31·60	3·21	28·38	1·19	3·57	3·04	1·02	3·33	11·14	5·09

* Determined in a hot water extract only, which gives a low value for the proteoses.

Enzymes present in the dried leaves. As was stated earlier, if the ground leaves are allowed to remain suspended in water for some hours (under a layer of toluene), proteolytic enzymes become active, with consequent increase of the simpler nitrogenous products in solution. Table VIII shows the analyses that have been made to illustrate this. A study of the figures shows that the action is complex, but the following points stand out fairly clearly

(1) There is at first a rapid increase in the water-soluble N tending to reach a limit with lapse of time.

(2) Monoamino N shows a continuous increase.

(3) Basic N shows an increase that is not maintained with lapse of time.

(4) Ammonia N shows at first a rapid decrease, tending to reach a limit. At the same time there is a nearly corresponding increase in the "amide N of asparagine."

(5) The sum ammonia N + "amide N of asparagine" shows a slow decrease, indicating re-synthesis of more complex bodies.

From (2) and (3) it is clear that an enzyme of the nature of a pepsin is present (the extract is slightly acid). But the chief point of interest is the rapid conversion of the ammonia N into the amide N of asparagine. In the leaf the origin and relationship between these two substances have not yet been definitely established. Under normal conditions the evidence is that the concentration of both of them is low. Whether the ammonia is due only to direct translocation from the root, or to protein degradation in the leaf itself or to both of these is not yet clear. Undoubtedly an enzyme capable of breaking down the protein into ammonia and amino acids exists in the dried leaves, but this may not be so in the normal fresh leaf. In a previous paper [Chibnall, 1922] there was described an experiment in which the living leaf was starved under conditions that precluded the translocation of the products of metabolism, and in spite of considerable protein degradation, no appreciable increase in either ammonia or "amide N of asparagine" was noted[1]. Suzuki [1897], by starving moist leaves for 48 hours, claims to have increased the "asparagine N" from 5 % to 13 % of the total N, but his results must be queried since he dried the leaves before analysis.

Asparagine itself cannot be a primary product of protein degradation, but must be a product of re-synthesis from the ammonia and amino acids. In very young plants direct translocation of it from the cotyledon reserve is possible, since Prianischnikoff [1896] showed in germinating seeds of *Vicia sativa* a steady increase of asparagine at the expense of the protein. Butkewitsch [1909], who experimented with seedlings of lupins and peas, considers, as does Suzuki [1894], that the formation of asparagine is to remove an injurious excess of free ammonia. This view, if modified to include an excess of monoamino-dicarboxylic acids also, is very likely, and the reaction "ammonia + aspartic acid $\rightleftharpoons$ asparagine" is probably due to an enzyme governed by the H ion concentration of the medium in which it is working. As far as is known, an enzyme of this nature, an asparaginase, has not yet been indicated, and it is therefore of great interest to observe that one appears to be present in the dried leaves used in this research. Furthermore, under the conditions of the present experiment, it appears to have marked synthetic activity.

It is hoped at an appropriate season to repeat these enzyme experiments with fresh leaves, and if an asparaginase should be indicated, to study the effect of change in the H ion concentration upon its action.

In conclusion the author would like to thank Prof. S. B. Schryver for his advice and encouragement throughout the research, which was made possible by a grant from the Department of Scientific and Industrial Research.

SUMMARY.

1. The effect of low-temperature drying on the nitrogenous bodies in the leaves of the runner bean has been investigated.

2. Protein autolysis takes place, with increase in the simpler water-soluble N products.

3. These products are chiefly ammonium salts, asparagine and amino acids.

4. The leaf proteins, whilst they are diminished in amount, are not appreciably changed in character.

5. Proteolytic enzymes are present in the dried leaves, and are activated by the addition of water.

[1] The season 1921 was abnormal, the bean plants bearing but few pods. At the time this experiment was carried out, 23 August 1921, no pods were in process of formation. Further starvation experiments are being carried out this season, and the results so far to hand from leaves picked in August 1922, when pod formation was active, indicate that considerable quantities of asparagine are formed by degradation of the leaf protein. Ammonia N, as before, is unchanged.

6. The presence of an asparaginase, activated by the addition of water, is indicated. Under the conditions of the present research it possesses marked synthetic activity.

7. The position of ammonia and asparagine in the N metabolism of the leaf is discussed.

REFERENCES.

Baumann and Bömer (1898). *Zeitsch. Untersuch. Nahr. u. Genussm.* 1, 106.
Boussingault (1860). Agronomie, chimie agricole et physiologie, 2nd ed.
Butkewitsch (1909). *Biochem. Zeitsch.* 16, 411.
Chibnall (1922). *Biochem. J.* 16, 344.
Couperot (1909). *J. Pharm. Chim.* 29, 100.
Emmerling (1887). *Land. Vers. Stat.* 34, 1.
Jodidi, Kellogg and True (1920). *J. Amer. Chem. Soc.* 32, 1061.
Kosutany (1897). *Land. Vers. Stat.* 48, 13.
Meunier (1880). *Ann. Agron.* 6, 275.
Müller (1887). *Land. Vers. Stat.* 33, 326.
Prianischnikoff (1896). *Land. Vers. Stat.* 46, 459.
—— (1899). *Land. Vers. Stat.* 52, 137, 347.
Suzuki (1894). *Bull. Coll. Agr. Tokyo*, 2, 409.
—— (1897). *Bull. Coll. Agr. Tokyo*, 3, 245.
Wassilieff (1901). *Land. Vers. Stat.* 55, 45.

LX. INVESTIGATIONS ON THE NITROGENOUS METABOLISM OF THE HIGHER PLANTS.

PART IV. DISTRIBUTION OF NITROGEN IN THE DEAD LEAVES OF THE RUNNER BEAN.

By ALBERT CHARLES CHIBNALL.

From the Biochemical Department, Imperial College of Science and Technology.

(*Received July 12th, 1922.*)

IN a former paper [Chibnall, 1922, 1], the seasonal variations in the N content of the leaves of the runner bean, *Phaseolus vulgaris* v. *multifloris* (Scarlet Champion), were demonstrated and discussed. The period examined covered the first twenty weeks of the plant's life, from the seedling stage, until the leaves were showing signs of chlorophyll degeneration. Eleven samples were analysed, numbered consecutively "series 1–11."

The plants died during the 24th week, following two nights' hard frost, and the present communication deals with the distribution of N found in the dead leaves, which will be denoted henceforth as series 12. As picked from the dead plants they were shrivelled up and dry, whilst their green colour was masked by a brownish-red stain, probably due to iron salts, since it was easily washed away with cold water.

The results given later for series 12 have in all cases been compared with the corresponding ones for series 9, 10 and 11, the four together covering the latter half of the plant's life, from the 11th to the 24th weeks. During this period the plants were fully grown and pod formation was complete, so that the general metabolism would be, presumably, on the down grade. The tables given later show how remarkably stable is the general equilibrium between the nitrogenous substances in the leaf during this period, the changes, even at death, being of quite a small order. There is no evidence of any great withdrawal of N from the leaves before death, nor is there any accumulation of asparagine. Miyachi [1896] assumed that asparagine accumulates in the aged leaves of *Paeonia albiflora* since old leaves from dying plants, if placed under water for 14 days, become rich in this substance. In the present author's opinion this experiment merely demonstrates the presence of active enzymes in the old leaves. (Compare Chibnall [1922, 2], where observations on dried leaves suspended in water are recorded.)

The methods of manipulation and analysis were those used for dried leaves, as described by Chibnall [1922, 2].

The results can be summarised under the following headings: (1) There is no great withdrawal of N from the leaves to the stems or roots when the plant becomes aged [*vide* Czapek, 1920]. (2) The protein- and proteose-free water-soluble N of series 12 shows a slight increase, due chiefly to ammonia (as ammonium salts or "amide N of asparagine") and monoamino N. From the author's experiments on the changes in the leaf N due to low-temperature drying [Chibnall, 1922, 2], it would appear that this increase has taken place during the period of dehydration of the leaf on the plant. Towards the latter end of the plant's life the nitric N has fallen, as one would expect from the ageing of the root system, otherwise the equilibrium between the water-soluble nitrogenous products has remained very steady, indicating a period of low metabolic activity in the leaf. (3) The proteins in the leaf have undergone slight amidisation, a change that must be ascribed to the ageing of the leaf, since it was not observed in the proteins of leaves dried at low temperatures. The ratio of protein to non-protein N in the leaf has suffered no appreciable change with age or death.

Table I. *Showing the percentage dry weight and total N in the leaves of series 9–12.*

Series number	Age above ground in weeks	% dry weight	% of N per fresh weight	% of N per dry weight
9	11	15·51	0·675	4·35
10	14	16·31	0·724	4·44
11	20	16·13	0·622	3·86
12	24	91·6	—	3·60

Table II. *Showing the amounts of protein and non-protein N in the dead leaves (series 12).*

(In percentages of total leaf N.)

Colloidal protein extracted	29·22
Protein extracted by 1 % HCl	29·49
N in residue (protein?)	6·62
*Protein coagulated by ammonium sulphate ...	0·77
Total protein N	66·10
Proteose	6·28
Proteose and protein-free water-soluble N ...	22·10
Soluble in alcohol and ether	2·15
Total N accounted for	96·63

* In estimating proteoses by saturation with ammonium sulphate, a small amount of N, probably due to traces of coagulated protein, is not resoluble in water.

Table III. *Showing the water-soluble N in series 9–12.*

(In percentages of total leaf N.)

Series number	Total water-soluble N	Proteose N	Coagulated protein N*	Protein- and proteose-free N
9	21·57	4·35	—	17·22
10	22·75	4·57	—	18·16
11	22·41	4·38	—	18·03
12	29·15	6·28	0·77	22·10

* As in footnote for Table II.

Table IV. *Showing the distribution of the water-soluble N in series 9–12.*

(In percentages of total leaf N.)

Series number	Total protein- and proteose-free N	Ammonia N	Amide N of asparagine	Nitric N	Humin N	Basic N	Mono-amino N	Other N
9	17·22	0·44	0·53	2·73	0·99	2·00	7·76	2·77
10	18·16	0·37	0·47	3·53	1·06	2·40	6·81	3·62
11	18·03	0·61	0·92	1·28	0·96	2·59	6·61	5·06
12	22·10	0·97	1·94	1·61	1·32	2·22	7·74	6·30

Table V. *Percentage distribution of N in the colloidal proteins of series 9–12.*

Series number	Amide N	Humin N	Basic N	Monoamino N	Other N
9	6·34	3·64	22·00	58·88	9·14
10	6·50	3·82	21·78	59·22	8·68
11	6·40	3·82	21·32	58·50	9·96
12	7·21	6·12	19·27	56·22	11·32
12*	8·81	6·67	19·27	51·90	13·35

* HCl extract of series 12.

REFERENCES.

Chibnall (1922, 1). *Biochem. J.* 16, 344.
—— (1922, 2). *Biochem. J.* 16, 595.
Czapek (1920). Biochemie der Pflanzen (Jena), 2, pp. 293–295.
Miyachi (1896). *Bull. Coll. Agr. Tokyo,* 2, 458.

LXI. NOTE ON THE NON-PROTEIN NITROGEN IN GOAT'S MILK.

By WILLIAM TAYLOR (*Carnegie Research Scholar*),

The Rowett Institute, Aberdeen.

(*Received June 28th, 1922.*)

As a result of an investigation into the factors controlling the percentage composition of milk, during which a goat was fed on various diets each of which was abnormally high in some one of the three energy-yielding constituents of the food, it was found that irrespective of the nature of the diet the percentages of protein, fat, lactose and ash all tended to vary with the daily volume of milk secreted, the percentages of protein, fat and ash tending to vary inversely and the percentage of lactose directly with the daily volume [Taylor and Husband, 1922]. So far as these constituents are concerned, therefore, the percentage composition of milk depends upon its rate of secretion rather than upon the percentage composition of the diet.

In the case of non-protein nitrogen no relationship could be traced between the percentage and the daily volume of the milk. The results of the following experiment show that the percentage of this constituent depends upon the nature of the diet.

EXPERIMENTAL.

A goat was fed on various diets of constant composition for periods varying from 8 to 16 days, and the total protein, caseinogen and albumin plus globulin contained in the milk were estimated daily. The figure for non-protein nitrogen was obtained from the difference between the percentage of total protein and the sum of the percentages of caseinogen, albumin and globulin. The figure for non-protein nitrogen is expressed in terms of milk protein, as the non-protein nitrogen was included in the estimation of total nitrogen from which the total protein was calculated.

The urine was collected and a daily estimation of the total excretion of urinary nitrogen was also carried out.

The following table shows, for each dietary period, the average daily total quantity of nitrogen in the urine and the average percentage of non-protein nitrogenous substances in the milk. The periods are arranged in the table according to the level of nitrogen excreted in the urine.

The table shows a definite correlation between the daily output of nitrogen in the urine and the percentage of non-protein nitrogen in the milk.

Average daily excretion of urinary nitrogen g.	Average percentage of non-protein nitrogenous substances in the milk per cent.	Nature of diet	Duration of experimental period Days
26·8	0·39	High protein	16
22·5	0·37	,, ,,	12
9·6	0·22	Potatoes	8
9·4	0·22	Grass	14
4·3	0·20	High fat	14
4·1	0·20	,, ,,	14
2·3	0·16	High carbohydrate	13

DISCUSSION OF RESULTS.

It has been shown that the non-protein nitrogenous substances in milk consist of amino-acids, creatine, creatinine, uric acid, and urea, the last named being present in greatest amount [Denis, Fritz Talbot and Minot, 1919]. The concentration of these in the blood seems to determine both the amount excreted in the urine and the percentage present in the milk. The mammary gland acts to some extent as an excretory organ, waste non-protein nitrogenous substances filtering through from the blood to the milk. The percentage in which these are found in milk seems to be determined by the degree of concentration in the blood of the end products of protein metabolism.

CONCLUSIONS.

1. In a lactating animal there is a correlation between the daily output of nitrogen in the urine and the percentage of non-protein nitrogen in the milk, both apparently being determined by the amount of protein in the food.

2. The mammary gland acts as an excretory organ passing through from the blood to the milk end products of protein metabolism.

REFERENCES.

Denis, Fritz Talbot and Minot (1919). *J. Biol. Chem.* **39**, 47.

Taylor and Husband (1922). *J. Agr. Sci.* **12**. 111.

LXII. SMELL.

By EDWIN ROY WATSON.

From the Dacca College, Dacca, Eastern Bengal.

(Received July 3rd, 1922.)

THE aim of this investigation was to find a method of accurately and scientifically describing smells. Sounds and colours can be accurately described by the frequencies of the vibrations of which they are composed, but no such method has yet been discovered for describing smells, and the best that can be done at present is to say that there is some kind of resemblance between the smells of certain substances though at the same time it is generally felt that such a description is very unsatisfactory and incorrect. Not only is it impossible to describe any smell satisfactorily but there is at present no criterion as to whether a smell is simple or complex. It is true we can sometimes detect more than one substance in a mixture by the smell but we are generally helped in such an analysis by the different volatility of the different constituents of the mixture, and in fact such analysis is very uncertain and liable to error.

The problem is difficult, because, so far as is known at present, smell is not due to any kind of vibration such as the air vibrations causing sound or the ether vibrations causing light. The fact that most and probably all odorous substances are volatile and that no odorous substance can be detected by its smell when enclosed in an air-tight receptacle made of any ordinary material has led to the general belief that contact between the odorous substance and the olfactory organ is necessary to cause the sensation of smell. At any rate temporarily the author has assumed that smell is due to chemical or physical reaction between the odorous substance and the olfactory organ. It is well known that certain odorous substances react chemically with proteins, *e.g.* the halogens, nitrous fumes, hydrochloric acid, formaldehyde. And as the olfactory organ is undoubtedly composed chiefly of proteins it seemed worth while first to see whether any chemical reaction could be detected between proteins and odorous substances in general. The method used by Bugarsky and Liebermann [1898] to prove chemical reaction between non-coagulated white of egg and hydrochloric acid seemed specially suitable. These investigators showed that the addition of egg-white to aqueous hydrochloric acid caused an elevation of the freezing-point, pointing to a decrease of the number of hydrochloric acid molecules in the solution. A similar experiment was therefore tried by the present investigator with aqueous ethyl acetate solution and white of egg but no elevation of the freezing point was observed. An experiment was also tried by shaking solid egg-albumin (Merck's) with aqueous ethyl acetate to see whether the amount of ethyl acetate was reduced. But the egg-albumin, although not readily soluble in water, was sufficiently soluble to make filtration very difficult. The experiment was therefore modified by using alcoholic solutions of various odorous substances, shaking them up with egg-albumin and ascertaining whether the amount of odorous substance in solution had decreased. At any rate it was expected that acids and bases would react with the egg-albumin. A decrease was observed in the case of hydrochloric acid but none in the case of ethyl acetate, acetic acid, pyridine or citronellal. It is obvious, therefore, that under these circumstances there is no chemical reaction between several typical odorous substances and the typical protein, egg-albumin.

In looking round for a physical reaction between odorous substances and the olfactory organ which might be the cause of smell the author was struck by the parallel between intensity of smell and depression of the surface tension of the aqueous solutions of various organic substances investigated by Traube. Physiologists give much attention to the problem of how chemical substances can enter the living cell, for it is obvious that no reaction can take place with the cell-substance unless entry can be obtained into the cell. Surface tension and adsorption are likely to play an important part in bringing substances into contact with the living cell, and it was therefore very interesting to notice in Traube's results that the strongest smelling substances produced the greatest depression of the surface tension in aqueous solutions.

But before much progress could be made in detecting a parallel between smell and depression of surface tension it was necessary to find a method of measuring the intensity of smell, and this seemed likely to be a very difficult matter as common experience seems to indicate that different animals have very different sensibilities as regards smells. Dogs by their sense of smell can obviously detect things which are quite inodorous to human beings, and one human being seems to have a much keener sense of smell than another. Moreover it seemed likely that the intensity of smell of a substance would depend among other things on its vapour pressure. Practical psychologists describe an instrument in which air is inhaled by the nostrils over a definite area of filter-paper surface moistened by a solution of the substance to be tested. But before experimenting with this instrument the simpler plan was tried of making aqueous solutions of different strengths of the substance to be tested and ascertaining the minimum strength of solution in which the substance could be detected by smelling the solution contained in an ordinary bottle at the ordinary temperature. Almost contrary to expectation it was found that consistent results could be obtained in this way, and that several persons obtained practically the same results. Details are given in the experimental part.

These determinations of the intensity of the odour of the substances examined by Traube confirmed the conclusion previously suspected, viz. that those substances had the strongest smell which produced the *greatest depression* of the surface tension in aqueous solution. The intensity of odour and also the surface tension of aqueous solutions of additional substances were determined, especially of essential oils and other substances well known for their strong odours, and with a few exceptions the above rule was found to hold. The constituents of the well-known perfumes, *e.g.* citral and geraniol, produced a very great depression of the surface tension. Out of 22 substances examined it was found that 17 would be arranged in exactly the same order whether they were arranged according to *increasing depression* of surface tension or increasing intensity of odour. The list is as follows: formic acid, methyl alcohol, acetic acid, ethyl alcohol, propionic acid, methyl acetate, methylamine, phenol, ethyl acetate, butyric acid, *iso*-amyl alcohol, quinoline, cinnamic aldehyde, citral, allyl sulphide, geraniol. The most noteworthy exceptions were ammonia, pyridine and ethyl mercaptan which did not depress the surface tension to the extent anticipated from their powerful odours.

The method used for the determination of surface tension was of the simplest character and it is well known that such determinations are liable to large experimental errors. Rather than spend more time on eliminating possible errors in surface tension determinations it was decided to make some determinations of adsorption. According to theory adsorption should run parallel with surface tension and it is obvious that from the point of view of contact of the odorous substance with the olfactory organ we are more interested in adsorption than in depression of surface tension which was merely used as an index of adsorption.

Arranged according to adsorption from aqueous solution by animal charcoal two of the exceptions above noted disappear, viz. pyridine and ethyl mercaptan, which have adsorptions in keeping with the intensity of their odours.

Arranged according to depression of surface tension	Arranged according to adsorption by charcoal	Arranged according to intensity of odour
Ammonia	Ammonia	Formic acid
Formic acid	Formic acid	Acetic acid
Acetic acid	Acetic acid	Propionic acid
Ethyl mercaptan	Propionic acid	Ammonia
Pyridine	Pyridine	Butyric acid
Propionic acid	Butyric acid	Pyridine
Butyric acid	Ethyl mercaptan	Ethyl mercaptan
	Citral	Citral

We may therefore enunciate the general rule that those substances have the strongest smell which are most readily adsorbed from aqueous solution.

It was thought that by replacing animal charcoal by a protein we might obtain a closer approach to the olfactory organ. Some experiments on adsorption by wool from aqueous solutions were tried but did not lead to any result of interest.

Although, apparently, no progress has been made in solving the problem outlined at the beginning of this paper, it is interesting to have found a connection between smell and physical properties capable of exact measurement, and it does not seem too much to hope that it may prove a first step in solving the problem which the author has set before him.

EXPERIMENTAL.

Determination of the minimum concentration of aqueous solution in which odorous substances can be detected by smell.

The solutions were all placed in similar bottles which were stoppered and shaken just before they were smelt by bringing the nose close to the mouth of the bottle. In order to avoid fatigue the more dilute solutions were presented to the observer before the more concentrated ones and in order to avoid the influence of suggestion he was not told what solutions were being presented to him. The results obtained by the author, two Indian colleagues (Dr A. C. Sircar and Babu S. C. Ganguli) and one Indian student (Babu J. K. Mazumdar) are recorded. It was more difficult to obtain consistent results with the fatty acids than with the other substances examined.

The substances are arranged in order of increasing intensity of odour.

Substance	Percentage strength in which detected by			
	E. R. W.	S. C. G.	J. K. M.	A. C. S.
Formic acid ...	0·1	1	1	—
Acetic acid ...	0·1 (0·3)	1 (0·1)	0·1	
Propionic acid ...	1 (0·1)	1 (0·03)	0·3	
Aniline	0·1	—	—	
Methyl alcohol ...	0·1 (0·01?)	0·1	0·01	—
Ethyl alcohol ...	0·1	0·1	0·1	
Methyl acetate ...	0·1	0·03	0·1	
Chloroform	—	0·1	—	—
Butyric acid ...	0·01	0·1 (0·003)	0·1 (?)	—
Ammonia	0·01	—	—	0·01
Methylamine ...	0·01	—	—	
Ethyl acetate ...	0·01	0·01	0·01	—
Phenol	—	0·01	—	
Pyridine	—	<0·001	—	. .
Quinoline	0·001	—	—	
Isoamyl alcohol ...	0·001	0·001	0·001	—
Cinnamic aldehyde ...	0·001	—	—	
Nitrobenzene ...	0·001	0·001	—	—
Ethyl mercaptan ...	0·0001	—	—	—
Citral	0·0001	0·0001	—	0·0001
Geraniol	—	0·0001	—	—
Allyl sulphide ...	0·0001	—	—	—
Phenyl isocyanide ...	0·00001	—		

Determination of surface tension of aqueous solutions of odorous substances.

This was done by observing the rise of the solution in a capillary tube.

In the following list the substances examined are arranged according to effect on surface tension, those having least effect being placed first:

1. Ammonia
2. Formic acid
3. Methyl alcohol
4. Acetic acid
5 and 6. Ethyl mercaptan / Ethyl alcohol
7. Pyridine
8. Propionic acid
9. Methyl acetate
10, 11, and 12. Methylamine / Phenol / Aniline
13. Nitrobenzene
14. Ethyl acetate
15. Butyric acid
16. Phenyl isocyanide
17. Isoamyl alcohol
18. Quinoline
19, 20 and 21. Allyl sulphide / Citral / Cinnamic aldehyde
22. Geraniol

Adsorption from aqueous solution by animal charcoal.

The determination was easy for acids and bases, which can readily be estimated by titration. An approximately normal solution (50 cc.) was shaken with 2 g. animal charcoal (Merck's, washed and dried) and allowed to stand with occasional shaking over two nights, then filtered and titrated. Similarly 100 cc. of an approximately decinormal solution were shaken with 1 g. animal charcoal.

Substance	Titration value (in cc. N soln.) of 10 cc.	
	Original solution	Final solution
Formic acid	21·7 × 0·939	19·9 × 0·939
	17·7 × 0·1108	16·1 × 0·1108
Acetic acid	16·8 × 0·939	14·9 × 0·939
	13·3 × 0·1108	11·55 × 0·1108
Propionic acid	12·3 × 0·939	10·25 × 0·939
	10·3 × 0·1108	8·15 × 0·1108
Butyric acid	8·7 × 0·939	6·6 × 0·939
	6·65 × 0·1108	4·35 × 0·1108
Ammonia	12·4 × 1·036	11·4 × 1·036
	12·3 × 0·11036	11·35 × 0·11036
Pyridine	9·7 × 1·036	7·6 × 1·036
	10·3 × 0·11036	7·3 × 0·11036

Citral. It was found that citral in very dilute aqueous solution such as required for these experiments could be estimated by extracting with ether and titrating the ethereal extract with Hübl's solution.

A solution of 0·3555 g. citral in 1 litre of water was shaken with 10 g. animal charcoal and allowed to stand for 24 hours. The smell of citral had entirely disappeared, so that concentration had been reduced to less than 0·000001.

A solution of 0·4 cc. citral dissolved in 1 litre was shaken with 2 g. charcoal and allowed to stand for 24 hours. The smell had almost entirely disappeared.

An estimation gave the approximate results:

$$\text{Concentration in water} = 0·00000256$$
$$\text{,,} \qquad \text{charcoal} = 0·1782$$

For the same concentration of butyric acid in charcoal we can calculate:

$$\text{Concentration in water} = 0·00201$$
$$\text{,,} \qquad \text{charcoal} = 0·1782$$

So that the adsorption of the citral is very much greater.

Ethyl mercaptan. An attempt was made to estimate ethyl mercaptan in aqueous solution by precipitating with lead acetate, filtering off, washing and weighing the lead mercaptide. It was not very satisfactory owing to the solubility of lead mercaptide.

Ethyl mercaptan (2 cc.) was dissolved in 250 cc. water.

(a) 40 cc. of this solution precipitated with 1 g. lead acetate dissolved in 15 cc. water; precipitate filtered off and washed with a measured quantity of water, 10 cc. at a time = 0·2800 g. (corresponds to 0·1055 mercaptan).

(b) 10 cc. of mercaptan solution diluted to 40 cc. and treated with 1 g. lead acetate solution, etc., exactly as above, using same quantity of wash water; lead mercaptide obtained = 0·0472 g.

(c) 50 cc. of mercaptan solution was shaken with 0·1 g. animal charcoal filtered and 40 cc. of filtrate treated exactly as above; lead mercaptide obtained = 0·0463 g.

(d) 50 cc. of mercaptan solution was shaken with 0·5 g. animal charcoal, filtered and 40 cc. filtrate treated exactly as above; lead mercaptide obtained = 0·0122 g.

Although the method of estimating mercaptan is unsatisfactory we can safely say that 40 cc. of the filtrate from (c) contained about the same quantity of mercaptan as 10 cc. of the original solution, so that

$$\text{Concentration in water} = 0·00166$$
$$\text{,,} \qquad \text{charcoal} = 2·502$$

From these data we can arrange the substances according to their adsorption by animal charcoal, those which are most adsorbed being placed last in the list: ammonia, formic acid, acetic acid, propionic acid, pyridine and butyric acid (bracketed together), ethyl mercaptan and citral (bracketed together).

The work described in the paper was carried out in the second half of 1919.

REFERENCE.

Bugarsky and Liebermann (1898). *Pflüger's Arch.* **72**, 51.

LXIII. STUDIES ON THE PITUITARY. I.

THE MELANOPHORE STIMULANT IN POSTERIOR LOBE EXTRACTS.

By LANCELOT THOMAS HOGBEN AND FRANK ROBERT WINTON.

(*Received July 4th, 1922.*)

INTRODUCTION.

To the manifold physiological responses evoked by extracts of the posterior lobe of the pituitary gland, it has lately become necessary to add that of inducing pigmental changes in those lower vertebrates (Fishes, Reptiles, Amphibia) which possess considerable capacity for temporary modification of bodily colour through the activity of a special type of effector organ, the chromatophore (melanophores, etc.). Recent work on the part played by endocrine organs in amphibian metamorphosis has in particular brought to light a number of interesting data relative to the factors which control pigment response in these organisms. One of the most characteristic consequences of pituitary removal in tadpoles noted by Allen [1917] and others was a condition of extreme pallor resulting from general contraction of the dermal melanophores (black pigment cells). More recently Swingle [1921] has shown that implantation of the pituitary in pale tadpoles induces a darkening of the skin, a result shown to follow injection of posterior lobe extracts into adult frogs by the writers [1922]. The present communication attempts to explore the use of frog melanophores as indicators of pituitary extracts and to emphasise the remarkable specificity of this reaction. Few people are aware how marked is the capacity for pigmental change possessed by the common frog (*R. temporaria*), the same individual varying from a light yellow to a coal black or dark grey tint within comparatively short time limits: the difference is mainly due to the "expansion" and "contraction" of the dermal melanophores. In general between temperatures of 0–25° frogs placed in bright illumination are pale if kept dry whether on a light or dark background. They may become dark if placed on a dark background in water. For the study of pigmental darkening by pituitary injection, it is best to leave the animals singly in dry glass dipping jars for a few hours in bright light on a white background to ensure maximal contraction of the melanophores. Observations can be supplemented by microscopic preparations of skin, fixed in Bouin's fluid, dehydrated, cleared and mounted without more elaborate treatment. In earlier experiments [Hogben and Winton, 1922] decerebrated animals were used. For injection Burroughs Wellcome's all glass tuberculin syringe graduated in 0·1 cc. was found to be most convenient: unless otherwise stated the intraperitoneal method was adopted.

In these experiments the frogs were stored in white porcelain tanks with glass tops, and were thus always ready for use. To facilitate observation individuals in which the skin is uniformly of a yellowish hue when pale, rather than the varieties with irregular markings, spotted or speckled aspect should be selected.

QUANTITATIVE EXPERIMENTS.

The following experiments were carried out to determine how far the melanophore response might be utilised to estimate the activity of posterior lobe extracts. The preparations employed were Burroughs Wellcome's 0·5 cc. phials of liquid sterile posterior lobe (21 %) extract ("infundin") stated to be standardised by action on the isolated mammalian uterus. Injection with 0·5 cc. of 1/1000 solution of this preparation invariably invoked the characteristic darkening of the skin, while 0·5 cc. of a 1/10,000 solution failed to do so in every case. The following preliminary experiments in which the animals employed were not weighed may be quoted to indicate the range of sensitivity of the frog's melanophores to pituitary extracts and the relative sensitivity with different methods of injection. The average temperature was 72° F.

Sample A. Three series of six frogs received an intraperitoneal injection equivalent to 0·00031, 0·00025, 0·00019, 0·00012, 0·00009 and 0·00006 cc. of infundin respectively. In two series the animals which received 0·00012 cc. and upwards underwent darkening of the skin and in the other series those which received 0·00019 cc. and upwards gave a positive reaction.

Sample B. Two series of six frogs were injected with 0·00025, 0·0002, 0·00015, 0·0001, 0·000075, 0·00005 cc. of infundin. In one series (intraperitoneal) the individuals which received 0·00015 cc. and upwards showed visible darkening of the skin accompanied, as seen on microscopic examination, with full expansion of the melanophores. The individual which received 0·0001 cc. displayed the melanophores in a slightly expanded (stellate) condition, while in the remaining frogs the melanophores were fully contracted. In the remaining series (dorsal lymph sac injection) microscopic examination revealed a precisely similar seriation, but the frog which received 0·0001 cc. showed a slight visible darkening which was not noticeable in the corresponding animal of the preceding group.

Sample C. Six frogs received an intraperitoneal and six an intravenous injection equivalent respectively to 0·0005, 0·0004, 0·0003, 0·00025, 0·0001 and 0·00005 cc. of B.W. liquid sterile extract. In both series the animals which received 0·0005, 0·0004 and 0·0003 cc. doses reacted with full expansion of the melanophores; a slight darkening of the frogs injected with 0·00025 and 0·0001 cc. was visible, but microscopic examination revealed the expansion of the melanophores in the former though not in the latter case. The animals injected with a dose of 0·00005 cc. did not react at all.

In these experiments summarised in Table I the frogs used were in no way selected, and it may be inferred that such methods are suitable for detecting the presence of posterior lobe extracts in quantities exceeding that equivalent to 5×10^{-5} g. of fresh gland (ox) substance, *i.e.* about 0·05 % of a clinical dose. In view of the relation between sensitivity and size indicated below it will be seen that by selecting sufficiently small individuals (or species of Amphibia), it should be possible to detect posterior lobe secretions in considerably smaller quantities than this and therefore in amounts that lie much beyond the limits of sensitivity of methods at present employed. We have found no evidence of a difference in the sensitivity of the reaction with the three methods of injection described.

Table I.

Sample	Method of injection	Minimal dose cc.	Subminimal dose cc.
A	Intraperitoneal	0·00012	0·00009
A	Intraperitoneal	0·00012	0·00009
A	Intraperitoneal	0·00019	0·00012
B	Intraperitoneal	0·00015	0·0001
B	Dorsal lymph sac	0·00015	0·0001
C	Intraperitoneal	0·00025	0·0001
C	Intravenous	0·00025	0·0001

Sample D. Two series, each of six frogs, were weighed and arranged in a heavy and a light series. Doses were injected so that pairs, one from each series, received the same dose per unit body weight—the results are shown in Table II, and demonstrate proportionality between body weight and the minimal effective dose.

Table II.

Approximate dose per 20 g. weight cc.	Heavy series			Light series		
	Wt. g.	Dose cc.	Response	Wt. g.	Dose cc.	Response
0·00025	31	0·00037	Darkening	16	0·0002	Darkening
0·0002	30	0·0003	,,	15	0·00015	,,
0·00015	25	0·0002	,,	15	0·00011	,,
0·00015	23	0·00018	,,	14	0·0001	,,
0·0001	23	0·00012	—	13	0·00007	—
0·00005	21	0·00005	—	12	0·00003	—

Sample E. (See Table III.) Eight frogs of 18–20 g. body weight received 0·0002, 0·00016, 0·00014, 0·00012, 0·0001, 0·00008, 0·00006, 0·00004 cc. infundin (intraperitoneal). The first five showed darkening: the rest remained pale, *i.e.* the minimal dose to produce maximal expansion of the melanophores lay between 0·0001 and 0·00008 cc. Six other frogs weighing 15–16 g. were injected *via* the peritoneum with 0·0002, 0·00015, 0·0001, 0·000075, 0·00005, 0·000025 cc. infundin. The first four only showed darkening of the skin, *i.e.* the minimal dose lay between 0·000075 and 0·00005 cc. A third set of eight frogs weighing 13–14 g. were tested from the same sample of infundin with the following doses (intraperitoneal): 0·0002, 0·00015, 0·0001, 0·00008, 0·00006, 0·00004,

0·00002, 0·00001 cc. The melanophores were fully expanded in the first three ,and contracted in the last three, while the frogs which received 0·00008 and 0·00006 cc. were found to display the pigment cells in a condition of partial expansion.

Sample F. (See Table III.) The following doses were administered to twelve individuals. A 1 (30 g.) and A 2 (25 g.), 0·00035 cc.; B 1 (30 g.) and B 2 (26 g.), 0·0003 cc.; C 1 (30 g.) and C 2 (26 g.), 0·00025 cc.; D 1 (29 g.) and D 2 (27 g.), 0·0002 cc.; E 1 (29 g.) and E 2 (28 g.), 0·00015 cc.; and F 1 (28 g.), and F 2 (28 g.), 0·0001 cc. infundiu. The pairs A, B, C, D, uniformly displayed an intense darkening of the skin after half an hour had elapsed. The two pairs E and F on the other.hand remained pale. This set of animals was deliberately chosen for large size; none of the females were carrying ripe eggs at the time.

The results obtained from the three foregoing experiments,' although obtained with three different samples of "infundin," and therefore heterogeneous, present sufficient uniformity to be instructive in connection with the problem of what order of error is introduced in the method. It will be seen that the interval between minimal and subminimal doses, representing the margin of uncertainty, can usually be reduced to 25 %–30 % of the minimum effective dose; occasionally however, as in the third series of experiment (*E*), unsuitability of frogs may extend this even to 60 %.

That the simple correlation between body weight and minimal dose does not invariably obtain, is illustrated by a further experiment which was undertaken with a view to determining the sensitivity of the reaction to a dose injected about two hours subsequent to a previous dose in the neighbourhood of the minimal dose.

Sample G. (See Table III.) A preliminary experiment with four frogs showed that this sample was somewhat more potent than some used previously, and suggested the doses described. 32 frogs were weighed, and divided into four groups.

Eight frogs (9–11 g. body weight) were injected respectively with 0·0002, 0·00015, 0·000125, 0·0001, 0·000075, 0·00005, 0·000025, 0·00001 cc. infundiu. Visible darkening of the skin occurred in the first five, the remaining three were unaffected. The minimal dose was in this case between 0·000075 and 0·00005 cc.

Eight frogs (12–13 g. body weight) were injected with 0·00025, 0·0002, 0·00015, 0·000125, 0·0001, 0·000075, 0·00005, 0·000025 cc. infundiu. The first three only showed visible darkening of the skin. The minimal dose was thus between 0·00015 and 0·000125 cc.

Eight frogs (14–15 g. body weight) were injected with the same series of doses. The first four in this instance displayed visible pigmental change, the dose necessary being between 0·000125 and 0·0001 cc.

Eight frogs (17–19 g. body weight) were injected with 0·00035, 0·0003, 0·00025, 0·0002, 0·00015, 0·0001, 0·000075 and 0·00005 cc. The first three only showed darkening of the skin, *i.e.* the minimum dose for this series was between 0·00025 and 0·0002.

Table III.

Sample	Actual dose			Dose per 20 g. body weight				Error in min.
	Wt. g.	Min. cc.	Submin. cc.	Wt. g.	Mean min. cc.	Min. cc.	Submin. cc.	%
D	23	0·00017	0·00012	23⎫		0·00015	0·0001	
D	14	0·0001	0·00007	13⎬ 0·00015				0
E	18	0·0001	0·00008	20⎫		0·0001	0·00008	− 17
E	15	0·000075	0·00005	16⎬ 0·00012		0·0001	0·00007	− 17
E	13	0·0001	0·00004	14⎭		0·00015	0·00006	+ 25
F	29	0·0002	0·00015	29⎫ 0·000145		0·00014	0·0001	− 3
F	27	0·0002	0·00015	28⎭		0·00015	0·00011	+ 3
G	19	0·00025	0·0002	17⎫		0·00025	0·00023	+ 14
G	15	0·000125	0·0001	14⎪		0·00017	0·00015	− 21
G	12	0·00015	0·000125	12⎪		0·00025	0·00021	+ 15
G	8	0·000075	0·00005	10⎬ 0·000215		0·00019	0·0001	− 11
G (repeat)	8	0·000075	0·00005	10⎪		0·00019	0·0001	− 11
G (repeat)	17	0·0002	0·00015	18⎭		0·00023	0·00017	+ 7

Mean deviation 11·8 %.

Effect of subsequent doses. After injection with concentrated extracts the condition of melanophore expansion may last 12 hours or more. After injection with minimal doses it passes off in about 1¾–2½ hours. After 1¾ hours following the first injection, the last series injected with sample *G* received corresponding doses of the same magnitude. The minimal dose was slightly lower, 0·0002–0·00015 cc. On injecting the first series two hours after the initial treatment the minimal dose was the same as before. It thus appears: first that given sufficient time for recovery the sensitiveness of the melanophores is not appreciably changed by previous treatment, second, that the effect of repeated doses is, in contrast with the action of the pressor principle [Howell, 1898], identical.

Like Dale and Laidlaw's original method[1] for estimating the uterine stimulant, this method is appropriate to comparison rather than absolute standardisation of extracts. Several factors which may lead to anomalous results might be considered, such as possible seasonal variation of sensitivity[2], and variations depending on the immediate previous history of the frogs with respect to moisture, temperature, etc. Again, different strains of frogs may respond with varying facility, and we have certainly noted that some are more and others less satisfactory in giving a sharply defined end point and well contrasted reaction. Furthermore, isolated frog's skin placed in Ringer shows the characteristic response to posterior lobe extracts, and may react with more constant sensitivity than an intact animal; these and other relevant considerations we have at present no opportunity to examine. The specificity,

[1] In Dale and Laidlaw's [1912] standard method for estimating the activity of pituitary extracts by their action on the isolated guinea pig's uterus, the minimal dose of infundin which produces a maximal contraction varies between ·003–·005 cc. which is diluted to about 100 cc.

[2] Compare the striking variation in the electrical phenomena of frog's skin which occurs during the breeding season—[Bayliss and Bradford, 1886].

certainty and delicacy of the melanophore response will recommend it for such biochemical and pharmacological purposes as do not require a degree of accuracy higher than that we have described, namely consistency of minimal dose, to the extent of a mean deviation of 12 %. In each of the foregoing experiments the actual number of frogs used was the minimal number requisite for demonstrating the end point in a series. It is evident that in testing samples on a large scale the degree of activity might be compared with greater confidence by employing a larger number of individuals. In any case a simple and direct method is provided for testing the retention of activity of pituitary products employed in the dispensary or hospital.

SPECIFICITY OF THE MELANOPHORE STIMULANT.

Concerning the mode of action of the melanophore stimulant there is no need to add anything further in this place to our previous observations except to emphasise the local response obtained on isolated strips of skin placed in Ringer's extracts. It is our intention to publish in the near future a comprehensive account of the reactions of frog melanophores to drugs; but it is of special interest to insert at this point a few observations under this heading to lay stress on the highly specific nature of the response evoked in amphibian melanophores by posterior lobe extracts. The physiological effects hitherto described as characteristic of posterior lobe extracts are equally characteristic of a group of other well-known drugs. In the case of the melanophore reaction described in this communication we have to deal with a property of the extract which is not precisely simulated by any of the more familiar pharmacological reagents, as the accompanying table signifies. The case of digitalis merits special comment. The water-soluble digitalin (Parke Davis) was employed. In a dosage of 1 mg. death followed 30–45 minutes after injection *via* the dorsal lymph sac and no colour change resulted. 0·5 mg. was generally fatal after a few hours and accompanied in one case by a partial expansion of the melanophores. Doses of 0·3 mg. (*i.e.* one half the minimum lethal dose)

Table IV.

Pale frogs were injected with doses of the drugs undermentioned.

(a) Pilocarpine	25 mg., 12·5 mg., 2·5 mg.	Remained pale
(b) Atropine	10 mg., 3 mg., 1 mg.	,,
(c) Strychnine	1 mg., 1/25 mg., 1/100 mg.	..
(d) Curare (decerebrated)	Curarised	
(e) Histamine	3·6 mg., ·36 mg., ·036 mg.	
(f) Ergatoxin	6·5 mg., ·65 mg., ·065 mg.	
(g) Tyramine	20 mg., 2 mg., ·02 mg.	
(h) Veratrine	1/4 mg., 1/20 mg., 1/100 mg.	
(i) Digitalis	Numerous doses of varying range below M.L.D. (cf. text)	
(j) Barium chloride ...	1 mg., 1/5 mg.	
(k) Sodium nitrite ...	5 mg.	..
(l) Caffeine	10 mg., 2 mg.	,,
(m) Apocodeine ...	2·5 mg.—·5 mg. without motor paralysis 10 mg. motor paralysis within 30 mins.	Slight darkening
(n) Nicotine	1/10 mg. no motor paralysis ... ... 1/5 mg.—1 mg. with motor paralysis ...	Remained pale Partial darkening

or less produced absolutely no effect on coloration. But no effect was obtained on the isolated skin. A partial expansion usually but not invariably occurs with nicotine and apocodeine only in dosage sufficient to produce motor paralysis without immediate lethal effects. In none of these cases is the reaction readily comparable with the pituitary reaction.

SOME CHEMICAL PROPERTIES OF THE MELANOPHORE STIMULANT.

Experiments were carried out to test how far the chemical properties of the pituitary melanophore stimulant agree with those of the other active constituents of the posterior lobe extracts.

Inactivation by boiling with dilute HCl. There appear to be at least two autacoid substances contained in extracts of the posterior lobe, since it is possible to destroy or reduce the pressor activity of "infundibular" extracts without diminishing in a corresponding manner their power to excite plain muscle. Dudley [1919] by extraction of dried and powdered infundibulum with acidified water, treatment of the solution with colloidal ferric hydroxide, and subsequent continuous extraction of the filtrate at reduced pressure with butyl alcohol, succeeded in separating a residue containing all the uterine stimulant ("oxytocic" principle) together with a portion of the pressor substance. The latter has been shown by Abel and Nagayama [1920] and Dale and Dudley [1921] to be rapidly destroyed by boiling with 0·5 % HCl. The oxytocic principle on the other hand is only slowly destroyed by such treatment. In this respect it appears that the melanophore stimulant is like the uterine principle and is not identical with the pressor substance.

The effect of continued boiling with dilute HCl was investigated as follows. A 0·5 % solution of the commercial extract was made up in 0·5 % HCl. This mixture was subjected to continuous boiling for five hours, a sample being removed at the end of thirty minutes. At the conclusion of the experiment a sample of the unboiled mixture, the portion which had been subjected to only half an hour's hydrolysis, and the residue were respectively neutralised with soda and diluted to a concentration approximately isotonic with frog's Ringer. From each of the three solutions, A (unboiled), B (boiled half an hour), C (boiled five hours), 0·5 cc. was injected into a pair of frogs whose pigment cells were fully contracted. The macroscopic and microscopic examination of the six animals at the conclusion of an hour revealed a marked contrast. The A and B pairs were dark and showed the typical pituitary reaction. The pair C remained pale. Microscopic preparations of the skin showed that in the C pair the melanophores were fully contracted, and the A and B pairs expanded. The result of the experiment indicates that pituitary extracts retain a considerable potency to induce melanophore response after half an hour's boiling with 0·5 % HCl; hence the melanophore stimulant is not identical with the pressor substance. According to Abel and Nagayama the oxytocic activity of the pituitary extracts after being reduced to about one-fifth by hydrolysis for 30 minutes resists further destruction. Dale and Dudley however did not find that the residue was stable to prolonged boiling with acid; and the behaviour of the melanophore stimulant in this experiment conforms to their experience of the uterine stimulant.

The oxytocic potency of the samples taken after 30 minutes' boiling with 0·5 % HCl was found by Dale and Dudley to be between a quarter and a fifth of the previous activity. The potency of pituitary extract to induce melanophore expansion after half an hour's hydrolysis appears to be reduced in somewhat similar proportions. In a further experiment an approximation to the amount of melanophore stimulant remaining after half an hour's boiling with 0·5 % HCl was investigated thus. Three portions A, B, C (15 cc.) were taken from a 2 % "infundin" solution. A was boiled for half an hour, after which 15 cc. 1 % HCl and sufficient 1 % soda to neutralise it were added. To B were added 15 cc. of 1 % HCl, which was neutralised before boiling for 30 minutes. C was acidified with 15 cc. of 1 % HCl, boiled for half an hour, and subsequently neutralised. Each solution thus contained originally 0·3 cc. of sterile extract. A and B were made up to two litres so that an injection of 0·3 cc. would be equivalent to 0·00015 cc. infundin. C was made up to half a litre: one-half of this (C 1), representing a dilution one quarter of A and B, was used for injection, the other half was diluted one in four (C 2). Twenty-eight frogs were injected as in the following table from which it appears, on making the correction necessary for body weight, that the samples A and B (controls) were about four times as potent as C, the solution which had been subjected to boiling with 0·5 % HCl.

The positive sign signifies darkening of the skin, the numbers in brackets refer to the body weight in grams, and the extreme left hand column gives the dosage in cc. injected.

Table V.

	A	B	C 1	C 2
0·8 cc.	+ (12)	+ (12)	+ (12)	+ (11)
0·7 ,,	+ (13)	+ (13)	+ (13)	− (11)
0·6 ,,	+ (14)	+ (13)	+ (14)	− (11)
0·5 ,,	+ (16)	+ (16)	+ (16)	− (11)
0·4 ,,	+ (17)	+ (18)	+ (17)	
0·3 ,,	− (18)	+ (19)	− (18)	
0·2 ,,	− (20)	− (20)	+ (20)	
0·1 ,,	− (23)	− (23)	− (23)	

Action of proteoclastic enzymes. Perhaps the most significant fact indicative of the type of chemical compounds to which the infundibular autacoids are allied is furnished by the action of proteoclastic ferments. Schafer and Herring [1906] stated that pepsin destroys the pressor activity of pituitary extracts leaving intact the diuretic principle. They denied further that trypsin affected either. Later work by Dale [1909] and Dudley [1919] indicates that trypsin rapidly destroys both the pressor and oxytocic principles. To test the action of proteoclastic ferments on the melanophore stimulant six solutions were made up: A (0·5 % infundibular extract in 0·2 % HCl and

0·5 % pepsin); B (0·5 % infundiu in 0·2 % HCl); C (0·2 % HCl and 0·5 % pepsin); D (0·5 % infundiu in 0·5 % saline trypsin); E (0·5 % saline trypsin); F (0·5 % infundin in 0·5 % *boiled* trypsin). After two hours' digestion A, B and C were neutralised, boiled and diluted till isotonic with Ringer. D, F were boiled. On injection into pairs of pale frogs, the pairs injected with A, B and F showed darkening of the skin with expansion of the melanophores. The rest remained pale with the melanophores contracted. Thus the melanophore stimulant is rapidly destroyed by trypsin and is not rapidly destroyed by pepsin. In this connection evidence is provided that the melanophore stimulant is not identical, as Abel and Kubota [1919] once believed the uterine principle to be, with histamine, since the latter (Dudley) is not destroyed by tryptic digestion. As a matter of fact histamine does not in any concentration cause melanophores to expand.

Relation to other pituitary autacoids. In view of the possible identity of the diuretic and pressor substances, and the likelihood that the oxytocic stimulant is responsible for the galactogogue action of posterior lobe extracts, it is not necessary to postulate the existence of more than two distinct autacoids secreted by the juxtaneural epithelium, It is possible that the melanophore stimulant is not identical with either the pressor or the oxytocic autacoid, but there is strong evidence that it is not identical with the former, namely (a) its slow rate of inactivation by boiling with dilute hydrochloric acid; (b) no reversal of response after successive action; (c) failure of drugs which agree with pituitary extracts in their effect on the blood pressure to evoke the same reaction. For the present therefore we may provisionally attribute the melanophore response to the oxytocic substance.

SOURCE OF THE PITUITARY MELANOPHORE STIMULANT.

In Swingle's [1921] experiments upon the implantation of the pituitary gland in amphibian larvae, the pars intermedia was found to be responsible for the darkening of the skin. The results obtained from injections are consonant with the conclusion that the juxtaneural epithelium is the source of the secretion which induces melanophore expansion.

The demarcation between the three parts of the gland in the case of ox pituitaries is very striking; and ox pituitaries secured within an hour of killing were therefore employed in the following experiment. The three portions of the gland were carefully separated: each portion was weighed, ground up with sand, and extracted with Ringer's solution at 35° for two hours. A 0·1 % and a 0·02 % solution were made up from each extract so prepared. A pair of frogs were injected with each of the six solutions. At the end of half an hour the two pairs injected respectively with the strong and weak anterior lobe extracts remained pale. The pair injected with a weak extract (0·02 %) of pars nervosa extract also remained pale. While the pairs injected with both weak and strong extracts of pars intermedia as well as the pair injected with the stronger pars nervosa extract were conspicuously darkened through, as microscopic examination displayed, the general expansion of the dermal melanophores. In view of the non-glandular character of the pars nervosa, and the higher concentration of the melanophore stimulant in the pars intermedia, it seems likely that the melanophore stimulant is secreted by the latter portion of the gland, diffusing rapidly into the former.

PHYLETIC DISTRIBUTION OF THE MELANOPHORE STIMULANT.

The following observations indicate the occurrence of the same melanophore stimulant in the pituitary gland of various classes of Vertebrata.

Mammalia. An adult female rabbit was decapitated, and its pituitary gland together with pieces of muscle, brain, ovary, pancreas, liver, suprarenals and spleen removed instantly. The tissues mentioned were severally ground with sand and extracted with Ringer in the usual way, after having been first weighed. A pair of frogs was injected with 0·5 cc. per individual of a 1 % solution of each extract. An additional pair received a corresponding amount of a putrid meat extract, while another pair received 0·5 cc. of a 0·1 % solution of the same pituitary extract. The two pairs injected with 1 % and 0·1 % pituitary extracts alone displayed a visible darkening of the skin, and, upon microscopic examination, the expansion of the melanophores.

Birds. Pituitaries were removed from eight ducks, weighed, ground up with sand, extracted and made up to 2 %, 0·4 %, and 0·08 % solutions which were administered respectively to pairs of pale frogs (0·5 cc. per individual). The animals injected with the two stronger solutions displayed the characteristic pigmentary response: the pair injected with the 0·08 % extract however remained pale.

Reptiles. Into each of a pair of frogs an injection of an extract of the pituitary of the lizard (*Lacerta viridis*) was made. The characteristic reaction was evoked. Two other frogs injected with an extract of brain tissue as controls, remained pale.

Amphibia. The experiment with frog's pituitary is of special interest as bearing on the rôle of the juxtaneural epithelium in normal pigmentary responses. Six frogs' pituitaries were removed and made up to 5 cc. in Ringer's solution. An injection of 0·5 cc. into each of a pair of pale frogs evoked an intense darkening of the skin. The remaining solution was diluted 1·5 in 10; and a pair of frogs injected with 0·5 cc. of this solution also underwent darkening of the skin. The 1·5/10 solution was again diluted 1/5. A pair of frogs injected each with 0·5 cc. of the latter showed in one case a darkening, while the other individual remained pale. On further dilution no response was induced by injection. It appears from the foregoing data that the amount of pituitary melanophore stimulant in the gland of one frog is not only sufficient to account for the darkening of the skin in one animal of the same species but in as many as fifty-six individuals.

Fishes. Pituitaries were removed from six cods (*Gadus*) and were made up to 2 %, 0·2 % and 0·04 % extracts. Two pairs of frogs injected with the usual amount of the two stronger solutions displayed the characteristic darkening of the skin. Only one of two individuals injected with 0·5 cc. of the 0·04 % extract underwent pigmentary darkening: a pair injected with a 5 % extract of cod's brain remained pale.

Tunicata. The widespread distribution of the pituitary melanophore stimulant in the Vertebrate series encouraged the search for a similar substance among the lower Chordata; and extracts of the so-called dorsal tubercle or subneural gland of the Tunicate, regarded by morphologists as the homologue of the Vertebrate hypophysis, were accordingly prepared with this end in view. *Ascidiella,* a solitary Ascidian about 1½ inches in length, was selected for the purpose. The dorsal tubercle in this form is very large and quite conspicuous through the translucent mantle after the gelatinous test has been removed. The organ in question was removed from twenty-four individuals and a 10 cc. extract was made. Injection into pale frogs was not followed by any pigmentary change.

REACTION OF CHROMATOPHORES TO PITUITARY EXTRACT.

The observations above mentioned refer to the behaviour of the dermal melanophores of the frog, and in their reactions to pituitary extracts these resemble those of other Amphibia, of which we have examined Hyla, Bombinator (Anura) and Amblystoma (Urodela). The frog possesses in addition at least three other types of pigmentary effector organs: (*a*) epidermal melanophores, responding to pituitary extracts in a manner similar to the dermal melano-phores, though their sensitivity may be different; (*b*) the dermal xantho-leucophores (yellow pigment cells) which respond *in the opposite sense, i.e.* contracting after pituitary injection—this property not being shared by Hyla in which the xantholeucophores appear to be non-contractile; (*c*) the retinal pigment cells. Fujita and Bigney [1918] have shown that these expand after treatment with adrenaline, an observation which we have confirmed. Pituitary injection has no effect upon them either in the expanded or contracted condition.

The fact that the frog itself possesses two types of chromatophores which differ diametrically in their mode of response to posterior lobe extracts is of interest in view of the fact that Spaeth [1917] describes the contraction of the melanophores of the fish (*Fundulus*) after action of the pituitary autacoid in certain circumstances. In view of the conformity of data derived from glandular extirpation and injection in the case of Amphibia, Spaeth's obser-vations require in our belief reconsideration, though the above-mentioned fact does not permit one to regard his conclusions as necessarily improbable.

SUMMARY.

1. The pars intermedia of the mammalian pituitary secretes a specific stimulant inducing expansion of the dermal melanophores in Amphibia: this property is shared by extracts from the pituitary of birds, reptiles, amphibia and fishes: it was not found in the subneural (hypophysial) gland of Tunicates. Sufficient can be extracted from the gland of one frog to induce intense visible darkening in more than fifty other pale individuals of the same species.

2. The action of pituitary extracts on melanophores is one of extreme delicacy, and can be used to detect with certainty the presence of posterior lobe secretions in quantities equivalent to 5×10^{-5} g. fresh glandular substance (*i.e.* about 1/2000th ordinary clinical dose) by their power to induce intense darkening of the skin in pale frogs.

3. A method of approximate quantitative estimation of potency of posterior lobe extracts is indicated by experiments on the sensitivity of the melanophore response of frogs, which further show within what limits of error such a procedure may be successful. The delicacy of response does not depend on the mode of injection, nor is it affected by previous injection of minimal doses of the extract; it is however correlated with body weight.

4. Whereas the physiological effects hitherto described as characteristic of posterior lobe extracts are equally characteristic of a group of other well-known drugs (*e.g.* digitalis, $BaCl_2$, histamine), the melanophore response described above is a property which is not precisely simulated by any of the more familiar pharmacological reagents. An effect in the same sense is indeed produced only by nicotine and apocodeine in doses sufficient rapidly to produce motor paralysis, and even here the darkening is much less conspicuous.

5. The melanophore stimulant in pituitary extracts is rapidly inactivated by tryptic but not by peptic digestion; its activity is diminished by boiling with 0·5 % HCl at about the same rate as is that of the uterine stimulant, but much more slowly than that of the pressor substance. The probable identity of the uterine and melanophore stimulants may be inferred.

6. In view of this probable identity of the uterine and melanophore stimulants, the accessibility of the materials, facility of manipulation, specificity and shortness of time implied by the method involving this response, it would seem to be serviceable for testing the activity of posterior lobe extracts in use in the laboratory and hospital.

The experiments were carried out in Prof. MacBride's laboratory at the Imperial College of Science, and the writers were advised on several points by Dr H. H. Dale, F.R.S., to whom grateful acknowledgment is made. The expenses of the research were defrayed by a grant made by the Government Grants Committee of the Royal Society.

REFERENCES.

Abel and Kubota (1919). *J. Pharm. Exp. Ther.* **13**, 243.
Abel and Nagayama (1920). *J. Pharm. Exp. Ther.* **15**, 347.
Allen (1917). *Biol. Bull.* **32**, 117.
Bayliss and Bradford (1886). *J. Physiol.* **7**, 216.
Dale (1909). *Biochem. J.* **4**, 427.
Dale and Dudley (1921). *J. Pharm. Exp. Ther.* **18**, 27.
Dale and Laidlaw (1912). *J. Pharm. Exp. Ther.* **4**, 75.
Dudley (1919). *J. Pharm. Exp. Ther.* **14**, 295.
Fujita and Bigney (1918). *J. Exp. Zool.* **27**, 391.
Hogben and Winton (1922). *Proc. Roy. Soc. B.* **93**, 318.
Howell (1898). *J. Exp. Med.* **3**, 2.
Schafer and Herring (1906). *Proc. Roy. Soc. B.* **77**, 571.
Spaeth (1918). *J. Pharm. Exp. Ther.* **11**, 209.
Swingle (1921). *J. Exp. Zool.* **34**, 119.

LXIV. ON THE SIGNIFICANCE OF VITAMIN *A* IN THE NUTRITION OF FISH.

By KATHARINE HOPE COWARD (*Beit Memorial Research Fellow*)
AND JACK CECIL DRUMMOND.

*From the Biochemical Laboratories, Institute of Physiology,
University College, London.*

(*Received July 8th, 1922.*)

It has recently been shown that the relatively large amounts of vitamin *A* found in the liver of the gadoid fishes is probably derived from their food and ultimately from the marine algae which synthesise this dietary factor. [Coward and Drummond, 1921; Hjort, 1922; Jameson, Drummond and Coward, 1922; Drummond and Zilva, 1922.] That the vitamin *A* probably fulfils some rôle in the development of the young fish is indicated by its presence in the gonads [McCollum and Davis, 1915; Hjort, 1922] and recently it has been shown that the amount stored in the reproductive cells of such a fish as the cod is extraordinarily large [Zilva and Drummond, 1922]. To study the fate of the vitamin in the fish ova it was decided to work with the brown trout which is easy to rear and maintain in the ordinary laboratory.

VITAMIN *A* IN TROUT EGGS.

Feeding tests on rats by the usual procedure demonstrated that trout eggs are very rich in vitamin *A*. A dose of two eggs per day per rat resulted in a resumption of normal growth, and it is to be regretted that tests were not made with smaller doses of the material, but none had been reserved (Fig. 2).

REARING OF YOUNG TROUT IN THE LABORATORY.

A thousand fertilised eggs were obtained from the Trout Fisheries, Stirling, and were allowed to hatch out in a single trough which we are describing because it enables trout to be reared with success for considerable periods of time in an ordinary laboratory, and we have encountered many cases where workers have failed to achieve this. All the eggs were placed in a flat boat-shaped receptacle (Fig. 1) 18″ long and cross section 5″ × 5″. The sides are of wood but the ends and bottom are made of narrow-mesh zinc gauze. Such receptacles are placed in an ordinary porcelain sink of which the outlet has been adjusted so as to maintain a depth of half an inch of water whilst a constant stream of water is supplied by means of a pipe leading from the tap. The incoming stream issues from a glass tube of half-inch diameter and is not directed centrally down the length of the "boat" but at an angle of about 20° to that axis. By this means the water stream does not pass directly out at the far end of the "boat" but tends to form a current round the sides. In this manner the unconsumed food, which must always be borne to the fish down stream, is carried round several times before being washed down the outflow pipe and wastage is thereby prevented.

The sinks and "boats" are kept scrupulously clean and the water current maintained without interruption. Partial shade for the trout is provided by placing a board crosswise over part of the sink. If these precautions are followed it is possible to rear the young trout almost without loss, at any rate for several months.

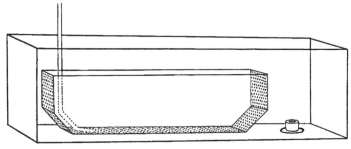

Fig. 1.

Vitamin *A* in Larval Trout.

The ova hatched out with few exceptions and the larvae were allowed to develop without any extra food being given for four weeks. The depth of water was increased to about one inch. It is difficult to obtain information as to the feeding habits of young fresh-water fish, and as far as we are aware there are available no careful studies analogous to Dr Lebour's investigations on marine fish [1918; 1919, 1, 2]. It is commonly believed that larval trout take no food during that stage of their development and that they are purely carnivorous when they begin to feed as soon as the yolk sac is nearly absorbed. Some specimens of young brown trout about one month old from a trout hatchery were examined by us and were found to contain definite remains of partially digested green microscopic plants in their alimentary tracts. It is possible that these were taken in accidentally by the fish or that they were previously ingested by minute crustaceans which the fish had swallowed. The similar presence of marine plants in young fish has been described by Lebour.

About half-way through the larval period our experimental trout were tested again for vitamin *A* and it was found that although the contents of the yolk sac were about half absorbed there yet remained considerable amounts of this dietary factor. At this stage one young fish per day was sufficient

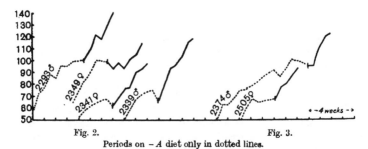

Fig. 2.   Fig. 3.

Periods on − *A* diet only in dotted lines.

Fig. 2. Resumption of growth of rats on feeding two trout eggs previous to basal diet each day.

Fig. 3. Resumption of growth of rats on feeding one larval trout with yolk sac half absorbed.

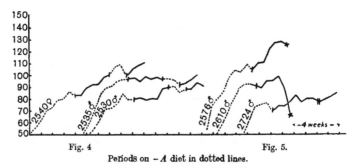

Fig. 4   Fig. 5.

Periods on − *A* diet in dotted lines.

Fig. 4. Very poor growth of rats on feeding one larval trout with yolk sac completely absorbed.

Fig. 5. Very poor growth of rats on feeding one trout that had been fed on cod muscle from the time of the absorption of yolk sac.

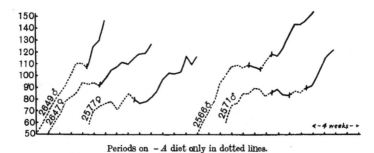

Periods on − *A* diet only in dotted lines.

Fig. 6. Resumption of growth of rats on feeding one trout that had been fed on liver and yolk of egg from the time of absorption of the yolk sac. Two short periods when the rats refused to eat the trout are seen in curves of rats 2566 and 2571. After a short time on − *A* only again the rats ate the trout and growth was rapid.

to cause a good resumption of growth in the test rats (Fig. 3). After about four weeks the yolk sacs were practically absorbed in every case and another test was made. From the results it was apparent that the vitamin A content of the fish had fallen very greatly and that actual utilisation of that originally present in the yolk sac had occurred (Fig. 4).

THE FEEDING OF POST-LARVAL TROUT.

Having found that the supplies of vitamin A present in the yolk sac of the newly hatched fish are apparently used up in the larval period, it was important to determine whether such fish are subsequently dependent on their food to maintain their supply of this factor.

Accordingly the remaining fish were divided up into three groups of about 250 each which were treated as follows:

Group I. Diet rich in vitamin A; fresh minced rats' liver and ground up yolk of hard boiled egg.

Group II. Diet deficient in vitamin A, freshly ground up cod muscle.

Group III. No extra food.

It was found necessary to feed Groups I and II several times a day by slowly dropping the finely minced food pulps into the stream of water in the "boats" so that the small particles were borne down to the fish on the current when the trout would immediately arrange themselves upstream and feed voraciously.

The influence of the feeding on the trout themselves was marked. Those in Group I grew rapidly and were very active, plump and well. Except by an accident late in the experiment when a large number of deaths followed the stoppage of the current of water, there was hardly a death in this group.

The fish on the vitamin A-deficient diet of cod muscle lived fairly well but did not grow much or show the vigour of those in Group I. Their weaker constitution was demonstrated by the increasing frequency of deaths as the experiment progressed. The group kept without food exhibited little or no growth but few died until about three weeks after the experiment had been in progress when deaths were suddenly very frequent.

INFLUENCE OF FOOD ON VITAMIN A RESERVES OF POST-LARVAL TROUT.

The feeding tests on rats demonstrated in a very striking manner the importance of the qualitative composition of the food supply for maintaining the vitamin A in the tissues of these fish. Very slight or no growth (Figs. 4 and 5) was given by supplements of the fish of Group III or those fed on the white fish muscle known to be deficient in vitamin A [Drummond, 1918].

The fish from Group I, however, served as an excellent source of the vitamin for rats and led to a prompt resumption of growth (Fig. 6).

Confirmation of this significance of the food supply was given by the survivors of Group II whose diet was later supplemented by an addition of egg-yolk pulp. They themselves immediately began to grow again and, when used as test material for the one rat surviving this test, showed that their tissues had once again become stored with the vitamin *A*. Similarly, those trout which had been previously starved were fed for ten days with egg; and the rats which had been used to test the starved trout were in the meantime fed only on the diet deficient in vitamin *A*. But when the trout now being fed were added to their diet, the rats immediately resumed growth.

The demonstration of the importance of this qualitative aspect of the diet of young fish may be of some interest in connection with the well-known critical period which coincides with the absorption of the yolk sac. Trout hatchers have assured us that this period is one of great uncertainty in the rearing of young trout, and it is possible that our experiments may suggest to them some modification of the usual existing methods of feeding which might lessen the loss which frequently occurs at that time. Some of our results, which are not fully recorded in this paper, tend to show us that mashed liver, the usual dietary of the fish in hatcheries, may not always supply sufficient of the vitamin *A* for their optimum development and well-being. The full absorption of the contents of the yolk sac is an equally critical stage in the development of marine fish. Fabre and Domergue [1900, 1905] found that certain young fish unable to obtain external food at this stage become anaemic and die, but that an addition of microscopical plant organisms to the water enables them to be reared successfully. The application of such observations to the practical question of sea fishery is well pointed out by Hjort [1914] who emphasises the importance of the immediate availability of suitable plankton food for the survival of larval marine fish.

SUMMARY.

1. The ova of the brown trout normally contain relatively large amounts of the vitamin *A*.

2. During the subsequent development of the young larval fish this substance is in some way utilised so that at or shortly after the stage at which the contents of the yolk sac are absorbed the supplies are almost exhausted, so far as our technique enables us to ascertain.

3. It is probable that this fact is one of the reasons why this stage is so critical a period in the development of the young fish.

4. If, in the post-larval period or even before the yolk sac is completely absorbed, the fish are given food rich in vitamin *A* their growth and development are satisfactory and they appear able to store that factor in their tissues.

5. On a diet containing adequate protein but deficient in the factor *A* no such storage occurs and growth is subnormal.

6. These experiments confirm the previous findings, referred to in the text, which show that the stores of vitamin *A* in the tissues of fish can be derived from the food.

We are indebted to Mr Peart, manager of the Chorley Wood Trout Hatcheries, for invaluable advice in the rearing of the trout, and also to the Medical Research Council for a financial grant which enabled the experiments to be made.

REFERENCES.

Coward and Drummond (1921). *Biochem. J.* **15**, 530.
Drummond (1918). *J. Physiol.* **52**, 95.
Drummond and Zilva (1922). *Biochem. J.* **16**, 518.
Fabre and Domergue (1900). *Bull. de la marine marchande*, Paris.
—— —— (1905). *Développement de la Sole*, Paris.
Hjort (1914). *Rapports, Conseil International pour l'exploration de la Mer*, 20.
—— (1922). *Proc. Roy. Soc. B.* **93**, 440.
Jameson, Drummond and Coward (1922). *Biochem. J.* **16**, 482.
Lebour (1918). *J. Marine Biol. Ass.* **11**, 434.
—— (1919, 1). *J. Marine Biol. Ass.* **12**, 9.
—— (1919, 2). *J. Marine Biol. Ass.* **12**, 22.
McCollum and Davis (1915). *J. Biol. Chem.* **20**, 641.
Zilva and Drummond (1922). *Lancet*, i. 1243.

LXV. NOTE ON KNOOP'S TEST FOR HISTIDINE.

By GEORGE HUNTER, *Strang-Steel Scholar.*

Institute of Physiology, University of Glasgow.

(*Received July 17th, 1922.*)

IN performing this test Knoop [1908] adds bromine water to the solution to be tested until there is just a very slight excess. On heating, the excess bromine disappears and in the presence of histidine a brownish red colour is developed; if the latter is in sufficient concentration a dark precipitate settles out. The solution to be tested should be slightly acid. The only substance other than histidine found to give this test is iminazolylethylamine or histamine. The test, according to Knoop, is given with histidine at a dilution of 1 : 1000.

When using this test the writer found that different intensities of colour were produced from solutions of the same concentration of histidine. This peculiar behaviour was accounted for by slight variations in the bromine excess. The colour produced is very soluble in excess of bromine and in dilute solutions of histidine an initial excess may prevent all colour development. The difficulty of adding just sufficient bromine especially to slightly coloured solutions has been overcome by adding a definite excess followed by washing the solution repeatedly with chloroform in a small separating funnel until the chloroform is no longer coloured. By this procedure even strongly coloured meat extracts and other fluids may be successfully tested by Knoop's method. The excess of bromine appears to oxidise the colouring matters and after washing with chloroform such solutions are almost colourless, so that on heating the brown colour due to histidine is readily detected. Most uniform results have been obtained when the washed solution is transferred to a test tube and set in a boiling water-bath.

That the chloroform does not extract any of the histidine compound was shown by adding to a series of test tubes equal amounts of 0·1 % histidine monohydrochloride and adding increasing small amounts of bromine water from a pipette until definite excess of bromine was present. By heating all the test tubes together on the water-bath it was found that there was first a rise, then a fall in colour values corresponding with insufficiency or excess of bromine. The maximum thus obtained was less than that obtained by the chloroform method.

The maximum colour appears to be developed when the histidine has absorbed just three atoms of bromine per molecule. A dilute solution of bromine water was standardised against $N/100$ sodium thiosulphate using a mixture of potassium iodide and starch as indicator on a tile. The bromine solution was $0.00785\ N$.

In one burette was placed $0.1\ \%$ solution of histidine monohydrochloride and in another the standard bromine water just before using it. At ordinary temperature it was found that 10 cc. of the histidine solution absorbed 18 cc. of the standard bromine water. When the solution was kept cold in ice the same quantity of histidine absorbed 16 cc. of the bromine solution, whilst at about $40°$ in a water-bath it absorbed 21.8 cc. of the bromine solution. (These amounts are not quite definite as on long standing more bromine is absorbed.)

The histidine-bromine solution was in each case transferred to a 100 cc. measuring flask which was filled up to the mark with water, so that the solution represented $0.01\ \%$ original histidine. About 10 cc. of each were put into three dry test tubes and heated on the water-bath. The colour developed was maximum when the bromine was absorbed at ordinary temperature.

It follows from this that one molecule of histidine monohydrochloride absorbs approximately three atoms of bromine at ordinary temperature.

To narrow the limits still further five lots of 5 cc. of $0.1\ \%$ histidine solution were taken and to these respectively were added 8, 8.5, 9, 9.5 and 10 cc. of standard bromine solution. On diluting to 50 cc. and heating portions, the colour was greatest where 9 cc. had been used. This is the same as the above. The maximum colour developed here was not greater than that got from solutions of the same concentration washed with chloroform.

By the modification above suggested Knoop's test can be obtained with certainty in solutions of histidine at a dilution of 1 : 10,000; a faint colour in pure solutions is still observable at 1 : 20,000.

The colour developed on heating is very variable according to the reaction of the liquid. In markedly acid solution it is a dull yellow brown. In approximately neutral solution, the shade is dark brown. In solutions made faintly alkaline with sodium hydroxide—after bromine has been added and the excess removed with chloroform—a bright pink colour develops which is less stable than that in neutral or acid solutions. If made very faintly alkaline with ammonia instead of caustic soda, a deep purple colour rapidly develops on heating. The presence of ammonia in just the requisite amount gives a more intense colour than in the other cases, but since it is difficult to regulate this amount, and further, since there is reason to believe that the specific value of the test may suffer—for carnosine gives a faint yellow by this treatment—the use of alkali cannot be recommended.

The colour developed from tryptophan by addition of bromine is readily distinguished from the colour due to histidine. The former develops in the cold and is easily extractable with amyl alcohol. The coloured substance in

Knoop's test is not extracted by any of the ordinary solvents and is peculiarly unaffected by reagents other than bromine.

SUMMARY.

By a modification of Knoop's test for histidine its delicacy is increased at least ten times and the test can be performed with greater certainty especially in coloured fluids. The proportion of three atoms of bromine per molecule of histidine is found to give the maximum colour on heating.

REFERENCE.

Knoop (1908). *Beiträge*, 11, 356.

LXVI. THE ESTIMATION OF CARNOSINE IN MUSCLE EXTRACT—A CRITICAL STUDY.

By GEORGE HUNTER, *Strang-Steel Scholar.*

Institute of Physiology, University of Glasgow.

(*Received July 17th, 1922.*)

In a recent communication [Hunter, 1921] a colorimetric method for the estimation of carnosine in muscle extract was generally outlined. This was based on a method of estimation of iminazoles devised by Koessler and Hanke [1919, 1]. This method depended on the coupling of carnosine with diazotised sulphanilic acid in alkaline solution to form an azo dye. The colour produced was measured against a standard solution of a mixture of Congo red and methyl-orange. Pure solutions of carnosine were found to react very satisfactorily, and in different extracts of the same muscles the colour values were consistent. The accuracy of the method could not be doubted, but the certainty that carnosine was responsible for the total colour production was called in question.

The problem—of the extent to which carnosine is responsible for the production of the azo colour in muscle extracts—may be attacked in two ways: 1. By the elimination of all interfering substances. 2. By a direct method of confirming the carnosine content.

I. THE ELIMINATION OF INTERFERING SUBSTANCES.

Potentially interfering substances. It is well known that aromatic amines and phenols readily couple with a diazotised aromatic amine and are thus potentially interfering substances with the reagent here employed for the estimation of carnosine.

The iminazoles form another well defined group—of which carnosine is only one member—giving the diazo reaction.

The purines which contain the iminazole ring are generally described as substances giving a positive diazo test—this applies at least to the members adenine, hypoxanthine, and xanthine [*v.* Plimmer, 1918].

There are various other substances diversely referred to throughout biochemical literature as giving the diazo test. Of these may be mentioned thymine [Thierfelder, 1908], bilirubin [Neubauer-Huppert, 1913], urochromogen and urobilinogen (Neubauer-Huppert), besides the "neutral-sulphur" compounds of urine (Neubauer-Huppert).

An exhaustive investigation into the behaviour of all substances—confined even to biochemical literature—towards the diazo reagent would probably be endless as well as fruitless. Attention will thus be confined only to those reagents ordinarily employed in the preparation of muscle aqueous extracts and to substances that may be present with a fair degree of probability in the extracts.

A. *Reagents.* Soluble chlorides, sulphates, nitrates, phosphates and sodium acetate, have been tested in relatively high concentrations and found to have no effect on the diazo reagent.

Ammonia and its salts not only interfere with the production of colour from other substances such as histidine [Koessler and Hanke, 1919, 2] but when added alone to the diazo reagent give a yellow colour.

Soluble sulphides give a colour in low concentration. The presence of sulphur in cystine probably accounts for the similar behaviour of that substance towards the diazo reagent.

Ethyl alcohol was found to give no colour with the reagent though Koessler and Hanke find that it inhibits colour production.

Hydrogen peroxide, formaldehyde and acetone all give yellow colours.

Tannic acid in very low concentration gives a marked colour. This is accounted for by its phenolic constituents.

All solutions in which carnosine is to be estimated by means of the diazo reagent should thus be free from ammonium salts, sulphides, and tannic acid. Formaldehyde or phenols, such as thymol, should not be used as preservatives in this connection. Muscle tissue may be preserved in alcohol and extracts by a layer of toluene.

B. *Muscle constituents.* Certain normal muscle constituents which may possibly act as interfering substances in the extracts require a more detailed examination, as these are much less within the control of the worker than substances which may be used in the preparation of the extracts.

The presence of bilirubin, urochromogen and urobilinogen in muscle extracts is too unlikely to claim further consideration for those substances. Thymine may also be dismissed on account of the difficulty with which it is liberated by hydrolysis from nucleic acid [v. Jones, 1920].

(1) *Phenols and aromatic amines.* The presence of these substances in protein-free extracts has not been shown by any of the tests used to detect them. Extracts from several types of muscle were shaken in acid solution with ether. The ether extracts were evaporated to dryness and the slight residues taken up in small quantities of water. These gave no diazo reaction nor has Millon's test been found positive either in the extracts themselves or in the ether fractions.

It is known that aromatic amines couple with the diazo reagent in acid solutions, whereas iminazoles and phenols require a weak alkaline medium before a colour is produced. Thus if a little 1. 2. 4-diaminotoluidine is added to the reagent without the addition of sodium carbonate, a strong orange colour is developed. If an iminazole or phenol is added under the same conditions no colour is produced[1].

Various extracts have been tested with the acid reagent but all with negative results. It may thus be concluded that there are no aromatic amines contributing to the azo colour developed in meat extracts[2].

(2) *Tyrosine.* This substance, which is not extractable by ether, has received considerable attention from the point of view of the diazo reaction. Totani [1915] devised a method to distinguish the azo colour developed by histidine from that developed by tyrosine. The colours were first reduced with zinc and hydrochloric acid, twice the volume of 25 % ammonia added and in both cases a golden yellow colour was obtained. In the case of histidine the addition of hydrogen peroxide changed the colour to lemon yellow whereas the colour was destroyed in the case of tyrosine. A dilution greater than 1 : 20,000 histidine was necessary to get the characteristic final colour.

Among the various substances considered by Totani no mention is made of iminazoles other than histidine, nor of purines, nor of aromatic amines nor of phenols other than tyrosine.

On repetition of Totani's procedure with histidine alongside carnosine, the two substances went through approximately the same colour changes. The final colour was not destroyed and so the presence of tyrosine could not be detected by this method.

Tyrosine gives a positive Millon's test; histidine and carnosine are negative to Millon's. Tyrosine gives a faint Millon's test at a dilution of 1 : 25,000.

A still more delicate test for tyrosine—on which is also founded a method of estimation—has recently been devised by Hanke and Koessler [1922]. This depends on a further modification of the diazo reaction and is approximately as delicate as that reaction is for carnosine. For quantitative purposes the procedure is the same as that adopted for iminazoles and phenols. The test cylinder is allowed to stand for exactly $5\frac{1}{2}$ minutes after the tyrosine solution has been added to the alkaline reagent. This gives rise to a primary yellow colour the intensity of which is not directly proportional to the amount of tyrosine used.

2 cc. of $3N$ sodium hydroxide solution are now added and the contents of the cylinder mixed. This gives rise to a colour intensification with a change of tint towards the red.

One minute after the addition of the sodium hydroxide 0·10 cc. of a 20 % solution of hydroxylamine hydrochloride is added and rapidly mixed. After

---

[1] The aromatic amines also give a colour in alkaline solutions.

[2] This test would also eliminate bilirubin which gives a blue colour in acid solution [Ehrlich, quoted by Neubauer-Huppert].

a latent period of 5 to 10 seconds an intense bluish red colour rapidly develops. This colour is stable and is directly proportional to the amount of tyrosine present.

Hanke and Koessler note that this intensification is also given by substances capable of a keto-enol tautomerism such as acetaldehyde, acetone and aceto-acetic acid.

It was decided to test this new modification on muscle extracts. It was first observed that the addition of sodium hydroxide and hydroxylamine hydrochloride in the manner above described to the azo colour developed by histidine or carnosine had the effect merely of a proportional dilution. Thus a solution of carnosine gave by the ordinary method a reading of 22·5 mm.; by the new method a reading of 18·3 mm., *i.e.* approximately the same as would be obtained by adding 2·1 cc. water to the cylinder. The following results were obtained on the extracts.

	Reading with diazo reagent mm.	Reading with diazo reagent, NaOH and $NH_4OH . HCl$ mm.	Calculated reading— assuming proportional dilution mm.	Intensification %
Ox muscle	27·2	21·4	21·6	None
Cat ,,	18·0	16·5	14·3	15·4
Rabbit muscle	10·8	11·3	8·5	33·0
Otter ,,	15·5	18·5	12·3	50·4
Salmon ,,	13·7	21·5	10·8	100·0

The percentage of intensification of these extracts is in the order of the yellowness of their azo colours. Thus ox muscle extracts give a colour which matches the carnosine colour standard. The colour developed from cat muscle extract is generally slightly more yellow; that from rabbit almost matches the histidine colour standard; that from otter is still more yellow and that from salmon is so yellow as to be almost unmatchable.

The intensification colours were pink. They did not show the purplish tinge given by tyrosine. No Millon's reaction was obtained in concentrated extracts of salmon muscle and it is concluded that tyrosine is not responsible for the intensification obtained in that tissue.

The specific cause of this intensification has not been determined.

(3) *Purines.* These are normally present in muscle extracts so that a consideration of the individual members is rendered necessary. Adenine and guanine were prepared from commercial plant nucleic acid. This was hydrolysed by suspending in methyl alcohol and passing dry hydrochloric acid gas according to the method employed by Levene [1921] for animal nucleic acid. This process was found to work very satisfactorily. The precipitated chlorides of adenine and guanine were filtered off and separated according to methods described by Jones [1920].

Hypoxanthine nitrate and xanthine were prepared from parts of the adenine and guanine respectively by deaminisation and subsequent purification.

Cytosine and uracil were also prepared from the residues freed from methyl alcohol by further hydrolysis in the autoclave at 160° with 25 % sulphuric acid for five hours according to Jones [1920].

Neither of these pyrimidines gave a colour with the diazo reagent.

0·05 % solutions were made of adenine sulphate, of guanine chloride, of hypoxanthine nitrate and of xanthine—all in 1·1 % sodium carbonate solutions.

1 cc. of each of these was tested in the ordinary way. Guanine and xanthine gave marked colours. The adenine solution showed only a slight reaction and there was no colour in the case of hypoxanthine. With a tenth of the above amounts guanine and xanthine were still strongly positive whilst adenine and hypoxanthine were entirely negative. The intensities of the colour were approximately the same in like concentrations of guanine and xanthine. With 0·025 mg. of guanine hydrochloride in the cylinder a reading of 8·5 mm. was obtained with the histidine colour standard. With 0·05 mg. the reading was 13 mm. Xanthine in the same amounts gave the respective readings 8 mm. and 12·5 mm.[1] The colour production is not directly proportional to the amounts of guanine and xanthine in the cylinder; nor do guanine and xanthine, mixed with a known amount of carnosine, give proportional colours. Thus 1 cc. of a mixture of 1 cc. 18·6 mm. per cc. carnosine solution with 1 cc. 13·5 mm. per cc. guanine solution gave a reading of 13·2 mm. with the test cylinder set at 20 mm. This is 2·8 mm. short of the calculated reading of 16 mm. A similar result was found with xanthine. Adenine and hypoxanthine do not give a colour or inhibit colour production at this concentration, but they tend to make the colour due to carnosine too yellow. Thus with 1 cc. of a mixture of carnosine solution of the above concentration with 1 cc. 0·05 mg. per cc. of adenine sulphate solution, the calculated reading of 9·3 mm. was obtained. Hypoxanthine behaves similarly.

Guanine and xanthine are about half as sensitive towards the diazo reagent as carnosine. If present in appreciable amounts in muscle extract they must seriously affect the estimation of carnosine.

Uric acid in excess gives a slight yellowness to the diazo reagent.

To what extent are purines likely to interfere with the estimation of carnosine in muscle extracts? On the assumption that there is present in extract of meat free purine nitrogen to the extent of 0·045 % as quoted by Lusk [1921] it is unlikely that more than 0·020 % nitrogen represents purines that give a colour with the diazo reagent. If the nitrogen be taken as representing 42 % of the purine molecule—an average figure from adenine, guanine, hypoxanthine, xanthine and uric acid—the purines affecting the diazo reagent amount to about 0·05 %. In muscles with a carnosine content of less than, say 0·1 % the presence of such a proportion of purines would make the results worthless.

[1] The readings are only approximate as the colours are much more yellow than the standard.

To determine the actual purine interference under the conditions of extraction previously recommended, it was then considered necessary to carry out some quantitative fractionations[1].

A quantity of ox muscle was extracted at 70° with small amounts of water until the filtrate no longer gave a positive diazo test. The proteins were precipitated with excess of lead acetate and the excess of lead was removed with disodium phosphate. The filtrate was neutralised with sodium hydroxide and its colour value measured with the carnosine colour standard. This was found to be 1050 mm. per cc.

50 cc. of the extract were taken in a small beaker, made just acid with nitric acid, and silver nitrate was added till a drop of the solution on a tile gave a brown colour with baryta. The test drops were washed back into the beaker.

After settling, the contents of the beaker were filtered by suction through washed asbestos in a small Hirsch funnel. The precipitate was carefully washed with water and after sucking dry was transferred with the asbestos to a small beaker, stirred up with a little water and hydrogen sulphide passed to precipitate the silver. The contents were again filtered through asbestos in the same way and the hydrogen sulphide removed by prolonged aeration with the help of a pump. The filtrate was neutralised. This is the purine fraction.

To the filtrate from the purine fraction excess of solid finely ground baryta was added. The precipitate was allowed to settle, filtered as above, and washed with saturated baryta water. The silver was removed with hydrogen sulphide and the barium with slight excess of sulphuric acid. The slightly acid solution was aerated to remove the hydrogen sulphide and the solution was then neutralised with sodium hydroxide. This is the carnosine fraction.

The filtrate from the carnosine fraction was freed from barium and hydrogen sulphide and neutralised as above. This is the final filtrate.

The exact volume of each of the three fractions was noted. The colour value per cc. of each of the fractions was then measured and their total colour values calculated. The sum total colour value was then obtained and compared with the original, i.e. 50 × 1050 or 52,500 mm.

---

[1] All the carnosine can be extracted from muscle with much smaller proportions of water than recommended by the writer in his previous paper on this subject. A filtrate of about 50 cc. per g. of muscle used was there recommended mainly because the proteins are easily precipitated by slight acidification with acetic acid and heat under those conditions. In concentrated extracts this simple process is not effective and the use of lead acetate or other precipitant is necessitated. The number of washings, however, should not be reduced. There is no need to reduce the proportion of water to muscle when the extract is to be tested only for colour value unless the carnosine content is less than 0·1 %. With that amount in muscle a reading of 12 mm. will still be obtained with the 50 cc. per 1 g. proportion. But for precipitation purposes and other tests it is desirable to keep the filtrate small rather than have to evaporate it to a low volume.

The distribution of colour value is shown by the total colour values in mm. of the three fractions:

Purine fraction	...		1,560
Carnosine fraction	...		47,587
Final filtrate	...	...	2,200
Sum total	...	...	51,347

The unrecovered colour value is thus 1153 mm. or approximately 2 %. The purine fraction accounts for about 3 % and the final filtrate for about 4·5 %. The carnosine fraction represents almost 91 % of the original colour value.

The relatively small error due to purines by the gentle method of extraction recommended is further seen in the table to follow.

Three portions were taken from the same muscle and two extracted as above. The third was extracted at a temperature just under boiling point.

Aliquot portions of the filtrates from the lead precipitates were taken for total nitrogen estimation and for further fractionation into purine and carnosine fractions. The final filtrates were rejected without evaluation. The total nitrogen was also estimated in the carnosine fractions. In the table carnosine, nitrogen and purine (calculated as carnosine) are shown as percentages of the original muscle. The colour values of the purine fraction are at best only a rough estimate on account of the difficulty in comparing with a much redder standard. The error due to the purines in ox muscle is however small.

Meat No.	Carnosine from lead acetate filtrate	Total N from lead acetate fraction	Carnosine from carnosine fraction	Total N from carnosine fraction	Colour value in purine fraction	Difference in % carnosine in lead acetate filtrate and in carnosine fraction
1	0·517	0·530	0·498	0·246	0·004	0·019
2	0·512	0·546	0·495	0·230	0·012	0·017
3	0·533	0·653	0·487	0·258	0·014	0·046

Before considering other substances likely to interfere with the estimation it may be noted that the more drastic extraction process of muscle No. 3 results in a higher percentage colour value. The total nitrogen in this extract is also raised. In the carnosine fraction the percentage falls again into line with the other two samples. The loss is not accounted for by a higher purine value. Though the colour value of the lead acetate filtrate is raised by the more rigorous extraction, the actual percentage of carnosine is lowered. The lowering of colour value becomes obvious even in the lead acetate filtrate if boiling is prolonged. Thus two samples of muscle treated in the usual way showed an average of 0·639 % carnosine in the lead acetate filtrate, whilst three samples of the same muscle after boiling for about 30 minutes showed an average value of only 0·614 % or a fall of 0·025 % reckoned as carnosine.

In muscles 1 and 2 approximately 96 % of the colour is recovered in the carnosine fraction. Calculated as carnosine this accounts for about 50 % of the nitrogen in the same fraction. There is no reason to doubt that carnosine is responsible for all the colour production in the carnosine fraction from ox muscle. In the case of rabbit muscle and notably salmon muscle, fractionation does not appreciably improve the match. Thus two portions of a rabbit muscle showed from the lead acetate filtrate colour values of 0·092 % and 0·099 % reckoned as carnosine. The carnosine fraction from the same muscles showed each a value of 0·072 % as carnosine. In the latter case the azo colour developed was still very yellow. The colour from salmon muscle extract is slightly more red in the carnosine fraction than in the lead acetate filtrate, yet it is far from satisfactory.

The precipitation process occupies much time and to obtain reliable quantitative results from it, great care must be exercised. With the additional possibility that the carnosine fraction may still contain other colour-giving substances, it would appear of little use to attempt the elimination of interfering substances by the silver-baryta precipitation except in very special cases.

(4) *Iminazoles.* Of this group only histidine and carnosine are known with certainty to occur in animal tissue. Histamine has been found in the intestinal mucosa by Barger and Dale where its presence is attributed to bacterial action [Barger, 1914]. Its isolation from the pituitary gland by Abel is accounted for by Hanke and Koessler through Abel's use of a commercial product. Hanke and Koessler [1920, 2] show that it is absent from fresh hypophysis cerebri.

In protein-free extract from fresh muscle there is no evidence that histidine is present provided the extraction process has been gentle. The ease with which carnosine may be hydrolysed (*v.* later), along with other factors to be considered, renders it important that there should be some ready means of detecting histidine in the presence of carnosine.

As far as the writer is aware there is no known method of separating these substances. They both precipitate in the carnosine fraction in the "arginine separation." Though histidine is more readily precipitated with ammonium-silver than carnosine, yet the latter also precipitates in such relatively dilute solutions as 0·5 % carnosine. Nor could any test be found in the literature to distinguish the two substances.

Various colour tests for histidine were performed on solutions of carnosine. The main differences were found in their behaviour towards the biuret and Knoop tests.

The biuret reaction was found to be entirely negative even in about 30 % solution of carnosine and after standing at least one hour. This difference in behaviour is however of little value for discrimination purposes as relatively concentrated solutions of histidine must be used to get a positive biuret. Thus 1 % solution of histidine monohydrochloride gives the reaction in the cold only after standing for about 1½ hours.

The writer has found Knoop's test to be the most useful way of detecting histidine in the presence of carnosine. Carnosine is quite negative to Knoop's test. By a modification of the test as explained in the preceding paper, the writer has been able to increase both its delicacy and certainty as a test for histidine. The presence of histidine may be detected with certainty at a dilution of 1 : 10,000.

Though this reaction has not yet the delicacy that might be desired, its very specific nature makes it valuable for work of this kind. Knoop [1908] states that the reaction is also positive for histamine. No other substance, as far as the writer is aware, gives the reaction.

Certain muscle extracts suspected of histidine, were then tested by the modified Knoop's test. Extracts of rabbit and salmon which give notably yellow colours were tested in sufficient concentration but without giving an indubitable Knoop's test. It is thus concluded that histidine is not a factor in causing the yellow colour of those extracts.

In various cases of purchased butcher's meat a positive Knoop's test has been obtained. If the meat is allowed to become just noticeably putrid a very marked reaction is given. Histamine may in part be responsible for the positive test in such cases.

The test may perhaps be used to most advantage as an indirect test for carnosine where the presence of that substance is doubtful in any tissue or fluid. If a protein-free extract is at first negative to Knoop's test, and after hydrolysis (v. later) is positive to the test, it appears necessarily to follow that the unhydrolysed liquid contains carnosine.

The test can also be applied to elucidate some other problems that had arisen in the course of this work. The effect of heat on carnosine solutions will first be considered.

The effect of heat on solutions of carnosine.

In the writer's preliminary communication on the estimation of carnosine, a slight fall was noted in the colour values of solutions of carnosine heated on the water-bath for one hour. Similar experiments have since been repeated but the period of heating has been extended to four hours. Thus a number of solutions of carnosine nitrate with an original colour value of 22 mm. per cc. showed after four hours on a boiling water-bath an average value of 19·4 mm. per cc. or a fall of 2·6 mm. per cc.

10 cc. of a 0·5 % solution of carnosine as nitrate was brought to dryness three times in an open basin on a steam-bath. The total period of heating was about four hours. It was finally taken up in about 2 cc. of water and Knoop's test applied. A decided brown colour was developed indicating that in the process the carnosine had been partly hydrolysed, giving rise to histidine. This observation confirms the fall in colour value.

Though four hours is a relatively long period of heating, the fact that carnosine is thus destroyed in pure neutral solution indicates that care should be exercised in the process of extraction from muscle.

The fall in colour value as estimated in carnosine from 0·639 % to 0·614 %

on prolonged extraction of the muscle, previously noted (p. 646), may be accounted for by the destruction of carnosine. The occurrence of histidine in muscle extracts will thus arise more likely from the hydrolysis of carnosine than from the hydrolysis of histidine-containing proteins. For if the latter process occurred to a marked degree the colour values would tend to rise rather than to fall.

It will thus be observed that any process of evaporation at atmospheric pressure must result in at least a partial destruction of the substance.

Solutions of carnosine whether pure or in muscle extracts may be evaporated *in vacuo* to dryness with very little destruction. Thus 10 cc. of a neutral dilute meat extract were evaporated to dryness at 20 mm. pressure of mercury from a 200 cc. distilling flask on a water-bath about 60°. Exactly 10 cc. of water were then pipetted into the flask and rinsed round to dissolve the dried residue. The colour value of the original solution was 25·6 mm. per cc. whilst that of the dried substance taken up in 10 cc. of water was 25 mm. per cc. (The volume of solid in the original 10 cc. was regarded as negligible.) Repetition of this experiment still showed a loss in colour value of about 2 %.

This experiment at the same time goes to show that a very small proportion, if any, of the colour-producing substances in muscle extract is volatile. It serves to confirm the results obtained by ether extraction and Millon's test.

(5) *Other substances.* The effects of cystine, leucine and arginine on the diazo reagent have been measured by Hanke and Koessler [1920, 1]. These workers find that there is no interference so long as the ratio of cystine to histidine does not exceed 6 : 1, which is higher, they note, than has yet been found in any protein. Cystine may thus be dismissed as a probable interfering factor in muscle extracts.

The effects of arginine and leucine are less marked than those of cystine.

Creatine in large amounts gives a slight yellowness to the reagent.

Creatinine, urea and lactic acid have been found to have no effect on the reagent.

Such a process of elimination as has here been attempted must yield mainly negative conclusions. None of the substances most likely to interfere with the colour values has any very marked effect. In the case of ox muscle it has been shown that most probably less than 5 % of the colour value is not due to carnosine. In cat and frog extracts, judging merely from the colour development, the non-carnosine colour value is likely to lie within the same limit, but of salmon little can be said except that it contains carnosine. Any elimination method must necessarily be unsatisfactory unless the substance can be obtained pure; and many factors militate against that possibility in the case of carnosine. Among those may be noted the high solubility of the substance in water, its facility for adsorption, the inadequacy of the methods for its fractional precipitation and the ease with which it is hydrolysed. It is upon this last property that the writer has sought a more direct method for confirming the values.

II. HYDROLYSIS EXPERIMENTS.

(1) *The stability of histidine.* Histidine has been observed by Koessler and Hanke [1919, 2] to be remarkably stable towards hot concentrated hydrochloric acid. They observed that when 2 cc. of a 1 % solution of histidine dihydrochloride were heated on a boiling water-bath for 10 hours with 25 cc. concentrated hydrochloric acid the histidine was quite unchanged, as shown by the colour values recovered. Over a shorter interval the writer has confirmed those findings as shown by the following experiment.

Into each of five numbered test tubes there was introduced 1 cc. of a solution of histidine along with 3 cc. concentrated hydrochloric acid. These were then set on a boiling water-bath and one removed every 15 minutes. As each test tube was removed from the water-bath it was cooled, about 1 sq. mm. of litmus paper introduced, and the contents neutralised with sodium hydroxide. The contents were again cooled and poured into a 25 cc. flask. The test tube was repeatedly washed into the measuring flask, then water was added up to the mark. 1 cc. of this was taken for measuring the colour value against the histidine colour standard. The original value of 23 mm. was obtained by directly diluting 1 cc. of the original to 25 cc. and taking 1 cc. for the test. The colour values after periods of boiling lasting 15, 30, 45, 60 and 75 minutes were found respectively to be 23·2, 23·1, 23, 22·9 and 23 mm. High concentrations of sodium chloride affect neither the tint nor the intensity of the colour developed.

(2) *Hydrolysis of carnosine—theoretical considerations.* It thus seemed very probable that if carnosine were hydrolysed with hydrochloric acid, the colour value would continue to fall until the reaction was complete. If the diazo value of the solution were measured before hydrolysis and again after hydrolysis, the ratio of the colour values thus obtained should be that of their molecular colour values. The molecular colour value of histidine was previously found to be 117 million mm. as compared with 114 million mm. determined by Koessler and Hanke. A revised determination of the molecular colour value of carnosine has led the writer—mainly from increased skill in matching the histidine and carnosine colour standards—to the conclusion that the molecular colour value of carnosine is somewhat greater than the 152 million mm. previously published, viz. 161 million mm. This value is checked by the following considerations.

The histidine colour standard is a mixture of 0·22 cc. methyl-orange with 0·20 cc. Congo red in 100 cc. water. For convenience call this S_H.

The carnosine colour standard is a mixture of 0·10 cc. methyl-orange with 0·25 cc. Congo red in 100 cc. water. Call this S_C.

These amounts were taken from 0·1 % stock methyl-orange and from 0·5 % stock Congo red [v. Hunter, 1921].

On comparing in the colorimeter it is found that 24 mm. S_C = 28·5 mm. S_H.

From the table previously published [Hunter, 1921] 0·040 mg. of carnosine in the test cylinder set at 20 mm. showed a reading of 24 mm. S_C.

0·04 mg. of carnosine should thus give a reading of 28·5 mm. S_H. $_{1n'}$

The molecular colour value of carnosine is thus

$$\frac{28·5}{4} \times 226 \times 10^5,$$

or approximately 161 million mm. S_H.

Now, carnosine yields by theory on hydrolysis 68·5 % of histidine, *i.e.* 0·04 mg. of carnosine yields on hydrolysis 0·0274 mg. of histidine.

0·04 mg. of histidine monohydrochloride gives a reading of 22·4 mm. S_H.

[N.B. The molecular colour value of histidine is thus

$$\frac{22·4}{4} \times 209 \times 10^5,$$

i.e. approximately 117 million mm. S_H.]

Histidine monohydrochloride contains 74 % histidine, *i.e.* 0·04 mg. of histidine monohydrochloride contains 0·0296 mg. of histidine.

0·01 mg. of histidine will thus give by calculation a reading of $\frac{22·4}{2·96}$ or 7·57 mm. S_H, and

0·0274 mg. of histidine will give 20·7 mm. S_H.

That is, a solution of carnosine giving an original reading of 28·5 mm. S_H will give after hydrolysis a reading of 20·7 mm. S_H.

The ratio of the final reading to the original reading should be the same as the ratio of the molecular colour value of histidine to the molecular colour value of carnosine.

Thus $$\frac{20·7}{28·5} = \frac{117}{161} = 0·73.$$

For convenience this figure will be termed the *hydrolysis quotient*.

(3) *Experimental.* The results obtained from the actual hydrolysis of carnosine solutions are less satisfactory than one might expect from the above considerations. Thus a series of carnosine solutions were treated in exactly the same way as that described for histidine. 0·1 % solution of carnosine was used. The original value of 1 cc. from 25 cc. dilution was 28·5 mm. S_H. After hydrolysis for periods of 15, 30, 45, 60 and 75 minutes 1 cc. showed the respective colour values 20·4, 19, 18·2, 18·1 and 18 mm. S_H. After 15 minutes' hydrolysis the colour was more yellow than S_H.

Carnosine is thus very readily hydrolysed and under the above conditions part of the histidine is destroyed. After the initial sudden fall from 28·5 mm. to 20·4 mm. continued boiling lowers the value at a much diminished rate. After 30 minutes' boiling the values remain almost constant but the colours are too yellow—indicating some destruction. It would appear that in the course of the hydrolysis the histidine passes through an unstable phase in which the iminazole ring is readily disrupted. The histidine that emerges from that hypothetical intermediate condition resists further boiling.

Other hydrolytic agents, such as sodium hydroxide and acetic acid, besides different conditions of temperature and concentration were tested, in .the attempt to overcome the difficulty. Finally an approximately constant-boiling mixture of hydrochloric acid was used and the hydrolysis conducted at 90°. Under these conditions a strictly quantitative conversion is not yet attained, but a comparison of the results from various extracts under standard conditions would appear to be of some weight. The extracts hydrolysed were of such a concentration that 1 cc. of the final dilution gave approximately the same values. To the amount of extract to be hydrolysed exactly half its volume of concentrated hydrochloric acid was added. The following are the results from a solution of pure carnosine and from muscle extracts of cat, of ox, and of salmon. This ox extract was slightly positive to Knoop's test. All the readings were taken with the histidine standard.

Time mins.	Carnosine mm.	Cat muscle extract mm.	Ox muscle extract mm.	Salmon muscle extract mm.
Original	28·5	29·0	28·5	27·0
15	25·4	24·3	26·0	26·5
30	23·3	23·0	23·5	24·0
45	21·5	21·8	22·7	22·5
60	21·0	21·2	22·0	22·1
75	20·9	20·7	21·7	21·4
Hydrolysis quotients after 75 mins.	0·73	0·71	0·76	0·79

The carnosine solution and the cat muscle extract agree very well in both their rate of fall and in their hydrolysis quotients. The ox muscle quotient is slightly high, probably owing to the presence of either histidine or histamine in the original. The salmon extract quotient is still higher, but it is remarkable in face of the very yellow azo colour given by the original that the rate of fall accords so well with the others.

Although this method of hydrolysis is not sufficiently sensitive to show that a definite percentage of the colour is due to carnosine, it at least gives one a sense of assurance in the use of a very unspecific reagent for the estimation of carnosine in such a complex solution as muscle extract. It would further seem to indicate that carnosine is responsible for a very high percentage of the colour as measured from fresh muscle extracts treated only with lead acetate.

The specific cause of the yellow colour produced in such cases as salmon has not been found. In the case of rabbit muscle, it has been observed that the colour is more red when the carnosine content of the muscle is high and yellow when low. Some cat and frog muscles with a low carnosine content also showed a yellow colour. This is not surprising as the chances of interference are greatly increased when the test portions are less dilute.

Though purines certainly give rise to a small error and exact quantitative results cannot be got when the colour is not of the right tint, yet the total error in the lead acetate filtrate is too small to eliminate. With a better hydrolytic agent it might yet be possible to evaluate the error.

Carnosine Content of some Muscles.

With the method of extraction and treatment of the extract as previously described, the following results have been obtained. Each result represents the percentage colour value reckoned as carnosine in the fresh skeletal muscle. Further, each result represents the average of at least two results obtained from different pieces of the same muscle.

For convenience the amounts found are given from lowest to highest carnosine contents. The contents for four members of each species of animal are given.

Rabbit muscles	0·026	0·064	0·090	0·101 % carnosine
Frog ,,	0·107	0·128	0·142	0·280 ,, ,,
Cat ,,	0·123	0·203	0·336	0·380 ,, ,,
Ox ,,	0·340	0·400	0·515	0·640 ,, ,,

The results show that the carnosine content of muscle varies not only with the species of animal but varies greatly in different animals of the same species. The highest values obtained in the one species are two to four times greater than the lowest values in the same species. This finding is at variance with that of Clifford [1921, 2]. This worker finds, for example, ox muscle to have an almost constant carnosine content of 1·1 %. But apart from the constancy of the results found, the writer is compelled to question their accuracy. In a previous paper [1921, 1] Clifford finds that 0·02 mg. of histidine gives a value of 10·75 mm. measured with the Koessler and Hanke histidine standard. With these data—and assuming the author means 0·02 mg. of histidine *dihydrochloride*—the molecular colour value of the histidine is

$$\frac{10\cdot75}{2} \times 228 \times 10^5,$$

which is approximately 122 million mm. S_H.

With the same colour standard Clifford records that 0·1 mg. of carnosine gives a reading of 30 mm. The molecular colour value of Clifford's carnosine is thus $30 \times 226 \times 10^4$ or approximately 68 million mm. As previously stated, the writer finds the molecular colour value of carnosine against this same colour standard to be 161 million mm. Assuming this figure represents 100 % pure carnosine, the carnosine employed by Clifford is thus only about 43 % pure. The carnosine content of the various muscles tested by Clifford should from this point of view be about 43 % of the values actually given.

Summary.

1. Extracts to be tested for carnosine should be free from ammonium salts, sulphides, phenols and aldehydes.

2. The degree of yellowness of the azo colours developed from muscle extracts is in the order of the percentages of the intensification of the azo colours as given by sodium hydroxide and hydroxylamine.

3. In ox muscle purines are responsible for about 3 % of the colour value reckoned as carnosine. In the same tissue there is probably about other

2 % of the colour not due to carnosine. The error in the method due to colour-producing substances other than carnosine is probably about 5 % in ordinary muscles.

4. Histidine under certain circumstances may be present in muscle extracts.

5. A test has been found to distinguish histidine from carnosine.

6. A more certain means than the diazo reagent is given for the identification of carnosine in any tissue.

7. The fall in colour value of hydrolysed meat extracts agrees with that of pure solutions of carnosine.

8. The carnosine content of muscle varies with the species of animal and with different members of the same species.

The writer is much indebted to Prof. Cathcart for his inspiring guidance throughout this work.

REFERENCES.

Barger (1914). Simple Natural Bases.
Clifford (1921, 1). *Biochem. J.* **15**, 400.
Clifford (1921, 2). *Biochem. J.* **15**, 725.
Hanke and Koessler (1920, 1). *J. Biol. Chem.* **43**, 527.
——— ——— (1920, 2). *J. Biol. Chem.* **43**, 557.
——— ——— (1922). *J. Biol. Chem.* **50**, 235.
Hunter (1921). *Biochem. J.* **15**, 689.
Jones (1920). Nucleic Acids.
Knoop (1908). *Beiträge*, **11**, 356.
Koessler and Hanke (1919, 1). *J. Biol. Chem.* **39**, 497.
——— ——— (1919, 2). *J. Biol. Chem.* **39**, 521.
Levene (1921). *J. Biol. Chem.* **48**, 177.
Lusk (1921). Science of Nutrition.
Neubauer-Huppert (1913). Analyse des Harns.
Plimmer (1918). Practical Organic and Biochemistry, 293–296.
Thierfelder (1908). Chemische Analyse, 164.
Totani (1915). *Biochem. J.* **9**, 385.

LXVII. IDENTIFICATION OF INULIN BY A MYCOLOGICAL METHOD.

By ALDO CASTELLANI AND FRANK E. TAYLOR.

From the London School of Tropical Medicine and King's College, University of London.

(Received June 1st, 1922.)

IN previous publications [1917, 1919, 1920] we have described a general mycological method, theoretically devised by one of us (C.) some years ago in Ceylon, which we have found useful in the identification of various carbohydrates and other carbon compounds. We propose in the present paper to describe briefly how this method can be applied to the determination of inulin.

It is generally stated that there is no organism which induces a complete fermentation of inulin, that is to say, fermentation with production of gas, but one of us (C.) has found a fungus which causes a complete fermentation of this carbohydrate with large production of gas. This fungus is *Monilia macedoniensis* Castellani and allied species, which ferment with production of gas in addition to inulin the following carbohydrates: glucose, levulose, galactose and saccharose.

By means of this fungus in conjunction with certain other fungi, it is possible to identify inulin, using a modification of the general mycological method we described some time ago for the identification of various sugars.

Technique. Let us suppose we have a substance about which we want to decide whether it is inulin or not. A sterile 1 % solution in sugar free peptone water is made and distributed into two tubes, No. 1 and No. 2, each containing a Durham's fermentation tube or similar appliance. The following procedure is then used:

(a) No. 1 tube is inoculated with *Monilia macedoniensis* Cast., No. 2 with *Monilia tropicalis* Cast. The two tubes are placed in an incubator at 35–37° for 72 hours. If after that time, No. 1 tube contains gas and No. 2 tube does not, we can come to the conclusion that the substance is inulin. This is easily understood by keeping in mind the fermentative reactions of the two monilias: *Monilia macedoniensis* ferments with production of gas, only the following carbon compounds: glucose, levulose, galactose, saccharose and inulin. *Monilia tropicalis* Cast. ferments with production of gas, only glucose, levulose, maltose, galactose and saccharose.

$$\left.\begin{array}{ll} \text{\textit{Monilia macedoniensis} Cast.} & + \\ \text{\textit{Monilia tropicalis} Cast.} & 0 \end{array}\right\} = \text{Inulin.}$$

43—2

(b) No. 1 tube is inoculated with *Monilia macedoniensis* Cast.; No. 2 with *Monilia rhoi* Cast. The two tubes are placed in an incubator at 35–37° for 72 hours. If after that time No. 1 tube contains gas and No. 2 does not we can come to the conclusion that the substance is inulin. This is easily understood remembering that *Monilia macedoniensis* ferments with production of gas, only glucose, levulose, galactose, saccharose and inulin, and *Monilia rhoi* ferments with production of gas, only glucose, levulose, galactose and saccharose.

$$\left. \begin{array}{l} \textit{Monilia macedoniensis} \text{ Cast. } + \\ \textit{Monilia rhoi} \text{ Cast.} \quad\quad 0 \end{array} \right\} = \text{Inulin.}$$

(c) No. 1 tube is inoculated with *Monilia macedoniensis*; No. 2 with *B. pseudocoli* or *B. neapolitanus*, or any other strain of the *communior* group of *B. coli* (ferment saccharose). The tubes are incubated at 37° for four days. If then tube No. 1 contains gas and tube No. 2 does not, we can again come to the conclusion that the substance is inulin, since glucose, levulose, galactose or saccharose would have been fermented also by *B. pseudocoli* or *B. neapolitanus* or any other strain of the *Coli communior* group.

$$\left. \begin{array}{l} \textit{Monilia macedoniensis} \text{ Cast.} \quad\quad\quad + \\ \textit{B. coli communior (B. pseudocoli} \text{ Cast.}, \\ \textit{B. neapolitanus} \text{ Emmerich, etc.)} \quad 0 \end{array} \right\} = \text{Inulin.}$$

(d) No. 1 tube is inoculated with *M. macedoniensis* Cast., No. 2 tube with *B. asiaticus* Cast. The two tubes are placed in an incubator at 37° for four days. If after that time No. 1 tube contains gas and No. 2 does not, we can come to the conclusion that the substance according to all probabilities is inulin. This is easily understood by remembering the fermentative reactions of the two organisms. *M. macedoniensis* ferments only glucose, levulose, galactose, saccharose and inulin with production of gas; whilst glucose, levulose, galactose and saccharose are also fermented by *B. asiaticus*; it must therefore be inulin.

$$\left. \begin{array}{l} \textit{Monilia macedoniensis} \text{ Cast. } + \\ \textit{B. asiaticus} \text{ Cast.} \quad\quad\quad 0 \end{array} \right\} = \text{Inulin.}$$

IDENTIFICATION OF INULIN WHEN PRESENT WITH SOME OF THE MORE COMMON FERMENTABLE SUBSTANCES.

If we suspect that a liquid contains inulin mixed with some of the more usual fermentable substances such as glucose, levulose, maltose, etc., we can find out the presence of inulin in the following manner. The mixture is fermented with *Monilia tropicalis* Cast. If, after exhaustion with *M. tropicalis*, the liquid can still be fermented with *M. macedoniensis* with production of gas, the inference is that the liquid contained inulin. Of course, the precaution should be taken of selecting strains of *M. tropicalis* and *M. macedoniensis* with approximately equal fermentative power on glucose, levulose, galactose and saccharose, which carbohydrates they both ferment.

Addendum.

For the reader's convenience we annex a table containing the fermentative characters of the various fungi and bacteria we use in our method, and we give also a list of the principal mycological formulae which we have devised and employed in the identification of various sugars and other carbon compounds. It is essential to use strains with permanent biochemical reactions. Acid fermentation without production of gas is not taken into account.

Table showing fermentation reactions of certain fungi and bacteria.

	Glucose	Levulose	Maltose	Galactose	Saccharose	Lactose	Mannitol	Dulcitol	Dextrin	Raffinose	Arabinose	Adonitol	Inulin	Sorbitol	Starch	Glycerol	Inositol	Salicine	Amygdalin	Isodulcitol
Monilia baleanica Cast.	G	0	0	0	0	0	0		0	0	0		0			0	0	0	0	0
M. Krusei Cast.	G	G	0	0	0	0	0		0	0	0		0			0	0	0	0	0
M. pinoyi Cast.	G	G	G	0	0	0	0		0	0	0		0			0	0	0	0	0
M. metalondinensis Cast.	G	G	G	G	0	0	0	**0**	0	0	0	**0**	0	**0**	**0**	0	0	0	0	0
M. tropicalis Cast.	G	G	G	G	G	0	0		0	0	0		0			0	0	0	0	0
M. rhoi Cast.	G	G		G	G	0	0		0	0	0		0			0	0	0	0	0
M. macedoniensis Cast.	G	G	**0**	G	G	0	0		0	0	0		G			0	0	0	0	0
Bacillus coli Escherich	G	G		G	0	G		G		G	G		0		0	G	0	G	0	G
B. pseudocoli Cast.	G	G		G	G	G		G		G	G		0		0	G	0	G	0	G
B. paratyphosus B var. M	G	G	**G**	G	0	0		G	G	0	G	**0**	0		0	G	0	G		G
B. paratyphosus A Schottmüller	G	G		G	0	0		G		0	G		0		0	0	0	0	0	G
B. asiaticus Cast.	G	G		G	G	0		G	0	G	G		0	G	0	G	0	0	0	G
B. pseudoasiaticus Cast.	G	G		G	G	0	**G**	G	G	G	G		0	**G**	0	G	0	G	0	G

G = gas; 0 = absence of gas. Simple acid fermentation is not taken into account.

Mycological Formulae.

Inulin.

Monilia macedoniensis Cast.	+ | = Inulin
M. tropicalis Cast.	0 |

M. macedoniensis Cast.	+ | = Inulin
M. rhoi Cast.	0 |

M. macedoniensis Cast.	+ | = Inulin
Bacillus coli communior (B. pseudocoli, B. neapolitanus)	0 |

M. macedoniensis Cast.	+ | = Inulin
B. asiaticus Cast.	0 |

Maltose.

M. tropicalis Cast.	+ | = Maltose
M. macedoniensis Cast.	0 |

M. metalondinensis Cast.	+ | = Maltose
M. macedoniensis Cast.	0 |

M. pinoyi Cast.	+ | = Maltose
M. krusei Cast.	0 |

M. pinoyi Cast.	+ | = Maltose
M. macedoniensis Cast.	0 |

Galactose.

M. metalondinensis Cast.	+ | = Galactose
M. pinoyi Cast.	0 |

M. metalondinensis Cast.	+ |
M. krusei Cast.	0 | = Galactose
M. macedoniensis Cast.	+ |

Galactose (continued).

M. tropicalis Cast.	...	...	...	...	...	+	= Galactose
M. bronchialis Cast.	...	...	...	...		0	

M. tropicalis Cast.	...	...	...	...	...	+	
M. macedoniensis Cast.	...	...	...	...		+	
M. krusei Cast.	...	...	...	...	...	0	= Galactose
B. paratyphosus B Schottmüller	...	...	...		+	

Saccharose.

M. tropicalis Cast.	...	...	...	...	...	+	= Saccharose
M. metalondinensis Cast. ...	...	...	...	...	0	

M. rhoi Cast.	...	...	...	...	...	+	= Saccharose
M. pinoyi Cast.	...	...	...	...	...	0	

M. tropicalis Cast.	...	...	...	...	...	+	= Saccharose
B. coli communis (sensu stricto) ...	...	...	...	0	

M. tropicalis Cast.	...	...	...	...	...	+	= Saccharose
B. paratyphosus B Schottmüller	...	...	...	0	

M. macedoniensis Cast.	...	...	...	...	...	+	
B. coli communis (sensu stricto) ...	...	...	...	0	= Saccharose
B. coli communior ...	...	...	...	...	...	+	

M. macedoniensis Cast.	...	...	...	...	...	+	
B. paratyphosus B Schottmüller	...	...	...	0	= Saccharose
B. coli communior ...	...	...	...	...	...	+	

B. coli communis Escherich (sensu stricto)	...	...	0	= Saccharose
B. neapolitanus Emmerich	...	...	...	...	+	

B. coli communis Escherich (sensu stricto)	...	...	0	= Saccharose
B. asiaticus	...	...	...	...	...	+	

Levulose.

M. krusei Cast.	...	...	...	...	...	+	= Levulose
M. pinoyi Cast.	...	...	...	...	...	0	

Glucose.

M. baleanica Cast.	...	...	...	...	...	+	= Glucose
M. krusei Cast.	...	...	...	...	...	0	

Inositol.

B. paratyphosus B var. M Schottmüller	...	...	+	= Inositol
B. paratyphosus A Schottmüller	...	...	...	0	

CHEMICO-MYCOLOGICAL FORMULAE.

Saccharose.

Fehling	...	...	...	...	...	0	= Saccharose
M. tropicalis Cast.	...	...	...	...	+	

Lactose.

Fehling	...	...	...	...	...	+	
B. paratyphosus B Schottmüller	...	...	...	0	= Lactose
B. coli communis Escherich	...	...	...	+	

Pentose.

Fehling	...	...	...	...	...	+	
M. tropicalis Cast.	...	...	...	...	0	Pentose
B. paratyphosus B Schottmüller	...	...	...	+	= (generally arabinose)
B. coli communis Escherich	...	...	...	+	

+ = gas; 0 = no gas; simple acid fermentation is not taken into account.

REFERENCES.

Castellani (1920). *Lancet,* i, 847
Castellani and Taylor (1917). *Brit. Med. J.* ii, 855.
—— —— (1919). *Brit. Med. J.* i, 183.

LXVIII. FEEDING EXPERIMENTS IN CONNECTION WITH VITAMINS *A* AND *B*.

III. MILK AND THE GROWTH-PROMOTING VITAMIN.

IV. THE VITAMIN *A* CONTENT OF REFINED COD-LIVER OIL.

By ARTHUR DIGHTON STAMMERS.

Work carried out at the Research Laboratory, Port Sunlight.

(*Received July 10th, 1922.*)

III.

DURING 1920 Osborne and Mendel [1920] published a paper entitled "Milk as a Source of Water-soluble Vitamin," in which they state that they are unable to produce results similar to those obtained by Hopkins in his classical experiments published in 1912 [Hopkins, 1912].

The administration of 2 cc. of milk per day in these experiments was sufficient to provide the accessory factors for growing rats, but the American authors did not obtain anything approaching normal growth until at least 16 cc. per day were supplied.

Osborne and Mendel's experiments included the use of winter milk (which was probably deficient in vitamins owing to stall-feeding of the cattle) and also milk from grass-fed animals. In each case comparable results were obtained. The authors also considered the possibility of an unsuspected inorganic substance in Hopkins' salt mixture, but experiments showed that this was not the cause of the discrepancy.

Subsequent to this, Hopkins [1920] repeated his experiments and confirmed his previous findings, without, however, being able to throw any light on the discrepancies of the American investigators. He suggests that a seasonal variation in the food value of milk may be a contributing cause, although an experiment upon goats' milk, secreted on winter and summer diets, planned to elucidate this point, failed to yield much evidence. He also suggested that there was a seasonal variation in the growth energy of rats.

In view of these conflicting results, the experiment now to be described may be of some interest, although the circumstances under which it was carried out were not comparable (as regards age of animals) to those in the experiments of Hopkins and of Osborne and Mendel referred to above.

The chief point of difference was that the animals used in the present work were the survivors from a previous experiment which had already lasted for 101 days, the results of which showed conclusively that the materials tested were almost entirely deficient in vitamin A. The average weight of the animals at the commencement of this experimental period was 57 g. and at the termination 80 g.—an increase of only 23 g. as compared with the normal for this period, which is 142.

There were ten of these survivors and their average age when selected for use in this new experiment was 150 days. Eight of them were showing marked signs of keratomalacia and general lack of condition and the other two were also out of condition but not to the same extent.

They received throughout the experiment 2 cc. of cow's milk per animal per day and the only other food supplied was entirely deficient in vitamins and composed as follows:

Basal dietary   ...   .   85 %
Steam distilled palm-kernel oil ...   15 %

[For method of preparation see Stammers, 1921.] No antineuritic or antiscorbutic was given, the milk supplied being the only source of these factors available to the animals.

The experiment was carried on for 111 days, and during this period the animals were weighed twice a week, first thing in the morning, before feeding.

It should be stated that the experiment was carried out during the months of December, January, February and March, and the milk used was obtained from stall-fed cows. Butter-fat obtained from the same source during this period showed the effect of stall feeding in giving growth far below the normal (see graph) when used as a control in other experiments, so that the milk may be considered to have been of fairly poor quality as regards its vitamin content.

RESULTS.

As already mentioned, the average weight of ten animals at the commencement of the experiment was 80 g. and their average age was 150 days. In 111 days the increase in average weight was 38 g., *i.e.* 80–118.

The normal increase for this period and age is 43 g., *i.e.* 202–245; hence there is a deficit of only 5 g. in average weight over a period of nearly sixteen weeks.

The animals improved out of all knowledge in general condition, the keratomalacia disappearing and the only difference from normal animals was in their weight and size. Otherwise, they appeared to be in excellent health.

A graph is appended (Fig. 1) which shows

A, growth from milk,
a, normal for A,
B, C, growth from winter butter-fat,
b, c, normals for B, C.

It will be noticed that the butter-fat curves are almost super-imposable.

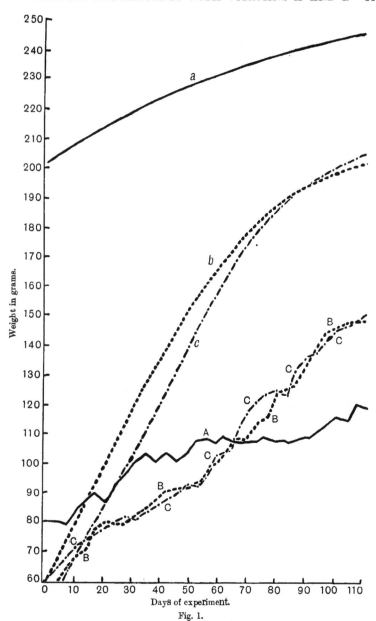

Fig. 1.
a, b, c, Donaldson's curves of normal growth.
A, growth from milk.   B, C, growth from butter-fat.

CONCLUSIONS.

While it is not maintained that this experiment either definitely confirms the results obtained by Hopkins, or disproves the statements of Osborne and Mendel, it seems to throw some additional weight of evidence in favour of the former, although, for the reasons already mentioned, the conditions under which the various experiments were carried out are not comparable.

It is considered, however, that the probabilities are that if an attempt had been made to duplicate Hopkins's experiments by making the conditions similar, the results, even with the sample of winter milk used, would have confirmed the findings of this worker.

If young animals had been used, the initial "growth momentum," as it were, derived from the food consumed prior to weaning, plus the 2 cc. of milk administered daily during the experimental period, would probably have sufficed to give normal growth, having regard to the seasonal variation; in other words the growth from milk would have compared favourably, if nothing more, with that from winter butter-fat.

It may be urged that an adult requires less vitamins to keep it in condition than does a young animal and an explanation of the results of this experiment may be found in this fact: on the other hand, this hypothesis is, in the writer's opinion, only applicable in the case of an animal which has received a normal diet up to the age of maturity and not to animals such as those used in this work, which had been fed on a vitamin-deficient diet since they were seven weeks old.

A further point of interest is now touched upon. When an animal has been fed upon a deficient diet and is suddenly placed upon one containing a greater or less quantity of the growth-promoting vitamin, a considerable stimulus is usually applied by the latter to the growth impulse and this can generally be well seen on the graphic record of the growth. It is quite common for animals in such a case to put on a large increase in weight in the first week, but if this occurs there is usually an almost corresponding drop within the next week or so. It will be noticed that, in this case, the effect of any stimulus applied by the administration of the milk was not reflected in a sudden jump in weight; in fact, during the first week there was a drop and if the curve is smoothed out, it will be seen that it remains almost parallel to the normal.

A possible explanation is that the deficiency disease caused by the previous experiment had suppressed the growth impulse to such an extent that the latter could not immediately respond to the stimulus applied. It is also possible that the greater the age of the animal the greater must be the magnitude of the stimulus to produce this effect, even though this may appear, under certain circumstances, to contradict the axiom that the older an animal is, the less vitamin it requires to keep it in health.

A possible criticism may be applied (owing to the advanced age of the animals used) to the conclusions drawn and therefore attention is invited at this stage to the next paper (IV).

It will be observed that in four of the experiments there described, the animals were also of a mature age (varying from 130 to 160 days) and in three of the experiments had been previously fed upon a diet which had failed to maintain adequate growth, thereby approximating to the conditions under which the present experiments were carried out.

Cod-liver oil is well known to be a rich source of vitamin *A*, but in spite of this fact its administration failed to raise the weights of the animals concerned to the normal, but the curves ran roughly parallel to that of normal growth (Donaldson) without any marked approach to the latter.

It might have been expected that cod-liver oil would be sufficiently potent to overcome the stagnation of the growth impulse and raise the weights to somewhere near the normal, but this did not take place even in any individual rat.

It therefore seems that age is the important factor and that once an animal has reached a certain age, either (*a*) the amount of vitamin which would have been adequate to give normal growth in a young animal, is insufficient, or (*b*) no amount of vitamin can fully restore the growth impulse.

The writer's opinion is that the latter is probably the correct explanation.

A study of these results therefore would seem to indicate that the curve of normal growth remains constant, within limits, for age and not for weight, and, if for any reason the growth impulse has been inhibited, an attempt at restoration of the impulse by however powerful a stimulus cannot succeed to an extent greater than that which would be expected from the age of the animal.

IV.

Six experiments are described in this paper and two varieties of cod-liver oil were used. It has already been established that crude cod-liver oil is exceedingly rich in vitamin *A* [Zilva and Drummond, 1921]. Of the two varieties used in the experiments now described, one, *A*, was an ordinary sample procured from a pharmacist and of a type commonly supplied for medicinal purposes; the other, *B*, was more crude and a brand suitable for administration to cattle. Sample *A* was used in Exps. 1, 2, 4, 5, and 6, and sample *B* in Exp. 3.

The animals in Exp. 1 were 160 days old at the commencement and had been fed (for the previous 110 days) upon butter-fat and used as a control in another set of experiments. Their average weight was 148 g. Those in Exp. 2 were also 160 days old and had previously been fed (also for 110 days) upon a partially deficient diet. Their average weight was 130 g. In Exps. 3 and 4 the animals were also of mature age (135 days) and had also been fed (for the previous 90 days) upon a partially deficient diet. Their average weights were 100 and 90 g. respectively.

Exps. 5 and 6 were confined to young animals. In Exp. 5 they were 43 and in Exp. 6, 33 days old. Their average weights were 77 and 49 g. respectively. The experimental technique was that usually adopted in this laboratory and has already been described in this journal [Stammers, 1921].

Exps. 1 and 2 were carried on for 101 days, Exps. 3 and 4 for 73 and Exps. 5 and 6 for 87 days.

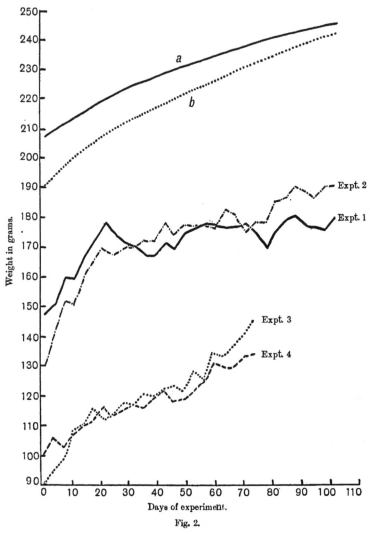

Fig. 2.

a, Normal for Expts. 1 and 2.   b, Normal for Expts. 3 and 4.

Expt. 1 ———   Expt. 2 –·–·–·   Expt. 3 ········   Expt. 4 – – – –

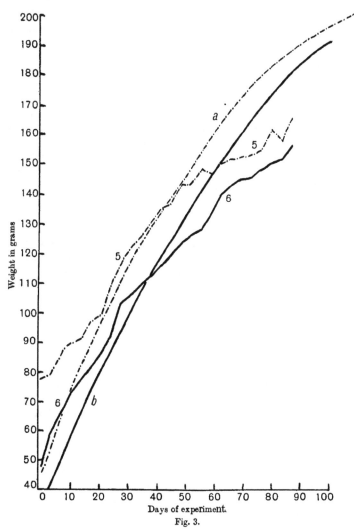

Fig. 3.

a, Normal for Exp. 5.   *b*, Normal for Exp. 6.
—·—·— Expt. 5 (cod-liver oil).   ———— Expt. 6 (cod-liver oil).

RESULTS.

The results are given in tabular form and graphs are also appended (Figs. 2 and 3) which show Donaldson's normal curves in addition to the growth recorded.

Exp.	Previous treatment	Age at commencement	Weight at commencement	Weight at conclusion	Actual increase	Normal increase (for age at commencement)
		days	g.	g.	g.	g.
1	Butter-fat	160	148	180	32	37
2	Partially deficient	160	130	191 ·	61	37
3	Partially deficient	135	100	133	33	41
4	Partially deficient	135	91	145	54	41
5	None	43	77	164	87	141
6	None	33	48	156	108	144

Taking the first four experiments, *i.e.* those in which mature animals were used, it will be seen that in Exps. 1 and 3 the growth was subnormal, while in Exps. 2 and 4, although it was supernormal (for age) the growth shown was not nearly sufficient to cause the curves to approximate to that of normal growth—in other words, although cod-liver oil is probably the richest source of vitamin *A* available, its administration in these cases was not attended by any very marked or maintained increase in weight. The less refined sample, used in Exp. 3, which, *ceteris paribus*, would be expected to give the best results, actually proved the least effective.

As regards Exps. 5 and 6 in which young animals just weaned were used, the growth was slightly subnormal in each case although the animals were considerably above normal weight at the commencement of the experiment: the curves in each case fell off gradually and crossed the normal between the seventh and eighth weeks. When the experiments were terminated, however, they were still on the up grade.

SUMMARY AND CONCLUSIONS.

Two samples of cod-liver oil, one more highly refined than the other, were tested upon rats of mature age, some of which had previously been fed on a diet partially deficient in vitamin *A*. The more highly refined oil was also tested on young animals.

In the latter case (Exps. 5 and 6) the growth registered was *slightly* subnormal and, in view of the potency of cod-liver oil as a source of vitamin *A*, expectations were hardly realised. The sample used was, however, highly refined and this may be responsible.

As regards the first four experiments, No. 3 may be considered first. Although the sample used in this case was fairly crude, it failed to restore deficiency animals to normal weight even after a period of 73 days, moreover there was no sharp initial rise in the growth curve as might have been expected.

In the other experiments an initial rise did take place, but after three weeks the curves flattened out and the growth recorded ran roughly parallel to the normal (for rats of that age). Exp. 2 showed the best results, but even after 101 days the animals averaged 54 g. below the normal.

The explanation offered for these results is (*a*) that the growth impulses of the animals concerned in these experiments had, owing to their ages (135 ·. and 160 days), been deprived to a great extent of their power to react to the stimuli applied even by a fat rich in vitamin *A* and consequently they failed to show substantially greater growth than the normal for their age, or (*b*) that the oil had in the process of refinement lost a large part of its original vitamin *A*.

Point is given to the former theory by the fact that in Exp. 1 the animals had previously been fed on a diet containing butter-fat, which would, in itself, not be expected to impair the growth impulse.

I am indebted to Messrs Lever Brothers, Limited, for permission to publish this research.

REFERENCES.

Hopkins (1912). *J. Physiol.* **44**, 425.
—— (1920). *Biochem. J.* **14**, 721.
Osborne and Mendel (1920). *J. Biol. Chem.* **41**, 15.
Stammers (1921). *Biochem. J.* **15**, 491.
Zilva and Drummond (1921). *Lancet*, ii, 753.

LXIX. ON THE CARDIAC, HAEMOLYTIC AND NERVOUS EFFECTS OF DIGITONIN.

By FRED RANSOM.

From the Pharmacological Laboratory of the London (Royal Free Hospital) School of Medicine for Women (University of London).

(*Received July 19th, 1922.*)

DIGITONIN, a sapo-glucoside found in *Digitalis purpurea*, has certain well-known properties—it causes haemolysis, forms a combination with cholesterol and some other sterols and has a marked action upon cardiac muscle.

The objects of the present investigations were (1) to compare the effects of certain allies of cholesterol upon the cardiac and haemolytic actions of digitonin and (2) to ascertain whether the activities of a digitonin solution are modified by shaking with an indifferent adsorbent.

The particular digitonin used was prepared and carefully purified by Mr J. A. Gardner, whom I desire to thank for his kind help.

In the first place it was necessary to measure both the cardiac and the haemolytic activity of the preparation.

Haemolysis. Graduated amounts of digitonin, in each case dissolved in 4 cc. of 0·95 % NaCl solution, were placed in a series of test tubes and 1 cc. of a 10 % dilution of defibrinated ox blood in physiological salt solution was added to each tube, so that all reactions took place in 5 cc. of fluid. The tubes were left standing in the laboratory for about 18 hours and then the condition of their contents was noted, *i.e.* whether haemolysed or intact. In Table I the results of such a test are set out. Repeated experiments showed that the minimum quantity of the digitonin which would, under the above conditions, completely lake 0·1 cc. of ox blood was from 0·0032–0·0036 mg., corresponding to a dilution of the digitonin of from 1 : 156,000 to 1 : 139,000. With cat's blood the minimum complete laking dose was rather less than 0·004 mg.

Table I. *Haemolytic power of Gardner's digitonin.*

Digitonin 0·004 %	NaCl 0·95 %	Ox blood 10 %	Dilution of digitonin	Amount of digitonin	Condition after 20 hrs. at room temp.
0·4 cc.	3·6 cc.	1 cc.	1 : 312,500	0·0016 mg.	Intact
0·5	3·5	1	1 : 250,000	0·0020	Very slight haemolysis
0·6	3·4	1	1 : 208,000	0·0024	Slight haemolysis
0·7	3·3	1	1 : 178,577	0·0028	Nearly haemolysed
0·8	3·2	1	1 : 156,250	0·0032	Very nearly haemolysed
0·9	3·1	1	1 : 139,000	0·0036	Completely haemolysed
1·0	3·0	1	1 : 125,000	0·004	,, ,,

Evaporation to dryness on a water-bath of a digitonin solution, whether in water or 90 % alcohol, did not affect the haemolytic power (Table II) but the presence of NaCl in hypertonic solution increased the resistance of the red cells to haemolysis by digitonin (Table III), as was previously observed by Luger [1921] for saponin.

Table II. *Digitonin solution evaporated to dryness and redissolved.*

Digitonin 0·004 %	NaCl 0·95 %	Ox blood 10 %	Condition after 18 hours
0·4 cc.	3·6 cc.	1 cc.	Trace of haemolysis
0·6	3·4	1	Partial „
0·8	3·2	1	Complete ..
0·9	3·1	1	„ ..

Table III. *Protective action of hypertonic NaCl.*

Digitonin	NaCl	10 % blood	Condition after 18 hours at room temp.
·004 mg.	4 cc. ·95 %	1 cc.	Complete haemolysis
·004	4 cc. 2·5 %	1	Partial „
·004	4 cc. 6·0 %	1	No „
·008	4 cc. 5·0 %	1	Complete

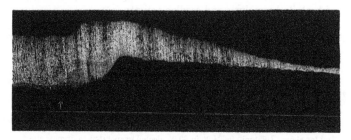

Fig. 1. Frog's heart perfused with Ringer's fluid. At arrow perfusion changed to Ringer + 0·004 % digitonin.

Cardiac action. Solutions of the digitonin in oxygenated Ringer's fluid were perfused through frogs' hearts *in situ* from a cannula inserted into the inferior cava, the fluid passing at a constant and moderate pressure through the heart and escaping by the cut aorta. A number of such perfusions showed that 0·0001 % of digitonin had very little or no effect; 0·001 % produced increase in the beat followed by rise of tone, diminution of diastole and finally stoppage or nearly so: 0·004 % produced the same effects somewhat more quickly (Fig. 1). Evaporation of a watery solution to dryness on a water-bath and redissolving in Ringer's fluid did not affect the activity. Perfusion of a frog's heart with a Ringer's fluid modified by omitting Ca quickly causes diminution of efficiency; if now, in the continued absence of Ca, 0·001 % digitonin is added to the perfusing fluid the normal efficiency is first partly restored and then the characteristic effect of digitonin is developed.

It is well known that digitonin treated with cholesterol loses its haemolytic power, it seemed therefore of interest to ascertain in how far bodies more or less closely allied to cholesterol such as phytosterol, β-cholestanol, coprosterol and pseudo-coprosterol affect the action of digitonin upon the blood and upon the heart.

EXPERIMENTS WITH SUBSTANCES ALLIED TO CHOLESTEROL.

Method. A solution containing 0·004 % (or other percentage) of digitonin in 90 % alcohol was mixed with an equal quantity of a solution of the same strength of one of the cholesterol allies, also in 90 % alcohol. The mixture was evaporated to dryness on a water-bath and the dry residue taken up in so much Ringer's fluid as would give a possible concentration of digitonin and the sterol equal to that of the original alcoholic solution. This solution of the dry residue was filtered and the filtrate tested for its haemolytic power and cardiac effect. The results obtained with phytosterol, β-cholestanol, and coprosterol were identical—in each case both the haemolytic effect and the cardiac action were either cut out or greatly diminished (haemolytic effect, Table IV). On the other hand, the mixture digitonin + pseudo-coprosterol retained both its haemolytic and cardiac activity.

Table IV. *Effect of phytosterol, β-cholestanol, coprosterol and ψ-coprosterol.*

Possible concentration of digitonin 0·004 %.

	Solution of residue	NaCl 0·95 %	Ox blood 10 %	Dilution of digitonin	Condition after 18 hrs. at room temp.
Phytosterol or	1 cc.	3 cc.	1 cc.	1 : 125,000	No haemolysis
β-Cholestanol	2	2	1	1 : 62,500	,,
or Coprosterol	4	—	1	1 : 31,250	,,
ψ-Coprosterol	1	3	1	1 : 125,000	Complete haemolysis

Control for above: Digitonin 0·004 % in 90 % alcohol evaporated to dryness and taken up in NaCl, so as to give a possible 0·004 % digitonin.

Solution of residue	NaCl 0·95 %	Ox blood 10 %	Dilution of digitonin	Condition after 18 hrs. at room temp.
0·6 cc.	3·4 cc.	1 cc.	1 : 208,333	No haemolysis
0·8	3·2	1	1 : 156,250	Partial haemolysis
1·0	3	1	1 : 125,000	Complete ,,

When stronger solutions of digitonin were mixed with correspondingly increased amounts of phytosterol, β-cholestanol or coprosterol the protection against haemolysis and cardiac action was still complete (Table V).

Table V. *Protective action of phytosterol, β-cholestanol or coprosterol.*

Possible concentration of digitonin 0·1 %.

Solution of residue	NaCl 0·95 %	Ox blood 10 %	Dilution of digitonin	Condition after 18 hrs. at room temp.
0·5 cc.	3·5 cc.	1 cc.	1 : 10,000	No haemolysis
1·0	3·0	1	1 : 5,000	,,
4·0	—	1	1 : 1,250	.

If the haemolytic test was carried out with mixtures containing a certainly haemolytic dose of digitonin and a gradually diminishing amount of phytosterol, β-cholestanol or coprosterol, the effect of these bodies in diminishing and finally extinguishing the haemolytic effect of the digitonin (Table VI) was perhaps still more strikingly in evidence.

Table VI. *Protective action of phytosterol.*

Digitonin 0·01 %		Phytosterol 0·01 %		Cat's blood 10 %	Condition after 20 hrs. at room temp.
0·5 cc.	+	0·5 cc.		1 cc.	No haemolysis
0·5	+	0·4	Evaporated	1	„
0·5	+	0·3	to dryness	1	„
0·5	+	0·2	and 4 cc. NaCl	1	„
0·5	+	0·1	added to	1	Some haemolysis
0·5	+	0·05	each, then	1	Complete haemolysis
0·5		—		1	„

In all the above experiments the solutions used were of equal percentage for the digitonin and the cholesterol allies; now the molecular weight of Gardner's digitonin is 1202, that of cholesterol and its isomer phytosterol is 386, so that when equal percentages by weight are used 1 phytosterol corresponds molecularly to 3·1 of digitonin. It was found that 0·31 cc. of a 0·01 % solution of digitonin (0·0031 mg.) will very nearly or sometimes completely haemolyse 0·1 cc. of ox blood. If the relationship between digitonin and cholesterol in the digitonin-cholesteride is monomolecular [Windaus, 1910], then 0·31 cc. of a 0·01 % solution of digitonin should be inactivated by 0·1 cc. of a 0·01 % solution of phytosterol or β-cholestanol or coprosterol (since the difference between the molecular weight of phytosterol, 386, and that of β-cholestanol or its isomer coprosterol, 388, is too small to be of importance in the conditions of the experiment); moreover multiples of these quantities should also be inactive. On the other hand, if the mixtures contain relatively an excess of digitonin more or less haemolysis should occur. In Table VII the results of an experiment carried out on these lines are shown. It will be seen that when molecular equivalents were mixed the neutralisation of the haemolytic action of the digitonin was complete, even when multiples of the minimum haemolysing dose were employed; when, however, the digitonin was relatively in excess haemolysis occurred and was most marked in the tubes containing the smallest relative proportion of the sterol.

As regards the cardiac activity it was found that the effect upon the heart ran parallel with that upon the red cells, the non-haemolytic mixtures had no cardiac effect and when the haemolytic action was not completely cut out there was also some cardiac effect. The results recorded in Table VII are strongly reminiscent of what occurs when diphtheria or other toxins are mixed with their respective antitoxins, and might perhaps be explained as simple adsorption phenomena (see below) were it not that other evidence points to a definite chemical union between digitonin and cholesterol. Moreover the fact that pseudo-coprosterol does not interfere with either the haemolytic or the cardiac action of digitonin strongly supports the chemical view. In view of the chemical relationship between cholesterol and cholalic acid a series of experiments was made with this substance on the same lines _ as those recorded above, using a 0·01 % solution of the acid with 0·01 % of digitonin, but no evidence was obtained which would indicate that cholalic acid has any influence either upon the haemolytic or the cardiac action of digitonin.

Table VII.

All reactions in 5 cc. of fluid.

0·01 % sol. of coprosterol	0·01 % sol. of digitonin	Cat's blood 10 %		Condition after 18 hrs. at laboratory temp.
0·1 cc.	0·31 cc.	1 cc.		No haemolysis
0·2	0·62	1	Molecular	,,
0·4	1·24	1	equivalents	,,
1·0	3·1	1		,,
0·05	0·31	1		No haemolysis
0·1	0·62	1	Digitonin	Trace of haemolysis
0·2	1·24	1	in excess	Partial ,,
0·5	3·1	1		Complete ,,
0·2	3·1	1	Digitonin in	Completely haemolysed
0·4	3·1	1	excess, constant,	,, ,,
0·6	3·1	1	coprosterol	Nearly ,,
0·8	3·1	1	varied	No haemolysis
—	0·1	1		No haemolysis
	0·2	1		Some ,,
	0·3	1	Controls for digitonin	Complete haemolysis
	0·4	1		,, ,,
	0·5	1		,, ,,

ADSORPTION EXPERIMENTS.

The importance of adsorption in many pharmacological problems suggested experiments to ascertain if the solutions of digitonin are affected by indifferent adsorbents. To this end experiments were carried out with animal charcoal, kaolin and starch.

Method. 50 cc. of 0·004 % solution of digitonin in distilled water were gently shaken with 2 g. of purified animal charcoal (washed and dried) or kaolin or starch for 30 mins., then filtered and to the filtrate salts to make Ringer's fluid were added; this isotonic solution was then tested as to its haemolytic and cardiac activities.

Charcoal. The digitonin solution before treatment with charcoal, but with salts for Ringer's solution added, caused haemolysis as usual (0·8 cc. completely); after charcoal 5 cc. had no haemolytic effect. As to cardiac action the solution which had been treated with charcoal had no effect, but when the perfusion fluid was changed to the digitonin solution untreated with charcoal the usual digitonin effect appeared at once. The charcoal remainder was shaken with 25 cc. 90 % alcohol for 30 mins., the alcohol evaporated and the residue taken up in 50 cc. Ringer's fluid, this solution caused haemolysis and had a marked cardiac action. It is evident that animal charcoal very effectively adsorbs digitonin. If the period during which the charcoal was applied or the amount of charcoal used was diminished the removal of the digitonin was not so complete.

Starch. In this case the digitonin solution was 0·001 %. This solution perfused through a frog's heart gave rise to increased systole and rise of tone, after treatment with starch the filtrate showed only a slight remnant of digitonin action. As to haemolysis, the results of the test are given in Table VIII and show that whereas before treatment with starch 1 cc. of the digitonin solution caused complete haemolysis, after starch 5 cc. did not do so. Digitonin then is adsorbed by starch and the filtrate from the starch is found to have lost both in haemolytic and cardiac activity.

Table VIII.

Before starch 1 cc. of 0·001 % digitonin gave complete haemolysis
After   „   1   „   „   „   no haemolysis
„   „   2   „   „   „   „   „
„   „   3   „   „   „   trace only
„   „   5   „   „   „   nearly complete haemolysis

Kaolin. In this case also the digitonin solution used was 0·001 %. Before treatment with kaolin 1 cc. completely haemolysed 0·1 cc. of ox-blood, after kaolin 5 cc. caused only partial haemolysis (Table IX). As to cardiac action, before kaolin the solution gave a characteristic digitonin effect, after kaolin there was still some evidence of digitonin effect but much less marked. Clearly, then, kaolin adsorbs digitonin from watery solution.

Table IX.

Before kaolin 1 cc. of digitonin sol. gave complete haemolysis
After   „   2   „   „   „   no haemolysis
„   „   4   „   „   „   trace of haemolysis
„   „   5   „   „   „   incomplete haemolysis

At first glance it would appear as though of the three adsorbents animal charcoal were the most effective but much stress cannot be laid upon this point, since no attempt was made to insure the same number and size of particles, *i.e.* surface area, in the amounts of each of the adsorbents. Nevertheless there are good grounds for believing that charcoal occupies a peculiarly favourable position among adsorbents [Michaelis and Rona, 1920].

Windaus founded his method of estimating cholesterol by means of digitonin upon his discovery that cholesterol forms with digitonin a stable compound which is not haemolytic and dissolves with difficulty in alcohol and upon the fact that cholesterol esters [Hausmann, 1905] do not form such a compound. He considered the digitonin-cholesteride to be an addition product and not an ester and gave its formula as $C_{55}H_{94}O_{28}$ (digitonin) + $C_{27}H_{46}O$ (cholesterol), regarding it as a combination of one molecule of cholesterol with one of digitonin without loss of water. The results obtained in the present investigation are confirmatory of this view, since they show that when phytosterol, coprosterol or β-cholestanol are mixed in monomolecular proportions with digitonin the latter loses completely its haemolytic power, whereas, if the mixtures contain less than this proportion of the sterol, more or less haemolysis takes place according to the degree of deficiency.

But now the very interesting question arises as to how the cardiac action of digitonin is affected. In the first place it may be noted that the effect of phytosterol, β-cholestanol or coprosterol upon the cardiac action runs strictly parallel to their effect upon the haemolytic power of digitonin—with monomolecular proportions both the haemolytic and the cardiac actions were cut out, whereas in mixtures containing a deficiency of sterol the more nearly the haemolytic action was neutralised the less was the cardiac effect and vice versa. Further, pseudo-coprosterol which does not prevent haemolysis, i.e. presumably does not combine with digitonin, did not affect the cardiac action either. These facts suggest that the action of digitonin upon the heart muscle depends upon the same factor as that which determines the haemolytic effect, viz. that there is present in cardiac muscle a body resembling or identical with cholesterol or coprosterol, etc., and that this sterol is at any rate in part free and not esterified. Moreover it has so important a function to perform in the muscle that its withdrawal as digitonin-sterol profoundly affects the physiological activity of the muscle cell. This supposition receives important confirmation from the fact that pseudo-coprosterol, which does not form a digitonide and does not affect haemolysis, has no effect upon the cardiac action of digitonin.

To show that the heart muscle probably does fix digitonin, the following experiment was carried out; a kitten was killed by pithing the brain and bleeding, the blood being collected and defibrinated. The heart was removed, washed free from blood, and the coronary vessels washed out with Ringer. The excised heart (7 g.) was pounded to as smooth a paste as possible, then well mixed with 50 cc. 0·002 % digitonin in Ringer and the mixture placed in the ice-chest. After 20 hours the mixture was centrifuged, the liquid poured off and divided into two parts, one of which was used to test haemolysis (Table X); with the other a frog's heart was perfused, with the result that both the haemolytic and the cardiac effects were markedly lessened.

Table X.

Heart emulsion + digitonin 0·002 %	Ringer	Kitten's blood 10 %	After 20 hrs.
0·8 cc.	4·0 cc.	0·2 cc.	Intact
1·0	3·8	0·2	,,
2·0	2·8	0·2	
5·0	—	0·2	
Digitonin sol. 0·002 %			
0·8 cc.	4·0 cc.	0·2 cc.	Partly haemolysed
1·0	3·8	0·2	Nearly completely
2·0	2·8	0·2	Completely

It may be recalled that many of the saponins, including digitonin, are powerful fish poisons, acting by paralysing the C.N.S., but that the cholesterol-saponide is non-toxic as well as non-haemolytic. In like manner saponin, *e.g.* agrostemma saponin, found in the corn cockle, when it causes poisoning in human beings does not kill by its action upon the blood but by its effect upon the C.N.S. and upon the heart. On the other hand, tetanus toxin, besides its characteristic action upon the C.N.S., has definite haemolytic powers, indeed tetanus toxin is usually regarded as a mixture of tetanospasmin and tetanolysin. Cholesterol neutralises tetanolysin without diminishing the specific action of tetanospasmin. So the question arises; is the central nervous effect of saponin due to the same factor as enables it to cause haemolysis and alteration in cardiac muscle or do the conditions resemble those in tetanus toxin? To throw light upon this point the following experiment was carried out. A young rabbit was killed by a blow on the neck, the carotids were then cut and the blood collected and defibrinated; brain removed (6·7 g.), rubbed

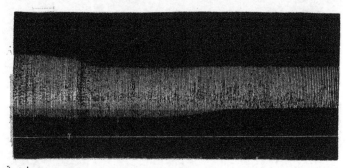

Fig. 2. Frog's heart perfused with Ringer's fluid. At arrow perfusion changed to emulsion of rabbit's brain in Ringer + 0·004 % digitonin. Compare with Fig. 1.

to a smooth paste in a mortar and then thoroughly mixed with 50 cc. of a 0·004 % solution of digitonin in Ringer's fluid; the emulsion was placed in the ice-chest for 20 hours; then well centrifuged, the supernatant fluid poured off and divided into two parts, one of which was used to perfuse a frog's heart, the other to test the haemolytic power. Table XI shows the results of the haemolytic test; the perfusion tracing is shown in Fig. 2. It will be seen at once that the digitonin-brain emulsion has no haemolytic power, at least not up to 5 cc. which is very much more than the ordinary haemolytic dose, and further the heart tracing shows merely a trace of digitonin action.

Table XI. *Protective action of brain substance.*

Possible concentration of digitonin, 0·004 %.

Brain + digitonin	Ringer	Ten rabbits' blood	After 18 hrs.
0·6 cc.	4·2 cc.	0·2 cc.	Intact
0·8	4·0	0·2	,,
1·0	3·8	0·2	
2·0	2·8	0·2	
5·0	—	0·2	
Digitonin 0·004 %			
0·8 cc.	4·0 cc.	0·2 cc.	Completely haemolysed

Considered in conjunction with the previous experimental evidence, these results appear to justify the conclusion that the action of saponin in causing narcosis in fishes and paralysis of the C.N.S. in man is an analogous phenomenon to digitonin haemolysis and digitonin cardiac action, *i.e.* there is free and unesterified in the brain a body allied to or identical with cholesterol, phytosterol, etc., which unites with digitonin. Moreover the function which this cholesterol-like body has to perform is of such importance that when it is inactivated by combination with digitonin (saponin) serious central nervous symptoms ensue.

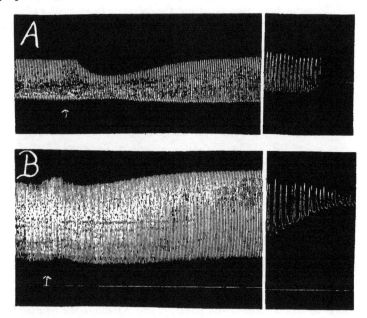

Fig. 3. Two frogs' hearts perfused with Ringer's fluid. *A*, at arrow perfusion fluid changed to emulsion of cat's brain in Ringer + 0·001 % strophanthin. *B*, at arrow perfusion fluid changed to Ringer + 0·001 % strophanthin. In each case the break in the tracing represents an interval of 15 mins.

Kobert [1904] has shown that the toxic action of saponin as well as the cardiac effects is abolished by the combination of saponin with cholesterol, and further that the extreme toxicity of sapotoxin solutions for fishes is entirely lost in the cholesterol-sapotoxin combination.

The effect of digitonin is also neutralised by adding milk to the digitonin solution.

It is interesting to note that strophanthin, which is a glucoside but not a saponin, does not lose its cardiac action when brain emulsion is added to the strophanthin solution (Fig. 3).

SUMMARY.

Monomolecular combinations of digitonin with phytosterol, coprosterol, β-cholestanol are non-toxic for red blood cells and for the frog's heart; any excess of digitonin beyond this proportion can be detected by the action of the solution upon red cells and upon the heart. Pseudo-coprosterol does not effect the activity of digitonin. Digitonin mixed with an emulsion of brain substance loses its toxicity for red cells and for the heart.

It is suggested that the toxic action of digitonin upon red blood cells, cardiac muscle cells and cells of the central nervous system (fishes) is essentially the same—it attacks a substance allied to or identical with cholesterol which is present free and not esterified in these various cells. This free sterol has in each case so important a function to perform that its inactivation by combination with digitonin leads to a profound modification of the physiological activities of the respective cells.

Digitonin is readily adsorbed from solution in water (*e.g.* 1 : 25,000) by animal charcoal, kaolin or starch; the adsorption by charcoal appears to be particularly complete.

Cholalic acid has no effect upon the action of digitonin.

REFERENCES.

Hausmann (1905). *Beiträge*, 6, 567.
Kobert (1904). Saponinsubstanzen, Stuttgart.
Luger (1921). *Biochem. Zeitsch.* 117, 145.
Michaelis and Rona (1920). *Biochem. Zeitsch.* 102, 268.
Windaus (1910). *Zeitsch. physiol. Chem.* 65, 110.

LXX. THE HEAT-COAGULATION OF PROTEINS.

By W. W. LEPESCHKIN.

From the Botanical Laboratory, the University of Kasan.

(Received July 21st, 1922.)

IT has been long pointed out that the heat-coagulation of proteins is not so simple a phenomenon as it at first may seem to be. Schmidt and his pupils, Aronstein [1874], Heynsius [1874], Kieseritzky [1882] and Rosenberg [1883], showed that albumin solutions from which the electrolytes had been removed by dialysis did not coagulate on heating; heat-coagulation could only be produced after the addition of electrolytes to such solutions; it took place also if a small amount of any salt was added to the cold albumin solutions which had been preliminarily heated. These authors therefore suggested the hypothesis that heating may transform the albumin into an unknown body, the solutions of which coagulate on the addition of a very small quantity of salts [Rosenberg, 1883, p. 34].

Hardy [1899, 1900, 1906] showed later that in solutions of protein altered by heating ultramicrons appear and that the protein can now be precipitated by adjusting the reaction so as to render the particles isoelectric with the solution. When the protein is made negative by adding alkali the valency of cations is of importance for the precipitation; on the contrary, the cations have no part in the precipitation if protein is made positive.

Then Michaelis came to the conclusion [Michaelis] that the properties of albumin solutions may be changed by heating from those of an "emulsoid" to those of a "suspensoid," whereas the high viscosity and the low surface-tension of solutions of albumin altered by heating point rather to their hydrophilic character. These solutions probably do not fall into either category of the colloidal state but form an intermediate group with some of the properties of both.

Chick and Martin [1913, 1] who investigated the heat-coagulation of proteins and the influence of various factors upon its course confirm the conclusion that this phenomenon consists of two distinct processes: (1) "denaturation" and (2) "agglutination" of the denaturated protein by electrolytes. They regard the former as a reaction between protein and hot water without determining this reaction more precisely.

Contradicting the results of the above investigations and of his own earlier experiments Pauli [1899, 1908] affirms that he has observed that protein solutions free from electrolytes ("amphoteres Eiweiss") coagulate on heating and that the addition of electrolytes to such solutions increases their coagulation temperature. Criticism and detailed consideration of this new opinion may be postponed till some experiments related to heat-coagulation have been described.

A number of scattered observations indicated that temperature is an accelerating factor in the process of heat-coagulation. Bnglia [1909] was the first who precisely determined the dependence of the velocity of heat-coagulation upon the temperature, and found this dependence to be logarithmic.

Chick and Martin [1910] found that the coagulation-rate increases 1·3–1·9 fold per 1° temperature-rise. Such a high temperature coefficient bears only on the denaturation, while the agglutination-velocity does not increase so rapidly with the temperature as the denaturation-rate, its temperature coefficient being equal to 2·5 per temperature-rise of 10° [Chick and Martin, 1913, 2]. These authors showed moreover that, if precautions are taken to keep the concentration of hydroxyl or hydrogen-ions constant during the process, the denaturation may proceed as a reaction of the first order.

(1) EXPERIMENTAL METHODS.

My experiments were made before I was acquainted with the papers by Chick and Martin, and the method used by me differs from theirs. I find however that it is more convenient for observing slight, and sometimes short, alterations of velocity in both the processes that are comprised in heat-coagulation.

In order to estimate alterations in the velocity of heat-denaturation and in coagulation ("agglutination") of the denatured protein under influence of various factors, in my experiments, the time required to produce a slight turbidity in the protein solutions like that of a ground glass (chosen for all the experiments) was noted. The solution to be tested was usually introduced by means of a capillary pipette into a small vessel made of two coverglasses (20 × 10 × 0·1 mm.) cemented with oil-lacquer (linseed-oil and minium) on both sides of a rectangular brass plate (0·7 mm. thick, 22 mm. long, 13 mm. broad) the middle of which was rectangularly cut out, so that the rim-breadth (at three sides) was equal to 3 mm.

A drawing of this vessel is given in one of my previous papers devoted to the reactions between starch and water brought about by heating[1]. There are also given a drawing and a photograph of the thermostat used.

---

[1] Lepeschkin, "Etude sur les réactions chimiques pendant le gonflement de l'amidon dans l'eau chaude." Le manuscrit est renvoyé à M. le professeur Chodat (Génève, Suisse) pour la publication dans l'*Archives des sciences naturelles*.

The above vessel with the protein solution to be tested was immersed in the water of the thermostat, in which the temperature variation did not surpass 0·03°. Close by the vessel, a similar rectangular brass plate (cut out in the middle) was placed; instead of the cover-glasses (see above) a piece of ground glass of the same size was cemented on one side of this plate. Both (viz. the vessel and the plate with ground glass) were observed through the front glass-wall of the thermostat box by means of a horizontal micro-scope of which the eye-piece was removed and the objective was substituted by a black plate having two rectangular apertures corresponding respectively to the front cover-glass of the vessel and to the ground glass. The protein solution and the ground glass were illuminated by an electric lamp (50 candles) through the back glass-wall of the thermostat, and the microscope-tube was situated obliquely to the front glass-wall (in order to observe most easily the turbidity produced in the protein solution by heating).

The water of the thermostat was uninterruptedly stirred by means of an electric stirrer.

The time-intervals were defined by means of a seconds pendulum.

The preliminary experiments showed that the liquid contained in the above vessel takes up the temperature of the thermostat water in about 20 seconds. This interval of time was always deducted from the time required to produce the above turbidity of the solution to be tested.

The material used was filtered egg-white, egg-albumin Kahlbaum from which electrolytes had been removed by dialysis, serum-albumin Kahlbaum similarly treated or purest recrystallised egg-albumin.

The first experiments are concerned with denaturation-velocity at various temperatures.

(2) The temperature coefficient of denaturation.

Preliminary experiments showed that the coagulation-velocity of the denaturated proteins in solutions containing a sufficiently great quantity of electrolytes is extraordinarily great in comparison with the denaturation-velocity at the same temperature. It could thus be assumed that the coagula-tion of denaturated protein proceeds instantaneously at the temperatures the influence of which upon the denaturation was investigated in my experi-ments. The appearance of the turbidity in the protein solutions containing a sufficient quantity of electrolytes marked, therefore, the denaturation pro-duced by heating (or is it truer to say the degree of denaturation corresponding with the turbidity of the solution like that of the ground glass). The results were the following.

Filtered egg-white. The temperature of water in the thermostat was in the experiments: 63·01°, 62·01°, 61·02°, 60·03°, 59·03°, 58·01°, 57·04° and 56·00°. The average (from six experiments) times (in seconds) passed from the moment of immersing the vessel with egg-white into the water to the appear-ing of the required turbidity of the solution T were found respectively:

40 secs., 97 s., 230 s., 595 s., 1535 s., 3720 s., 9520 s., and 22,600 s. The temperature coefficients per rise of 1° thus were found to be: 2·4, 2·4, 2·5, 2·6, 2·4, 2·6, 2·4 or on the average 2·5.

3 % *solution of egg-albumin Kahlbaum.* This had been dialysed during 5 days and no longer coagulated on heating; in order to make it coagulable 0·4 % ammonium sulphate was added. Temperatures of the thermostat: 70·04°, 69·00°, 68·00°, 67·04°, 66·04°, 65·06°. The average (from five experiments) times T: 91 secs., 136 s., 207 s., 306 s., 452 s., 690 s. The average temperature coefficient per rise of 1° = 1·5.

2·5 % *solution of serum-albumin Kahlbaum.* This had been dialysed during 20 hours and no longer coagulated on heating. 0·7 % potassium chlorate was added. The temperatures were: 75·00°, 72·00°, 70·00°, 68·00°. The average times T: 59 secs., 205 s., 470 s., 1298 s. The average temperature coefficient = 1·6.

1 % *solution of recrystallised egg-albumin* (not coagulating on heating). 1·5 % ammonium sulphate was added. The temperatures were: 70·00°, 69·00°, 68·00°, 67·00°. The average times T: 154 secs., 331 s., 784 s., 1860 s. The average temperature coefficient = 2·3.

The coefficient of denaturation thus is equal to 1·5–2·5 *per* 1°, *viz.* 58–9540 *per* 10°. A similar coefficient was found by me for the chemical reaction between starch and water during the swelling of starch in hot water. See footnote p. 679.

(3) On the nature of denaturation. The influence of hydrogen- and hydroxyl-ions upon the denaturation-rate.

Chick and Martin's researches [1910] show that water is of great significance in the denaturation process of proteins, and reasoning from analogy with the above mentioned reaction between starch and water, I should incline, in accordance with Chick and Martin, to regard this process as a chemical reaction between protein and water. Further we shall see indeed, that the chemical properties of denaturated protein differ considerably from those of native protein.

In my cited paper it was suggested that during the swelling of starch in hot water no hydrolysis takes place but the starch forms a compound with water. Indeed, the velocity of the reaction between starch and water increases not at all proportionally to the hydrogen-ion concentration but depends upon the nature of the acid added, *e.g.* sulphuric acid, in weak concentrations, even lowers the reaction-rate. So then the hypothesis, that starch preliminarily forms a compound with acid before it reacts with water, must be adopted. The preliminary formation of compounds with potassium and sodium hydroxide which react more easily with water than free starch is also certain.

The fact that denaturation proceeds as a reaction of the first order (monomolecular reaction) indicates that the protein reacts with the solvent, but leaves it uncertain whether the denaturation is a hydrolysis, like the inversion of sugar, or is accompanied by the formation of a compound between protein and water. It is important in this connection to know the effect of acids and alkalies upon the denaturation-velocity.

Chick and Martin [1913] who found the denaturation-rate to be increased with increasing concentration of hydrogen- or hydroxyl-ions in protein solutions suggest that the cause of this phenomenon lies in the fact that protein forms salts with acids and alkalies, and that these salts are in a more intimate association with water.

The researches of Hoffmann [1889, 1890], however, had already shown that slight quantities of acid which are added to a pure protein solution form no salts with protein, and this result is in accordance with the recently found weakly acid nature of proteins (about 0·00001 N hydrogen-ion concentration). Nevertheless my experiments showed that a very remarkable increase of denaturation-rate is observed even when albumin solutions contain 0·00016 mol. nitric acid per litre. The increasing effect of such small concentrations of acid upon the denaturation-rate cannot be explained except by an acceleratory influence of hydrogen-ions like that observed in the case of hydrolysis of carbohydrates, viz. by their catalytic properties. The denaturation-velocity was however found in my experiments not to be proportional to the concentration of hydrogen-ions present in the solution when strongly different concentrations of the same acid had been tested. It increased more slowly than the hydrogen-ion concentration when the latter was not greater than 0·002 mol. per litre, but it began to increase more rapidly when the hydrogen-ion concentration attained 0·01 mol. per litre.

However, if various acids were tested at about the same molecular concentrations, the denaturation-velocity was found to depend upon the hydrogen-ion concentration, so that contrary to the case of the reaction between starch and water, the nature of the acid has here no influence upon the denaturation-velocity.

The material used in my experiments was a solution of egg-albumin Kahlbaum which had been dialysed during many days. The solution was diluted partly with water, partly with nitric acid so that the albumin-content in all solutions was 0·66 %. In order to make the solutions obtained coagulable 0·8 % $(NH_4)_2SO_4$ was added. When the concentration of nitric acid was 0 N, 0·00016 N mol., 0·00032 N, 0·002 N, 0·016 N, the average time of coagulation at 63° was found respectively to be 2440 secs., 1550 s., 1220 s., 315 s., 11 s. In the case of different acids the average coagulation-time at 53° was found to be equal to 167 secs. in the presence of 0·016 N HNO_3 (viz. hydrogen-ion concentration = 0·015 N), 161 secs. in the presence of 0·015 N HCl (hydrogen-ion concentration = 0·014 N), 86 secs. in the presence of 0·018 N H_2SO_4 (hydrogen-ion concentration = 0·024 N) and 1275 secs. in the presence of 0·016 N acetic acid (hydrogen-ion concentration = 0·0016 N).

The absence of proportionality between the denaturation-velocity and the concentration of hydrogen-ions in the case of strongly different concentrations of the same acid, is not unexpected. Indeed a strict proportionality between the hydrogen-ion concentration and the reaction-rate also does not exist in the case of the hydrolysis of sugar by acids, particularly when salts are present in the sugar solution.

Arrhenius [1889] who investigated the influence of salts upon the inversion-velocity showed that the latter departs from its normal value almost proportionally to the salt-content of the solution; moreover, the weaker the acid concentration, the more considerable is the influence of salt; so for instance, the inversion-rate increased by 35–40 % on adding KBr to the concentration of 0·4 N, when hydrobromic acid was 0·002 N, and it increased by 200–240 % when the acid was 0·0005 N.

At the same time, it was necessary, in my experiments, to add ammonium sulphate to the albumin solutions to bring about the coagulation of denaturated albumin. No doubt the salt increased the catalytic power of the hydrogen-ions normally contained in the albumin solution, the presence of which is due to the electrolytic dissociation of the protein molecules (hydrogen-ion concentration of about 0·00001 N); then, the salt induced thereby an extraordinarily strong increase of the denaturation-rate[1]. However, this increasing effect of salt upon the catalytic power of hydrogen-ions strongly diminished on the further increase of the concentration of these ions by adding nitric acid so that the denaturation-rate increased only by 1·6 fold, while the hydrogen-ion concentration increased by 15 fold. By virtue of the same cause the denaturation-velocity continued to increase more slowly than the concentration of hydrogen-ions on the addition of acid. When, however, this concentration became more than 0·002 N the denaturation-rate began to increase more rapidly than the hydrogen-ion concentration (an increase of the latter of 8·3 fold was accompanied by an increase of the former of 28 fold). The same was found by Chick and Martin [1912].

The more rapid increase of the denaturation-rate by stronger concentrations of acid could not be explained except by the formation of compounds between albumin and acid which react more easily with water than the unaltered albumin, the acid concentration 0·016 N being evidently sufficient for the formation of such compounds.

As the increasing effect of salts upon the catalytic power of hydrogen-ions varies only with a very considerable alteration of concentration of these ions (Arrhenius, see above), it is very comprehensible that the denaturation-rate remained in my experiments almost proportional to the hydrogen-ion concentration in the case where denaturation proceeded in solutions containing similar concentrations of various acids.

In view of the importance of hydrogen-ions in the reaction between protein and water, it seems to be very probable that the denaturation is a slight hydrolysis of protein; such hydrolysis cannot be discovered by chemical

[1] It is uncertain whether this increase of denaturation-rate is due to the formation of an active albumin, analogically to the active sugar of Arrhenius.

analysis, the protein molecule being too large, and it is not surprising that Chick and Martin could obtain no evidence of hydrolysis by denaturation of pure proteins in the chemical way.

We go on now to consider the influence of alkali upon denaturation.

From Chick and Martin's experiments [1913] it follows that in the presence of very weak alkali concentration (hydroxyl-ion concentration equal to $10^{-3\cdot8} N$) the denaturation proceeds too slowly and only on addition of more alkali (hydroxyl-ion concentration $10^{-2\cdot7} N$) does the denaturation-rate increase so as to be comparable with that in the original, slightly acid solution. These experiments confirm therefore the above hypothesis that the denaturation of albumin, free from electrolytes, is accelerated by its own hydrogen-ions present in the solution in concentration of about $10^{-5} N$: after the neutralisation of albumin by a small amount of alkali the accelerating effect of these ions is abrogated and the denaturation-rate strongly diminished. The strong increase of the latter on the addition of more alkali could then be explained by the formation of albumin salts with alkali (alkali albuminate) which react more easily with water than the unaltered albumin.

Unfortunately it was only possible for Chick and Martin to institute a comparison over a small portion of the curves expressing the dependence of the denaturation-rate upon the hydroxyl-ion concentration (namely, from a concentration equal to $10^{-3\cdot8}$ to $10^{-2\cdot7}$). Nevertheless it seemed to me to be desirable to investigate the effect of stronger concentrations of alkali upon the denaturation-velocity. My experiments made for this purpose showed that an increase of the hydroxyl-ion concentration beginning from $10^{-2\cdot2} N$ only influence the denaturation-rate very little. Thus an increase of alkalinity of over sixty-fold was accompanied by a decrease in denaturation-rate of 27 %.

In my experiments a solution of serum-albumin Kahlbaum dialysed during 6 weeks and containing 0·49 % albumin was used. This solution was diluted with an equal volume of a solution of potassium hydroxide of known concentration and in order to keep it coagulable ammonium sulphate (concentration of 0·2 N, or 0·4 N when the alkali concentration was great) was added to it. When the hydroxyl-ion concentrations were 0·006 N (=$10^{-2\cdot2}$), 0·012 N, 0·067 N, 0·150 N, 0·400 N, the average denaturation-times at 70° were found to be equal respectively to 270 secs., 285 s., 300 s., 320 s., 370 s.

The study of the influence of alkalies and acids upon the denaturation showed therefore that the above hypothesis concerning the nature of this reaction is true: the denaturation is a weak hydrolysis of protein.

This hydrolysis can be accompanied also by the splitting off of any amino-acids containing sulphur from the protein molecule [Moll, 1904] and by an increase in the amount of carboxyl group, for, as will be shown below, the acid properties of albumin increase after the denaturation.

Concerning the known increase of alkalinity of a solution of albumin, after denaturation, Chick and Martin showed that the diminution of acidity during heat-coagulation takes place only in acid solutions, while in alkaline solutions a diminution of alkalinity is observed [Chick and Martin, 1913, 1]. It has already been pointed out that protein can unite with acids and alkalies before denaturation; the coagulation following the latter would then evidently remove not only protein from the solution but also a quantity of acid or alkali so that the acidity or the alkalinity of solution would be diminished during the heat-coagulation.

(4) Influence of salts upon the denaturation-velocity of protein.

As mentioned above, salts increase the catalytic power of hydrogen-ions and produce thereby an increase of inversion-rate of sugar. Salts probably also accelerate the denaturation of protein for they increase the catalytic power of hydrogen-ions whose presence in the electrolyte-free protein solution is due to the electrolytic dissociation of the protein molecules which are more acid than alkaline (see above, hydrogen-ion concentration of about $10^{-5} N$). Moreover, Arrhenius found the accelerating effect of salts upon inversion to increase with the increasing salt concentration. On the other ha.. ., as is known, salts diminish the electrolytic dissociation of electrolytes, and they may thus diminish that of protein; moreover the researches of Lillie [1907] showed that salts considerably diminish the osmotic pressure of albumin solutions; that is to say they lower their degree of dispersion and, therefore, the particle-surface of the protein; so that the reaction between protein an. ` water may be retarded.

We should therefore expect that an increase of salt concentration would accelerate the denaturation of albumin when the salt concentration is small, it would however diminish it when the salt concentration is great and it would leave it unaltered if the salt concentration is intermediate.

My experiments showed that this suggestion is right. But, unfortunately, they could not render clear the effect of very small concentration of salts upon the denaturation-velocity, the method employed having required the presence in the protein solution of an amount of salt sufficiently great to produce the coagulation of the denaturated albumin (see above).

The material used in my experiments was egg-albumin Kahlbaum, which had been dialysed many days (content 0·66 %). When the concentration of added ammonium sulphate was 0·03 N, 0·06 N, 0·12 N, 1·1 N and 1·32 N, the average times of denaturation at 70° were found respectively to be 932 secs., 185 s., 185 s., 270 s. and 605 s. These results are in accordance with the results of Chick and Martin, who investigated only the concentrations from 0·1 to 1 N and above.

As is known, salts accelerate the inversion of sugar by acid in different degree, according to their lyotropic properties, which are chiefly due to the effect of anions: $SO_4 < Cl < Br < SCN$ [Spohr, 1888]. It was therefore to be expected that salts would accelerate the denaturation of acid solutions of protein in the same series of anions. Indeed, my experiments showed that the denaturation-velocity of protein in acid solutions is incomparably more increased by potassium thiocyanate than by potassium sulphate; potassium chlorate takes the middle position but it is, in its effect, nearer to potassium sulphate.

The material used was egg-albumin Kahlbaum, the 2 % solution of which had been dialysed during 5 days. The solution obtained was diluted with water and 0·1 N HNO$_3$, so that the content of albumin in all the solutions was 0·41 %, and the content of acid 0·01 mol. per litre. To the solutions such quantity of various salts was added as to render the salt-content of each solution equal to 0·083 mol. per litre. In the presence of KCNS, KCl, K$_2$SO$_4$ the average time of denaturation at 50° was found respectively to be 15 secs., 3000 s. and 4650 s.

If albumin solutions contain no free hydrogen-ions but are slightly alkaline, the presence of salts could not evidently accelerate the denaturation (there are no hydrogen-ions in the solution the hydrolytic power of which could be increased), and it is to be expected that the nature of the anion would have no influence upon the denaturation-velocity.

Indeed, my experiments showed that the denaturation-rate in this case is almost equal in the solutions containing salts of various anions. A slightly alkaline solution of egg-albumin Kahlbaum had been dialysed until it no longer coagulated on heating. When 0·1 N K$_2$SO$_4$, KCl, KI and KOCOCH$_3$ were added, it coagulated at 75° respectively in 212, 220, 225 and 227 secs.

Concerning the increasing influence of various cations upon the catalytic power of hydrogen-ions in the case of denaturation, it is almost the same for various alkali metals. The acceleration of denaturation by salts of alkaline earths and heavy metals is however so strong, as to make necessary the suggestion that the salts in question first enter into chemical unions with protein which are soluble in water and react with it more easily than the free protein, and therefore are more easily denaturated. The other suggestion that alkaline earths and heavy metals form insoluble compounds with native protein (which are therefore no longer capable of denaturation by heating), the formation of which is accelerated by high temperature as are all other chemical reactions, is less probable, for the temperature coefficients of the reaction were found in my experiments to be too high: 1·28–1·38 per rise of 1°.

A solution of egg-albumin Kahlbaum, after dialysis during 6 days, was diluted with water and salt solutions so that all the solutions to be tested contained 0·3 % albumin. In the presence of 0·1 N KCl the average time of denaturation was found, at 75°, to be 2095 secs., while in the presence of 0·1 N BaCl$_2$ denaturation proceeded instantaneously at the same temperature; in the presence of 0·003 N BaCl$_2$ it required 95 secs., 0·003 N MgCl$_2$ 120 secs., 0·0001 N HCl 52 secs., while it proceeded instantaneously in the presence of 0·003 N HCl.

The study of the influence of salts upon denaturation showed, therefore, that there was no contradiction with the foregoing hypothesis.

(5) THE CAUSE AND THE TEMPERATURE COEFFICIENT OF THE COAGULATION OF DENATURATED PROTEIN.

As mentioned above, it has long been known, that protein solutions made heat-incoagulable by dialysis acquire the property to coagulate on heating when some electrolyte is added to them. They coagulate also on the addition of electrolytes when preliminarily heated and cooled. The presence of electrolytes is therefore required to produce the second process comprised in heat-coagulation. In order to render this process clear it was first of all necessary to answer the questions, whether the coagulation of denaturated protein is a pure physical process or, at the same time, also a chemical reaction; whether this process is, in its nature, like the other processes of coagulation of suspension-colloids by electrolytes.

In this connection it was interesting to estimate the temperature coefficient of coagulation of denaturated protein and to compare it with that of chemical reactions and of other coagulations.

My experiments showed firstly that the coagulation of denaturated protein proceeds at a certain temperature considerably more rapidly than the denaturation at the same temperature if the protein solution contains a sufficiently great amount of salt.

Thus, in one of my experiments, the average time required to produce the necessary turbidity of 0·57 % solution of dialysed egg-albumin Kahlbaum to which ammonium sulphate was added to the concentration of 0·05 N at 65° was found to be 2220 secs., while the average time required to produce the same turbidity of the same albumin solution which had been boiled and to which after cooling the same amount of ammonium sulphate was added (*i.e.* the time to produce the coagulation) was found to be 10 secs. at 65°.

Further, in accordance with Chick and Martin, my experiments showed that the average temperature coefficient of coagulation of denaturated albumin is usually equal to 1·1 (it varies from 1·08 to 1·2) per temperature-rise of 1°. But sometimes it increases to 1·3. The results of four of my experimental series are the following:

I. Solution of egg-albumin Kahlbaum made non-coagulable on heating (by dialysis) and containing 0·57 % protein was boiled. After cooling $(NH_4)_2SO_4$ was added to it to the concentration of 0·025 N. The average times of coagulation at 80°, 75°, 70°, 65°, and 60° were found to be equal respectively to 165 secs., 405 s., 943 s., 1922 s., 2800 s. The temperature coefficient = 1·15.

II. Solution of egg-albumin Kahlbaum. Content of albumin = 0·41 %. Content of ammonium sulphate added after boiling and cooling the solution = 0·05 N. The average times of coagulation: at 65° 10 secs., at 60° 40 secs., at 55° 138 secs., at 53° 251 secs., at 50° 687 secs., at 47° 1630 secs. The temperature coefficient = 1·33.

III. 0·14 % *solution of recrystallised egg-albumin*, made non-coagulable on heating by dialysis. Content of ammonium sulphate added after boiling and cooling to the solution = 0·1 N. The average times of coagulation were: at 45° 72 secs., at 40° 152 secs., at 35° 425 secs., at 30° 1200 secs. The average temperature coefficient = 1·2.

IV. *Egg-white* had been dialysed till the absence of heat-coagulation, diluted with water fivefold, boiled and cooled. $(NH_4)_2SO_4$ was then added to the concentration of 0·1 N. The average times of coagulation were: at 60° 35 secs., at 50° 297 secs., at 45° 687 secs., at 40° 1325 secs. The average temperature coefficient = 1·18.

As known, the denaturated protein is regarded as a suspension-colloid (see above). It was therefore interesting to compare the temperature coefficients found with those of coagulation of other suspension-colloids. But there are yet no investigations dealing with the dependence of coagulation upon temperature. Scattered up and down a number of observations only pointed out that colloidal solutions are generally less stable at higher temperatures [see, for example, Zsigmondy, 1912].

I therefore made some experiments with colloidal solutions of arsenic trisulphide and lecithin by means of the method above described, which, in the case of arsenic, was a little modified; the vessel for the colloidal solution was observed with the intact horizontal microscope (small magnification) and the time-intervals were noted which were required to produce a complete coagulation. This was marked by the appearance of orange particles of arsenic trisulphide, all through the liquid, on the pure white ground.

A 0·5 % colloidal solution of arsenic trisulphide (obtained by passing sulphuretted hydrogen into water containing powdered As_2O_3), from which H_2S had been removed by a current of air, was centrifuged until particles visible under the microscope were absent. Then ammonium sulphate was added to the resulting colloidal solution to the concentration of 0·01 mol. per litre.

The average times of coagulation were found to be at 85° 228 secs., at 75° 401 secs., at 65° 601 secs., at 55° 1110 secs. The temperature coefficient = 1·05 (per rise of 1°).

In another experimental series where 0·4 % solution of arsenic trisulphide was used and the concentration of added salt (NaCl) was 0·04 mol. per litre, the average times of coagulation were: at 60° 19 mins., at 50° 29 mins., at 40° 39 mins., at 30° 56 mins. and at 17·3° 86 mins. The temperature coefficient was therefore 1·04 (per rise of 1°).

1 % solution of lecithin in water was made by mixing an ethereal solution of lecithin with water and removing ether by blowing air through the solution. Twenty-four hours after the preparation ammonium sulphate was added to the solution (concentration 0·08 N). The average times of coagulation were at 90° 106 secs., at 80° 500 secs., at 70° 910 secs., at 60° 1960 secs. The average temperature coefficient was therefore 1·10. When the concentration of

lecithin was 0·5 %, the times of coagulation were: at 90° 376 secs., at 80° 606 secs., at 70° 1356 secs., at 60° 3290 secs. The average temperature coefficient was therefore 1·07 (per rise of 1°).

From the experiments cited it may be seen that the temperature coefficient of coagulation of denatured albumin is greater than that of arsenic trisulphide or lecithin but it is nearer to that of lecithin, which is known to occupy an intermediate position between hydrophilic and hydrophobic colloids (resp. emulsoids and suspensoids). In the case of arsenic trisulphide the temperature coefficient of coagulation is like that of the diffusion of salts. The temperature-rise only increases the quantity of salt acting upon the colloidal particles. The temperature effect, in the case of lecithin coagulation, is however more complex. The temperature influences in this case not only the salt quantity acting upon the colloidal particles, but also the degree of dispersion and therefore the reacting surface of the colloidal particles, and the coefficient is greater.

The most complex effect of temperature is that upon the coagulation of denatured protein. The temperature influences probably, in this case, not only the dispersion of the colloid and the acting salt quantity, but also the rate of reaction between salt and protein. For, a formation of loose chemical compounds ("adsorption compounds") of salts with denatured protein during the coagulation is very probable (see below).

(6) Influence of Acids and Alkalies upon the Coagulation of Denatured Protein.

Denatured proteins are known to be soluble in alkalies and in strong acids [Cohnheim, 1904], so that the formation of compounds of denatured proteins with alkalies and acids is undoubted. From the solution of denatured albumin in weak potassium hydroxide the protein can be precipitated by ammonium sulphate. The precipitate is however easily soluble in a stronger solution of this alkali, but a greater quantity of salt precipitates protein anew. This precipitate is also soluble in more alkali, etc. The increase of alkali concentration produces therefore a continuous change of colloidal properties of denatured albumin owing evidently to the formation of alkali compounds of protein, and the greater the alkali concentration of the solution the more alkali these compounds contain.

We should therefore expect that alkalies and acids would have a considerable influence upon the coagulation-velocity of denatured albumin. My experiments showed, indeed, that, generally, acid strongly increases and alkali strongly diminishes the coagulation-rate of denatured albumin. Moreover 0·00005 N HNO$_3$ increases this rate more strongly than 0·0001 and even 0·0002 N acid so that the increasing effect cannot be explained by the influence of hydrogen-ions; it is evidently due to the formation of acid compounds of denatured protein the colloidal properties of which change with the amount of acid contained in them. The same result was also found for the alkali effect. The decrease of coagulation-velocity is in this case not at all proportional to the hydroxyl-ion concentration: $0.001\,N$ hydroxyl-ion concentration, in the case of ammonia, produces a similar diminution of velocity as $0.011\,N$ hydroxyl-ion concentration in the case of potassium hydroxide. Conversely the like molecular concentrations of ammonia and potassium hydroxide bring about a like decrease of coagulation-velocity so that the decreasing effect of alkali cannot be explained except by the formation of alkali compounds of denaturated albumin having other colloidal properties than the free albumin.

The experiments were made with egg-albumin Kahlbaum, the solutions of which had been dialysed during many days and no longer coagulated on heating.

I. A solution of albumin (containing 2·07 % protein) was boiled and, after cooling, diluted fivefold partly with water, partly with nitric acid. To the solutions obtained potassium sulphate was added to the concentration of 0·083 mol. per litre. The average time of coagulation was found at 75°, in the absence of acid, to be 2700 secs., whilst it was found with $0.00005\,N$ acid to be 102 secs., with $0.00010\,N$ acid 236 secs., $0.00020\,N$ acid 178 secs. But in the presence of $0.01\,N$ acid the coagulation proceeded instantaneously.

II. A solution of albumin (containing 3·2 % protein) was boiled and, after cooling, diluted fivefold partly with water, partly with solutions of potassium hydroxide or of ammonia of certain concentrations. Ammonium sulphate was added to the concentration of $0.06\,N$. The coagulation proceeded at 70°, in the absence of alkali, instantaneously, while the average time of coagulation, with $0.0018\,N$ ammonia (H-ion concentration $= 0.00031\,N$), was 370 secs., with $0.018\,N$ ammonia (H-ion concentration $= 0.00100\,N$), 760 secs., of $0.2\,N$ ammonia (H-ion concentration $= 0.00400\,N$), 2200 secs. In the presence of $0.018\,N$ KOH (H-ion concentration $= 0.01100\,N$) it was 770 secs.

My experiments showed further, that the temperature coefficient of coagulation is in the presence of acid still less than that of coagulation of denaturated albumin in the presence of alkali and even than that of coagulation of arsenic trisulphide. So in a series of experiments, in which dialysed egg-albumin Kahlbaum in the presence of 0·0001 mol. nitric acid per litre was tested, the coagulation time was found to be equal to 237 secs. at 75°, to 400 secs. at 65°, to 720 secs. at 55°. So that the average temperature coefficient was 1·05. In another series, in which the acid concentration was 0·0002 mol. per litre, the coagulation time was: 760 secs. at 75°, 1597 secs. at 55° and 1720 secs. at 50°. The average temperature coefficient of coagulation was, therefore, 1·02. In the presence of 0·018 mol. ammonia per litre this coefficient was found to be 1·12 and in the presence of 0·18 mol. ammonia per litre 1·19.

We might thus suggest, that the colloidal state of acid compounds of denaturated albumin is nearer to that of typical suspension colloids, while that of alkali compounds is nearer to the state of emulsion colloids. The free denaturated albumin seems to occupy an intermediate position between the alkali compounds and the acid compounds.

Finally, my experiments showed that the acid compounds of denaturated albumin are not identical with the denaturated acid compounds of native albumin, formed in the presence of the same concentration of acid. Its colloidal state is different.

One portion of solution of dialysed egg-albumin containing 2·07 % protein and not coagulating on heating was diluted two and a half-fold with water and boiled up. After cooling, to this protein solution a solution of potassium sulphate and diluted nitric acid were added, altogether in such quantities that the original albumin solution was diluted fivefold. The concentration of acid was 0·0002 mol. per litre, that of potassium sulphate 0·083 mol. The average time of coagulation at 65° was found to be 240 secs., at 55° 365 secs.

The second portion of the same solution of egg-albumin was directly diluted fivefold with diluted nitric acid so that the acid concentration was also 0·0002 mol. per litre. Subsequently the solution was boiled up and cooled; afterwards potassium sulphate was added to the concentration of 0·083 mol. per litre. The average time of coagulation at 65° was found to be 320 secs., at 55° 660 secs.

(7) INFLUENCE OF THE DEGREE OF DISPERSION OF DENATURATED ALBUMIN UPON ITS COAGULATION-VELOCITY.

My experiments relating to the comparison of the time required to produce a certain degree of denaturation of an albumin solution with the time required to bring about the coagulation of the same albumin denaturated by boiling led to an interesting observation which is in accordance with the fact cited above (found by Lillie) that salts lower the degree of dispersion of albumin solutions.

My experiments showed, namely, that the appearance of the definite turbidity of a 0·57 % solution of dialysed egg-albumin Kahlbaum, containing 0·1 mol. potassium chlorate per litre, required on an average 220 secs. at 75°, and 1630 secs. at 71°. However, the same turbidity appeared in 1380 secs. at 75° and in 2020 secs. at 71°, when before adding potassium chlorate the same albumin solution had been boiled and cooled. In both cases the turbidity marked the coagulation of denaturated albumin, but in the first case this coagulation, following the denaturation, proceeded in a lapse of 220 and 1630 secs. (or still more rapidly), while in the second case this coagulation required 1380 and 2020 secs.

It is evident that the colloidal properties of denaturated albumin formed in the solution containing 0·1 mol. potassium chlorate were unlike those of denaturated albumin formed in the solution containing no potassium chlorate, for the former albumin required less time to be coagulated than the latter.

This phenomenon can only be explained by a greater degree of dispersion of the denaturated albumin formed in absence of potassium chlorate. Lillie showed that the degree of dispersion of albumin in the presence of salts is smaller than in the absence of it, even before the denaturation takes place. The denaturation of an albumin of a smaller degree of dispersion leads therefore to the formation of a denaturated albumin which also has a smaller degree of dispersion.

In agreement with the results obtained by Lillie, my experiments showed that the presence of potassium chlorate even in concentration of 0·01 mol. per litre affects the colloidal properties of albumin. So in one of my experiments a 1 % solution of dialysed egg-albumin Kahlbaum (not coagulating on heating) was boiled and, after cooling, diluted with an equal volume of 0·2 N KCl. This albumin solution coagulated at 75° on an average in 1380 secs. On the other hand, to the same solution of albumin Kahlbaum KCl was first added to the concentration of 0·01 mol. per litre and the solution was then boiled and after cooling diluted with an equal volume of 0·19 N KCl. In this case the protein solution coagulated at 75° on an average in 225 secs., though the concentration of KCl was in both cases the same, namely 0·1 mol. per litre.

The coagulation-velocity of denaturated albumin formed in the presence of potassium chloride was found therefore to be almost sixfold as great as the coagulation-velocity of denaturated albumin formed in the absence of salt. The result obtained corresponds precisely with the value of diminution of osmotic pressure of albumin solution by potassium chloride found by Lillie.

(8) INFLUENCE OF SALTS UPON THE COAGULATION OF DENATURATED ALBUMIN.

In the preceding part of this paper it has several times been pointed out that protein solutions from which salts have been removed by dialysis do not coagulate on heating, but they become coagulable when salts are added to them. Conversely Pauli and his pupil Handovsky affirm that after a very long dialysis the serum (viz. serum-albumin) acquires the property to coagulate on heating without adding salts. Salts added to such serum bring about a rise of its "coagulation temperature," and thiocyanates and iodides make it even heat-uncoagulable [Pauli, 1908; Pauli and Handovsky, 1908, 1909].

These authors call the protein contained in their serum "amphoteric protein," pointing out that it conducts the electrical current not more than distilled water and shows no cataphoresis. It is however a pity that the authors did not define the salt content of their protein, as for instance, by determining the quantity of ash, for salts might not have been dissolved in the water of the serum but adsorbed on the protein particles. Then after the denaturation of albumin they might be liberated anew and produce the coagulation.

At all events, it is impossible to regard the small salt content of "ampho-teric protein" of Pauli, as the cause of the possibility of its heat-coagulation, and to suppose that the absence of heat-coagulation of dialysed protein solutions which has been observed by all other authors is due to the incom-pleteness of the dialysis. Indeed, the addition of salts to the "amphoteric protein" could cause only a rise of coagulation temperature, and only thio-cyanates and iodides (normally absent in natural liquids) made, in the experiments of Pauli, the heat-coagulation of serum impossible.

In order to render the influence of salts upon the coagulation of denaturated protein clear, I made experiments with egg-albumin and serum-albumin Kahlbaum. We consider first the influence of salts upon the coagulation of egg-albumin.

The solutions of egg-albumin (5–10 %) were usually centrifuged and filtered. They had a slight alkaline reaction (litmus) and were almost water-clear. After drying a part of the solutions in vacuum and afterwards in the drying-stove the dry albumin was found to contain about 3 % ash, of which the most part (2 %) was water-soluble (KCl, NaCl, K_2SO_4, K_2CO_3). The solutions coagulated on heating.

The dialysis was executed by means of the thinnest parchment paper which could be purchased. If the dialysis was prolonged an antiseptic was added to the solutions tested and the dialyser was placed in a covered vessel. The solution layer in the dialyser never exceeded $\frac{1}{2}$ cm. in depth.

After 24 hours of dialysis (distilled water was changed each two hours) the protein solutions usually became neutral (litmus) in reaction and then lost the property of coagulating on heating (if the protein-content was relatively great, e.g. 10 %) or they lost this property only after two or three days of dialysis (if the protein-content was small).

The dialysed solutions were usually centrifuged and filtered anew till they became water-clear. The content of protein and of mineral substances in the solutions (which did not coagulate on heating) is set out below.

Number of tested solutions	Original concentration of albumin	Duration of dialysis	Concentration of albumin after dialysis	Content of mineral substances in dry albumin after dialysis	Content of soluble mineral substances in g. per 100 cc. of solution after dialysis
	%	days	%	%	
1	10	1	7·5	1·8	0·05
2	4·9	2	3·29	1·1	0·02
3	4·1	5	2·07	1·2	0·01

In order to answer the question whether solutions of egg-albumin which no longer coagulate on heating would acquire the property to coagulate anew by a still more prolonged dialysis, solutions 1 and 3 were employed for a further dialysis.

To solution No. 1 an excess of chloroform was added as an antiseptic. After 10 days of dialysis at 20° this solution showed heat-coagulation anew.

But it acquired an unpleasant scent and, in spite of the chloroform, contained many bacteria. The heat-coagulation was evidently due in this case to organic acids or salts which had been formed by the bacteria. Indeed, after a dilution of the solution with water the heat-coagulation disappeared anew, but it reappeared if some ammonium sulphate was added to it.

To solution No. 3 an emulsion of melted thymol in water was added as an antiseptic. The dialysis lasted three weeks. The solution contained after such dialysis 0·99 % albumin. The dry albumin (drying in vacuum) was found to contain 0·8 % ash of which about a half was soluble in water. The solution contained 0·008 % dissolved mineral substances and coagulated on heating. An unpleasant scent and bacteria were however absent.

The albumin solution thus obtained showed the required degree of coagulation (turbidity like that of the ground glass, see above) at 75° in 27 secs. After dilution of the solution with water to double the volume, the same degree of coagulation was observed at 75° in 700 secs., at 80° in 370 secs., at 70° in 1260 secs. The striking increase of coagulation-time after a relatively inconsiderable dilution with water indicates that the very small amount of salts contained in the albumin, which was sufficient to bring about the coagulation of protein in 27 secs. at 75°, after the dilution with water, could no longer produce the coagulation in this time. At the same time, the small temperature coefficient—1·12 and 1·14 per temperature-rise of 1°—showed that the coagulation-velocity, after the dilution with water, was smaller than the denaturation-velocity.

When this albumin solution was diluted fivefold with water, it ceased to coagulate even on boiling, whereas it coagulated easily if some ammonium sulphate was first added to it. Similarly it coagulated if after boiling and cooling this salt was added to it.

It seemed, therefore, that the heat-coagulation of denaturated egg-albumin has two optimal concentrations of salts present in the solution. Namely the coagulation on heating takes place when this concentration is more than 0·1 % (viz. about 0·01 N), it ceases to be observed when this concentration is between 0·01 and 0·05 % (viz. about 0·001 and 0·006 N), it appears anew when this concentration diminishes to 0·008 % (viz. about 0·0008 N), and, finally, it ceases again to appear when the salt-concentration becomes about 0·001 % (viz. about 0·0001 N).

Similarly, two most favourable salt-concentrations, viz. two coagulation-zones, were observed also on the other colloids, as for example, on lecithin and cholesterol [Porges and Neubauer, 1908]. Nevertheless, my experiments could not confirm the existence of any coagulation-zones in the case of heat-coagulation of protein.

They showed that a gradual increase of salt-content produces, instead of a retardation or a cessation of heat-coagulation, an acceleration of this coagulation of protein. In my experiments, the above solution of albumin containing 0·008 % dissolved mineral substances was diluted with an equal volume of solution of potassium chlorate of various concentrations and the heat-coagulation time was determined by the previous method.

When potassium chlorate was substituted by potassium thiocyanate, the phenomenon remained the same, and only very strong concentrations of this salt (2·5 mol. per litre) brought about a diminution of coagulation-velocity. A suppression of heat-coagulation by adding KCNS could never be observed. At a concentration of KCl of 0·003 N the temperature coefficient was found to be equal to 1·17; and only at the concentration of 0·01 N did this coefficient attain the value of the normal denaturation coefficients 1·35–1·40, and the denaturation began to proceed more slowly than the coagulation of the denaturated albumin. The same was also observed, when a sufficient amount of KCNS (0·1 mol. per litre) was added to the solution. On the other hand an excess of this salt (2·5 mol. per litre) retards the coagulation anew and the denaturation begins to proceed more rapidly than the coagulation (the temperature coefficient = 1·04).

The albumin contained in the solution which had been dialysed during four weeks is therefore not identical with that contained in the solution which had been dialysed during 5–7 days. After denaturation the former albumin coagulates more easily under the influence of salts than the latter. The degree of dispersion of the former is probably smaller than that of the latter. At all events the very prolonged dialysis evidently alters the original albumin. Both albumins are however "amphoteric" (see Pauli), for neither the one nor the other showed in my experiments cataphoresis by an electrical current of 110 volts; but both migrated to the anode after denaturation (according to the increase of alkalinity of the solution).

We pass now to the experiments with serum-albumin.

A 5 % solution of serum-albumin Kahlbaum was centrifuged and filtered. The slightly alkaline solution obtained was very faintly opalescent, coagulated on heating and contained 3·07 % dry residue of which 0·25 % was mineral matter.

The dry protein therefore contained 8·4 % mineral substances. The soluble part of the latter consisted of sodium chloride, potassium chloride, sodium carbonate, etc., the insoluble part was also insoluble in acids (SiO_2?).

The albumin solution obtained was dialysed during 20 hours. After the dialysis it was centrifuged and filtered anew. It reacted neutral (litmus) and no longer coagulated on heating, but only became opalescent. It now contained 2·4 % dry residue, of which 0·07 % was mineral and 0·05 % water-soluble. The dry albumin contained now about 3 % mineral substances.

After adding salts to this albumin solution heat-coagulation could again be produced on heating.

The high temperature coefficient of heat-coagulation (= 1·55) after adding potassium chloride showed that 0·1 mol. per litre of this salt was already sufficient to make the coagulation of the denatured albumin proceed more swiftly than the denaturation.

In fact, heating the solution of dialysed serum-albumin after a preliminary boiling and cooling and the subsequent addition of potassium chloride to a concentration of 0·1 mol. per litre, showed in my. experiments that the coagulation of the denatured albumin at 70° required on the average only 35 secs., at 68° 50 secs., at 65° 90 secs., etc. On the other hand the same concentration of potassium thiocyanate only brought about a coagulation which proceeded slower than the denaturation, and the temperature coefficient was found in this case to be low. Only between 65° and 70° did the denaturation-velocity begin to be greater than the coagulation-velocity and the temperature coefficient increased, while the coagulation-time became equal to that observed in the presence of potassium chloride.

Further dialysis renders the solution of serum-albumin heat-coagulable again, and even after 6 weeks of dialysis the solution coagulated on heating. At the same time even after 5 days of dialysis the filtered solution contained 1·28 % dissolved solids, of which 0·02 % was mineral; the dry albumin therefore now contained 1·4 % mineral substances (half of which was soluble in water). Further dialysis however altered this salt-content of the albumin solution very little.

Similarly to the solutions of egg-albumin those of serum-albumin, when dialysed during a long time, had a low temperature coefficient of heat-coagulation, the denaturation proceeding evidently more rapidly than the coagulation of the denatured albumin. The heat-coagulation at 70° required in my experiments on the average 840 secs. and at 80° 510 secs.; the temperature coefficient per rise of 1° was thus found to be equal to 1·05.

Further, a very small concentration of salts (KCl or KCNS) was found in my experiments to accelerate the coagulation of denatured serum-albumin, and the increase of salt concentration accelerated it so strongly that its velocity became greater than the denaturation-velocity, and the temperature coefficient was found high again and like that observed for solutions of serum-albumin which had not been dialysed (1·69).

The results obtained in my experiments on serum-albumin are therefore like those obtained on egg-albumin. The difference is only quantitative. The solution of serum-albumin loses its mineral substances considerably more readily by dialysis than the solution of egg-albumin. After two days of dialysis the solution of serum-albumin has lost the greater part of its mineral matter; and after 5 days only one-twelfth of this remains in the solution.

Further, the disappearance of heat-coagulation of serum-albumin is observed when the salt-content of the solution is between 0·03 and 0·07 %, while the absence of heat-coagulation of egg-albumin takes place when this content is between 0·01 and 0·05 %. On the other hand the addition of salts (even of potassium thiocyanate) to the solutions of both albumins, made coagulable on heating by a very prolonged dialysis, cannot render them incoagulable on heating anew.

We have thus to conclude that in both cases albumin is altered by a very prolonged dialysis in such a manner that, after denaturation it shows a greater susceptibility to salts than before. In what however this alteration consists, is still unknown. It is probable that the degree of dispersion of both albumins became smaller by lapse of time, or that a small part of the salts, contained in albumin solutions, are chemically united with the native albumin and the prolonged dialysis decomposes such salt compounds. Free albumin, after denaturation, is in this latter case more susceptible to salts than its unions with salts.

It was formerly pointed out that Pauli and Handovsky affirm that they have found the addition of salts to "amphoteric albumin" to bring about an increase of its "coagulation-point." This observation contradicts the results of my experiments which showed, as mentioned, an increasing effect of salts upon the coagulation-velocity of albumins. But nevertheless, according to Pauli and Handovsky, my experiments showed that potassium thiocyanate and iodide in strong concentration (more than 1 mol. per litre) diminish the coagulation-velocity of denaturated albumin and in the case of serum-albumin can even make it incoagulable on heating when they are present in almost saturated solution. This phenomenon can be explained by the formation of chemical compounds of these salts with denaturated serum-albumin. This albumin is still soluble in a strong and boiling solution of potassium thiocyanate [Pauli and Handovsky, 1908][1].

Moreover, the results of my experiments concerning the influence of dilution with water upon the coagulation-velocity of denaturated albumin and particularly upon that of denaturated serum-albumin also contradict the result obtained by Pauli and Handovsky. These authors point out that an increasing concentration of protein produces a considerable rise of the coagulation-point. My experiments show however that dilution of heat-coagulable albumin solutions which had been very long dialysed diminishes the coagulation-velocity and a sufficiently strong dilution makes it so small as to render it inaccessible to observation.

What is the cause of all these contradictions? Could it not lie in that conjuncture which had taken place in the experiments of Pauli and Handovsky, who, before determining the coagulation-point of serum dialysed 6 weeks, left it to settle during 3–5 months [1908, pp. 416–8]? It is possible that during such a long time the serum-albumin was altered under the influence of water and the oxygen of the air. In one of my experiments, at least, a solution of serum-albumin, having been dialysed during 6 weeks and coagu-

[1] The degree of dispersion of compounds of albumin with thiocyanates and iodides is probably only greater than that of free denaturated albumin, a sufficient amount of ammonium sulphate making them coagulable anew.

lating on heating lost this property and became slightly alkaline after standing during one year (thymol as antiseptic).

We now pass to the experiments on the influence of salts upon the coagulation-velocity of denaturated albumin dialysed during some days. In these experiments albumin solutions were dialysed until they no longer coagulated on heating without the addition of salts.

We have seen above that increase of salt concentration accelerated the coagulation of denaturated albumin which had been very thoroughly dialysed and coagulated on heating. By virtue of low temperature coefficients we have, indirectly, concluded that in this case of heat-coagulation we have been dealing with the coagulation of denaturated albumin. To investigate the influence of salt concentration upon the coagulation of denaturated albumin not coagulating on heating, without the addition of salts, is evidently much simpler. The solutions of dialysed albumin were in my experiment first boiled up and, after cooling, mixed with salt solutions of a certain concentration; the albumin solutions obtained were then tested at certain temperatures in the thermostat. I will cite here an example of my experiments.

A solution of egg-albumin Kahlbaum containing 2·85 % protein and not coagulating on heating (owing to dialysis during 5 days) was boiled up, and, after cooling, diluted fivefold with solutions of ammonium sulphate of various concentrations so that the albumin solutions obtained contained 0·025, 0·05, 0·1 and 0·15 mol. ammonium sulphate per litre. At 47° the times required to produce the standard turbidity were found to be respectively equal (on an average) to 10,600 secs., 1620 s., 40 s. and 17 s.

The experiment cited shows that the dependence of the coagulation-rate upon the salt concentration of the solution is not like the dependence of the adsorption of salt upon the salt concentration. The coagulation is therefore not a simple adsorption of electrolytes whose ions electrically discharge the colloidal protein particles.

According to the well-known adsorption-isotherm $\frac{x}{m} = ac^{\frac{1}{n}}$, the adsorbed quantity of dissolved substance increases more slowly than the concentration of this substance in the solution. In my experiment however the increase of the salt concentration of twofold was accompanied by an increase of coagulation-velocity of ten- or fourteen-fold.

As mentioned above Hardy showed that the denaturated protein can be precipitated by adjusting the reaction so as to render the particles iso-electric with the solution; but when the protein is made negative the valency of the cation is of importance for its precipitation. It has already been pointed out that the colloidal particles in the solution of egg-albumin Kahlbaum, which has been dialysed for some days (4–6 days), are iso-electric with the solution (no cataphoresis was observed) and that they become negative after denaturation. According to Hardy we should therefore expect that the cation only would be important in the coagulation of denaturated protein of the albumin solution in question. My experiment showed however that not only cations but also anions are of importance in the process. The results of experiments are set out below.

I. The material used was egg-albumin Kahlbaum, a solution of which had been dialysed for 6 days (albumin-content 2·8 %). It was boiled and after cooling diluted with solutions of salts (fivefold). The average times of coagulation at 75° were found to be 2100 secs. in the presence of 0·1 N KCl, 1450 secs. at 0·1 N NH$_4$Cl, 1180 secs. at 0·1 N NaCl, 1057 secs. at 0·1 N LiCl, 1287 secs. at 0·003 N BaCl$_2$, 1320 secs. at 0·003 N MgCl$_2$, 1080 secs. at 0·1 N FeCl$_3$, 2540 secs. at 0·1 N AlCl$_3$. In the presence of 0·02–0·1 N BaCl$_2$ or MgCl$_2$ the coagulation proceeded instantaneously. In the presence of 0·003 N FeCl$_3$ very slowly.

II. The same solution of egg-albumin. The average times of coagulation at 75° were found to be 215 secs. in the presence of 0·1 N K$_2$SO$_4$, 826 secs. at 0·1 N potassium tartrate, 1070 secs. at 0·1 N KNO$_3$, 1363 at 0·1 N KCl, 1406 secs. at 0·1 N KBr, 1826 secs. at 0·1 N KI and 3826 secs. at 0·1 N KCNS.

Therefore, the cations act upon the coagulation-velocity of denaturated albumin not at all proportionally to their valency; the tervalent cations (Al, Fe) produce an acceleration of this velocity almost equal to that produced by the univalent ions. The bivalent cations (Ba, Mg) are however strikingly active, producing in the concentration of 1/33 of the univalent ions an almost equal acceleration of coagulation-velocity. In the single series of cations of equal valency the accelerating effects are as follows: Li > Na > NH$_4$Cl > K; Ba > Mg and Fe > Al. In all cases the temperature coefficients are relatively great (1·1), and then the suggestion emerges that the different effects of cations are due to some chemical influence produced by these cations. The latter could, for example, form adsorption compounds with denaturated protein. The yellow colour of the precipitate brought about by ferric chloride directly confirms this suggestion. At all events, the observed effect of cations shows that the coagulation of denaturated albumin cannot simply be regarded as a process of electrical discharge of protein particles by the ions. The same results also from the observed effect of various anions.

In spite of the slight negative charge carried by the particles of denaturated albumin the anions employed can be placed in the following series:

$$SO_4 > \text{tartr.} > NO_3 > Cl > Br > I > \text{Thiocyanate.}$$

Both the series of ions (that of univalent cations and that of anions) are the well-known lyotropic series of ions which were found for all precipitations of emulsion-colloids.

The observed difference in effect of single anions becomes still sharper when the solution of denaturated albumin is made more alkaline by adding potassium hydroxide. In this case, as already mentioned, protein certainly forms compounds with alkali, which are definitely nearer to the emulsion-colloids than the free denaturated albumin (see above). In my experiments the coagulation of denaturated albumin proceeded at 75° instantaneously when the solution contained 0·001 mol. per litre potassium hydroxide and 0·5–1 mol. K_2SO_4; whilst the average time of coagulation was 180 secs. in the presence of the same quantity of alkali and 1 mol. KCl, it was 1220 secs. with 1 mol. KNO_3. In the presence of 1 mol. of KI or KCNS the coagulation did not appear at all. KCNS produces a coagulation only when it saturates the solution.

As already mentioned, dialysed albumin shows, after denaturation, a cataphoresis to the anode, so that it is probable that the reaction of the solution becomes in this case slightly alkaline (see above, Hardy). But it was pointed out that even very slight concentrations of alkali can cause the formation of compounds with denaturated protein. It is therefore possible that the lyotropic series of ions found is observed only when the coagulation of alkali compounds of protein is investigated. In order to study the influence of ions upon the coagulation of alkali-free albumin, I added to the albumin solution, after denaturation, nitric acid to the concentration of 0·0002 mol. per litre. This concentration is not sufficient for the formation of acid-compounds of albumin (see above). Nevertheless the lyotropic series of anions was perceptible even in this case.

If in my experiments the acid concentration increased to 0·004 mol. per litre and above, the denaturated albumin formed compounds with acid (see above) and the lyotropic series of anions became indefinite, but the valency of the anions came to the foreground. Potassium sulphate, for instance, was found to accelerate the coagulation considerably more strongly than potassium chloride; on the other hand, according to its lyotropic property (the series is in this case inverse) potassium thiocyanate had a still greater effect upon the coagulation than potassium sulphate. The results obtained by Hardy were confirmed only in part.

The study of the influence of salts upon the coagulation-velocity showed generally that the process of coagulation of denaturated albumin is not simply a pure physical phenomenon of discharge of colloidal protein particles owing to an adsorption of electrolytes (ions), but that it is, at least partly, a chemical phenomenon, in which not only electrical properties but also chemical pro-pertics of salts are significant.

REFERENCES.

Aronstein (1874). *Pflüger's Arch.* 8, 75.
Arrhenius (1889). *Zeitsch. physikal. Chem.* 4, 239.
Buglia (1909). *Kolloidzeitsch.* 5, 291.
Chick and Martin (1910). *J. Physiol.* 40, 413.
——— ——— (1912). *J. Physiol.* 43, 2.
——— (1913, 1). *J. Physiol.* 45, 61.
——— (1913, 2). *J. Physiol.* 45, 288.
Cohnheim (1904). Chemie der Eiweisskörper, 132.

Hardy (1899). *J. Physiol.* **24**, 158.
—— (1900). *Zeitsch. physikal. Chem.* **33**, 385.
—— (1906). *J. Physiol.* **30**, 251.
Heynsius (1874). *Pflüger's Arch.* **9**, 514.
Hoffmann (1889). *Zentr. klin. Med.* 793.
—— (1890). *Zentr. klin. Med.* 521.
Kieseritzky (1882). Die Gerinnung d. Faserstoff. etc., Dissertation, Dorpat.
Lillie (1907). *Amer. J. Physiol.* **20**, 127.
Michaelis. *Phys. Chemie der Kolloide, Richter-Koranyi's Handb.* **2**, 391
Moll (1904). *Beiträge*, **4**, 563.
Pauli (1899). *Pflüger's Arch.* **78**, 315.
—— (1908). *Kolloidzeitsch.* **3**, 2.
Pauli and Handovsky (1908). *Beiträge*, **11**, 415.
—— —— (1909). *Biochem. Zeitsch.* **18**, 340.
Porges and Neubauer (1908). *Biochem. Zeitsch.* **7**, 154.
Rosenberg (1883). Vergleichende Untersuchungen betr. d. Alkalialbuminate, etc., Dissertation, Dorpat.
Spohr (1888). *Zeitsch. physikal. Chem.* **4**, 237.
Zsigmondy (1912). Kolloidchemie, 66.

LXXI. THE SYNTHESIS OF GLYCINE FROM FORMALDEHYDE.

By ARTHUR ROBERT LING AND DINSHAW RATTONJI NANJI.

From the Department of Biochemistry of Fermentation,
University of Birmingham.

(Received July 25th, 1922.)

FOR the purpose of some work on which we are engaged, it became necessary to secure a considerable quantity of glycine. The published methods for the preparation of this compound are unsatisfactory and difficult to carry out. Glycine was first obtained by Braconnot [1820] by the hydrolysis of gelatin with dilute sulphuric acid or baryta. Perkin and Duppa [1858] prepared it by treating bromoacetic acid with ammonia. It has also been prepared by treating chloroacetic acid with ammonia or ammonium carbonate. This method of preparation is dealt with in subsequent papers by Heintz [1862], Nencki [1883], Mauthner and Suida [1888, 1890]. When prepared by either of these methods, the glycine must be isolated by means of the copper salt, and the yield never exceeds 20 % of the theoretical.

Gabriel and Kroseberg [1889] obtained an almost theoretical yield of glycine by hydrolysing ethylphthalylglycine with hydrochloric acid. This method from an economic standpoint is unsuitable for the preparation of large quantities of glycine.

Eschweiler [1894] states that he obtained glycine in almost theoretical yield by treating methylene cyanohydrin with a large excess of ammonia. Here the difficulty is the preparation of the cyanohydrin.

Attempts have been made by us to devise a direct method for the synthesis of glycine from formaldehyde. Methylene-aminoacetonitrile can be obtained in a yield of 60 % of the theoretical by the condensation of formaldehyde (2 mols.) with ammonium cyanide as shown by Klages [1903], thus:

$$2H . CHO + NH_4CN = CH_2 : N . CH_2 . CN + 2H_2O.$$

Klages' method is carried out as follows. Finely powdered ammonium chloride (360 g.) is added to 40 % formaldehyde (1000 g.), in a wide-necked glass jar, cooled to 5° in a freezing mixture, the solution being stirred by means of a strong electric turbine. Potassium cyanide (440 g.), dissolved in water (600 cc.) is slowly run in during 3 hours, the temperature being kept during this time below 10°. It may be mentioned that commercial 96 % cyanide, which consists of a mixture of potassium and sodium cyanide, may be employed. When half the cyanide solution has been added, the ammonium chloride will have completely dissolved. The remainder of the cyanide and at the same time glacial acetic acid (250 cc.) is then dropped in. When the whole of the acetic acid has been added methylene-aminoacetonitrile commences to separate in glistening white crystalline flocks. The solution is now stirred with a turbine for 2 hours and the crystalline mass collected on a Buchner funnel. After washing with cold water and drying on a porous plate, the yield is about 280 g. or 60 % of the theoretical. It melts at 129°.

When the nitrile is hydrolysed with hydrochloric acid, the plan adopted by Klages, considerable difficulty is encountered in separating the glycine from the ammonium chloride produced simultaneously. It was found, however, that the nitrile could be easily hydrolysed by boiling it with a concentrated (40 %) solution of barium hydroxide.

The nitrile (10 g.) is added in small portions at a time to a boiling 40 % solution of barium hydroxide (100 cc.) in an open beaker. The mixture is boiled until no more ammonia is evolved, the volume of the liquid being kept approximately constant by the addition of water from time to time. The total period of boiling is 3 hours. From the solution, which now contains the methylene derivative of glycine, the barium is precipitated with sulphuric acid, the filtrate and washings from the barium sulphate are acidified until 3 % of sulphuric acid is present, and the liquid is boiled in an open beaker until no more formaldehyde is given off, which usually requires about 4 hours. The reaction proceeds as follows:

$$CH_2(NCH_2)COOH + H_2O = CH_2(NH_2)COOH + HCHO.$$

The sulphuric acid is removed by the addition of barium hydroxide, and the filtrate and washings containing the glycine are decolorised by boiling with norit. Traces of barium are removed by the addition of the requisite quantity of standard sulphuric acid. The filtrate is concentrated on a water-bath and the glycine which separates is recrystallised from alcohol. The yield is 90 % of the nitrile employed or taking the yield of the nitrile as 60 % of the theoretical, the yield of glycine from formaldehyde is 54 % of the theoretical.

This is the highest yield of glycine yet recorded as a result of its direct synthesis from formaldehyde. Our procedure has many advantages over the methods previously published, one being that the glycine is obtained uncontaminated with inorganic salts which are very troublesome to separate. Indeed methods of separating these salts have formed the subject matter of patents [Farbw. Meister, Lucius and Brüning, 1903; Siegfried, 1907].

REFERENCES.

Braconnot (1820). *Ann. Chim. Phys.* [2], **13**, 114.
Eschweiler (1894). *Annalen*, **278**, 229.
Farbw. Meister, Lucius and Brüning (1903). *Zeitsch. angew. Chem.* **16**, 527. D.R.P. 141,976.
Gabriel and Kroseberg (1889). *Ber.* **22**, 426.
Heintz (1862). *Annalen*, **122**, 257.
Klages (1903). *Ber.* **36**, 1506.
Mauthner and Suida (1888). *Monatsh.* **9**, 732.
—— —— (1890). *Monatsh.* **11**, 373.
Nencki (1883). *Ber.* **16**, 2827.
Perkin and Duppa (1858). *Quart. J. Chem. Soc.* **11**, 22.
Siegfried (1907). *Chem. Zentr.* **11**, 1466. D.R.P. 188,005.

LXXII. AN INVESTIGATION OF THE CHANGES WHICH OCCUR IN THE PECTIC CONSTITUENTS OF STORED FRUIT.

By MARJORIE HARRIOTTE CARRÉ.

Department of Plant Physiology and Pathology, Imperial College of Science and Technology, London.

(Received July 27th, 1922.)

AN accurate method of estimating pectin in dilute solution by precipitating as calcium pectate, has recently been established. The method has been successfully applied to the estimation of pectin in some fruit juices and by the application of this method it was proved that a complete extraction of the soluble pectin of apples can be effected by a process of continuous washing out with water.

As a result of this preliminary work it was found possible to follow the changes which take place in the pectic constituents of apples kept in cold and ordinary storage and to determine if the improved keeping properties of the apples in cold storage is accompanied by any marked difference in pectin content.

The soluble pectin probably develops from an insoluble pectic substance contained in the cell wall, which is left behind in the pulp residues after the aqueous extraction of the soluble form. This insoluble pectin corresponds to the protopectin of Fellenberg [1918], and to the pectose of earlier investigators.

Estimations of the soluble pectin were systematically carried out at regular intervals throughout the period of storage. The apples chosen for the experiments were very different, Lane's Prince Albert, a hard and acid cooking apple, and the soft, sweet Cox's Orange Pippin. Equal quantities were stored in cupboards at the ordinary temperature, and in a refrigerator at 60° C.

In view of the possibility that the time of harvesting might influence the keeping properties of the fruit, the estimations were carried out on three pickings, which were subsequently kept in cold and ordinary store. Batches were gathered three weeks earlier than the usual harvesting, at the normal time in early September, and the third picking was left on the trees till October.

Estimation of the Soluble Pectin.

The following is a summary of the process of extraction and estimation of the soluble pectin; a more detailed account of the method has recently been published [Carré and Haynes, 1921].

Samples of ten apples were cut up, freed from skin and core, and thoroughly mixed. Fifty grams were used for each estimation throughout the season.

After killing the cells, by freezing the weighed portion of the material in an efficient freezing mixture, the extraction of the soluble pectin was carried out by washing with water and pressing out in a small hand press. These processes were repeated until all the pectin was extracted. The completeness of the extraction was verified in all cases by testing the last washings of the pulp for pectin by precipitation as calcium pectate.

The dilute juice was filtered through fluted papers, to remove insoluble matter and disintegrated cell substance. It was then boiled to destroy enzyme action, and after cooling, the solution was made up to a known volume.

Aliquot portions were taken for each estimation, the quantity used depending on the concentration of the pectin solution. It was found advisable to make a rough preliminary determination of the amount of pectin contained in the juice solution, since, as the fruit approached maturity, the development of the soluble pectin from an initial negligible quantity to as much as 0·8 % necessitated a careful adjustment of the conditions of estimation.

The best results were obtained by using a volume of solution yielding· 0·02–0·03 g. of calcium pectate.

For this weight of pectin, 100 cc. $N/10$ soda were used for the hydrolysis, and after 24 hours' standing the pectin was precipitated as calcium pectate by acidification with 50 cc. of $N/1$ acetic acid, and subsequent addition of 50 cc. $M/1$ calcium chloride. During the later period of storage in the case of both Cox and Lane apples, it was observed that, although the pectin content did not appreciably differ and the conditions of experiment were exactly as before, the coagulation of the calcium pectate became ill defined, and the precipitate was slimy and difficult to wash and filter. It was found however that a satisfactory coagulation could again be obtained by using double the amount of $N/10$ soda previously found adequate for hydrolysis. The larger amount of soda was found to be necessary as long as the supply of fruit was available, and it seems reasonable to conclude from these observations that some change occurs in the state of aggregation of the pectin sol as the fruit develops, or possibly an alteration may take place in the constitution of the pectin.

THE DEVELOPMENT OF SOLUBLE PECTIN.

The accompanying graphs (Figs. 1 and 2) are representative of the results obtained from the estimations. Fig. 1 illustrates the development of soluble pectin in cold and ordinary storage, and Fig. 2 shows the very close similarity in the development of the soluble pectin in all three pickings of apples.

The experiments indicate that in the early stages of the maturation of the apple there is no soluble pectin, but it gradually develops as ripening proceeds till it attains a maximum amount when the fruit reaches its fully ripe condition. The Cox apples ripened earlier than the Lane, and in both cases the fully ripe state was found to be coincident with the greatest amount of pectin developed. The maximum pectin content was maintained in all cases for about 4 weeks, during which the apples were in their prime condition. This period was followed by a condition of general softening and a sudden drop in the pectin content was found to accompany the change of condition in the fruit. This softening process is the concluding phase in the apple's life, and it leads eventually to a state of physiological breakdown or of general decay brought about by fungal or bacterial action. During the whole softening

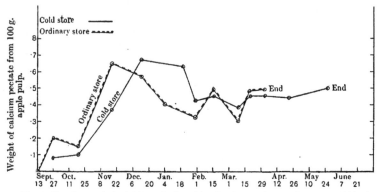

Fig. 1. Development of soluble pectin in cold and ordinary storage (Lane—1st picking).

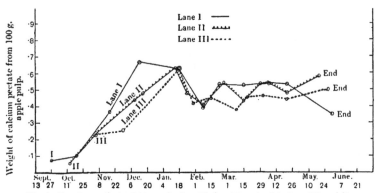

Fig. 2. Development of soluble pectin in cold stored Lane's picked at different times.

period the soluble pectin content showed considerable variations. In order to ascertain if these variations could be accounted for by differences in the condition of the fruit estimations of pectin were made on sets of apples in different states of softness, each set being carefully selected to be as uniform as possible. The results obtained show that there is a marked decrease in pectin as the fruit softens. The following experiments illustrate this point:

	Hard sample	Soft sample
Weight of soluble pectin from 100 g. ...	0·51 g.	0·42 g.

It may therefore be concluded that the fluctuating values obtained for the soluble pectin are to be attributed to the great variability which was observed in the state of maturity of the individual apples taken for each estimation, and it follows that in samples of ten taken for each separate determination a large sampling error might easily result. It is proposed during the next season to carry out estimations on each of the individual apples and to calculate the probable error due to sampling.

The routine estimations showed that apples contained varying but appreciable amounts of pectin as long as supplies of fruit were available, and in no case where sound apples were examined did the pectin wholly disappear. On the other hand when diseased apples were examined there was found in most cases to be a complete absence of soluble pectin.

EFFECT OF COLD STORAGE ON THE DEVELOPMENT OF SOLUBLE PECTIN.

The curves showing the development of pectin in cold and ordinary stored fruit (Fig. 1) are very similar, the only effect of the low temperature being to prolong the period of ripening. The maximum point of soluble pectin content occurs six weeks to two months later in the cold stored fruit.

After the maximum was passed and during the breakdown of the fruit the pectin content of the cold and ordinary stored fruit was found to be practically identical as long as samples of both were available for comparison.

It was possible to carry out these comparisons up to the end of April on the cold and ordinary stored Lane's, but the Cox's Orange were so badly affected by Bitter Pit that the supply of ordinary stored fruit was exhausted by the end of December, soon after the maximum point had been reached. The greatly prolonged life of the cold stored Cox's as compared with those kept in ordinary storage was due to the effect of cold in retarding the progress of disease in the fruit.

If the results of the estimations of soluble pectin developed in cold and ordinary store during the normal life of the apple are added together, the quantity is found to agree very closely for apples of the same kind whether they have been picked at different stages of maturity or kept under different conditions of temperature.

Total of (12) estimations during six months.

			Cold stored	Ordinary stored
Lane I[1]	...	...	4·05 g.	4·07 g.
Lane II	...	...	3·99	4·03

Total of (7) estimations during three months.

Cox I ...	...	...	3·01 g.	3·07
Cox II	...	...	2·77	2·82

			Picking		
			1	2	3
Lane	...	...	4·29	4·11	4·45
Cox	...	...	4·95	4·70	4·64

[1] = the numerals indicate the different pickings of the apples.

This agreement is too close to be merely fortuitous, but it is impossible to attempt any complete interpretation at present. It may however be inferred from the foregoing results that a definite amount of soluble pectin is produced during the process of ripening and that it is subject to secondary change during the later stages of maturation.

Estimations of protopectin which are described below, throw some further light on the subject.

Estimation of Protopectin.

With a view to examining the pectin changes in the apple more thoroughly it became necessary to devise a quantitative method of estimating the insoluble pectin constituent or protopectin[1], as it will be called in the following account, and so to obtain some idea of the relations between the soluble and insoluble forms.

The method adopted was to treat the residue after extraction of the soluble pectin with 100 cc. $N/20$ HCl in an autoclave for about an hour at a temperature of 110°. The material became thoroughly disintegrated during the process and the protopectin was converted into the soluble form, presumably by a process of hydrolysis. It was easily washed out with very little pressing, and the extract was estimated by the calcium pectate method. That the treatment removes all the insoluble pectic substances capable of being transformed by the acid was proved by repeating the autoclaving with a similar amount of $N/20$ HCl in which case no soluble pectin was found in the second extract. Subsequent treatments with greater concentrations of acid $N/5$ and $N/1$ also failed to produce any further trace of pectin.

To test the accuracy of the method equal amounts of apple pulp were taken, and the soluble pectin in them was extracted and estimated. The residues were treated as described above in order to convert the protopectin into soluble pectin, and the extracts thus obtained were also estimated. The following results illustrate the measure of agreement in the amounts obtained from two samples of uniform material. Both protopectin and soluble pectin are estimated in terms of the calcium pectate obtained from them.

Weight (g.) obtained from 100 g. of pulp.

	Soluble pectin	Protopectin	Total pectin (Protopectin and pectin)
(a)	0·43	0·36	0·79
(b)	0·385	0·45	0·835

It has been found possible to estimate the total pectin directly from the apple pulp without preliminary washing out of the soluble form.

---

[1] The adoption of this name for the hydrolysable pectin present in the cell wall implies that this is the precursor of soluble pectin. While there is much reason to believe that this is so, the fact cannot be regarded as finally established, and the present nomenclature must therefore be considered provisional.

Apple Sample I from Lane.

(Weight obtained from three similar samples of 100 g. of apple.)

	Picking		
	1	2	3
Soluble pectin ...	—	—	0·348
Protopectin ...	—	—	0·094
Total ...	0·404	0·418	0·442

Apple Sample II from Cox.

Soluble pectin ...	—	—	0·52
Protopectin ...	—	—	0·24
Total pectin	0·71	0·71	0·76

The amounts estimated in this way are in fairly close agreement with the total arrived at by adding the weights obtained by separate estimation of the soluble and insoluble pectin. The slightly lower values obtained in both these examples where the total pectin is extracted from the pulp may be entirely due to experimental error, but it is possible that some change may be brought about by the acids and salts contained in the juice.

ESTIMATION OF PROTOPECTIN OF APPLES KEPT IN COLD AND ORDINARY STORE.

During the latter half of the period of storage this method was systematically used in a preliminary examination of the protopectin content of apples. Estimations were carried out on the residue of the weighed sample from which all the soluble pectin had been extracted. The residue was then autoclaved with acid as described above, and the resulting soluble pectin precipitated as calcium pectate and weighed. In this way the weight of protopectin and soluble pectin was determined in a given sample of apple.

The results obtained are of necessity incomplete, and the following account must be regarded as in the nature of a preliminary survey.

The protopectin determinations were made after the maximum soluble pectin content had been passed and the protopectin shows the tendency to fluctuate which was observed in the case of the soluble pectin.

The following estimations of the protopectin content of hard and soft apples show that there is less in the softer fruit. As the samples contained individuals differing greatly in their degree of maturity these fluctuations may be attributed, as in the case of the soluble pectin, to the variation of individual apples in each set taken for estimation.

Weight of protopectin in 100 g. of samples	Hard sample	Soft sample
A. Lane II ordinary stored	0·45	0·35
B. Lane II cold stored	0·28	0·18

A series of estimations of pectin and protopectins carried out at regular intervals on the same apples, showed that a very definite relationship exists between them and that the changes in the two constituents tend to be equal and opposite in amount. The accompanying graph (Fig. 3) is typical of the results obtained:

The foregoing results show that after the maximum soluble pectin content is passed the quantity of soluble pectin tends to decrease and that the decrease is roughly proportional to a simultaneous increase in the insoluble pectin. On the other hand both the protopectin and soluble pectin have been shown to decrease markedly in soft over-ripe fruit as compared with harder less ripe fruit.

These observations seem to indicate that there is a balance between the protopectin and soluble pectin during the normal life of the apple, but that in the later stages of breakdown the total pectin tends to decrease. On the other hand there is some evidence to show that the total pectin content tends to *increase* as the apple ripens. The following estimations were carried out

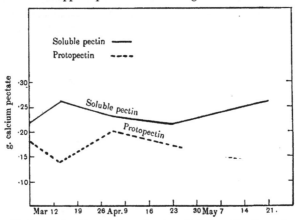

Fig. 3. Relationship of pectin and protopectin in Lane II ordinary store.

on cold stored "Cox's" when the apples were in the early stages of their development and during the period of fluctuation when the fruit was beginning to get over-ripe.

	Soluble pectin	Protopectin	Total
Early unripe state (Sept. 20) ...	0·05	0·48	0·53
Fully ripe state (Feb. 16–Mar. 16)[1]	0·65	0·36	1·01

[1] These figures represent the mean of the weights obtained during this period.

These results need confirmation but appear to show that in the few weeks following the picking of the apples the total pectin increases rapidly owing to the marked development of soluble pectin. The change in the protopectin content is much less pronounced and is insufficient to account for the increase of the soluble form. It may therefore be inferred that the total increase in pectin is due to a third source of pectin, which either gives rise to soluble pectin directly, or is transformed into it through the intermediate state of protopectin.

THE FACTORS CONTROLLING THE DEVELOPMENT OF SOLUBLE PECTIN.

The following investigations were made to obtain some information as to whether the production of soluble pectin is due to enzyme activity.

Samples of apple which had been previously killed by freezing were crushed with sand and thoroughly washed and pressed free from soluble pectin. The residues were left with water and a little thymol added to prevent bacterial action. At the end of a week the residues were again pressed out and tested for soluble pectin. In all cases examined pectin was found to have developed. The samples were put back in water and the process repeated every week till no more pectin was found to have developed. The following examples carried out on ordinary stored "Cox's" illustrate the results so obtained.

WEIGHT OF SOLUBLE PECTIN DEVELOPED FROM 100 G. OF APPLE PULP.

	A Cox III Ordinary stored	B Cox II Ordinary stored
1st week	0·22	0·30
2nd week	0·35	0·05
3rd week	0·04	0·05
4th week	0·00	0·06
5th week	0·00	0·02
6th week	0·00	0·00

A series of experiments on similar lines was carried out on swedes previously killed by ether and washed free from soluble pectin. As with the experiments on apple material a sample left standing with water showed a steady development of soluble pectin. Another portion was boiled with water for three hours and then pressed out and examined. Much soluble pectin was found in the extract. The residue was washed free and then again boiled for another three hours. No soluble pectin was obtained in this case. The residue was divided into two parts for different treatment. One portion was autoclaved with a solution of $N/20$ HCl at 110° and subsequent examination showed that a great deal of pectin had been produced in the process. The second portion was kept in water for one month but no soluble pectin was developed. It was then autoclaved with acid and a great deal of pectin was produced.

The foregoing observations show that a certain amount of pectin appears, if a residue previously washed free from pectin is left with water for a time, but that a steady state is finally reached when no more pectin is produced however long the sample is kept. Similar results are obtained by boiling; a certain amount of pectin is brought into solution but subsequent boiling for long periods has no further effect and no more pectin is developed after prolonged standing with water. On the other hand in both these cases large quantities of pectin are produced by autoclaving with acid. A great deal of further investigation is necessary before any definite conclusions can be made, but such facts as have already been ascertained support the theory that the production of soluble pectin is due to the action of an enzyme which must be supposed to be present in the cell walls in a form which it was not possible to extract.

SUMMARY.

An account is given of the development of soluble pectin in different kinds of apples kept in cold storage and at ordinary temperature. It is shown that the pectin reaches a maximum during the process of ripening and then gradually falls as the apple becomes over-ripe. The date of picking of the fruit has no effect on the development of the pectin in either cold or ordinary store.

A method is described for the quantitative estimation of protopectin by hydrolysis with weak acid. Preliminary work carried out by the application of this method suggests that there is a definite relationship between the amounts of soluble and insoluble pectin constituents. The experiments indicate the possibility of a third source of pectin.

Preliminary experiments tend to show that the development of the soluble pectin may be attributed to enzyme activity.

This investigation has been carried out for the Food Investigation Board of the Scientific and Industrial Research Department. The author wishes to express her gratitude to Dr Haynes of this Department for her valuable criticism and advice.

REFERENCES.

Carré and Haynes (1921). *Biochem. J.* 15, 60.
Fellenberg (1918). *Biochem. Zeitsch.* 85, 118.

PHYSIOLOGICAL ABSTRACTS

Monthly, price 5s. net.   Annual Subscription, post free, 42s.

This Journal is issued by the PHYSIOLOGICAL SOCIETY, acting in co-operation with numerous physiological organisations in Great Britain, America, and other countries. The Editor, Professor HALLIBURTON, is assisted by a staff of competent abstractors both at home and abroad.

The Journal aims at issuing promptly abstracts of the papers published throughout the world, in physiological and allied sciences (including Plant Physiology).

A number is published at the beginning of each month.

Volume VII. began with the April (1922) issue.

Subscriptions are only taken for April to March of the next year, and are payable in advance.

Back numbers, excepting some which are out of print, may be obtained from the Publishers, or any Bookseller. An extra charge is made for some numbers the stock of which is nearly exhausted. Prices on application.

Binding Cases, price 2s. net, are supplied by the Publishers, who can also arrange for binding; prices on application.

Published by H. K. LEWIS & CO. LTD.

136 GOWER STREET, LONDON, W.C. 1

Publishing Office, 28 GOWER PLACE, LONDON, W.C. 1

Subscriptions may be paid through any Bookseller

Cambridge University Press

Practical Plant Biochemistry. By M. WHELDALE ONSLOW, formerly fellow of Newnham College, Cambridge, and Research Student at the John Innes Horticultural Institution. Royal 8vo. 16s. *net*

Basic Slags and Rock Phosphates. By GEORGE SCOTT ROBERTSON, D.Sc. (Dunelm.), F.I.C., with a Preface by Sir E. J. RUSSELL, D.Sc. (Lond.), F.R.S. With 8 plates and a map. Small Royal 8vo. 14s. *net.* Cambridge Agricultural Monographs.

A Course of Practical Chemistry for Agricultural Students. By H. A. D. NEVILLE, M.A., F.I.C., and L. F. NEWMAN, M.A., F.I.C. Demy 8vo. Vol. I, 10s. 6d. *net*; Vol. II, Part I, 5s. *net.*

A Course of Practical Work in Agricultural Chemistry for Senior Students. By T. B. WOOD, M.A., F.R.S., Drapers Professor of Agriculture in the University of Cambridge. Demy 8vo. Paper covers. 2s. *net.*

A Course of Practical Physiology for Agricultural Students. By J. HAMMOND, M.A., and E. T. HALNAN, M.A. Demy 8vo. With blank pages interleaved. Cloth, 6s. 6d. *net*; paper covers, 4s. 6d. *net.*

Physiology of Farm Animals. By T. B. WOOD, C.B.E., M.A., F.R.S., and F. H. A. MARSHALL, Sc.D., F.R.S. **Part I, General.** By F. H. A. MARSHALL. Demy 8vo. 16s. *net.*

Insect Pests and Fungus Diseases of Fruit and Hops. A Complete Manual for Growers. By P. J. FRYER, F.I.C., F.C.S. Crown 8vo. With 24 plates in natural colours and 305 original photographs and diagrams. 45s. *net.*

Fungoid and Insect Pests of the Farm. By F. R. PETHERBRIDGE, M.A. With 54 illustrations. Crown 8vo. 5s. 6d. *net.* Cambridge Farm Institute Series.

Manuring for Higher Crop Production. By Sir E. J. RUSSELL, D.Sc., F.R.S. Second edition, revised and extended. Demy 8vo. With 17 text-figures, 5s. 6d. *net.*

The Fertility of the Soil. By Sir E. J. RUSSELL, D.Sc., F.R.S. With 9 illustrations. 16mo. Cloth, 3s. *net*; lambskin, 4s. *net.* Cambridge Manuals Series.

A Student's Book on Soils and Manures. By Sir E. J. RUSSELL, D.Sc. Second edition. Revised and enlarged. Large crown 8vo. With 41 illustrations. 8s. *net.*

Inorganic Plant Poisons and Stimulants. By WINIFRED E. BRENCH-LEY, D.Sc., F.L.S., Fellow of University College, London. Royal 8vo. With 19 illustrations. 9s. *net.*

Plants Poisonous to Live Stock. By HAROLD C. LONG, B.Sc. (Edin.), of the Board of Agriculture and Fisheries. Royal 8vo. With frontispiece. 8s. *net.*

Brewing. By A. C. CHAPMAN. With 14 illustrations. Royal 16mo. 2s. 6d. *net* and 3s. *net.* Cambridge Manuals Series.

Chemistry for Textile Students. By B. NORTH and N. BLAND. Demy 8vo. 30s. *net.* Cambridge Technical Series.

Cambridge University Press

Fetter Lane, London, E.C. 4: C. F. Clay, Manager

ii

PHYSIOLOGICAL REVIEWS

PUBLISHED BY THE

AMERICAN PHYSIOLOGICAL SOCIETY

UNDER THE EDITORIAL DIRECTION OF

W. H. HOWELL, Baltimore   J. J. R. MACLEOD, Toronto
REID HUNT, Boston   LAFAYETTE B. MENDEL, New Haven
F. S. LEE, New York   H. GIDEON WELLS, Chicago
D. R. HOOKER, *Managing Editor*, Baltimore

Containing short but comprehensive articles dealing with the recent literature in Physiology, using this term in a broad sense to include Bio-Chemistry, Bio-Physics, Experimental Pharmacology and Experimental Pathology.

CONTENTS. VOLUME III, 1923

PUBLISHED   SUBSCRIPTION $6.00 DOMESTIC
QUARTERLY   IN ADVANCE ONLY $6.50 FOREIGN

Subscriptions should be sent to

Dr D. R. HOOKER,
1222 ST PAUL STREET,
BALTIMORE, Md., U.S.A.

LXXIII. THE PRESENCE OF THE ANTINEURITIC AND ANTISCORBUTIC VITAMINS IN URINE.

By N. VAN DER WALLE.

From C. Eijkman's Laboratory of Hygiene, Utrecht, Holland.

(Received July 29th, 1922.)

FUNK [1914], SEIDELL [1921] and others applied chemical methods in order to determine the structure of the antineuritic vitamin. Their efforts to isolate these mysterious substances were not perfectly successful and therefore we do not wonder that nothing is known as yet of the metabolism of these factors. However, recently some experiments have been made to examine the excretion of antineuritic vitamins from the organism.

Cooper [1914] succeeded in curing polyneuritis-pigeons with an alcoholic extract of faeces.

Muckenfuss [1918] tried to find antineuritic vitamins in ox-bile, human urine and saliva by shaking up these substances with fuller's earth. The fuller's earth, treated in this way, purified with water and alcohol and then dried, proved to be efficient in curing pigeons with polyneuritis.

Gaglio [1919] at Rome also investigated the result of administering urine to polyneuritis-pigeons. For this purpose he concentrated the urine and found that even 3-4 cc. were able to cure the birds. Urine of rabbits that had fasted for 15-20 days was likewise active, though less than that of normal rabbits which were fed on cabbage leaves and bran. The ash of urine proved to be without action.

Gagliò [1919] as well as Curatolo [1920] are convinced that these cures are not to be attributed to the action of vitamins but to the influence of non-specific products of metabolism. On this supposition is founded the research of di Matteï [1920], who tried to cure pigeons with coffee. Indeed, 5 % of mocha proved to have a strong curative power, but to his astonishment he perceived that pure caffeine (in non-toxic doses) had only a slight effect.

Funk [1912] had previously demonstrated that some purine and pyrimidine derivatives had a more or less favourable influence on polyneuritis.

It has also been proved that a mixture of NaCl and KCl may have a good, though temporary effect on the disease [Eijkman, van Hoogenhuyze and Derks, 1922].

Still more remarkable is the observation made by Theiler who described the favourable result of subcutaneous injection of distilled water [1921].

It is difficult to explain this effect from a stimulation of vitamins, still present in the organism. Theiler attributed this action to "*spontaneous recovery.*"

Jansen [1920], in the Medical Laboratory at Weltevreden (Java), confirmed the observations of Eijkman and Theiler.

Yet, the results described by Gaglio and Curatolo are so striking, that we doubt whether such a lasting and absolute recovery may be attributed to the influence of non-specific products of metabolism.

Funk informs us that the recovery obtained by treatment with hydantoin lasted only nine days, whereas a pigeon treated with thymus-nucleic acid lived on for a fortnight. Allantoin, adenine and other pyrimidine derivatives were less active. From the administration of uric acid he did not obtain any effect.

It occurs to us that much is to be said in favour of the opinion of Muckenfuss who believes antineuritic vitamins to be present in urine.

By a series of experiments on pigeons we tested the opinions of these investigators and tried to find the most probable explanation.

First, 12 pigeons (Nos. 1–12) were fed on polished rice and water till they showed symptoms of polyneuritis. Then 5 cc. of urine that had been concentrated *in vacuo* at 45° were administered *per os.* In six cases this had undoubtedly a favourable influence.

In most cases the symptoms had much improved one day after the administration; sometimes even the influence was evident after a few hours. The body weight, having diminished before the appearance of the clinical symptoms, went up again in all cases of improvement. Yet we did not succeed in regaining the initial weight.

Eight pigeons (Nos. 13–20) were treated prophylactically with the administration of concentrated urine from the very first. The result was that the average incubation period increased from 22 to 42 days.

To enhance, if possible, the curative power of the urine the latter was dried in large open Petri dishes in the incubator at 37°. After about 16 hours a thin layer of tough, semi-solid substance was found which was dried in an exsiccator at room temperature; 1 g. of dried substance was found to be equivalent to 25 cc. of urine with a s.g. of 1·030.

Seven polyneuritis-pigeons (Nos. 21–27) were treated with this dried urine, dissolved in autoclaved milk and administered in quantities of 1·5–2 g. per day. Four of these pigeons were cured rapidly and absolutely. After stopping the administration of dried urine, some days passed before the birds showed further symptoms of the disease. We succeeded in curing a second attack in two pigeons.

Three pigeons (Nos. 28–30) were treated by subcutaneous injection of dried urine, dissolved in sterilised milk. These however died after a few days.

It is well known that strong heating destroys the antineuritic vitamin. We therefore tried the action of urine which had been heated for one hour at 130°. After sterilisation the urine was filtered, neutralised with hydrochloric acid and then concentrated *in vacuo* to a specific gravity of 1·050.

Ten polyneuritis-pigeons (Nos. 31–40) were treated with 15 cc. of this liquid daily. In one case the disease lasted only seven days, whereas the medicament had not the least effect on the other birds. This proves that by strong heating the curative power of the urine is decreased.

We also investigated the action of urine ash. For this purpose dried urine was incinerated, by which process it lost about 75 % of its weight. Ten pigeons (Nos. 45–54) were treated *per os* with this ash, suspended in water, and it was found to be almost inactive.

Chamberlain and Vedder [1911] observed that the antineuritic factor was removed by filtering through animal charcoal. We made use of this discovery and shook up urine with purified animal charcoal, 200 cc. of urine being shaken up for 45 minutes with 5 g. of animal charcoal; afterwards the urine was centrifuged and the charcoal shaken up anew with 200 cc. of fresh urine. This process was repeated three times. The charcoal was then washed once with distilled water and dried at 37°.

Six polyneuritis-pigeons (Nos. 60–65) were treated with this "*carbo cum urina*," 2 g. of this substance suspended in a little water being administered per day. All the birds recovered within a short time and we kept them healthy for at least a fortnight. The second attack too was cured. The smallest active dose proved to be 1 g.

Crude (non-purified) charcoal, shaken up with urine and treated in the same way, was far less active (Table I).

Table I. *Treatment of polyneuritis-pigeons with carbo cum urina.*

Laboratory no. of pigeon	Incubation period (days)	Period of illness* (days)	Effect of treatment	Dose of *Carbo cum urina*, g.	Remarks	Weight (g.)	
						Beginning of experiment	End of experiment
					Crude charcoal		
55	16	1	0	2	Died	240	200
56	21	2	0	2	,,	265	175
57	45	1	0	2	,,	240	175
58	22	5	±	2	Tempor. improved	350	225
59	16	12	±	2	,, ,,	260	200
					Carbo anim. puriss.		
60	24	18	+	2	Much improved	295	175
61	17	18	+	2	,, ,,	265	175
62	35	—	+	2	,, ,,	280	220
63	22	30	+	2	,, ,,	380	315
64	14	30	+	2	,, ,,	270	?
65	32	—	+	2	,, ,,	265	?
66	14	1	+	1	,, ,,	?	?
67	18	15	±	1	Slightly improved	355	260
68	17	19	±	0·5	,, ,,	330	290

* The period of illness is counted from the moment the treatment begins until death or, in case of recovery, until the end of experiment.

Another experiment on 12 pigeons (Nos. 69–80) showed that the non-treated animal charcoal had absolutely no effect.

Normal urine, shaken up with a surplus of charcoal, then filtered and concentrated *in vacuo*, had no effect (pigeons Nos. 81-89).

Then eight polyneuritis-pigeons (Nos. 90-98) were treated with animal charcoal, which had been shaken up with urine, heated previously for three hours at 120°. We expected this charcoal to be ineffective. On the contrary, two pigeons recovered and walked normally for ten days: six other pigeons were not cured. Without a doubt the curative effect of this charcoal is far inferior to that of charcoal shaken up with fresh urine. However, it must be admitted that there are some substances present in urine, possessing a curative power, and capable of being adsorbed by charcoal, and yet not bearing the character of the so-called vitamins. We tried to find out whether these substances have an organic or an inorganic character. For this purpose a mixture of inorganic salts was made, approximately in the same proportion as those found in human urine:

NaCl	9·2 %	Na_2HPO_4	5·0 %
K_2SO_4	2·7	$MgCl_2$	1·5
$(NH_4)_2SO_4$	1·5	$CaH_4(PO_4)_2$	0·25
KH_2PO_4	2·8		

This solution was shaken up with purified animal charcoal and ten pigeons (Nos. 99-108) were treated with this charcoal, which proved to have a very small effect as was expected beforehand, Rona and Michaëlis [1919] having demonstrated that animal charcoal will only slightly adsorb inorganic salts, the cations and anions present in urine being least adsorbed.

We may therefore safely assume that sterilised urine has a definite curative power owing to the presence of non-specific organic substances.

Finally we made an investigation as to whether the bacilli present in urine had any influence. The urine used was fresh, but not sterile. We did not expect to have a positive result here as the quantity of bacilli as far as weight was concerned, was very small.

Besides it has been already shown that *B. coli* does not contain any antineuritic vitamin even when cultivated in an extract of rice bran [Eijkman, van Hoogenhuyze and Derks, 1922], whilst Damon [1921] demonstrated that the B-vitamin is absent from *B. paratyphosus B., B. coli* and *B. subtilis*. Fresh urine was kept at 37° for 24 hours. The bacilli out of this urine were cultivated on agar, then suspended in distilled water and afterwards heated at 100° for 30 minutes. This suspension of dead bacilli was then filtered through animal charcoal. After this process the latter was dried at 37°. As expected this charcoal proved to have scarcely any curative power (Table II).

Finally the fact that yeast is only active in a medium that contains antineuritic vitamin or its components was made use of [Eijkman, van Hoogenhuyze and Derks, 1922]. Baker's yeast was cultivated at 28° for 24 hours in fresh urine, the acid reaction of which had been diminished and to which had been added 5 % of glucose. This yeast was centrifuged, washed once with distilled water and dried at 37°. It proved to have a strong curative action on polyneuritis of pigeons, 2 g. of yeast being sufficient to ensure the birds an absolute and lasting cure. This experiment makes it probable that antineuritic vitamins or their components are present in fresh, normal urine.

Table II. *Treatment with bacilli-charcoal.*

Laboratory no. of pigeon	Incubation period (days)	Period of illness (days)	Effect of treatment	Dose of bacilli-charcoal, g.	Remarks	Weight (g.)	
						Beginning of experiment	End of experiment
109	27	5	±	2	Slightly improved	330	230
110	31	6	±	2	,, ,,	360	350
111	19	4	0	2	Died	300	190
112	20	5	0	1	,,	300	250
113	28	4	0	0·5	,,	340	205
114	28	10	±	0·25	Slightly improved	350	200
115	15	7	±	0·25	,, ,,	325	260
116	19	4	0	0·25	Died	?	?
117	19	4	0	0·25	,,	?	?
118	19	4	0	0·25	,,	305	240

In order to trace whether diet has any influence on the activity of urine, the urine of a dog also was examined. Charcoal shaken up with this urine proved to possess as curative an action as human urine. Two g. per day were quite sufficient, but one pigeon even recovered from the use of 0·25 g. per day. As long as we were experimenting with its urine, the dog was fed on potatoes, polished rice and meat. This food contains a sufficient quantity of anti-neuritic vitamin. Then the diet was altered by autoclaving all the dog's food (bread, rice, meat and potatoes) during three hours. Only three weeks after the dog had been taking the vitaminless diet, the urine was again examined every day and shaken up with charcoal. The urine proved to have lost all its curative power (Table III).

Table III. *Treatment with charcoal shaken up with urine of a dog.*

A. The dog was fed on normal diet.

Laboratory no. of pigeon	Incubation period (days)	Period of illness (days)	Effect of treatment	Dose of charcoal, g.	Remarks	Weight (g.)	
						Beginning of experiment	End of experiment
123	13	—	+	2	Much improved	280	230
124	23	—	+	1·5	,, ,,	340	285
125	28	—	+	1	,, ,,	380	300
126	25	—	+.	1	,, ,,	380	250
127	42	1	0	1	Died	300	195
128	33	8	±	0·5	Slightly improved	310	290
129	31	1	0	0·5	Died	320	190
130	37	—	+	0·25	Improved	290	245
131	35	8	0	0·25	Died	?	?

B. The dog was fed on a vitamin-free diet.

132	23	6	0	2	Died	320	260
133	28	2	0	2	,,	300	220
134	22	4	0	2	,,	315	200
135	21	2	0	2	,,	360	?
136	26	5	0	2	,,	330	?
137	22	6	±	2	Slightly improved	350	250
138	17	2	0	2	Died	300	240
139	26	5	0	2	,,	340	205

It became more and more obvious that the favourable result must at least partly be attributed to the presence of the antineuritic vitamin, yet we went on to treat some fowls with urine because up till now it has never been demonstrated that non-specific substances have any influence on the polyneuritis of fowls. We found that fresh, concentrated urine (s.g. 1·050), as well as dried urine, when administered in large quantities is able to cure polyneuritis-fowls. Still more efficient was the action of animal charcoal, shaken up with fresh human urine. This however has to be administered in doses of 15 g. (equivalent to 2·5 litres of urine!) a day. Even then it required ten days to cure the fowl whilst we did not always succeed in curing the symptoms of paralysis absolutely. By a check experiment on six fowls it was shown that the charcoal itself did not exercise any curative action, and a fowl treated with charcoal that had been shaken up with urine, heated previously at 130° for one hour, did not recover.

The result of the experiments on fowls strengthened our opinion as to the presence of a small quantity of antineuritic vitamin in urine.

Antiscorbutic vitamin.

We then examined whether the antiscorbutic vitamin was also present in the urine.

For this purpose some guinea-pigs were put on a scorbutic diet receiving in addition fresh urine *per os*. The standard diet selected was a mixture of oats and bran, which together with water was given *ad lib*. Great care was taken to give the animals the best conditions possible. Following the example of Delf [Delf and Tozer, 1918] five guinea-pigs (Nos. 1–5) (Fig. 1) were given about 60 cc. of milk, previously autoclaved for an hour at 120°. Two other animals (Nos. 6 and 7), kept on the same diet, received daily 10 cc. of urine (which had been diluted to a specific gravity of 1·020). All these guinea-pigs developed scurvy and succumbed after some weeks. By post-mortem histological examination of the costochondral junctions we found the abnormalities described by Tozer [Delf and Tozer, 1918] in the severe stage of scurvy: irregularity of the junction, disorder of the rows of cartilaginous growing cells, usually a well-developed reticular zone, haemorrhages etc.

The average loss of weight of the guinea-pigs which were getting urine came to 29 % as against 19 % of the former group, the average lifetime being 28 days against 27 of the first group.

Evidently the administration of urine had no influence on the severity of the microscopical changes and on the average lifetime, whereas the loss of weight had increased.

(Though we made a histological examination of the ribs of all the guinea-pigs we are conscious that we have to be cautious in judging the results. It is known that Tozer [1921] demonstrated that the lack of A-vitamin may bring about departures from the normal which closely resemble those which are present in chronic cases of scurvy.)

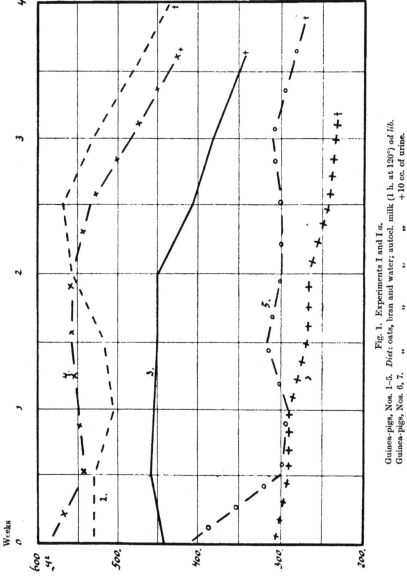

Fig. 1. Experiments I and I *a.*

Guinea-pigs, Nos. 1-5. *Diet:* oats, bran and water; autocl. milk (1 h. at 120°) *ad lib.*
Guinea-pigs, Nos. 6, 7. " " " " +10 cc. of urine.

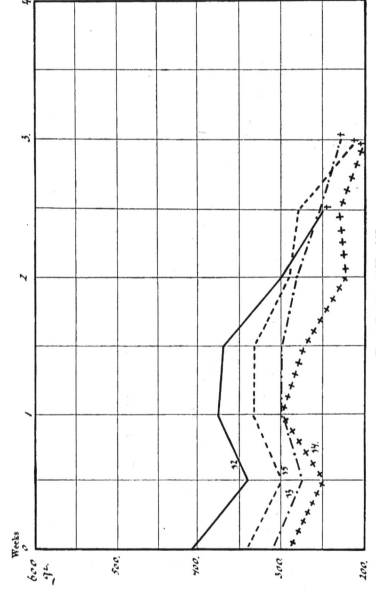

Fig. 2. Experiments II and II a.

Guinea-pigs, Nos. 12–15. *Diet:* oats, bran, water, 10 cc. of autocl. milk.

Guinea-pigs, Nos. 8–11.   "   "   "   "   +10 cc. of urine.

It appeared to us that the guinea-pigs were loth to take the autoclaved milk and it was necessary to give an additional quantity with a syringe, which took a great deal of time.

Experiment II. Four guinea-pigs (Nos. 12-15) were given with a syringe 10 cc. of autoclaved milk besides the usual scorbutic diet. Just as had been demonstrated by Delf, it was found that the loss of weight had increased (about 34 %). The average lifetime was about the same (25 days), and the microscopical changes were very similar in both experiments (Fig. 2).

Experiment II a. Another series of animals (Nos. 8-11), fed on the same standard diet, were also given 10 cc. of fresh urine daily. The average loss of body weight was 31 % and the average lifetime was 26 days. The microscopical abnormalities were not less than those in the above experiments.

Experiment II b. In this experiment the amount of urine was increased; 20 cc. being administered instead of 10. The average loss of weight of this group of animals (Nos. 16-19) was 26 %, and the average lifetime was 23 days (Fig. 3). On histological examination we found marked disorganisation of the normal structure. In this experiment too the urine proved to be without antiscorbutic power. (Although the administration of a large quantity of autoclaved milk had a favourable influence on the weight and the general state of the guinea-pigs, we thought it advisable to administer only 10 cc. of autoclaved milk to the animals in the following experiments, by which method it was still possible to make a comparison with the former groups.)

As we did not think it advisable to administer a greater dose of urine, we tried to find out in the following experiments whether urine was able to increase the antiscorbutic power of a small quantity of green cabbage.

Experiment III. Four guinea-pigs were given besides the diet of oats and bran 1 g. of green cabbage. These animals (Nos. 20-23) died in about 33 days, the average loss of weight being 32 %. The macroscopical as well as the microscopical changes were far less than in the former experiments (Fig. 4).

Experiment III a. 1 g. of green cabbage was evidently not quite enough absolutely to prevent the onset of scurvy. We then gave four guinea-pigs (Nos. 24-27) in addition 15 cc. of fresh urine. It was found that the average lifetime of these last animals, as compared to that of those in Expt. III, had increased only two days, the average loss of weight being 24 %. The result of the microscopical examination was similar to that of Group III. (We think the decrease of the average loss of body-weight was only owing to guinea-pig No. 23, the effect of the green cabbage being extraordinarily favourable in this case.) Without a doubt this experiment shows that the antiscorbutic power of 1 g. of green cabbage was great (a well-known fact), but that a dose of 15 cc. of urine was unable to exercise any curative effect.

Experiment IV. A further group of animals (Nos. 28-32) was given, besides the standard diet, 2 g. of green cabbage. By this treatment the average loss of weight was only 17 % whereas the average lifetime increased to 69 days. The influence of this quantity of cabbage proved to be much greater than the

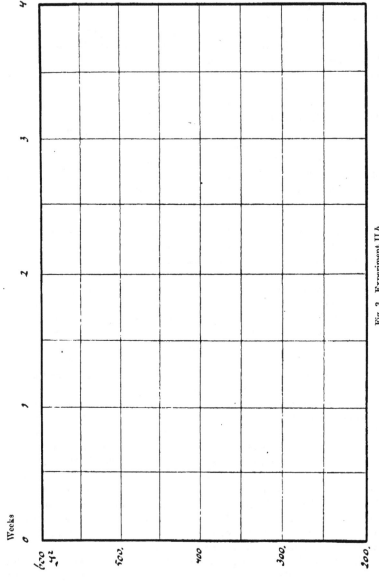

Fig. 3. Experiment II b.

Guinea-pigs, Nos. 16-19. Diet: oats, bran, water + 10 cc. of autocl. milk + 20 cc. of urine.

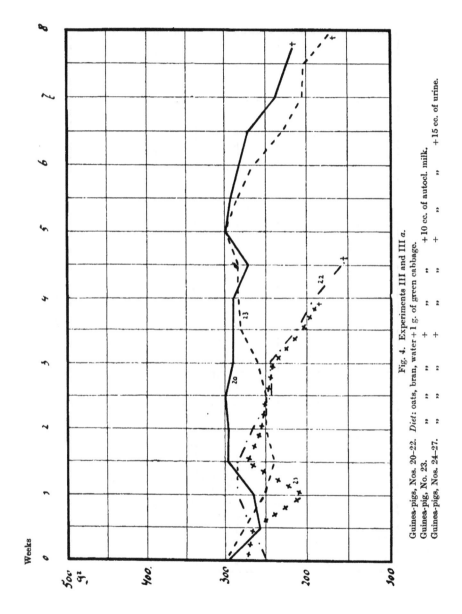

Fig. 4. Experiments III and III *a*.

Guinea-pigs, Nos. 20–22. *Diet:* oats, bran, water + 1 g. of green cabbage.
Guinea-pig, No. 23.   „   „   „   „   + „   „   „   „   „   + 10 cc. of autocl. milk.
Guinea-pigs, Nos. 24–27. „   „   „   „   + „   „   „   „   „   + „   „   „   „   „   + 15 cc. of urine.

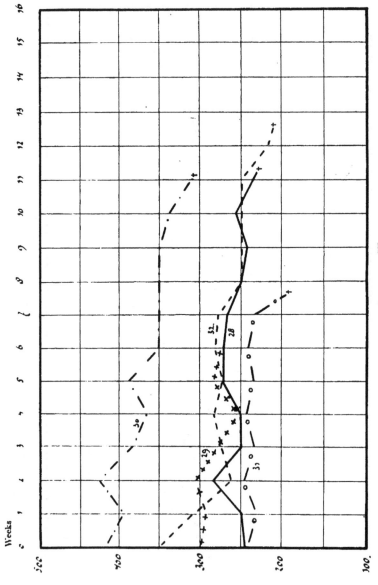

Fig. 5. Experiment IV.

Guinea-pigs, Nos. 28–31. *Diet*: oats, bran, water + 2 g. of green cabbage.
Guinea-pig, No. 32.   „   „   „   +   „   „   „   + 10 cc. of autocl. milk.

effect of giving a combination of 1 g. of cabbage plus 15 cc. of urine (Fig. 5). The microscopical changes found post-mortem were approximately those described by Tozer in the incipient stage.

In none of these experiments were we able to detect any antiscorbutic power of the urine. Of course it is possible that the urine might prove to be effective if given in greater quantities. This was practically impossible. Theoretically one might try to concentrate the urine just as Harden and Robison [1920] concentrated orange-juice and Bassett-Smith lemon-juice. We did not think it advisable to do this with urine because of the unknown influence of the other substances in urine on the labile antiscorbutic vitamins. Harden and Zilva [1918] demonstrated that charcoal does not adsorb the antiscorbutic vitamin and so we could not use this method of procedure, which proved so successful in the case of the antineuritic vitamin.

CONCLUSIONS.

1. It was shown that concentrated, normal urine may exercise a favourable influence on polyneuritis of pigeons.

2. Urine, dried at 37° for 16 hours, was likewise active.

3. By heating the urine at 130° for one hour, the curative power was destroyed.

4. The ash of urine was inactive.

5. *Carbo anim. puriss.*, shaken up with urine, centrifuged, washed with distilled water and then dried, had a strong curative power.

6. Urine, treated with a surplus of charcoal and then filtered, had lost its curative power.

7. Charcoal, shaken up with urine, which had been heated previously for three hours at 120°, was far less active.

8. Charcoal not shaken up with urine proved to be quite inactive.

9. Charcoal, shaken up with a solution of inorganic salts, containing the cations and anions present in urine, had only a slight effect.

10. The curative power exercised by urine may not be attributed to the presence of bacilli.

11. Yeast, cultivated in urine, to which had been added 5 % of glucose, exercised a strong curative action.

12. The urine of a dog had as strongly curative an action as that of man, but when the dog was fed on a vitamin-free diet, the urine lost its curative power.

13. It was shown that in some cases fresh, normal urine or dried urine, when given in large quantities, may have a good influence on polyneuritis of fowls.

14. Charcoal shaken up with normal urine was likewise active.

15. The charcoal itself had no curative influence on the polyneuritis of fowls.

16. Charcoal treated with urine heated previously for one hour at 130° had no effect.

17. It was pointed out that a small quantity of antineuritic vitamin is present in the urine, as proved by the above experiments.

18. We were unable to detect the presence of the antiscorbutic vitamin in urine.

LITERATURE.

Chamberlain and Vedder (1911). *Philippine J. Sci.* **6**, 395.
Cooper (1914). *J. Hygiene*, **14**, 20.
Curatolo (1920). *Il Policlinico.*
Damon (1921). *J. Biol. Chem.* **48**, 379.
Delf and Tozer (1918). *Biochem. J.* **12**, 416.
Eijkman and van Hoogenhuyze (1913). *Archiv Schiffs- und Tropenhygiene*, **17**, 328.
Eijkman, van Hoogenhuyze and Derks (1922). *J. Biol. Chem.* **50**, 311.
Funk (1912). *J. Physiol.* **45**, 489.
—— Die Vitamine (Bergmann, 1914).
Gaglio (1919). *Il Policlinico.*
Harden and Robison (1920). *Biochem. J.* **14**, 171.
Harden and Zilva (1918). *Biochem. J.* **12**, 260.
Jansen (1920). *Mededeelingen uit het geneesk. Laboratorium te Weltevreden*, 3ᵒ serie, 23.
di Mattei (1920). *Il Policlinico.*
Muckenfuss (1918). *J. Amer. Chem. Soc.* **40**, 1606.
Rona und Michaëlis (1919). *Biochem. Zeitsch.* **94**, 240.
Seidell (1921). *J. Ind. Eng. Chem.* **13**, 1111.
Theiler (1915). *The third and fourth reports of the Director of Veterinary Research*, Pretoria, Nov.
Tozer (1921). *J. Pathol. Bact.* **24**, 306.

LXXIV. THE SUCTION PRESSURE OF THE PLANT CELL.

A NOTE ON NOMENCLATURE.

By WALTER STILES.

From the Department of Botany, University College, Reading.

(*Received August 4th, 1922.*)

As a result of the work of Pfeffer and de Vries during the last thirty years of the nineteenth century, the vacuolated cell came to be regarded as a system consisting of a solution containing osmotically active substances (the cell sap) surrounded by a semi-permeable membrane (the protoplasm) permeable to water but impermeable to the solutes in the cell sap, the whole being enclosed by an elastic cellulose envelope (the cell wall), supposed to be permeable both to water and dissolved substances. How far such a view really represents the facts is not in question here.

When such a system is immersed in water, it is to be expected that water will enter the cell on account of the osmotic pressure of the cell sap. This must result in the development of a hydrostatic pressure and consequent stretching of the wall, which, being elastic, exerts an inwardly directed pressure on the cell contents. At any time, therefore, there is a pressure, the osmotic pressure, tending to send water into the cell, and the inwardly directed component of the pressure exerted by the wall tending to compress the cell contents and so send water out. The hydrostatic pressure of the water in the cell is called the turgor pressure, to which the inwardly directed wall pressure must be equal and opposite. If this pressure is denoted by T and the osmotic pressure by P, the net pressure sending water into the cell when this is surrounded by water is thus $P - T$, which varies from P when the cell is completely deturgid to zero when it has absorbed so much water that $T = P$ and the cell can be regarded as saturated with water.

Just as it is convenient to have the term osmotic pressure to denote the pressure developed when a solution is separated from pure water by a semi-permeable membrane, so it is convenient to have a term to denote the actual net pressure sending water into the cell when this is immersed in pure water. For this quantity Ursprung and Blum [1916] have used the term "suction force" (Saugkraft) and there appears to be a tendency for other continental writers to adopt this. The chief objection to this term is that the quantity is a pressure and not a force, and is equated with a pressure (osmotic pressure). In this country Thoday [1918] has proposed the term "water absorbing power' of the cell for this quantity. This term appears particularly unfortunate, for not only is it perhaps a trifle clumsy, but if the quantity in question is not a force, still less is it a power, a rate of doing work. Nevertheless there is a tendency for the term to be accepted by English writers, one of whom indeed gives an equation in which a quantity defined as a surface tension (a force per unit length) is put equal to a quantity defined as a pressure (a force per unit area) less a quantity defined as a power (work per unit time).

To avoid such anomalies in nomenclature, I propose for this quantity, already described as a force and a power, but which is in reality a pressure, the term "suction pressure." This gives an accurate description of the quantity, is not clumsy, and practically embodies the term used by Ursprung and Blum, whose expression, other things being equal, is entitled to precedence on the ground of priority.

REFERENCES.

Thoday (1918). *New Phyt.* **17**, 108.
Ursprung and Blum (1916). *Ber. deut. bot. Ges.* **34**, ϋ39.

LXXV. RESPIRATORY EXCHANGE IN FRESH WATER FISH.

PART IV.

FURTHER COMPARISON OF GOLD-FISH AND TROUT.

By JOHN ADDYMAN GARDNER AND GEORGE KING.

From the Physiological Laboratory, University of London, South Kensington.

(*Received August 21st, 1922.*)

IN Part II of this series of papers [1914, 2] the behaviour of trout at various temperatures under reduced oxygen tension was described, and measurements were recorded of the quantity of oxygen in the water at the asphyxial point for the various temperatures. It was found that the oxygen content of the water at the asphyxial point was much larger at higher temperatures than at lower, a result to be expected since, as was shown in Part I [1914, 1], the oxygen consumption of trout in fully saturated water increases with the rise in temperature—within the ordinary limits of temperature of natural water consumption doubles for rise of 10°—and consequently the fish would have to pump a much larger volume of water through their gills at a higher temperature than at a lower to obtain the necessary amount of oxygen.

Gold-fish as we showed in Part III [1922] require much less oxygen than trout and live at an altogether lower plane of metabolism, though their oxygen requirements are also proportional to temperature. It seemed of interest to measure the oxygen tension at the asphyxial point at various temperatures in the case of these animals.

METHOD.

The apparatus described in Part II proved unsuitable for gold-fish, as we found it impossible to reduce the oxygen in the water to the asphyxial tension within a reasonable time. We therefore used a much smaller apparatus based on the same principle, and filled with water which had previously been heated in order to get rid of the bulk of the dissolved gas and allowed to cool in a closed bottle. A six litre vacuum desiccator with slightly domed lid served for the water tank. This was filled to the edge of the ground flange with the water so that the gas space above the water was only that of the shallow dome of the lid. The tubulure of the lid was closed by a rubber stopper carrying (1) a tube leading to the bottom of the vessel and ending in a perforated bulb through which nitrogen could be sprayed through the water, (2) an exit tube for nitrogen bent over so as to dip into a beaker of water and (3) a capillary tube fitted with a tap through which samples of the water could be drawn off as required. The desiccator stood in a thermostat. The fish were placed in the tank and the air in the water was gradually displaced by a current of nitrogen, which also served to stir the water. The compressed commercial nitrogen used contained about 0·5 % of oxygen; this quantity of oxygen was too much, and the gas was purified by passing through several wash bottles containing strong alkaline pyrogallol. In this way the percentage of oxygen in the nitrogen could easily be reduced to 0·14. A sample of the water was withdrawn for analysis at the asphyxial point, which was taken as that point at which the fish rolled over on their sides and ceased to breathe. The sample was withdrawn by displacement of mercury and analysed by the method fully described in previous papers.

Between temperatures of, say, 10° and 27° the oxygen content of the water could fairly easily be reduced to the asphyxial tension by means of a current of nitrogen, but at low temperatures it was not found possible to reduce the oxygen sufficiently by this method. In experiments at low temperatures the oxygen was displaced as far as possible by a current of nitrogen, and the apparatus was then closed and the fish made to use up the remaining oxygen themselves.

EXPERIMENTAL.

Experiment at high temperature.

Two fish, A weighing 53 g. and B weighing 58 g., were placed in the partially boiled out water at a temperature of about 22° and the tank put in the thermostat at 28°; nitrogen was then bubbled through the water. At the beginning the fish were very active, A being particularly violent for a gold-fish, and attempting to jump out of the water. The respirations of both animals were about 117 per minute. As the oxygen tension decreased the respiration became more spasmodic, irregular and slow. The movements of the fish gradually became less vigorous, and towards the end of the experiment the animals took short intervals of rest with scarcely any perceptible movements of the gills, alternating with short rushes round the vessel mainly near the surface. After about 90 minutes A remained on his side without any obvious sign of life and B was nearly in the same condition. A sample of water was now withdrawn for analysis, the temperature being 27·2°. The fish were now removed from the experimental vessel and placed in a trough of tap water at laboratory temperature. After two minutes B recovered his normal position, while A began to breathe and showed slight movements after five minutes, but still remained on his side. The fish were then placed in the store tank in running water and after two to three hours A recovered. Next day both fish appeared to be perfectly normal and healthy. On analysis 1 litre of the water was found to contain the following quantities of gas measured at 0° and 760:—free and combined CO_2 12·25 cc., oxygen 0·422 cc. and nitrogen

8·287 cc.. Taking Winkler's value 0·02727 as the absorption coefficient of oxygen at 27·2° (Landolt, Börnstein, Meyerhoffer's tables) the partial pressure of the dissolved oxygen was therefore 1·547 % of one atmosphere.

Experiment at temperature 11–12°.

Two fish, C 51 g. and D 40 g., were placed in six litres of partially boiled out water as before and purified nitrogen bubbled through the water. Both animals were fairly active at first and their respirations 80–90 per minute, the normal in fully aerated still water being between 40–50 per minute. During the first two hours of the experiment, as the oxygen gradually decreased, the fish became less continuously active, the number of respirations decreased to 40–50 per minute, and the breathing, if we can use this term, became shallower. A noticeable feature was that under these low oxygen tensions the fish indulged in spells of backward swimming. After 135 minutes the nitrogen current was stopped and the fish reduced the oxygen content of the water naturally. The breathing became shallow and very spasmodic and the respiration rate was reduced to 20–30 per minute. At the end of the fourth hour it was noticed that the fish remained inert for periods of 1–1½ minutes. This was followed by a few respirations then another period of inertness and so on. After six hours both fish were on their sides and seemed to have ceased to respire, a sample of water was therefore withdrawn for analysis. The temperature was 11·4°. One litre of the water contained 10·732 cc. free and combined CO_2, 0·397 cc. oxygen and 15·362 cc. nitrogen measured at 0° and 760. Taking the coefficient of absorption for oxygen in water at 11·4° as 0·03686, the partial pressure of the oxygen was 1·077 % of one atmosphere. The fish gradually recovered on being restored to the store tank.

Experiment at low temperature.

It was not found possible at the low temperature of the ice chest to reduce the oxygen tension sufficiently, by means of a current of nitrogen, to incommode, let alone asphyxiate, the fish. After reducing the tension as far as possible by this method we had to rely on the fish to use up the remaining oxygen, and this proved a very slow process.

Two fish, E 52 g. and F 42 g., were placed in 5½ litres of boiled out water and a rapid current of nitrogen bubbled through for two hours on March 27th. The nitrogen inlet and outlet taps were then closed and the vessel placed in the ice chest and left for a week. On the second day the fish appeared to be quite normal and during the week they remained very quiet and sluggish sometimes near the surface and sometimes at the bottom of the vessel. On April 5th they showed a rolling gait symptomatic of distress. On the morning of April 6th F was more or less on his side and in the afternoon appeared to be dead or nearly so. E was still breathing occasionally in a spasmodic manner and had a rolling gait as though finding difficulty in keeping upright. The temperature recorded by the thermometer in the upper part of the chest varied from 3-5° during the week. A sample of water was withdrawn for analysis at 2.35 p.m. on April 6th; its temperature was 2·6°. The quantity of oxygen in the water was less than 'the limit of the experimental error. On analysis one litre of water was found to contain the following volumes of gas reduced to 0° and 760:—free and combined CO_2 96·96 cc., oxygen nil, nitrogen 22·97 cc.

Comparison with trout.

Trout normally use very much larger quantities of oxygen per kilo per minute than gold-fish, and at a diminishing oxygen tension behave quite differently. The behaviour of trout was described in detail in Part II [1914, 2, p. 595]. As the oxygen diminished the respiratory movement gradually increased in rapidity and force and acquired a dyspnoeic character. The animals became more active, rushing about and jumping into the air space above the water. These periods of activity alternated with periods of rest at the bottom with the dyspnoeic respiration mentioned. The periods of rest gradually became longer than the periods of activity, and as the oxygen approached the asphyxial tension they remained nearly all the time at the bottom and ultimately quietly rolled over on their backs. In this condition they died in two or three minutes, but if oxygen were at once sprayed into the water they recovered in the course of a few minutes and afterwards appeared to suffer no ill effects. The asphyxial condition in the case of trout takes place within quite small limits of oxygen tension; at or below this tension they very quickly die but slightly above they recover. Gold-fish on the other hand could stand much lower oxygen tensions for any given temperature and are much less sensitive, so that the precise asphyxial tension is difficult to ascertain. This difference is still more marked at low temperatures. In Table I we give a comparison of the measurements for trout and gold-fish.

Table I.

Temperature	8" trout, weight 80-90 g. Oxygen in water at asphyxial point*		5" to 6" gold-fish, weight 45-55 g. Oxygen in water at asphyxial point		Normal water	
	cc. of oxygen measured at 0° and 760 per litre of water	partial pressure of oxygen in per cent. of 1 atmosphere	cc. of oxygen measured at 0° and 760 per litre of water	partial pressure of oxygen in per cent. of 1 atmosphere	cc. of oxygen at 0° and 760 per litre	partial pressure in per cent. of 1 atmosphere
2·6°	—	—	nil	nil	9·49	
6·4°	0·79	1·89	—	—	8·60	
9·5 to 10°	0·81	2·19	—	—	8·00	
11·4°	—	—	0·397	1·077	7·9	
17°	1·37	4·17	—	—	6·75	app. 21 %
18°	1·49	4·62	—	—	6·61	
24°	1·97	6·83	—	—	5·89	
25°	2·40	8·49	—	—	5·78	
27·2°	—	—	0·422	1·547	5·52	

* Gardner and Leetham [1914, 2].

GLYCOGEN CONTENT OF TROUT AND GOLD-FISH UNDER VARIOUS CONDITIONS.

In Part I [1914, 1] of this series it was shown that at low temperature, *i.e.* below about 6°, trout appeared to be in a condition, for a time at any rate, akin to that of hibernating animals, and their respiratory quotients were very low. These low respiratory quotients could not be directly connected with nitrogen metabolism, and it was suggested that a possible explanation might be found in the fat. If at low temperature the animals were in a state of hibernation or starvation they might be living on their fat and partially converting it into glycogen and sugar. Some such process might result in a low respiratory quotient, as indicated in the following equation:

$$2C_3H_5(C_{18}H_{33}O_2)_3 + 64O_2 = 16C_6H_{12}O_6 + 18CO_2 + 8H_2O$$
$$CO_2/O_2 = \tfrac{18}{64} = 0.281.$$

In the hope of throwing light on this point comparative estimations of the glycogen content of starving trout at various temperatures were made. The glycogen was estimated by Pflüger's method. The precipitated glycogen, after solution and reprecipitation two or three times, was hydrolysed and the sugar estimated by Fehling. If in sufficient quantity the cuprous oxide was weighed, otherwise it was dissolved in ferric sulphate and the ferrous salt produced estimated by permanganate. The first experiments were done on four-inch trout which however proved too small for the purpose.

EXPERIMENTAL.

Experiments with four-inch trout.

(1) *Control.* Four fish weighing 44·65 g. were kept in running water in a large tank with a gravel bottom. The average temperature of the water was 19°. The fish were not fed though they may have picked up some food. After treatment by Pflüger's method and hydrolysis of the glycogen 0·0013 g. Cu_2O was obtained. Percentage of glycogen was therefore 0·0052.

(2) Four fish weighing 40·6 g. were placed in an open vessel in the ice chest and the water kept well oxygenated by a spray of oxygen. At the end of the first day the temperature of the water was 7°, at the end of the second day 6° and the third day below 5°. The fish were very inert and sluggish and scarcely moved when touched. On the afternoon of the third day they were killed and analysed. 0·0032 % of glycogen was found.

(3) Five fish weighing 57·5 g. were kept in the ice chest under the same conditions as in experiment 2 for six days. On the last day two died and as soon as this was observed the rest were killed and the whole analysed. The final temperature of the water was 3·5°. The percentage content was 0·0022.

(4) Three fish weighing 27·8 g. were kept in the ice chest for two days at reduced oxygen tension. They were then analysed but no glycogen could be detected.

As the fish were too small to give a suitable weight of material without using an inconvenient number of fish the experiments were repeated using eight-inch animals.

Experiments with eight-inch trout.

(5) *Control experiments.* The fish were kept in a large outside tank in running water for several days. They were not fed and the temperature of the water was about 16°. On analysis, fish A weighing 89·1 g. gave no trace of glycogen, fish B weighing 109 g. was found to contain 0·0012 %.

(6) Two fish were placed in separate earthenware bowls in the ice chest and a spray of oxygen passed through the water by means of circular lead pipes punctured with small holes. The fish were kept at 2·6° for two days; they were very sluggish but otherwise appeared healthy. Fish C weighing 114·3 g. contained no trace of glycogen nor did fish D weighing 134·3 g.

(7) Two fish were kept under similar conditions to those in experiment (6) for five days. Final temperature of the water was 1·7°. On the fifth day one of the animals turned over and they were therefore cut up and analysed. Fish E weighing 101·3 g. gave 0·1179 g. of Cu_2O and contained 0·0469 % glycogen. Fish F weighing 108 g. gave 0·0798 g. Cu_2O and hence contained 0·0294 % glycogen.

We do not know what is the limit of accuracy of Pflüger's method for estimating glycogen, but as the conditions—proportion of alkali, duration of hydrolysis, volume of alcohol—were kept as nearly as possible the same in experiments 5, 6 and 7 it is clear that the fish in experiment 7 contained a very much larger percentage of glycogen than the ntrols ... to fit in with our hypothesis as to the cause of the low respiratory quotients at low temperature. At the same time we cannot pretend that the experiments are altogether conclusive.

GLYCOGEN CONTENT OF GOLD-FISH.

It seemed of interest to ascertain how glycogen content of gold-fish, which did not show abnormal quotients at low temperatures, compared with that of trout.

(8) *Control experiments.* Two gold-fish were taken from the running water tank in which they had been living many months. The tank contained plenty of food. Two fish weighing 73 g. analysed together were found to contain 0·86 % of glycogen. Temperature of the tank when the fish were caught 13–14°. Two fish weighing 90·5 g. taken from water at about 16° contained 1·57 %. Two fish weighing 61·2 g., temperature of tank 18°, contained 1·235 % glycogen.

(9) Two fish weighing together 73·3 g. were kept in an open vessel in the ice chest for seven days; the water was aerated by a current of oxygen and the final temperature was 2–3°. The fish were not fed, and on analysis they contained 1·21 %.

(10) In this experiment two fish weighing 81·8 g. were placed in the apparatus used for determining the asphyxial point and the oxygen tension considerably reduced by bubbling nitrogen through the water. The apparatus

was then placed in the ice chest and kept at about 2° for seven days. Percentage of glycogen was found to be 0·52.

(11) Two fish total weight 61·6 g. were placed in a trough of ordinary tap water through which oxygen was bubbled occasionally and kept in the ice chest for eleven days; the fish were not fed. Percentage of glycogen was found to be 0·54.

(12) Two fish weighing 60·45 g. were placed in six litres of partially boiled out tap water in the asphyxial point apparatus referred to above. Pure nitrogen was passed through the water for half an hour when the nitrogen tubes were closed and the apparatus placed in the ice chest and left for several days. The average temperature of the water was 3·7°. On the morning of the fifth day one fish was found on its side but the others were still alive and in normal position. They were killed and the whole analysed and found to contain 0·024 % glycogen.

These results appear to be in conformity with the low level of metabolism of these fish compared with trout.

We take this opportunity of expressing our thanks to the Government Grants Committee of the Royal Society for aid in carrying out this work; and also to Mr A. R. Peart of Hungerford who supplied the trout.

REFERENCES.

Gardner and Leetham (1914, 1). *Biochem. J.* **8**, 374.
—— (1914, 2). *Biochem. J.* **8**, 591.
Gardner, King and Powers (1922). *Biochem. J.* **16**, 523.

LXXVI. RESPIRATORY EXCHANGE
IN FRESH WATER FISH.

PART V. ON EELS.

By JOHN ADDYMAN GARDNER AND GEORGE KING.

From the Physiological Laboratory, University of London,
South Kensington.

(*Received August 25th, 1922.*)

EELS are said to be peculiarly averse to cold, and the fact that the temperature of the brackish water of estuaries is usually higher than that of unmixed salt or fresh water has been advanced as one of the reasons for their seaward migration on the approach of winter. During the cold of winter these fish lose their appetite and become torpid, large numbers of them congregating together for the sake of the additional warmth thus obtained, and burying themselves to a depth of many inches in places where the receding tide leaves them more or less dry.

In the eel the bronchial openings are small and lead into a sac, from which another sac is given off. The gills are thus exposed only slightly to the drying influence of the atmosphere, and it is believed that it is owing to this and to the slimy condition of the skin, that eels are able to exist for a very considerable time, compared with other fish, out of water. They can evidently exist at a low plane of metabolism, but like most animals which pass the winter in a torpid condition, are relatively voracious during the summer months.

For these reasons it seemed of interest to study the respiratory exchange of these animals under various conditions of temperature, and compare them with other fish, particularly trout, which, as we have shown, are also sluggish, for a time at any rate, on exposure to cold.

The fish used in these experiments were the common *Anguilla vulgaris* and were 15 to 18 inches in length. They were kept for several days in a large tank with gravel bottom in running water before being used for experiment. The tank contained a number of small worms and the like, which the eels could have picked out of the gravel, but whether they did so we have no means of judging. The fish however were in good condition.

The method adopted was that fully described in Part I [1914] and Part III [1922] of this series.

Experimental.

At high temperatures.

(1) Six fish, weight 940 g., were ·placed in a tank of water at the temperature of the outside tank—about 15°—and gradually warmed up to 20° before being placed in the experimental bottle. The initial temperature of the water in the bottle was 21·1° and the final temperature 23·5°; the duration of the experiment was 5 hours 4 minutes, and during this period 275 cc. of commercial oxygen were added (measured at 16·5° and 760 mm.) to the air above the water. Analyses of the air and water at the beginning and end of the experiment gave the following results in cc. reduced to standard temperature and pressure.

	Initial vol.	Final vol.	Difference
Free and combined CO_2	1827·4	2023·0	+ 195·6
Oxygen	4102·3	3810·2	– 292·1
Nitrogen	14693·1	14721·4	+ 28·3

Nitrogen error + 0·1990 %.

At the end of the experiment the water (24619 cc.) had a strong faecal smell, and contained a good deal of dejecta. The eels at this high temperature were very active and difficult to catch by hand.

At medium temperatures.

(2) Five fish, weight 650 g. Initial temperature of water 16·5°, final temperature 16·5°. Duration of experiment 5 hours 45 minutes. Oxygen was added as before. The fish were lively.

	Initial vol.	Final vol.	Difference
Free and combined CO_2	1821·0	1950·1	+ 129·1
Oxygen	4147·5	3981·5	– 166·0
Nitrogen	15982·2	15981·8	– 0·4

Nitrogen error – 0·0025 %.

(3) Five fish; weight 725 g. Initial temperature 11·7°, final temperature 8·4°. Duration 3·44 p.m. to 12.10 p.m. = 20 hours 36 minutes. Oxygen was added, and pumping was continuous except for a few hours owing to a slipped belt.

	Initial vol.	Final vol.	Difference
Free and combined CO_2	1552·9	1828·8	+ 275·9
Oxygen	4868·8	4548·4	– 320·4
Nitrogen	17594·7	17607·8	+ 13·1

Nitrogen error + 0·074 %.

(4) Four fish, weight 680 g. This experiment was commenced at 11.37 a.m. and continued until 12 noon on the following day—24 hours 23 minutes. The initial temperature of the water in bottle was 16·6° and the air 16·7°. After about three hours the water in the thermostat was packed with ice, and the temperature of the thermometer in the air of the bottle was noted at intervals, the final temperature of the water was 6·4°. It was estimated

that the average temperature was between 9° and 10°. Oxygen was added during the experiment as before. During the night the pumping was interrupted for a few hours owing to the slipping of a belt, probably in the early morning owing to change over of the dynamos in the power station.

	Initial vol.	Final vol.	Difference
Free and combined CO_2	1650·1	1949·8	+ 299·7
Oxygen	4655·4	4371·7	− 283·7
Nitrogen	16875·8	16903·4	+ 27·6

Nitrogen error +0·16 %.

At low temperatures.

(5) For this purpose the eels were kept in well oxygenated water in the ice chest for 24 hours before the experiment, the average temperature being between 3° and 4°. The animals were very inert and sluggish and scarcely moved when handled. Five fish, weight 725 g. Initial temperature of water 6·5°, final temperature 5·3°. Duration of experiment 24 hours 8 minutes. Oxygen was added during the experiment and the pumping was continuous.

	Initial vol.	Final vol.	Difference
Free and combined CO_2	1343·5	1414·3	+ 70·8
Oxygen	5559·5	5397·2	− 162·3
Nitrogen	20211·6	20211·5	− 0·1

Nitrogen error −0·00049 %.

In all the above experiments except that at high temperatures the water remained clear and no unpleasant smell was noticed. The fish remained perfectly healthy after the experiments. The results are summarised in the following table.

No. of experiment	Average temperature	Average weight of fish used in g.	Oxygen absorbed per fish per hour in cc. at 0° and 760	Oxygen absorbed per kg. per hour at 0° and 760	CO_2 evolved per fish per hour at 0° and 760	CO_2 evolved per kg. per hour at 0° and 760	Respiratory quotient
(1)	22·2°	157	9·60	61·29	6·43	41·04	0·67
(2)	16·5°	130	5·77	44·41	4·49	34·54	0·78
(3)	10°	145	3·11	21·45	2·68	18·47	0·86
(4)	about 9°	170	2·91	17·11	3·07	18·08	1·06
(5)	5·9°	145	1·35	9·28	0·59	4·05	0·44

It will be noticed that the oxygen consumed per fish, and per kg. of fish per hour is proportional to the temperature and if the oxygen is plotted against temperature the values are approximately on a straight line.

On comparison with the figures given in Part I [1914] for 8-inch brown trout it is evident that eels live at a much lower plane of metabolism than trout. At medium temperatures, about 16°, trout use about four times as much oxygen as eels, and at low temperatures 10 to 12 times as much.

The respiratory quotients in the first four experiments are approximately normal but at low temperature, at which the animals are in a torpid condition, the value falls much below normal. In this respect their behaviour recalls that of trout.

REFERENCES.

Gardner and Leetham (1914). *Biochem. J.* **8**, 374.
Gardner, King and Powers (1922). *Biochem. J.* **16**, 523.

LXXVII. ON A SERIES OF METALLO-CYSTEÏN DERIVATIVES. I.

By LESLIE JULIUS HARRIS.

From the Biochemical Laboratory, Cambridge.

(*Report to the Committee of Scientific and Industrial Research.*)

(*Received September 14th, 1922.*)

WHEN a stream of air is blown through an alkaline solution of cysteïn, the oxidation is generally accompanied by the production of a violet coloration; an observation that will be familiar to workers who have had occasion to prepare cystine in this manner. Although the phenomenon has been long known its cause has remained obscure, and the present investigation was undertaken primarily with the object of ascertaining its nature.

The fact that cysteïn gives with ferric chloride and ammonia an intense coloration, which becomes darker on shaking, was first noticed by Andreasch [1884]. Arnold [1911], describing a number of tests by which cysteïn may be recognised, distinguishes between the colour produced with ferric chloride and that with ammonia. "If a solution of cysteïn is made weakly alkaline a violet colour is produced which is more permanent than that given by sodium nitroprusside...." There can be little doubt however that the coloration is in reality due not so much to the action of the alkali upon the cysteïn, as would appear from Arnold's account, as to the presence of traces of a metallic (generally ferric) salt, which produce the typical coloration in alkaline solution. Taking care to ensure the complete absence of a metallic ion I have been unable to observe the production of the slightest colour.

An important feature of the reaction is its impermanent character. Arnold observes that the colour fades after a while, but can be renewed by shaking. The nature of this curious alternation of fading and restoration of colour receives no explanation from Arnold.

The Reversible System: Cysteïn—Iron—Ammonia.

When cystine is isolated by the use of mercuric sulphate reagent, the Hg-cystine compound which is precipitated, yields, on decomposition with H_2S, a solution of the reduced cysteïn. This can be re-oxidised, after removal of H_2S, by making alkaline with ammonia and blowing a stream of air through the solution. Under these circumstances there is generally enough iron present as impurity to produce a noticeable violet coloration in the solution as soon as the active oxidation is started. Often a state of delicate equilibrium is set up in which the colour disappears upon the air current being stopped, and immediately reappears upon its continuation. The necessary conditions may be reproduced artificially by adding to a test tube containing (say) 4 cc. of a 5 % cysteïn solution a single drop of weak $FeCl_3$ solution. Upon making alkaline with weak NH_4OH, a violet colour appears, only to fade almost immediately. The colour reappears if the tube be shaken, or if air (or, better, oxygen) be otherwise brought into intimate contact with the solution. Bubbling hydrogen, carbon dioxide or nitrogen through the liquid has no effect.

Variation in the concentrations has a visible effect on the course of the phenomena. The depth of colour produced is directly proportional to the quantities of both cysteïn and ferric chloride present together. Again, the greater the concentration of ferric chloride relative to that of cysteïn, the longer does the violet colour survive the cessation of the air current, or, to look at it in another way, increasing the proportion of cysteïn in the mixture, brings about a more rapid fading when the air current is stopped. The solution remains violet (or alternatively capable of yielding a violet colour on aeration) so long as unoxidised cysteïn is present; the presence of the latter can be demonstrated by the extremely delicate nitroprusside test[1]. When oxidation is complete the solution consists of cystine together with a minute quantity of $Fe(OH)_3$, derived from the added iron salt. The effect of increasing the alkalinity is to yield a solution which less readily decolorises in the absence of the air current, but in which complete and permanent loss of colour (or ability to produce it on addition of the iron salt) occurs more rapidly, owing apparently to the greater rapidity of the transformation cysteïn $\longrightarrow$ cystine in the more alkaline solution.

Numerous ferric salts, also suspended ferric hydroxide, have been investigated as substitutes for $FeCl_3$. All were equally efficient. Potassium ferricyanide was without action.

Explanation of the Phenomena.

Several explanations might be adopted to account for the observations described above. At first sight it would seem likely that the colour is due to the formation of an unstable compound in a state of oxidation intermediate between cysteïn and cystine; its presence being made manifest as a result of its power to yield a coloration with ferric salts.

Secondly, it may be assumed that direct oxidation of cysteïn to a body of the cystine peroxide type occurs, and that this hypothetical substance is reactive to $FeCl_3$. The consecutive reaction is between cysteïn and the peroxide to yield the end product, cystine. There are theoretical grounds for postulating the possible existence of such a peroxide.

---

[1] It is possible to fix approximately the completion of oxidation by the disappearance of the violet coloration.

In view, however, of similar reactions between cysteïn and salts of certain other metals, notably copper, the following explanation would appear to fit the facts more closely than either of these.

Cysteïn is an acid, and it may be supposed that in the presence of $Fe^{...}$ ions, and in alkaline solution, it forms a violet-coloured ferric derivative. The main characteristic of cysteïn is its reducing power:

$$2COOH—CH(NH_2)—CH_2—S\boxed{H + O} \longrightarrow (COOH—CH(NH_2)—CH_2—S—)_2 + H_2O.$$

If excess is present and the supply of air be limited the cysteïn is oxidised to cystine at the expense of the ferric ion, which is thereby reduced to the ferrous state. The latter yields no coloured derivative with cysteïn, and the solution therefore fades. If now air be blown through the mixture, the $Fe^{..}$ is rapidly re-oxidised to $Fe^{...}$, $Fe^{..}$ as is well known being unstable in alkaline solution when available oxygen is present. The colour is therefore regenerated provided some unoxidised cysteïn remains to form the anion of the coloured body, and it will be retained until the process of oxidation is complete. In the presence of abundance of free oxygen the cysteïn will be oxidising continuously, and the iron will be maintained in the ferric condition.

This explanation accounts for the following experimental observations:

(1) The intensity of colour on making alkaline with ammonia is proportional to the amounts of cysteïn and Fe salt present together (*i.e.* the amount of cysteïn iron derivative produced).

(2) Excess of cysteïn causes a more transient coloration (since the $Fe^{...}$ is more rapidly reduced in the presence of excess of the reducing agent).

(3) Excess of iron causes a more stable coloration.

(4) Excess of alkali causes the coloration to be less *transient* (since $Fe^{..}$ is increasingly unstable in alkaline solution).

Apparently the reduction potential of cysteïn does not develop with increasing alkalinity so rapidly as that of $Fe(OH)_2 \longrightarrow Fe(OH)_3$.

(5) With increasing alkalinity the *total duration* of the colour phenomena is less, since the cysteïn disappears more rapidly. (Cystine becomes more readily autoxidisable with increase of p_H.)

(6) It is found that the addition of *free cysteïn* to the coloured solution aids the (temporary) decolorisation. (Explanation as in (2).)

(7) It is a well known experimental fact that iron acts as a catalyst in accelerating the oxidation of cysteïn to cystine. This is readily explained on the basis of the above theory: the iron acts as a carrier; in the ferrous state it is a more efficient oxygen acceptor than the cysteïn, and oxidised it functions as an oxygen donator.

(8) The most convincing support to these views is afforded by the behaviour of copper (*vide infra*). It is found moreover that reversible colorations analogous with that given by iron could be obtained only by those metals which exist in more than one state of oxidation, and in which change of valency from the one form to the other can occur with tolerable ease.

With regard to the catalytic action of iron (see (7) above) mention should be made of the work of Mathews and Walker [1909] on the velocity of spontaneous oxidation of cysteïn in the presence of salt solutions. This paper was published prior to that of Arnold and I was unaware of the explanation therein advanced of the mechanism of the catalysis when the work described in the present paper was carried out.

"An explanation is given" by Mathews and Walker "of the action of different metals based on their solution pressure." These authors account for the catalytic action by assuming the momentary formation of an intermediate metallic compound

"MeO + Cysteïn = Cysteïn . MeO.
Cysteïn . MeO = Cystin + Me + H_2O."

It would seem that this view fails to account for the experimental observation that the coloured derivative is only instantly decomposed when excess of free cysteïn is present. The pure cupric cysteïn derivative can be isolated and shows no tendency for internal spontaneous scission such as is suggested.

The action of cysteïn upon the cupric and other coloured metallic derivatives, can leave little doubt that the mechanism of the reaction consists in the reduction of the oxidised metallic atom of the derivative by the action of excess of cysteïn.

The formation of various metallic derivatives of cysteïn.

Fe'''. The behaviour of iron is explained above. The deep coloration obtained when (excess of) ferric chloride is added to cysteïn was first described by Andreasch, and then by Arnold. The latter author distinguished between the colour given by $FeCl_3$ and by "ammonia." With sufficiency of iron the colour remains until oxidation of cysteïn to cystine is complete, and cannot then be regenerated by aeration.

Fe''. Under ordinary conditions $Fe(OH)_2$ cannot be precipitated sufficiently free from $Fe(OH)_3$ to prevent the production of the violet ferric compound. Investigations have been started under anaerobic conditions.

Double salts (ferric alum, etc.) in which iron functions as Fe''' act like $FeCl_3$ or $Fe(OH)_3$. When iron functions in the complex anion as in K_4FeCy_6 or K_3FeCy_6 it is without action on the cysteïn anion.

Mn''' and **Mn''.** If a few drops of manganous sulphate be added to cysteïn (in HCl solution) and the liquid made alkaline with ammonia a fine *green* coloration (Mn-cysteïn cpd.) appears. It rapidly fades provided sufficient cysteïn be present. On blowing air through the faded solution the colour is regenerated. The general behaviour of manganese towards cysteïn resembles that of iron described at length above.

It appears probable that the green body is the Mn''' derivative and the decolorised solution contains Mn''. The addition of alkali to a manganous salt yields $Mn(OH)_2$, which however rapidly goes to $Mn(OH)_3$ or Mn_2O_3, H_2O

unless air be excluded. When cysteïn is also present the green colour is given in air, while in absence of air the solution remains colourless (Mn··). When oxidation of cysteïn has reached completion brown Mn_2O_3, H_2O remains. The addition of fresh cysteïn to this yields the green derivative and the sequence can then be repeated afresh.

Thunberg [1913] has found that in the presence of $MnCl_2$ the autoxidation of various thio-compounds is accelerated, a fact which is readily accounted for by a cycle such as the above, in which the salt functions as an oxygen carrier. According to Mathews and Walker [1909] however $MnCl_2$ has no effect on the velocity of oxidation of cysteïn.

Mn^{vi}. Dilute $KMnO_4$ added to cysteïn solution suffers decolorisation. On making alkaline with ammonia a *rose-red* colour appears, fading on standing and reappearing on aeration. The depth of colour is increased with increasing concentration of the two substances; excess of cysteïn producing more rapid fading on stoppage of the air current, coupled with a less readily regenerated colour on aeration, while increasing the amount of $KMnO_4$ gives a more stable coloration. Here again the behaviour is parallel with that of iron salts.

The rose colour may be due to sexvalent manganese, corresponding with MnO_3, and the colourless reduced solution may contain quadrivalent Mn— (corresponding with MnO_2) formed from the MnO_3 by the reducing action of excess of cysteïn.

The first stage would be the production of MnO_2

$$2KMnO_4 + 6H.S—CH(NH_2)_2—COOH = 2MnO_2 + 3(S—CH_2—CH(NH_2)—COOH)_2$$
$$+ 2KOH + 2H_2O.$$

In alkaline solution however Mn^{vi} is stable,

e.g.
$$Mn_2O_7 \xrightarrow{\text{alkali}} 2MnO_3 + O$$

$$3K_2Mn^{vi}O_4 + 2H_2O \underset{\text{in acid solution}}{\overset{\text{in alkaline solution}}{\rightleftharpoons}} 2KMn^{vii}O_4 + MnO_2 + 4KOH$$

hence we may have Mn^{vi} in the alkaline aerated cysteïn mixture. In the reduced solution MnO_2 is almost certainly present. The action of $KMnO_4$ is being further investigated.

Cu·· and **Cu·**. The existence of a copper derivative of cysteïn has been known for some time [Suter, 1895]. The addition of excess of a cupric salt to cysteïn in neutral solution produces a blue-black precipitate of a Cu··- cysteïn compound. The latter is almost completely insoluble in neutral or slightly acid reaction, and may advantageously be employed for the separation and estimation of cysteïn [Embden, 1901]. When the solution is somewhat acid it is better to employ moist freshly precipitated $Cu(OH)_2$ in place of the copper salt, following the procedure used by Hopkins [1921] in the separation of glutathione. The blue-black precipitate dissolves on addition of ammonia forming an intense dark brown (sepia) solution. It is reprecipitated on reducing the p_H to about 7 and dissolves to an almost colourless solution in excess of acid.

· Interesting results have been obtained by the use of excess of cysteïn. The addition of cysteïn to the alkaline (brown) solution of the cysteïn-copper derivative causes rapid decolorisation. It will be shown that this is due to the reduction of the copper to the cuprous condition, and that the decolorised solution contains the cuprous-cysteïn derivative. If the decolorised solution be now shaken the dark brown colour reappears: the Cu˙ has been re-oxidised by the air in the alkaline solution.to Cu˙˙. On standing, the solution fades once more; and these changes can be repeated so long as sufficient unoxidised cysteïn remains. By the prolonged action of oxygen or air on the brown solution the cysteïn is completely oxidised to cystine and a cuprammonium colour persists.

Copper is singular among the metals so far examined in that *both* its cysteïn salts (cuprous and cupric) are insoluble in neutral solution. As mentioned above an alkaline solution of the cuprous derivative is formed when the ammoniacal cupric—*i.e.* the brown—solution reduces spontaneously in the presence of excess of cysteïn. On neutralisation of the decolorised solution by the addition of dilute HCl the cuprous cysteïn compound is precipitated as a highly characteristic voluminous white precipitate. It may be obtained in a variety of ways. Add to a few cc. of a 5 % cysteïn solution one drop of $CuSO_4$aq. A blue-black precipitate of the cupric derivative first appears but is rapidly transformed into the white cuprous derivative. The Cu˙˙ has become reduced to Cu˙ by excess of cysteïn. The white precipitate is readily soluble in excess of acid or ammonia to a colourless solution. The alkaline solution readily darkens in presence of air owing to oxidation of Cu˙. The blackened alkaline solution will of course again yield the cupric cysteïn derivative on neutralisation, while the air-free clear alkaline solution will give the white precipitate.

The cuprous derivative has also been obtained directly by the use of cuprous iodide. To a cysteïn solution a little well washed freshly precipitated cuprous iodide was added and the solution made alkaline with ammonia. An almost colourless solution resulted. This was neutralised and yielded the characteristic white precipitate. The strongly alkaline solution readily darkens in air—owing to oxidation of the cuprous ion—especially when the amount of copper is large. The colour can be seen extending downwards from the exposed surface of a large tube.

Co. The addition of a drop of cobalt acetate to cysteïn solution followed by ammonia resulted in the production of a dark yellow colour. No appreciable intensification resulted on prolonged passage of air through the mixture. The intensity of colour was increased by addition of further Co solution.

Ni. Cysteïn solution containing a little $NiSO_4$ assumed an orange colour on addition of ammonia. The colour deepened to a reddish brown (mahogany) appearance when some more cysteïn and nickel sulphate were added. The colour was retained so long as cysteïn remained in the solution.

Cr. On adding cysteïn solution to a weak, acidified solution of $K_2Cr_2O_7$

an intense chrome yellow colour appeared. On making alkaline with ammonia the usual greenish chromate colour was observed; it was not formed so readily however on the direct addition of ammoniacal $K_2Cr_2O_7$ to cysteïn. Potassium dichromate is peculiar in giving the coloration in *acid* solutions. Potassium chromate, which also contains the hexavalent chromium atom, behaves in the same way.

Sn^{iv}, Sn$^{..}$. The formation of a tin derivative of cysteïn may be demonstrated in the following manner. Have two test tubes each containing 3 cc. of water, and to one add also about 30 mg. of cysteïn. Add a little $SnCl_4$ to both tubes. Now make both tubes alkaline by the addition of ammonia and then acid by means of sulphuric acid. The presence of a tin cysteïn derivative is shown by its ready solubility in acid and more especially in ammoniacal solution. The tube containing the cysteïn will be seen to become perfectly clear and colourless in presence of ammonia or larger excess of acid; while a bulky precipitate remains in the blank tube under the same circumstances.

It might be thought that the presence of $SnCl_4$ would favour oxidation of the cysteïn. There seems no doubt however that the production of the clear solution is due to unoxidised cysteïn, since its presence is shown by the nitroprusside reaction. Indeed the presence of $SnCl_4$ seems to inhibit the oxidation. In one experiment a merest trace of cysteïn was added to a bottle containing a large excess of $SnCl_4$ and water, and the mixture was made alkaline with ammonia. It still showed a nitroprusside reaction after the lapse of five days.

Tin has a particularly strong affinity for sulphur and no doubt it protects that atom in the cysteïn molecule from oxidation.

With $SnCl_2$ the results observed were in every respect identical with those described above for $SnCl_4$.

As$^{..}$, As$^{...}$. Arsenic appears to be too electro-negative to afford any ready evidence of functioning as the basic part of a cysteïn derivative. As_2S_2 and As_2S_3 were examined but without result.

Pb. The addition of $Pb(NO_3)_2$ to cysteïn solutions failed to yield a coloration.

Hg$^{..}$. If insufficiency of "$HgSO_4$ reagent' is added to cysteïn the white precipitate first formed disappears. The same result can be obtained by addition of excess of cysteïn to the mercury precipitate. The phenomenon appears to be due to the reduction of the mercury from the bivalent to the univalent state, Hg$^{.}$, which is inefficient as a precipitant. Analogous results may be obtained with other mercuric derivatives. If cysteïn be added in excess to a suspension (1) of *mercuric iodide*, or (2) of the *double ammonium compound* which is precipitated when ammonia is added to acidified mercuric iodide, the precipitate will be seen to dissolve in both cases.

When sufficient mercuric salt is added, however, precipitation occurs. Mercuric nitrate, chloride and acetate all yield insoluble compounds with cysteïn. These substances have no precipitant action on cystine.

Bi. Add cysteïn to an ammoniacal suspension of bismuth hydroxide (obtained from bismuth nitrate or subnitrate). The precipitate readily dissolves with the production of a yellow solution, indicative of a $Bi^{...}$ cysteïn compound.

The action of BiO_4 and Bi_2O_5 remains to be investigated.

It will be seen that the observations recorded above open up considerable ground for further inquiry. A number of metals have still to be investigated. In view of the fact that considerable supplies of cysteïn will be required for continuation of the research and in view of the calls of other research work it has been thought advisable to give an account of the observations so far made.

I should like to express my sincere gratitude to Professor Hopkins for the help and encouragement he has given me and the interest he has shown in this work.

SUMMARY.

(1) The so-called "ammonia" test for cysteïn is only effective in the presence of traces of a metallic compound.

(2) The following metals—viz. $Fe^{...}$, $Mn^{...}$, $Mn^{vi(?)}$, $Cu^{..}$, $Hg^{..}$—can form part of a system which is reversible so long as unoxidised cysteïn is present; the controlling factors being the relative concentrations of cysteïn, metal salt and atmospheric oxygen. In alkaline solution and in presence of oxygen, the reduced metal is equivalent to an oxygen acceptor, and in absence of oxygen the oxidised metal is equivalent to a donator of oxygen, the cysteïn forming the oxidisable body.

(3) The fading of alkaline solutions containing a metal *plus* cysteïn and the recoloration on aeration are explained on these grounds.

(4) Coloured metallic derivatives of cysteïn have been formed from $Fe^{...}$, $Mn^{...}$, $Mn^{vi(?)}$, $Cu^{..}$, Co, Ni, Cr, Bi. $Cu^{.}$ gives a characteristic white derivative insoluble in neutral solutions. Sn and Hg derivatives are also discussed.

(5) Cysteïn exhibits greater readiness to form metallic derivatives than is the case with cystine.

REFERENCES.

Andreasch (1884). *Maly's Jahresber.* 76.
Arnold (1911). *Zeitsch. physiol. Chem.* **70**, 317.
Embden (1901). *Zeitsch. physiol. Chem.* **32**, 94.
Hopkins (1921). *Biochem. J.* **15**, 290.
Mathews and Walker (1900). *J. Biol. Chem.* **6**, 306.
Suter (1895). *Zeitsch. physiol. Chem.* **20**, 575.
Thunberg (1913). *Skan. Arch. Physiol.* **30**, 293.

LXXVIII. THE INFLUENCE OF FAT AND CARBOHYDRATE ON THE NITROGEN DISTRIBUTION IN THE URINE.

By EDWARD PROVAN CATHCART.

Institute of Physiology, University of Glasgow.

(Received September 22nd, 1922.)

IN previous papers it has been shown that the complete withdrawal of carbohydrate from the diet produces well marked changes in the metabolism. Landergren [1903] in his original paper showed clearly that (1) on an exclusively carbohydrate diet the output of total nitrogen steadily fell and (2) when the diet was changed to one composed exclusively of fat the output of total nitrogen rose. This finding I confirmed and extended [1909]. The series of experiments detailed in the present paper were intended to extend further our knowledge of the subject. They were all carried out previous to July 1914 on Mr R. Lang and it was intended to complete the series with the effects of such limited diets on the capacity of the individual to do muscular work, but unfortunately Mr Lang on his demobilisation was no longer available and so far no other subject, either willing or suitable to undergo the necessary privation, has been found.

METHODS.

Previous to the ingestion of the various experimental diets the subject on each occasion attempted for two or three days to consume roughly the same diet so that the base line would be approximately the same. Unfortunately owing to circumstances over which the subject had no control this was not always possible. The life led by the subject was uniform throughout.

The oil used was the finest olive oil procurable. It was always emulsified before taking by shaking, after the addition of about 1 g. of potassium carbonate dissolved in water. In this form the oil was fairly readily consumed and during the three days of the duration of the experiment it gave rise to no untoward digestive disturbance or diarrhoea. As a matter of fact in one only of the three day experiments did a single movement of the bowels take place. Mr Lang found it impossible to carry on for more than three days on account of the nausea induced by the mere sight and smell of the emulsified oil. The addition of even 150 g. of glucose did not materially lessen this disgust although the subject always declared that a feeling of "tiredness" which was prominently associated with the ingestion of pure oil was distinctly reduced when sugar was added even in comparatively small amounts. The sugar used was chemically pure anhydrous dextrose. The calorie values of the various diets were kept approximately constant; they varied between 3102 C. with pure oil and 2978 C. with oil plus 150 g. sugar. The analytical methods employed were: total nitrogen, Kjeldahl; ammonia, Folin; urea, Folin and urease; uric acid, Hopkins-Folin; creatinine and creatine, Folin.

Results.

Output of Nitrogen.

Diet	Day of experiment	In g.						In percent. of T.N.				
		Total	Urea	Ammonia	Total creatinine	Uric acid	Undetermined	Urea	Ammonia	Total creatinine	Uric acid	Undetermined
323 g. olive oil	1	10·00	7·31	0·32	0·50	0·094	1·76	73·1	3·3	5·0	0·94	17·6
	2	14·24	11·08	0·66	0·62	0·031	1·84	77·8	4·6	4·4	0·22	12·9
	3	[10·75]	7·75	1·07	0·63	0·033	1·27	72·1	9·9	5·8	0·31	11·8
323 g. olive oil	1	10·95	7·88	0·29	0·49	0·123	2·16	72·0	2·7	4·5	1·12	19·7
	2	14·35	11·05	0·56	0·52	0·059	2·17	77·0	3·9	3·6	0·41	15·1
	3	14·18	10·44	1·13	0·59	0·029	1·99	73·6	8·0	4·2	0·20	14·0
310 g. olive oil	1	9·41	6·94	0·21	0·58	0·136	1·53	73·7	2·2	6·2	1·44	16·2
30 g. dextrose	2	11·72	7·48	0·44	0·57	0·087	3·15	63·8	3·7	4·9	0·74	26·7
	3	10·23	6·60	0·58	0·60	0·066	2·38	64·5	5·7	5·9	0·64	23·2
297 g. olive oil	1	9·78	7·36	0·26	0·52	0·142	1·50	·75·3	2·6	5·3	1·45	15·3
60 g. dextrose	2	9·31	7·07	0·31	0·49	0·119	1·32	75·9	3·4	5·3	1·28	14·2
	3	8·60	5·86	0·38	0·50	0·108	1·76	68·1	4·4	5·8	1·25	20·4
279 g. olive oil	1	7·97	5·38	0·26	0·48	0·133	1·73	67·5	3·2	5·9	1·67	21·7
100 g. dextrose	2	7·75	5·24	0·25	0·54	0·118	1·60	67·6	3·3	7·0	1·52	20·6
	3	7·12	4·79	0·18	0·54	0·145	1·47	67·3	2·5	7·6	2·03	20·6
257 g. olive oil	1	10·18	7·66	0·37	0·56	0·121	1·47	75·2	3·6	5·5	1·18	14·4
150 g. dextrose	2	9·95	7·28	0·24	0·56	0·152	1·68	73·2	2·4	5·6	1·52	16·8
	3	7·37	5·05	0·16	0·56	0·143	1·46	68·5	2·2	7·6	1·94	19·8

Total nitrogen.

The table clearly shows that with increasing amounts of carbohydrate there is a general tendency for the output of total nitrogen to decrease. On the pure oil, although in each instance there is a rise in the output of total nitrogen on the second day of the experiment, there is not the further rise noted on the third day of the experiment which both Landergren and I found in previous experiments. This may be due to the fact that in this series, in contradistinction to the earlier ones, olive oil emulsified by alkali was used in place of butter fat or cream. There is no doubt that the olive oil and alkali is much more readily tolerated by the intestine than either cream or butter. Instead of experiencing the acute diarrhoea which usually results from the ingestion of these products when taken alone, as above noted Mr Lang was constipated throughout the experiments.

Zeller [1914] has also carried out a series of observations, both on dog and man, on the effect of varying the fat and carbohydrate values of practically protein-free diets. It is to be regretted that his experimental periods on the fat-rich diets were so short, two days with the 75 % fat diet and only a single day with 100 % fat. Experiments of such very short duration do not permit of the metabolism adjusting itself to the new conditions. The effect of the fat-rich diet had however the usual result, viz. a marked rise in the output of total nitrogen both in dog and man.

Urea.

The urea output resembles very closely that of total nitrogen. It is interesting to note that when the percentage outputs of urea and ammonia nitrogen are compared there is a very marked fall in the output of urea after the sugar, even with the smallest dose, although at the same time there is also a fall in the output of ammonia, whereas it might have been expected that with the decrease in the ammonia output there would have been a rise in the urea output. In the case of the experiments with oil alone there does seem to be a balance between the two outputs, a rise in the ammonia output being associated with a fall in the urea.

When the outputs of both ammonia and urea are taken together it may be definitely stated that when oil is given alone there is a definite and steady rise in the excretion of these nitrogenous materials, whereas in the experiments in which sugar is added there seems to be just as general a tendency for the united output to fall during the three days of the experiment.

Zeller in his series also found that the urea output followed very closely the curve for the output of total nitrogen. His percentage output of urea reached however a far lower level than any that I obtained. Thus when a pure carbohydrate diet was given immediately after a pure fat diet he found in one instance that urea only formed on the average 49·7 % of the total nitrogen and in the other it reached the very low level of 39·7 %.

Ammonia.

In each case with the pure oil diet a very definite steady rise both of the absolute and the percentage output of ammonia nitrogen occurs on the three days of the experiment. A slight rise is also noted on the third day when 30 and 60 g. of sugar are given although, even with these small amounts of sugar, there is a definite reduction in the total ammonia output. When the larger amounts of sugar are given there is a steady fall in the output of ammonia during the three days of the test.

Zeller too in his experiments found, like other observers, that the giving of fat-rich diets led to a marked rise in the output of ammonia. In all of his experiments, just as he found a much lower percentage output in the case of urea, he obtained much higher outputs of ammonia, both with the carbohydrate-rich and carbohydrate-poor diets, than those given in the present paper. His maximum percentage output of ammonia was 15·4 % with the 100 % fat diet.

Uric acid.

The variation in the excretion of uric acid as the result of the alterations in the composition of the diets is very striking and most interesting. As the table shows it may generally be stated that when carbohydrate is completely removed from the diet there is a well marked reduction in the output and that this output gradually and steadily rises as the amount of glucose added is increased. It was shown in a previous paper [1909] that the output of uric acid varied with the carbohydrate intake; that the output was high when the subject was on a carbohydrate-rich diet and low when on a fat-rich, carbohydrate-poor diet. Graham and Poulton [1913] who studied the influence of carbohydrate and fat on the output of endogenous purine also found that a diet in which fat predominated was associated with a low output and a diet in which carbohydrate predominated with a high output of uric acid. Later Umeda [1915] working in my laboratory confirmed, both in the case of man and the dog, the variation in uric acid output as the result of alteration of the diet. He found in man that, although there was a definite fall in the output of total purines, the output of purine bases on a fat diet was higher than on a carbohydrate one. Graham and Poulton had also noted that when the carbohydrate content of the diet was reduced there was a tendency for the purine base output to rise. In the case of the dog, Umeda found that the allantoin output behaved like uric acid in its relation to the nature of the diet. Zeller also investigated the influence of carbohydrate and fat diets both on the output of uric acid and the purine bases. He too found that there was a reduction in the output of uric acid on a 100 % fat diet but there was little or no influence on the output of the purine bases. As already noted Zeller's experiments were of exceptionally short duration.

In an interesting and suggestive paper Ackroyd and Hopkins [1916] showed very definitely that (1) when both arginine and histidine were removed from the diet the amount of allantoin excreted in the urine of the rat was much decreased; (2) it was somewhat diminished when only one of these amino acids was present and (3) when both were restored to the diet the allantoin excretion returned to normal. Their conclusion that arginine and histidine probably constitute the most readily available raw material for the synthesis of the purine ring in the animal body would seem to be justified. Still, although these two amino acids may be considered to supply the raw material for the synthesis, this does not dismiss the probability that the presence of carbohydrate is necessary as in the experiments of Ackroyd and Hopkins the carbohydrate supply was abundant. Further, the well known experiment of Knoop and Windaus [1905] in which they demonstrated the formation of the iminazole ring when a solution of dextrose was acted upon by ammonia in the presence of zinc and sunlight, shows, at least, that a synthesis from very simple compounds is possible. If the rest of the diet played but a minor part in the synthesis of the purine it would have been expected that the output of purines would have risen when a general increase in the breakdown of the protein molecule took place, as is the case when oil alone is the food material supplied, i.e. when there would be in all probability an increased amount of amino acids including arginine and histidine free in the tissues. Further the diminution of the output of uric acid on the fat diet is not due to delayed oxidation of other purine bodies because reference to the output of undetermined nitrogen, which would include purine bases, shows that the excretion on the fat diet is actually lower than when sugar is given. It is probable then that, although arginine and histidine may be regarded as the actual nitrogenous source of the purine, before the synthesis can take place carbohydrate in some form or other must be present.

Creatinine and Creatine.

Inasmuch as many of the estimations of these substances were made without the precautions suggested by Graham and Poulton the total creatinine output only is given. It may be stated that in two other experiments when every precaution was taken the presence of creatine in the urine was definitely shown. The work of Underhill and Baumann [1916] showed very conclusively that acidosis alone cannot account for the presence of creatine in the urine.

Undetermined nitrogen.

Although the absolute amount of undetermined nitrogen daily excreted does not vary very greatly in any of the experiments, yet, when calculated on a percentage basis, it is found that there is a very definite rise when carbohydrate is added to the diet. Zeller in one of his pure fat experiments found the highest percentage output whereas in the other the output was low, if not the lowest of the series. Zeller in addition to the substances referred to in this paper also determined amino and peptide nitrogen. The output of these substances rose when the pure fat diet was given.

DISCUSSION.

The consideration of the question of isodynamic replacement and the inferences to be drawn from the consideration of experiments on complete replacement such as those described, irrespective of the question as to whether such experiments fall within the normal capacities of the tissues, open up very wide issues. The mere statement of the isodynamic law virtually upholds the thesis that the basis of nutrition is the exchange of energy and not the exchange of material or, at the very least, that *Kraftwechsel* predominates over *Stoffwechsel*. Rubner [1883] who enunciated the hypothesis that fat and carbohydrate are mutually replaceable in a diet in isodynamic amounts, although in a much more recent paper he definitely stated that it was impossible to replace completely any of the proximate principles, definitely selected the term isodynamic after the consideration of such less specific terms as *gleichwertig*.

The work of Rubner is generally stated to be substantiated by the work of Atwater and Benedict [1903]. The experiments of these workers do undoubtedly support it but the type of experiment they adopted could not be.

expected to determine the ultimate degree of replacement, as only variations in a mixed diet on the human subject under approximately normal conditions were employed in contradistinction to Rubner's method of a rigorous administration of a single food stuff. It is undoubtedly true that within limits fat and carbohydrate may replace one another in the diet and it is obvious that in the average diet such an arrangement is automatically adopted. The statement of the case in the form of a general isodynamic law is simply untenable. At present the evidence is clear and convincing in support of the statement that it is impossible to replace carbohydrate completely by fat but so far the evidence available in support of the view that fat as such is a necessary constituent of a diet is scanty and unsatisfactory.

No one will of course seriously maintain that nutrition can ultimately be reduced merely to the satisfying of the energy demands: the calorie factor may be regarded as strictly secondary to the supply of material. We do not live on calories, yet all our general estimates of food requirements are quite properly for the most part made in terms of calories. Calorie value is simply a very convenient physical standard for the assessment of diets, but merely because such a standard has proved of great utilitarian value there is no real justification for placing this standard as the foundation stone of hypotheses framed to offer an explanation of cellular activity. Many writers are obsessed with the idea of the calorie, forgetting that the organism is certainly not a heat engine. It is perfectly true that calories are a measure of heat, but it must not be forgotten that we do not consume actual heat units but only potential heat-giving substances which can eventually be degraded to the form of heat and be measured as such. The thermal aspect of nutrition is unduly stressed, for, while heat may be a necessary product of tissue activity, it is, after all, a by-product.

The use of the term isodynamic in connection with problems of nutrition should be strictly limited. One can undoubtedly speak of isodynamic quantities of various substances but it does not follow that they are of equal, or indeed of any, value to the organism. When dealing with foodstuffs we ought to keep constantly in view that the material side is fully as important as the energy side. Therefore one ought not to stress so much the equality in energy as the equality in sparing or preventing tissue breakdown, the isoeconomic or, as I prefer to call it, the *isotamieutic* (Gk. tamieuo = to husband or to spare) value. Such a value is more physiological than isodynamic as it covers all phases of cellular activity. At present the data available do not suffice to permit of any adequate explanation of metabolic phenomena. Considerations such as those on which Carl Voit based his theory of metabolism, so actively rebutted by Pflüger, are not dismissed by the more modern hypothesis of Folin. Folin simply dealt with entirely superficial results and within these limitations the hypothesis is admirable. He did not attempt to elucidate the causal factors which lie beneath the phenomena which he correlated in his papers. The old question discussed so energetically by Voit and Pflüger as to whether the newly ingested material becomes an integral part of the living molecule before utilisation is still unanswered.

CONCLUSIONS.

1. The output of total nitrogen, urea, and ammonia rises on a fat diet and falls on the addition of carbohydrate.

2. The output of uric acid is low on the fat diet and increases on the addition of carbohydrate.

3. The output of total creatinine is but little affected by the change of diet. Small amounts of creatine are excreted on a carbohydrate-free diet.

4. The output of undetermined nitrogen is greater on diets containing carbohydrate than on those from which carbohydrate is absent.

REFERENCES.

Ackroyd and Hopkins (1916). *Biochem. J.* **10**, 551.
Atwater and Benedict (1903). *U.S. Dept. Agr. Bull.* 136.
Cathcart (1909). *J. Physiol.* **39**, 311.
Graham and Poulton (1913). *Quart. J. Med.* **7**, 13.
Knoop and Windaus (1905). *Beiträge*, **6**, 392.
Landergren (1903). *Skan. Arch. Physiol.* **14**, 112.
Rubner (1883). *Zeitsch. Biol.* **19**, 312.
Umeda (1915). *Biochem. J.* **9**, 421.
Underhill and Baumann (1916). *J. Biol. Chem.* **27**, 127 *et seq.*
Zeller (1914). *Arch. Physiol.* 213.

LXXIX. ON THE INFLUENCE OF THE SPLEEN UPON RED BLOOD-CORPUSCLES. I.

By NICOLAAS ALBERT BOLT AND PIETER ANTON HEERES.

From the Physiological Laboratory, University of Groningen, Holland.

(*Received October 3rd, 1922.*)

SINCE the researches of Hunter [1892] on the destruction of red blood-corpuscles, an active rôle in this action has been ascribed to the spleen. The appearance of phagocytosis and haemolysis in the spleen had already been discovered by Koelliker [1847]. From this discovery resulted the researches of Virchow, Quincke and others on the transformation of blood pigments, in which for the rest the spleen was only considered as an organ of accumulation.

But it was Hunter who pointed out that the cause of the blood destruction partly lies in an action of the spleen itself. This view was based specially upon this author's researches on toluylene-diamine poisoning. As is known the action of this poison is much stronger in normal than in splenectomised animals, so that the opinion is justified that the spleen plays an interfering part in toluylene-diamine poisoning.

To determine the haemolytic power of the spleen the following four methods have been used:

1. Microscopical examination of spleen pulp in normal, experimentally modified and pathological conditions. This morphological method is not well adapted for learning the mechanism of the process observed.

2. Study of the consequences of splenectomy. We do not propose to discuss all the experiments made in this direction. The most generally accepted result of this operation is the increased resistance of the red blood-corpuscles of the animal against hypotonic salt solutions, a fact which has mainly been settled by the Dutch investigator Pel Jr. [1911].

3. Study of the action of splenic extracts. The results of this method, which appears to us as a very unreliable and inaccurate one, are moreover very contradictory.

4. Comparative examination of blood from the splenic vein and splenic artery. The latter may of course be replaced by arterial blood in general.

Destructive action of the spleen being admitted, it was obvious to investigate traces of this action in the blood of the vena lienalis. There are three principal qualities of this blood, which may be examined:

(*a*) the number of erythrocytes in a cubic millimetre;
(*b*) the amount of decomposition products of red cells, especially haemoglobin and bilirubin;
(*c*) the resistance of red blood-cells to haemolytic agents.

The number of erythrocytes in a cubic millimetre of the blood of the splenic vein, as compared with blood from the splenic artery, is not a very exact measure of the function of the spleen. In the first place the difference which might be expected would be so small as to be within the limits of technical error. If the lifetime of a red cell is estimated at 20 days (Quincke, Bénard; Rubner consider it to be at least 70–90 days, Dekhuysen still longer) and if one supposes that the total quantity of blood of the body passes through the spleen about thirty times a day [Burton-Opitz, 1912], it is clear that of each volume of blood which passes through the organ only $1/20 \times 1/30 = 1/600$ is destroyed. So one might expect to find numbers like 6,000,000 and 5,990,000 for splenic artery and vein. This difference is too small to be determined with sufficient accuracy. Larger differences might be found, if this action of the spleen were intermittent, but there is no evidence that this is so.

Moreover the operation which is necessary for the puncture of the splenic vessels will cause a stagnation of the blood which has a great influence upon the number of erythrocytes.

Of late this method has been used again by Frey [1920]; his values for the difference between the blood from an artery of the abdominal wall and the splenic vein are so large (on an average 800,000 in a cubic millimeter) that in our opinion they must be due to some error in technique.

An increased quantity of bilirubin in the splenic vein, as compared with that in the blood from other vessels, might possibly be due to the formation of bilirubin in the spleen (Hymans van den Bergh and Snapper, Ernst and Szappanyos), it would not be conclusive proof of a haemolytic function of that organ.

The best method in our opinion is the investigation of the resistance of the red cells against hypotonic salt solutions, the so-called osmotic resistance. It is Eppinger[1920], who also stated that the osmotic resistance of the erythrocytes was decreased in the spleen, asserting that this fact is "eigentlich die einzig fassbare Tatsache die im Sinne einer zerstörenden Tätigkeit der Milz spricht, vorausgesetzt dass die Resistenzherabsetzung als ein einleitender Vorgang der Erythrozytenzerstörung aufgefasst werden kann."

French authors were the first to use the resistance determination in hypotonic salt solutions, originating from Hamburger [1904], to estimate the condition of the red blood-cells in several diseases. The value of this method however has been greatly increased by the modifications and extensions given by Brinkman [1922]. This author showed in the first place that the haemolysis in hypotonic NaCl solution is not only due to the difference of osmotic pressure, but also to the lyotropic effect of pure NaCl, the liquefaction of the cellular membrane-colloids by the Na-ions. Especially when the erythrocytes are washed several times with an isotonic NaCl solution, as is done by many authors, the cells are injured, as is shown by the diminished osmotic resistance (Snapper). This lyotropic effect may be abolished by the addition of Ca-ions to the solution in a very definite concentration. It would lead us too far to discuss the modified Ringer solution, which is used by Brinkman; we give here only the composition:

NaCl	...	...	0·7 %	
NaHCO₃	...	...	0·2 %	$[H^{\cdot}] = 0·45 \times 10^{-7}$
KCl	...	...	0·01 %	
CaCl₂6aq.	...	...	0·02 %	

Hypotonic solutions are obtained by diminishing only the concentration of NaCl.

· If the red blood-cells are washed with this isotonic solution, the result is an increase of the osmotic resistance. This is due, as has been shown by Brinkman, to the removal of an auto-haemolytic substance from the surface of the erythrocytes, a phosphatide, perhaps lecithin[1]. This same substance, if you make an emulsion of it in modified Ringer solution in an appropriate concentration, added to red blood-cells, may dissolve them.

In the surface of the blood-cells the haemolytic action of the phosphatide is balanced by the cholesterol, which is kept in solution by the first. The osmotic resistance has been shown to be dependent on the proportion cholesterol/phosphatide.

It must still be mentioned, that the modified Ringer liquid may be substituted by a mixture of primary and secondary phosphate, containing 16·30 g. Na_2HPO_4. aq. and 2·18 g. KH_2PO_4 per litre. The freezing-point of this liquid in which the erythrocytes have the same volume as in their plasma, is −0·46°. Without any theoretical consideration, the equivalence of this solution with the modified Ringer solution has been proved empirically. Its advantages are the easiness of its preparation and its stability, combined with the physiological properties of the modified Ringer solution. It was necessary to mention this liquid, because we used it in our resistance-determinations. For further particulars reference must be made to the publications of Brinkman.

After this digression we will discuss briefly the results obtained by others in the determination of the resistance of red blood-cells from the splenic vein in comparison with that of erythrocytes from arterial blood.

Gabbi [1893] finds like Hammarsten an increased resistance in the splenic vein.

Pugliese and Luzatti [1900] observed however a lowered resistance of the red blood-cells leaving the spleen.

Chalier and Charlet [1913] once more found that both minimum and maximum resistance of the blood-cells from the splenic vein (*i.e.* respectively the resistance of the weakest and of the strongest erythrocytes in regard to haemolysis by hypotonic salt solution) were higher than those of erythrocytes from the splenic artery. From their observations they conclude that the spleen only lets the strongest erythrocytes pass and so exerts a haemolytic action upon the blood. This conclusion would be more justified, if only the minimum resistance had increased, the weakest erythrocytes entering the spleen having been destroyed. How is the increase of the maximum resistance, however, from this point of view to be explained, *i.e.* the fact that the most resistant blood-corpuscles which enter the spleen, leave the organ still more resistant?

The remaining investigators, Strisower and Goldschmidt [1914], Widal and Abrami [cited by Eppinger, 1920], Banti and Furno [cited by Eppinger, 1920], Polak Daniëls and Hannema [1916], Eppinger [1920], all found a *diminished* resistance of the erythrocytes leaving the spleen. We have already quoted Eppinger's opinion on the importance of this fact.

We also have tried to study the haemolytic function of the spleen by investigating its influence upon the resistance of the red blood-cells passing through the organ. For practical reasons we took our materials from sheep at the abattoir[2]. Immediately after the killing of the animal, some of the blood spurting out of the carotid artery was collected in a glass tube and defibrinated by shaking it with glass beads. Then as soon as possible the

---

[1] It may be obtained by evaporating the washing fluid and extracting the phosphatide from the dry residue, or, by direct extraction of the erythrocytes with chloroform, after they have been extracted with light petroleum, in which cholesterol and neutral fats dissolve.

[2] We were able to do so by the kindness of the authorities of the abattoir in Groningen.

spleen was removed, care being taken not to injure the capsule and the splenic vessels. The organ was then rapidly conveyed to the physiological laboratory. By introducing into the splenic artery a glass canula, which was connected with a reservoir, containing the modified Ringer solution, placed about 175 cm. above the working table, it was possible to obtain from the splenic vein the blood which had remained in the organ, diluted with the salt solution. The suspension of erythrocytes obtained in this manner was centrifuged immediately. These red corpuscles, which had been for some time in the spleen, longer than they are normally, but too short a time for autolysis to take place, seemed to us very well adapted for the investigation of an eventual resistance-decreasing power of the spleen. By the dilution of the blood from the spleen with modified Ringer solution, the resistance of the erythrocytes is practically not changed. The effect of washing red corpuscles with this solution, which we have already described, fails practically to be evident, if the total blood is "washed" with the liquid. To produce the same conditions, however, we have also diluted the arterial blood from the carotis and centrifuged it to get the corpuscles.

Both kinds of the corpuscles were then put in a series of tubes containing the phosphate mixture in geometrically decreasing concentrations, viz.:

9/10 isotonic (obtained by adding one vol. aq. dest. to 9 vols. of an isotonic solution).

$(9/10)^2$ isotonic (obtained by adding again one vol. of aq. dest. to 9 vols. of the 9/10 isotonic solution).

Etc.

Ten minutes after the addition of the erythrocytes, the latter were centrifuged and the degree of haemolysis read by Arrhenius' method. In 20 determinations we always reached the same result, that is to say, the blood from the spleen was *less resistant* than that of the carotis. Both the beginning and the complete haemolysis of the erythrocytes of the spleen occurred in a salt solution of higher concentration than those of the blood-corpuscles of the carotis. In a hypotonic salt solution of a definite concentration the degree of haemolysis of the blood from the spleen was always larger than that of erythrocytes from the carotis.

We do not think that *in vivo* the difference of the resistance of red blood-corpuscles from art. and vena lienalis will be as large as in our determinations, because in the latter the erythrocytes had remained for some time in the spleen. Eppinger observed *e.g.* that if the blood stagnated for 10 minutes in the spleen the resistance of erythrocytes from the vena lienalis was distinctly lower than before the stagnation. There is, however, evidence that *in vivo* also stagnation in the organ is possible.

Table I contains the determinations:

Table I.

No.	Kind of blood	Isotonic	0·9	0·9²	0·9³	0·9⁴	0·9⁵	0·9⁶
1	Spleen	0	0	10	30	70	100	—
	Carotis	0	0	0	0	0	0	—
2	Spleen	0	0	0	20	50	90	—
	Carotis	0	0	0	0	0	0	—
3	Spleen	0	20	40	90	100	100	—
	Carotis	0	0 -	0	0	30	75	—
4	Spleen	0	5	10	50	70	100	—
	Carotis	0	0	0	0	5	10	—
5	Spleen	0	0	5	20	50	90	—
	Carotis	0	0	0	0	5	—	—
6	Spleen	0	0	0	50	90	95	100
	Carotis	—	—	—	0	5	60	90
7	Spleen	0	0	5	50	100	100	100
	Carotis	—	—	0	5	30	50	70
8	Spleen	—	—	10	30	50	100	100
	Carotis	—	—	0	10	50	80	100
9	Spleen	—	5	5	20	50	80	100
	Carotis	—	0	0	5	10	50	70
10	Spleen	—	0	5	10	40	60	90
	Carotis	—	0	0	0	5	30	90
11	Spleen	—	5	20	40	60	100	100
	Carotis	—	0	5	10	30	50	70
12	Spleen	—	0	0	10	50	70	100
	Carotis	—	0	0	0	10	50	70
13	Spleen	—	0	0	10	30	60	100
	Carotis	—	0	0	0	5	20	50
14	Spleen	—	0	5	10	40	80	100
	Carotis	—	0	0	5	30	50	70
15	Spleen	—	0	5	20	50	100	100
	Carotis	—	0	0	0	30	70	90
16	Spleen	—	0	5	10	30	70	100
	Carotis	—	0	0	0	30	70	90
17	Spleen	—	5	10	30	50	80	100
	Carotis	—	0	0	0	10	40	80
18	Spleen	—	5	15	30	80	100	100
	Carotis	—	0	0	10	30	70	100
19	Spleen	—	5	10	40	70	90	100
	Carotis	—	0	0	10	40	100	100
20	Spleen	—	5	20	50	100	100	100
	Carotis	—	0	0	10	60	100	100

If one calculates for every concentration the average degree of haemolysis one obtains the following table:

Table II.

Salt concentration	% haemolysis Carotis	% haemolysis Spleen
0·9 × isotonic	0	0
0·9² × ,,	0	5
0·9³ ×	0	30
0·9⁴ ×	20	60
0·9⁵ ×	50	85
0·9⁶ ×	75	100
0·9⁷ ×	100	100

One may consider this as an accidental case. Fig. 1 shows the "resistance-curves" of this case (the left one of the corpuscles of the spleen, the other of the carotis corpuscles):

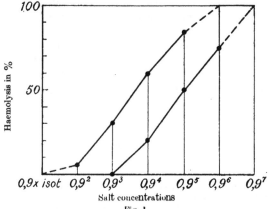

Fig. 1.

To deduce from the decreased osmotic resistance of the erythrocytes leaving the spleen conclusions as to the haemolytic function of the organ, it is necessary to investigate how far the phenomenon is specific. Of course it is not necessary that only the erythrocytes from the vena lienalis should have a decreased resistance; it might be also the case with blood-corpuscles leaving the liver and perhaps other organs in so far as these are concerned with blood destruction. We have contented ourselves for the present with showing that an organ like the kidney to which no haemolytic function is assigned fails to show a resistance-diminishing power. Table III shows our determinations:

Table III.

No.	Kind of blood	0·9 isotonic	0·9^2	0·9^3	0·9^4	0·9^5	0·9^6
			% haemolysis Salt concentrations				
1	Kidney	0	0	5	50	75	100
	Carotis	0	0	5	60	75	100
2	Kidney	0	5	10	50	80	100
	Carotis	0	5	10	30	50	70
3	Kidney	0	0	10	50	80	100
	Carotis	0	0	10	30	70	100

We have already mentioned the opinion of Chalier and Charlet and of Gabbi, that the spleen would exert a destroying action upon the weakest blood-corpuscles, so that the result is an increased resistance of the erythrocytes in the splenic vein, as was found by these authors, and we have observed that it is difficult to explain the increase of the maximum resistance from this point of view. From our researches, however, it would follow that the spleen exerts a haemolytic influence upon all blood-corpuscles, passing through the

organ, as is proved by the diminished minimum and maximum resistance of
erythrocytes from the vena lienalis. The parallel course of the curves in Fig. 1
gives a clear image of this fact. If one makes *e.g.* the assumption that the
spleen produces an internal secretion, which causes an alteration of all erythrocytes passing through the organ—and perhaps also has some influence upon
the bone marrow—the mode of action of the spleen becomes more intelligible.
We also point to the fact that after splenectomy a general increase of the resistance of the blood-corpuscles has been found, *i.e.* an increase of the minimum
and maximum resistance whilst the "breadth of resistance" (the difference
in concentration between the solutions in which beginning and complete
haemolysis is observed) remains the same. Pel gives *e.g.* the following table
as the average of a large number of experiments:

Table III

	Normal dog %	Splenect. dog %
Average concentration of NaCl, in which beginning · haemolysis is observed	0·42	0·35
Average concentration of NaCl, in which just com- plete haemolysis is observed	0·30	0·23
Breadth of resistance	0·12	0·12

This phenomenon would also be explained in the most simple way by the
assumption that after removal of the spleen the production of an internal
secretion ceases, that has normally a harmful action upon the erythrocytes,
which is measured by the diminished resistance.

We have tried to ascertain more particulars about the resistance-diminishing power of the spleen. Assuming the action of an internal secretion, it
was obvious that in the first place the surface layer of the erythrocytes would be
altered by such an action. From the researches of Brinkman and Miss van Dam
[1921] we know however the importance of the proportion cholesterol/phos-
phatide in the surface layer for the osmotic resistance. A decrease of this
proportion means a diminution in the osmotic resistance. To test the assumption that only or principally the surface layer of the erythrocytes is altered
by the action of the spleen it is only necessary to investigate the resistance of
the blood-corpuscles of the vena lienalis, when this surface layer has been
removed by washing the erythrocytes with modified Ringer solution. We have
made these determinations by washing the blood-corpuscles obtained in the
manner described on p. 757 and also the erythrocytes from the art. carotis
three times with the modified Ringer solution and then determining the resistance in the same way as described above. In all our determinations we
found a considerable difference between the two kinds of corpuscles, before
washing, which disappeared however almost entirely after this process.
Table IV contains the values of these determinations:

Table IV.

No.[1]	Kind of blood. Three times washed with modified Ringer	Isotonic	0·9	0·9²	0·9³	0·9⁴	0·9⁵	0·9⁶
1	Spleen	0	0	0	0	5	20	—
	Carotis	0	0	0	0	0	5	—
2	Spleen	0	0	0	0	20	50	—
	Carotis	0	0	0	0	0	0	5
3	Spleen	0	0	0	0	5	20	—
	Carotis	0	0	0	0	0	5	—
4	Spleen	—	0	5	10	30	50	100
	Carotis	—	0	0	10	30	50	100
5	Spleen	—	0	0	0	5	10	40
	Carotis	—	0	0	0	5	10	40
6	Spleen	—	0	0	10	40	80	100
	Carotis	—	0	0	10	40	80	100
7	Spleen	—	0	0	20	50	70	100
	Carotis	—	0	5	20	30	70	100
8	Spleen	—	0	0	10	30	60	90
	Carotis	—	0	0	20	30	50	90
9	Spleen	—	0	0	0	10	40	100
	Carotis	—	0	0	0	10	30	70
10	Spleen	—	0	0	10	40	80	100
	Carotis	—	0	0	20	50	80	100

Table V contains the average values which are represented also in Fig. 2.

Table V.

Salt concentration	% haemolysis Washed spleen Corpuscles	Washed carotis
0·9 × isotonic	0	0
0·9² × ,,	0	0
0·9³ × ,,	5	5
0·9⁴ × ,,	20	15
0·9⁵ × ,,	45	35
0·9⁶ × ,,	90	85

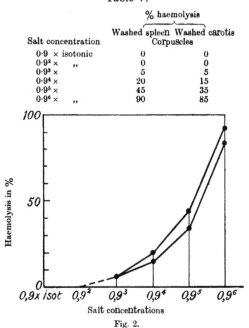

Fig. 2.

[1] The resistance of the unwashed corpuscles of these cases may be seen respectively in Nos. 1, 2, 3, 4, 11, 13, 14, 15, 16, 17, 18 of Table I.

The resistance of the arterial erythrocytes does not increase in the same degree as that of the blood-corpuscles of the spleen, so that after the washing the difference between the two kinds of blood-corpuscles has been reduced considerably, as is shown by Table V and Fig. 2.

There remains a small difference; this is perhaps partly due to the defibrinating of the blood from the art. carotis (the resistance is increased hereby somewhat); moreover arterial erythrocytes are always somewhat more resistant than blood-corpuscles from venous blood (Hamburger). The same difference exists between the resistance of erythrocytes from the vena renalis and art. carotis.

From our experiments we draw the conclusion that the influence exerted by the spleen upon the red blood-cells is confined to the adsorbed surface layer of the latter, *i.e.* to the layer which may be removed by washing with a modified Ringer solution. As we have not found any conception in the literature about the mode of alteration of the erythrocytes themselves in the spleen, except the following of Jacobsthal [1921], it seemed worth while to quote it. "Ich habe mir die Vorstellung gebildet, dass die Abnahme der Blutkörperchenresistenz beim haemolytischen Ikterus sich so entwickelt, dass die die Milz passierenden Blutscheiben jedesmal nicht haemolysiert, sondern *an ihrer Lipoidhülle sozusagen nur angenagt werden.*"

The three following alterations in the surface layer of the red blood-corpuscles in the spleen are possible:

1. Amount of phosphatide the same; amount of cholesterol decreased.

2. Amount of phosphatide increased; amount of cholesterol the same.

3. Amount of phosphatide and cholesterol both increased, the former however in the highest degree.

In all these cases the proportion cholesterol/phosphatide would diminish and this means a decrease of osmotic resistance. It seems that of these three possibilities the third one holds true. It has appeared to us that the erythrocytes leaving the spleen contain more cholesterol than the erythrocytes from the art. carotis[1]. It is therefore necessary that the phosphatide should undergo a considerable augmentation. We want to point out here already that this increase may be as well of a qualitative as of a quantitative nature. The specific influence of the phosphatide and cholesterol of the surface of the erythrocyte upon its resistance is undoubtedly due to the physico-chemical properties of these substances. If one makes an emulsion of lecithin *ex ovo* in an isotonic modified Ringer solution, this emulsion has a low surface tension and a great haemolytic power. If, however, cholesterol is added in increasing small amounts, the surface tension and the viscosity of this emulsion rise gradually and at the same time the haemolytic power decreases. The haemolytic power of an emulsion of lecithin is dependent on the *quantity* of the substance per cc. of the solvent, but apparently small *qualitative* alterations of the lecithin also have an enormous influence upon its haemolytic power.

[1] Publication of these investigations will follow.

This has been shown by Delezenne and Fourneau [1914], who withdrew oleïc acid from lecithin by means of cobra venom and in this way obtained a substance, called by them "désoléolécithine" or "lysocythine," which had a very strong haemolytic power. It is possible that *in vivo* also analogous chemical changes play a part in haemolytic processes. Perhaps it will be possible to settle the nature of the actual quantitative or qualitative changes in the phosphatides in the erythrocytes in the spleen and elsewhere by further researches.

To judge the value of our opinions, developed after investigations on the sheep's spleen, for normal and pathological physiology (Eppinger's Hypersplenie) of the human spleen, it will be necessary to repeat the determinations of the resistance described by us in man. This is possible where splenectomy is performed. Till now only one case has been investigated in this respect. For these figures we are indebted to Dr H. H. de Zoo de Jong, curator of the clinic of Prof. L. Polak Daniëls, University Hospital, Groningen. Splenectomy was performed in the case of a haemolytic anaemia. Some blood was taken from the vena lienalis and the red blood cells compared with erythrocytes from the peripheral blood in regard to their osmotic resistance before and after washing with a modified Ringer solution. Table VI contains the determinations:

Table VI.

		% haemolysis			
		Before washing		After washing	
Salt concentration		Peripheral	Spleen	Peripheral	Spleen
½ isotonic		20	30	0	0
0·9 ×	,,	35	65	40	40
0·9² ×		70	80	70	70
0·9³ ×		80	85	80	80
0·9⁴ ×	..	95	95	90	95
0·9⁵ ×	,,	100	100	100	100

The difference between peripheral blood and blood from the vena lienalis would perhaps have been more evident if the resistance had also been determined in solutions stronger than ½ × isotonic. Then the beginning of haemolysis would also have been determined.

For the rest it is clear that in this case the investigations on the sheep's spleen are wholly confirmed. It will, however, be necessary to repeat these determinations in regard to man if possible.

CONCLUSIONS.

The spleen has the power of *diminishing* the osmotic resistance of the erythrocytes. These are prepared hereby for haemolysis, which partially takes place in the organ itself.

Erythrocytes from the vena lienalis, which have been washed with an equilibrated salt solution, *do not show* this decreased osmotic resistance. This fact proves that the point of attack of the haemolytic power of the spleen lies in the removable surface layer of the erythrocytes. The alteration in this layer must consist in a decrease of the proportion cholesterol/phosphatide.

REFERENCES.

Brinkman (1922). Résistance osmotique et phosphatides du sang, Groningen.
Brinkman and Frl. E. v. Dam (1921). *Biochemische Zeitsch.* **108.**
Burton-Opitz (1912). *Pflüger's Archiv,* **146.**
Chalier and Charlet (1913). *J. Phys. Path. Gén.* **13.**
Delezenne and Fourneau (1914). *Bull. Soc. Chim.* (4), **15.**
Eppinger (1920). Die hepatolienalen Erkrankungen, Berlin.
Frey (1920). *Deutsch. Arch. klin. Med.* **133.**
Gabbi (1893). *Beiträge u. allgem. Path.*
Hamburger (1904). Osmotischer Druck und Ionenlehre, Wiesbaden.
Hunter (1892). *Lancet.*
Jacobsthal (1921). *Verhandl. deutschen path. Gesell.* **66.**
Koelliker (1854). Micr. Anatomie oder Gewebelehre des Menschen, Leipzig.
Pel Jr. (1911). Onderzoekingen by miltlooze dieren, Amsterdam.
Polak Daniëls and Hannema (1916). *Folia microbiologica* **4.**
Pugliese and Luzatti (1900). *Arch. Ital. Biologie.*
Strisower and Goldschmidt (1914). *Zeitsch. gesammt Med.* **14.**

LXXX. THE ENZYMES OF THE LATEX OF THE INDIAN POPPY (*PAPAVER SOMNIFERUM*).

By HAROLD EDWARD ANNETT.

Agricultural College, Cawnpore.

(*Received October 3rd, 1922.*)

THIS paper must be considered in the nature of a preliminary note. The work done has simply been qualitative and pressure of other work has prevented a more detailed treatment hitherto. In view of certain interesting results obtained and as it is possible the work may have to be discontinued it is deemed worth while to put it on record.

When poppy capsules are lanced for opium[1] the latex which immediately exudes varies considerably in colour from capsules in the same field and from the same pure race of poppy and even at times from capsules on the same plants. It may be pure white, smoky grey, light pink or deep pink in colour. It rapidly darkens to some shade of brown, either light chestnut or deep mahogany and at times it is almost black. When this opium is dried at air temperature and powdered the resulting powder may vary from a pale straw colour to almost black.

Moreover in a field sown with a single pure race of poppy, samples of opium collected on the same day at the same time under apparently identical conditions, on being subsequently air dried and powdered vary considerably in colour, *i.e.* from pale straw colour to almost black.

It would appear probable that the change is due to oxidation and one would expect to find powerful oxidising enzymes in the latex.

That the change is due to oxidation seems proved by the following experiments.

A. Some of the fresh latex diluted slightly with water to make it more liquid was drawn into the bulb of a Lunge nitrometer over mercury. The tap was turned off, thus leaving some of the latex exposed in the cup of the nitrometer. The remainder of the latex was effectually sealed up by the mercury out of contact with air. The latex exposed to air rapidly darkened. That sealed up after three months still shows no sign of darkening.

B. Portions of fresh latex were placed in small distillation flasks which were exhausted with a Geryk pump and sealed off. After three months these samples still show no sign of darkening whereas latex in similarly sealed flasks which were afterwards cracked rapidly darkened.

---

[1] For details of method of lancing see Annett [1921].

Latex kept in closed vessels over alkaline pyrogallol darkens because the complete absorption of oxygen by this reagent is a slow process unless the vessel is vigorously shaken and this is not feasible in the presence of the latex.

These preliminary experiments would appear to show that the darkening of the poppy latex is due to an oxidation process. It therefore appeared of interest to examine the latex for the presence of oxidising enzymes.

(1) Some fresh latex was shaken with water and divided into portions (a) and (b). The flask containing (a) was placed in boiling water for two minutes. (a) and (b) were then both filtered. Next day the filtrate from (b) was much darker than the filtrate from (a), thus indicating that the boiling had destroyed the oxidising enzyme.

(2) Latex mixed with water and fresh guaiacum tincture gave no blueing but when the test was repeated in presence of H_2O_2 a strong blue colour was immediately produced. It would therefore appear that a peroxidase is present in the latex. Spence [1908] working with rubber latex obtained a similar result but on dialysis with water for 24 hours he obtained both oxidase and peroxidase.

(3) Some fresh latex was immediately tested for oxidising enzymes with the following results:

(a) Latex and water + guaiacum tincture gave an immediate green colour but when the test was repeated in presence of H_2O_2 no reaction was obtained.

(b) A similar test with benzidine gave a deep colour immediately in absence of H_2O_2, but when the test was repeated in presence of H_2O_2 no reaction was obtained.

(c) A similar test with tyrosine suspension in absence of H_2O_2 gave darkening within 15 minutes and next day the liquid was very dark. On repeating the test in presence of H_2O_2 no darkening was obtained.

(d) Pyrogallol gave similar results, i.e. oxidation in absence of H_2O_2 but no reaction in its presence.

This result is directly opposed to that obtained under (2) above. The latex used under (3) was, however, used within 20 minutes of the lancing of the capsules while that used under (2) was probably two hours old. Some of the same latex used for (a), (b), (c) and (d) above tested about an hour later gave similar results except in the case of benzidine with which reagent a strong reaction was obtained both with and without H_2O_2.

In these cases controls were done with boiled liquids and all gave negative results.

This would indicate that the relative freshness of the two lots of latex used had nothing to do with the different results under (2) and (3).

(4) 10–15 g. of latex, collected for experiments in (3) above, were rubbed up in a mortar with distilled water and poured into a cleaned goat's stomach. This was allowed to diffuse in a vessel through which a slow stream of tap water was passed. The tap water was passed through thymol before reaching the diffusing vessel. Next day the walls of the goat's stomach were quite

black, indicating the presence of tyrosinase in the latex. 100 cc. of the dialysed liquor, the total volume of which was about 300 cc., were removed for enzyme tests. It was very faintly acid in reaction to litmus paper. 40 cc. of the liquid were boiled in a water-bath for five minutes. The table sets out the results of experiments then performed with the boiled and unboiled dialysed liquid. 3 cc. of the liquid were taken for the test in each case: "–" indicates no reaction, "+" a reaction, and "++" a very powerful reaction. Where H_2O_2 was used one drop of the 10 volume reagent was employed.

	Unboiled dialysed liquid		Boiled dialysed liquid	
Reagent	No H_2O_2	With H_2O_2	No H_2O_2	With H_2O_2
Guaiacum	+ +	–	–	–
Benzidine	+ +			
Pyrogallol	{ + + + next day	–		
Tyrosine	{ + + + next day	–		

After another two days more dialysed liquid was removed from the goat's stomach. It was then quite neutral to litmus. The above tests were repeated with exactly similar results except that the unboiled liquid gave a very faint reaction with benzidine in presence of H_2O_2.

Another portion of the dialysed liquid was filtered and tested as above. Almost identical results were obtained. That is to say reactions were obtained with guaiacum, benzidine and tyrosine in absence of H_2O_2. In the case of guaiacum and tyrosine no such reaction was obtained in presence of H_2O_2, but in the case of benzidine a very faint reaction was obtained in the presence of H_2O_2.

(5) Some latex was collected and tested within 10 minutes of lancing with guaiacum and a deep green colour was immediately obtained. The test when repeated in presence of H_2O_2 gave an even more intense reaction. This is again a contradictory result.

The rest of this latex was shaken with water and divided into two portions A and B. A was rendered faintly alkaline with $NaHCO_3$ and one-half of it, A', was boiled in a water-bath for five minutes. One-half of B was boiled similarly and is called B'. A, A', B and B' were then tested with guaiacum, benzidine, and tyrosine with and without addition of H_2O_2.

The boiled solutions gave no reaction in any case. The reaction with guaiacum in case of both A and B solutions was very doubtful probably owing to the latex having been too largely diluted.

With benzidine both in the case of A and B a very strong reaction was obtained in absence of H_2O_2, the reaction being distinctly stronger in the alkaline solution. In the presence of H_2O_2 a faint but positive reaction was obtained with both solutions A and B but the reaction was much weaker than in absence of H_2O_2.

With tyrosine a practically similar result was observed namely a strong reaction in absence of H_2O_2 in both solutions A and B, but it was not more marked in the alkaline solution. In presence of H_2O_2 a very faint but positive reaction was obtained in the case of both A and B solutions.

(6) Some of the latex collected for tests under section (3) was immediately after collection rubbed up with water and poured into a tall cylinder. Toluene was poured on the surface and the cylinder allowed to stand undisturbed. The insoluble portions of the latex settled out at the bottom of the cylinder. It will be remembered that when tested under section (3) a strong reaction was given with both guaiacum and benzidine in absence of H_2O_2 but there was no reaction in presence of H_2O_2.

After the cylinder had stood for eight days some of the liquid was drawn up from the bottom with a pipette. It contained pieces of the insoluble matter of the latex. This liquid reacted with guaiacum both in presence and absence of H_2O_2 the reaction being slightly stronger in the latter case. It reacted with benzidine in absence of H_2O_2 but not when H_2O_2 was present.

(7) It appeared of interest to test old powdered opium for oxidising enzymes.

Some opium powder two years old was rubbed up with water and dialysed. The dialysed liquid was then tested with guaiacum and benzidine in presence and absence of H_2O_2. Guaiacum gave no reaction in either case; benzidine reacted strongly in absence of H_2O_2 but in presence of H_2O_2 there was no reaction.

Tests for other enzymes in the latex.

We have been unable to find any reference to work on the enzyme content of poppy latex, with the exception of a statement that it contains only a small amount of protease [Czapek].

Fifty capsules were lanced and the latex which exuded was immediately scraped off. It was macerated with sand and 100 cc. water. After straining through coarse cloth the filtrate was divided into two portions A and B. The latter was boiled. The usual tests for enzymes were then carried out on the following substances: 1 % starch solution, 1 % cane sugar solution, 1 % α-methyl glucoside solution, 1 % urea solution and white flour extract (10 %). The result of these tests indicated the complete absence of amylase, invertase, maltase and urease. The test for protease indicated the possible presence of protease in small quantity as a very faint positive result was obtained.

Emulsin was tested for with amygdalin as in Armstrong and Horton's test, and a negative result obtained.

There was one interesting observation we made in all these tests, namely that the liquids in the tests with unboiled latex all darkened distinctly with reference to the control experiments carried out with boiled latex. This is another indication of the presence of oxidising enzymes.

Conclusions.

1. The darkening of the latex of the Indian Opium Poppy is a process of oxidation.

2. The latex has a powerful oxidising action on guaiacum tincture, pyrogallol, benzidine and tyrosine in the absence of H_2O_2. In almost all the experiments tried it was found that H_2O_2 inhibited these reactions.

Chandat and Staub [1907] found that hydrogen peroxide retarded the action of tyrosinase.

Bach [1906] and also von Fürth and Jerusalem [1907] have found that though hydrogen peroxide in certain amounts retards the action of tyrosinase, yet minute quantities accelerate the action of the enzyme.

3. The dialysed latex both before and after filtration also oxidises guaiacum tincture, pyrogallol, benzidine and tyrosine in absence of H_2O_2 but the reaction is inhibited in the presence of this reagent.

4. The actions on benzidine and tyrosine were particularly powerful. In the case of tyrosine the darkening always appeared as a surface film on the liquid and gradually diffused downwards.

5. Opium powder stored for three years has been shown to possess an oxidising enzyme which acts on benzidine.

The author has recently shown that dry opium powder loses a considerable proportion of its morphine on storage. This may be due to the action of oxidising enzymes.

6. The following enzymes were tested for in the latex and not found, viz. amylase, invertase, maltase, emulsin and urease. There were indications of a weak proteolytic activity.

REFERENCES.

Annett (1921). *Mem. Dept. Agric. India*, **6**, Nos. 1 and 2.
Bach (1906). *Ber.* **39**, 2126.
Chaudat and Staub (1907). *Arch. Sci. Phys. Nat.* **24**, 172–191 (*vide* "The Oxidases," J. H. Kastle, p. 85).
Czapek. Biochemie der Pflanzen, **3**, 716.
von Fürth and Jerusalem (1907). *Beiträge*, **10**, 131.
Spence (1908). *Biochem. J.* **3**, 165, 351.

LXXXI. THE BLOOD OF EQUINES.

By CHRISTIAN PETRUS NESER.

Preliminary Communication[1] *from the Division of Veterinary Education and Research, Onderstepoort, Union of South Africa.*

(*Received October 3rd, 1922.*)

Technique. Without offering details of technique it may be briefly stated that blood, drawn from the jugular vein except where otherwise mentioned, was collected in 10 cc. bottles containing a measured adequate amount of sodium citrate solution. From this were determined, (*a*) percentage volume of erythrocytes by centrifuging in specially prepared uniform tubes of 2 cc. capacity, blown out slightly at the sealed end to facilitate rapid sedimentation; (*b*) corpuscle count after diluting 1 in 200 with Hayem's fluid for erythrocytes, and 1 in 10 with 0·5 % acetic acid for leucocytes, the Bürker chamber being used in both cases; (*c*) haemoglobin content by Sahli's method.

Smears were made by Craandyk's [1918] method and also by a modification of Ehrlich's method which is so simple and effective as to merit brief description. A small drop of blood is quickly transferred to a slide by means of a suitable platinum loop, and a cover slip, as broad as the slide, immediately lowered over it. The droplet at once spreads as a thin circular layer, and on drawing the slip lengthwise over the slide all the blood is left upon the latter as a thin film in which the distribution of leucocytes is remarkably uniform, and eminently suitable for differential count after Giemsa staining in the usual way.

Limits of error. From numerous observations expressly designed to determine the degree of accuracy of methods in conventional use, it was concluded that (*a*) the centrifuge gives very accurate and consistent data, (*b*) red counts are liable to an error of up to 10 % even with the improved Bürker chamber, (*c*) white counts are less liable to error, (*d*) the Sahli reading cannot be relied upon to within five scale units, (*e*) smears made by the described modification of Ehrlich's method show a very regular distribution of leucocytes.

THE RED CORPUSCLES.

Influence of work. Comparative study of the blood of different horses showed a remarkable difference in percentage volume of erythrocytes from individual to individual, but a fairly constant figure for any given individual over short intervals of time. The outstanding difference between the blood of different horses was at first very puzzling. Since all were fed alike, diet

---

[1] The full paper, giving detailed protocols, will appear in the forthcoming "Report of the Director of Veterinary Education and Research" (P.O. Box 593, Pretoria).

was not a factor. Age was excluded by statistical comparison of data. That sex played no part was evidenced by similar variation in mares and geldings. Since all the animals under observation had been purchased for the horse-sickness experiments of the institution, they were a mixed lot whose histories were generally unknown, and no mere inspection sufficed to explain the observed variations.

The clue, however, was given by two animals (laboratory Nos. 11144 and 11775) one of which was permanently stabled and showed a percentage volume of 23, while the other was used as a saddle horse and showed a figure of approximately 40. This remarkable difference at once suggested that *work* was the deciding factor in determining the percentage volume of erythrocytes in horse blood. Following up this clue, horses were grouped according to the work performed, as "fast working," "permanently stabled," and "other horses." It was at once noted, *without exception*, that the blood of fast-working horses showed a high percentage volume of erythrocytes, or high red count, while in horses stabled for six months or more the figures were correspondingly low. With the third group the data were variable, but it is probable that if the previous histories of these miscellaneous animals had been known, many of them would have been classified in one of the other groups. At a later date it was found possible to procure the blood of a few race-horses in training, and their high volume of red corpuscles bore out the earlier conclusions in most striking fashion. Graph I represents a summary of the data from over 200 animals from all sources, Graph II the data from three race-horses in various stages of training, while Graph III illustrates the change actually occurring during hard training of three young race-horses.

These data leave no doubt that hard fast work brings about a marked increase in the percentage volume of red blood corpuscles in the horse. The contrast shown in Graph II, for three race-horses from the same stable, is particularly striking, the figure for the fully trained animal showing that 52 %, or more than half the total volume of blood, may be made up of erythrocytes. This corresponds to a count of approximately 12 million as against an average count of about 8 million for the ordinary slow working horse, and less than 6 million for permanently stabled laboratory horses.

These already conclusive observations are further clinched by reference to Graph III, which shows a steady increase in numbers of red corpuscles during the actual course of hard training. In five weeks the count increased from less than 7 million to over 9 million in one case, and from 6·3 million to 8·8 million in a second case. In the third case training was interrupted owing to injury about the third week, the rapidly rising count then remaining stationary. Of special interest is the further fact that although the high count follows so rapidly in the wake of hard training, the reverse effect, or diminution of red blood corpuscles with rest, is a slow process. Thus in two young horses a decrease of two millions occurred only after three months, while in several old animals the decrease after five months was hardly noticeable at all.

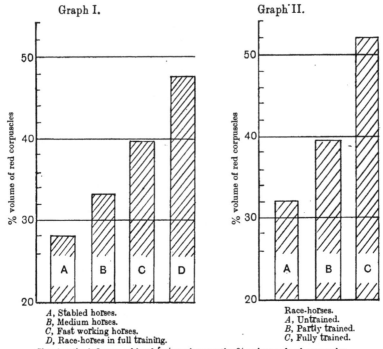

Graph I.

Graph II.

A, Stabled horses.
B, Medium horses.
C, Fast working horses.
D, Race-horses in full training.

Race-horses.
A, Untrained.
B, Partly trained.
C, Fully trained.

Showing the influence of hard fast work upon the % volume of red corpuscles.

Graph III.

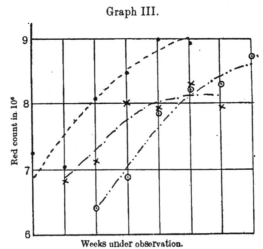

Weeks under observation.
Showing the influence of time upon the red count in three
young race-horses, during hard training.

Relationship between volume, count, and measurement of erythrocytes. In numerous cases, the percentage volume as determined by the centrifuge, the actual count in the Bürker chamber, and the diameter of the cells as measured in smears, were compared for the same horses.

The *diameter* of individual corpuscles varied between 4μ and 8μ: but, with one exception, the average of 200 cells remained fairly constant at 5.5μ. In respect, therefore, to average size of erythrocytes, different horses are very uniform.

The *ratio* between percentage volume and numerical count of red blood corpuscles varied between $4k$ and $4.7k$, with an average of $4.35k$—k being a constant dependent upon the units of measurement used. From the relatively constant average ratio, the observed liability of the red count to an error of 10 %, and the observed constancy of centrifugal readings, it may be concluded that the variations observed are mainly due to errors in counting and not to real variations in the average size of the corpuscles. These considerations led to the conclusion that a count calculated from percentage volume on an assumed ratio of $4.35k$ is even more correct than an actual count; at least for the blood of healthy horses. It is quite certain that the percentage volume and the red count give the same information, and that the observed increase of the former with work is due to a real increase in the number of normal red cells per unit volume of blood; normal also in respect to haemoglobin content as determined by the Sahli method. This increase in erythrocytes is probably absolute, and a genuine response of the blood-forming tissues to the increased demands made upon the oxygen-carrying capacity of the blood.

Diurnal variations. Quite remarkable differences in percentage volume of erythrocytes were observed from day to day, and even from hour to hour. Numerous experiments in which horses were worked, fasted and then fed, allowed to go thirsty and then watered, all failed to account for the variations observed in jugular blood. The data obtained indicated that any change in the concentration of the blood, arising from these factors, is at most very transient. The real explanation of the variations, however, was at once found on tapping at different points in the circulation. In comparing blood from the jugular vein with that from the ear, it was found, with very few exceptions, that the latter gave a higher red count than the former, and was also subject to far greater variation. The difference between ear blood and jugular blood was greatest in animals feeding at rest, and least in animals excited or at work; an increase in one count being generally associated with a decrease in the other.

These observations indicated the mechanical state of the circulation as the important factor in determining distribution of erythrocytes throughout the body, and the few simple experiments recorded in Graph IV provided direct evidence of its operation.

Graph IV.

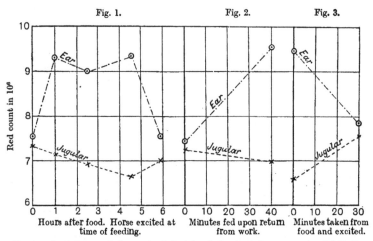

Fig. 1.   Fig. 2.   Fig. 3.

Hours after food. Horse excited at time of feeding.   Minutes fed upon return from work.   Minutes taken from food and excited.

Showing the influence of the mechanical state of the circulation, upon the distribution of the red cells.

Fig. 1 of Graph IV represents the changes in red corpuscle count in an animal excited just prior to feeding; Fig. 2 those of an animal rested and fed immediately upon return from work; Fig. 3 those of an animal taken from the manger and threatened for 30 minutes by exhibition of a familiar whip. In all these cases the counts on peripheral and systemic blood converge most closely when the circulation is most active, and disagree most widely when the circulation is more sluggish. A less active circulation thus results in a concentration of the red corpuscles in the peripheral capillaries, possibly owing to relative increased lymph formation and possibly owing to sedimentation.

On many occasions it was also noticed that both ear and jugular red counts had increased or decreased together; observations which may be explained by assuming concentration or dilution in some other part of the circulation.

The general results, however, show that in horses the jugular blood is far more representative of the average circulation than is blood taken by ear puncture; that factors beyond the control of the observer may give very misleading results when ear blood is studied; but that even jugular blood is not free from variation.

Pregnant and nursing mares. The results obtained from a few animals of this class may be summarised by stating that no influence of gestation or lactation, as such, could be established. The red corpuscle counts varied greatly between individual mares, being always highest in those with a past history of active life, and such differences obscured possible variations centring around sex.

Foals. In foals the observed red count was very high just after birth, but in two months had decreased considerably. Two cases will serve to show that the most rapid fall occurs in stabled foals at rest:

	Percentage volume of red corpuscles	
Foal	At birth	After 2 months
(a) Permanently stabled	48	30
(b) Running with mother	42	36

Again the influence of physical activity is manifest.

Donkeys and Mules. The red cell picture of donkey and mule blood differs in an interesting way from that of the horse. The average diameter of the erythrocytes is approximately 6.2μ for both donkey and mule as against 5.5μ for the horse. For the donkey the ratio of percentage volume and numerical count was generally found greater than $5k$, while for mules it approximated $4.35k$, the figure for the horse. The inference is that the corpuscles of the mule are thinner than those of donkey and horse.

THE LEUCOCYTES.

The classification of the leucocytes adopted in this work is that which is generally accepted to-day, as fully discussed by du Toit [1917] in his article on bovine blood. Equine leucocytes are not unlike those of bovines, as studied in stained smears, with the exception of the eosinophiles and basophiles. The granules of the latter cells are very large and form a characteristic feature of equine blood.

The *number* of leucocytes in the blood of healthy horses is very variable, not only as between individuals, but also in the same animal from time to time. The general range is from 5 to 20 thousand per cubic millimetre. In any individual animal the hourly and daily variations are not great, although exceptions are not uncommon.

Several experiments were undertaken to ascertain the influence of ordinary non-pathological factors upon the white count. Results may be summarised as follows:

(a) Moderate variations in supply of food and water showed no definite effects, and the jugular blood certainly reflected no evidence of any "digestive leucocytosis."

(b) Exercise always increased the white count of jugular blood, in some cases by over three thousand per cmm. On cessation of exercise the numbers often decrease again very rapidly, even to a figure below the pre-exercise count.

(c) Simultaneous observations upon ear and jugular blood yielded no conclusive results, except with exercise. In this case the results obtained indicate that the distribution of leucocytes also depends largely upon the mechanical state of the circulation.

To ascertain the extent to which technique of sampling influenced the white count, variations were made upon the usual procedure of first slapping the ear and collecting from a large puncture. A small puncture was made and blood obtained by squeezing at the base and at the apex. The following data represent one case:

	Leucocyte count
Ear, ordinary technique	$11 \cdot 3 \times 10^3$
Ear, squeezed at base	$7 \cdot 4 \times 10^3$
Ear, squeezed at apex	$14 \cdot 3 \times 10^3$
Jugular blood for comparison	$12 \cdot 4 \times 10^3$

From these marked variations it may be concluded that the white cells cling to the walls of the capillaries and the wound, especially if the blood flow is slow. When the circulation is rapid, or when the blood is forced out under considerable pressure, the leucocytes become detached, or are prevented from clinging, with consequent increase in apparent numbers. This explains why the ear white count is usually lower than the jugular.

Differential Counts. Differential leucocyte counts, obtained by the useful method (modified Ehrlich) already described, showed the following general figures:

Differential count of equine leucocytes	Lympho-cytes	Mono-cytes	Neutro-philes	Eosino-philes	Baso-philes
Horses, average of 7 each:					
(a) Stabled	36	4	54	5	1
(b) Medium	39	4	52	4	1
(c) Fast working	40	5	50	4	1
(d) General average	38	4	53	4	1
Normal variations	45–30	8–2	60–45	9–3	3–0
Extreme variations in 200 cases	50–25	9–0	62–54	15–1	3–0
Donkeys, average	53	4	34	8	1
Mules:					
(a) Clinically healthy	41	4	49	6	1
(b) Slight injuries	32	3	61	4	—
(c) Inoc. earlier against anthrax	52	3	39	6	1

The most interesting feature of this comparative table is the curious reversal of the figures for lymphocytes and neutrophiles, as between horses and donkeys. The horse shows lymphocytes 38 and neutrophiles 53 as general average, while the donkey shows neutrophiles 34 and lymphocytes 53. The count for mules approximates that of horses, when clinically healthy animals are considered. The slightly higher neutrophile count in group (b) mules may perhaps be accounted for by the slight injuries from which they were suffering. The high lymphocyte count in group (c) mules may perhaps be due to the fact that these animals had been inoculated against anthrax some months earlier.

Breeding mares and foals.

To these data may be added a few observations upon mares and foals. Taken just before or after parturition, the mares showed a very high neutrophile count, even over 70 %. An even higher neutrophile count, about 80 %, was characteristic of foals shortly after birth. After about two months this neutrophilia had disappeared, provided the animals remained healthy.

Influence of other factors.

Various minor deviations from the usual mode of life did not appear to influence the differential count in any regular way; except for the eosinophiles, which were generally increased by water-drinking after a period of thirst or fasting. Graph V offers a typical example of transient eosinophilia following large consumption of water after dry feeding.

Graph V.

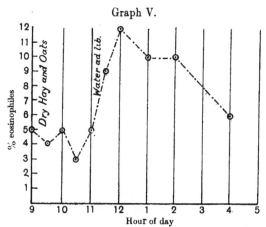

Showing the influence of food and water upon the % of eosinophiles in one horse.

GENERAL DISCUSSION.

It is of interest to compare the findings recorded in this paper with the limited data already available in the same direction. References to healthy equine blood in the general haematological literature appear to be very scanty, and it is to be regretted that South African library facilities do not always allow of consultation of such references as can be tracked by abstract. Burnett quotes various authors as obtaining erythrocyte counts ranging from 6·3 to 8·5 million, and leucocyte counts varying between 5·6 and 15 thousand, for adult horses. The white count is approximately that recorded in this paper, but the red counts quoted by Burnett suggest that other workers have contented themselves too readily with animals of one class. The data now recorded show a much wider range, considerably below 6·3 million for permanently stabled horses and even up to 12 million for race-horses in hard training.

Frei [1909], working in this laboratory, compared the centrifuge and the counting chamber, obtaining 4·7k for the ratio between percentage volume and numerical count, and preferring the volumetric method for determining the proportion of erythrocytes. His ratio is thus about 8 % higher than the one now offered on a more extensive series of observations. The horses available to Frei belonged to the same miscellaneous class as our own original lot, and showed variations from⁻22 to 43 for corpuscular volume. The astonishing thing is that no serious attempt was made to elucidate the nature of these differences, and the extraordinary effect of regular work seems to have wholly escaped his notice.

In regard to corpuscular dimensions, some authors give the average size of erythrocytes as $5 \cdot 5 \mu$, but others give figures as high as $5 \cdot 8 \mu$. The former figure is the average now established upon several hundred horses.

The differential counts quoted by Burnett for horses show higher neutrophile and lower lymphocyte percentages than those recorded in this paper. Stephens and co-workers [1921] have called attention to the variation in distribution of leucocytes in different parts of smears made in the ordinary way, and since our own technique has overcome this difficulty, the data now recorded would seem the more reliable.

By far the most striking observations recorded by us concern the remarkable influence of regular exercise upon the proportion of erythrocytes in horse blood. The available recent works upon horse blood all deal with pathological conditions, and nowhere does any reference seem to be made to the fundamental influence of normal work. In the various standard text-books upon Physiology and Haematology, the same omission is conspicuous. Dealing, as they do, largely with the human subject, in which normal variations are less noticeable, no stress is laid upon this factor, which plays so remarkable a rôle in the physiology of the horse. It is generally accepted that slight variations occur with mode of life, but whether or not these are specifically related to exercise is hardly discussed, although factors such as "altitude" are almost unduly stressed—perhaps because immediately detectable. Some workers on human blood have found that "robust" people have a higher, and "obese" people a lower, red count. The numbers are also lower for women, and it has been recorded that school children show a higher count at the end of a long vacation, than at the beginning. It is true that the variations generally recorded in healthy human individuals are registered in hundreds of thousands as against the tremendous variation of 4 million to 12 million between the stabled laboratory animal and the fully trained race-horse, but is it not possible that the minor differences in the human are due to differences in the amount of exercise rather than to other factors?

For the horse, the facts now recorded are perhaps not so surprising when considered in the light of its evolutionary history. The ancestor of the modern horse depended upon speed and endurance for its very existence; two qualities conditioned by a highly efficient circulatory system, and the demand for a high oxygen-carrying capacity of the blood. Is it then astonishing that the modern horse is born with a potential capacity for high speed and great endurance, and that during hard training his store of haemoglobin should be increased as fast as his muscles develop, so that finally more than half the total blood volume of a "finished race-horse" consists of erythrocytes?

In how far does this phenomenon of increase of red corpuscles with exercise, so strikingly shown here for the horse, apply to other animals? The problem at least seems a promising one for future investigation.

REFERENCES.

Burnett. Clinical Pathology of the Blood of Animals (Taylor and Carpenter, Ithaka).
Craandyk (1918). *Folia Haematologica,* **23,** Heft 2.
Frei, W. (1909). Physical Chemical Investigations into South African diseases: *Transvaal Department of Agriculture. Report of the Government Veterinary Bacteriologist for the year* 1907–8.
Stephens, J. W. W., Yorke, W., Blacklock, B., Macfie, J. W. S., Cooper, C. Forster, and Carter, Henry F. (1921). *Annals of Trop. Med. and Parasit.*
du Toit, P. J. (1917). *Archiv wissenschaft. prak. Tierheilkunde,* **43,** Heft 2 and 3.

LXXXII. THE FOOD VALUE OF MANGOLDS AND THE EFFECTS OF DEFICIENCY OF VITAMIN A ON GUINEA-PIGS.

By ELLEN BOOCK and JOHN TREVAN.

From the Wellcome Physiological Research Laboratories.

(*Received October 11th, 1922.*)

OUR colleagues Glenny and Allen [1921] presented to the Pathological Section of the Royal Society of Medicine an account of an investigation into an epizootic amongst a stock of guinea-pigs. They presented conclusive evidence that the epizootic could be entirely controlled by alteration of the diet. The diet which resulted in the outbreak consisted of bran, oats, water and mangolds. Substitution of the mangolds by cabbage, grass or lucerne stopped the epizootic. When guinea-pigs in adjacent runs were fed on the two diets, those on mangolds were attacked, whilst those with grass remained healthy, although no precautions whatever were taken to prevent infection of the healthy animals. Isolation was not carried out, the runs were not disinfected, and the attendants handled both groups of animals indiscriminately. These results are of such importance that the following attempt was made to determine the factors or deficiencies in the mangold which rendered the guinea-pigs sensitive to the epizootic. A few experiments were done by Glenny in which an alcoholic extract of carrot was given to the guinea-pigs with a view to supplying a possible deficiency of vitamin A. These indicated that the supply of vitamin A, although it reduced the incidence of the disease, did not stop it, and the guinea-pigs did not put on weight.

The following experiments were begun after the epizootic had been controlled by dietary means, and the effect of the diet was re-investigated.

Chart I shows the weight curves of four guinea-pigs fed on unlimited bran, oats, mangolds and water, the animals eating about 40 g. of mangold each per diem, and Chart II those of guinea-pigs kept on a control diet of bran, oats, autoclaved milk and greenstuff. These latter gained weight rather faster than Miss Hume's [1921] "standard" guinea-pigs (43·7 % in 25 days, as against 34·2 % in Miss Hume's experiments).

It was found that all the guinea-pigs placed on the mangold diet eventually succumbed within a period of about two months. In Chart I, the guinea-pigs were about 200 g. weight when started. In Chart III, they were younger—about 150 g.—and the younger guinea-pigs succumbed earlier than the older ones.

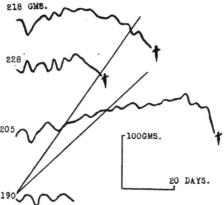

INTUSSUSCEPTION. KILLED.

The charts are drawn to different scales indicated in each case. The oblique straight lines on each chart represent the greatest and least rate of growth shown in the experiment with a normal diet represented in Chart II. The abscissa for each curve is at a different level; the figure at the beginning of each curve shows the initial weight of the guinea-pig in grams.

Chart I. Effect of basal diet of bran, oats, and mangolds. One of the animals had an intussusception and was killed. We have met this condition fairly frequently in animals under 200 g. in weight which have been deprived of greenstuff.

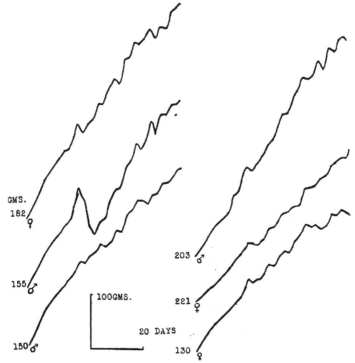

Chart II. "Normal" diet consisting of bran, oats, lucerne, grass and autoclaved milk. This diet was used as a control for some experiments on scurvy. We have obtained since writing this paper even better results than these, with the "synthetic" diet described later.

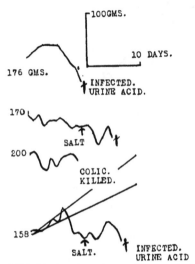

Chart *III*. Effect of basal diet of bran, oats, and mangolds on younger guinea-pigs. It will be seen that these animals died sooner than those in experiment 1.

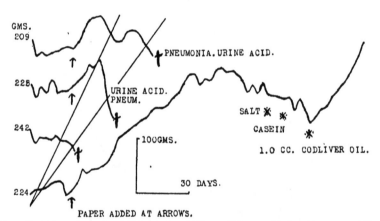

PAPER ADDED AT ARROWS.

Chart *IV*. Guinea-pigs on basal diet as in I and III, to which was added packing paper at the arrows. One of the animals died the same day, the others started to grow again, one growing for over two months before decline set in. As the result of experiments which had been going on in the meanwhile, salt mixture (calcium lactate and sodium chloride), caseinogen and cod-liver oil were added to the diet (see further charts). It will be seen that the limiting factor in this guinea-pig was the absence of something supplied by cod-liver oil—presumably vitamin A.

It was noticed whilst weighing the animals on the deficient diet, that they eagerly nibbled at any pieces of paper or cardboard which they could reach; so clean white packing paper was placed in the cages. The guinea-pigs each consumed about ·5 g. of paper every day. Chart IV indicates the effect on the growth, and it will be seen that out of four guinea-pigs, one died on the first day that the paper was given, but the other three gained in weight as a result, and one of the three continued to gain for over two months before a decline set in. Suggestions are put forward below as to the explanation of this.

Post-mortem examinations revealed several abnormalities in the guinea-pigs whilst on the mangold diet.

As in Glenny's experiments, there was constantly some sign of an infection present, either

(a) Alkaline stomach contents.

(b) Pneumonic consolidation or abscess of the lung.

(c) Purulent pericarditis.

(d) Enteritis.

When the urine was tested during life and post-mortem, it was always found to be strongly acid, and therefore quite free from the normal precipitate of phosphates which renders the urine of the healthy guinea-pig turbid. This observation led to the suggestion that the mangold, bran and oats diet is deficient in certain basic constituents, and following the indication of McCollum's feeding experiments on rats [1920] (from which he concludes that roots and seeds, besides other deficiencies, lack the three inorganic elements calcium, sodium and chlorine), it was decided to add a salt mixture to the bran, oats and mangold diet. A mixture of calcium lactate and sodium chloride was made and mixed as thoroughly as possible with the bran and oats, about 1·5 g. of calcium lactate and 1·0 g. of sodium chloride per guinea-pig per diem being given. This resulted in a return of the urine to the normal alkalinity, when examined during life and post-mortem, and improved the growth curves. (First part of Chart V.)

In connection with this, we made an examination of the ash, by drying mangold at 100°, and then incinerating in a muffle furnace. Magnesium, sodium, potassium and iron were found to be present, but no qualitative test for calcium could be obtained, and in attempting to estimate calcium by Cahen and Hurtley's method [1916], no potassium permanganate was used up in the final titration.

The beneficial effect of the salt mixture is therefore probably due to its supplying the extra calcium. The packing paper also helps in the same direction, for analyses of the ashed paper showed that there is a certain amount of calcium present:

Percentage of ash in paper ...	... 1·7 %	1st determination.
,, ,, ,, ,, ...	... 1·6 %	2nd	,,
,, ,, calcium in paper	... 0·126 %	1st	,,
,, ,, ,, ,,	... 0·190 %	2nd	,,

Using Sherman and Gettler's tables [1912], we have calculated that the calcium supplied by the 5 g. of paper eaten per guinea-pig per diem would be about equal to that supplied by 6 cc. of cow's milk per diem, and is about equal to half the calcium in 40 g. of cabbage (Miss Hume's standard amount).

Besides supplying this small amount of calcium, it is possible that the paper may act as "roughage" in the diet, the alimentary tract of the guinea-pig being only suited to an extremely bulky diet, and requiring a comparatively large amount of ballast in order to secure normal intestinal movements. It is, however, very difficult to believe that a guinea-pig devouring 40 g. of mangold per diem is in need of any further cellulose, and we are inclined to the opinion that the paper acts chiefly by partially supplying the calcium deficiency. Mangold, though deficient in calcium, contains about the same percentage of potassium as cabbage.

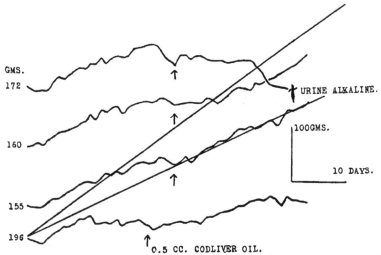

Chart V. Guinea-pigs on basal diet as in I and III, plus salt mixture and paper. 0·5 cc. cod-liver oil per guinea-pig per diem administered at the point marked by the arrows. One guinea-pig only then approached a normal rate of growth, the deficiency of protein probably being the limiting factor in the case of the other three guinea-pigs, one of which ultimately succumbed.

An important point in connection with the first part of Chart V arises in consideration of the changes in the acidity of the urine.

On the mangold diet, the urine is acid. On a greenstuff diet, the urine is alkaline. 40 g. of cabbage, according to Miss Hume's results, is sufficient for normal growth in guinea-pigs. If, however, a comparison is made between the ash of mangold and cabbage, it will be found that whereas 40 g. of ashed mangold give sufficient base to neutralise 35·2 cc. $N/10$ acid (this is obtained from a direct titration of the ash—is probably therefore too high because of the loss of sulphur, but the error is only of the order of 2 or 3 cc.), 40 g. of ashed cabbage will neutralise 42·6 cc. $N/10$ acid (Sherman and Gettler, calculated), not a significant difference. There are two explanations that offer themselves:

(1) The potassium salts of the mangold are not absorbed by the intestinal mucous membrane in the absence of calcium salts to balance their toxic effects. It is a commonplace that potassium poisoning is very difficult or impossible to produce by oral administration of potassium salts. As a result, the acid-base balance is disturbed—the animal's only source of salts is the bran and oats, the ash of which is acid.

(2) The deficiency of calcium may lead to a disturbance of the metabolism, possibly secondarily to an infective process, which leads to excessive breakdown of body protein, and the consequent production of abnormal amounts of sulphuric and phosphoric acids in excess of the neutralising value of the food salts absorbed.

The experiments so far done are in favour of the first explanation, for animals that died as a result of other deficiencies in diet, but with a good supply of calcium lactate, mostly had an alkaline urine, whereas animals fed on the diet without mangold, and with no other source of vitamin C, although supplied with calcium lactate, developed an acid urine.

This is only one of many indications of the need for further investigation of the physiological nature of the disturbances of the animal economy by a deficient diet, and is one which we are further examining.

In Chart V, four guinea-pigs were started on an initial diet of mangolds, bran and oats, salt mixture and paper, and it will be seen that the salt mixture and paper did not sufficiently supplement the diet to give continued normal growth curves, and after about a month, all the guinea-pigs started to decline in weight. The similarity of the post-mortem lesions to those found in rats fed on diets deficient in vitamin A suggested the addition of some source of vitamin A. This had been attempted before with carrot extract prepared by Zilva's method, but without much success, for other deficiencies of the mangold were not made up. We now tried cod-liver oil instead of the carrot extract and 0·5 cc. per guinea-pig per diem was administered by hand. This was found·to cause a great improvement, and the guinea-pigs, with only one exception, resumed a fair rate of growth.

It follows, therefore, that the mangold, bran and oats diet, besides being deficient in calcium and perhaps roughage, is also lacking in the fat-soluble vitamin A. Our impression is that the dose of 0·5 cc. of the oil used in this experiment for our guinea-pigs was marginal, which may account for the failure of one of the four guinea-pigs to survive on the oil-supplemented diet. We have frequently noted that rats kept for a prolonged period on a diet deficient in vitamin A seem to undergo some permanent change, which renders them unresponsive to any treatment with cod-liver oil. It is possible that guinea-pigs also develop a similar condition of unresponsiveness, when deprived of the vitamin for prolonged periods of time.

But further feeding experiments with another set of guinea-pigs on a diet

of bran and oats, mangolds, salt mixture, paper and oil, revealed a further
deficiency in the diet, viz. protein.

Chart VI shows the weights of two guinea-pigs which were started with
a diet of bran, oats, mangolds, paper and salt mixture, and then when a
decline in weight, due to deficiency of fat-soluble A, occurred, oil was given,
with a result that growth was again resumed. After about a month of this
supplemented diet, however, a fresh drop in the weights of both guinea-pigs
occurred, and one of them succumbed, but the other guinea-pig quickly
picked up weight again on adding caseinogen to the diet. This guinea-pig
continued a normal rate of growth, and was still continuing to grow at the
time of writing, so that it is concluded that a fully supplemented diet has
been attained. (See also the latter part of Chart IV.) The question arises
whether the caseinogen acts beneficially by:

(a) Supplementing the phosphorus intake.

(b) Supplementing deficient protein, either a deficiency in certain amino-
groupings, or a deficiency in total amount.

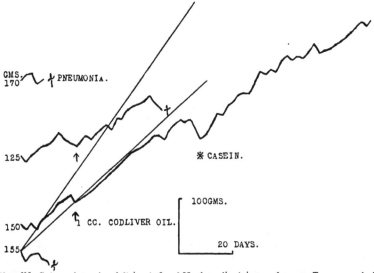

Chart VI. Guinea-pigs on basal diet as in I and II, plus salt mixture and paper. Two succumbed
before administration of oil was commenced (at arrows). Of the two survivors, one eventually
succumbed to protein deficiency, but the other quickly resumed a normal rate of growth on
the addition of caseinogen to the diet (at star).

Osborne and Mendel [1918] have shown with rats, that a shortage of
phosphorus leads to a considerable slowing in growth followed by a fall in
weight, but on adding caseinogen to the phosphorus-free diet, an improvement was made, and still greater improvement when inorganic phosphorus
was added. The addition of edestin (a phosphorus-free protein), however, only
led to complete cessation of growth and a decline.

The 12·5 g. of bran and oats eaten in a day contains 0·05 g. of phosphorus [Sherman and Gettler, 1912], whereas 40 g. of cabbage only adds 0·013 g. of phosphorus, an amount which could be easily covered by a small extra consumption of bran and oats, such as often takes place. We are led to conclude that, since cabbage can fully supplement a bran and oats diet, the addition of phosphorus is probably not the influencing factor, but that both caseinogen and cabbage supply a certain amino-acid or acids not present in bran and oats or in mangolds. This is a rather remarkable conclusion, considering the small amount of nitrogen in greenstuff. It is paralleled by the observations of Thomas that the N of potato is especially effectual in supplying the N requirements of the body.

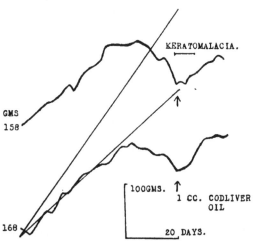

Chart VII. Guinea-pigs on diet of bran, oats, salt mixture, caseinogen and mangolds. After definite symptoms of vitamin A deficiency had become apparent, cod-liver oil was administered, at the arrows. Duration of the keratomalacia marked by straight line.

Two guinea-pigs, Chart VII, were started on a diet of bran and oats, caseinogen, salt mixture, paper and mangolds, and it will be seen that this diet was complete with one exception, viz. fat-soluble A. After a good initial growth for over a month, both guinea-pigs dropped in weight considerably, and one of them developed keratomalacia. There was a clouding of the cornea of one eye, which gradually became opaque. No haemorrhagic discharge was seen, and no conjunctivitis. In the Report of the Medical Research Committee [1919], it is stated that in rats it is the swelling of the eyelids and conjunctivitis which appears first, and this, if untreated, leads to thickening and clouding of the cornea, and ultimate blindness. With guinea-pigs, however, we have found that the corneal clouding always appears first, and several times we have received guinea-pigs from stock with completely opaque corneas, which have gone on to panophthalmia. We have never observed any haemorrhagic or purulent discharge preceding the keratomalacia in guinea-pigs. In rats, we have sometimes observed the uncomplicated corneal change first, but more frequently conjunctivitis precedes the corneal cloudiness.

When the guinea-pigs had each lost about 50–100 g., and the keratomalacia had developed unmistakably in the one, 1 cc. of cod-liver oil was given to each guinea-pig every day, which resulted in an almost immediate increase in weight, and a resumption of a normal rate of growth. There was a complete disappearance of the eye trouble after about one week's feeding with the oil.

We have confirmed the absence of fat-soluble A from mangold by feeding experiments with rats (Chart VIII).

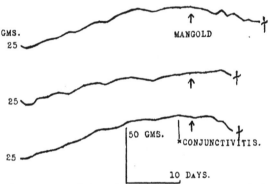

Chart VIII. Rats on purified synthetic diet deficient in vitamin A. Conjunctivitis developed at point marked by cross, and here mangold was added to the diet. No improvement in the condition and weight of the rats resulted, and they all eventually died.

The fact that our guinea-pigs developed no symptoms of scurvy whilst on the mangold diet supplemented as above, for a period extending over more than two months, indicates that the mangold supplies an adequate amount of the anti-scorbutic vitamin C. Vitamin B is supplied in sufficient quantity by the bran and oats mixture, and is also present in the mangold.

Our experiments therefore lead to the conclusion that mangold is deficient in the following: (1) calcium, (2) fat-soluble A, (3) protein, (4) ? roughage; but contains a sufficiency of the anti-scorbutic vitamin to keep guinea-pigs free from scurvy. This confirms the rule laid down by McCollum for seeds, roots and tubers generally.

The only essential dietary substance added by the mangold to bran and oats is vitamin C, although there is of course some energy value in the various other constituents, and by feeding experiments on rats we have shown that there is a fair quantity of vitamin B present.

Once again, as other observers have noted, vitamin A deficiency accompanies a low calcium content, so that it can almost be laid down as a general rule that where there is vitamin A deficiency, calcium deficiency probably runs parallel with it; and conversely, where there is a good supply of the fat-soluble vitamin, there is often a high percentage of calcium, as the following figures taken from Sherman and Gettler's tables show:

		Vitamin A	Calcium content %
I.	Egg yolk	+	0·143
	Cow's milk	+	0·124
	Beans (dried)	+	0·165
	Nuts	+	0·270 (almonds)
II.	Egg white	–	0·011
	Fish (white)	–	0·022
	Potatoes	–	0·006
	Rice	–	0·006–0·01

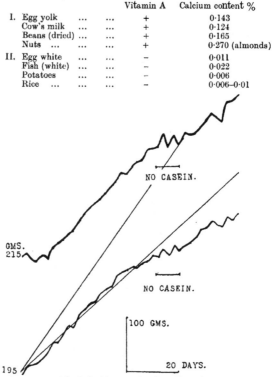

Chart IX. Guinea-pigs on complete diet of bran and oats, salt mixture, orange juice, cod-liver oil and caseinogen. Flattening of the growth curves when caseinogen was withdrawn from the diet for a period marked by the horizontal line.

We have made various efforts to feed guinea-pigs on a basal artificial diet similar to that used in feeding experiments on rats, but made up according to calculations based on analyses of bran and oats, milk and grass, with the idea of supplementing this basal diet with the three vitamins A, B and C in turn. All our experiments, however, have been brought to an untimely conclusion by the fact that the guinea-pigs refuse to eat any of the artificial diet.

Starting with bran and oats as a basal diet, and supplementing this with heated caseinogen, salt mixture as above, and paper, one can, however, study the effects of the vitamins A and C in the guinea-pig in a more or less uncomplicated way.

Two guinea-pigs were placed on this "synthetic" diet, as shown in Chart IX, bran and oats, heated caseinogen, salt mixture and paper being given, together

with 3 cc. lemon juice and 0·3 cc. cod-liver oil per guinea-pig per diem. After
a time, however, 5 cc. orange juice was substituted for the lemon, as it was
found that the guinea-pigs were averse to the sour taste of the lemon, whereas
they drank the orange juice eagerly. It will be seen from the curves that a
very satisfactory rate of growth can be maintained on this diet. A flattening
of the growth curves occurred when caseinogen was withdrawn from the diet
for a short period, thus emphasising the need of supplementing the protein
of the bran and oats. It is, of course, of great importance that the temperature
of the animal house in which the guinea-pigs are kept should be kept as
constant as possible.

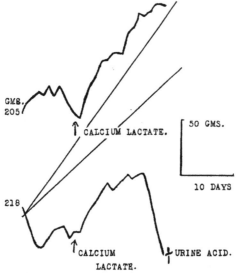

Chart X. Showing effect of calcium deficiency. Guinea-pigs on bran and oats, mangold and
cod-liver oil. (Paper and caseinogen were omitted because of the possibility of their containing small quantities of calcium.) A rapid decline set in, from which both animals
quickly recovered on the addition of calcium lactate to the diet. One guinea-pig later
succumbed to protein deficiency.

We are unable to confirm Miss Hume's statement [1921] that guinea-pigs
have an intolerance for unemulsified fat, for we have performed a large number
of experiments in which we have fed guinea-pigs with cod-liver oil, and in
every case the oil has had a beneficial effect provided other essential food
factors are not missing. We would suggest that Miss Hume's failure to obtain
growth on oil was probably due to the basal diet of bran and oats and orange
juice being inadequate, so that when the test substance, e.g. butter-fat or oil,
was added to the diet, the beneficial effect of the addition of vitamin A was
masked by the protein and inorganic deficiencies. For example, in Chart X,
guinea-pigs were put on a diet of mangolds, bran and oats and cod-liver oil,
and very soon a decline in weight occurred, which might have been ascribed

to an intolerance for oil on the part of the guinea-pig. That this was not the case was shown by supplying the deficient calcium of the dietary, when a quick recovery of the weight of the guinea-pigs was obtained, and they continued growing satisfactorily for some time, the oil still being continued, until finally, the deficiency of protein became apparent.

It is obvious that the need of the guinea-pig for an adequate supply of calcium is an urgent one, and the effects of its absence show themselves earlier than those of the vitamins. Vitamin A deficiency takes about 40–50 days to become apparent, and scurvy takes about three weeks to develop. On a diet containing inadequate calcium, however, guinea-pigs lose weight in a few days (see Chart X).

These experiments fully confirm Miss Hume's conclusion that guinea-pigs require a large amount of vitamin A, and they have the advantage that the addition of vitamin A to the diet was made in the form of cod-liver oil— much less admixed with other essential substances than in the case of the greenstuff and milk studied by her. The effect of deprivation of vitamin A is much more regularly obtained in guinea-pigs than in our own stock of rats, several of which grow for long periods on a deficient diet and even continue to grow whilst developing obvious eye changes. The larger dose of oil necessary for the guinea-pigs also suggests that they may be more suitable than rats for the estimation of vitamin A, and we are making some experiments with that in view.

CONCLUSIONS.

(1) The deficiencies of a diet of mangold, bran and oats and water which was the controlling factor in an epidemic amongst guinea-pigs have been investigated.

(2) Vitamin A and calcium salts have been shown to be deficient, and the protein to be deficient in quantity or composition or both.

(3) Keratomalacia in the guinea-pig as a result of vitamin A deficiency has been observed and cured by the administration of cod-liver oil.

(4) The administration of cod-liver oil to guinea-pigs has been shown to be an adequate means of administering vitamin A, and to be well tolerated by guinea-pigs if the diet is otherwise satisfactory.

(5) Direct confirmation of the necessity for vitamin A in the diet of guinea-pigs, inferred by Miss Hume, has been obtained.

(6) Another instance is provided of a foodstuff in which a deficiency of vitamin A is accompanied by a deficiency of calcium.

REFERENCES.

Cahen and Hurtley (1916). *Biochem. J.* **10**, 308.
Glenny and Allen (1921). *Lancet*, ii, 1109.
Hume (1921). *Biochem. J.* **15**, 47.
McCollum (1920). The Newer Knowledge of Nutrition, 36, 49, 64.
Osborne and Mendel (1918). *J. Biol. Chem.* **34**, 131.
Report of the Medical Research Committee (1919). Special reports, **38**, 17.
Sherman and Gettler (1912). *J. Biol. Chem.* **11**, 328.

LXXXIII. THE CATALYTIC DESTRUCTION OF CARNOSINE IN VITRO.

By WINIFRED MARY CLIFFORD.

Physiology Department, Household and Social Science Department, King's College for Women, Kensington.

(*Received October 20th, 1922.*)

IN a previous paper [Clifford, 1922] an account was given of the destruction of carnosine in beef and rat muscle kept in cold storage. Since the temperature at which the destruction occurred was at or below zero, it was improbable that it was due either to enzyme or bacterial action. The time taken however was prolonged—6–10 months—and therefore a slow change due to an enzyme or to bacteria might have been the cause. In order to exclude these possibilities experiments were carried out at 100°. A similar though far more rapid disappearance of carnosine took place at this temperature pointing to a simpler catalyst than an enzyme, while bacterial life is also inconsistent with a temperature range of 0°–100°.

EXPERIMENTAL RESULTS.

(1) *Destruction of carnosine in beef at 100°.*

From 3–5 g. of minced lean beef were weighed into test-tubes and 10 cc. water added. The tubes were then plugged with cotton wool and placed in a water-bath kept at 100°. When necessary more water was put in the tubes to prevent drying of the meat.

At intervals of 1–4 days a tube was removed, the contents made up to 95 cc. with distilled water and 5 cc. of 20 % metaphosphoric acid added to precipitate proteins. The carnosine present in the filtrate was estimated colorimetrically by the method described in a previous paper [Clifford, 1921, 1].

The numerical results of four experiments were as follows:

Days at 100°	8	15	29	29 A
Start	0·96 %	0·90 %	1·05 %	1·05 %
1	—	0·80	—	—
2	0·70	—	0·90	0·89
4			0·80	0·74
5	0·60	0·66	—	—
7	0·54	0·60	0·68	0·65
10			0·60	0·64
12	0·53	—		
14	—	0·60	0·55	0·57
16	0·30	—	—	—
17		0·40	0·43	0·40
21	—	0·33	—	—

These results, Fig. 1, indicate the presence of a catalyst which removes carnosine from muscle. This removal is considerably hastened[1] by a rise of temperature and is therefore probably a chemical and not a physical change. On comparing the results with those previously published [Clifford, 1922] it is found that for the percentage of carnosine to fall to the same level 9–10 months is necessary at 0° as against 21 days with a temperature of 100°.

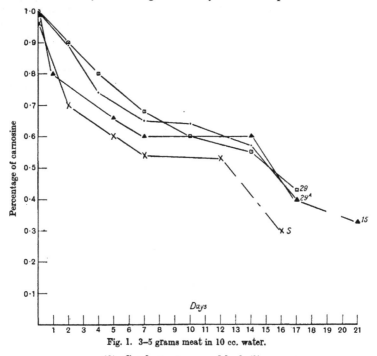

Fig. 1. 3–5 grams meat in 10 cc. water.

(2) *Catalyst not removed by boiling.*

Experiments were performed to find whether the catalyst were present in an extract made by boiling beef with water. Minced lean beef was put into a flask with water and boiled in a water-bath for 30–60 minutes. It was then filtered, and the residue again boiled with water for a similar period. The filtration was repeated and the two filtrates united. An estimation of the carnosine percentage of the extract was made, and portions of 10 cc. put in test-tubes plugged with cotton wool. The tubes were then placed in a water-bath at 100°.

In five experiments lasting 13–26 days no change took place in the carnosine content of the extract. Similar results were obtained in two other 13 day experiments, one kept at 60°, and the other at 37° after sterilisation by boiling.

[1] The coefficient is 1·5 for 10° rise whilst for physical actions it is generally 1·03–1·05.

Since the carnosine content of the extract in these experiments remained unchanged, the catalyst cannot be removed from beef by extraction with water for 1–2 hours at 100°.

(3) Catalyst remains in beef after boiling.

A series of test-tubes was taken containing 10 cc. of beef extract prepared as above together with 3 g. of boiled minced beef which had been washed in cold running water for 15 minutes to remove traces of extractives. The result as expected was identical with that obtained with fresh beef in water.

(4) Catalyst removed by prolonged washing of boiled meat.

Further experiments were carried out to find the result of prolonged washing of the boiled meat.

Five experiments with series of tubes containing 10 cc. of extract and 3 g. of boiled meat washed 3, 18, 24, 48, 48 hours respectively were kept at 100° with the following results:

	21 3 hours' washing	20 18 hours' washing	19 24 hours' washing	30 48 hours' washing	9 48 hours' washing
Days.	%	%	%	%	%
Start	1·0	1·0	1·05	0·99	0·98
3	0·9	1·0	1·0	0·99	0·98
6	0·9	0·9	1·0	0·99	0·98
8	0·45	0·6	0·77	0·99	0·98
10	0·45	0·6	0·70	0·99	0·98
13	0·40	0·55	0·65	0·99	0·98
23	0·35	0·35	0·50	0·99	0·98
27	0·30	0·33	0·48	0·99	0·98

From these results it appears that the catalyst present in boiled beef which destroys carnosine can be removed by prolonged washing in running water, probably by separation of an adsorption compound from the muscle protein.

In one experiment the preliminary boiling of the meat was omitted and fresh unboiled minced beef was washed in running water for 24 hours. Three g. of this washed meat was then added to each of a series of tubes containing 10 cc. extract at 100°. The result was a disappearance of carnosine similar to that given by boiled beef washed for three hours.

Start	1·0 %	10 days	0·40 %
3 days	0·95 %	13 ,,	0·40 %
6 ,,	0·80 %	23 ,,	0·25 %
8 ,,	0·45 %	27 ,,	0·25 %

These results, Fig. 2, indicate that the catalytic destroyer of carnosine is present in some physical combination with the protein of beef muscle and may be removed by prolonged washing in cold running water. This removal is facilitated by boiling the meat before washing, though mere boiling for 1–2 hours does not extract the catalyst from the muscle.

The curve of disappearance of the base is peculiar and it is difficult to account for the delay in the start of the reaction with washed meat and extract as compared with fresh meat and water. In all the curves there is a sharp fall between the 6th and 8th days followed by a stationary period and then a slower steady fall.

This three-step curve is indicated in experiments on fresh meat and water (Fig. 1) and is strongly marked in experiments with cod muscle and with liver. A paper previously published [Clifford, 1922] shows a similar curve from beef kept in cold storage.

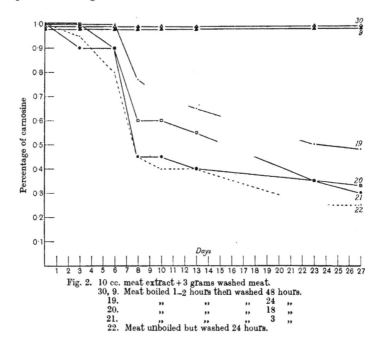

Fig. 2. 10 cc. meat extract + 3 grams washed meat.
    30, 9. Meat boiled 1_2 hours then washed 48 hours.
    19.   ,,   ,,   ,, 24 ,,
    20.   ,,   ,,   ,, 18 ,,
    21.   ,,   ,,   ,,   3 ,,
    22. Meat unboiled but washed 24 hours.

(5) *Presence of catalyst in muscles of white fish.*

Carnosine has been shown [Clifford, 1921, 2] to be absent from the muscles of white fishes. A series of experiments was performed using 3 g. of fresh minced cod muscle in beef extract in place of 3 g. of boiled and washed beef. Unexpectedly it was found that the carnosine disappeared giving a similar curve to that with washed beef in its peculiar three-stepped nature, though the actual position of the stationary period varies (Fig. 3).

Three experiments were carried out at 60° after preliminary sterilisation by boiling and three were kept constantly at 100°.

Days	P 60° %	H 60° %	7 60° %	E 100° %	14 100° %	24 100° %
Start	0·96	0·99	0·99	0·99	1·04	1·0
1	—	—	—	—	0·80	—
2	0·90	0·90	—	0·90	—	—
3	—	—	—	—	—	1·1
4	—	—	0·95	—	—	—
5	0·85	—	—	—	0·80	—
6	—	—	—	—	—	0·75
7	0·58	0·80	0·72	0·90	0·55	—
8	—	—	—	—	—	0·60
9	—	—	—	—	—	—
10	—	0·80	—	0·88	—	0·46
11	—	—	0·60	—	—	—
13	0·58	—	—	—	—	0·46
14	—	0·66	0·55	0·50	0·50	—
16	—	0·66	—	0·44	—	—
17	0·52	—	—	—	—	—
18	—	0·44	0·55	—	0·45	—
20	—	—	—	—	—	—
21	—	0·40	0·50	0·40	0·35	—
23	—	0·35	—	—	—	—
25	—	0·30	—	—	—	0·30
28	—	—	0·40	—	—	—

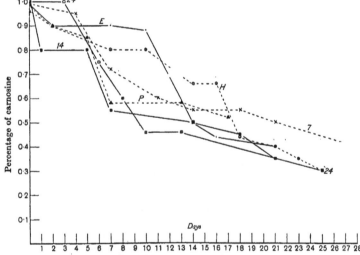

Fig. 3. 10 cc. meat extract + 5 grams cod muscle.
Unbroken lines 100° C. constantly.
Broken lines 60° C. constantly after sterilisation by boiling.

(6) *Presence of catalyst in liver.*

An experiment was performed using tubes with 10 cc. of beef extract and 3 g. of calf liver kept constantly at 100°. Carnosine estimations again gave a stepped curve of disappearance of the base.

A second experiment in which 1 g. of freshly killed rat's liver was used in place of the 3 g. of calf liver, and in which the temperature was 60° in place of 100°, also showed a loss. As was expected from the lower temperature and smaller amount of tissue taken, the rate of disappearance was slower than in the first experiment (Fig. 4).

Days of experiment	16 100° and 3 g. liver. %	Q 60° and 1 g. liver. %
Start	1·0	0·96
1	0·80	—
2	—	0·90
5	0·55	0·77
7	0·55	0·77
13	—	0·64
14	0·50	—
17	0·42	0·60
21	0·25	—

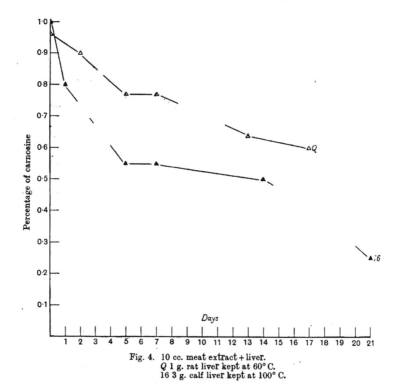

Fig. 4. 10 cc. meat extract + liver.
Q 1 g. rat liver kept at 60° C.
16 3 g. calf liver kept at 100° C.

A third experiment lasting 18 days, in which 3 g. of frozen calf liver was used as a source of the catalyst, showed no loss of the base. These findings point to the presence of the carnosine-destroying catalyst in liver as well as in muscle, whilst prolonged keeping in cold storage destroys the catalyst itself.

(7) Absence of catalyst from kidney.

Since skeletal muscle and liver had both been shown to cause the disappearance of carnosine from muscle extract, it was possible that the catalyst was present in all cells. Therefore an experiment was carried out using 1 g. of freshly killed rat's kidney, in place of liver or muscle, in 10 cc. of beef extract. The tubes were kept at 60°. No change in carnosine content took place till the 21st day, when a very slight fall took place. This fall however lay well within the limits of experimental error.

Similar results were obtained using 3 g. of sheep's kidney, 3 g. of bullock's kidney and 3 g. of frozen bullock's kidney, all kept at 100° for at least 20 days.

There is, therefore, no carnosine-destroying catalyst in the kidney, and consequently the possibility of existence of this catalyst in all cells cannot be maintained.

(8) Absence of catalyst from invertebrate muscle.

A series of test-tubes containing 10 cc. of beef extract together with 3 g. of fresh lobster muscle was left in a bath at 100°. No change in the carnosine content of the tubes was observed up to the end of 46 days, and therefore the catalyst was not present in the lobster muscle.

Oysters were next used as a further type of the invertebrata. Here from the first day onwards no pink colour could be obtained on diazotising the protein-free filtrate. However a deep yellow colour resulted and if estimated as carnosine showed no loss of base up to the 23rd day.

In all other experiments, e.g. with liver and muscle, or in simple keeping of meat in cold storage, disappearance of the base coincided with a less depth of colour, and not an alteration of hue. Therefore the lack of red colour on diazotising the oyster-beef extract was probably due to the presence of an inhibitory substance similar to that shown to exist in salmon muscle [Clifford, 1921, 2].

Unfortunately there was not enough of the beef extract with oyster muscle to use the precipitation method of Dietrich [1914] to show whether this were the case or not, but experiments are in progress to determine this point.

The definitely negative result with lobster muscle and the probably negative findings with oyster point to a lack of carnosine-destroying catalyst in the invertebrata.

DISCUSSION OF RESULTS.

The experiments described in this paper indicate the presence of a substance which is capable of destroying carnosine in muscle extracts, and owing to its temperature range it must be a relatively simple catalyst.

It is not known whether the absence or lessening of red colour on diazotising filtrates from meat kept several days in water at 100° is due to actual destruction of the iminazole ring, or to a synthesis which prevents coupling of this ring with the diazo-reagent. Experiments are in progress with the object of elucidating this problem. In either case, whether the loss of carnosine is due to a synthesis or to a breaking up of the base, it will be of interest to know the fate of the β-alanine portion of the molecule, as this is the only known β-amino acid in the animal organism.

The stepped nature of the curves of loss of carnosine is peculiar and totally unlike any enzyme curve. It may indicate that not one but two agents are active in the change, and that they act in an analogous way to enzymes and co-enzymes. On the other hand the catalyst may be a single substance and the reaction take place in two stages separated by an inactive period. Neither of these explanations can account for the initial delay shown when using washed meat as a source of catalyst, since after this delay the three-stepped curve appears.

The catalyst is present in ox, rat, and cod muscle and therefore is probably found in vertebrate skeletal muscle generally, but experiments with lobster and oyster muscle indicate its absence from invertebrates. The livers of the rat and ox also contain this catalyst, which however is absent from frozen ox-liver. This indicates another connection between hepatic and muscular activity already seen in glycogen metabolism and also with urea, creatine and creatinine.

The catalyst is not present in all cells since experiments have proved it to be absent from the kidney of the rat, ox, and sheep. The existence of this catalyst may account for the non-appearance of ingested carnosine in the urine.

Earlier experiments in which carnosine was shown to be absent from the striped muscles of the white fishes and the finches appeared to suggest that carnosine is a merely accidental substance, somewhat arbitrary in its presence or absence. But the experiments detailed above, particularly those that show the presence of a carnosine-removing catalyst in the muscle of white fish, suggest that it is an intermediary product of metabolism and that its appearance and its percentage in muscle are determined by the rate at which it is formed and the rate at which it is removed in the different types of muscle.

The expenses of this research were defrayed by a grant from the Medical Research Council.

Thanks are due to Professor V. H. Mottram for the interest shown by him in the work and for his helpful criticism.

REFERENCES.

Clifford (1921, 1). *Biochem. J.* **15**, 400.
—— (1921, 2). *Biochem. J.* **15**, 725.
—— (1922). *Biochem. J.* **16**, 341.
Dietrich (1914). *Zeitsch. physiol. Chem.* **92**, 212.

LXXXIV. ON THE VITAMIN D.

By TREVOR BRABY HEATON.

From the Department of Pharmacology, Oxford University.

(*Received October 25th, 1922.*)

THE well-known fact, first described by Wildiers, that yeast cells when in low concentration fail either to ferment sugar or to grow, but may be induced to do so by the addition of "bios," is as yet imperfectly understood. The need of yeast cells for "bios" depends clearly on dilution; for yeast, when in adequate initial concentration, is able to synthesise this substance indefinitely from the ingredients of a simple medium. It is an organic substance, soluble in water or alcohol, dialysable, and thermostable; it appears therefore to belong to the group of the vitamins. It seems, however, to be more thermostable than the vitamin which promotes the growth of rats [Souza and McCollum, 1920]; it is more thermostable also than the co-enzyme of zymase [Tholin, 1921]. The capacity for activating yeast in this manner has been used nevertheless by Williams [1919] and others [see Sherman, 1921], as a measure of the richness of any material in the vitamin B; and there is no doubt that in many cases, in yeast itself for instance, this vitamin and "bios" are closely associated.

Now the comparative value of the various organs of animals, for preventing the onset of polyneuritis in pigeons fed on polished rice, has been determined by Cooper [1914], who places them in the following order: liver, heart, cerebrum, cerebellum, muscle; 0·5 g. of yeast being equivalent for this purpose to 0·9 g. of liver, 1·2 g. of brain, or 5·0 g. of muscle (dried weights in each case). In respect of promoting the growth of young rats, Osborne and Mendel [1918] place the tissues in a similar order. The distribution of "bios" among the organs of animals, therefore, seemed of interest as a test of the closeness of association between it and this vitamin; and a comparison in this respect between the organs of normal animals and those of animals suffering from vitamin-B deficiency, as an indication how far this deficiency can be attributed to a shortage of "bios." Such an investigation forms the subject of the experiments to be described.

METHOD.

Two methods have been adopted for measuring "bios," or the substance which activates yeast; the one depends upon the rate of multiplication of yeast cells, the other upon the rate of fermentation, as indicated by the evolution of CO_2. The majority of workers have preferred to study growth, as being a less erratic phenomenon than that of CO_2 production. The number and complexity of the factors which influence growth, however, are likely to be no less great than those which influence fermentation; and the fermentation method has been used in the present experiments, since it has the advantages of convenience and speed, and allows therefore many estimations to be made and an average drawn.

The medium employed has been Nägeli's solution, omitting ammonium nitrate. Ammonium nitrate of course is necessary for the growth of yeast, and leads also to more rapid fermentation; but where the rate of fermentation by a given quantity of yeast is alone in question, a factor which only promotes growth is a source of error; and results without it, while less sensitive, seemed likely to be more accurate than in its presence. The composition of the medium used, then, was:

Cane sugar	10 g.
Potassium biphosphate ...	0·5 g.
Magnesium sulphate	0·25 g.
Calcium phosphate	0·05 g.
Distilled water to	·100 cc.

To this medium is added 0·01 % of fresh brewer's yeast (1 cc. of a 1 % suspension to 100 cc.), a quantity in itself insufficient to produce any fermentation whatever in 24 hours: this constitutes the standard yeast-Nägeli medium. To 10 cc. in each of a series of tubes is added a graduated quantity of a watery extract of the substance under investigation. The mixtures are transferred with a serum-syringe to a series of 3 cc. glass ampoules, such as are used for vaccines, two ampoules from each tube. The experiments are performed in duplicate, so that four determinations are made for each concentration of the activating substance. The ampoules are then inverted over a receptacle, and incubated at 33° for 24 hours. As fermentation occurs, fluid is displaced, and drips out of the neck of the ampoule into the receptacle. Each ampoule is weighed empty (A), full (B), and at the end of the experiment (C). The difference, $B - C$, represents the cc. of CO_2 produced in $B - A$ cc. of solution; from which the percentage CO_2 formation may be calculated.

Results found to be most reliable are obtained when the fermentation amounts to about 5–10 %; the yeast tends to adhere to the glass as the fluid recedes, vitiating results when larger amounts of fermentation have occurred. And the progressive emptying of the ampoule constantly diminishes the quantity of fluid in which the recorded fermentation is taking place. Other influences, moreover, which may affect the rate of fermentation, such as the presence of amino-acid or of phosphate, cause an increasingly greater percentage error. The suggestion has indeed been made that such influences completely vitiate the test. At these concentrations of yeast, however, the error that they produce is not large. Table I shows the fermentation produced in the medium, in the absence of a specific activating substance, by various influences of this kind.

Table I. *Showing absence of fermentation in standard yeast-Nägeli medium, in absence of specific activating substance.*

Added to 10 cc. yeast-Nägeli medium ... ·05 ·1 ·15 ·2 cc.

				Percentage fermentation			
Sodium phosphate 2 %	...	...	...	0	0	1	1
Commercial glucose 2 %	...	...	...	1	1	1	1
Dried blood 2 % extract	...	...	...	1	1	1	2

Table II shows, further, that when an extract of dried yeast, which is an efficient activating substance, is added to the medium, the further addition of phosphate makes very little difference.

Table II. *Showing absence of effect of sodium phosphate on fermentation produced in activated yeast-Nägeli medium.*

10 cc. yeast-Nägeli medium, with 0·075 cc. of 2 % extract of dried yeast	Fermentation
Without sodium phosphate	4·5
With sodium phosphate	4·25

A more searching test, however, as to whether the activation of minimal quantities of yeast is or is not the effect of a single substance, and whether the fermentation thereby produced is or is not a satisfactory measure of such substance, is afforded by a comparison of the fermentation curves obtained when the activating substance is derived from various sources. The following have been used for this purpose: cow's milk, a 10 % solution of "Glaxo," a 2 % extract of the desiccated spleen of the calf (Armour), and a 2 % extract of dried yeast. Table III represents the average of a large number of determinations of the fermentation produced in the standard medium, after addition of these four substances in increasing quantity.

Table III. *Average fermentation obtained in standard yeast-Nägeli medium, after addition of increasing amounts of various activating substances.*

Amount added to 10 cc. yeast-Nägeli medium ...	·05	·075	·1	·125	·15	·175	·2	·225	·25	·3 cc.
Substance added	Percentage fermentation in 24 hours									
Cow's milk	2·5	4·6	8·6	11·8	16·5	16·9	23·4	—	—	—
10 % Glaxo	—	—	4	—	7	—	8·5	—	15	24
2 % dried yeast ...	1·5	3·3	7·7	11·6	16·2	24·2	—	—	—	—
2 % desiccated spleen ...	—	2·5	4·4	6·5	12	14·5	18·1	21·5	—	—

It is clear from this table that while the activating power of cow's milk is almost exactly the same as that of 2 % yeast extract, it is considerably greater than that of either 10 % Glaxo or 2 % spleen. If, however, the activating powers be calculated from the above figures, of 17 % Glaxo and of 2·6 % spleen respectively, and the results plotted on a curve, it is found that such fermentation curves, for the four activating substances used, almost exactly coincide (Fig. 1). This seems to indicate that the activating substance is a single and definite entity, and that the method is a fair one for estimating its quantity in any material.

In the following experiments the curve shown in Fig. 1 has been taken as a standard, indicating the fermentation produced by a 2 % extract of dried yeast. The organs of various animals have been air-dried at 33°, and powdered, and their activating power determined in 2 % watery extract. By comparison with Fig. 1 (or Table III) it is possible to say how much dried yeast is the equivalent in activating power of unit weight of the material investigated. This, for convenience, I have called the "yeast-equivalent" of the substance. From Table III, for instance, it will be seen that the yeast-equivalent of this preparation of desiccated spleen is 0·8.

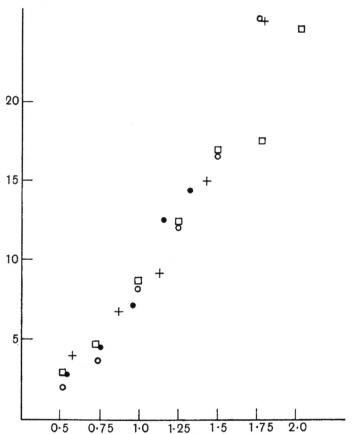

Fig. 1. Curve of fermentation in yeast-Nägeli medium, induced by addition of various activating substances.

Abscissae: Percentage addition to the medium of the various substances. Ordinates: Percentage fermentation in 24 hours at 33°. Crosses: 17 % Glaxo[1]. Squares: Cow's milk. Hollow circles: 2 % dried yeast. Solid circles: 2·6 % desiccated spleen[1].

[1] Calculated from observations with 10 % Glaxo and 2 % spleen respectively.

RESULTS.

(1) *Pigeons*. The organs were taken and dried under the same conditions from a series of birds, being (1) normal pigeons, and (2) pigeons which on an exclusive diet of polished rice had developed avian polyneuritis, and had either died from this condition, or had been killed at the point of death. It was thought that if the substance which activates yeast were the same as that whose absence causes polyneuritis, the organs of polyneuritic birds should show a deficiency of this substance; and this deficiency should be most marked in the brain, where the symptoms are so nearly exclusively localised. Funk [1912], as is well known, has shown a deficiency in the N- and P-content of the brains of birds in this condition.

Because the symptoms of avian polyneuritis are markedly cerebellar in type, the cerebral hemispheres and cerebellum were dried and estimated separately. In order to illustrate the degree of consistency of the results, the figures obtained in the case of the brain (cerebral hemispheres) are given in detail in Table IV. Table V is a summary of results from all organs, the average alone being given in each case, together with the calculated "yeast-equivalent."

Table IV. *Details of experiments showing percentage fermentation obtained in yeast-Nägeli medium by addition of 2 % extracts of dried cerebral hemispheres of pigeons.*

Pigeon	1	2	3	4	Av.	1	2	3	4	Av.
I. Polyneuritic										
P	2·5	6	2·5	7·5	4·9	11	11	11·5	9	10·5
Q	4·5	3·5	4	4·5	4·1	9	7	8	4·5	7·1
O	3	3	1·5	1·5	2·2	5	5	8·5	6	6·1
Y	8	4	3·5	5·5	5·2	8	9	10	10	9·2
Z	4	8	4	5	5·2	9	11	10	12	10·5
B	3	2	1·5	3·5	2·5	6	9	10	6	7·7
X	3·5	3	4·5	4	3·8	8	5	7·5	4	6·1
Average	4·1	4·2	3·1	4·9	4·0	8	8·1	9·4	7·4	8·2
II. Normal										
D	5	3	5·5	4	4·4	8	11·5	11	6	9·1
H	5	3	5	6	4·5	9	12	10	5	9·0
A	4	5	6	4	4·7	11	10	9	11	10·2
Average	4·7	3·7	5·5	4·3	4·5	9·3	11·2	10	7·3	9·5

Tables IV and V show, firstly, that the power to activate yeast is approximately the same in all the organs of pigeons that were examined, and that this power is not very much less than that of dried yeast itself: the liver, which is slightly more active than other organs, being scarcely at all inferior to yeast. This offers a striking contrast with the figures of Cooper, already referred to, showing the relative efficiency of the organs in preventing the development of polyneuritis.

So marked a contrast seems to show quite clearly, not only that the yeast-activating substance is not identical with the antineuritic vitamin, but that the two are not associated together in constant relative proportions.

Table V. *Showing average fermentation produced in standard yeast-Nägeli medium by addition of 2 % extracts of dried organs of pigeons, and "yeast-equivalents" calculated therefrom.*

Organ ...	2 % cerebrum			2 % cerebellum			2 % liver			2 % kidney			2 % heart			2 % muscle		
Cc. added to 10 cc. yeast-Nägeli medium ...	·1	·15		·1	·15		·1	·15		·1	·15		·1	·15		·1	·15	
Pigeon	% Ferm.		Yeast equiv.	% Ferm.		Yeast equiv.	% Ferm.		Yeast equiv.	% Ferm.		Yeast equiv.	% Ferm.		Yeast equiv.	% Ferm.		Yeast equiv.
I. Polyneuritic																		
P	5	10·5	·8	1·5	4	·4	9·5	14·5	1·0	7	12	·9	8	12	·95	5	11	·8
Q	4	7	·7	5	12	·85	4·5	12	·8	4	8	·75	5	9	·75	5·5	11	·85
O	2	6	·5	4	9	·75	8·5	14	1·0	4·5	10	·8	4	8	·75	4	6	·7
Y	5	9	·8	5·5	11	·85	7	14	·9	—		·85	—			—		—
Z	5	10·5	·8	—		—	8	9·5	·9	—			—			—		
B	2·5	7·5	·6	—		—	8	12	·9	6	10·5	·85	6	8	·8	5·5	9·5	·8
X	4	6	·6	3·5	8	·7	6·5	8·5	·8	—			—			—		
Average	4	8·2	·7	4	9	·75	7·5	12	·9	5·4	10	·8	5·7	9	·8	5	9·5	·8
II. Normal																		
D	4·5	9	·8	5	10	·8	8·5	25	1·1	4	8	·75	6·5	11	·85	4	5	·6
H	4·5	9	·8	4	8	·75	—		—	—			—			—		—
A	4·5	10	·8	—		—	5	7·5	·7	8	14	1·0	4·5	9	·8	5	8	·7
Average	4·5	9·3	·8	4·5	9	·8	6·7	16·3	·9	6	11	·85	5·5	10	·8	4·5	6·5	·7

The tables show, moreover, secondly, that the activating power of the organs of polyneuritic pigeons is only very slightly less than that of the normal organs. This fact serves to strengthen the conclusion that deficiency of the substance which activates yeast is not the deficiency which causes polyneuritis.

There might indeed be a possibility that the organs of polyneuritic animals, while retaining their percentage of this activating substance, suffered a loss in their total content, by diminution in their whole weight. However, as Abderhalden has shown [1921], such a diminution in the weight of individual organs in polyneuritis occurs only in the case of the liver and voluntary muscles, not in that of the brain. This observation was confirmed in the foregoing experiments, as is seen in Table VI.

Table VI. *Showing weights of normal and of polyneuritic pigeons, and of their organs.*

Pigeon	Total body weight	Weights of organs		
		Cerebrum	Cerebellum	Liver
I. Polyneuritic				
P	150	0·98	0·22	4·52
Q	165	1·05	0·20	7·78
O	145	1·13	0·22	3·15
Y	150	1·12	0·23	5·22
Z	165	1·01	0·26	6·10
B	224	1·11	0·27	8·67
X	199	0·94	0·27	5·85
Average	171	1·05	0·24	5·90
II. Normal				
D	270	1·08	0·28	6·39
H	—	1·03	0·25	—
A	323	1·15	0·24	6·86
Average	296	1·09	0·26	6·68

(2) *Rats.* It follows from the above experiments, that the yeast-activating substance cannot be identified with the antineuritic vitamin, its amount being very nearly as great in the organs of polyneuritic as in those of normal pigeons. Similar experiments were performed on rats, the organs of normal animals being compared with those of others, which had been either killed when moribund, or had died, as the result of a diet free from water-soluble vitamin; and with those of a third series which, as the result of such a diet, had ceased to grow for several weeks, but had been killed before any other symptoms had manifested themselves. The diet consisted the usual mixture of purined caseinogen, starch, sugar, butter, and salts. Upon this diet growth ceased at once; but if the initial weight of the animals was not less than about 70 g., they remained active and in good condition for many weeks; eventually dying with or without irregular nervous symptoms. The course of this condition, as has often been observed, is in striking contrast with the regular and rapid onset and constant symptoms of polyneuritis in birds. Table VII gives the summary of the average results of experiments on the content of the organs of these rats in the substance which activates yeast.

It will be seen that in the case of normal rats, the "yeast-equivalents" approximate very closely to those found in pigeons; but that in the rats which had died as the result of deprivation of water-soluble vitamin, this "equivalent" has diminished considerably. The rats which were killed after a shorter period (six weeks) of vitamin starvation occupy in this respect an intermediate position.

Table VII. *Showing average percentage fermentation produced in standard yeast-Nägeli medium by addition of 2 % extracts of dried organs of rats, with "yeast-equivalents" calculated therefrom.*

Organ ... Cc. added to 10 cc. yeast-Nägeli medium Rat number	2 % brain				2 % liver				2 % kidney				2 % heart			
	·1	·15 % fermentation	·2	Yeast equiv.	·1	·15 % fermentation	·2	Yeast equiv.	·1	·15 % fermentation	·2	Yeast equiv.	·1	·15 % fermentation	·2	Yeast equiv.
I. Normal rats																
11	5·5	13	—	0·85	5·5	17	—	0·9	14	28·5	—	1·3	5	10	16	0·8
6	5	11	19·5	0·8	6	14	—	0·9	7	17	—	0·9	5·5	9	15	0·8
7	8	20	—	1·0	3·5	6	—	0·6	5·5	16	—	0·9	5	7	16	0·75
Average	6	15	—	0·9	5	12	—	0·8	9	20·5	—	1·0	5	9	16	0·75
II. Vitamin-free diet six weeks. No symptoms																
8	4·5	6	13	0·7	5	10	15	0·8	6	9	14	0·8	3·5	4·5	6	0·6
9	2	3	9	0·5	4	9	13	0·75	7·5	9	16	0·85	1·5	2·5	6	0·45
10	2	4	8	0·5	2·5	4	8	0·5	6	11	14	0·8	3	7	10	0·6
Average	3	4·3	10	0·6	4	8	12	0·7	6·5	9·7	15	0·8	2·7	4·7	7	0·5
III. Vitamin-free diet till death or moribund																
1	1	—	8·5	0·5	1·5	—	11	0·6	2·5	—	7·5	0·5	1	—	2·5	0·3
2	2	—	10·5	0·55	1	—	5	0·4	1	—	5	0·4	2	—	7·5	0·5
3	1·5	—	10	0·55	1·5	—	7·5	0·45	4	—	11	0·55	1	—	6	0·45
4	1·5	—	6·5	0·45	3	—	7·5	0·5		3	8	0·5	—	1	6	0·4
5		6	9	0·55	—	7	9	0·55	9·5	15·5	0·8			3	4·5	0·4
Average	1·5	6	9	0·55	1·5	—	8	0·55	2·5	6	9·5	0·6	1·5	2	5·3	0·4

There is therefore a considerable difference between the condition of these rats and that of polyneuritic pigeons. This is due, no doubt, to the long duration of the diet; for pigeons on a diet of polished rice are invariably dead from polyneuritis within 28 days. But it shows at any rate, that rats suffering from deficiency of water-soluble vitamin are depleted of a substance which polyneuritic pigeons still possess; since death occurs before this second de-.ency is manifested.

These findings afford support to the conclusions of Funk and Dubin [1921], that there are two water-soluble vitamins present in yeast: it is the absence of one of these, vitamin B, which causes polyneuritis in birds; while the other, which these authors call vitamin D, is the "bios" of Wildiers, necessary to the activity of low concentrations of yeast cells; both of them are necessary for the proper growth of rats. The prevailing uncertainty as to the identity of the antineuritic and the growth-promoting vitamins is due, therefore, to the fact that the latter includes the antineuritic substance, but contains also another essential factor, the vitamin D.

So far as yeast is concerned, the "bios" effect indicates that the vitamin D alone is necessary for the activation of low concentrations of yeast cells, and leads not only to fermentation but also to growth. This must not be taken to imply, however, that fermentation and growth are identical phenomena; nor that vitamin D is the only requisite for the manifestation of either or both of them. The fact that vitamin B is present in yeast cells is a proof to the contrary; and other vitamins, it may be, are responsible for various functions in the life-history of the yeast-plant. The "bios" effect signifies merely that vitamin D is one of the substances necessary to the life of the cell, and that it differs from most other such substances in undergoing dilution by diffusion into the surrounding medium.

SUMMARY.

(1) The substance which activates minimal concentrations of yeast is a definite chemical entity, though obtainable from many sources. It may be measured fairly satisfactorily by the fermentation to which it gives rise.

(2) It is present, in approximately the same amount, in the following organs, both of pigeons and rats: cerebrum, cerebellum, liver, kidney, heart, and voluntary muscle; its amount in any of these is about 80 % of its amount in dried yeast. This distribution is contrasted with that of the antineuritic vitamin.

(3) In the organs of pigeons rendered polyneuritic by a diet of polished rice, it is present in the same amount as in those of normal pigeons.

(4) In the organs of rats fed on a diet purified of water-soluble vitamin, its amount progressively diminishes, and when death occurs is only about one-half the normal.

(5) For these reasons this activating substance cannot be identified with

the antineuritic vitamin B. The deprivation of water-soluble vitamin from rats is held to involve a double deficiency; a deficiency of vitamin B in the first place, and of the yeast-activating substance in the second. This is the substance which Wildiers called "bios," and Funk the vitamin D.

The author wishes to express his cordial thanks and obligation to Dr J. A. Gunn, Professor of Pharmacology in the University of Oxford, for facilities to carry out the above work in his laboratory.

REFERENCES.

Abderhalden (1921). *Pflüger's Archiv*, **193**, 355.
Cooper (1914). *J. Hygiene*, **14**, 12.
Funk (1912). *J. Physiol.* **44**, 50.
Funk and Dubin (1921). *Proc. Soc. Exp. Biol.* **19**, 15.
Osborne and Mendel (1918). *J. Biol. Chem.* **34**, 17.
Sherman (1921). *Physiol. Reviews*, i, 598.
Souza and McCollum (1920). *J. Biol. Chem.* **44**, 113.
Tholin (1921). *Zeitsch. physiol. Chem.* **115**, 235.
Williams (1919). *J. Biol. Chem.* **38**, 465.

LXXXV. A NEW PHOSPHORIC ESTER PRODUCED BY THE ACTION OF YEAST JUICE ON HEXOSES.

By ROBERT ROBISON.

From the Biochemical Department of the Lister Institute.

(*Received October 30th, 1922.*)

THE effect of sodium phosphate in increasing the fermentative power of yeast juice was first observed by Wroblewski [1901] and again by Buchner [Buchner, E. and H. and Hahn 1903] by whom it was attributed to the alkalinity of the salt. That such was not the true explanation followed from the work of Harden and Young [1905; 1906, 1 and 2; 1908, 1 and 2; 1909; 1910, 1 and 2; 1911] whose extended investigations revealed the important rôle played by phosphates in alcoholic fermentation. It was shown by these authors that when a suitable amount of a soluble phosphate is added to a fermenting mixture of yeast juice and glucose, fructose or mannose, the rate of fermentation rapidly increases but after a short period again falls to a constant rate, which is only slightly greater than that of the original yeast juice and sugar. During this period of increased fermentative activity the total evolution of carbon dioxide is increased by an amount which is equivalent, molecule per molecule, to the phosphate added, while the latter undergoes a transformation into a form that is no longer precipitable by magnesium citrate mixture.

From such solutions Young [1907; 1909] succeeded in isolating an ester of phosphoric acid by precipitation in the form of its insoluble lead salt. He showed [1909; 1911] that this compound possessed the constitution of a hexosediphosphoric acid, $C_6H_{10}O_4(PO_4H_2)_2$, and that the same substance was formed from glucose, fructose or mannose when fermented with yeast juice or zymin in the presence of phosphate. The acid was slightly dextrorotatory ($[\alpha]_D = + 3\cdot2°$) and on hydrolysis yielded phosphoric acid and a laevorotatory reducing substance from which fructose was isolated, although Young was not satisfied that the latter was the only sugar produced. A number of salts and derivatives were described, including the phenylhydrazine salt of a phenylhydrazone in which both phosphoric acid groups were retained, and an osazone in the formation of which one molecule of phosphoric acid had been split off. The evidence brought forward by Young was, however, insufficient to decide fully the constitution of the ester.

About the same time this compound was also discovered by Ivanov [1905; 1907; 1909], who attributed to it the constitution of a triosemonophosphoric ester, and at a somewhat later date was examined by Lebedev [1909; 1910] who at first supposed it to be a hexosemonophosphoric ester but afterwards [1911] came to the same conclusion as Young on this point.

During the course of some experiments carried out in 1913 by the author in conjunction with Professor A. Harden indications were noted that hexose-diphosphoric acid is not the only compound of this type produced during the fermentation of sugar by yeast juice in presence of phosphate. It was observed that during the preparation of hexosediphosphoric acid from fructose and after the precipitation of its lead salt by the addition of lead acetate, the filtrate always contained very appreciable amounts of phosphorus in organic combination, and that a further precipitate containing most of this phosphorus was thrown down on the addition of basic lead acetate. On examining the acid solutions obtained by decomposing such basic lead precipitates with hydrogen sulphide it was found that the ratio of the reducing power, as determined by Bertrand's method, to the phosphorus content was much higher than that given by solutions of hexosediphosphoric acid, while polarimetric observations gave evidence of the presence of a substance much more strongly dextrorotatory than the latter.

The barium salt of this new acid was prepared, and proved to be readily soluble in both hot and cold water, differing in this respect markedly from barium hexosediphosphate. The salt was amorphous and was obviously contaminated with substances derived either from the yeast juice or formed during the fermentation of the fructose, but the results of analyses appeared to indicate that the compound was the barium salt of a hexosemonophosphoric acid, $C_6H_{11}O_5PO_4Ba$. These facts were published in a preliminary note [Harden and Robison, 1914] and the investigation of the compound was continued. Its purification, however, proved unexpectedly difficult and was not completed when the work was interrupted by the war. It was resumed in 1919 but was again held up by the difficulty experienced in obtaining yeast juice of reasonable fermentative power. The variability in the activity of juice prepared from the same type of yeast and apparently under similar conditions has been frequently observed, but no explanation could be found for the long run of inactive or feebly active juices obtained during 1919 unless it were the disheartening effect of the war beer on the yeast. That difficulty no longer exists and improved methods for the isolation and purification of the new compound have been worked out. Sufficient evidence has now been obtained to confirm the opinion already stated as to its constitution. It is, however, not identical with the hexosemonophosphoric acid prepared by Neuberg [1918], by partial hydrolysis from hexosediphosphoric acid. A specimen of barium hexosemonophosphate prepared according to the method described by Neuberg was found to be optically almost inactive but to yield a strongly laevorotatory product on further hydrolysis. The compound iso-

:lated from the products of fermentation is, on the other hand, strongly dextrorotatory, and on hydrolysis yields a dextrorotatory reducing substance from which glucosazone has been obtained. The ratio of the rotatory power of this hydrolysis product to its reducing power as estimated by Bertrand's method is, however, much lower than that given by solutions of pure glucose. In this and in other respects the behaviour of the ester suggests that it may possibly be a mixture of isomeric hexosemonophosphoric acids, while the substance described by Neuberg may represent one or more other isomers. The isomerism may be due to the hexose molecule itself or to the position of the phosphoric acid radicle.

So far, only one salt of hexosemonophosphoric acid, that of brucine, has been obtained in a crystalline condition and the hope that this salt might be used to effect the purification of the acid, and its possible resolution into two or more isomers was for long disappointed owing to the difficulty with which crystallisation was effected. Within the last few weeks, however, more success has been obtained and it is hoped that the question may yet be solved in this way.

With phenylhydrazine, hexosemonophosphoric acid yields the phenylhydrazine salt of an osazone in which the phosphoric acid radicle is retained. This compound is not identical with the osazone of the same empirical formula prepared by Young, Ivanov and Lebedev from hexosediphosphoric acid.

Hexosemonophosphoric acid is only very slowly hydrolysed by boiling with mineral acids and a considerable portion of the sugar produced is destroyed during the operation. It is also hydrolysed by emulsin, yielding free phosphoric acid and a dextrorotatory reducing substance. Emulsin likewise hydrolyses hexosediphosphoric acid but, as with acid hydrolysis, the reducing substance produced is strongly laevorotatory. It is therefore improbable that hexosemonophosphoric acid is formed during the fermentation by partial hydrolysis—enzymic or otherwise—of the diphosphoric ester.

The behaviour of hexosemonophosphate towards yeast, yeast juice and zymin is being studied in conjunction with Dr Harden and is yielding interesting results that are, however, difficult to interpret. The alkali salts are readily fermented by yeast juice and by zymin and the initial rate is relatively high, even approximating to the "phosphate rate" for glucose, but rapidly falls to a lower level, which nevertheless is decidedly higher than that for hexosediphosphate.

The same compound is apparently formed during the fermentation of either fructose or glucose in presence of phosphate. It is true that some differences were observed in the specific rotations of the various preparations, but such differences were not greater between preparations made from glucose and those from fructose than between different preparations made from the same sugar. This may perhaps be taken as further evidence that all specimens were mixtures of isomeric compounds in somewhat varying proportions.

Owing to the many processes employed in the different attempts to obtain the ester in a pure condition, and the fact that even the best of these methods involves some loss of the substance, the record of yields obtained does not at present give very trustworthy evidence as to the most favourable conditions for the production of the monophosphoric ester or for determining the relative amounts of this and the diphosphoric ester that are formed during fermentation.

Calculations based on the values of the ratio

$$\frac{\text{Reducing power (as glucose)}}{\text{Phosphorus}}$$

as determined for the acid solutions derived from both normal and basic lead precipitates would appear to indicate the presence of a much higher proportion of the monophosphoric ester than has ever been actually isolated. It is just possible that in spite of the thorough washing to which they were subjected, these precipitates still held some adsorbed sugar which would affect the above ratio. The possibility of other reducing substances being present must also be considered.

The best yields of the hexosemonophosphoric acid actually obtained from fructose and from glucose correspond with only 15 % and 7 % respectively of the sodium phosphate added during the fermentation. These figures are based on the weight of barium salt of at least 90 % purity; the total amounts formed would certainly be higher and might well correspond with double these percentages.

The question as to what rôle, if any, is played by hexosemonophosphoric acid in alcoholic fermentation must be left unanswered until more information is gained on the various points indicated above.

EXPERIMENTAL.

A typical series of values of the ratios

$$\frac{\text{Reducing power (as glucose)}}{\text{Phosphorus}} \quad \text{and} \quad \frac{\text{Rotation in 4 dm. tube}}{\text{Reducing power (as glucose) in 100 cc.}}$$

as determined for the acid solutions derived from normal and basic lead precipitates, is given below. They are taken from one of the first experiments, which led to the discovery of the monophosphoric ester, and are quoted because of their bearing on the relative amounts of the two esters formed during the fermentation.

The reducing power was estimated by Bertrand's method and calculated as glucose; the phosphorus was estimated by Neumann's method.

Fermentation: 1220 cc. juice from 4 kilos of fresh pressed brewery yeast was allowed to ferment 122 g. fructose, 131 g. (0·4 Mol.) of $Na_2HPO_412H_2O$ being added gradually, as indicated by the rate of evolution of CO_2, during $1\frac{1}{2}$ hours. Fermentation was stopped and protein coagulated by blowing steam through the solution. After neutralisation of the filtered solution with caustic soda and removal of free phosphate by precipitation with magnesium acetate, a slight excess of lead acetate was added to the filtrate. The precipitate was removed by centrifuging and washed nine times with water. The filtrate and washings were treated with basic lead acetate and the precipitate again separated by centrifuging and thoroughly washed. These precipitates were decomposed by hydrogen sulphide and the excess of the latter removed by means of a current of air. The acid solutions were each neutralised to phenolphthalein by the addition of caustic soda and were again treated successively with normal and with basic lead acetate, any traces of free phosphate being first removed. These operations were repeated as shewn below (Table I).

The acid solutions derived from the various lead precipitates gave the following ratios:

Table I.

No. of sol.	Description of precipitate from which solution was derived	Reducing power as glucose P	Rotation in 4 dm. tube / Reducing power as glucose in 100 cc.	$[a]_D$ of acid calculated on P-content
1	First precipitate by normal lead acetate (crude lead hexosediphosphate)	2·50		
2	Solution 1 precipitated a second time by normal lead acetate	2·25		
3	Solution 2 precipitated a third time by normal lead acetate. Precipitations carried out in two fractions: 1st fraction	1·50		
4 a	Solution 3 precipitated a fourth and fifth time with normal lead acetate, the fifth time in two fractions: 1st fraction	1·21	0·647	Calculated for $C_6H_{10}O_4(PO_4H_2)_2$ $+3·56°$
4 b	2nd ,, ...	1·67	—	—
5	Solution 4 a precipitated a sixth time with normal lead acetate (pure lead hexosediphosphate)	1·16	0·623	$+3·26°*$
6	First precipitate by basic lead acetate after removal of first normal lead precipitate	5·50		
7 a	Solution 6 precipitated a second time with normal then with basic lead acetate: Normal lead precipitate ...	—		
7 b	Basic ,, ,, ...	4·24		—
8 a	Solution 7 b precipitated a third time with normal then with basic lead acetate: Normal lead precipitate ...	3·96	1·60	Calculated for $C_6H_{11}O_5(PO_4H_2)$ $+19·0°$
8 b	Basic ,, ,,	4·36	2·40	$+31·2°$
9 a	Solution 8 b precipitated a fourth time with normal then with basic lead acetate: Normal lead precipitate ...	4·13	2·18	$+25·8°†$
9 b	Basic lead precipitate Amount too small for determination of ratios	—	—	

* $[a]_D$ of hexosediphosphoric acid $= +3·2°$ (Young).

† $[a]_D$ of hexosemonophosphoric acid $= +25·0°$ (p. 817).

Solutions 9 a and 9 b were each treated with baryta until neutral to phenolphthalein and the barium salts precipitated by the addition of an equal volume of alcohol. They were purified by redissolving in water and reprecipitating with alcohol and were then dried at 100° to constant weight and analysed:

Barium salt from solution 9 a. Ba = 34·33 %; P = 7·07 %.

„ „ „ „ 9 b. Ba = 34·86 %; P = 7·22 %.

Calculated for $C_6H_{11}O_5PO_4Ba$. Ba = 34·75 %; P = 7·85 %.

It seemed probable therefore that the chief constituent of the basic lead precipitates was the salt of a hexosemonophosphoric acid, but that this was contaminated with salts of other acids containing less phosphorus or none at all. Whether the differences in the ratios for solutions Nos. 6 to 9 are due entirely to phosphorus-free compounds, or whether they indicate the presence of yet another phosphoric ester, e.g. of a disaccharide, is impossible at present to say.

In the case of the solutions derived from the normal lead precipitates a steady decrease in the ratio Reducing power/P occurs as the purification proceeds. If the final ratio of 1·16[1] be accepted as correct for pure hexosediphosphoric acid, then it is obvious that solution 1 must have contained a considerable proportion of some other reducing substance. The presence of sugar would seem to be precluded by the thorough washing of the precipitate. That the precipitate contained some hexosemonophosphate is certain since, although this ester gives in concentrated neutral solutions no precipitate with normal lead acetate, a copious precipitate, probably of basic lead salt, is formed in dilute solutions.

If solution No. 1 contained no reducing substance or phosphorus compound other than the two hexosephosphoric acids, then the ratio 2·50 would indicate that nearly 50 % of the phosphorus was present in the form of the mono-ester, in addition to the amount remaining in the filtrate and precipitated by basic lead acetate. As already stated, the yields of hexosemonophosphoric acid actually isolated in the form of its barium salt have never reached this proportion and have usually been much lower.

The purification of the ester proved a problem of considerable difficulty and much time was spent in solving it. Attempts to prepare salts in a crystalline condition were unsuccessful except in the case of the brucine salt, and fractional precipitation of the metallic salts by alcohol effected only slight improvement in the analytical figures. Fractional precipitation of the syrupy acid from its alcoholic solution by ether, was also of little use. Repeated extraction of the aqueous solution of the acid with ether or ethyl acetate did remove a small proportion of crystalline substance which proved to be succinic acid, but the phosphorus content of the barium salt was thereby only slightly raised. Further fractional precipitation by lead acetate, normal and basic, gave a purer product but involved too rapid diminution in the amount of

[1] The values for this ratio calculated from Young's data lie between 2·1 and 2·3. They are, however, not comparable with those shown above as the reducing power was determined by a different method(Pavy).

substance to be practicable. The greatest improvement was effected by the use of mercuric acetate which precipitates a considerable proportion of the impurities and by combination of this with other processes, a product giving satisfactory analytical figures was at length obtained.

Preparation and purification of the hexosephosphoric esters.

The methods employed in the preparation of the esters have been considerably modified as experience suggested. Heating the fermentation mixture to destroy the enzymes and coagulate the protein was discontinued on account of the risk of causing partial hydrolysis of the diphosphoric acid. Preliminary separation of the two esters by means of their lead salts was given up owing to the large proportion of the monophosphoric acid brought down in such dilute solutions by lead acetate. The method now used is as follows:

Juice prepared from fresh pressed English mild ale yeast is mixed with 10 % of its weight of fructose or glucose and warmed to 26° in a water-bath. Fermentation having commenced a 20 % aqueous solution of $Na_2HPO_4 12H_2O$ is added from time to time in such quantity as to produce the maximum rate of evolution of carbon dioxide. The addition of phosphate solution is best regulated by means of a guide experiment carried out with 25 cc. juice in the apparatus described by Harden, Thompson and Young [1910] (v. "Alcoholic Fermentation" by A. Harden, 1923 edition, p. 28). When the total volume of phosphate solution added exceeds half the original volume of juice a further quantity of sugar should be dissolved in the fermentation mixture. When the further addition of phosphate no longer causes any considerable rise in the rate of evolution of carbon dioxide, solid barium acetate in amount equal to the total weight of crystalline sodium phosphate that has been added, is dissolved in the reaction mixture which is then rendered just alkaline to phenolphthalein with baryta, and treated with an equal volume of alcohol. The barium salts of the two hexosephosphoric acids and of any excess of free phosphoric acid are thus precipitated together with the protein of the yeast juice. The precipitate is filtered off on a large Buchner funnel, and thoroughly washed with 70 % alcohol. It is then treated with boiling absolute alcohol and allowed to remain in contact with the alcohol over night. In this way the protein present is denatured and rendered insoluble. The crude barium salts are then dried in an evacuated desiccator over sulphuric acid or more quickly by exposure to a current of warm air. They are next thoroughly ground up with 10 parts of cold water which dissolves the barium hexose-monophosphate, but scarcely any of the hexosediphosphate. The residue is washed twice with small quantities of water (being preferably removed from the Buchner funnel and again ground up) and is then extracted with 200 parts of water which dissolves the barium hexosediphosphate, leaving behind the insoluble barium phosphate etc. After removal of any traces of mineral phosphate by precipitation with magnesium acetate, the hexosediphosphoric acid is precipitated from the filtered solution in the form of its lead salt by the addition of lead acetate. The precipitate is filtered or centrifuged off and washed. It is then suspended in water, decomposed by a current of sulphuretted hydrogen, the clear filtrate is freed from sulphuretted hydrogen by a current of air and finally neutralised to phenolphthalein with caustic soda. The precipitation of the lead salt is repeated several times [v. Young, 1909].

The first aqueous extract of the crude barium salts contains the barium hexosemonophosphate together with a little hexosediphosphate and other barium salts of acids derived from the yeast juice or formed during the fermentation. It is treated with basic lead acetate and the insoluble basic lead salt is filtered and washed. It is then decomposed with sulphuretted hydrogen as described above and the acid solution neutralised to phenolphthalein with a hot saturated solution of baryta. The solution is filtered and the precipitation with basic lead acetate repeated, after which the barium salt is again formed and is precipitated from its solution by alcohol, filtered off, washed with absolute alcohol and dried in an evacuated desiccator over sulphuric acid. It is dissolved in 10 parts of water to which 10 % of alcohol is finally added. The filtered solution is treated with mercuric acetate so long as this causes any precipitate to form, and is allowed to stand for some hours before filtering. The basic lead salt is then reprecipitated from the clear filtrate and once more converted into the barium salt, which is finally purified by repeatedly dissolving it in 10 % alcohol, filtering and reprecipitating with an equal volume of alcohol. In presence of a small proportion of water the salt readily forms a syrup, and it should therefore be washed several times with absolute alcohol and rapidly dried in an evacuated desiccator. If the results of analysis are not satisfactory the purification by mercuric acetate etc. is repeated. The mercury precipitate and a sparingly soluble barium salt, which is obtained in very small amounts during the final stage of the purification, both contain phosphorus and are being submitted to further investigation.

Hexosemonophosphoric acid and its salts.

Barium hexosemonophosphate, prepared by the above method, is a colourless, amorphous substance very readily soluble in hot and cold water, but insoluble in organic solvents. Dried over sulphuric acid it still retains about 4·5 % moisture which is given up at 100°, the salt at the same time taking on a brownish tint. The percentage of water corresponds approximately with one molecule (calculated for $C_6H_{11}O_5PO_4BaH_2O$, $H_2O = 4·36$ %) but is not constant. The anhydrous salt takes up about the same amount of moisture with great avidity at ordinary temperatures even when kept in a closed vessel over calcium chloride.

For analysis the salt was dried to constant weight in an evacuated vessel (2 mm. pressure) at 78° over phosphorus pentoxide, only slight discoloration taking place under these conditions. Table II gives the results of analyses of specimens prepared from glucose and fructose, together with the specific rotation of the barium salt and of the acid prepared from it by the addition of the exact amount of sulphuric acid.

The results obtained with a specimen of barium hexosemonophosphate prepared by hydrolysis of hexosediphosphoric acid (Neuberg's method) are also shown in the same table for purposes of comparison.

Table II.

No.	Sugar from which salt was prepared	Ba	P	C	H	$[\alpha]_D^{20}$ of salt	$[\alpha]_D^{20}$ of acid
1	Fructose	35·35	7·61	—	—	—	—
2	,,	34·89	7·86	—	—	$+12\cdot5°\ (c=3\cdot9\ \%)$	$+24\cdot9°\ (c=2\cdot1\ \%)$
3	Glucose	35·12	7·60	—	—	—	—
4	,,	35·05	7·66	—	—	$+14\cdot4°\ (c=14\ \%)$	—
5		34·82	7·45	—	—	$+12\cdot7°\ (c=17\ \%)$	—
6		34·83	7·73	—	—	$+15\cdot6°\ (c=11\cdot2\ \%)$	$+28\cdot8°\ (c=2\cdot7\ \%)$
7	,,	{34·87	7·82	18·37	2·93}	$+12\cdot2°\ (c=3\cdot7\ \%)$	$+25\cdot0°\ (c=3\ \%)$
		{35·09	7·85	18·16	2·99}		
	Calculated for $C_6H_{11}O_5PO_4Ba$	34·75	7·85	18·20	2·81	—	—
8	Hexosemonophosphoric acid obtained by partial hydrolysis of hexosediphosphoric acid (Neuberg's method)	33·26	7·58	—	—	$+0\cdot38°\ (c=5\cdot2\ \%)$	$+1\cdot49°\ (c=3\cdot3\ \%)$
	Calculated for $C_6H_{11}O_5PO_4BaH_2O$	33·22	7·51	—	—	—	—

Molecular weight determination by the cryoscopic method yielded the following results:

1. 0·0849 barium salt in 13·17 g. water gave $\Delta = 0·062°$, whence apparent Mol. weight = 198.

2. 0·3037 barium salt in 13·17 g. water gave $\Delta = 0·192°$, whence apparent Mol. weight = 228.

$C_6H_{11}O_5PO_4Ba$ requires Mol. weight = 395·5; if 100 % dissociated, apparent Mol. weight = 197·8.

$C_{12}H_{20}O_9(PO_4Ba)_2H_2O$ requires Mol. weight = 791; if 100 % dissociated, apparent Mol. weight = 263·7.

The experimental results are therefore consistent with the first formula but not with the second.

The brucine salt was prepared by dissolving the calculated amount of the base in an aqueous solution of hexosemonophosphoric acid prepared from the barium salt. On evaporation of the solution at a low temperature a crystalline residue was obtained. This was very soluble in water, moderately soluble in warm methyl alcohol but sparingly soluble in ethyl alcohol. Recrystallisation was not readily effected but on treating the concentrated aqueous solution with acetone and cooling to 0°, a semi-crystalline precipitate was obtained. After twice repeating this operation the salt was dried to constant weight at 78° and 2 mm. pressure over phosphorus pentoxide, no discoloration taking place.

Phosphorus estimated by Neumann's method: found P = 2·73 %.
Calculated for $C_6H_{11}O_5PO_4H_2(C_{23}H_{26}N_2O_4)_2$; P = 2·96 %.
Specific rotation of salt in water (conc. = 6·6 %); $[\alpha]_D^{19} = -23\cdot4°$.

The investigation of the brucine salt is being continued.

The normal lead and copper salts are also soluble in concentrated solutions but on diluting these, precipitates, presumably of the basic salts, are formed. Hexosemonophosphoric acid itself was obtained as a colourless, very viscous liquid by evaporating its aqueous solution *in vacuo*. It is moderately soluble in alcohol but very sparingly soluble in ether or acetone. Whether prepared from fructose or glucose it reduces Fehling's solution and gives a red coloration when heated with resorcinol and hydrochloric acid (Selivanov's reaction), though the colour is not so intense as that given by the equivalent amount of fructose. With water it yields a strongly acid solution which can be titrated sharply with alkalis (two equivalents) using phenolphthalein as indicator. All attempts to obtain the acid in crystalline condition have failed.

Osazone of hexosemonophosphoric acid.

Attempts to prepare a phenylhydrazone of hexosemonophosphoric acid were not successful but on warming the solution with excess of the base, a crystalline derivative was obtained. 10 g. barium hexosemonophosphate, prepared from fructose, were treated with the exact amount of sulphuric acid and the barium sulphate removed by filtration. The clear solution was mixed with 15 g. phenylhydrazine dissolved in acetic acid, and was heated on the water-bath during 30 minutes. It was then cooled to 0° and the yellow precipitate filtered off, washed with chloroform, and dried.

Weight of product = 8 g. M. pt. 135°–137°.

Recrystallisation was effected by dissolving the compound in boiling ethyl-alcohol, adding an equal volume of hot chloroform and cooling to 0°. Short, pale yellow needles were formed. M. pt. 139° with decomposition. Moderately soluble in methyl alcohol, less soluble in ethyl alcohol, still less in chloroform and insoluble in light petroleum. Very sparingly soluble in water but readily soluble in dilute sodium hydroxide.

Analysis of substance dried at 66° *in vacuo* over phosphorus pentoxide: N = 15·33 %; P = 5·60 %.

Calculated for $C_6H_5NHNH_2H_2PO_4 . C_4H_5(OH)_3C(N_2HC_6H_5)CHN_2HC_6H_5$; N = 15·38 %; P = 5·68 %.

The compound therefore appeared to be the phenylhydrazine salt of the osazone of hexosemonophosphoric acid. It is, however, not identical with the compound of the same empirical formula obtained by Young from hexose-diphosphoric acid, with elimination of one molecule of phosphoric acid, since the latter melted at 151°.

The chloroform extract from the crude osazone yielded, on evaporation, a small quantity of a crystalline compound readily soluble in ethyl alcohol, which, by recrystallisation from ethyl alcohol and chloroform, was obtained in yellow plates, melting with decomposition at 190°. The substance contained phosphorus and was readily soluble in dilute sodium hydroxide, though very sparingly soluble in water. Unfortunately the quantity was insufficient for analysis.

Hydrolysis of hexosemonophosphoric acid by acids.

Hexosemonophosphoric acid is very resistant to hydrolysis by acids at ordinary temperatures.

A solution of the acid in N sulphuric acid, after being kept for four months at room temperature contained only a trace of free phosphoric acid. When the solution of the acid is boiled either alone or with the addition of mineral acid, free phosphoric acid is slowly liberated and a reducing substance formed. Table III gives the data of an experiment in which 5·4 g. of hexosemonophosphoric acid prepared from glucose, was heated with 250 cc. of $N/5$ sulphuric acid at 97°, the rotation and free phosphoric acid being estimated from time to time. The total phosphorus = 0·643 g.

Table III.

Time of heating, hours	Rotation in 1 dm. tube	P present as free PO_4, g.	P present as free PO_4 in percentage of total P
0	+0·625°	0	0
1	+0·568°	—	—
3	+0·544°	—	—
16½	+0·507°	0·170	26·4
36	+0·427°	0·274	42·6
80	+0·327°	0·374	58·2

After 80 hours the solution had become dark brown in colour and the rotation was difficult to measure. The hydrolysis was accordingly stopped and the solution neutralised with baryta. The precipitate, which consisted entirely of barium phosphate and sulphate, was filtered off and the clear filtrate treated with an equal volume of alcohol, which precipitated the barium salt of the unchanged hexosemonophosphoric acid. The phosphorus content and specific rotation of this salt agreed with those of the original barium hexosemonophosphate. The alcoholic filtrate reduced Fehling's solution and contained only a trace of phosphorus. After distilling off the alcohol the reducing power of the solution was estimated by Bertrand's method and its rotation measured.

Reducing power calculated as glucose = 0·083 g. per 100 cc. equivalent to 0·98 for total original solution.

Rotation in 4 dm. tube = + 0·097°.

$$\frac{\text{Rotation in 4 dm. tube}}{\text{Reducing power as glucose in 100 cc.}} = 1\cdot17.\ \text{Calculated for glucose} = 2\cdot11.$$

The amount of sugar formed during the hydrolysis, calculated from the amount of phosphoric acid liberated is equal to 2·17 g. or more than twice the quantity actually estimated in the solution. It would seem therefore that a considerable proportion of the sugar is destroyed during the boiling. This solution yielded an osazone which, after being twice recrystallised from dilute pyridine, melted at 205°–206° and did not lower the melting point of glucosazone when intimately mixed with the latter.

In another experiment, Table IV, a solution of hexosemonophosphoric acid prepared from laevulose was boiled under a reflux condenser, the reducing power and the rotation being measured from time to time.

Table IV.

Time in hours	Rotation in 4 dm. tube	Reducing power as glucose in 100 cc.	$\dfrac{\text{Rotation}}{\text{Reducing power}}$
0	+1·339°	0·59	2·27
2	+1·064°	0·64	1·66
4	+1·024°	0·64	1·60
10	+0·896°	0·64	1·40
16	+0·788°	0·65	1·21

At the end of 16 hours' boiling the free phosphate amounted to 0·0795 g. P per 100 cc. or 53 % of the total phosphorus (0·1499 g. per 100 cc.). After precipitating the unchanged ester by basic lead acetate the filtrate was freed from lead by sulphuretted hydrogen and the reducing power etc. determined. The value found for the ratio $\dfrac{\text{Rotation in 4 dm. tube}}{\text{Reducing power as glucose in 100 cc.}}$ was equal to 0·90.

Hydrolysis of hexosemonophosphoric acid by emulsin.

Hexosemonophosphoric acid is hydrolysed by emulsin yielding also in this case, free phosphoric acid and a dextrorotatory substance from which glucosazone can be obtained.

A solution of 1·876 g. anhydrous barium hexosemonophosphate in 24 cc. water was treated with 2 cc. of an extract of emulsin[1] prepared from sweet almonds, and was incubated at 37° in presence of toluene. At intervals the rotation was measured, after filtering off the precipitated barium phosphate. Table V gives the results of these determinations. On the 10th day a further 1 cc. emulsin extract was added but the rotation has been calculated for the original volume of the solution.

Table V.

Time of incubation, days	Rotation in 2 dm. tube at 20°	Percentage of salt hydrolysed as measured by free PO_4
0	+1·680°	—
1	+1·579°	
4	+1·376°	—
10	+1·334°	—
13	+1·339°	66

On the 13th day the remaining solution was neutralised to phenolphthalein with baryta and the total free phosphate estimated. The clear filtrate was treated with 1½ volumes of alcohol and the precipitate filtered off, dried at 78° *in vacuo*, and examined.

Weight = 0·370 g. equivalent to 0·53 g. for the total original solution.
Found Ba = 32·8 %; P = 6·87 %; $[\alpha]_D^{20} = +11\cdot3$ %.

[1] This extract contained a slight trace of reducing substance.

The substance probably consisted chiefly of unchanged barium hexose-monophosphate. The filtrate and washings from this precipitate were made up to 100 cc. They contained a reducing substance equivalent to 0·240 g. glucose as estimated by Bertrand's method. This amount is equivalent to 0·347 g. for the original solution.

Rotation in 4 dm. tube = + 0·350°.

$$\frac{\text{Rotation in 4 dm. tube}}{\text{Reducing power as glucose in 100 cc.}} = 1\cdot46.$$

On evaporation this solution yielded a syrup which could not be made to crystallise but with phenylhydrazine yielded an osazone. After recrystallisation from dilute pyridine this melted at 204°–205°, and did not lower the melting point of glucosazone when intimately mixed with the latter.

Hydrolysis of barium hexosediphosphate by emulsin.

The hydrolytic action of emulsin on hexosediphosphate was established by Harding [1912] who estimated the phosphoric acid set free but did not further examine the products.

In view of the possibility that the optically active substances produced under these conditions might differ from those produced by acid hydrolysis it was thought advisable to investigate this point.

1 g. of barium hexosediphosphate was incubated with 1 cc. emulsin extract and 10 cc. water at 37°, in presence of toluene. The salt only dissolved to a slight extent in the water. After 14 days the solution was filtered from the solid matter, which consisted of unchanged barium hexosediphosphate and barium phosphate, and examined polarimetrically.

$$\alpha^{18} \text{ in 1 dm. tube} = -0\cdot530°.$$

Estimation of free and total phosphate in both precipitate and solution indicated that about 60 % of the phosphoric acid had been liberated. On neutralising the solution with baryta and adding alcohol a small amount of soluble barium salt was precipitated while the solution contained the laevo-rotatory substance.

It appears therefore that the hydrolysis of hexosediphosphoric acid by enzymes results in products similar to those obtained by acid hydrolysis. Further experiments are, however, being carried out.

The hexosemonophosphoric acid produced by partial hydrolysis of hexosediphosphoric acid.

A quantity of this ester was prepared by Neuberg's [1918] method in order to compare it with the compound obtained by the fermentation of sugar.

9·3 g. of crude barium salt was thus obtained from 19 g. barium hexose-diphosphate by boiling the latter with a solution of 9·5 g. oxalic acid in 150 cc. water. After purification by repeatedly dissolving it in 10 % alcohol and precipitating the clear solution with alcohol, the product (5·3 g.) was dried at 78° in vacuo over phosphorus pentoxide and analysed.

Found Ba = 33·26 %; P = 7·58 %; $[a]_D^{19}$ in 5·2 % aqueous solution

$$= + 0·38°.$$

Calculated for $C_6H_{11}O_5PO_4BaH_2O$; Ba = 33·22 %; P = 7·51 %.

According to Neuberg the barium salt holds one molecule of water which is not given up at 105° and the above analyses are in agreement with this view.

The specific rotation of the acid, prepared from the solution of the barium salt by adding the calculated amount of oxalic acid was also determined:

$$[a]_D^{20} \text{ in } 3·3 \text{ % aqueous solution} = + 1·49°.$$

With such small polarimetric readings, however, the possible error in the above values is relatively large. The solution of the acid was boiled with such excess of oxalic acid as brought the total acidity equal to N.

The solution rapidly became laevorotatory, as is shown in Table VI.

Table VI.

Time of boiling (hours)	Rotation in 2 dm. tube	Percentage of salt hydrolysed as measured by free phosphate
0	+0·094	0
½	−0·176	—
1½	−0·481	—
4½	−0·822	68

The behaviour of fermentation hexosemonophosphoric acid on hydrolysis is thus very different from that of hexosediphosphoric acid, and of the monophosphoric acid obtained from it by partial hydrolysis, both of which yield laevorotatory products. Table VII gives the results of experiments with each of the above compounds for comparison, the figures for the diphosphoric ester (Nos. 7–9) being those of Young [1909].

Table VII.

Substance	Method of hydrolysis	Percentage of phosphate liberated	Rotation in 4 dm. tube Before hydrolysis	After hydrolysis
1. Hexosemonophosphoric acid from glucose	Solution of acid in $N/5$ H_2SO_4 at 97° 80 hours	58·2	+2·50°	+1·31°
2. Do. Do.	Solution of acid boiled 27 hours	73	+0·374°	+0·234°
3. Do. from fructose	,, ,, 16 ,,	53	+1·339°	+0·788°
4. Do. Do.	,, ,, 10½ ,,	42	+0·816°	+0·532°
5. Barium hexosemonophosphate from glucose	Emulsin; 13 days at 37°	66	+1·680°*	+1·339°*
6. Hexosemonophosphoric acid from hexosediphosphoric acid by partial hydrolysis (Neuberg's method)	Solution of acid in N oxalic acid boiled 4½ hours	68	+0·094°	−0·822°
7. Hexosediphosphoric acid from glucose (Young)	Solution of acid boiled 27 hours	91		
8. Do. Do.	,, ,, 14 ,,	—	+0·161°	−0·658°
9. Do. from fructose (Young)	,, ,, 10 ,,	—	+0·416°	−1·514°
10. Barium hexosediphosphate	Emulsin; 14 days at 37°	60	(+0·30† calc.) (salt not in solution)	−0°·530†

* 2 dm. tube. † 1 dm. tube.

It would seem improbable therefore that the hexosemonophosphoric acid found in the products of fermentation is formed from hexosediphosphoric acid by hydrolysis, enzymic or otherwise. It may conceivably be an intermediate stage in the formation of the diphosphoric ester and indeed its behaviour with yeast juice and zymin lends some weight to such a view. On the other hand it may be formed by the synthetic power of the yeast enzymes but have no part in the main cycle of reactions which constitutes alcoholic fermentation.

The analytical data so far obtained are all in agreement with the formula assigned to the compound, that of a hexosemonophosphoric ester, but it is admitted that further criteria of purity and homogeneity are desirable before this formula can be considered as established. The formation of an osazone retaining the phosphoric acid radicle indicates that the latter is not attached either to the terminal aldehydic carbon atom or that adjacent to it, but its position in the molecule cannot be definitely stated at present. The nature of the hexose is also undecided. The reducing substance produced on hydrolysis is dextrorotatory but less so than glucose. The ester itself whether produced from glucose or fructose gives Selivanov's reaction for fructose but not so strongly as an equivalent quantity of the latter sugar. These facts and the formation of glucosazone from the products of hydrolysis may perhaps indicate that the compound is a mixture of the monophosphoric esters of glucose and fructose. More cannot be said until further work has been done.

SUMMARY.

1. When fructose or glucose is fermented by yeast juice in presence of suitable amounts of a soluble phosphate, an ester having the composition of a hexosemonophosphoric acid is formed in addition to the hexosediphosphoric acid described by Young, Ivanov and Lebedev.

2. A method is described for the separation of this new phosphoric ester from the other products of the fermentation and for the purification of its barium salt.

3. Hexosemonophosphoric acid is strongly dextrorotatory; $[\alpha]_D^{20} = + 25 \cdot 0°$ in aqueous solution.

The metallic salts with the exception of the basic salts of the heavy metals, are all readily soluble in water and are amorphous. A crystalline brucine salt has been obtained.

4. The phenylhydrazine salt of the osazone of hexosemonophosphoric acid has been obtained. It is not identical with the compound of the same empirical formula prepared by Young and Lebedev from hexosediphosphoric acid.

5. On hydrolysis by acids or by emulsin, hexosemonophosphoric acid yields free phosphoric acid and a dextrorotatory reducing substance from which glucosazone has been obtained. The rotatory power of this product is, however, less than that of pure glucose.

6. By its specific rotation and its behaviour on hydrolysis the compound is sharply distinguished from the hexosemonophosphoric acid prepared by Nenberg by partial hydrolysis of hexosediphosphoric acid.

7. The alkali salts of hexosemonophosphoric acid are readily fermented by yeast juice and zymin.

8. It is improbable that hexosemonophosphoric acid is produced by the hydrolysis of the diphosphoric ester during fermentation. It may perhaps be an intermediate stage in the formation of the latter compound.

In conclusion I wish to thank Professor A. Harden for the interest which he has taken in this investigation.

REFERENCES.

Buchner, E. and H. and Hahn (1903). Die Zymasegärung, München.
Harden and Robison (1914). Proc. Chem. Soc. 30, 16.
Harden, Thompson and Young (1910). Biochem. J. 5, 230.
Harden and Young (1905). Proc. Chem. Soc. 21, 189.
—— —— (1906, 1). Proc. Roy. Soc. B. 77, 405.
—— —— (1906, 2). Proc. Roy. Soc. B. 78, 369.
—— —— (1908, 1). Proc. Roy. Soc. B. 80, 299.
—— —— (1908, 2). Proc. Chem. Soc. 24, 115.
—— —— (1909). Proc. Roy. Soc. B. 81, 336.
—— —— (1910, 1). Centr. Bakt. Par. Abt. II, 26, 178.
—— —— (1910, 2). Proc. Roy. Soc. B. 82, 321.
—— —— (1911). Biochem. Zeitsch. 32, 173.
Harding (1912). Proc. Roy. Soc. B. 85, 418.
Ivanov (1905). S. Travaux de la Soc. des Naturalistes de St Petersburg, 34.
—— (1907). Zeitsch. physiol. Chem. 50, 281.
—— (1909). Centr. Bakt. Par. Abt. II, 24, 429.
Lebedev (1909). Biochem. Zeitsch. 20, 114.
—— (1910). Biochem. Zeitsch. 28, 213.
—— (1911). Biochem. Zeitsch. 36, 248.
Neuberg (1918). Biochem. Zeitsch. 88, 432.
Wroblewski (1901). J. pr. Chem. (2), 64, 1.
Young (1907). Proc. Chem. Soc. 23, 65.
—— (1909). Proc. Roy. Soc. B. 81, 528.
—— (1911). Biochem. Zeitsch. 32, 178.

LXXXVI. MAMMARY SECRETION. IV.

THE RELATION OF PROTEIN TO OTHER DIETARY CONSTITUENTS.

By GLADYS ANNIE HARTWELL.

From the Physiological Laboratory, Household and Social Science Department, King's College for Women, Kensington, London.

(*Received October 24th, 1922.*)

I. INTRODUCTORY AND HISTORICAL.

IT is generally held that the diet of a lactating animal should contain a large amount of protein, and it seems quite obvious that a maximal growth of the young cannot be obtained if the mother's protein intake is low. On the other hand it has been demonstrated [Hartwell, 1921, 2] that large amounts of protein fed to the mother, result in harmful effects to the young. It has, however, been found [Hartwell, 1922] that large quantities of whole milk, and large amounts of yeast extract added to the mother's excess protein diet, prevent the bad symptoms in the litters. The obvious, though possibly not the correct explanation of such results, was that the important factor was vitamin B, since this is present in both milk and yeast. Also it was shown [Hartwell, 1922] that lactose, butter, milk-ash and calcium lactate were ineffective in improving the bad condition. Consequently the milk effect was not due to any obvious dietary factor.

The experiments described in this paper were started in the hope of throwing some light on the relation of vitamin B to protein in the metabolism of the lactating animal.

That there is a possible relation between vitamin B and protein is to be seen from the work of Karr [1920]. He found that the dog refused food unless vitamin B was included in the diet, and since this animal is a carnivore, it may be that the vitamin is in some way essential for protein metabolism.

In the experiments to be detailed here the mother rats were fed on an excess protein diet, to which were added various fruit and vegetable juices and extracts made from other foods said to contain vitamin B. From the results described, it appears that the addition to the mother's protein-rich diet of juices of, or extracts from, foods reported to contain vitamin B, is effective in rendering the nature of the milk more normal.

At the present time, the published results as to presence or absence of vitamins in foods show discrepancies. This is probably due to

(1) The different methods used for testing the presence of the vitamins.

(2) The lack of quantitative estimations.

It appears to be generally accepted that vitamin B is widely distributed in the plant kingdom, although there are differences of opinion as to the amount present.

Osborne and Mendel [1920, 2] state that the edible part of an orange contains vitamin B and that the potency is similar to that of comparable volumes of cow's milk. Grape juice they consider to be less potent than orange; apples contain the vitamin, but are not rich in it. These observers [1920, 1] also find that tomatoes and carrots contain appreciable amounts of the water-soluble vitamin; potato is quite a good source, but not so good as tomato. Beetroot contains less of the substance. The occurrence of vitamin B in animal tissues is possibly less extensive, but it may be present in smaller quantities. Cooper [1913] prepared a curative solution of the antineuritic substance from horse flesh and from beef.

According to Drummond [1919] animal tissues (with one or two exceptions) are deficient in the water-soluble antineuritic factor. This observer fed young rats on protein-free extracts of muscle from herring, cod and salmon and found that his solutions contained no vitamin B. When the extract was made from the whole herring, he obtained a slight growth and concludes that the source of the vitamin in this case was the ova and glands, shown to contain the antineuritic vitamin by Chick and Hume [1917]. Drummond does not explain how he prepared the protein-free solutions and it is possible that any vitamin present was adsorbed when the protein was precipitated. If the factor investigated in the experiments below be vitamin B the results obtained contradict those of Drummond.

II. METHODS.

The technique employed has been fully discussed in a previous paper [Hartwell, 1921, 1]. At least three mother rats were fed on the various diets and each attempted to rear six baby rats. One typical curve is shown as representative of the three. In some experiments where two or three members of one litter survived and another litter all died (e.g. cucumber juice, Fig. 1) two growth curves are given as showing the best and worst growth on the given diet.

The basal diet was:

(i) *White bread,*

(ii) *Butter* for fat and vitamin A,

(iii) *Lemon juice* for vitamin C,

(iv) *Caseinogen*
(v) *Salt mixture* } [Hartwell, 1922],

to which was added one of the following as potential source of vitamin B:

(a) *Fruit* or *vegetable juices*. These were made as follows: the soft fruits were placed in muslin and the juice squeezed out by hand. The harder vegetables (*e.g.* potato, carrot) were grated and then squashed in a hand press. The juice was filtered through muslin and allowed to stand in a cylinder so that the starch (if present) settled at the bottom of the vessel and the juice was decanted. All juice was freshly prepared, just before being fed to the animals.

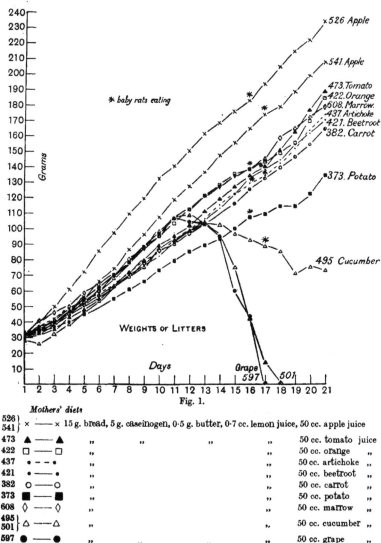

Fig. 1.

(b) *Meat or fish extracts.* The meat was freed from fat as far as possible and minced finely. The fish muscle was removed from the bones (in the herring the backbone was taken out, but probably a few small bones were left) and minced. No roe was included.

The minced meat and fish were weighed, placed in a bottle, warm distilled water added (1 cc. water to 1 g. flesh) and the whole shaken on a mechanical shaker for $1\frac{1}{2}$ hours. The contents of the bottle were then transferred to a flask, which was placed in a boiling water-bath for some of the protein in the solution to coagulate. The whole was filtered through paper and the filtrate fed to the animals.

(c) *Wheat germ extract.* The wheat germ was added to warm distilled water in the proportion of 3 g. wheat to 50 cc. water. This mixture was shaken for $1\frac{1}{2}$ hours and then filtered through paper. The volume of the filtrate was measured and the liquid boiled to coagulate some of the protein. When cold the volume was made up with distilled water to what it was before boiling and the protein filtered off. This filtrate was fed to the animals.

(d) *Extract rich in vitamin B made from crude "marmite."* 250 g. "marmite" were dissolved in 400 cc. distilled water and transferred to a Winchester quart bottle. 1600 cc. absolute alcohol were added and the whole shaken for 3 hours. This liquid was decanted (filtered if necessary) and distilled *in vacuo* at 40°–50° until all the alcohol had been distilled off. The remainder was left to cool, filtered, and the volume made up to 250 cc. with distilled water. It was boiled for 4 hours to sterilise.

(e) *Egg yolk solution.* The egg yolk was separated from the white as much as possible, 50 cc. warm distilled water added and the whole well mixed.

The diet given.

In all the experiments to be described in this paper the proportion of white bread to protein was the same as that previously used as an excess protein diet, *i.e.* 15 g. bread; 5 g. protein (caseinogen).

The initial basal ration consisted of: 15 g. bread, 5 g. protein (caseinogen), 0·7 g. salt mixture, 0·5 g. butter, 0·7 cc. lemon juice.

To this were added 50 cc. of fruit or vegetable juice, meat extract, or egg yolk solution etc.

When wheat germ or soya bean was given as solids, 3 g. were added to the above ration and water was used to mix the dry constituents.

All the constituents were increased proportionately when necessary.

III. Results.

(i) *Vegetable and fruit juices.*

Jerusalem artichoke, carrot, tomato and potato gave excellent results. The baby rats were normal in all respects and were extraordinarily fit and healthy. The litters belonging to the mothers which were given tomato juice in their diets, were, perhaps, just fitter than the others and had very thick coats. Representative growth curves are shown in Fig. 1, 437, 382, 473, 373.

Apple, orange, beetroot and vegetable marrow juices were also effective in preventing very bad symptoms in the young, but the babies were not so healthy and fit as were those whose mothers received artichoke, carrot, tomato or potato juice, although the "apple babies" grew at a greater rate.

In all these experiments slight spasms were seen in the young, but the condition of the babies was never very bad. The spasms were generally noticed for one day only, and the growth curve was not interfered with, *i.e.* no loss of weight was observed. In the case of the "apple babies" the rate of growth was practically maximal. From the 18th–20th day screaming fits [Hartwell, 1921, 2] were noticed, but in spite of this the babies seemed well and gained weight. All were successfully weaned at the 21st day (Fig. 1, 526, 541, 422, 421, 608).

Cucumber juice when fed to the mother was of no effect in relieving the bad symptoms in the young. All three litters suffered badly, the spasms were severe and only 4 out of the 18 young survived. These were weakly for about two weeks after weaning (Fig. 1, 495, 501).

Grape juice, gave even worse results than cucumber juice. All three litters suffered badly and none of the young survived (Fig. 1, 597).

(ii) *Wheat germ and soya bean.*

These two foods were tried because both are reported to be rich in vitamin B.

(*a*) *Wheat germ* 3·0 g. *and wheat germ extract* 50 cc. (preparation see p. 828) were most effective. The litters did well and showed none of the bad symptoms. In the first series of experiments, it is interesting to note that the mother was receiving still more protein, since wheat germ itself contains 24·3 % of protein [Plimmer, 1921] (Fig. 2, 405, 480).

(*b*) *Soya bean.* When 3 g. were added to the mother's basal diet the litters suffered severely; all the babies had bad spasms; 11 out of 16 survived, but were weakly for some time after weaning.

Soya bean contains 33·7 % protein [Plimmer, 1921] and therefore it seemed possible that the bad results of the above experiments might be explained by the increase of protein in the mother's diet. Accordingly three more experiments were tried in which the soya bean was increased from 3 to 6 g. in proportion to the 15 g. bread, 5 g. protein etc. In this case the babies all survived and were practically normal. Slight spasms were noticed, but only for one day. The results were comparable to those obtained when the mother received apple juice (Fig. 3).

(iii) *Fish and meat extracts. Egg yolk. Whey.*

Lean muscle (beef) and herring extracts when added to the mother's diet entirely protected the young from any harmful effects (Fig. 4, 390, 433).

Cod extract was nearly as good, but not quite. No spasms were noticed, but the baby rats were not quite so lively as normal animals. However, three days after weaning (on a bread and whole milk diet) they had recovered entirely (Fig. 4, 447).

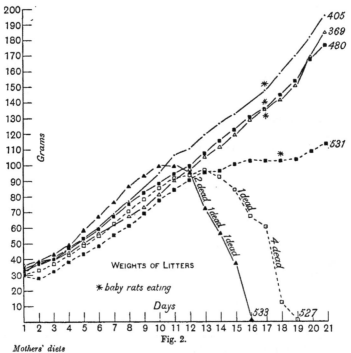

Fig. 2.

Mothers' diets

405 •——• 15 g. bread, 5 g. caseinogen, 0·5 g. butter, 0·7 g. salt mixture, 0·7 cc. lemon juice, 3·0 g. *wheat germ.*

480 ■——■ 15 b., 5 c., 0·5 b., 0·7 salt, 0·7 lemon (see above), *50 cc. wheat germ extract.*

369 △——△   „   „   „   *3 cc. extract from marmite.*

531 ■– –■ } 15 b., 5 c., 0·5 b., 0·7 salt, 0·7 lemon (see above), *50 cc. wheat extract previously*
527 □– –□ }   *boiled with charcoal.*

533 ▲——▲ 15 b., 5 c., 0·5 b., 0·7 salt, 0·7 lemon (see above), *50 cc. marmite extract previously boiled with charcoal.*

Egg yolk was extraordinarily good. The growth curve of the young was equal to that obtained on any diet (*i.e.* maximal) and the babies were absolutely fit in all respects. Their coats were especially thick and long. This is interesting because egg yolk itself contains 15 % of protein (Fig. 4, 464).

Whey. It has previously been shown that the whey from 100 cc. whole milk when added to the mother's excess protein diet was adequate in safeguarding the young. In these experiments 50 cc. were tried in order to compare these results with those obtained when the mothers received 50 cc. of fruit juice, etc.

The babies suffered badly; 13 out of 24 survived, but these were miserable specimens. They were weakly and their coats were thin and dirty (a typical sign that an animal is not fit). They were put on a bread and milk diet on weaning, but it was over a fortnight before they could be considered normal (Fig. 4, 515).

(iv) *Extract rich in vitamin B made from crude "marmite."*

(Preparation given on p. 827.)

3 cc. of the extract were added to the basal ration, and proved most effective. All three litters did well and were normal in all respects (Fig. 2, 369).

The results of the above experiments are tabulated on p. 832. The distribution of vitamin B as given in the report of the Medical Research Council [1919] and by Eddy [1921] is included.

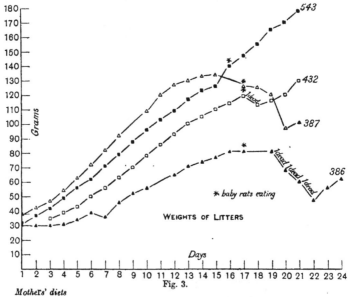

Fig. 3.

Mothers' diets

543 ■ —— ■⎫ 15 g. bread, 5 g. caseinogen, 0·5 g. butter, 0·7 g. salt mixture, 0·7 cc. lemon
432 □ —— □⎭ juice, *6 g. soya bean.*
386 ▲ --- ▲⎫ 15 g. bread, 5 g. caseinogen, 0·5 g. butter, 0·7 g. salt mixture, 0·7 cc. lemon
387 △ —— △⎭ juice, *3 g. soya bean.*

(v) *Experiments with a diet physiologically incomplete.*

The rats were fed on a diet of bread and caseinogen, as before, in the proportion of 15 : 5 and to this was added tomato juice, wheat germ or marmite extract. Such diets were incomplete physiologically, since salt mixture and fat were omitted. Also vitamin A would be poorly represented, if present at all. As explained in a previous paper [Hartwell, 1922] the caseinogen used was not pure and might therefore contain a small amount of vitamin A, as would also the wheat germ and tomato juice.

Three mothers were fed on each diet and in all nine experiments they remained fit and well. The young had a good growth curve and no spasms were seen. The baby rats appeared perfectly fit and healthy, and were quite equal to those whose mothers were getting a more complete diet (Fig. 5, 396, 519, 553).

Fruit and vegetables.

50 cc. juice	M.R.C. on 3 + system	Eddy on 4 + system	Condition of litters
Apple[1]	Not given	—. + +	Slight spasms. Growth good
Artichoke	,,	Not given	No spasms. Growth good
Beetroot[2]	,,	+	Spasms. Babies weakly, but all 18 survived
Carrot[3]	+	+ + +	No spasms. Growth good
Cucumber	Not given	Not given	Bad spasms. 4 out of 18 young survived
Grapes[4]	,,	+	Bad spasms, all young died
Orange[1]	,,	+ + +	Slight spasms. Growth good
Potato	+	+ + +	No spasms. Slight jerky type of walking. Growth good
Soya bean (3 g.)	+ +	+ + +	Bad spasms, 11 out of 16 young survived, but were weakly
,, ,, (6 g.)	+ +	+ + +	Very slight spasms. Babies normal and healthy
Tomato[2]	Not given	+ + +	No spasms. Growth good. Babies coats very thick. Young very fit
Vegetable marrow	,,	Not given	Slight spasms, but quite good growth
Wheat germ[5] 3 g.	+ + +	+ + +	No spasms. Very good growth
Wheat germ extract	—	—	,, ,, ,,
Yeast 3 g.[5]	+ + +	+ + + +	,, ,, ,,
Yeast extract "marmite"	+ + +	+ + +	,, ,, ,,

Meat, fish, eggs, milk.

Cod[6]	White fish very slight if any	+	No spasms. Good growth. Not *quite* fit
Egg yolk[6]	Eggs + + +	Eggs + +	No spasms. *Very* good growth, extraordinarily fit babies
Herring	Very slight if any	+ +	No spasms. Quite good growth
Meat extract[6] 1 g. 1 cc. water	Lean meat +	0	No spasms. Good growth
Whey	Milk +	+ + +	Bad spasms. 13 out of 24 young survived, but all were weakly

[1] Osborne and Mendel [1920, 2].
[2] ——— [1919].
[3] Sugiura and Benedict [1918].
Denton and Kohman [1918].
[4] Osborne and Mendel [1920, 1]. (Commercial grape juice.)
[5] Chick and Hume [1917, 1, 2].
[6] Cooper [1913].

From such a series of experiments it seems justifiable to conclude that protein and the substance supplied in tomato, marmite, wheat germ etc. (? vitamin B) are of primary importance in the diet of a lactating animal.

(vi) *Experiments demonstrating adsorption by charcoal of the accessory factor.*

From the experiments just described it is clear that the protective substance has a distribution similar to that of vitamin B. It has been shown by Chamberlain and Vedder [1911] that the active principle (prepared from rice polishings) which prevents polyneuritis in fowls is adsorbed by filtering through animal charcoal.

Cooper [1913] found that a curative solution prepared from horse flesh lost its activity to the extent of 30 % when filtered six times through a bed of animal charcoal. It was therefore decided to try if the protective substance shared this property of vitamin B. Accordingly the following experiments were made in which the mother rats were fed on marmite and wheat germ extracts which had previously been boiled with animal charcoal.

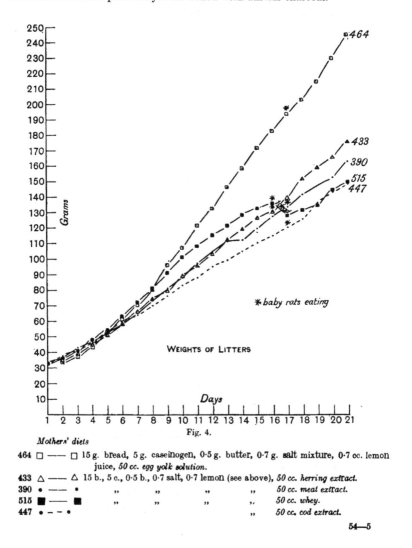

Fig. 4.

Mothers' diets

464 □ —— □ 15 g. bread, 5 g. caseinogen, 0·5 g. butter, 0·7 g. salt mixture, 0·7 cc. lemon juice, *50 cc. egg yolk solution.*
433 △ —— △ 15 b., 5 c., 0·5 b., 0·7 salt, 0·7 lemon (see above), *50 cc. herring extract.*
390 • —— •   „   „   „   „   *50 cc. meat extract.*
515 ■ —— ■   „   „   „   „   *50 cc. whey.*
447 • – – •   „   „   „   *50 cc. cod extract.*

54—5

Preparation of extracts.

(a) *Wheat germ.* The extract was prepared as previously described (p. 828). Animal charcoal was added and the whole boiled. When boiling, a little of the liquid was taken in a spoon; if the charcoal settled rapidly and left a clear liquid on top, the whole was allowed to cool. If the charcoal remained suspended in the liquid, more charcoal was added and the boiling continued until the clear liquid settled out when tested in the spoon. When cool the volume was made up with distilled water to what it was before boiling, and the whole filtered. This clear filtrate was added to the diet.

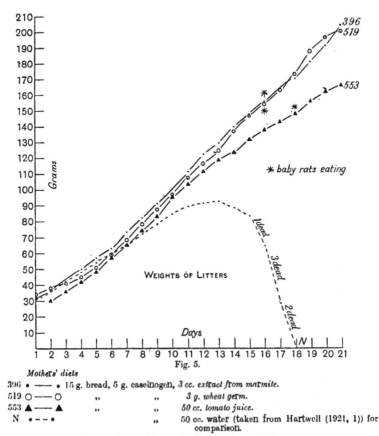

Fig. 5.

Mothers' diets

396	•——•	15 g. bread, 5 g. caseinogen,	3 cc. extract from marmite.
519	O——O	,,	,,	3 g. wheat germ.
553	▲——▲	,,	,,	50 cc. tomato juice.
N	•--•	,,	,,	50 cc. water (taken from Hartwell (1921, 1)) for comparison.

(b) *"Marmite" (yeast extract).* 6 % marmite solution was made with hot water. This was boiled with charcoal (as described above), cooled, the volume made up and filtered. This process was repeated (usually three times) until a clear, colourless filtrate was obtained. This was used for the feeding.

The diet was the same as that used in the preceding experiments, *i.e.* 15 g. bread, 5 g. protein, 0·5 g. butter, 0·7 g. salt mixture, 0·7 cc. lemon juice + 50 cc. of one of the above extracts.

Three experiments were made with each diet.

Results. When the mothers received the marmite extract which had been boiled with animal charcoal, the babies suffered badly and none survived. A typical curve is shown in Fig. 2 with the yeast "charcoal" extract, two litters died and one survived, though all had bad spasms. The young which survived were weakly and were not normal for some time after weaning.

The slight difference in these results may be accounted for by the fact that the marmite extract was boiled three times with charcoal and the wheat extract, only once, so that in the former case there might be more complete adsorption (Fig. 2, 527, 531, 533).

IV. COMMENTS.

Many natural foods or extracts made from them, contain a factor which protects the suckling rats from the harmful effects of excess of protein in the mother rat's diet.

The distribution of this factor is similar to that of vitamin B. The foods reported to be especially rich in this vitamin, *e.g.* egg yolk, wheat germ, when added to the mother's diet not only prevent the bad symptoms, but provide for good growth in the young. Grape juice when fed to the mother was of no effect in relieving the symptoms in the baby rats; all the 18 young (3 litters) died. According to Eddy the vitamin B content of grapes is represented by + (see Table, p. 832).

There are, however, discrepancies between my findings, the work of other physiologists, and the tables published by the Medical Research Council and Eddy.

Drummond [1919] finds that there are no appreciable amounts of the water-soluble factor in the muscles of cod and herring. Eddy gives the vitamin B content of cod as + and herring + +. According to the experiments described in this paper, the extracts from cod and herring muscle had appreciable protective properties, since no spasms were seen in the young, when such extracts were added to the mother rat's diet.

Again Eddy represents orange and tomato as equal and containing + + + and apple as + +. My experiments show tomato to be very superior to orange and that apple and orange are about equal. When the mothers received a ration of 50 cc. orange juice per day, the litters developed spasms, while a ration of 25 cc. tomato juice added to the mother's diet was sufficient to give complete protection to the young.

Soya bean gave interesting results (Eddy + + +, M.R.C. Report + +). When 3 g. were fed to the mother rats, the babies developed spasms and were in a very poor condition; but with 6 g. instead of 3 g., the litters were practically normal, only slight spasms were noticed and the growth was good. Soya bean contains 33·7 % of protein and hence adding 3 g. soya bean to the mothers' diet also increased the protein ration. With 6 g., however, the vitamin content of the diet was nearly doubled, while the percentage of protein was very little different from that of the diet with 3 g. protein. On the whole my results agree more closely with those of Eddy than with those published in the Medical Research Council's Report.

It is possible that the disagreement may be due to:

1. Method of investigation—some observers find the curative dose, while others merely add a definite amount of the food under investigation to a diet deficient in vitamin B.

2. The method of preparation of extracts used, since vitamins easily form adsorption compounds.

Further the protective factor is similar to vitamin B in its solubility and its property of adhering to charcoal. Consequently there is presumptive evidence that the protective factor is vitamin B. This is interesting as vitamin B is thus definitely linked with protein metabolism; the one gland that synthesises protein in large amount functions badly in the absence of vitamin B when there is excess of amino acids in the blood. The lactating animal, when there is excess of protein in the diet, needs much more vitamin B than normal, not for herself, but to regulate the activity of the mammary glands.

SUMMARY.

1. When 50 cc. tomato, artichoke or carrot juices are added to the mother's basal high protein diet, the litters do well and show no harmful effects, such as would otherwise be produced by such a diet.

3 g. wheat germ or 50 cc. wheat germ extract give equally good results. Soya bean, 3 g., are ineffective, but 6 g. added to the above diet greatly improve the condition of the litters.

2. Apple and orange juices give good results, but not quite equal to those obtained with potato etc. The litters show good growth, but slight spasms are noticed. These do not last long and the growth curve is not impaired.

3. Wheat germ extract and marmite extract after being boiled with animal charcoal have no beneficial effect.

4. Beef muscle, herring, and cod extracts, when added to the mother's diet have also the property of protecting the young from any bad effects. Egg yolk has a similar protective action and the growth curve of the young is exceptionally good.

5. The substance in foods which exercises this protective action resembles vitamin B in its distribution, its solubility and the ease with which it is adsorbed by charcoal.

I wish to thank Prof. V. H. Mottram for his criticism and advice. Thanks are also due to the Medical Research Council for a grant which defrayed the expenses of this research.

REFERENCES.

Chamberlain and Vedder (1911). *Philippine Journal of Science,* **6**, 395.
Chick and Hume (1917, 1). *J. R.A.M.C.* **29**, 121.
—— —— (1917, 2). *Proc. Roy. Soc.* B. **90**, 44.
Cooper (1913). *Biochem. J.* **7**, 268.
Denton and Kohman (1918). *J. Biol. Chem.* **36**, 249.
Drummond (1919). *J. Physiol.* **52**, 95.
Eddy (1921). The Vitamine Manual, 59.
Hartwell (1921, 1). *Biochem. J.* **15**, 140.
—— (1921, 2). *Biochem. J.* **15**, 563.
—— (1922). *Biochem. J.* **16**, 78.
Karr (1920). *J. Biol. Chem.* **44**, 255.
Medical Research Committee (1919). Special Report Series, No. 38, 50.
Osborne and Mendel (1919). *J. Biol. Chem.* **39**, 29.
—— —— (1920, 1). *J. Biol. Chem.* **41**, 451.
—— —— (1920, 2). *J. Biol. Chem.* **42**, 465.
Plimmer (1921). Analyses and Energy Values of Foods.
Sugiura and Benedict (1918). *J. Biol. Chem.* **36**, 171.

INDEX